VOLUME SEVENTY EIGHT

Current Topics in MEMBRANES

Na Channels from Phyla to Function

CURRENT TOPICS IN MEMBRANES, VOLUME 78

VOLUME SEVENTY EIGHT

Current Topics in MEMBRANES

Na Channels from Phyla to Function

Edited by

ROBERT J. FRENCH

Department of Physiology and Pharmacology and the Hotchkiss Brain Institute; Cumming School of Medicine University of Calgary, Calgary, Alberta, Canada

SERGEI YU. NOSKOV

University of Calgary, Centre for Molecular Simulation, Department of Biological Sciences, Calgary, Canada

ELSEVIER

Amsterdam • Boston • Heidelberg • London
New York • Oxford • Paris • San Diego
San Francisco • Singapore • Sydney • Tokyo

Academic Press is an imprint of Elsevier

Academic Press is an imprint of Elsevier
50 Hampshire Street, 5th Floor, Cambridge, MA 02139, United States
525 B Street, Suite 1800, San Diego, CA 92101-4495, United States
125 London Wall, London EC2Y 5AS, United Kingdom
The Boulevard, Langford Lane, Kidlington, Oxford OX5 1GB, United Kingdom

First edition 2016

ISBN: 978-0-12-805386-7
ISSN: 1063-5823

Publisher: Zoe Kruze
Acquisition Editor: Alex White
Editorial Project Manager: Helene Kabes
Production Project Manager: Vignesh Tamil
Cover Designer: Greg Harris

Typeset by TNQ Books and Journals

DEDICATION

We dedicate this volume to the memory of Harry A. Fozzard (1931–2014), an extraordinary physician-scientist, colleague, collaborator, and friend. Harry made landmark contributions to the understanding of cardiac physiology in general, and of voltage-gated sodium channels, in particular.

CONTENTS

CONTRIBUTORS

K. Aimar
Department of Biomedical Physiology and Kinesiology, Simon Fraser University, Burnaby, BC, Canada

T.W. Allen
RMIT University, Melbourne, VIC, Australia; University of California Davis, Davis, CA, United States

C. Barbosa
Indiana University School of Medicine, Indianapolis, IN, United States

C. Boiteux
RMIT University, Melbourne, VIC, Australia

E.D. Brodie, III
University of Virginia, Charlottesville, VA, United States

V. Carnevale
Temple University, Philadelphia, PA, United States

C.E. Clancy
University of California, Davis, Davis, CA, United States

R.B. Clark
University of Calgary, Calgary, AB, Canada

T.R. Cummins
Indiana University School of Medicine, Indianapolis, IN, United States

N. D'Avanzo
Université de Montréal, Montréal, QC, Canada

P.G. DeCaen
Children's Hospital Boston, Boston, MA, United States; Harvard Medical School, Boston, MA, United States; Northwestern University, Chicago, IL, United States

K.R. DeMarco
University of California, Davis, Davis, CA, United States

C. Domene
King's College London, London, United Kingdom; University of Oxford, Oxford, United Kingdom

S.C. Dudley, Jr.
The Warren Alpert Medical School of Brown University, Providence, RI, United States

S. Furini
University of Siena, Siena, Italy

V.S. Gawali
Medical University of Vienna, Vienna, Austria

S. Geffeney
Utah State University Uintah Basin, Vernal, UT, United States

R. Ghadiry-Tavi
Department of Biomedical Physiology and Kinesiology, Simon Fraser University, Burnaby, BC, Canada

M-R. Ghovanloo
Department of Biomedical Physiology and Kinesiology, Simon Fraser University, Burnaby, BC, Canada

W.R. Giles
University of Calgary, Calgary, AB, Canada

D. Granata
Temple University, Philadelphia, PA, United States

B.R. Green
University of Utah, Salt Lake City, UT, United States; Monash University, Parkville, VIC, Australia

C. Hanifin
Utah State University Uintah Basin, Vernal, UT, United States

C. Ing
Hospital for Sick Children, Toronto, ON, Canada; University of Toronto, Toronto, ON, Canada

L.L. Isom
University of Michigan Neuroscience Program, Ann Arbor, MI, United States; University of Michigan Medical School, Ann Arbor, MI, United States

B. Joos
University of Ottawa, Ottawa, ON, Canada

M.A. Kasimova
Temple University, Philadelphia, PA, United States

M. Liu
The Warren Alpert Medical School of Brown University, Providence, RI, United States

E.G. Moczydlowski
California Northstate University, Rancho Cordova, CA, United States

C.E. Morris
Ottawa Hospital Research Institute, Ottawa, ON, Canada

V. Oakes
King's College London, London, United Kingdom

B.M. Olivera
University of Utah, Salt Lake City, UT, United States

R. Pomès
Hospital for Sick Children, Toronto, ON, Canada; University of Toronto, Toronto, ON, Canada

P.C. Ruben
Department of Biomedical Physiology and Kinesiology, Simon Fraser University, Burnaby, BC, Canada

D.B. Tikhonov
Russian Academy of Sciences, St. Petersburg, Russian Federation

H. Todt
Medical University of Vienna, Vienna, Austria

G. Toledo
University of Virginia, Charlottesville, VA, United States

T.N. Vien
Tufts University, Boston, MA, United States

J.J. Winters
University of Michigan Neuroscience Program, Ann Arbor, MI, United States

K.-C. Yang
The Warren Alpert Medical School of Brown University, Providence, RI, United States

A. Yu
Department of Biomedical Physiology and Kinesiology, Simon Fraser University, Burnaby, BC, Canada

B.S. Zhorov
McMaster University, Hamilton, ON, Canada; Russian Academy of Sciences, St. Petersburg, Russian Federation

PREFACE

Sodium Channels: Structure, Function, and Beyond Emerging Views in the 21st Century

In 2012, the entire field of membrane biophysics and physiology celebrated the 60th anniversary (Catterall, 2012) of seminal papers published by Hodgkin and Huxley revealing a selective, gated Na^+ conductance as an essential and vital part of the action potential propagation in excitable cells (Hodgkin & Huxley, 1952a, 1952b; Hodgkin, Huxley, & Katz, 1952). We now know this conductance to be mediated by members of the family of voltage-gated sodium channel proteins (Nav). Nav channels have been the focus of numerous recent reviews, including Ahern, Payandeh, Bosmans, and Chanda (2016), Catterall (2014a, 2014b), Payandeh and Minor (2015), which were published in the last 2 years. In the past six decades, progress in the study of ion transport by Na^+ channels and various roles played by them in virtually all tissues have been accelerated due to a number of advancements in experimental techniques ranging from conventional X-ray crystallography to cryo-EM, spectroscopic techniques including rapidly evolving FRET and LRET, low-noise electrophysiology, and ultrahigh resolution microscopy to name a few of many (Ahern et al., 2016; Lacroix, Campos, Frezza, & Bezanilla, 2013; Payandeh & Minor, 2015)!

The legacy of Hodgkin and Huxley instilling marriage between physics of ion transport and experimental protocols led to an intimate connection between experimental studies and mathematical modeling of the action potential where detailed models of sodium channel function were tightly coupled to electrophysiological recordings of selective ion transport, activation and deactivation dynamics and its perturbation by mutations, drugs, and other environmental factors.

With the determination of the first crystal structure at high resolution (Payandeh, Scheuer, Zheng, & Catterall, 2011), the field received fresh impetus enabling detailed mapping of the molecular machinery of gating, location of drug-binding pockets, and providing molecular bases for various roles played by the sodium channel in disease-linked mutations (Catterall, 2014a; McCusker et al., 2012).

Parallel to the growth in experimental insights and sophisticated mathematical models of Na^+ channel function, the biomolecular simulations of ion channels in model membranes have made significant contributions to our understanding of selective ion transport, drug binding, and gating mechanisms. As it is all but impossible to observe the atomic details responsible for specific transport processes with existing experimental approaches, computational studies provide an invaluable window onto these processes, establishing a basis for formal theories and providing testable predictions. Therefore, we are in a unique position to obtain detailed molecular insights into the properties of membrane proteins and directly link experimental observations to molecular simulations at different scales. We can now apply a combination of modeling and experimental techniques to studies of Na^+ channels to provide a multiscale platform linking chemical transitions in single-ion channels to the control of processes such as the dynamic variation of the heart's rhythm at the cellular or organ level. The union of molecular simulations driving and/or driven by experimental explorations into membrane proteins has already produced remarkable results, including the discovery of fundamental kinetic and thermodynamic mechanisms underlying selective transport critical in cellular signaling and the elucidation of key mechanisms of protein function regulation by lipids, to name a few.

Of particular note, experimental and theoretical studies over recent decades have revealed converging evolutionary mechanisms by which biophysical function is coupled to protein structure. Hence, methods that link slow conformational dynamics of proteins with membrane dynamics such as bending are critical to making progress in our study of biomembranes. This volume provides a comprehensive discussion on the development of novel methodologies, multiscale simulations and experimental studies of Na channels in various cells and tissues, and apparent implications to normal heart and brain function. The well-orchestrated opening and closing of Na channels is an essential control mechanism for electrical excitation and relaxation in brain and heart tissues. Abnormalities resulting from inherited mutations and/or pharmacological blockade of Na channels are tightly associated with aberrations in action potentials, leading to the development of severe arrhythmias and neurological disorders such as neonatal epilepsy.

The first structures of Na channels opened an avenue for detailed computational studies of selective transport mechanisms providing unprecedented insights into the bases of Nav channel function and roles.

This present issue of *Current Topics in Membranes* entitled *Na-Channels, From Phyla to Function* presents a collection of articles that summarizes the cutting edge of investigations into structure and function for this remarkable class of membrane proteins from the excitable cells. Although there are numerous conceptual cross-links among the different sections, we have arbitrarily organized the remaining chapters as follows.

In **Chapter 1**, Moczydlowski (2016) provides a whimsically deft view of historical ideas and studies, which sets the stage for our growing understanding of Nav channel proteins.

Chapters 2–4: Evolutionary Adaptations of Channels and Toxins

In these three chapters, the authors illustrate evolutionary adaptations among Nav channels and the toxins that target them (Green & Olivera, 2016; Toledo, Hanifin, Geffeney, & Brodie, 2016; Vien & DeCaen, 2016). In delving into these topics, the different groups have greatly enriched our understanding of the molecular bases of ligand–sodium channel interactions and their interdependence during molecular evolution and adaptation.

Chapters 5–10: Computational Studies and Modeling

Bridging from themes introduced in the last section is the chapter by Zhorov and Tikhonov (2016) on "Computational structural pharmacology and toxicology of voltage-gated sodium channels". The three following chapters (Boiteux & Allen, 2016; Ing & Pomes, 2016; Oakes, Furini, & Domene, 2016) have major foci on properties of the ion-conducting pore, including ion permeation and drug interactions: Boiteux and Allen, "Understanding sodium channel function and modulation using atomistic simulations of bacterial channel structures"; Oakes, Furini, and Domene, "Voltage-gated sodium channels: mechanistic insights from atomistic molecular dynamics simulations," with a comparison of conduction and selectivity in Nav channels and related K channels; Ing and Pomès examine a variety of approaches with their chapter "Simulation studies of ion permeation and selectivity in voltage-gated sodium channels."

The interest shifts to gating properties and their functional implications in chapters by Kasimova, Granata, and Carnevale (2016), "Voltage-gated sodium channels: evolutionary history and distinctive sequence features" with particular emphasis on phylogenetic analysis and Nav channel gating; and DeMarko and Clancy (2016) "Cardiac Na channels: structure to function" where structure of these channels were linked to its function by cellular simulations of cardiac electrical activity highlighting the physiological and pathological roles of Nav1.5.

Chapters 11–15: Functional Studies—Modulation by Other Molecules and Roles of Inactivation and Prolonged Currents

Direct functional modulation of function and regulation of expression are discussed by Winters and Isom (2016), in the chapter "Developmental and regulatory functions of Na^+ channel non-pore-forming β subunits"; some of these protein subunits also have specific functions when expressed independently of the Nav1 α subunits. The chapter by D'Avanzo (2016) explores the modulatory effects of the membrane lipid environment, "Lipid regulation of sodium channels." In their chapter "Mechanism of inactivation in voltage-gated Na channels," Gawali and Todt (2016) provide a detailed account of inactivation processes ranging from fast to ultraslow in their kinetics, and their likely underlying mechanisms. Clark and Giles (2016), in their chapter "The current–voltage relation for the late Na current in mammalian ventricular myocytes," present new experimental data on the prolonged component of Nav current in heart. Increases in the late Nav current may increase the likelihood of arrhythmias both by direct electrical effects and secondary changes in cytoplasmic sodium and calcium contractions. The chapter by Ruben (2016), "Physiology and pathophysiology of inactivation in voltage-gated sodium channels," examines changes in both fast and slow Nav inactivations associated with naturally occurring mutations in cardiac Nav channels and the clinical implications of those channels.

Chapters 16–18: Sodium Channels Under Stress and in Disease

Although several different Nav channel isoforms have been identified in heart, Nav1.5 is generally recognized as the cardiac sodium channel. Liu, Yang, and Dudley (2016) pose the question, "Cardiac Sodium Channel Mutations: Why So Many Phenotypes?" They suggest that clinically significant Nav1.5 phenotypic variations can arise from "...other genetic variations," and "alterations in the life cycle of $Na_V1.5$ such as gene transcription, RNA processing, translation, post-translational modifications, trafficking, complex assembly, and degradation." On another level of organization, Joos and Morris (2016), explore the consequences of mechanical damage to excitable membranes in their chapter, "Nav channels in damaged membranes," and conclude that "Smart Nav inhibitors designed for sick excitable cells would target bleb-damaged membrane, buying time for cell-mediated removal or repair of Nav-bearing membrane." Finally, as an example of the control of pathological hyperexcitability, Cummins and Barbosa (2016) discuss "Unusual voltage-gated sodium currents as targets for pain."

R.J. French and S. Noskov
University of Calgary, Calgary, AB, Canada

REFERENCES

Ahern, C. A., Payandeh, J., Bosmans, F., & Chanda, B. (2016). The hitchhiker's guide to the voltage-gated sodium channel galaxy. *Journal of General Physiology, 147*(1), 1–24. http://dx.doi.org/10.1085/jgp.201511492.

Boiteux, C., & Allen, T. W. (2016). Understanding sodium channel function and modulation using atomistic simulations of bacterial channel structures. In R. J. French, & S. Noskov (Eds.), *Current topics in membranes: Na channel* (Vol. 78). Academic Press.

Catterall, W. A. (2012). Voltage-gated sodium channels at 60: structure, function and pathophysiology. *Journal of Physiology-London, 590*(11), 2577–2589.

Catterall, W. A. (2014a). Sodium channels, inherited epilepsy, and antiepileptic drugs. *Annual Review of Pharmacology and Toxicology, 54*, 317–338. http://dx.doi.org/10.1146/annurev-pharmtox-011112-140232.

Catterall, W. A. (2014b). Structure and function of voltage-gated sodium channels at atomic resolution. *Experimental Physiology, 99*(1), 35–51.

Clark, R. B., & Giles, W. R. (2016). Current-voltage relationship for late Na^+ current in adult rat ventricular myocytes. In R. J. French, & S. Noskov (Eds.), *Current topics in membranes: Na channel* (Vol. 78). Academic Press.

Cummins, T. R., & Barbosa, C. (2016). Unusual voltage-gated sodium currents as targets for pain. In R. J. French, & S. Noskov (Eds.), *Current topics in membranes: Na channel* (Vol. 78). Academic Press.

D'Avanzo, N. (2016). Lipid regulation of sodium channels. In R. J. French, & S. Noskov (Eds.), *Current topics in membranes: Na channel* (Vol. 78). Academic Press.

DeMarko, K., & Clancy, C. E. (2016). Cardiac Na channels: structure to function. In R. J. French, & S. Noskov (Eds.), *Current topics in membranes: Na channel* (Vol. 78). Academic Press.

Gawali, V. S., & Todt, H. (2016). Mechanism of inactivation in voltage-gated Na^+ channels. In R. J. French, & S. Noskov (Eds.), *Current topics in membranes: Na channel* (Vol. 78). Academic Press.

Green, B. R., & Olivera, B. M. (2016). Venom peptides from cone snails: pharmacological probes for voltage-gated sodium channels. In R. J. French, & S. Noskov (Eds.), *Current topics in membranes: Na channel*. Academic Press.

Hodgkin, A. L., & Huxley, A. F. (1952a). Currents carried by sodium and potassium ions through the membrane of the giant axon of Loligo. *Journal of Physiology, 116*(4), 449–472.

Hodgkin, A. L., & Huxley, A. F. (1952b). Movement of sodium and potassium ions during nervous activity. *Cold Spring Harbor Symposia on Quantitative Biology, 17*, 43–52.

Hodgkin, A. L., Huxley, A. F., & Katz, B. (1952). Measurement of current-voltage relations in the membrane of the giant axon of Loligo. *Journal of Physiology, 116*(4), 424–448.

Ing, C., & Pomes, R. (2016). Simulation studies of ion permeation and selectivity in voltage-gated sodium channels. In R. J. French, & S. Noskov (Eds.), *Current topics in membranes: Na channel* (Vol. 78). Academic Press.

Joos, B., & Morris, C. E. (2016). Nav channels in damaged membranes. In R. J. French, & S. Noskov (Eds.), *Current topics in membranes: Na channel* (Vol. 78). Academic Press.

Kasimova, M. A., Granata, D., & Carnevale, V. (2016). Voltage-gated sodium channels: evolutionary history and distinctive sequence features. In R. J. French, & S. Noskov (Eds.), *Current topics in membranes: Na channel* (Vol. 78). Academic Press.

Lacroix, J. J., Campos, F. V., Frezza, L., & Bezanilla, F. (2013). Molecular bases for the asynchronous activation of sodium and potassium channels required for nerve impulse generation. *Neuron, 79*(4), 651–657.

Liu, M., Yang, K.-C., & Dudley, S. C. (2016). Cardiac sodium channel mutations: why so many phenotypes?. In R. J. French, & S. Noskov (Eds.), *Current topics in membranes: Na channel* (Vol. 78) Academic Press.

McCusker, E. C., Bagneris, C., Naylor, C. E., Cole, A. R., D'Avanzo, N., Nichols, C. G., & Wallace, B. A. (2012). Structure of a bacterial voltage-gated sodium channel pore reveals mechanisms of opening and closing. *Nature Communications, 3*.

Moczydlowski, E. G. (2016). The natural and unnatural history of the voltage-gated Na^+ channel. In R. J. French, & S. Noskov (Eds.), *Current topics in membranes: Na channel* (Vol. 78). Academic Press.

Oakes, V., Furini, S., & Domene, C. (2016). Voltage-gated sodium channels: mechanistic insights from atomistic molecular dynamics simulations. In R. J. French, & S. Noskov (Eds.), *Current topics in membranes: Na channel* (Vol. 78). Academic Press.

Payandeh, J., & Minor, D. L., Jr. (2015). Bacterial voltage-gated sodium channels (BacNa(V) s) from the soil, sea, and salt lakes enlighten molecular mechanisms of electrical signaling and pharmacology in the brain and heart. *Journal of Molecular Biology, 427*(1), 3–30. http://dx.doi.org/10.1016/j.jmb.2014.08.010.

Payandeh, J., Scheuer, T., Zheng, N., & Catterall, W. A. (2011). The crystal structure of a voltage-gated sodium channel. *Nature, 475*(7356), 353–358.

Ruben, P. C. (2016). Physiology and pathophysiology of inactivation in voltage-gated sodium channels. In R. J. French, & S. Noskov (Eds.), *Current topics in membranes: Na channel* (Vol. 78). Academic Press.

Toledo, G., Hanifin, C., Geffeney, S., & Brodie, E. D. (2016). Convergent evolution of tetrodotoxin resistant sodium channels in predators and prey. In R. J. French, & S. Noskov (Eds.), *Current topics in membranes: Na channel*. Academic Press.

Vien, T. N., & DeCaen, P. G. (2016). Biophysical adaptations within the family of prokaryotic voltage gated sodium channels. In R. J. French, & S. Noskov (Eds.), *Current topics in membranes: Na channel*. Academic Press.

Winters, J. J., & Isom, L. L. (2016). Developmental and regulatory functions of Na^+ channel non-pore-forming subunits. In R. J. French, & S. Noskov (Eds.), *Current topics in membranes: Na channel* (Vol. 78). Academic Press.

Zhorov, B., & Tikhonov, D. (2016). Computational structural pharmacology and toxicology of voltage-gated sodium channels. In R. J. French, & S. Noskov (Eds.), *Current topics in membranes: Na channel* (Vol. 78). Academic Press.

ACKNOWLEDGMENTS

Many people have worked hard and long to make this volume possible. In particular, we express our heartfelt thanks to:

- The authors, who have found time and motivation, in already frenetic lives and professions, to produce a collection of chapters which provide a remarkably broad and original overview of contemporary studies on voltage-gated sodium channels from both prokaryotes and eukaryotes.
- Sid Simon, for the invitation to embark on this task, and for his perceptive, and tireless work editing and improving the manuscripts.
- Helene Kabes and Vignesh Tamilselvvan, our frontline supporters at Elsevier, for their persistent and remarkably patient support throughout the assembly of the volume.
- The numerous unnamed colleagues who have for decades, by their free sharing of encouragement, criticism, ideas, constructs, software and data, have made the study of voltage-gated sodium channels and their relatives, the captivating and satisfying endeavor that it has been, and continues to be. This has allowed us, like scientists of yore, to clamber toward new insights supported on the backs of giants.

PREVIOUS VOLUMES IN SERIES

Current Topics in Membranes and Transport

Volume 23 Genes and Membranes: Transport Proteins and Receptors★ (1985)
Edited by Edward A. Adelberg and Carolyn W. Slayman

Volume 24 Membrane Protein Biosynthesis and Turnover (1985)
Edited by Philip A. Knauf and John S. Cook

Volume 25 Regulation of Calcium Transport across Muscle Membranes (1985)
Edited by Adil E. Shamoo

Volume 26 Na^+-H^+Exchange, Intracellular pH, and Cell Function★ (1986)
Edited by Peter S. Aronson and Walter F. Boron

Volume 27 The Role of Membranes in Cell Growth and Differentiation (1986)
Edited by Lazaro J. Mandel and Dale J. Benos

Volume 28 Potassium Transport: Physiology and Pathophysiology★ (1987)
Edited by Gerhard Giebisch

Volume 29 Membrane Structure and Function (1987)
Edited by Richard D. Klausner, Christoph Kempf, and Josvan Renswoude

Volume 30 Cell **Volume** Control: Fundamental and Comparative Aspects in Animal Cells (1987)
Edited by R. Gilles, Arnost Kleinzeller, and L. Bolis

Volume 31 Molecular Neurobiology. Endocrine Approaches (1987)
Edited by Jerome F. Strauss, III, and Donald W. Pfaff

Volume 32 Membrane Fusion in Fertilization, Cellular Transport, and Viral Infection (1988)
Edited by Nejat Düzgünes and Felix Bronner

Volume 33 Molecular Biology of Ionic Channels★ (1988)
Edited by William S. Agnew, Toni Claudio, and Frederick J. Sigworth

Volume 34 Cellular and Molecular Biology of Sodium Transport★ (1989)
Edited by Stanley G. Schultz

Volume 35 Mechanisms of Leukocyte Activation (1990)
Edited by Sergio Grinstein and Ori D. Rotstein

★Part of the series from the Yale Department of Cellular and Molecular Physiology

Volume 36 Protein–Membrane Interactions* (1990)
Edited by Toni Claudio

Volume 37 Channels and Noise in Epithelial Tissues (1990)
Edited by Sandy I. Helman and Willy Van Driessche

Current Topics in Membranes

Volume 38 Ordering the Membrane Cytoskeleton Trilayer* (1991)
Edited by Mark S. Mooseker and Jon S. Morrow

Volume 39 Developmental Biology of Membrane Transport Systems (1991)
Edited by Dale J. Benos

Volume 40 Cell Lipids (1994)
Edited by Dick Hoekstra

Volume 41 Cell Biology and Membrane Transport Processes* (1994)
Edited by Michael Caplan

Volume 42 Chloride Channels (1994)
Edited by William B. Guggino

Volume 43 Membrane Protein–Cytoskeleton Interactions (1996)
Edited by W. James Nelson

Volume 44 Lipid Polymorphism and Membrane Properties (1997)
Edited by Richard Epand

Volume 45 The Eye's Aqueous Humor: From Secretion to Glaucoma (1998)
Edited by Mortimer M. Civan

Volume 46 Potassium Ion Channels: Molecular Structure, Function, and Diseases (1999)
Edited by Yoshihisa Kurachi, Lily Yeh Jan, and Michel Lazdunski

Volume 47 Amiloride-Sensitive Sodium Channels: Physiology and Functional Diversity (1999)
Edited by Dale J. Benos

Volume 48 Membrane Permeability: 100 Years since Ernest Overton (1999)
Edited by David W. Deamer, Arnost Kleinzeller, and Douglas M. Fambrough

Volume 49 Gap Junctions: Molecular Basis of Cell Communication in Health and Disease
Edited by Camillo Peracchia

Volume 50 Gastrointestinal Transport: Molecular Physiology
Edited by Kim E. Barrett and Mark Donowitz

Volume 51 Aquaporins
Edited by Stefan Hohmann, Søren Nielsen and Peter Agre

Volume 52 Peptide–Lipid Interactions
Edited by Sidney A. Simon and Thomas J. McIntosh

Volume 53 Calcium-Activated Chloride Channels
Edited by Catherine Mary Fuller

Volume 54 Extracellular Nucleotides and Nucleosides: Release, Receptors, and Physiological and Pathophysiological Effects
Edited by Erik M. Schwiebert

Volume 55 Chemokines, Chemokine Receptors, and Disease
Edited by Lisa M. Schwiebert

Volume 56 Basement Membranes: Cell and Molecular Biology
Edited by Nicholas A. Kefalides and Jacques P. Borel

Volume 57 The Nociceptive Membrane
Edited by Uhtaek Oh

Volume 58 Mechanosensitive Ion Channels, Part A
Edited by Owen P. Hamill

Volume 59 Mechanosensitive Ion Channels, Part B
Edited by Owen P. Hamill

Volume 60 Computational Modelling of Membrane Bilayers
Edited by Scott E. Feller

Volume 61 Free Radical Effects on Membranes
Edited by Sadis Matalon

Volume 62 The Eye's Aqueous Humor
Edited by Mortimer M. Civan

Volume 63 Membrane Protein Crystallization
Edited by Larry DeLucas

Volume 64 Leukocyte Adhesion
Edited by Klaus Ley

Volume 65 Claudins
Edited by Alan S. L. Yu

Volume 66 Structure and Function of Calcium Release Channels
Edited by Irina I. Serysheva

Volume 67 Advances in Adrenergic Receptor Biology
Edited by Qin Wang

Volume 68 Membrane Fusion
Edited by Leonid V. Chernomordik and Michael M. Kozlov

Volume 69 Metal Transporters
Edited by Svetlana Lutsenko and José M. Argüello

Volume 70 Co-Transport Systems
Edited by Mark O. Bevensee

Volume 71 Store-Operated Calcium Channels
Edited by Murali Prakriya

Volume 72 Functional Organization of Vertebrate Plasma Membrane
Edited by Vann Bennett

Volume 73 Exchangers
Edited by Mark O. Bevensee

Volume 74 Thermal Sensors
Edited by León D. Islas and Feng Qin

Volume 75 Lipid Domains
Edited by Anne K. Kenworthy

Volume 76 Basement Membranes
Edited by Jeffrey H. Miner

Volume 77 Dynamic Plasma Membranes: Portals Between Cells and Physiology
Edited by Vann Bennett

SECTION ONE

Historical Overview

CHAPTER ONE

On the Natural and Unnatural History of the Voltage-Gated Na^+ Channel[1]

E.G. Moczydlowski
California Northstate University, Rancho Cordova, CA, United States
E-mail: edward.moczydlowski@cnsu.edu

Contents

Abstract

This review glances at the voltage-gated sodium (Na^+) channel (NaV) from the skewed perspective of natural history and the history of ideas. Beginning with the earliest natural philosophers, the objective of biological science and physiology was to understand the basis of life and discover its intimate secrets. The idea that the living state of matter differs from inanimate matter by an incorporeal spirit or mystical force was central to vitalism, a doctrine based on ancient beliefs that persisted until the last century. Experimental electrophysiology played a major role in the abandonment of vitalism by elucidating physiochemical mechanisms that explained the electrical excitability of muscle and nerve. Indeed, as a principal biomolecule underlying membrane excitability, the NaV channel may be considered as the physical analog or surrogate for the vital spirit once presumed to animate higher forms of life. NaV also epitomizes the "other secret of life" and functions as a quantal transistor element of biological intelligence. Subplots of

[1] This article contains material excerpted from a book in progress by the same author, entitled *The Electrobiology Paradigm*.

Current Topics in Membranes, Volume 78
ISSN 1063-5823
http://dx.doi.org/10.1016/bs.ctm.2016.06.002

this incredible but true story run the gamut from electric fish to electromagnetism, invention of the battery, venomous animals, neurotoxins, channelopathies, arrhythmia, anesthesia, astrobiology, etc.

1. INTRODUCTION TO NaV AS SURROGATE VITAL SPIRIT

My first encounter with the voltage-gated sodium (Na^+) channel (NaV) might best be described as a stochastic event—not more than a casual textbook affair. Its unique physiological significance is perhaps what enhanced the affinity. Back then, NaV was still in its premolecular incarnation, but it eventually won me over as the most fascinating and captivating thing imaginable (as a scholarly pursuit). Today, NaV hardly turns heads anymore. It is merely another ho-hum signaling component of the human proteome. However, back in the 1970s, NaV came into its own as a special object of intellectual allure.

Physiology textbooks of that era explained what little was known about electrical excitability of muscle and nerve. So-called excitable cells of creatures as diverse as squid and humans are identified by their unique ability to "fire" an action potential—a voltage impulse that compels muscles to move and hearts to beat. Moreover it was obvious that the action potential's transient flicker or "spike" has a larger significance that is often left unstated. It serves as the symbolic element of a physiological code—an electrical script of sensory perceptions of the external world conducted along nerve fibers to the highest cognitive centers of the brain.

It did not take much imagination for a naïve but enamored student to hyperbolize the action potential as the unitary signal of biological intelligence. Beyond the well-documented role of enabling blips of voltage that trigger release of neurotransmitters across the synaptic crevasse, it seemed that the action potentials might also be understood as informational bits flashing within the brain's network microarchitecture. In effect, the brain's connectome and its real-time electrical signals could be idealized as a logical operating system that underlies a program of thoughts, consciousness, and inner will running within the computationally active mind—perhaps even, the complex system from which "self" emerges.

Especially intriguing in those formative years of membrane biophysics was the unassailable experimental evidence, first advanced in squid by

Hodgkin and Huxley (1952a, 1952b, 1952c, 1952d) and later extended to vertebrates, which deconstructed the nerve action potential into separate currents of Na^+ and K^+ ions flowing, respectively, inward and outward across the cell membrane. The physical entities responsible for this vital function were unknown and mysterious, but hypothesized to be two different kinds of voltage-sensing channels that spanned the oleaginous cell membrane and conducted an aqueous phase current of Na^+ or K^+ in preference to one another. Given its leading role in action potential initiation, the NaV can be said, as I shall argue, to be the fundamental biological device that makes us "alive" in body (and spirit)—essentially, a surrogate for the elusive (or illusory, if you prefer) concept of "vital spirit."

Indubitably, NaV is a key piece of the puzzle of sentience—a critical part of biological hardware that enables us to interrogate the external world and act upon our brain's receipt of overwhelming sensory data. In the 1970s, the out-sized significance of NaV for nervous system function was already well supported by compelling evidence; eg, tetrodotoxin, a paralyzing poison extracted from the self-inflating, spine-covered pufferfish was found to specifically abolish the action potential and the signature Na^+ current along with it (Narahashi, Moore, & Scott, 1964). As happened to me and many other biochemically intrigued, but biophysically challenged scientists, who came of age in the pregenome era, NaV became a major obsession. But nobody knew exactly what it was.

To appreciate how understanding of NaV ascended from obscure historical origins to its current heights of molecular intricacy, first consider that our *Homo sapiens* species has long sought to find the thing that makes us alive. Many ancient cultures share a similar concept of *spirit* or *soul* that is closely allied with the life principle. For example, ancient Egyptians went so far as to entomb funerary food offerings with their dead to nourish a hypothetical vital essence called *ka* that was presumed to leave the body after death. A similar notion of life spirit called *qi* is central to ancient Chinese thought and medicine. The corresponding Sanskrit word for life energy in the Hindu tradition is *prana*. Finally, but by no means exhaustively, interpretations of the early Greek concept of soul, also known as *psyche*, encompass the idea of a "breath of life" that animates the living. These many cross-cultural examples of the spirit concept of life are evidence of a universal human desire, in the words of the late pioneer of DNA structure, Sir Francis Crick, to know the "borderline between the living and nonliving" (Crick, 1988, p. 17).

But what do unscientific metaphysical notions of ancestral civilizations have to do with the discovery and modern biological conception of the

NaV channel? Proposition: the science of physiology emerged from the concerted attempt by early philosophers to elucidate the life principle. Aristotle (384–322 BCE), who is often hailed as the first natural historian, devoted much effort to interrogate the nature of soul in his philosophical inquiries. As described by Aristotle, the soul is "the cause and first principle of the living body" (Ackrill, 1987, p. 171). His quest to expose the soul's true nature led him to propose certain physiological correlates of the life force. He distinguished three kinds of soul particular to the hierarchy of living species; ie, the lowest soul of plants is characterized by growth, nutrition, and reproduction; the higher soul of animals also exhibits sensation and purposive movement; whereas, the human soul is uniquely defined by capacity of reason and mind. The following passage from Aristotle's discourse *On the Soul (de anima)* demonstrates his attempt to rationalize functional manifestations of soul in physiological terms:

> *...that which has soul is distinguished from that which has not by life. But life is spoken of in many ways, and we say that a thing lives if but one of the following is present—intellect, perception, movement, rest in respect of place, and furthermore the movement involved in nutrition, and both decay and growth.*
>
> ***Ackrill (1987, p. 167)***

The Aristotelean emphasis on active processes of the living state, particularly those functions concerned with metabolism and intelligence, presaged efforts of later generations of biologists to emphasize physical (as opposed to metaphysical) mechanisms that animate life—molecular events that actually make life work at cellular and multicellular levels. Theology, of course, appropriated the soul concept for its ecclesiastical utility. However, within the intellectual tradition of science, the notion of soul and its correlate, vital spirit, had a distinctly biological connotation.

It may be hard to fathom in this day and age, but until sometime quite recently in the middle of the last century, physiologists labored under the lingering cloud of an alternate theory of life called *vitalism*. Its key tenet basically held that the essence of life is a mystical life force, a *vital spirit* not directly accessible in the laboratory. Such ideas undoubtedly sprang from the innate human propensity for magical thinking and contrived myth in the absence of readily available knowledge of physical cause and effect. Before the scientific method emerged as a trusted protocol for solving seemingly intractable problems, myths provided surrogate explanations of natural phenomena. Biology was virtually the last domain of science to reach the functional maturity of an applied technology simply due to the fact that foundational results from mathematics, chemistry, and physics were necessary for cogent interpretation of many biological observations and experiments.

The significance and influence of the now-defunct vitalist tradition in biology is illustrated in *Man on His Nature*, a book by pioneer neuroscientist Sir Charles Sherrington based on the Gifford lectures he delivered at the University of Edinburgh in 1937–38 (Sherrington, 1940). The organizing theme of these endowed lectures was Natural Theology, a topic meant to bridge the philosophical divide between science and religion. However, Sherrington used his academic license to analyze how "brain collaborates with psyche" and more generally, how scientific understanding of living body and mind emerged from the ancient concept of spirit.

As background to his larger exposition, Sherrington discussed at length the antiquated notions of Jean Francois Fernel (1497–1588), physician to King Henry II of France, who authored a three-part medical treatise entitled *Universa Medicina*. Fernel, also a Renaissance scholar, is widely recognized as a major figure in the genesis of modern physiological science. According to Sherrington, Fernel was "the earliest to draw together into one discipline physiology, calling it for the first time by that name, and holding it to be the necessary introduction to scientific medicine" (Sherrington, 1940, p. 6). Fernel's conception of life and biology was fundamentally vitalist:

> *Our sixteenth-century physician-philosopher, Jean Fernel, supposed in the body a something incorporeal. The material body did not work itself. The body was tenanted by a principle which made it 'live.'...Fernel supposes this not only of man but of animate creation generally. In other words, there is throughout animate Nature a living principle which informs its bodies with life.*
>
> ***Sherrington (1940, p. 34)***

By way of illustration from his medical physiology text, Fernel actually described the mechanism of saliva secretion as a sequence of causational events beginning with the "volitional faculty (the highest subdivision of the rational soul, incorporeal)" (Sherrington, 1940, p. 41). Essentially then, modern physiology began with the assumption of an imaginary substance as the ultimate effector of any physiological cascade, as one might say in the modern parlance of signal transduction.

Given that science in Fernel's time slowly progressed within a Renaissance culture pervaded by astrology, superstition, occultism, and biblical interpretation of the natural world, it is not surprising that modern physiology started from vitalist assumptions. To purge such ingrained notions of an ineffable life substance, scientists had to produce palpable evidence of a physical force resident within the body that served the same purpose. In effect, they needed to find a surrogate vital spirit at work in the real world of animal bodies.

Sherrington saw fit to link the onset of vitalism's demise to a culinary event that occurred in the laboratory of Luigi Galvani (1737–98), a professor of medicine at the University of Bologna. In reference to the supposed animal spirits:

> *Their death-knell rang when Lucia Galvani told her husband Luigi that the frogs' legs prepared for the meal seemed alive on the copper wire. Slowly and surely the ensuing century's analysis resolved this aspect of the spirits of the anima into transient electric potentials travelling the fibers of the nervous system. They were no longer 'spirit', but were become a physical event describable under energy.*
>
> ***Sherrington (1940, p. 248)***

The key breakthrough needed for biology to graduate from unnatural epistemology to natural science was indeed Galvani's discovery of animal electricity; ie, the electrophysiological basis of nervous system function. By explaining the bioelectric phenomena of membrane potential, action potentials, and synaptic potentials in terms of the ordinary physics of electricity, the subdiscipline of electrophysiology actually made it possible to fully comprehend the living world from a material perspective. Galvani's electrical experimentation on partially dissected frogs' legs led to a paradigm shift of major significance.

But how did this *Electrobiology Paradigm*, as I like to call it, refine understanding of the so-called "borderline between the living and nonliving," or the conditional state of being alive versus not alive? Instead of being quickened by an incorporeal spirit, the living condition is now recognized by certain physiological parameters or 'vital signs' of a metabolically active cell, multicellular organism, or human person. Physiology has identified key parameters (Na^+, K^+, Ca^{2+}, membrane potential, glucose, ATP, respiration, heart rate, blood pressure, etc.) that must reside within homeostatic norms for the life state of matter to continue to operate in a self-sustaining fashion. In turn, these vital metrics are maintained by complex metabolic processes ultimately driven by the electromagnetic force operating at the level of chemical reactions, atomic interactions, and electrical potentials within a cell. From a reductionist perspective, the elusive thing that ancient philosophers were actually seeking in their contemplation of soul or spirit was electromagnetism, the fundamental force that underlies practically everything from the origin and composition of matter to molecular processes that animate cell physiology. The molecular transducers of electromagnetism's "life spirit" are nothing more than familiarities of the modern biochemistry textbook: nucleic acids, proteins, enzymes, metabolites, inorganic ions, oxygen, phospholipids, etc. Any one of these latter molecules might qualify as a surrogate vital spirit required for the living state as we now know it.

However, for purposes of this review, consider the NaV channel to rank near the highest echelon of surrogate vital spirits. For what other molecular transducer is more emblematic of vital spirit than the protein device responsible for the signature electrical event of the nervous system? To buttress this point, if a certain number of our own NaV channels were to be blocked, say by a potent neurotoxin, we would be rendered *instantly* paralyzed and insensate, a condition soon to be followed by tissue death from asphyxiation resulting from the electrical silencing of key respiratory nerves and muscles. There are few other molecular transducers, perhaps some additional ion channels that one might propose, whose blockage would instantaneously result in a sudden death syndrome. Inhibition of many other enzymes would be similarly fatal, but it would take an extended interval of time for morbidity, paralysis, and brain death to ensue. In fact, NaV does reside at the literal borderline of life and death. The many advances in NaV channel research documented in this volume and elsewhere (Ahern, Payandeh, Bosmans, & Chanda, 2016) only comprise the recent chapter of an amazing story written by inquisitive souls, Aristotle, Fernal, Galvani, Sherrington, and many other contributors to physiological science who helped to banish the fallacy of vitalism from the domain of serious thought.

2. NaV AS PHYSICS LESSON: ELECTROBIOLOGY EXPERIMENTS LED TO ELECTROMAGNETISM

Today's electrophysiology laboratories would be largely dysfunctional without sophisticated electronic instruments such as patch clamp amplifiers to measure ionic current and computers to program voltage protocols and analyze data. Modern understanding of ion channels and the methodology used to study them assumes basic knowledge of electromagnetic theory. By the same token, breakthroughs that led physicists to discover the laws of electromagnetism were enabled and accelerated by the work of electrobiologists who pioneered the study of muscle and nerve.

The origins of electromagnetism as a science can be traced to ancient curiosity about two materials of nature: lodestone (or magnetite) and amber (Halliday & Resnick, 1978). The Greek philosopher-mathematician known as Thales of Miletus (c. 624–546 BCE) is cited as a "Father of Science" in the western tradition for being one of the earliest thinkers to eschew supernatural explanations of natural phenomena (Sagan, 1980). He was particularly interested in materials that appeared to possess rudimentary forms of intelligent behavior such as the through-space attraction of iron to lodestone and the similar electrostatic attraction of bits of straw and feather to rubbed amber. Many centuries of dogged investigation were required to eventually establish that the electromagnetic force underlying all charge–charge interactions is fundamentally the same "vital force" that enlivens cells and empowers biological intelligence.

Besides attracting iron, a magnet in the form of a free-floating needle aligns with the earth's magnetic field. A similar principle allowed Danish physicist Hans Christian Oersted (1777–1851) to detect the magnetic field that surrounds an electric current flowing through a copper wire with a simple compass. Amber, on the other hand, is a natural product of biology, fossilized resin from conifer trees that lived approximately 300 million years ago. Ancient fascination with the charge-collecting property of amber as a dielectric material led to the adoption of the Greek word for amber, *elektron*, as the etymological root for many terms ranging from electron to electricity to electrophysiology. Specifically, in the Renaissance period, physician-scientist William Gilbert (1544–1603) first used the word *electricus* to refer to the attractive property of rubbed amber. He also had a major role in instigating the science of electromagnetism by inventing the first instrument to detect electric charge. His device, known as a versorium electroscope, consisted of a polarizable metal needle mounted on a pivot that tilted in the direction of a charged object such as amber or glass rubbed with fur or wool. Following upon Gilbert, succeeding generations of physicists performed many entertaining experiments with electrostatics. These experiments involved friction machines and electrostatic generators used to transfer charge to and from various dielectric materials. In 1745, invention of the Leyden Jar, basically a large capacitor, gave experimentalists the ability to store electric charge produced by a hand-cranked friction apparatus and explore its effect on biological preparations such as the human body. In a communication to the Paris Academy of Science, one of the inventors of the Leyden Jar, Pieter van Musschenbroek (1692–1761), wrote in 1746 that upon discharging the jar by touching its conductor leads, "my right hand was hit with so much commotion that my body was rattled as if struck by a thunderbolt" (Finger & Piccolino, 2011, pp. 166–167). Subsequently, Benjamin Franklin, founding-father of the United States and pioneer membrane biophysicist (Wang, Steiglitz, Marden, & Tamm, 2013), demonstrated that lightning is actually a form of electricity by flying a kite in a thunderstorm to charge a Leyden Jar with the assistance of his young son William (Priestly, 1794, p. 159). Franklin also used his electrical knowledge to invent the lightning rod but admirably declined to patent it to further the greater good. These and many other imaginative experiments with biological preparations consisting of bodies and body parts of humans and animals proved to be crucial for ground-breaking advances in both physics and physiology.

In particular, the study of electric fish such as the *Torpedo* ray and *Electrophorus* eel, two unique species capable of generating ~220 and ~600 V, respectively, greatly contributed to our understanding of electricity and bioelectricity. A fascinating account of the enormous impact that electric fish have had on science and medicine is found in *The Shocking History of Electric Fishes*, an engaging book by two historians of science, Stanley Finger and Marco Piccolino (2011). Their story recounts many interesting (and often bizarre) experiments and demonstrations that have been conducted in efforts to understand the exact nature of the torpor-producing ability of such fish. Such experiments led to the realization that electrocution by an electric fish is quite similar to the shock received from a Leyden Jar. Thus, the plausible hypothesis emerged that certain fish have the ability to discharge electricity equivalent to that produced by devices that are sources of voltage and produce sparks; eg, a Taser gun and Van de Graaf generator. A landmark experiment performed by naturalist John Walsh on August 1, 1776 in the presence members of the Royal Society of London demonstrated that the electrical discharge of a healthy *Electrophorus* eel can be detected as the flash of a visible spark across a metallic gap in a darkened room (Finger & Piccolino, 2011, p. 282). By consensus of 18th century electrician-scientists, the observation of a spark discharge was the *sine qua non* of electricity.

Experiments with high-voltage fish eventually led physiologists to wonder whether less shocking animals such as frogs and humans also have electricity. The answer to this question emerged from what is perhaps the most famous debate in the annals of science—the extraordinary war of words over the existence of animal electricity waged between Luigi Galvani (1737–98), professor of medicine at the University of Bologna, and Alessandro Volta (1745–1827), professor of physics at the University of Pavia. The furor that accompanied the clash of their intellects and ideas is documented in *The Ambiguous Frog* by Marcello Pera (1992). As Pera notes, it is difficult to ascertain the real substance of the argument between the two scientists since they often seemed to be using different verbiage to say the same thing. In essence, Galvani favored a biological origin for events underlying the contraction of dissected frogs' legs in response to electrical stimuli; whereas, Volta preferred a physical explanation more closely aligned with his own work on electrical contact behavior of metals. Furthermore, Volta never quite abandoned the vitalist notion of a nonphysical life sprit. By contrast, Galvani clearly suspected that the supposed animal spirits assumed by ancient philosophers were in reality a corporal form of bioelectricity. The following two excerpts from their writings give a flavor of the Galvani-Volta controversy:

> *If it will be so, then the electric nature of animal spirits, until now unknown and for a long time uselessly investigated, perhaps will appear in a clear way. Thus being these things after our experiments, certainly nobody would, in my opinion, cast doubt on the electric nature of such spirits…*
>
> ***Luigi Galvani (from the conclusion to* De viribus electricitatus*, published in 1791. (As quoted by Finger & Piccolino, 2011, p. 321)***

> *Here then, after having done and written so much to disprove the alleged animal Electricity—the one that is excited by means of metals in severed limbs, etc.—I too admit a genuine animal Electricity in the Torpedo, the shaking Eel and other shock-imparting fishes. And I am also inclined to attribute a similar one, in the manner explained above, to all animals. This Electricity is all the more truly animal as it depends on the soul [anima] and therefore obeys the will, and has little or no actual extension outside its locus.*
>
> ***Alessandro Volta (in a letter to Luigi Brugnatelli on October 19 1798, written 46 days before Galvani's death (As quoted by Pera, 1992, p. 151)***

The truth is that neither Volta nor Galvani could possibly articulate how nerves and muscles interact with external sources of voltage since knowledge of electricity during their lifetimes was barely rudimentary. However, the attention attracted by Volta's and Galvani's often less-than-collegial disagreement prompted a surge of research in electrophysiology and the physics of electricity. Volta's own experiments involving the use of bimetallic strips (eg, silver and zinc) to stimulate contraction of frog's legs set him on a path that led to the invention of the battery.

As originally described, Volta's battery was a cylindrical apparatus consisting of a repeating stack of silver, zinc, and moist paper (Volta, 1800). This simple electrochemical contraption known as a Voltaic pile or Galvanic cell was the first device capable of charging a condenser to produce a spark discharge or impart a shock to the human finger without requiring frictional rubbing of a dielectric material to first accumulate static charge. Interestingly, the original name that Volta gave to his battery was *organe électrique artificiel* or artificial electric organ. He also directly compared its construction to the columnar cellular microarchitecture of the torpedo electric organ (Finger & Piccolino, 2011, pp. 335–338). This highly abbreviated bit of science history illustrates that major advances in physics and electromagnetism were intimately linked to biological science. It also reminds us that the technological breakthrough that opened the door to practical applications of electromagnetism—a stable voltage source of electricity in the form of a battery—was based on the biomimetic principle; ie, a device that mimics nature in the form of an electric fish.

3. NaV AS DISCOVERY CHANNEL: FINDING OUR INNER ACTION POTENTIAL

Physicists must have been terribly frustrated in the era when the most sensitive voltmeter available to them was a chopped-up animal in the form of Galvani's "frog preparation"—a pair of frog hind legs crudely dissected to retain a piece of motor nerve cut from the lower vertebrae. Similarly, progress in physiological science of the early 19th century was greatly hampered by the lack of appropriate instruments to measure electrical signals from nerves, muscles, and electric organs. Oersted's discovery of the magnetic field component of electric current led to development of simple galvanometers that could be used to investigate the electrical behavior of biological preparations. Carlo Matteucci (1811–68), a professor of physics at the University of Pisa extended Galvani's work by discovering so-called "injury currents" or "demarcation currents" that developed in cut muscle fibers of frogs and other animals (Finger & Piccolino, 2011; McComas, 2011). This signal was detected when one lead of a galvanometer was placed on the intact surface of a muscle and the other lead was placed near the transversely cut or "injured" surface of the same muscle. The current that deflected Matteucci's galvanometer needle in this configuration corresponds to the flow of ions along the surface of a piece of cut muscle that is electrically polarized due to a resting membrane potential of about −80 mV, inside negative. Neither Matteuci nor anyone else at the time had a clue that selective permeability of the resting cell membrane to K^+ was the underlying basis of the injury current.

The next set of major advances in electrophysiology were led by German physiologists, many of whom were trained in the laboratory of Johannes Peter Müller (1801–58) at the University of Berlin. One of Müller's students, Emil Du Bois-Reymond (1818–96), is credited with discovery of a signal called the negative *Schwankung* or (negative variation) which essentially corresponds to the summation of action potentials fired simultaneously by many individual fibers of a whole muscle or nerve (Finkelstein, 2003, 2013). Curiously, Müller, who is considered one of the founders of modern experimental physiology along with Claude Bernard (1813–78) of France, was an ardent vitalist who presented the life force (*Lebenskraft*) as scientific fact in his influential *Handbuch der Physiologie* textbook. Du Bois-Reymond had few qualms in criticizing his mentor on this point, describing Müller's *Lebenskraft* as an idea in which "reason has been put to sleep" (Derig, 2002). He also once vowed it his life's mission to "demonstrate the truth that no forces operate in living organisms except physico-chemical ones" (Derig, 2002; Finkelstein, 2013, p. 64).

Acting on his convictions, Du Bois-Reymond used a sensitive galvanometer to measure current from skeletal muscle preparations before and after tetanic stimulation. Repetitive stimulation of a motor nerve at high frequency results in the summation of unitary twitch contractions of individual muscle fibers to a maximally contracted state known as tetanus. He discovered that the galvanometer reading reversed direction and decreased in magnitude during tetanus relative to the outward injury current of a resting, unstimulated muscle (Finger & Piccolino, 2011; McComas, 2011). With this protocol, the galvanometer responded to the difference in electrical polarity between K^+ current flowing outward from the intact muscle membrane in the resting hyperpolarized state versus the opposite (depolarizing) inward current that flows in response to the electrical activation of NaV channels. He also detected a similar signal from a length of excised peripheral nerve subjected to repetitive electrical stimulation. Although he was not able to explain how the excitation of muscle and nerve gave rise to such electrical changes, Du Bois-Reymond realized that the negative *Schwankung* signal was a finding of major significance. He wrote that he had achieved "the 'hundred years' dream of physicists and physiologists, to wit, the identity of the nervous principle with electricity." (Schuetze, 1983) Appropriately, Du Bois-Reymond is credited with discovery of the action potential and, correspondingly, the first intimation of the electrical function of voltage-gated Na^+ and K^+ channels.

By this time, the existence of animal electricity was no longer in doubt with the technological means to measure electric current from excited biological tissue. But what could be said about the actual mechanism responsible for such current? Major advances leading to the theory of ionic permeability of the cell membrane were made by Julius Bernstein (1839–1917). Bernstein's academic pedigree can also be traced back to Müller since his doctorate degree was supervised by Du Bois-Reymond, who succeeded Johannes Müller as Chair of Physiology in Berlin. Bernstein himself eventually became Chair of Physiology at Martin Luther University at Halle-Wittenberg where he did much of his seminal research. His first important contribution was partial resolution of the time course of the underlying waveform of the negative *Schwankung* signal with the use of an ingenious instrument known as the differential rheotome or "current slicer" (Bernstein, 1868; Seyfarth, 2006). This device consisted of a rotating wheel used to control both the timing of electrical stimulation and the duration of a brief sampling window on the order of 0.3 ms used to record the galvanometric response of the nerve. Bernstein further improvised a signal amplification technique that involved repetitive trials of stimulation monitored at specific time points of the nerve impulse. Using this strategy, he was able to reconstruct the approximate time course of the compound nerve impulse displayed as a brief wavelike current on the order of a millisecond in duration that rapidly rose to a maximum and returned to baseline. This result published in 1868 (see Figure 5 of Schuetze, 1983) is recognized as the first graphical representation of the nerve action potential as a plot of amplitude versus time.

Not only did Bernstein resolve the approximate time course of the action potential, but near the end of his career he finally hit upon the correct line of reasoning needed to explain its ionic basis. These ground-breaking ideas were first outlined in his landmark *Membrantheorie* paper of 1902 (Bernstein, 1902; Seyfarth, 2006) and later elaborated in his *Elektrobiologie* book (Bernstein, 1912).

Like many major advances in science, Bernstein's membrane theory relied upon a previous set of developments in a different but related field, in this case, physical chemistry (De Palma & Pareti, 2011). These developments began with the theory of solution osmotic pressure derived by Jacobus Henricus van't Hoff (1852–1911), a Dutch chemist and recipient of the first Nobel Prize in Chemistry (1901). This was followed by the ionic theory of solutions describing the dissociation of neutral salts such as NaCl and KCl in aqueous solution into electrically active cations

(Na^+, K^+) and anions (Cl^-) as proposed by Swedish physicist, Svante Arrhenius (1859–1927), also a Chemistry Nobel laureate (1903). Another Nobelist (1920), Walther Nernst (1864–1941), contributed the essential theoretical tool needed to interpret bioelectric potentials in the form of the Nernst Equation which originally described the dependence of the electrical potential of electrochemical cells (eg, batteries) on the aqueous concentrations of the reactants. He and others realized that basic principles of solution chemistry and electrochemistry suggested an origin for the resting voltage associated with biological membranes. These concepts were consolidated by a fourth physical chemist and Nobel awardee (1909), Wilhelm Ostwald (1853–1932), who had the following flash of insight in 1890: the principles of osmosis, solution chemistry of ions, and electrochemical potential elucidated by van't Hoff, Arrhenius, and Nernst provided a quantitative framework to interpret bioelectrical phenomena. In a commentary published in *Zeitschrift für physikalische Chemie*, Ostwald proposed that selective diffusion of ions across a semipermeable membrane would result in a separation of charge (Ostwald, 1890). Bernstein considered this insight so significant that he quoted from Ostwald in his own 1902 *Membranetheorie* paper (Bernstein, 1902):

> *It is perhaps not too daring to express here the hypothesis that, through the properties of the semi-permeable membranes here discussed, it would be possible to find an explanation not only for the currents in muscles and nerves, but also—and particularly—for the enigmatic effects of electric fish.*
>
> **Ostwald (1890); as quoted in Finger and Piccolino (2011, p. 386)**

Bernstein took Ostwald's suggestion to the next level. He specifically proposed that the resting voltage of muscle and nerve is due to a microscopic separation of charge, an *elektrische Doppelschicht* (double electric layer), maintained across a semipermeable cell membrane that separates unequal physiological solutions of an ion such as K^+ (Finger & Piccolino, p 390). He proposed that the voltage difference (ΔV) across such a membrane at equilibrium would obey a version of the same equation that Nernst used to describe the behavior of a concentration cell (ie, two half cells, electrically connected by a salt bridge, consisting of the same electrochemical reduction–oxidation reaction but containing two different concentrations of the same electrolyte):

$$\Delta V = E_K = \frac{RT}{zF} \ln \frac{[X^+]_{out}}{[X^+]_{in}}$$

where ΔV is equivalent to E_K, the equilibrium diffusion potential for K^+, R is the ideal gas constant, T is absolute temperature, $z = +1$ valence of a cation X^+, F is the Faraday constant, and $[X^+]$ refers to the concentration of X^+, on either side of the membrane.

While Bernstein correctly proposed that selective permeability to an ion such as K^+ might be the basis for the resting voltage of muscle and nerve cell membranes, he did not manage to fully explain how the directionality or sign of the transmembrane voltage changes polarity during the transient time course of the action potential. He did suggest that the negative *Schwankung* or *Actionstrom* (translated as "action current," first used by Hermann Helmholz which gave us the modern term, action potential (Finger & Piccolino, 2011)), might be due to an increase in permeability of certain ions that are normally blocked from carrying current across the membrane in the resting state. Such a transient loss of K^+-selectivity of the resting membrane could result from an increase in membrane permeability to all ions; in effect, a physical breakdown of membrane integrity. This plausible but rather inelegant idea of a nonselective "membrane breakdown" plagued electrophysiology for many years. Only after the decisive experiments of Hodgkin and Huxley was it proved certain that the action potential involves two distinct changes in ionic permeability consisting of a transient increase in permeability to Na^+ followed by a delayed increase in permeability to K^+ (Hodgkin & Huxley, 1952a, 1952b, 1952c, 1952d).

But well before Hodgkin and Huxley voltage-clamped their first squid axon, another major discovery concerning the role of Na^+ in muscle excitability was made by Charles Ernest Overton (1865–1933). English by birth, Overton obtained his doctorate degree at the University of Zürich where he began work in cellular physiology with studies of plant cells. He moved to the University of Würzburg in Germany in 1901 to commence his postgraduate academic career but later moved to the University of Lund, Sweden, in 1907 where he held long tenure as Chair of Pharmacology. In 1902, Overton published a remarkable set of findings in the same volume of the journal (*Pflügers Archiv*) in which Julius Bernstein's pivotal *Membranetheorie* paper appeared (De Weer, 2000; Overton, 1902).

The principal observation was that frog muscle became insensitive to electrical stimulation and lost the ability to contract when transferred from a saline solution of similar composition to extracellular fluid (eg, ~100 mM NaCl) to a sugar solution of the same osmotic strength (eg, isotonic glucose or sucrose). Muscle excitability was restored by changing the solution from sucrose back to NaCl, demonstrating facile reversibility of the effect. In contrast to the specific action of Na^+ in supporting electrical excitability, external solutions with K^+ as the sole inorganic cation strongly inhibited muscle excitation. The paralyzing effect of high K^+ on muscle contraction is due, of course, to its depolarizing effect on membrane potential (ie, a positive change in voltage due to an increase in the Nernst equilibrium potential for K^+). Overton's simple but profoundly revealing experiments demonstrated that particular inorganic cations (eg, Na^+, K^+, and Ca^{2+}) have special roles in excitability. In commenting on the significance of Overton's work, Andrew Huxley is said to have once modestly remarked that "If people had listened to what Overton had to say in 1902, the work of Alan (Hodgkin) and my own would have been obsolete" (Kleinzeller, 1999, p. 16). As an aside, Overton is also known for his role in development of the Meyer-Overton hypothesis of general anesthesia which established the correlation between the lipid solubility and anesthetic potency of general anesthetic molecules. Today we recognize that a variety of ion channel proteins including NaV are molecular targets of these marvelous drugs used to reversibly suppress consciousness during major surgery (see Section 6 below).

The final part of the story of how humanity found her inner action potential and was forced to abandon the improbable notion of incorporeal vital spirit is more widely known (Ashcroft, 2012). The science part is largely contained within the pages of four seminal papers by Alan Hodgkin and Andrew Huxley (1952a, 1952b, 1952c, 1952d) published in the *Journal of Physiology*. Today virtually every student who studies neurophysiology is introduced to the Hodgkin—Huxley analysis via any respectable textbook on this subject (Hille, 2001). An account of the broader human endeavor is found in Hodgkin's autobiographical memoir, *Chance and Design*, which elaborates upon aspects of his family upbringing, education at Cambridge University, work on radar air defense technology during World War II, and collaborative interactions with prominent neuroscientists such as Herbert Gasser, Joseph Erlanger, Kenneth S. Cole, Andrew Huxley, and many other fellow physiologists (Hodgkin, 1992). The tangible outcome of Hodgkin and Huxley's deconstruction of the action potential by their novel voltage-clamp analysis was the lucid explanation of how it was mediated by two independent channel-like conductors of ionic current, one selective for Na^+ and one for K^+. Without the benefit of any biochemical information on the nature of the macromolecules involved, Hodgkin and Huxley's current—voltage data and mathematical theory correctly inferred all the functional properties of ion channels that we recognize today: a pore of atomic dimensions, ionic selectivity, activation and inactivation gating, and voltage dependence.

Before pausing the remote on the NaV discovery channel, it would be remiss not to acknowledge the roles of many other scientists who had a part in the production. Among Nobel laureates, the field of ion channels is indebted to both the founder of quantum theory, Max Planck (1858–1947), and relativity theorist, Albert Einstein (1870–1955). Planck is noted for his work on electrodiffusion memorialized by the namesake Nernst–Planck equation (Hille, 2001, see p 318; Planck, 1890) that describes ionic current as a sum of diffusional and electromotive components. Einstein's (1905) classic work on the analysis of Brownian motion as the random walk of molecules gave molecular meaning to the diffusion coefficient (D) as a prominent term in the Nernst–Planck equation and ultimately inspired the powerful approach of computational molecular dynamics to simulate the flow of ions through the pores of channels as well as the gating motions of channel proteins. Two other key contributions acknowledged by Hodgkin and collaborators (Hodgkin, 1992) include the invention of the glass capillary microelectrode by Ralph Gerard, Gilbert Ling, and Judith Graham (Graham & Gerard, 1946; Ling & Gerard, 1949) and the invention and application of the voltage clamp technique by George Marmont and Kenneth S. Cole that allowed measurement of cell membrane currents under conditions of constant voltage (Cole, 1949; Marmont, 1949).

Of course, a historically compete account of credits for virtual movies on the NaV discovery channel can never be fully compiled. Annotating the full fabric of such science would require the unraveling every significant thread of insight contributed by thinkers extending back to human origins. What astonishes, though, is how a mysterious entity that biophysicists first imagined as a special kind hole in the nerve cell membrane has materialized from mere idea to protein biomolecule in the brief span of 32 years from the 1952 papers of Hodgkin and Huxley to publication of the primary sequence of the *Electrophorus* NaV channel (Noda et al., 1984). The following excerpt from the seminal paper of David E. Goldman that gave rise to the Goldman–Hodgkin–Katz equation widely used to assess the ion selectivity of channels via permeability ratios (Hille, 2001; see p. 448) is a reminder of NaV's embryonic origin as a hypothetical concept to be pursued with further experiments: "If, then, the membranes are to be regarded as having a sieve-like structure, the 'holes' must be almost of molecular dimensions...For many biological membranes this assumption is made in the absence of evidence to the contrary" (Goldman, 1943, p. 50).

4. NaV AS BIOMOLECULE: THE OTHER SECRET OF LIFE

Within the cloistered community of academic biologists, the cellular capacity for electrical signaling is properly regarded as a wonder close to the heart of what makes us alive. But ion channels, let alone NaV, are not the first thing that comes to mind when practical achievements of biology are touted in public discourse. The universal metaphor for biology is DNA. The latter acronym for deoxyribonucleic acid has acquired a distinctly nonchemical connotation that is routinely used to denote anything of primary essence. Whatever behavior, tendency, or exceptionalism is thought to be inborn or instinctive, the near universal off-handed explanation is that "it is in our DNA." But our immediately purposive behavior could just as readily be expressed by attributions such as "it is the way our ion channels work" or "my NaV channel made me do it." The default ascription of all things biological to DNA only reinforces the sad reality that electrophysiologists have done a terrible job of educating the public on their discoveries.

James Watson and Francis Crick are said to have joked about discovering the "secret of life" in their minicelebration held in the Eagle pub soon after they had derived the structure of DNA (Watson & Berry, 2003, p. 54). Granted, DNA is the molecular basis of genes, proteins, and evolutionary change, but omission of ion channels from the secrets of life seems overly dismissive. This kind of chauvinism is bluntly stated in *DNA: The Secret of Life*, a popular commentary on the history of molecular biology and biotechnology (Watson & Berry, 2003). Its authors claim that the "discovery of the double helix sounded the death knell for vitalism" (Watson & Berry, 2003, p. 61) and further assert:

> *In retrospect we know that Dr. Darwin's* exploration of what was called 'galvanism' was a red herring; the secret of life remained a secret until 1953. Only with the discovery of the double helix and the ensuing genetic revolution have we had grounds for thinking that the powers held traditionally to the exclusive property of the gods might one day be ours. Life, we know now, is nothing but a vast array of coordinated biochemical reactions. The "secret" to that coordination is the breathtakingly complete set of instructions inscribed, again chemically, in our DNA.*
>
> ***(Watson & Berry, 2003, pp. 395–396)*** ****[*****a reference to Erasmus Darwin, the physician and naturalist-philosopher grandfather of Charles Darwin*****]***

Putting aside issues of personal ego that sometimes obscure holistic perspective, it seems appropriate in a volume documenting the impact of NaV channel research to revisit the wisdom of Erwin Schrödinger, a pioneer of quantum mechanics, who is often credited with initiating the molecular biology revolution with his provocative little book entitled *What is Life?* (Schrödinger, 1944). In it, he made a coherent argument for the existence of a chemical code required to perpetuate life, but he also clearly understood that the code itself would not define the living state of matter. Specifically, he argued that the condition of being alive is determined by active metabolism which we know requires not merely DNA, but proteins to do the work. Here is the decisive quote from Schrödinger:

> *What is the characteristic feature of Life? When is a piece of matter said to be alive? When it goes on 'doing something', moving, exchanging material with its environment, and so forth, and that for a much longer period than we would expect an inanimate piece of matter to 'keep going' under similar circumstances.*
>
> ***Schrödinger (1944, p. 69)***

The idea distilled here is that a cell with DNA alone would be quite dead without protein molecules to transduce energy from the environment, power metabolism, and conduct the electrochemical signaling that actually makes it alive. Luigi Galvani's animal electricity, also known as galvanism, cannot be easily dismissed as a red herring with regard to the secret of life. At the very least, in terms of the defining feature of life as "doing something" (ie, acting alive), transmembrane voltage and action potentials mediated by ion channels, in particular NaV, merit more than equal footing with DNA as the *Other Secret of Life.*

This brings us back to the issue of why ion channels such as NaV have not risen to prominence in the public consciousness. Unfortunately, one has to point to the much maligned educational system. The topic of ion channels ought to be introduced much earlier in biological science education, perhaps as early as elementary school. Every educated person needs to know that cells are electrically active and not merely sacks of DNA. Ion channel proteins should be introduced as extraordinary biomolecular devices that do amazing things such as automating the heart and firing electrical signals along nerve fibers. The basics of how cells talk to each other electrically ought to be as commonly known as other important "facts of life."

In the context of an educational field trip or science video, the experience of seeing a live electric eel in an aquarium tank light a light bulb with its high voltage discharge naturally excites the inquisitive mind. As a graduate and postdoctoral student in membrane biochemistry, the electric organs of *Electrophorus* and *Torpedo* had a similarly stimulating impact on my own education. I was fortunate to be an active researcher during exciting times when ion channel proteins from fish electric organs were first purified, sequenced, and eventually cloned via complementary DNA (Noda et al., 1984, 1983). These breakthroughs generated much excitement in the field of molecular physiology. The thrill of exploring newly discovered primary sequences of the *Torpedo* nicotinic acetylcholine receptor channel and the *Electrophorus* NaV channel for clues of function was to physiological science as the first step on the moon was to space exploration. In the years since ion channels of excitability were first revealed as protein molecules, nearly all the closely held secrets of NaV and other ion channels have similarly come to light: subunit structures, voltage-sensing elements, ion-selectivity filters, gating mechanisms, and drug-binding sites (Ahern et al., 2016; Hille, 2001). Vital spirit happens to be a real biomolecule.

5. NaV AS BIOTRANSISTOR: ELEMENTARY ON–OFF SWITCH OF NEURONAL INTELLIGENCE

In commenting on the crystal structure of KvAP that provided a structural glimpse of the K^+ channel voltage sensor (Jiang, Lee, et al., 2003; Jiang, Ruta, Chen, Lee, & MacKinnon, 2003), Fred Sigworth (2003) insightfully compared voltage-gated ion channels to field-effect transistors composed of silicon semiconductor materials. Field-effect transistors are unitary on–off switching elements of electronic circuits that essentially function as voltage-gated channels for electrons. In comparing the steepness of their voltage-response functions, Sigworth further noted that typical voltage-sensitive ion channels are actually more sensitive to voltage than the transistor elements in digital logic circuits of computers. Since integrated circuits of microprocessors with vast arrays of transistors effectively comprise the physical basis of machine intelligence, is it too much of an exaggeration to define neuronal ion channel proteins such as NaV (and their evolutionary channel precursors in primordial cells) as the elementary particles (ie, the electrical quanta) of biological intelligence?

Functional comparison of the biological mind to the artificial intelligence of the silicon chip computer is a fascinating topic that will probably remain impenetrable until a major breakthrough occurs in understanding the neural basis of human thought. Meanwhile, exponentially advancing progress in computer technology is soon predicted to approach a "singularity" moment when computational power of the smartest machine exceeds and merges with that of the human brain (Kurzweil, 2006). In pursuing its relentless goal of emulating and exceeding the human brain, computer science has always been closely wedded to neuroscience. This platonic relationship pervades the literature of artificial intelligence. A well-known example is Alan Turing's landmark paper on *Computing Machinery and Intelligence* (1950) which proposed the famous operational Turing Test for whether machines can think: if a machine-executed program can fool a human into believing its output was produced by a human, then the machine passes the test.

As another example, John von Neumann (who together with Turing was a principal architect of the modern computer and first predicted the singularity moment more recently elaborated by Kurzweil (2006)) set forth a basic framework for the mind—machine comparison in his unfinished manuscript on *The Computer and the Brain* written for the Silliman Lecture at Yale University (Von Neumann, 1958). Von Neumann recognized both the analog and digital nature of signaling networks of brain neurons. On one level he described individual neurons as idealized input—output devices that function as a binary elements of a computational network by either firing an action potential equivalent to a binary "1" digit output, or not-firing equivalent to a "0" digit. But he also recognized that brain neurons have many additional complexities such as a voltage firing threshold and an inordinate number of simultaneous excitatory and inhibitory synaptic inputs integrated through branches of the dendritic tree. Taking account of these and other biological realities, he conceded that the analogous computing structure of the brain is more accurately represented by a massively parallel, mixed analog—digital model.

Even before the advent of Colossus in 1944 (the first programmable digital computer constructed from vacuum tubes), the psychiatrist-mathematician team of McCulloch and Pitts (1943) attempted to describe logical operations of the brain by simple digital networks of idealized neurons. Their theoretical paper on *A Logical Calculus of Ideas Immanent in Nervous Activity* inspired the neural network approach to artificial intelligence (McCulloch & Pitts, 1943; Piccinini, 2004). They showed that simple networks of "all-or none" or two-state model neurons readily perform computations of discrete mathematics (eg, AND, OR, and NOT Boolean logic operations). Similar logic gates comprised of transistors are the basis of the high-precision mathematical calculations performed by any modern electronic calculator or computer.

Insofar as the science of artificial intelligence and neural network theory attempt to infer how the brain works from a reverse engineering perspective, it is relevant to consider how ion channels and NaV, in particular, fit into the mind—computer analogy. Molecular biophysics implies that a single NaV channel functions something like an elementary nanoswitch. Clusters of voltage-activated NaV channels in a given cell respond both to their own positive feedback and the simultaneous electrical switching of other ion channels in a given cell. In this scenario, the firing of a single action potential is something like a touchdown in football, a scoring event that infrequently occurs after many smaller plus or minus skirmishes that precede it. Models that represent the firing of an action potential by a single neuron as the equivalent to a transistor's switching to the on-state greatly underestimate the complexity of the brain's "biological computer." Given all that is known about the numbers and diverse types of channels in a single cell, their mechanisms and modulations, the function of single neuron is much more elaborate than a single transistor. A neuron, or any cell for that matter, with its sophisticated network of chemical and electrical signal transduction elements is more closely analogous to a small microprocessor composed of many transistors. By this analogy, higher-order networks of brain neurons appear to function as an integrated cluster of many single-celled computers capable of remodeling themselves in real time, rather than as a reduced network of two-state transistors. To speculate beyond the usual bounds of moderation, if the analogy of a single NaV channel as the relevant quantal transistor unit of neuronal circuitry is correct, this implies that neural network models will need to be greatly expanded before they can successfully simulate higher cognitive function. Perhaps the rapidly approaching singularity moment in the history of computing will be delayed a bit more into the future.

In pursuit of the physical basis of extraordinary human intelligence, the preserved brain of physicist Albert Einstein has been subjected to extensive analysis in the hope of discovering differentiating microanatomical features that might explain his ability to derive the theory of relativity (Kremer, 2015). Would it not make more sense to analyze the channelome (the complete sequences of all ion channel genes (Shcherbatyy et al., 2015)) of mathematical prodigies for evidence of unusual ion channel polymorphisms that correlate with higher intelligence? In any case, comprehensive understanding of ion channel biophysics may certainly be expected to help decipher the biological basis of higher intelligence and finally establish the mind—brain connection.

6. NaV AS A PHARMACOLOGICALLY TUNABLE ON–OFF SWITCH OF PAIN AND CONSCIOUSNESS

As if being a surrogate vital spirit biomolecule, the other secret of life, and a biotransistor of intelligence were not enough, Nature has endowed the NaV channel with additional responsibilities. Among these, the role of being a drug target with major biomedical implications looms large. The first observation is that a single NaV channel gene is physiologically inadequate. The human genome has nine different functional NaV channel genes located on four different chromosomes (Ahern et al., 2016): NaV1.1 (gene SCN1A), NaV1.2 (SCN2A), NaV1.3 (SCN3A), NaV1.4 (SCN4A), NaV1.5 (SCN5A), NaV1.6 (SCN8A), NaV1.7 (SCN9A), NaV1.8 (SCN10A), and NaV1.9 (SCN11A). These nine NaVs are further diversified by possible coassociation with four different accessory beta subunits coded by human genes: SCN1B, SCN2B, SCN3B, and SCN4B. Of the nine homologous NaVs, NaV1.4 and NaV1.5 primarily function in the initiation of action potentials in skeletal muscle and the heart, respectively. The other NaVs have an analogous role in various neurons of the central and peripheral nervous system. Gene redundancy may be seen as a failsafe evolutionary mechanism for survival when one member of a large gene family is genetically defective. It also allows for functional specialization and expansion of differential mechanisms of regulation among various members of the gene family. Much research in the NaV channel field has been devoted to defining structural and functional differences of the nine NaVs. An important goal of such work is to develop strategies for selective targeting or selective delivery of drugs that control conditions of hyperexcitablity such as pain, muscle myotonia, heart arrhythmia, convulsions, and epileptic seizures (eg, Jensen, Grunnet, & Bastlund, 2014; Kwong & Carr, 2015).

Given their commanding role in excitation, action potentials are fine in moderation. However, firing too many action potentials above the normal frequency or firing at the wrong time is often physiologically problematic. The presence of approximately 78 different K^+ channels in the human genome may be seen as the evolutionary outcome of the necessity for suppression of excessive cellular excitation within tolerable limits. The natural endogenous feedback inhibitors of excitation, K^+-selective and Cl^--selective ion channels, are often insufficient to contain the depolarizing excesses of NaV. When subjected to a barrage of physiological hyperactivity abetted by external stimuli, humans have always sought relief from neuroactive drugs.

The widespread use of addictive substances such as ethanol (a positive modulatory enhancer of the inhibitory $GABA_A$ Cl^- channel) and opiates (suppressors of excitation of nociceptive neurons) may be interpreted as a basic human inability to withstand an overproduction of action potentials. Even though many neuroactive drugs do not target NaV channels directly, the ulterior goal of using and administering such medications is often to depress the frequency of NaV and CaV channel openings in a pharmacological attempt to achieve physiological quietude.

More to the point, NaV channels are the direct target of many important drugs. The historical observation that the plant natural product, cocaine, has a locally acting anesthetic effect led to development of many synthetic drugs of the local anesthetic class (procaine, lidocaine, etc.) that block NaV channels (Dos Reis, 2009), often with a use-dependent action linked to repetitive gating of the channel (Hanck et al., 2009). Other uses of NaV-targeting drugs include treatment of certain types of heart arrhythmias (class I antiarrhythmic agents such as lidocaine and flecainide) and epilepsy (eg, phenytoin). Recent advances in understanding the specific role of certain NaV channel isoforms in the initiation and relay of action potentials involved in pain perception (eg, NaV1.7, 1.8, and 1.9) have prompted renewed efforts to develop new classes of analgesic drugs (Kwong & Carr, 2015).

Certain general anesthetics that silence perception of conscious pain are now known to act by hyperpolarizing central neurons via activation of the $GABA_A$ receptor (eg, propofol) or two-pore K^+ channels (eg, halothane) (Franks, 2006). The sweet bliss of general anesthesia enjoyed in the course of invasive surgery is the end result of activating such inhibitory hyperpolarizing ion channels that suppress NaV-driven action potentials in brain regions underlying the state of consciousness. It is not at all presumptuous then to state that NaV functions not only as a quantal unit of intelligence but also plays a pivotal accessory role in the pharmacological on–off switching of consciousness.

7. NaV AS TRAGEDY: A LOCUS OF NERVOUS SYSTEM MALADIES

We humans are not always lucky enough to inherit a perfectly functioning set of NaV channel genes. The list of genetic diseases termed NaV channelopathies linked to inheritance of a mutation in one of the nine NaV channels is unfortunately very long. Such mutations can affect virtually every aspect of NaV channel function resulting in diverse pathologies of hyperexcitablity or hypoexcitability.

The genetic disorder of hyperkalemic periodic paralysis (HyperPP) gained notoriety as an autosomal dominant mutation of the muscle isoform NaV1.4 of the American Quarter Horse, a classic breed of race horse. Apparently over 4% of quarter horses in the United States inherited this disorder through the bloodline of a prolific sire named Impressive (Jurkat-Rott & Lehmann-Horn, 2007; Rudolph et al., 1992). Affected horses are subject to muscle twitching, weakness, and paralysis precipitated by hyperkalemia (high plasma K^{+}) or extremes of temperature. In humans, five different types of NaV channelopathies affecting skeletal muscle are recognized by different inherited mutations in all four internally repetitive Domains I–IV of NaV1.4: HyperPP, K^{+}-aggravated myotonia, paramyotonia congenita, hypokalemic periodic paralysis, and a form of congenital myasthenic syndrome (Jurkatt-Rott, Holzherr, Fauler, & Lehmann-Horn, 2010). Such mutations typically affect the kinetics of NaV channel activation and inactivation resulting in a propensity toward muscle stiffness (myotonia) or paralysis.

A potentially lethal channelopathy of the NaV1.5 cardiac channel gene is responsible for a type of Long-QT syndrome classified as LQT3 (Wang et al., 1995). This syndrome is characterized by an abnormally long-duration cardiac action potential due to a small increase in the probability of late openings of NaV1.5 caused by defective inactivation gating. Such a defect has the potential to trigger the onset of sudden death from ventricular fibrillation. People born with this condition are often unaware of it until it is too late with the onset of fatal arrhythmia. Mutations in the NaV1.7 channel expressed in nociceptor (pain sensing) neurons of the dorsal root ganglion can either result in acute hypersensitivity to pain (erythromelalgia, paroxysmal extreme pain disorder) or the virtual lack of peripheral pain sensation (channelopathy-associated insensitivity to pain) (Drenth & Waxman, 2007) The paradoxically opposite symptomology of these maladies is explained by inheritance of either a dominant missense mutation that interferes with normal NaV channel inactivation gating resulting in hyperexcitability; or, a recessive nonsense mutation that prevents expression of a functional Na1.7 channel protein.

A large number of epilepsy channelopathies are also linked to various mutations in the genes for human NaV1.1 and 1.2 channels that are widely expressed in brain neurons (Ito et al., 2002; Kamiya et al., 2004). The available treatments for such conditions are often less than satisfactory. However, with the advent of whole-genome sequencing and personalized genetic medicine, it is becoming increasingly possible to identify and diagnose the physiological implications of such mutations. New therapies of the future involving stem cells and advanced methods of genetic modification may one day enable medicine to correct some of Nature's NaV channel mistakes.

8. NaV AS THE FAVORITE TARGET OF NATURE'S VENOMS

As the preeminent surrogate for life's vital spirit (as justified by arguments given above), wild Nature has not been kind to the NaV channel. The competition for survival among living forms has naturally singled out the NaV channel as a favorite physiological target for pharmacological attack. For what better way is there to incapacitate one's attacker or immobilize one's prey than to interfere with the ability to fire action potentials? Many of the most toxic and venomous species of earth's nefarious menagerie—ranging from dinoflagellates to the pitohui birds of New Guinea—harbor toxins that target NaV. Virtually every available site on the NaV channel protein that can serve as a crevice or a peptide thread for molecular contact with a toxin molecule is targeted as a binding site. In terms of formal nomenclature, the major recognized toxin binding sites of NaV include:

Site 1: Residues located in the outer vestibule of the ion conduction pore bind guanidinium toxins such as tetrodotoxin and saxitoxin as well as cone snail venom peptides such as μ-conotoxins (Moczydlowski, 2013).

Site 2: Residues of the four S6 helical transmembrane segments of internally homologous Domains I–IV line the inner pore and chemically interact with a diverse array of lipophilic and alkaloidal small molecule activators such as batrachotoxin, veratridine, grayanotoxin, and aconitine (Wang & Wang, 2003).

Site 3: Residues of external loops of the S1–S4 voltage sensor subdomain within Domains I and IV bind a variety of α-scorpion and sea anemone peptide toxins tarantula venom peptide toxins (Rogers, Qu, Tanada, Scheuer, & Catterall, 1996).

Site 4: Residues of the S3–S4 loop of Domain II bind certain peptide β-scorpion toxins and tarantula peptide toxins (Cestele et al., 1998; Xiao et al., 2008).

Comprehensive studies involving chimera construction and mutagenesis further indicate that all four NaV voltage sensors of Domains I, II, III, and IV may potentially serve as target sites of peptide toxins that affect activation and inactivation gating. Some toxins may interact with multiple voltage sensors and differentially target multiple NaV gene isoforms (Bosmans, Martin-Eauclaire, & Swartz, 2008).

Other molecules that target NaV channels include plant insecticides such as pyrethrins of chrysanthemums and lipophilic cyclic polyethers (brevetoxins and ciguatoxins) that are produced by dinoflagellates and bioaccumulate in fish. Similar to NaV channelopathy mutations, NaV channel toxins can disable the production of action potentials by preventing activation of the channel or by directly blocking flow of Na^+ through the channel. Alternatively, particular toxins may depolarize the cell membrane and cause repetitive firing by shifting the voltage dependence to a lower threshold range or by disabling or slowing inactivation gating of the channel. The steroidal alkaloid, batrachotoxin, contained within the skin of poison dart frogs of the *Phyllobates* genus and in skin and feathers of the hooded pitohui bird, has a remarkably unique triple-play mechanism of action that includes shifting the range of voltage-activation gating by approximately −50 mV, virtual removal of inactivation gating, and altering the ionic selectivity for permeant cations by lowering the normally high permeability ratio for Na^+ over K^+ (Moczydlowski & Schild, 1994).

While suffering this chemical onslaught, the NaV targets of Nature's toxins have not been idle in the evolutionary arms race between toxin-producing species and their victims. Many species that coevolve with toxin producers have responded by developing genetically based chemical immunity to NaV channel toxins. For example, NaV channels of toxic newts and pufferfish that accumulate tetrodotoxin within their tissues have characteristic mutations in the NaV outer pore region that renders the channel resistant to the toxin. Certain species of garter snakes (eg, *Thamnophis sirtalis*) that feed on tetrodotoxin-containing newts (*Taricha granulosa*) are likewise toxin resistant and have similar mutations in NaV channel genes (for review, see Moczydlowski, 2013).

An extraordinary example of toxin-channel tit-for-tat is the case of the grasshopper mouse (*Onychomys torridus*) and the bark scorpion (*Centruroides sculpturatus*). The sting of this scorpion is intensely painful to the ordinary house mouse via the action of venom peptides that activate the NaV1.7 channel of nociceptive neurons causing prodigious firing of action potentials propagated to interpretive pain centers in the brain. However, the grasshopper mouse of the Arizona desert resides comfortably within the same territory inhabited by the *Centruroides* bark scorpion due to the benefit of mutations in a companion nociceptive channel NaV1.8 that results in a gain-of-function block of this channel by other components in the venom (Rowe, Xiao, Rowe, Cummins, & Zakon, 2013). This mouse has managed to exploit evolution to its benefit by transforming (over eons of evolutionary time) in the fashion of a comic superhero from scorpion-victim into scorpion-eater.

As chemical methods for peptide isolation, identification, and sequencing become increasingly more powerful and sensitive, the list of known venom neurotoxins has expanded proportionally. With the availability of such data for bioinformatic analysis, it would be interesting to determine which ion channel or receptor is the most toxin-targeted protein biomolecule in nature. Chances are high that NaV is Nature's favorite target, although voltage-gated K^+ channels and nicotinic acetylcholine receptor channels also deserve an honorable mention.

9. NaV EXOBIOLOGY: WHY Na^+?

Extrapolating from knowledge of earth-based biology, it seems reasonable to conclude that ion channels are required for life; or, at the very least, ion channels are required for higher intelligence. The supporting arguments are relatively straightforward. Metabolically autonomous life forms whether unicellular or multicellular require H_2O as an essential solvent for solution chemistry. All cells are bounded by an impermeable plasma membrane composed of lipids that prevents vital biomolecules from being infinitely diluted into the surrounding void. As such, all cells are osmotically sensitive and have adopted various strategies for dealing with this contingency. Cells face an existential threat of swelling and bursting in a hypoosmotic environment; ie, an extracellular solution with lower osmolarity than the intracellular compartment. Some cells (eg, bacterial and plant) are partially protected mechanically by cell walls that mitigate osmotic stress, but the basic problem remains. Porelike channels that facilitate diffusive fluxes of inorganic ions and water across the cell membrane are thought to have originated to satisfy this fundamental requirement for osmotic regulation (Armstrong, 2005).

Indeed, modern versions of ancestral genes coding for K^+ channels and Cl^- channels appear to be present in all microorganisms (Hille, 2001). Ca^{2+} and Na^+ channels are ubiquitous in higher organisms, but they are also found in some microbes suggesting that these two classes of cation channels arose later in evolution along with the emergence of increasingly intelligent behavior. The ion-selectivity preference of channels at the origin of primordial life would be dictated by the ionic composition of the aqueous environment. Na^+ and Cl^- are the predominant inorganic ions of earth's salty oceans. The earliest cells needed Cl^- channels to permit the escape of intracellular Cl^- as required to maintain electroneutrality when anionic biomolecules such as nucleic acids, acidic metabolites, and proteins came to predominate over cationic biomolecules inside the cell. K^+-selective channels fulfill the need for a diffusive pore for a charge-compensating cation that counterbalances extracellular Na^+, as well as serving as a battery in the form of a K^+ diffusion potential, inside negative. Voltage-gated Ca^{2+} channels and intracellular Ca^{2+}-release channels arose to fulfill a major role in Ca^{2+}-dependent cellular signaling (eg, signal transduction, activation of contractile proteins, exocytotic vesicle fusion) (Cai, Wang, Patel, & Clapham, 2015). NaVs appeared most recently in the evolution of excitable cells of higher invertebrates and vertebrates with the emergence of electrical signaling of neurons and muscle cells (Hille, 2001). ATP-driven Ca^{2+} and Na^+ pumps were correspondingly required to exclude these latter ions from the cell interior and maintain the integrity of their signaling functions.

While all cells on planet earth appear to require ion channels, it is not inconceivable that other chemistries and molecular strategies for solving osmotic and ionic quandaries may have arisen in other worlds of life within the universe. One of the newest fields of biological science called astrobiology or exobiology leads the effort to seek evidence of extraterrestrial life. The number of known exoplanets orbiting a sunlike star at a Goldilock's distance compatible with the liquid state of water as identified by ongoing astronomical research is ever increasing. A major question is how to determine whether there is life on these distant planets. Was life on earth seeded by meteorites from distant galaxies that transported dormant cells encapsulated within robust spores? Or did it arise spontaneously here on earth through an extraordinary series of fortuitous chemical reactions? At the other end of the problem from the perspective of SETI (Search for Extraterrestrial Intelligence) research, various initiatives have attempted to detect nonrandom electromagnetic radiation from distant planets that would be likely to correspond to radio communications of intelligent alien civilizations (Shostak, 2015).

If such intelligent life does exist elsewhere in the universe, it is euphemistically a "no-brainer" to expect that it utilizes some kind of ion channels or biotransistors for the electrical activity expected to underlie higher cognitive functions of communication and critical reasoning. The question then becomes: what ions, what channels? Space explorations seeking to advance the frontier of astrobiology would be well advised to explore the inorganic chemistry of other planets for aqueous pools containing Na^+, K^+, Ca^{2+}, Cl^-, and other ions that could potentially mediate bioelectric signaling. In many lower invertebrates, electrical signaling within nervous systems is based on Ca^{2+}-action potentials. Excitable cells of certain plants with unusual electromechanical abilities exhibit slow Cl^--driven action potentials (Samejima & Sibaoka, 1982). Should we start taking bets on the preferred ion for alien action potentials? Na^+, anyone?

10. NaV ENDGAME

How much more research will be required before we can be reasonably certain to know everything important there is to know about the NaV channel? The discovery of tetrameric bacterial Na^+ channels and recent determinations of their crystal structures have succeeded in casting new light on the molecular underpinnings of ionic selectivity and unique features of channel gating (Payandeh & Minor, 2015). The next major breakthrough in structural understanding may come with the crystal structure of a pseudo-tetrameric NaV channel from a eukaryote. Beyond that, we can continue to expect the unexpected of NaV from diverse fields of research spanning cell physiology, neuroscience, neuropharmacology, molecular biophysics, and unnatural history.

REFERENCES

Ackrill, J. L. (1987). *A new Aristotle reader*. Princeton, New Jersey: Princeton University Press. From "On the Soul," Book I, (Chapter 4). Translation: D. W. Hamlyn (Clarendon Aristotle Series, 1968); Text: W. D. Ross (Oxford Classical Texts, 1956).

Ahern, C. A., Payandeh, J., Bosmans, F., & Chanda, B. (2016). The hitchhiker's guide to the voltage-gated sodium channel galaxy. *Journal of General Physiology, 147*, 1–24.

Armstrong, C. (2005). Channels and pumps early in evolution. In F. V. Sepulveda, & F. Bezanilla (Eds.), *Pumps, Transporters, and ion channels: Studies on their structure, function, and cell biology* (pp. 1–10). New York, NY: Kluwer Academic/Plenum Publishers.

Ashcroft, F. (2012). *The spark of life: Electricity in the human body*. New York: W. W. Norton & Co.

Bernstein, J. (1868). Über den zeitlichen Verlauf der negativen Schwankung des Nervenstromes. *Pflügers Archiv, 1*, 173–207 (Title translated as "On the time course of the negative variation of the nerve current.").

Bernstein, J. (1902). Untersuchungen zur Thermodynamik der bioelektischen Ströme. *Pflugers Archiv, 92*, 521–562 (English translation in: Boylan, J. W. (ed) 1971 Founders of experimental physiology. München: J. F. Lehmanns Verlag, pp. 258–299).

Bernstein, J. (1912). *Electrobiologie. Der Lehre von den elektrischen Vorgängen im Organismus auf moderner Grundlage dargesstellt.* Braunschweig: Vieweg & Sohn.

Bosmans, F., Martin-Eauclaire, M.-F., & Swartz, K. J. (2008). Deconstructing voltage sensor function and pharmacology in sodium channels. *Nature, 456*, 202–208.

Cai, X., Wang, X., Patel, S., & Clapham, D. E. (2015). Insights into the early evolution of animal calcium signaling machinery: a unicellular point of view. *Cell Calcium, 57*, 166–173.

Cestele, S., Qu, Y., Rogers, J. C., Rochat, H., Scheuer, T., & Catterall, W. A. (1998). Voltage sensor-trapping: enhanced activation of sodium channels by beta-scorpion toxin bound to the S3-S4 loop in domain II. *Neuron, 21*, 919–931.

Cole, K. S. (1949). Dynamic electrical characteristics of the squid axon membrane. *Arch Sci Physiol, 3*, 253–258.

Crick, F. (1988). *What mad pursuit: A personal view of scientific discovery*. New York, NY: Basic Books.

De Palma, A., & Pareti, G. (2011). Bernstein's long path to membrane theory: radical change and conservation in nineteenth-century German electrophysiology. *Journal of the History of Neuroscience Basic Clinical Perspectives, 20*, 306–337.

De Weer, P. (2000). A century of thinking about cell membranes. *Annual Review of Physiology, 62*, 919–926.

Derig, S. (2002). *Apollo's laboratory. (An essay on E. Du Bois-Reymond)*. The Virtual Laboratory. Max Planck Institute for the History of Science. http://vlp.mpiwg-berlin.mpg.de/essays/data/art8.

Dos Reis, A., Jr. (2009). Sigmund Freud (1856–1939) and Karl Koller (1857–1944) and the discovery of local anesthesia. *Revista Brasileira de Anestesioligia, 59*, 244–257.

Drenth, J. P. H., & Waxman, S. G. (2007). Mutations in sodium-channel gene SCN9A cause a spectrum of human genetic pain disorders. *Journal of Clinical Investigation, 117*, 3603–3609.

Einstein, A. (1905). On the movement of small particles suspended in a stationary liquid demanded by the molecular kinetics theory of heat. *Annals of Physics, 17*, 549–560. Republished in translation in Einstein, A., 1956, Investigations on the Theory of Brownian Movement. Dover Publications, New York, 1–18.

Finger, S., & Piccolino, M. (2011). *The shocking history of electric fishes: From ancient epochs to the birth of modern neurophysiology*. Oxford, UK: Oxford University Press.

Finkelstein, G. (2003). M. du Bois-Reymond goes to Paris. *British Journal for the History of Science, 36*, 261–300.

Finkelstein, G. (2013). *Emil du Bois-Reymond: Neuroscience, self, and society in nineteenth century Germany*. Cambridge, MA: The MIT Press.

Franks, N. P. (2006). Molecular targets underlying general anesthesia. *British Journal of Pharmacology, 147*, S72–S81.

Goldman, D. E. (1943). Potential, impedance, and rectification in membranes. *Journal of General Physiology, 27*, 37–60.

Graham, J., & Gerard, R. W. (1946). Membrane potentials and excitation of impaled single muscle fibers. *Journal of Cellular and Comparative Physiology, 28*, 99–117.

Halliday, D., & Resnick, R. (1978). *Physics: Parts I and II combined* (3rd ed.). New York, NY: John Wiley & Sons.

Hanck, D. A., Nikitina, E., McNulty, M. M., Fozzard, H. A., Lipkind, G. M., & Sheets, M. F. (2009). Using lidocaine and benzocaine to link sodium channel molecular conformations to state-dependent antiarrhythmic drug affinity. *Circulation Research, 105*, 492–499.

Hille, B. (2001). *Ion channels of excitable membranes* (3rd ed.). Sunderland, MA: Sinauer Associates.

Hodgkin, A. (1992). *Chance & design: Reminiscences of science in war and peace*. Cambridge, UK: Cambridge University Press.

Hodgkin, A. L., & Huxley, A. F. (1952a). Currents carried by sodium and potassium through the membrane of the giant axon of *Loligo*. *Journal of Physiology (London), 116*, 449–472.

Hodgkin, A. L., & Huxley, A. F. (1952b). The components of membrane conductance in the giant axon of Loligo. *Journal of Physiology (London), 116*, 473–496.

Hodgkin, A. L., & Huxley, A. F. (1952c). The dual effect of membrane potential on sodium conductance in the giant axon of Loligo. *Journal of Physiology (London), 116*, 497–506.

Hodgkin, A. L., & Huxley, A. F. (1952d). A quantitative description of membrane current and its application to conduction and excitation in nerve. *Journal of Physiology (London), 117*, 500–544.

Ito, M., Nagafuji, H., Okazawa, H., Yamakawa, K., Sugawara, T., Mazaki-Miyazaki, E., … Kaneko, S. (2002). Autosomal dominant epilepsy with febrile seizures plus missense mutations of the (Na^+)-channel alpha subunit gene, SCN1A. *Epilepsy Research, 48*, 15–23.

Jensen, H. S., Grunnet, M., & Bastlund, J. F. (2014). Therapeutic potential of NaV1.1 activators. *Trends in Pharmacological Sciences, 35*, 113–118.

Jiang, Y., Lee, A., Chen, J., Ruta, V., Cadene, M., Chait, B. T., & MacKinnon, R. (2003). X-ray structure of a voltage-dependent K^+ channel. *Nature, 423*, 33–41.

Jiang, Y., Ruta, V., Chen, J., Lee, A., & MacKinnon, R. (2003). The principle of gating charge movement in a voltage-dependent K^+ channel. *Nature, 423*, 42–48.

Jurkatt-Rott, K., Holzherr, B., Fauler, M., & Lehmann-Horn, F. (2010). Sodium channelopathies of skeletal muscle result from gain or loss of function. *Pflügers Archiv – European Journal of Physiology, 460*, 239–248.

Jurkat-Rott, K., & Lehmann-Horn. (2007). Genotype-phenotype correlation and therapeutic rationale in hyperkalemic periodic paralysis. *Neurotherapeutics: Journal of the American Society for Experimental Neuro Therapeutics, 4*, 216–224.

Kamiya, K., Kaneda, M., Suguwara, T., Mazaki, E., Okamura, N., Montal, M., … Yamakawa, K. (2004). A nonsense mutation of the sodium channel gene SCN2A in a patient with intractable epilepsy and mental decline. *Journal of Neuroscience: The Official Journal of the Society for Neuroscience, 24*, 2690–2698.

Kleinzeller, A. (1999). Charles Ernest Overton's concept of a cell membrane. *Current Topics in Membranes, 48*, 1–22.

Kremer, W. (2015). The strange afterlife of Einstein's brain. *BBC News Magazine*. April 18, 2015 http://www.bbc.com/news/magazine-32354300.

Kurzweil, R. (2006). *The singularity is near: When humans transcend biology*. New York, NY: Penguin Books.

Kwong, K., & Carr, M. J. (2015). Voltage-gated sodium channels. *Current Opinion in Pharmacology, 22*, 131–139.

Ling, G., & Gerard, R. W. (1949). The normal membrane potential of frog sartorius fibers. *Journal of Cellular and Comparative Physiology, 34*, 383–396.

Marmont, G. (1949). Studies on the axon membrane. I. A new method. *Journal of Cellular Physiology, 34*, 351–382.

McComas, A. J. (2011). *Galvani's spark: The story of the nerve impulse*. New York, NY: Oxford University Press.

McCulloch, W. S., & Pitts, W. (1943). A logical calculus of the ideas immanent in nervous activity. *Bulletin of Mathematical Biophysics, 5*, 115–133.

Moczydlowski, E., & Schild, L. (1994). Unitary properties of the batrachotoxin-trapped state of voltage-sensitive sodium channels. In C. Peracchia (Ed.), *Handbook of membrane channels: Molecular and cellular physiology* (pp. 137–160). Orlando, FL: Academic Press.

Moczydlowski, E. G. (2013). The molecular mystique of tetrodotoxin. *Toxicon, 63*, 165–183.

Narahashi, T., Moore, J. W., & Scott, W. R. (1964). Tetrodotoxin blockage of sodium conductance increase in lobster giant axons. *Journal of General Physiology, 47*, 965–974.
Noda, M., Shimizu, S., Tanabe, T., Takai, T., Kayano, T., Ikeda, T., ... Numa, S. (1984). Primary structure of *Electrophorus electricus* sodium channel deduced from cDNA sequence. *Nature, 312*, 121–127.
Noda, M., Takahashi, H., Tanabe, T., Toyosato, M., Kikyotani, S., Furutani, Y., ... Numa, S. (1983). Structural homology of *Torpedo californica* acetylcholine receptor subunits. *Nature, 302*, 528–532.
Ostwald, W. (1890). Elektrische Eigenschaften halbdurchlässiger Scheidewande. *Zeitschrifte für physikalische Chemie, 6*, 71–82.
Overton, E. (1902). Beiträge zur allegemeinen Muskel- und Nervenphysiologie. II. Mitteilung. Über die Unentbehrlichkeit von Natrium- (oder Lithium-) Ionen für den Contractionsact des Muskels. *Pflügers Archiv, 92*, 346–386.
Payandeh, J., & Minor, D. L., Jr. (2015). Bacterial voltage-gated sodium channels (BacNavs) from the soil, sea, and salt lakes enlighten molecular mechanisms of electrical signaling and pharmacology in the brain and heart. *Journal of Molecular Biology, 427*, 3–30.
Pera, M. (1992). *The ambiguous frog: The Galvani-Volta controversy on animal electricity*. Translated by Johnathan Mandelbaum. Princeton, NJ: Princeton University Press.
Piccinini, G. (2004). The first computational theory of mind and brain: a close look at McCulloch and Pitts's "logical" calculus of ideas immanent in nervous activity. *Synthese, 141*, 175–215.
Planck, M. (1890). Ueber die Potentialdifferenz zwischen zwei verdünnten Lösungen binärer Elektrolyte. *Ann Phys Chem Neue Folge, 40*, 561–576.
Priestly, J. (1794). *The history and present state of electricity with original experiments* (5th ed.) (London, UK: printed for J. Johnson et al. Reproduction from Harvard University Houghton Library. Gale ECCO Print Editions).
Rogers, J. C., Qu, Y., Tanada, T. N., Scheuer, T., & Catterall, W. A. (1996). Molecular determinants of high affinity binding of alpha-scorpion toxin and sea anemone toxin in the S3-S4 extracellular look in domain IV of the Na^+ channel alpha subunit. *Journal of Biological Chemistry, 271*, 15950–15962.
Rowe, A. H., Xiao, Y., Rowe, M. P., Cummins, T. R., & Zakon, H. H. (2013). Voltage-gated sodium channel in grasshopper mice defends against the bark scorpion toxin. *Science, 342*, 441–446.
Rudolph, J. A., Spier, S. J., Byrns, G., Rojas, C. V., Bernoco, D., & Hoffman, E. P. (1992). Periodic paralysis in quarter horses: a sodium channel mutation disseminated by selective breeding. *Nature Genetics, 2*, 144–147.
Sagan, C. (1980). *Cosmos. 2013*. New York, NY: Ballantine Books.
Samejima, M., & Sibaoka, T. (1982). Membrane potentials and resistances in excitable cells in the petiole and main pulvinus of *Mimosa pudica*. *Plant & Cell Physiology, 23*, 459–465.
Schrödinger, E. (1944). *What is life? The physical aspect of the living cell with mind and matter & autobiographical sketches*. Cambridge, UK: Cambridge University Press Edition 1992.
Schuetze, S. M. (1983). The discovery of the action potential. *Trends in Neurosciences, 6*, 164–168.
Seyfarth, E.-A. (2006). Julius Bernstein (1839–1917): pioneer neurobiologist and biophysicist. *Biological Cybermatics, 94*, 2–8.
Shcherbatyy, V., Carson, J., Yayalaoglu, M., Jäckle, K., Grabbe, F., Brockmeyer, M., ... Eichele, G. (2015). A digital atlas of ion channel expression patterns in the two-week old rat brain. *Neuroinformatics, 13*, 111–125.
Sherrington, C. (1940). *Man on his nature*. London, UK: Cambridge University Press.
Shostak, S. (2015). Searching for clever life. *Astrobiology, 15*, 949–950.
Sigworth, F. J. (2003). Life's transistors. *Nature, 423*, 21–22.
Turing, A. (1950). Computing machinery and intelligence. *Mind, 59*, 433–460.

Volta, A. (1800). On the electricity excited by the mere contact of conducting substances of different species. *Philosophical Transactions of the Royal Society, 90*, 403–431.
Von Neumann, J. (1958). *The computer and the brain. 3rd edition (2012).* New Haven, CT: Yale University Press.
Wang, D.-N., Steiglitz, H., Marden, J., & Tamm, L. K. (2013). Benjamin Franklin, Philadelphia's favorite son, was a membrane biophysicist. *Biophysical Journal, 104*, 287–291.
Wang, Q., Shen, J., Splawski, I., Atkinson, D., Li, Z., Robinson, J. L., ... Keating, M. T. (1995). SCN5A mutations associated with an inherited cardiac arrhythmia, long QT syndrome. *Cell, 80*, 805–811.
Wang, S. Y., & Wang, G. K. (2003). Voltage-gated sodium channels as primary targets of diverse lipid-soluble neurotoxins. *Cell Signaling, 15*, 151–159.
Watson, J. D., & Berry, A. (2003). *DNA: The secret of life*. New York, NY: Alfred A. Knopf.
Xiao, Y., Bingham, J. P., Zhu, W., Moczydlowski, E., Liang, S., & Cummins, T. R. (2008). Tarantula huwentoxin-IV inhibits neuronal sodium channels by binding to receptor site 4 and trapping the domain II voltage sensor in the closed configuration. *Journal of Biological Chemistry, 283*, 27300–27313.

SECTION TWO

Evolutionary Adaptations of Channels and Toxins

CHAPTER TWO

Biophysical Adaptations of Prokaryotic Voltage-Gated Sodium Channels

T.N. Vien* and P.G. DeCaen[§,¶,||,1]

*Tufts University, Boston, MA, United States
[§]Children's Hospital Boston, Boston, MA, United States
[¶]Harvard Medical School, Boston, MA, United States
[||]Northwestern University, Chicago, IL, United States
[1]Corresponding author: E-mail: Paul.DeCaen@childrens.harvard.edu

Contents

Abstract

This chapter describes the adaptive features found in voltage-gated sodium channels (Na_Vs) of prokaryotes and eukaryotes. These two families are distinct, having diverged early in evolutionary history but maintain a surprising degree of convergence in function. While prokaryotic Na_Vs are required for growth and motility, eukaryotic Na_Vs selectively conduct fast electrical currents for short- and long-range signaling across cell membranes in mammalian organs. Current interest in prokaryotic Na_Vs is stoked by their resolved high-resolution structures and functional features which are reminiscent of eukaryotic Na_Vs. In this chapter, comparisons between eukaryotic and prokaryotic

Current Topics in Membranes, Volume 78
ISSN 1063-5823
http://dx.doi.org/10.1016/bs.ctm.2015.12.003

Na_Vs are made to highlight the shared and unique aspects of ion selectivity, voltage sensitivity, and pharmacology. Examples of prokaryotic and eukaryotic Na_V convergent evolution will be discussed within the context of their structural features.

1. INTRODUCTION

The earliest forms of life likely evolved in conditions analogous to oceanic thermal vents. Progenitors of these organisms resembled modern day anaerobic thermohalophilic marine archaea (Cooper, 2000). This ecological niche is a high salt ($\approx$1 OsM) and temperature (>41°C) environment. To survive under these extreme osmotic and atmospheric pressures, primitive prokaryotic organisms possessed at least one cell wall which provides rigidity to the internal lipid bilayer (comprised of special lipids) and adapted ion channels for chemiosmosis of salt ions. In all prokaryotes, chloride and potassium channels evolved first and are essential for chemiosmosis (Fig. 1A) (Anderson & Greenberg, 2001). Uncontrolled influx of sodium or calcium would quickly reach saturation due to an abundance of these ions in seawater (550 mM and 10 mM, respectively) and the hyperpolarized potential ($\Psi = -140$ to -180 mV) of the prokaryotic membrane (Armstrong, 2015). Yet voltage-gated sodium channels (Na_Vs) from terrestrial and marine halophilic bacteria have been identified and characterized, demonstrating high selectivity for sodium ions, a feature shared by their eukaryotic counterparts (Fig. 1B) (Ren et al., 2001). These prokaryotic organisms utilize Na^+ to drive flagella stators and have adapted Na^+ transporters and pumps which reduce Na^+ accumulation (Fujinami, Terahara, Krulwich, & Ito, 2009; Ito et al., 2004). Discussed in other chapters of this book, human Na_Vs are essential for a variety of physiological processes including neuronal and cardiac electrical transmission and are targeted by many therapeutically useful drugs. Thus, how Na_Vs can rapidly and selectively conduct Na^+ ions in response to changes in the electrical membrane potential is a crucial biophysical question rooted in the evolution of life on earth. In this chapter, we will discuss the structural and biophysical determinants of eukaryotic and prokaryotic Na_V channels and highlight the adaptive features over the course of evolution which have enabled their survival in extreme environments. We will discuss recently solved prokaryotic Na_V structures and evaluate how these findings extend our understanding of more complex eukaryotic Na_V proteins. In the Section 4, we will compare and contrast the pharmacophore of eukaryotic and prokaryotic Na_V antagonists and determine which receptor sites are likely to be shared across the two families.

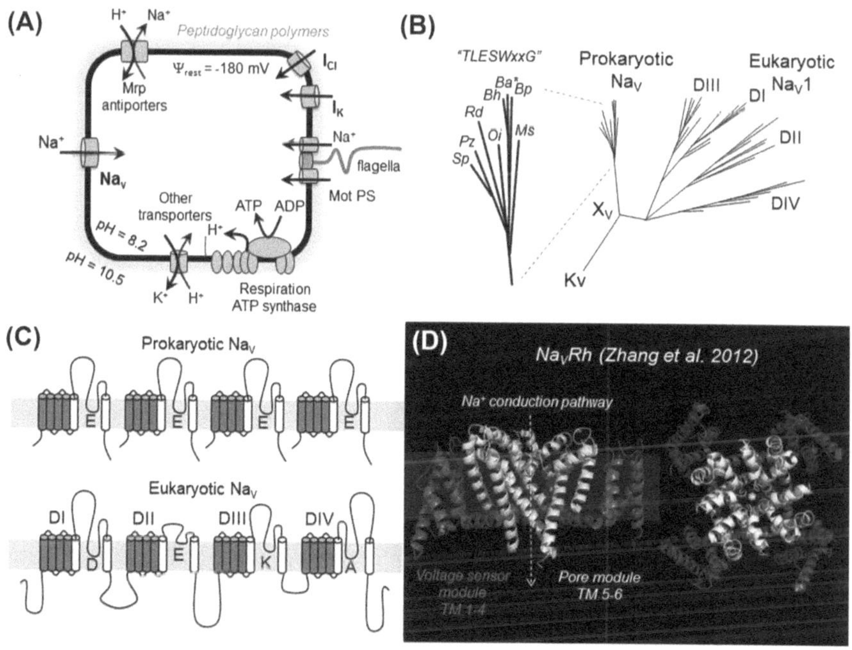

Figure 1 ***Relatedness and defining structural features among prokaryotic and eukaryotic Na_V families.*** (A) Prototypic cation cycle of a thermohalophilic prokaryote depicting the membrane proteins involved in the cation cycle. The cation/proton antiporters, including Mrp antiporters, catalyze net proton accumulation in the cytoplasm in cells that are extruding protons during respiration. Na^+ reentry in support of pH homeostasis is achieved by Na^+ solute symporters and through members of the prokaryotic Na_V channel family. (B) Phylogenetic analysis of the prokaryotic and eukaryotic Na_Vs using methods *(Adapted from Liebeskind, B.J., Hillis, D.M., & Zakon, H.H. (2013). Independent acquisition of sodium selectivity in bacterial and animal sodium channels. Current Biology: CB, 23, R948–R949.* http://dx.doi.org/10.1016/j.cub.2013.09.025.*)*. (C) Transmembrane topology of prokaryotic and eukaryotic Na_Vs. Transmembrane helices that compose the voltage-sensing model are represented as gray cylinders while those from the pore module are in white. (D) Transmembrane and extracellular views of the Na_VRh structure (Zhang et al., 2012).

1.1 Origins of Na_V channels

Potassium leak and voltage-dependent K^+ (Kv) channels appeared 3 billion years ago in bacteria and are found in all organisms (Anderson & Greenberg, 2001; Strong, Chandy, & Gutman, 1993). Kv channels are the oldest of the family of ion-permeating channels whose basic structure is a protein of six transmembrane helices (TM1-6) which associate as tetramers to form a channel. At some point in early prokaryotic evolution, a novel singular domain channel of unknown ion selectivity likely evolved from common ancestor shared by Kv channels (Fig. 1B) (Strong et al., 1993). Based on phylogenetic analysis, the lineage of prokaryotic Na_V likely diverged from a progenitor of eukaryotic Na_Vs and voltage-gated calcium channels (Ca_Vs) (Liebeskind, Hillis, & Zakon, 2013; Zakon, 2012). At this time, the prokaryotic gene duplicated, giving rise to a protein with two domains which dimerize to form a four-domain structure (Liebeskind, Hillis, & Zakon, 2012). The Ca^{2+} and Na^+ permeable eukaryotic two-pore channel family that localizes in endosome and lysosome compartments is proposed to be related to these prokaryotic genes (Galione et al., 2009). A second round of gene duplication occurred creating four domain channels, which provided the topological blueprint for eukaryotic Na_Vs (and Ca_Vs) (Cai, 2012). Thus analysis of the pore-forming sequences from eukaryotic genes led to the proposal that Na_Vs and Ca_Vs are derived from a common progenitor (Yu & Catterall, 2004). When considering that prokaryotic and eukaryotic Na_V channels diverged early in evolution, it should be expected that differences are present between the two channel families. In the Subsections 1.2-1.3, we will identify which features of eukaryotic and prokaryotic Na_V function differ (or are likely to) especially as they relate to recent advances in the structural biology of prokaryotic Na_Vs.

1.2 Topology and structural elements of prokaryotic and eukaryotic Na_V

In eukaryotes, the alpha subunits of Na_Vs and Ca_Vs are four-domain pseudo-heterotetramers where each domain contains 6 transmembrane helices for a total of 24 (Fig. 1C) mammalian alpha Na_Vs and Ca_Vs complex with auxiliary subunits. Expression of the Na_V alpha subunit is sufficient to generate measurable Na^+ current, but coexpression with beta subunits enhances current density and hypopolarizes the voltage dependence of gating (Patino & Isom, 2010) (see Chapter 11: Developmental and Regulatory Functions of Na^+ Channel Beta Subunits for more details). The alpha subunits of Ca_Vs complex with a unique intracellular beta subunit, as well as a membrane-associated alpha2delta disulfide-linked complex, and a transmembrane gamma subunit, which regulate current density and modulates channel function (Catterall, 2011). These large (260 kDa) and complex (multiple subunits with heteromeric stoichiometry) transmembrane proteins have posed a major challenge for structural biologists. Prokaryotic Na_Vs have a single domain containing a voltage-sensing module (TM1-4) and pore-forming module (TM5-6) which oligomerize as a homotetramer (like Kv channels) which makes it a more tractable target than mammalian Na_Vs for structural analysis using X-ray crystallography (Ren et al., 2001). Each subunit or monomer of prokaryotic Na_Vs is viewed as analogous to the four domains of eukaryotic Na_Vs (Scheuer, 2014; Tang et al., 2014). Thus, the structure and biophysical features of prokaryotic Na_Vs have generated considerable interest. To date, three full-length (Na_VCt, Na_VAb, and Na_VRh) (Payandeh, Gamal El-Din, Scheuer, Zheng, & Catterall, 2012; Payandeh, Scheuer, Zheng, & Catterall, 2011; Tsai et al., 2013; Zhang et al., 2012) and two pore-only (Na_VMs and Na_VAe) (McCusker et al., 2012; Shaya et al., 2014) prokaryotic Na_V structures have been resolved with resolutions up to 3 Å. All full-length structures demonstrate that four identical subunits form a functional channel where the first four transmembrane segments form the voltage sensor module (VSM) and flank pore module (PM) which assembles along a central ion conducting pathway (Fig. 1D). Some of the structural features responsible for voltage dependence and ion selectivity will be discussed in the Sections 1–4 but usually in the context of adaptations. There is one eukaryotic Na_V structure available from the electric eel *Electrophorus electricus* determined by single particle cryo-electron microscopy (Sato et al., 2001). Although, the structure is at low resolution (19 Å), some coarse features of the folded protein were resolved including a bell-shaped outer surface and several inner cavities connected to four small holes and eight orifices close to the extracellular and cytoplasmic membrane surfaces.

Considering the recent advances that in single particle cryo-electron microscopy which enables near atomic resolution structure (<4 Å) of integral membrane proteins in a manner not requiring crystals, it seems likely that refined structures of multiunit eukaryotic Na_V1 (and Ca_Vs) will soon be within reach (Liao, Cao, Julius, & Cheng, 2014). Readers interested in further details of the structural features of prokaryotic Na_Vs solved using X-ray crystallography are encouraged to view several comprehensive reviews (Catterall & Zheng, 2015; Payandeh & Minor, 2015).

1.3 Proposed function of Na_Vs in prokaryotes

While the need for K^+ and Cl^- channels in bacteria membranes for osmotic regulation is forthright, the need of a Na^+ channel is not. Thus the discovery of the prokaryotic Na_Vs family was unexpected (Ren et al., 2001). Most bacteria require the ion-motive forces provided by H^+ or Na^+ ions that move through the MotAB or MotPS stator to power the rotary motor proteins to drive the motion of the flagella (Fig. 1A) (Fujinami et al., 2009; Ito et al., 2004). In soda lakes and other extremely H^+-poor (pH 9–12; 1 nM–1 pM [H^+]) and Na^+-rich environments (up to 500 mM), alkaliphilic prokaryotes couple motility and ion transporters to Na^+ rather than H^+. Alkaliphilic bacteria growth and motility is impaired when extracellular sodium is removed or if the gene encoding Na_V is ablated. Thus it appears that a sodium conductance(s) was adapted in early prokaryotes to provide a metabolic and charge substrate to support cell division and drive flagella-mediated locomotion. Given that eukaryotic and prokaryotic Na_Vs have unique structural topology and different physiological functions, what can aspects of these divergent families are shared? Molecular determinates and adaptive selectivity in Na_Vs will be compared in the Section 2. The mechanisms and conditions which mediate Na_V opening and inactivation are addressed in the Section 3. In the Section 4, eukaryotic and prokaryotic Na_V pharmacology will be compared in the light of recent structural advances in the field.

2. ADAPTATIONS IN CATION SELECTIVITY IN Na_V CHANNELS

2.1 Marine conditions remain influential in human Na_V and Ca_V channels

As discussed in the introduction, the eukaryotic Na_Vs and calcium channels (Ca_Vs) likely share a common progenitor that evolved in conditions analogous to marine thermal vents. Evidence of this primordial marine environment from which these channels evolved is still present in human Na_V and Ca_Vs and is reflected in the ion concentration by which their conduction is optimized (B_{max}, see Fig. 2A). Human Na_V1.1 conducts sodium with the highest efficiency at 454 mM, which approximates 470 mM sodium found in seawater (specifically, "salinity 35" which comprises the saline content found in 95% of global oceans) (Gruber & Sarmiento, 2006). Members of the prokaryotic sodium channels (NaChBac, Na_VMs, and NS_VBa) have the same efficiency ($B_{max} = 492–534$ mM [Na^+], Fig. 2A). Analysis comparing domains of bacteria and mammalian Na_V channels indicates that Na^+ selectivity was independently acquired in these channel families (Liebeskind et al., 2013). These data suggest that both acquisitions occurred in similar sodium conditions. The R-type hCa_V2.1 channel can conduct both Na^+ and Ca^{2+} ions but exhibits a strong selectivity for Ca^{2+} ($P_{Ca}/P_{Na} < 100$). Remarkably, the B_{max} for human Ca_V2.1 is 9 mM Ca^{2+} and 598 mM Na^+ which reflects the approximate conditions for both cations in seawater ([Ca^{2+}] = 12 mM, Fig. 2A). Like Ca_V2.1, the B_{max} for Ca_V3.2 is 20 mM and approximates the Ca^{2+} concentration of seawater but does not conduct Na^+

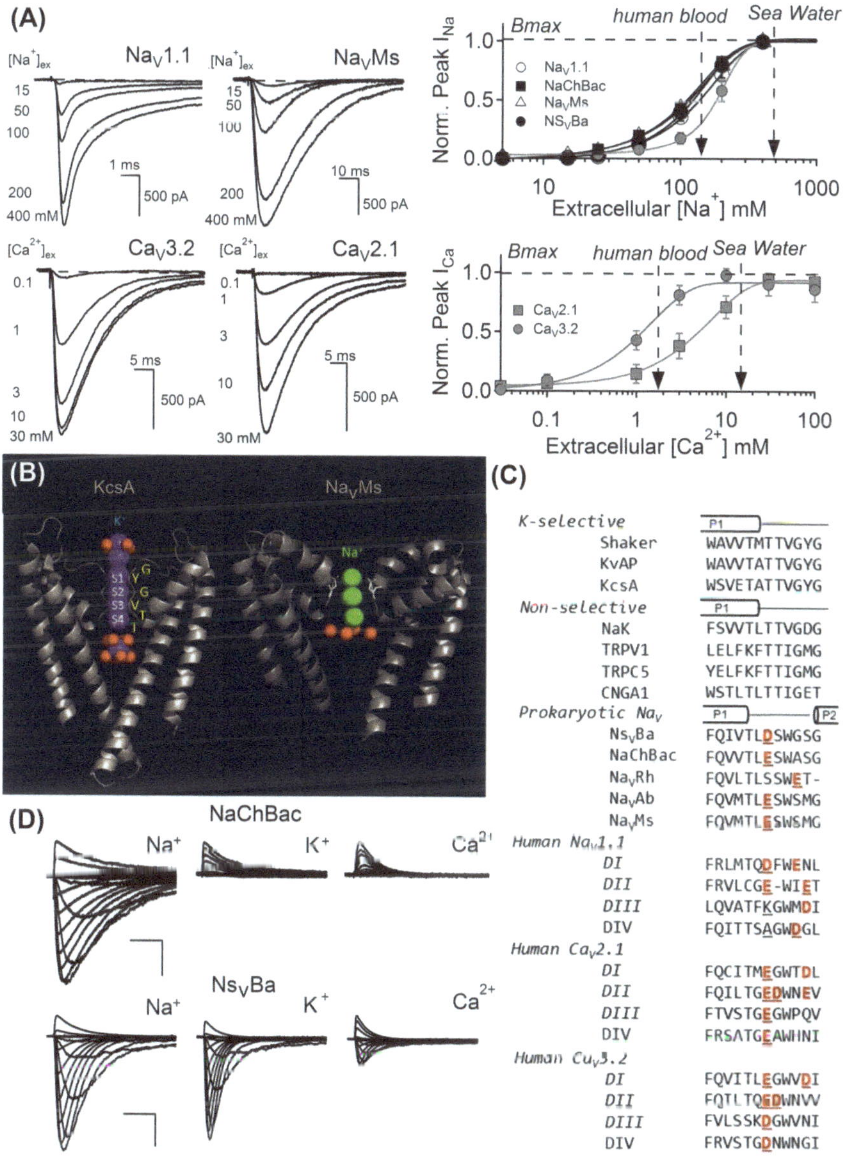

Figure 2 ***Ion selectivity among prokaryotic and eukaryotic Na_Vs.*** (A) Left, Na^+ or Ca^{2+} currents (human neuronal $Na_V1.1$; Na_VMs prokaryotic Na_V from *Magnetococcus marinus*; human P/Q type $Ca_V2.1$ and human T-type $Ca_V3.2$.) recorded under increasing cation concentrations. Right, Resulting saturation curves are plotted to estimate B_{Max}. (B) Transmembrane views of the KcsA K^+ channel (Zhou, Morais-Cabral, Kaufman, & MacKinnon, 2001) and Na_VMs (McCusker et al., 2012). (C) An alignment of the selectivity filters from various ion channels. (D) Voltage-dependent currents generated from NaChBac and NS_VBa prokaryotic channels. NaChBac only generates inward Na^+ currents but not K^+ and Ca^{2+} because it is a sodium selective channel, whereas the nonselective channel NS_VBa can conduct all three cations.

ions as found in $Ca_V2.1$. When considering that in mammals the blood concentrations of Na^+ and Ca^{2+} are 145 mM and 1.8 mM, respectively, these channels do not conduct ions at maximum efficiency in the environment which they currently reside. Rather, their B_{max} reflects the likely marine environment in which their cation selectivity evolved. Evidently, the level of selectivity and efficiency is sufficient for the eukaryotic Na_V and Ca_V to carry out the significant functions in excitable cells. Additionally, it is not surprising that many molecular dynamic simulation conditions required 500 mM of Na^+ to sample enough ion conduction events, owing to the weak association energy (3−4 kcal/mol as calculated from the apparent Kd) of sodium and the Na_V filter (Ke, Timin, & Stary-Weinzinger, 2014; Ulmschneider et al., 2013; Zhang et al., 2013). How the Na_V filter discriminates Na^+ among other ions available (termed "ion selectivity") and examples of adaptations to Na_V selectivity are discussed in the Subsection 2.2.

2.2 Convergent evolution of Na^+ selectivity in prokaryotic and eukaryotic Na_Vs

The first prokaryotic Na_V, cloned from *Bacillus halodurans C-125*, was called NaChBac (Na$^+$ Channel of Bacteria, also called Na_VBh) (Ren et al., 2001). This channel exhibited high Na^+ selectivity ($P_{Na}/P_K = 171$) as observed in eukaryotic Na_Vs. Since then, all full-length channels exhibit Na^+ selectivity (Irie et al., 2010; Ito et al., 2004; Koishi et al., 2004; Ren et al., 2001; Ulmschneider et al., 2013) with the exception of a homologue cloned from the alkaliphilic prokaryote, *Bacillus alcalophilus AV* called NS_VBa (Nonselective voltage-gated *Bacillus alcalophilus*) (Fig. 2D) (DeCaen, Takahashi, Krulwich, Ito, & Clapham, 2014). Among the first alkaliphilic extremophiles described in the literature, the gram-positive, rod-shaped *B. alcalophilus AV* (Vedder, 1934) was initially isolated from human feces and thus are part of the human microbiome. Growth and motility of *B. alcalophilus AV* is better supported in K^+ rather than Na^+ conditions (Terahara, Sano, & Ito, 2012). In *B. alcalophilus AV*, either ion can drive the flagellar motor, and a single mutation in the MotS protein was identified that converted the naturally nonselective MotPS (Na^+ or K^+) stator into a Na^+-selective stator. Presumably, prokaryotic Na_V channels provide a source of cations that drives the stators and maintains ion homeostasis required for growth. NS_VBa conducts Na^+, K^+, and Ca^{2+}, as opposed to Na^+ alone as found in other Na_V family members cloned from soil and marine dwelling alkaliphilic bacteria. Although the human GI concentration of Na^+ is relatively low (100−120 mM) and high for K^+ (12−20 mM), it is unclear if this was the environment or the selective pressure in which K^+ was coopted by *B. alcalophilus* MotPS and Na_V proteins.

Although the sequence of NS_VBa is homologous to other prokaryotic Na_V channels, the selectivity filter sequence (TLDSWGSG) deviates at the high field strength site ($Site_{HFS}$) with an aspartate residue replacing glutamate (D186) and a flexible glycine at an extracellular position (G189). When aligned, G189 and G191 form a "GxG motif" commonly found in K^+-selective and nonselective ion channels (eg, several members of the transient receptor potential (TRP) and cyclic nucleotide gated channel families; Fig. 2B and C). Instead of reflecting a nonselective progenitor of voltage-gated ion channels, phylogenetic analysis suggests that this change represents a convergent adaptation to the selectivity filter. Mutagenesis of the NaChBac and NS_VBa channel genes demonstrates that the prokaryotic selectivity filter is adaptable to support nonselective (K^+, Na^+, and Ca^{2+}) and Ca^{2+}- or Na^+-selective conductance with one to three nucleic acid substitutions (DeCaen et al., 2014). The heterotetrameric Na_V channel homologs from the sea anemone *Nematostella vectensis* ($NvNa_V2.1$) contains a noncanonical selectivity site (DEEA) that is nonselective among K^+ and Na^+ ions (Gur Barzilai et al., 2012). Thus it appears that the homomeric prokaryotic NS_VBa and heteromeric eukaryotic $NvNa_V2.1$ channels are examples of evolution of Na^+ selectivity to nonselectivity that occurred independently within prokaryotes and metazoans. These findings highlight the high degree of adaptive versatility of Na_V selectivity filters and the importance of Na^+ selectivity in biological systems ranging from prokaryotes to advanced eukaryotes.

2.3 Structural basis for Na^+ conductance and selectivity in Na_V channels

While the structural and biophysical determinants of K^+ selectivity are well described (Alam & Jiang, 2009; Doyle et al., 1998; Jiang et al., 2002, 2003; Yellen, 1987), those of Na^+ and Ca^{2+} remain unresolved. The selectivity for Na^+ ions in vertebrate Na_V channels is attributed to an asymmetric ring of four amino acids (Asp, Glu, Lys, and Ala: DEKA) contributed by each of the pore-lining loops of the four domains (Sun, Favre, Schild, & Moczydlowski, 1997). Ca_Vs are thought to achieve Ca^{2+} selectivity by a symmetric ring of four glutamate residues (EEEE), each contributed by one domain of the polypeptide (Hess, Lansman, & Tsien, 1986; Kim, Morii, Sun, Imoto, & Mori, 1993). The EEEE ring is present in all Ca_V1/Ca_V2 channels and EEDD is present in all Ca_V3 channels found in every known animal species to date, down to the single cell choanoflagellates (Stephens, Guan, Zhorov,

& Spafford, 2015). In contrast, prokaryotic Na_Vs achieve Na^+ selectivity from an apparent fourfold symmetry of acidic residues EEEE, each from a subunit in the tetramer. In previous studies, the Clapham laboratory demonstrated that the Na^+-selective NaChBac (TLESWASG) filter could be converted into one that prefers the divalents Ba^{2+} and Ca^{2+} by introducing acidic residues into three positions in the filter sequence (TLDDWADG) (Yue, Navarro, Ren, Ramos, & Clapham, 2002). This filter sequence also was grafted into the Na_VAb channel (called Ca_VAb) which rendered the Na_VAb channel Ca^{2+}-selective, and the high-resolution structure was determined (Tang et al., 2014). These studies identified three sites occupied by Ca^{2+} ions within the mutated selectivity filter. To date, no structure of a Na_V has been published where the locations of Na^+ ions have been resolved in sufficient detail to describe the occupancy sites. Nonetheless, Na^+ occupancy sites ($Site_{HTS}$, $Site_{CEN}$, $Site_{IN}$) within the Na_VAb structure has been proposed based on the locations of the side chain and backbone carbonyl oxygen atoms that line the filter and imitate a square array of four water molecules (Fig. 2B and C) (Payandeh et al., 2011). The selectivity filter is supported by two pore helices (P1 and P2) from each subunit, as opposed to a singular helix found in Kv channels. P1 and P2 direct an electronegative dipole directed at the filter, and the four acidic residues create an electronegative region that attracts cations. Microsecond molecular dynamics simulations of Na^+ conduction through the selectivity filter of Na_VAb, Na_VRh, and Na_VMs suggest a "knock-on" mechanism of Na^+ permeation where occupancy of extracellular sites influence the displacement of a second Na^+ occupying intracellular sites generating a net flux of sodium into the cell (Ulmschneider et al., 2013; Zhang et al., 2013) (see Chapters 14: Computational studies of permeation and selectivity in voltage-gated Na^+ channels and 17: Learning about function and modulation from simulations of bacterial sodium channels for more details). From the structure of the prokaryotic Na_V channels, selectivity for Na^+ among cations is believed to be achieved (in part) by favoring small monovalent cation (Na^+ or Li^+) occupancy at the high field strength site ($Site_{HTS}$) (Boiteux, Vorobyov, & Allen, 2014; Chakrabarti et al., 2013) as initially proposed in Eisenman's selectivity theory (Eisenman, 1961). Some features from these models of prokaryotic Na_V sodium conduction and selectivity were initially postulated in Hille's single-ion pore model (Hille, 1972), where Na^+ is partly dehydrated by an anion residue (like $Site_{HTS}$) found in the filter during conduction. However, the molecular arrangement that enables Na^+ selectivity in prokaryotes might differ from human Na_Vs. As mentioned previously, eukaryotic Na_V1 channel selectivity filter includes a protonatable lysine (K) from the third repeat and a neutral alanine (A) from the fourth repeat (the DEKA ring). Thus, it is not clear if structural features that determine ion selectivity in homotetrameric prokaryotic Na_Vs reflect the heterotetrameric eukaryotic Na_V1s.

3. ADAPTATIONS IN VOLTAGE-DEPENDENT PROCESSES IN THE PROKARYOTIC Na_V FAMILY

3.1 Prokaryotic Na_Vs as a model of voltage-dependent activation

Voltage-gated ion channels are characterized by their ability to turn on and off the flow of ionic current by opening and closing their conductive pore in response to changes in membrane potential. The opening and closing of the pore is mechanically linked to outward movement of highly conserved positively charged residues (called "gating charges," R1–R4) found in the fourth transmembrane helix (TM4) of voltage-gated ion channels. Movement of the gating charges precedes the opening and flow of the ionic current and can be measured in Na_V and Kv channels and are calculated to be equal to $\approx 3-4e^-$ per VSM (Armstrong & Bezanilla, 1974; Keynes, Rojas, & Rudy, 1974; Kuzmenkin, Bezanilla, & Correa, 2004). In addition to the gating charges, there are highly conserved negatively charged amino acid residues in transmembrane helices (TM1-3) which have been proposed to serve as counter charges to stabilize the TM4 gating charges (called "helical screw" or "sliding helix" models of voltage sensor motion) (Catterall, 1986; Guy & Seetharamulu, 1986). Prior to the solved structure of prokaryotic Na_V channels, using a disulfide cross-linking method, over 100 mutations of the NaChBac channel were tested and demonstrated discrete and sequential voltage-dependent ion-pair interactions (DeCaen, Yarov-Yarovoy, Sharp, Scheuer, & Catterall, 2009; DeCaen, Yarov-Yarovoy, Zhao, Scheuer, & Catterall, 2008; DeCaen, Yarov-Yarovoy, Scheuer, & Catterall, 2011; Yarov-Yarovoy et al., 2012). Seven disulfide interactions were resolved as a function of time and membrane potential. The differences in onset of interaction led to the proposal that the VSM has at least two activated states (Fig. 3A; activated state 1 and activated state 2) (Yarov-Yarovoy et al., 2012). In addition, these results predicted that the secondary structure was in a partial 3_{10} helix rather than an alpha helix. A 3_{10}-helical conformation during activation would allow outward movement of gating charges and exchange of ion-pair partners without helical rotation, as opposed to the "sliding helix" or "helical screw" models of voltage sensing which required 30–90° turn. In conjunction with structural modeling of the disulfide locking data, TM4 is expected to move in and out of the plane of the membrane

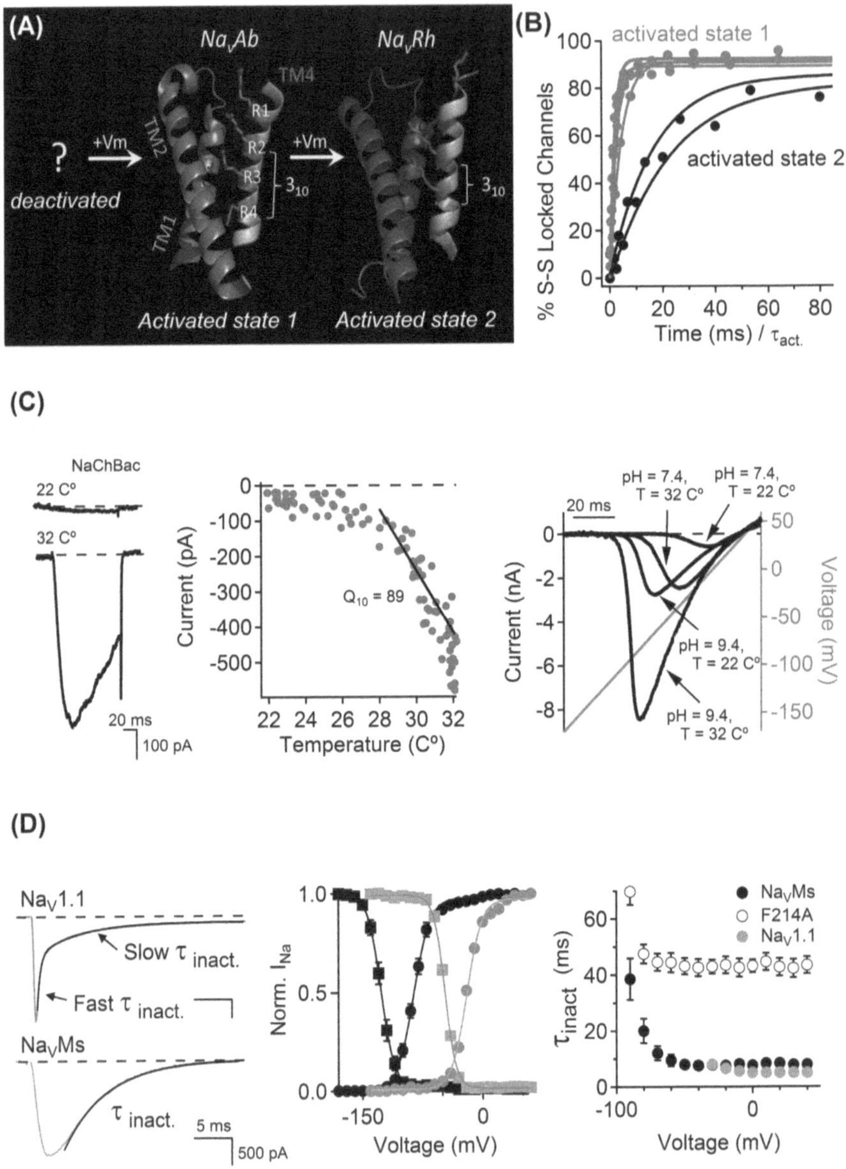
(A)
NaVAb
NaVRh
TM4
TM2
TM1
R1
R2
R3
R4
3_{10}
+Vm
?
deactivated
Activated state 1
Activated state 2
(B)
activated state 1
activated state 2
% S-S Locked Channels
Time (ms) / $\tau_{act.}$
(C)
NaChBac
22 C°
32 C°
20 ms
100 pA
Current (pA)
Temperature (C°)
Q_{10} = 89
pH = 7.4, T = 32 C°
pH = 7.4, T = 22 C°
pH = 9.4, T = 22 C°
pH = 9.4, T = 32 C°
Current (nA)
Voltage (mV)
(D)
NaV1.1
Slow τ inact.
Fast τ inact.
NaVMs
τ inact.
5 ms
500 pA
Norm. I_{Na}
Voltage (mV)
NaVMs
F214A
NaV1.1
τ_{inact} (ms)

of 8–13 Å. Subsequently, the structures of Na_VAb and Na_VRh provided two snapshots of the activated forms of the VSM, where the TM4 has translated 3 Å between the two states (Payandeh et al., 2011; Zhang et al., 2012). In a separate study, cysteine substitutions of gating charges generated voltage-dependent "omega currents" emanating from the VSM of NaChBac (Gamal El-Din, Scheuer, & Catterall, 2014). Omega currents (also called "gating pore currents") are independent from currents conducted by the central pore and are caused by mutations in the VSM which create an aberrant leak ion conduction pathway. Indeed, omega currents have been observed in mutations which encode for gating charges of skeletal muscle Na_V1.4 channel in inherited hypokalemic periodic paralysis (Sokolov, Scheuer, & Catterall, 2007). The stepwise formation of the omega currents in the

Figure 3 ***Prokaryotic Na_Vs as models of voltage-dependent gating and inactivation.*** (A) The voltage sensor modules (VSMs) from prokaryotic Na_Vs captured in two activated states (Payandeh et al., 2011; Zhang et al., 2012). (B) Onset of disulfide interactions between cysteine pairs engineered to track the movement of gating charges (R1–R4) within VSM of the NaChBac channel in the whole cell patch clamp assay *(Summarized from DeCaen, P.G., Yarov-Yarovoy, V., Zhao, Y., Scheuer, T., & Catterall, W.A. (2008). Disulfide locking a sodium channel voltage sensor reveals ion pair formation during activation. Proceedings of the National Academy of Sciences of the United States of America, 105, 15142–15147.* http://dx.doi.org/10.1073/pnas.0806486105; *DeCaen, P.G., Yarov-Yarovoy, V., Sharp, E.M., Scheuer, T., & Catterall, W.A. (2009). Sequential formation of ion pairs during activation of a sodium channel voltage sensor. Proceedings of the National Academy of Sciences of the United States of America, 106, 22498–22503.* http://dx.doi.org/10.1073/pnas.0912307106; *DeCaen, P.G., Yarov-Yarovoy, V., Scheuer, T., & Catterall, W.A. (2011). Gating charge interactions with the S1 segment during activation of a Na+ channel voltage sensor. Proceedings of the National Academy of Sciences of the United States of America, 108, 18825–18830.* http://dx.doi.org/10.1073/pnas.1116449108; *Yarov-Yarovoy, V., DeCaen, P.G., Westenbroek, R.E., Pan, C.Y., Scheuer, T., Baker, D., & Catterall, W.A. (2012). Structural basis for gating charge movement in the voltage sensor of a sodium channel. Proceedings of the National Academy of Sciences of the United States of America, 109, E93–E102.* http://dx.doi.org/10.1073/pnas.1118434109.*)*. Two groups of interacting pairs approximate the activated states found in the Na_VAb and Na_VRh crystal structures. (C) Left, steady-state effects of temperature of NaChBac sodium currents activated by −60 mV depolarization from a −140 mV holding potential. Middle, the resulting temperature coefficient (Q_{10}) estimated by the linear slope of the temperature-sodium current relationship measured between 28 and 32°C. Right, alkaline pH and temperature independently or additively shift the voltage dependence of activation toward negative potentials. (D) Inactivation processes in Na_VMs and hNa_V1.1 channels. Left, Na_V1.1 currents inactivate on two times courses whereas prokaryotic Na_Vs inactivate on a singular time course. Middle, the voltage dependence of activation (*circles*) and inactivation (*squares*) for Na_VMs and Na_V1.1. Right, The voltage dependence of inactivation from Na_V1.1 (slow $\tau_{inact.}$), Na_VMs and the Na_VMs mutant channel F214A.

NaChBac channel supports a mechanism where outward and sequential interaction of TM4 gating charges and negative charged residues found within the VSM (Gamal El-Din et al., 2014). Since the membrane potential of neurons and bacteria ranges between −180 and −90 mV and these structures were crystallized at zero millivolts, additional translation of TM4 is expected for the deactivated state. Thus with the structure of a deactivated Na_V VSM is at large and a complete analysis of the VSM conformations is incomplete.

3.2 Voltage dependence of prokaryotic Na_Vs is modulated by heat and proton

Voltage-gated ion channels respond to changes in membrane potential by opening and closing their ion conducting pathway across the cell membrane in a process called "gating." Gating is reversible, occurs at the millisecond time scale, and is sensitive to a narrow range of voltages. In excitable eukaryotic tissues such as nerves and muscle, the half-activation threshold ($V_{1/2}$) of Na_Vs is within ~80 mV of their cellular resting membrane potential (Ψ_{rest} ~ −60 to −90 mV). This appears to contrast with the 140 mV difference between the Ψ_{rest} and $V_{1/2}$ of alkaliphilic *Bacillus* (see legend of Fig. 3). With such a large difference between Ψ_{rest} and $V_{1/2}$, under what condition will the prokaryotic membrane potential be depolarized enough to activate the Na_V channel? Since these channel evolved in high temperature and pH conditions, we tested these conditions on the voltage dependence of activation of *Bacillus* and human Na_Vs. First, we tested temperature on steady-state currents by depolarizing the cell to potentials below the threshold of activation (−80 mV for NaChBac), a method previously used to assess the Shaker Kv channel temperature sensitivity (Chowdhury, Jarecki, & Chanda, 2014). Here, current amplitudes that were extremely small (<60 pA) at 22°C became ≈100 times larger at 32°C (Fig. 3C, left). The estimated 10-degree temperature coefficient (Q_{10}) using this method is high (≈90), placing *Bacillus* Na_Vs alongside other temperature sensitive channels such as some members of the TRP family (Patapoutian, Peier, Story, & Viswanath, 2003). However, when measuring the peak current during a voltage ramp, the Q_{10} is much smaller (3.5−5) for *Bacillus* Na_Vs. The reason for this apparent difference is that temperature can shift prokaryotic Na_V $V_{1/2}$'s by approximately −45 mV, allowing the channels to open at potentials where they would otherwise be closed (Fig. 3C, right). A similar effect on the voltage dependence occurs when pH is increased from 7.4 to 9.4. Thus, when combining alkaline pH and heat, the effects on $V_{1/2}$ is an additive −100 mV shift in voltage dependence. If the sodium current sensitivity is extrapolated to 60°C (as found near marine thermal vents), the voltage dependence would likely be well within 80 mV of the most negative bacterial resting membrane potential. Thus the evolutionary origins of prokaryotic Na_Vs can be characterized by the modulation voltage dependence by heat and alkaline pH, reflecting the thermohalophilic environment from which they evolved. Since bacterial Ψ_{rest} is a function of metabolic state (Na^+/H^+ pumping) and bacteria occupy more variable ionic environments than neurons, the larger difference between "optimal" Ψ_{rest} and $V_{1/2}$ makes physiological sense. We suspect that the fidelity of neuronal firing in metazoans, which depend on their Na_Vs, led to evolution of less sensitivity to pH and temperature, perhaps by selection of the gating charge residues that become exposed to waters within the VSM (Chowdhury et al., 2014). In contrast, we propose that alkaliphilic bacteria (eg, *B. halodurans* and *B. alcalophilus*), which thrive in high temperature (35–60°C) and high pH (9–11) environments, employ homomeric Na_V channels for an entirely different purpose, that of rapidly adjusting internal sodium concentration for metabolic control. This would thus impact growth rates of these *Bacillus* species as observed (DeCaen et al., 2014; Ito et al., 2004). Recently, waves of membrane depolarizations have been observed during the growth phases of *Bacillus subtilis* biofilms through changes in voltage-sensitive and extracellular potassium–sensitive dye fluorescence (Liu et al., 2015; Prindle et al., 2015). *Bacillus subtilis* membrane depolarizations are initiated by low concentrations of internal glutamate (indicative of low nutrient availability and slower metabolism) which triggers the opening of the yugO K-channel and allows potassium to flow out of the cell in a concentration-dependent manner (K: $[300\ mM]_{in}$ vs $[5\ mM]_{out}$). This generates a localized and propagating wave of increased concentrations of extracellular K^+ as each bacterium's membrane is depolarized which reduces the uptake of glutamate (Glu^-) and triggers additional extracellular K^+ release by opening of the yugO channel. Although *B. subtilis* is not an alkaliphile, the oscillations in membrane potential and K^+ extrusion triggered by nutrient availability in biofilms may alter the gating of Na_Vs found in other gram-positive *Bacillus* halophiles. Likewise, it is possible that other *Bacillus* species use a similar strategy as found in *B. subtilis* to coordinate their communal growth. Evidently, the function of prokaryotic ion channels is not limited by the state of the solitary bacterium, as the work by Prindle et al. (2015) demonstrates that some are used for cross-cellular communication (ie, quorum sensing).

3.3 Voltage-dependent inactivation in eukaryotic and prokaryotic Na_Vs

Following the activation process, a second process called inactivation ensues which blocks the central pore from conducting ions despite the depolarized state of the cell membrane. Compared to the activation process, the structural mechanism of inactivation is poorly understood (see Chapters 10: Mechanism of inactivation in voltage-gated Na channels and 13: Physiology and Pathophysiology of Inactivation in Voltage-Gated Sodium Channels for details). In humans, Na_V1s are inactivated by both fast and slow processes which involve distinct locations within the channel (Fig. 3D). The fast inactivation process occurs on a 1–2 ms time course and recovers within a few milliseconds after the membrane repolarization to facilitate the rapid frequency action potentials in nerve and muscle cells (Hodgkin & Huxley, 1952). Based on mutagenesis experiments, the intracellular loop connecting domains III and IV of the eukaryotic Na_V alpha subunit is proposed to fold into the cytoplasmic side of the pore module to block the ion conductance (Smith & Goldin, 1997). In human Na_Vs, slow inactivation is 5–7 times slower than fast inactivation (Fig. 3D). The structural motif responsible for slow inactivation is proposed to be located within the pore domain of these channels. Prokaryotic Na_V channels have only one kinetic (see Fig. 3 legend) upon entry into inactivation, depending on the homologue (8–230 ms) (Gamal El-Din et al., 2014; Irie et al., 2010; Ito et al., 2004; Koishi et al., 2004; Ren et al., 2001). Prokaryotic Na_V inactivation is proposed to be analogous to eukaryotic slow inactivation, which is believed to be caused by a molecular rearrangement of the pore. While this mechanism is vague and not well understood, structural and functional evidence supports the filter and pore as sites involved in the inactivation process. Removal of the C-terminus of the NaChBac channel does not alter the rate of inactivation (Pavlov et al., 2005). However, some prokaryotic Na_Vs have a cytosolic C-terminal coiled-coil motif (one per subunit) which oligomerizes as a tetramer (Shaya et al., 2014). Missense and nonsense mutations in the C-terminal linker and the coiled-coil motif of Na_VSp and Na_VMs channels cause a slower inactivation rate (12–42 times slower than wild-type channels) (Bagneris et al., 2013; Irie et al., 2010; Irie, Shimomura, & Fujiyoshi, 2012). However, it is unlikely that multidomain Na_Vs assemble using coiled-coil motifs analogous to those found in some prokaryotic Na_V channels, thus it is unclear if slowing inactivation effects in this motif have any implications on human Na_V biophysical properties. Several mutation sites within TM6 of the pore module or in the selectivity filter of prokaryotic Na_Vs delay the inactivation process (Pavlov et al., 2005; Zhao, Yarov-Yarovoy, Scheuer, & Catterall, 2004). Additional prokaryotic Na_V (Na_VRh, Na_VCt, and Na_VAb) structures that have been determined are proposed to be in a state which underlies eukaryotic Na_V1 and/or Ca_V channel inactivation (Payandeh et al., 2012; Tsai et al., 2013; Zhang et al., 2012). In both structures, the TM6 and P-loops of the pore domains are asymmetrically collapsed, causing distortion of the selectivity filters which would disfavor Na^+ conductance. Evidently, structural characterization of eukaryotic Ca_V or Na_V1 channels in a native inactivated state is obtainable since their purification condition neutralizes ($\Psi = 0$ mV) the cell membrane potential and would favor this conformational change. While the pore of prokaryotic and eukaryotic Na_V families share a modest degree of homology, a detailed functional comparison of would be required to validate the proposed inactivated states found in Na_VRh, Na_VCt, and Na_VAb structures.

4. DIVERGENCE IN THE EUKARYOTIC AND PROKARYOTIC Na_V PHARMACOPHORES

Na_Vs are the targets of clinically used pharmaceuticals for the management of pain and several neurological and cardiovascular diseases (see Chapters 15: Sodium channels alterations and disease, 16: Unusual voltage-gated sodium currents as targets for pain, and 19: The Current -Voltage Relation for the late Na Current in Mammalian Ventricular Myocytes for details). Pharmaceutical investment in targeting Na_V1s has expanded as emerging clinical data suggest that subtype specificity can be achieved to avoid unintended side effects (Bagal et al., 2014; Nardi, Damann, Hertrampf, & Kless, 2012). Early site-directed mutagenesis studies revealed a common receptor site for local anesthetics and anticonvulsants comprised of specific residues in domains I, III, and IV in cardiac and neuronal Na_V1s (Ragsdale, McPhee, Scheuer, & Catterall, 1994, 1996; Vedder, 1934). Recently solved structures of prokaryotic Na_Vs and the cocrystallization of the Na_VMs channel with small molecule antagonists may provide insight for rational drug design (Bagneris et al., 2014). However, as described in Section 4, the pharmacophore shared between these families is limited.

Lipids modulate prokaryotic Na_V activity primarily from interacting with the VSM (D'Avanzo et al., 2013) Since the genes which encode Na_V proteins are evolved to support membrane interactions, it is not surprising that lipids modulate their function (see Chapter 12: Lipid Regulation of Sodium Channels for details). However, the structures from prokaryotic Na_VAb and Na_VRh channels present a unique type of lipid interaction not observed in previously crystallized ion channels which has implications for therapeutic Na_V1 antagonists (Payandeh et al., 2011; Zhang et al., 2012). The Na_VAb

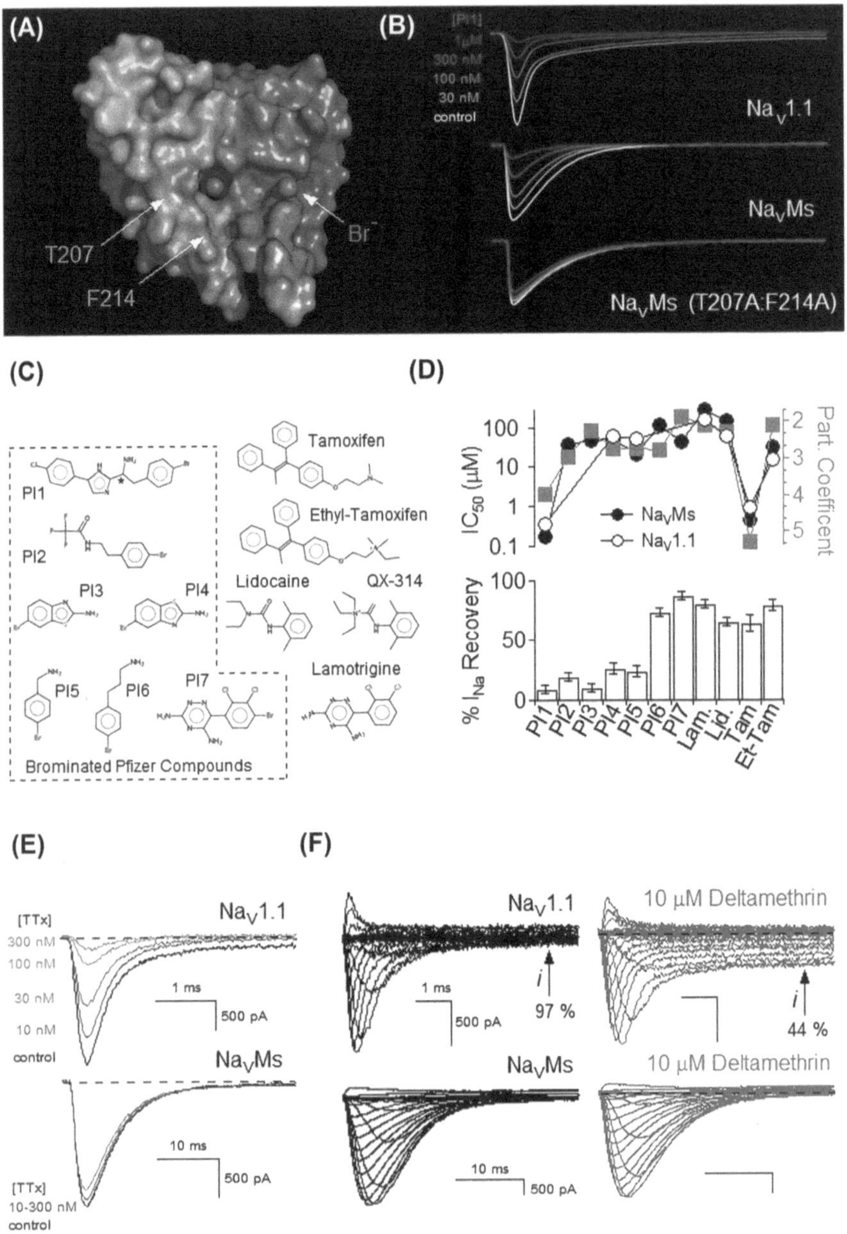
(A)
T207
F214
Br⁻
(B)
[PI1]
1 μM
300 nM
100 nM
30 nM
control
Na$_V$1.1
Na$_V$Ms
Na$_V$Ms (T207A:F214A)
(C)
PI1
PI2
PI3
PI4
PI5
PI6
PI7
Brominated Pfizer Compounds
Tamoxifen
Ethyl-Tamoxifen
Lidocaine
QX-314
Lamotrigine
(D)
IC$_{50}$ (μM)
Part. Coefficent
Na$_V$Ms
Na$_V$1.1
% I$_{Na}$ Recovery
PI1
PI2
PI3
PI4
PI5
PI6
PI7
Lam.
Lid.
Tam
Et-Tam
(E)
[TTx]
300 nM
100 nM
30 nM
10 nM
control
1 ms
500 pA
10 ms
10-300 nM
(F)
97 %
44 %
10 μM Deltamethrin

pore module reveals four fenestrations that open within the plane of the membrane which house opposing acyl molecules which penetrate through the water vestibule and bisect the ion conduction pathway. The residues which line the fenestrations in the Na_VAb and Na_VRh structures align to the local anesthetic drug binding sites initially observed in eukaryotic heart and cardiac Na_V1s. Subsequent cocrystallization of the pore module of the Na_VMs channel with brominated small molecules supports this hypothesis (Bagneris et al., 2014). *In silico* docking aided by the anomalous signature of the bromine atom identified T207 and F207 as critical drug interaction residues located within the fenestrations (Fig. 4A and B). Several brominated and unbrominated analogs were screened for drug antagonism of the human Na_V1.1 and Na_VMs and their potency correlated with increasing hydrophobicity, suggesting that access to the receptor site was limited by the compound solvation into the cell membrane (Fig. 4C and D). Use-dependent block of sodium currents from hNa_V1.1 and Na_VMs channels by lamotrigine (a phenyltriazine) suggests that this antiepileptic gains access to the receptor site in open/inactivated channel state using a mechanism that is conserved across eukaryotic and prokaryotic Na_Vs (Bagneris et al., 2014; Ragsdale, Scheuer, & Catterall, 1991). Subsequently, a lipid acyl chain was also observed in the central cavity in the TRAAK mechanosensitive K-channel crystallized in a nonconducting conformation (Brohawn, Su, & MacKinnon, 2014). Thus, the lipid occupied fenestrations observed in TRAAK, Na_VAb, and Na_VAp channel structures also provide an important glimpse into ion channel evolution within the membrane environment.

Thus, despite millions of years of divergent evolution between the eukaryotic and prokaryotic Na_Vs genes, the pharmacophore appears to be similar. However, some differences have emerged in the mechanism of drug block between prokaryotic and eukaryotic Na_Vs. While lidocaine inhibits NaChBac current with a comparable potency to eukaryotic Na_V1s,

Figure 4 ***Pharmacology of prokaryotic Na_VMs and eukaryotic Na_V1.1 channels.*** (A) A transmembrane view of two subunits of the tetrameric Na_VMs channel where the bromine atom of the Na_V antagonist PI-1 is found near the fenestration binding site (Bagneris et al., 2014). (B) Sodium current antagonism by PI-1 compound on the Na_VMs and Na_V1.1 channels (note the same concentrations did not block currents from the binding site mutant T204A:F214A channel). (C) Small molecules used in the drug screen. (D) Results from the drug screen from Na_V1.1 and Na_VMs graphing the half maximal inhibitory concentration; Oil–water partition coefficient and the percent sodium current recovery after drug block (Bagneris et al., 2014). (E) Sodium current antagonism by tetrodotoxin on the Na_VMs and Na_V1.1 channels. (F) The effect of deltamethrin on Na_V slow inactivation. Percent of total current inactivate (*I*) is indicated by the *arrow*.

the drug-induced enhancement in NaChBac inactivation kinetics suggests a different mechanism of channel antagonism (Lee, Goodchild, & Ahern, 2012a, 2012b). In addition, NaChBac and Na_VMs appear to be insensitive to tetrodotoxin (TTX) (Fig. 4E). TTX is a metabolic product of symbiotic *Vibrio* species of bacteria that is absorbed in the tissues some species of snails, crabs, fish and salamanders (see Chapter 4 for details). These animals use TTX mainly as protection against ingestion by predators and have evolved insensitivity to this neurotoxin (Feldman, Brodie, Brodie, & Pfrender, 2009; Jost et al., 2008; Kaneko, Matsumoto, & Hanyu, 1997; Lee et al., 2000). The TTX-receptor site is located above the selectivity filter on the extracellular mouth of eukaryotic Na_V1 channels. Sensitivity to TTX among SCN gene splice variants (which encodes the alpha subunits $Na_V1.1-9$) is enhanced by the presence of a glutamate residue in domain IV at the junction of the pore helix and the filter of the channel (Terlau et al., 1991). Adaptive resistance has been observed in SCN from garter snakes where the pore aspartic and glutamic acids are not present (Feldman, Brodie, Brodie, & Pfrender, 2012). An acidic residue at this position is not present in any of the prokaryotic Na_V that have been functionally characterized, thus would explain the lack of potent sodium current antagonism by TTX.

Another example of pharmacological divergence between prokaryotic and eukaryotic Na_Vs is the pyrethroid ester insecticides (eg, deltamethrin, see Fig. 4). These compounds are proposed to diffuse through the cell membrane and bind at a site near the inner mouth of the channel and prevent slow inactivation (Tan et al., 2005). This causes neurotoxicity in insects due to the hyperexcitability of the neurons resulting from continuous sodium current (Soderlund, 2012). Resistance has been observed in several insect species with sequence polymorphisms associated with pyrethroid sensitivity in TM6 of domain 1 genes which encode this receptor site (Soderlun & Lee, 2001). Although insect Na_V1 homologs are more sensitive to pyrethroids than human Na_V1s, the effect of deltamethrin on the slow inactivation process can still be observed when the drug concentration is increased (Fig. 4F). Deltamethrin does not block the inactivated state of prokaryotic Na_Vs, NaChBac, and Na_VMs (up to 100 μM). Thus the pharmacophores of eukaryotic and prokaryotic Na_Vs are not identical and their relevance to human Na_V1 pharmacology will likely diminish once the eukaryotic Na_V1 structures are solved. However, this raises the possibility that eukaryotic and prokaryotic Na_V divergence can be utilized as antimicrobials for industrial or medical purposes. Concentrations of lidocaine, nifedipine, and tamoxifen that inhibit heterologously expressed bacterial Na_V channels (Na_VBp and NS_VBa) also block the growth and motility of native *Bacillus* species (*Bacillus pseudofirmus* and *B. alcalophilus*) (DeCaen et al., 2014; Ito et al., 2004). These findings demonstrate that block of Na^+ entry, via Na_V, which disrupts the sodium cycle, can impair *Bacillus* motility and growth. In the light of the rising frequency of drug resistance to clinically important antibiotics (eg, vancomycin and linezolid), alternative prokaryotic targets are desired in the treatment of gram-positive pathogens (Soderlun & Lee, 2001; Zumla et al., 2014). Given the differences in the pharmacophore of eukaryotic and prokaryotic Na_Vs, one could envision specifically targeting voltage-gated channels found in pathogenic *Bacillus* (eg, *Bacillus cereus* and *Bacillus anthracis*) or other gram-positive strains. Biofilms, which contain billions of packed bacterium, cause chronic infections and show increased tolerance to antibiotics as well as resisting phagocytosis by human immune cells (eg, *Pseudomonas aeruginosa* and *Staphylococcus aureus*) (Hoiby, Bjarnsholt, Givskov, Molin, & Ciofu, 2010). As discussed previously (Prindle et al., 2015), growth of some *Bacillus* in biofilms is electrically regulated. Thus voltage-gated ion channels which control opportunistic and pathogenic bacteria membrane potentials might be an important drug target for future antibiotics.

REFERENCES

Alam, A., & Jiang, Y. (2009). Structural analysis of ion selectivity in the NaK channel. *Nature Structural & Molecular Biology, 16*, 35–41. http://dx.doi.org/10.1038/nsmb.1537.

Anderson, P. A., & Greenberg, R. M. (2001). Phylogeny of ion channels: clues to structure and function. *Comparative Biochemistry and Physiology. Part B, Biochemistry & Molecular Biology, 129*, 17–28.

Armstrong, C. M. (2015). Packaging life: the origin of ion-selective channels. *Biophysical Journal, 109*, 173–177. http://dx.doi.org/10.1016/j.bpj.2015.06.012.

Armstrong, C. M., & Bezanilla, F. (1974). Charge movement associated with the opening and closing of the activation gates of the Na channels. *The Journal of General Physiology, 63*, 533–552.

Bagal, S. K., Chapman, M. L., Marron, B. E., Prime, R., Storer, R. I., & Swain, N. A. (2014). Recent progress in sodium channel modulators for pain. *Bioorganic & Medicinal Chemistry Letters, 24*, 3690–3699. http://dx.doi.org/10.1016/j.bmcl.2014.06.038.

Bagneris, C., Decaen, P. G., Hall, B. A., Naylor, C. E., Clapham, D. E., Kay, C. W., & Wallace, B. A. (2013). Role of the C-terminal domain in the structure and function of tetrameric sodium channels. *Nature Communications, 4*, 2465. http://dx.doi.org/10.1038/ncomms3465.

Bagneris, C., DeCaen, P. G., Naylor, C. E., Pryde, D. C., Nobeli, I., Clapham, D. E., & Wallace, B. A. (2014). Prokaryotic NavMs channel as a structural and functional model for eukaryotic sodium channel antagonism. *Proceedings of the National Academy of Sciences of the United States of America, 111*, 8428–8433. http://dx.doi.org/10.1073/pnas.1406855111.

Boiteux, C., Vorobyov, I., & Allen, T. W. (2014). Ion conduction and conformational flexibility of a bacterial voltage-gated sodium channel. *Proceedings of the National Academy of Sciences of the United States of America, 111*, 3454–3459. http://dx.doi.org/10.1073/pnas.1320907111.

Brohawn, S. G., Su, Z., & MacKinnon, R. (2014). Mechanosensitivity is mediated directly by the lipid membrane in TRAAK and TREK1 K^+ channels. *Proceedings of the National*

Academy of Sciences of the United States of America, 111, 3614–3619. http://dx.doi.org/10.1073/pnas.1320768111.

Cai, X. (2012). Ancient origin of four-domain voltage-gated Na^+ channels predates the divergence of animals and fungi. *The Journal of Membrane Biology, 245*, 117–123. http://dx.doi.org/10.1007/s00232-012-9415-9.

Catterall, W. A. (1986). Molecular properties of voltage-sensitive sodium channels. *Annual Review of Biochemistry, 55*, 953–985. http://dx.doi.org/10.1146/annurev.bi.55.070186.004513.

Catterall, W. A. (2011). Voltage-gated calcium channels. *Cold Spring Harbor Perspectives in Biology, 3*, a003947. http://dx.doi.org/10.1101/cshperspect.a003947.

Catterall, W. A., & Zheng, N. (2015). Deciphering voltage-gated Na(+) and Ca(2+) channels by studying prokaryotic ancestors. *Trends in Biochemical Sciences, 40*, 526–534. http://dx.doi.org/10.1016/j.tibs.2015.07.002.

Chakrabarti, N., Ing, C., Payandeh, J., Zheng, N., Catterall, W. A., & Pomes, R. (2013). Catalysis of Na^+ permeation in the bacterial sodium channel Na(V)Ab. *Proceedings of the National Academy of Sciences of the United States of America, 110*, 11331–11336. http://dx.doi.org/10.1073/pnas.1309452110.

Chowdhury, S., Jarecki, B. W., & Chanda, B. (2014). A molecular framework for temperature-dependent gating of ion channels. *Cell, 158*, 1148–1158. http://dx.doi.org/10.1016/j.cell.2014.07.026.

Cooper, G. M. (2000). *The cell: A molecular approach*. Sinauer Associates.

D'Avanzo, N., McCusker, E. C., Powl, A. M., Miles, A. J., Nichols, C. G., & Wallace, B. A. (2013). Differential lipid dependence of the function of bacterial sodium channels. *PLoS One, 8*, e61216. http://dx.doi.org/10.1371/journal.pone.0061216.

DeCaen, P. G., Takahashi, Y., Krulwich, T. A., Ito, M., & Clapham, D. E. (2014). Ionic selectivity and thermal adaptations within the voltage-gated sodium channel family of alkaliphilic Bacillus. *eLife, 3*. http://dx.doi.org/10.7554/eLife.04387.

DeCaen, P. G., Yarov-Yarovoy, V., Scheuer, T., & Catterall, W. A. (2011). Gating charge interactions with the S1 segment during activation of a Na^+ channel voltage sensor. *Proceedings of the National Academy of Sciences of the United States of America, 108*, 18825–18830. http://dx.doi.org/10.1073/pnas.1116449108.

DeCaen, P. G., Yarov-Yarovoy, V., Sharp, E. M., Scheuer, T., & Catterall, W. A. (2009). Sequential formation of ion pairs during activation of a sodium channel voltage sensor. *Proceedings of the National Academy of Sciences of the United States of America, 106*, 22498–22503. http://dx.doi.org/10.1073/pnas.0912307106.

DeCaen, P. G., Yarov-Yarovoy, V., Zhao, Y., Scheuer, T., & Catterall, W. A. (2008). Disulfide locking a sodium channel voltage sensor reveals ion pair formation during activation. *Proceedings of the National Academy of Sciences of the United States of America, 105*, 15142–15147. http://dx.doi.org/10.1073/pnas.0806486105.

Doyle, D. A., Morais Cabral, J., Pfuetzner, R. A., Kuo, A., Gulbis, J. M., Cohen, S. L. … MacKinnon, R. (1998). The structure of the potassium channel: molecular basis of K^+ conduction and selectivity. *Science, 280*, 69–77.

Eisenman, G. (1961). On the elementary atomic origin of equilibrium ionic specificity. In *Symposium on membrane transport and metabolism*. New York: Academic Press.

Feldman, C. R., Brodie, E. D., Jr., Brodie, E. D., 3rd, & Pfrender, M. E. (2009). The evolutionary origins of beneficial alleles during the repeated adaptation of garter snakes to deadly prey. *Proceedings of the National Academy of Sciences of the United States of America, 106*, 13415–13420. http://dx.doi.org/10.1073/pnas.0901224106.

Feldman, C. R., Brodie, E. D., Jr., Brodie, E. D., 3rd, & Pfrender, M. E. (2012). Constraint shapes convergence in tetrodotoxin-resistant sodium channels of snakes. *Proceedings of the National Academy of Sciences of the United States of America, 109*, 4556–4561. http://dx.doi.org/10.1073/pnas.1113468109.

Fujinami, S., Terahara, N., Krulwich, T. A., & Ito, M. (2009). Motility and chemotaxis in alkaliphilic Bacillus species. *Future Microbiology, 4*, 1137–1149. http://dx.doi.org/10.2217/fmb.09.76.

Galione, A., Evans, A. M., Ma, J., Parrington, J., Arredouani, A., Cheng, X., & Zhu, M. X. (2009). The acid test: the discovery of two-pore channels (TPCs) as NAADP-gated endolysosomal Ca(2+) release channels. *Pflugers Archiv: European Journal of Physiology, 458*, 869–876. http://dx.doi.org/10.1007/s00424-009-0682-y.

Gamal El-Din, T. M., Scheuer, T., & Catterall, W. A. (2014). Tracking S4 movement by gating pore currents in the bacterial sodium channel NaChBac. *The Journal of General Physiology, 144*, 147–157. http://dx.doi.org/10.1085/jgp.201411210.

Gruber, J. L., & Sarmiento, N. (2006). *Ocean biogeochemical dynamics*. Princeton University Press.

Gur Barzilai, M., Reitzel, A. M., Kraus, J. E., Gordon, D., Technau, U., Gurevitz, M., & Moran, Y. (2012). Convergent evolution of sodium ion selectivity in metazoan neuronal signaling. *Cell Reports, 2*, 242–248. http://dx.doi.org/10.1016/j.celrep.2012.06.016.

Guy, H. R., & Seetharamulu, P. (1986). Molecular model of the action potential sodium channel. *Proceedings of the National Academy of Sciences of the United States of America, 83*, 508–512.

Hess, P., Lansman, J. B., & Tsien, R. W. (1986). Calcium channel selectivity for divalent and monovalent cations. Voltage and concentration dependence of single channel current in ventricular heart cells. *The Journal of General Physiology, 88*, 293–319.

Hille, B. (1972). The permeability of the sodium channel to metal cations in myelinated nerve. *The Journal of General Physiology, 59*, 637–658.

Hodgkin, A. L., & Huxley, A. F. (1952). The dual effect of membrane potential on sodium conductance in the giant axon of Loligo. *The Journal of Physiology, 116*, 497–506.

Hoiby, N., Bjarnsholt, T., Givskov, M., Molin, S., & Ciofu, O. (2010). Antibiotic resistance of bacterial biofilms. *International Journal of Antimicrobial Agents, 35*, 322–332. http://dx.doi.org/10.1016/j.ijantimicag.2009.12.011.

Irie, K., Kitagawa, K., Nagura, H., Imai, T., Shimomura, T., & Fujiyoshi, Y. (2010). Comparative study of the gating motif and C-type inactivation in prokaryotic voltage-gated sodium channels. *The Journal of Biological Chemistry, 285*, 3685–3694. http://dx.doi.org/10.1074/jbc.M109.057455.

Irie, K., Shimomura, T., & Fujiyoshi, Y. (2012). The C-terminal helical bundle of the tetrameric prokaryotic sodium channel accelerates the inactivation rate. *Nature Communications, 3*, 793. http://dx.doi.org/10.1038/ncomms1797.

Ito, M., Xu, H., Guffanti, A. A., Wei, Y., Zvi, L., Clapham, D. E., & Krulwich, T. A. (2004). The voltage-gated Na^+ channel NaVBP has a role in motility, chemotaxis, and pH homeostasis of an alkaliphilic Bacillus. *Proceedings of the National Academy of Sciences of the United States of America, 101*, 10566–10571. http://dx.doi.org/10.1073/pnas.0402692101.

Jiang, Y., Lee, A., Chen, J., Cadene, M., Chait, B. T., & MacKinnon, R. (2002). The open pore conformation of potassium channels. *Nature, 417*, 523–526. http://dx.doi.org/10.1038/417523a.

Jiang, Y., Lee, A., Chen, J., Ruta, V., Cadene, M., Chait, B. T., & MacKinnon, R. (2003). X-ray structure of a voltage-dependent K^+ channel. *Nature, 423*, 33–41. http://dx.doi.org/10.1038/nature01580.

Jost, M. C., Hillis, D. M., Lu, Y., Kyle, J. W., Fozzard, H. A., & Zakon, H. H. (2008). Toxin-resistant sodium channels: parallel adaptive evolution across a complete gene family. *Molecular Biology and Evolution, 25*, 1016–1024. http://dx.doi.org/10.1093/molbev/msn025.

Kaneko, Y., Matsumoto, G., & Hanyu, Y. (1997). TTX resistivity of Na^+ channel in newt retinal neuron. *Biochemical and Biophysical Research Communications, 240*, 651–656. http://dx.doi.org/10.1006/bbrc.1997.7696.

Ke, S., Timin, E. N., & Stary-Weinzinger, A. (2014). Different inward and outward conduction mechanisms in NaVMs suggested by molecular dynamics simulations. *PLoS Computational Biology, 10*, e1003746. http://dx.doi.org/10.1371/journal.pcbi.1003746.

Keynes, R. D., Rojas, E., & Rudy, B. (1974). Proceedings: demonstration of a first-order voltage-dependent transition of the sodium activation gates. *The Journal of Physiology, 239*, 100P–101P.

Kim, M. S., Morii, T., Sun, L. X., Imoto, K., & Mori, Y. (1993). Structural determinants of ion selectivity in brain calcium channel. *FEBS Letters, 318*, 145–148.

Koishi, R., Xu, H., Ren, D., Navarro, B., Spiller, B. W., Shi, Q., & Clapham, D. E. (2004). A superfamily of voltage-gated sodium channels in bacteria. *The Journal of Biological Chemistry, 279*, 9532–9538. http://dx.doi.org/10.1074/jbc.M313100200.

Kuzmenkin, A., Bezanilla, F., & Correa, A. M. (2004). Gating of the bacterial sodium channel, NaChBac: voltage-dependent charge movement and gating currents. *The Journal of General Physiology, 124*, 349–356. http://dx.doi.org/10.1085/jgp.200409139.

Lee, M. J., Jeong, D. Y., Kim, W. S., Kim, H. D., Kim, C. H., Park, W. W. … Kim, D. S. (2000). A tetrodotoxin-producing Vibrio strain, LM-1, from the puffer fish *Fugu vermicularis radiatus*. *Applied and Environmental Microbiology, 66*, 1698–1701.

Lee, S., Goodchild, S. J., & Ahern, C. A. (2012a). Molecular and functional determinants of local anesthetic inhibition of NaChBac. *Channels, 6*, 403–406. http://dx.doi.org/10.4161/chan.21807.

Lee, S., Goodchild, S. J., & Ahern, C. A. (2012b). Local anesthetic inhibition of a bacterial sodium channel. *The Journal of General Physiology, 139*, 507–516. http://dx.doi.org/10.1085/jgp.201210779.

Liao, M., Cao, E., Julius, D., & Cheng, Y. (2014). Single particle electron cryo-microscopy of a mammalian ion channel. *Current Opinion in Structural Biology, 27*, 1–7. http://dx.doi.org/10.1016/j.sbi.2014.02.005.

Liebeskind, B. J., Hillis, D. M., & Zakon, H. H. (2012). Phylogeny unites animal sodium leak channels with fungal calcium channels in an ancient, voltage-insensitive clade. *Molecular Biology and Evolution, 29*, 3613–3616. http://dx.doi.org/10.1093/molbev/mss182.

Liebeskind, B. J., Hillis, D. M., & Zakon, H. H. (2013). Independent acquisition of sodium selectivity in bacterial and animal sodium channels. *Current Biology: CB, 23*, R948–R949. http://dx.doi.org/10.1016/j.cub.2013.09.025.

Liu, J., Prindle, A., Humphries, J., Gabalda-Sagarra, M., Asally, M., Lee, D. Y. … Suel, G. M. (2015). Metabolic co-dependence gives rise to collective oscillations within biofilms. *Nature, 523*, 550–554. http://dx.doi.org/10.1038/nature14660.

McCusker, E. C., Bagneris, C., Naylor, C. E., Cole, A. R., D'Avanzo, N., Nichols, C. G., & Wallace, B. A. (2012). Structure of a bacterial voltage-gated sodium channel pore reveals mechanisms of opening and closing. *Nature Communications, 3*, 1102. http://dx.doi.org/10.1038/ncomms2077.

Nardi, A., Damann, N., Hertrampf, T., & Kless, A. (2012). Advances in targeting voltage-gated sodium channels with small molecules. *ChemMedChem, 7*, 1712–1740. http://dx.doi.org/10.1002/cmdc.201200298.

Patapoutian, A., Peier, A. M., Story, G. M., & Viswanath, V. (2003). ThermoTRP channels and beyond: mechanisms of temperature sensation. *Nature Reviews Neuroscience, 4*, 529–539. http://dx.doi.org/10.1038/nrn1141.

Patino, G. A., & Isom, L. L. (2010). Electrophysiology and beyond: multiple roles of Na^+ channel beta subunits in development and disease. *Neuroscience Letters, 486*, 53–59. http://dx.doi.org/10.1016/j.neulet.2010.06.050.

Pavlov, E., Bladen, C., Winkfein, R., Diao, C., Dhaliwal, P., & French, R. J. (2005). The pore, not cytoplasmic domains, underlies inactivation in a prokaryotic sodium channel. *Biophysical Journal, 89*, 232–242. http://dx.doi.org/10.1529/biophysj.104.056994.

Payandeh, J., Gamal El-Din, T. M., Scheuer, T., Zheng, N., & Catterall, W. A. (2012). Crystal structure of a voltage-gated sodium channel in two potentially inactivated states. *Nature, 486*, 135–139. http://dx.doi.org/10.1038/nature11077.

Payandeh, J., & Minor, D. L., Jr. (2015). Bacterial voltage-gated sodium channels (BacNa(V)s) from the soil, sea, and salt lakes enlighten molecular mechanisms of electrical signaling and pharmacology in the brain and heart. *Journal of Molecular Biology, 427*, 3–30. http://dx.doi.org/10.1016/j.jmb.2014.08.010.

Payandeh, J., Scheuer, T., Zheng, N., & Catterall, W. A. (2011). The crystal structure of a voltage-gated sodium channel. *Nature, 475*, 353–358. http://dx.doi.org/10.1038/nature10238.

Prindle, A., Liu, J., Asally, M., Ly, S., Garcia-Ojalvo, J., & Suel, G. M. (2015). Ion channels enable electrical communication in bacterial communities. *Nature*. http://dx.doi.org/10.1038/nature15709.

Ragsdale, D. S., McPhee, J. C., Scheuer, T., & Catterall, W. A. (1994). Molecular determinants of state-dependent block of Na^+ channels by local anesthetics. *Science, 265*, 1724–1728.

Ragsdale, D. S., McPhee, J. C., Scheuer, T., & Catterall, W. A. (1996). Common molecular determinants of local anesthetic, antiarrhythmic, and anticonvulsant block of voltage-gated Na^+ channels. *Proceedings of the National Academy of Sciences of the United States of America, 93*, 9270–9275.

Ragsdale, D. S., Scheuer, T., & Catterall, W. A. (1991). Frequency and voltage-dependent inhibition of type IIA Na^+ channels, expressed in a mammalian cell line, by local anesthetic, antiarrhythmic, and anticonvulsant drugs. *Molecular Pharmacology, 40*, 756–765.

Ren, D., Navarro, B., Xu, H., Yue, L., Shi, Q., & Clapham, D. E. (2001). A prokaryotic voltage-gated sodium channel. *Science, 294*, 2372–2375. http://dx.doi.org/10.1126/science.1065635.

Sato, C., Ueno, Y., Asai, K., Takahashi, K., Sato, M., Engel, A., & Fujiyoshi, Y. (2001). The voltage-sensitive sodium channel is a bell-shaped molecule with several cavities. *Nature, 409*, 1047–1051. http://dx.doi.org/10.1038/35059098.

Scheuer, T. (2014). Bacterial sodium channels: models for eukaryotic sodium and calcium channels. *Handbook of Experimental Pharmacology, 221*, 269–291. http://dx.doi.org/10.1007/978-3-642-41588-3_13.

Shaya, D., Findeisen, F., Abderemane-Ali, F., Arrigoni, C., Wong, S., Nurva, S. R.… Minor, D. L., Jr. (2014). Structure of a prokaryotic sodium channel pore reveals essential gating elements and an outer ion binding site common to eukaryotic channels. *Journal of Molecular Biology, 426*, 467–483. http://dx.doi.org/10.1016/j.jmb.2013.10.010.

Smith, M. R., & Goldin, A. L. (1997). Interaction between the sodium channel inactivation linker and domain III S4-S5. *Biophysical Journal, 73*, 1885–1895. http://dx.doi.org/10.1016/S0006-3495(97)78219-5.

Soderlund, D. M. (2012). Molecular mechanisms of pyrethroid insecticide neurotoxicity: recent advances. *Archives of Toxicology, 86*, 165–181. http://dx.doi.org/10.1007/s00204-011-0726-x.

Soderlun, D. M., & Lee, S. H. (2001). Point mutations in homology domain II modify the sensitivity of rat Nav1.8 sodium channels to the pyrethroid insecticide cismethrin. *Neurotoxicology, 22*, 755–765.

Sokolov, S., Scheuer, T., & Catterall, W. A. (2007). Gating pore current in an inherited ion channelopathy. *Nature, 446*, 76–78. http://dx.doi.org/10.1038/nature05598.

Stephens, R. F., Guan, W., Zhorov, B. S., & Spafford, J. D. (2015). Selectivity filters and cysteine-rich extracellular loops in voltage-gated sodium, calcium, and NALCN channels. *Frontiers in Physiology, 6*, 153. http://dx.doi.org/10.3389/fphys.2015.00153.

Strong, M., Chandy, K. G., & Gutman, G. A. (1993). Molecular evolution of voltage-sensitive ion channel genes: on the origins of electrical excitability. *Molecular Biology and Evolution, 10*, 221–242.

Sun, Y. M., Favre, I., Schild, L., & Moczydlowski, E. (1997). On the structural basis for size-selective permeation of organic cations through the voltage-gated sodium channel. Effect of alanine mutations at the DEKA locus on selectivity, inhibition by Ca^{2+} and H^+, and molecular sieving. *The Journal of General Physiology, 110*, 693–715.

Tan, J., Liu, Z., Wang, R., Huang, Z. Y., Chen, A. C., Gurevitz, M., & Dong, K. (2005). Identification of amino acid residues in the insect sodium channel critical for pyrethroid binding. *Molecular Pharmacology, 67*, 513–522. http://dx.doi.org/10.1124/mol.104.006205.

Tang, L., Gamal El-Din, T. M., Payandeh, J., Martinez, G. Q., Heard, T. M., Scheuer, T. … Catterall, W. A. (2014). Structural basis for Ca^{2+} selectivity of a voltage-gated calcium channel. *Nature, 505*, 56–61. http://dx.doi.org/10.1038/nature12775.

Terahara, N., Sano, M., & Ito, M. (2012). A Bacillus flagellar motor that can use both Na^+ and K^+ as a coupling ion is converted by a single mutation to use only Na^+. *PLoS One, 7*, e46248. http://dx.doi.org/10.1371/journal.pone.0046248.

Terlau, H., Heinemann, S. H., Stuhmer, W., Pusch, M., Conti, F., Imoto, K., & Numa, S. (1991). Mapping the site of block by tetrodotoxin and saxitoxin of sodium channel II. *FEBS Letters, 293*, 93–96.

Tsai, C. J., Tani, K., Irie, K., Hiroaki, Y., Shimomura, T., McMillan, D. G. … Li, X. D. (2013). Two alternative conformations of a voltage-gated sodium channel. *Journal of Molecular Biology, 425*, 4074–4088. http://dx.doi.org/10.1016/j.jmb.2013.06.036.

Ulmschneider, M. B., Bagneris, C., McCusker, E. C., Decaen, P. G., Delling, M., Clapham, D. E. … Wallace, B. A. (2013). Molecular dynamics of ion transport through the open conformation of a bacterial voltage-gated sodium channel. *Proceedings of the National Academy of Sciences of the United States of America, 110*, 6364–6369. http://dx.doi.org/10.1073/pnas.1214667110.

Vedder, A. (1934). Bacillus alcalophilus n. sp.; benevens enkele ervaringen met sterk alcalische voedingsbodems. *Antonie Van Leeuwenhoek, 1*, 143–147.

Yarov-Yarovoy, V., DeCaen, P. G., Westenbroek, R. E., Pan, C. Y., Scheuer, T., Baker, D., & Catterall, W. A. (2012). Structural basis for gating charge movement in the voltage sensor of a sodium channel. *Proceedings of the National Academy of Sciences of the United States of America, 109*, E93–E102. http://dx.doi.org/10.1073/pnas.1118434109.

Yellen, G. (1987). Permeation in potassium channels: implications for channel structure. *Annual Review of Biophysics and Biophysical Chemistry, 16*, 227–246. http://dx.doi.org/10.1146/annurev.bb.16.060187.001303.

Yu, F. H., & Catterall, W. A. (2004). The VGL-chanome: a protein superfamily specialized for electrical signaling and ionic homeostasis. *Science's STKE: Signal Transduction Knowledge Environment*, re15. http://dx.doi.org/10.1126/stke.2532004re15.

Yue, L., Navarro, B., Ren, D., Ramos, A., & Clapham, D. E. (2002). The cation selectivity filter of the bacterial sodium channel, NaChBac. *The Journal of General Physiology, 120*, 845–853.

Zakon, H. H. (2012). Adaptive evolution of voltage-gated sodium channels: the first 800 million years. *Proceedings of the National Academy of Sciences of the United States of America, 109*(Suppl. 1), 10619–10625. http://dx.doi.org/10.1073/pnas.1201884109.

Zhang, X., Ren, W., DeCaen, P., Yan, C., Tao, X., Tang, L. … Yan, N. (2012). Crystal structure of an orthologue of the NaChBac voltage-gated sodium channel. *Nature, 486*, 130–134. http://dx.doi.org/10.1038/nature11054.

Zhang, X., Xia, M., Li, Y., Liu, H., Jiang, X., Ren, W. … Gong, H. (2013). Analysis of the selectivity filter of the voltage-gated sodium channel Na(v)Rh. *Cell Research, 23*, 409–422. http://dx.doi.org/10.1038/cr.2012.173.

Zhao, Y., Yarov-Yarovoy, V., Scheuer, T., & Catterall, W. A. (2004). A gating hinge in Na^+ channels; a molecular switch for electrical signaling. *Neuron, 41*, 859–865.

Zhou, Y., Morais-Cabral, J. H., Kaufman, A., & MacKinnon, R. (2001 Nov 1). Chemistry of ion coordination and hydration revealed by a K+ channel-Fab complex at 2.0 A resolution. *Nature, 414*(6859), 43–48.

Zumla, A., Memish, Z. A., Maeurer, M., Bates, M., Mwaba, P., Al-Tawfiq, J. A. … Hui, D. S. (2014). Emerging novel and antimicrobial-resistant respiratory tract infections: new drug development and therapeutic options. *The Lancet Infectious Diseases, 14*, 1136–1149. http://dx.doi.org/10.1016/S1473-3099(14)70828-X.

CHAPTER THREE

Venom Peptides From Cone Snails: Pharmacological Probes for Voltage-Gated Sodium Channels

B.R. Green[*,§] **and B.M. Olivera**[*,1]
[*]University of Utah, Salt Lake City, UT, United States
[§]Monash University, Parkville, VIC, Australia
[1]Corresponding author: E-mail: olivera@biology.utah.edu

Contents

Abstract

The venoms of cone snails provide a rich source of neuroactive peptides (conotoxins). Several venom peptide families have been identified that are either agonists (ι- and δ-conotoxins) or antagonists (μ- and μO-conotoxins) of voltage gated sodium channels (VGSCs). Members of these conotoxin classes have been integral in identifying and characterizing specific neurotoxin binding sites on the channel. Furthermore, given the specificity of some of these peptides for one sodium channel subtype over another, conotoxins have also proven useful in exploring differences between VGSC subtypes.

Current Topics in Membranes, Volume 78
ISSN 1063-5823
http://dx.doi.org/10.1016/bs.ctm.2016.07.001

This chapter summarizes the current knowledge of the structure and function based on the results of conotoxin interactions with VGSCs and correlates the peptides with the phylogeny of the *Conus* species from which they were derived.

List of Abbreviations

Na_V1 α-Subunit of the voltage-gated sodium channel
$Na_V\beta$ β-Subunit of the voltage-gated sodium channel
VGSC Voltage-gated sodium channel

1. INTRODUCTION

Neuroactive peptides from the venoms of predatory organisms have been integral in understanding the structure and function of voltage-gated sodium channels (VGSCs). Of particular interest are the venoms of predatory marine snails of the genus *Conus* (ie, cone snails), which are extremely rich in compounds that act on molecular targets in nervous systems.

Analyses of cone snail venoms suggest that each species has a repertoire of c. 100 neuroactive peptides, with most of these tailored to modulate the function of a specific ion channel, receptor, or transporter in the nervous system of their prey, predators, or competitors. Among these targets are VGSCs. Five known classes of conotoxins (ι-, δ-, μ-, μO-, and μO§-conotoxins—see Fig. 1) act on VGSCs either to promote or inhibit their activity in conducting Na^+ ions across the membrane, thereby directly affecting the initiation and propagation of action potentials.

Groups of conotoxins have been found to act in a concerted manner to elicit a desired physiological endpoint; these are known as *toxin cabals*. For example, the peptides comprising the so-called *lightning-strike cabal* cause a dramatic and immediate immobilization of an envenomated animal (Terlau & Olivera, 2004; Terlau et al., 1996). Conversely, peptides that act to elicit a complete inhibition of neuromuscular transmission belong to the *motor cabal* (Terlau & Olivera, 2004). In this review, we discuss the role of VGSC-targeted venom peptides from cone snails as major components of different cabals.

The generalized structure of the mammalian VGSC consists of a pore-forming α-subunit (Na_V1), roughly 260 kDa in size, and one or more auxiliary β-subunits ($Na_V\beta$) (Catterall, 1995). The activities of the α-subunit are influenced by nature-derived neurotoxins that elicit their effects on either ion permeation or voltage-dependent gating. Traditionally, pharmacological assessment of the effects of such compounds on VGSC activity was

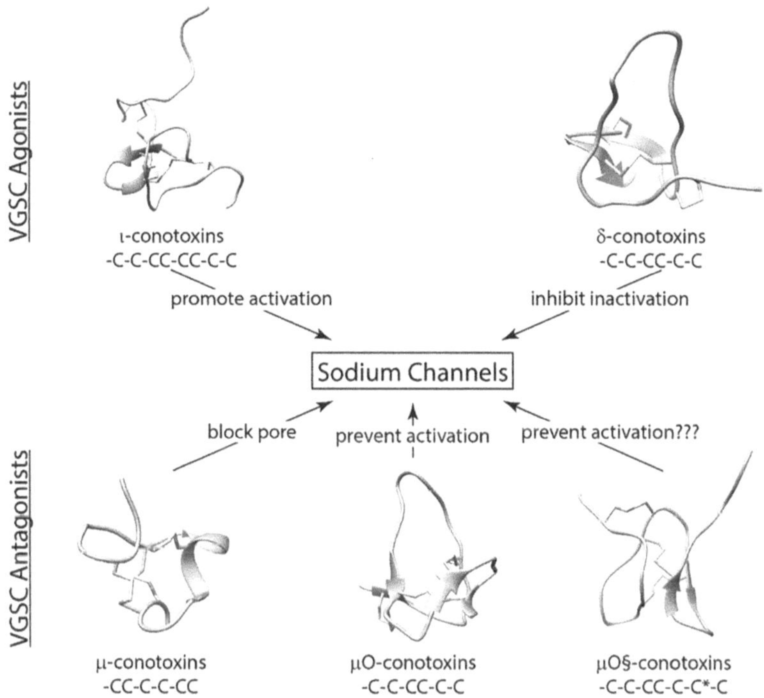

Figure 1 Voltage-gated sodium channel (VGSC)-modulating peptide classes from cone snail venoms. To date, five classes of toxins that modulate VGSC activity have been described. These conotoxins can be broadly categorized as either VGSC agonists (ι- and δ-conotoxins) or antagonists (μ-, μO-, and μO§-conotoxins) that bind to sites either within the sodium-conductance pore or at sites on the external surface of voltage sensors of the channel. NMR structures of representative members of each class were created using UCSF Chimera using coordinates available on the RCSB Protein Database (www.rcsb.org). PBD IDs: 2JRJ (ι-RxIA), 1G1P (δ-EVIA), 1TCK (μ-GIIIA), 1RMK (μO-MrVIB), and 2N8H (μO§-GVIIJ).

performed on Na_V1-subtypes as this transmembrane protein has been shown to be the minimal requisite structure to carry out the processes of a fully functional channel (Catterall, 2012). Perhaps less well understood are the roles of $Na_V\beta$ subunits on channel activity. These protein subunits have been shown to modulate the kinetics and voltage dependencies of Na_V1 as well as play a role in cellular adhesion, migration, aggregation, and interactions with the cytoskeleton (Isom, 2001). Beyond their roles in these important cellular functions, coexpression of these "accessory" proteins with Na_V1 has also been shown to alter the effects of conotoxins on the channel (Wilson et al., 2011, 2015; Zhang, Wilson, Azam, et al., 2013). This chapter will emphasize the interactions of conotoxins with the α- and β-subunits and will describe the integral role that these peptides have played in understanding the structure–function of VGSCs.

2. OVERVIEW OF PHYSIOLOGICAL ROLES OF VOLTAGE-GATED SODIUM CHANNEL–TARGETED *CONUS* VENOM PEPTIDES

The different superfamilies of sodium channel–targeted *Conus* venom peptides are summarized in Table 1A. In this section, the physiological roles of different Na^+ channel–targeted conopeptide superfamilies are discussed in the context of the biology of the specific lineages of *Conus* that evolved these peptides. The genus *Conus* comprises ~700 species. In the latest taxonomic treatise on cone snails (Puillandre et al., 2014), some species that used to be included in the genus *Conus* have been separated based on molecular data—three additional genera of cone snails are recognized, *Conasprella, Californiconus*, and *Profundiconus*; together these comprise c. 100 species, all of which are included in the family Conidae, but not in the genus *Conus*. The VGSC-targeted venom peptides in these other genera of the family Conidae are largely uncharacterized, except for work carried out on *Californiconus californicus* by Gilly and coworkers (Elliger et al., 2011; Gilly et al., 2011), but a biochemical characterization of VGSC-targeted toxins has not yet been reported.

Within the genus *Conus*, three broad classes of species are recognized, ie, fish-hunting, worm-hunting, and mollusk-hunting species. Most of our focus will be on the fish-hunting species, comprising over 100 different *Conus* that have been divided into eight subgenera. The subgenera of *Conus* shown to express the different superfamilies of VGSC-targeted peptides (as of April 2016) are summarized in Table 1B and are shown in Fig. 2A and B. The venoms of worm-hunting and mollusk-hunting *Conus* species are much less well characterized; the worm-hunting and mollusk-hunting species that have been analyzed are shown in Fig. 2C.

The first characterized superfamily of *Conus* venom peptides targeted to VGSCs was the μ-conotoxins, which are pore blockers. To date these have only been found in fish-hunting *Conus* species. Because these inhibit sodium-ion flux through the pore, not surprisingly they elicit paralysis when injected by the cone snail into their fish prey. The μ-conotoxins are part of the "motor cabal," a set of peptides found in fish-hunting cone snail venoms that block neuromuscular transmission. Together with the ω-conotoxins that block presynaptic Ca^{2+} channels, and the α-conotoxins

Table 1 *Conus* venom peptide families targeted to Na^+ channels

(A) Summary of VGSC-modulating peptide classes from *Conus* venoms

Pharmacological family	Super family	Cysteine scaffold (framework)	Pharmacological activity	References
μ	M	CC-C-C-CC (III)	VGSC antagonist (Site 1)	Cruz et al. (1985) Gray et al. (1988)
δ	O1	C-C-CC-C-C (VI/VII)	VGSC agonist (Site 6)	Shon et al. (1994) Hasson et al. (1995)
μO	O1	C-C-CC-C-C (VI/VII)	VGSC antagonist (Site 4)	McIntosh et al. (1995) Zorn et al. (2006)
μO§	O1	C-C-CC-C-C*-C (VI/VII)	VGSC antagonist (Site 8)	Gajewiak et al. (2014)
ι	I1	C-C-CC-CC-C-C (XI)	VGSC agonist (Site 4)	Jimenez et al. (2003)

(B) Phylogenetic range of conotoxin families targeted to Na^+ channels

Subgenera of *Conus*	Families of VGSC-targeted toxins
Fish hunters	
Pionoconus	μ, δ
Chelyconus	μ, δ
Textilia	μ, δ
Gastridium	μ, μO
Afonsoconus	μ
Phasmoconus	ι
Snail hunters	
Conus	μO
Cylindrus	δ
Worm hunters	
Tesselliconus	δ
Kalloconus	δ

VGSC, voltage-gated sodium channel; *C** denotes S-cysteinylated cysteine.
Biochemically, no direct evidence is available, but the mechanism is consistent with other Site 4 toxins.

that block postsynaptic nicotinic acetylcholine receptors, μ-conotoxins, most of which preferentially target VGSCs in the muscle plasma membrane thereby abolishing muscle action potentials, the motor cabal peptides, comprise a powerful set of pharmacological agents that together effect an

Figure 2 *Conus* species from which voltage-gated sodium channel (VGSC)-targeted venom peptides have been characterized. The majority of the 21 species shown are fish-hunting *Conus* (14 out of 21). The 21 species belong to 11 different subgenera of *Conus*. Shown are the shells of species that have been analyzed to date. In A and B, the fish-hunting species are shown. (A) The seven species in the subgenus *Pionoconus* are illustrated. (B) Seven other fish-hunting species belonging to five different subgenera (*Gastridium, Textilia, Chelyconus, Phasmoconus*, and *Alfonsoconus*) are illustrated. The species shown in panel (C) are nonpiscivorous *Conus* from which VGSC-targeted venom peptides have been elucidated and characterized. These include species form three mollusk-hunting clades (*Conus, Cylindrus*, and *Leptoconus*) and from two worm-hunting clades (*Tesselliconus* and *Kalloconus*).

irreversible paralysis in the envenomated fish. It is likely that there are μ-conotoxins expressed in the venoms of all fish-hunting cone snails. As will be discussed below, μ-conotoxins from different subgenera of fish-hunting *Conus* have different selectivity for neuronal VGSCs.

Importantly, μ-conotoxins have not been found outside the fish-hunting subgenera of *Conus*. A potent peptide that inhibits VGSCs was characterized from the mollusk-hunting species *Conus marmoreus*—these are the μO-conotoxins, which belong to a different superfamily of cone snail peptides from the μ-conotoxins (the O1- vs the M-superfamily—see Fig. 1). So far, these have only been characterized from the subgenus *Conus*. In contrast to the μ-conotoxins, which are pore blockers, the μO-conotoxins prevent the movement of voltage sensors thereby inhibiting the activation of the channel and consequent pore opening. It seems likely that μO-conotoxins are part of a mollusk-specific "motor cabal."

In contrast, the δ-conotoxins act as agonists of VGSCs by inhibiting inactivation. These peptides are much more widely distributed phylogenetically than the families described previously (see Table 1B), having been characterized from fish-hunting, snail-hunting, and worm-hunting *Conus*. For two subgenera of fish-hunting cone snails, the physiological role of δ-conotoxins has been clearly elucidated. To effectively capture fish, a "taser-like effect" of the venom has been evolved in species of the subgenera *Pionoconus* and *Chelyconus*. In essence, δ-conotoxins bind to axons close to the venom injection site and, by inhibiting Na^+-channel inactivation, elicit a massive depolarization of axonal membranes. By keeping Na^+ channels open and in combination with κ-conotoxins that block voltage-gated K channels on the same axons, the two major pathways to repolarization of the axonal membrane after depolarization has occurred are completely inhibited. Almost immediately after venom injection, because of these two toxins, the fish undergoes a tetanic paralysis and is almost instantly immobilized. The group of peptides that efficiently cause this rapid immobilization is known as the "lightning-strike cabal," and δ-conotoxins are key components.

The δ-conotoxins are also found in snail-hunting *Conus* belonging to the subgenus *Cylindrus*. In this case, the physiological role of δ conotoxins is apparently to cause an uncontrollable seizure-like repetitive motion in the envenomated snail. The purpose is to prevent the snail from withdrawing into its shell; snail-hunting cones face the challenge of keeping envenomated prey from withdrawing deep into their shell; even if the snail were immobilized by the venom, if it withdrew into its shell, the soft parts of the prey would be inaccessible (cone snails have no mechanical weaponry for breaking the shell of the prey). The δ-conotoxins cause a hyperactivity of the nervous system that prevents the prey snail from withdrawing into its shell, therefore making it vulnerable to further attack. Species in the subgenus *Cylindrus* (see Fig. 2) will inject prey multiple times, and it is likely that the venom composition changes as the envenomated gastropod becomes more and more disabled. After a prey snail is attacked successfully by *Conus textile*, nothing remains but an empty shell; the entire soft body of the mollusk is devoured by the predatory cone. δ-Conotoxins have also been found in venoms of worm-hunting *Conus* species. Some of these are highly potent on mammalian VGSCs, and it has been postulated that these serve as deterrents to fish competitors of worm-hunting cone snails, since they would presumably elicit pain. However, worm-hunting species in the subgenus *Tesselliconus* may also use δ-conotoxins as an offensive weapon to opportunistically immobilize fish, their secondary prey. Although they are not specialized for fish hunting, they have been observed to sting weakened or compromised fish prey (Aman et al., 2015).

In contrast to the three VGSC-targeted conopeptide families described previously, the ι-conotoxins and the μO§-conopeptide are less well defined in their physiological roles. It is probable that the ι-conotoxins play a role similar to the role of δ-conotoxins in the lightning-strike cabal of fish-hunting cone snails, but this has not been demonstrated directly.

Although *Conus* is one of the venomous animal lineages that has been most extensively analyzed, the studies that have been carried out have only scratched the surface, and there is an enormous untapped potential for the discovery of novel VGSC-targeted venom peptides. Only 21 out of the c. 700 species of *Conus* have been analyzed at all, and of these, one-third of the analyses have been carried out on a single subgenus of *Conus*, *Pionoconus*. To put this in context, *Pionoconus* is only one of 57 different subgenera proposed within the genus *Conus*; 41 of these subgenera have not been examined at all.

Two-thirds of the species that have been analyzed are fish hunting, half from the subgenus *Pionoconus* and the other half distributed among five other clades of fish-hunting cone snails (see Fig. 2 and Table 1B). Many of the novel peptides that have been characterized from fish-hunting cone snails have come from the analysis of clades (or subgenera) from which a single species has been analyzed (eg, for μ-conotoxins from the *Textilia* clade, *Conus bullatus*). Because the analysis that has been carried out so far is so incomplete, generalizations that we provide must be regarded as tentative and are subject to revision after a more thorough analysis of *Conus* venoms has been completed.

It appears, however, that there are some VGSC-targeted conopeptides that are very widely distributed phylogenetically, such as the δ-conotoxins. This group of peptides that inhibit VGSC inactivation have been found in fish-hunting, mollusk-hunting, and worm-hunting species. Thus, this is a conopeptide family that is widely distributed phylogenetically. In contrast, three of the VGSC-targeted conopeptide families have been found in a single subgenus: μO-conotoxins have only been found in the subgenus *Conus*, ι-conotoxins have only been found in the subgenus *Phasmaconus*, and μO§-conotoxins have been detected only in *Gastridium*. Intermediate between these two extremes are the μ-conotoxins that have been found in five subgenera of fish-hunting cone snails, but not in any species of nonpiscivorous *Conus*. Thus, the overall picture that emerges with regard to the phylogenetic distribution of VGSC-targeted conopeptides is that while some may be restricted to a single species or subgenus within the genus *Conus*, others may be more broadly distributed. The phylogenetic distribution of different VGSC-targeted subfamilies has implications with regard to their origin and evolutionary history.

3. CONOTOXIN INTERACTIONS WITH Na_V1

In the absence of a model of the mammalian VGSC, neurotoxins such as conotoxins have been integral in mapping discrete binding sites on the α-subunit of VGSCs. To date, eight neurotoxin binding sites have been proposed on the α-subunit (sites 1–8). Both agonists (δ- and ι-conotoxins) and antagonists (μ-, μO-, and μO§-conotoxins) have been shown to elicit their pharmacological effects through binding either deep within the sodium-conductance pore, or on the external surface of the channel.

3.1 Na_V1 agonists

VGSCs briefly open in response to local depolarization events to allow Na^+ ions to flow into the cell. Thus far, two classes of excitatory peptides targeted to sodium channels have been identified from *Conus* venoms.

3.1.1 δ-Conotoxins

To date, 22 members of the δ-conotoxin family have been characterized (www.conoserver.org). These six cysteine–containing peptides possess in a Type VI/VII cysteine framework (C-C-CC-C-C) macrocyclized by three

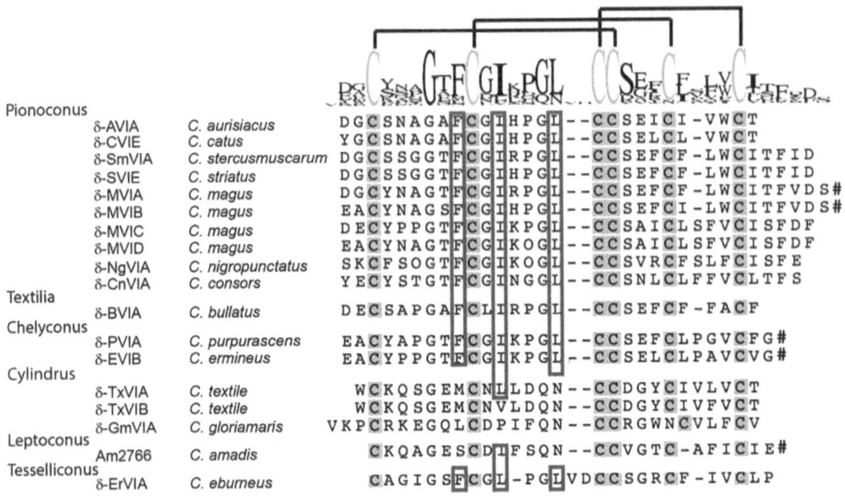

Figure 3 Sequences of δ-conotoxins from *Conus*. Alignments were constrained by the cysteine framework. Hydrophobic peptide residues deemed important for δ-conotoxin interactions with hydrophobic residues in the channel at neurotoxin binding site 6 are indicated by the *red boxes* (light gray in print versions). # denotes C-terminal amidation of the peptide. Sequence alignment and positional conservation of amino acids were constructed using Geneious software.

disulfide bridges in an inhibitor cysteine knot (ICK) motif (Fig. 3). Representatives of the δ-conotoxin family have been found in both fish- and worm-hunting *Conus* species (Bulaj et al., 2001), and it has been proposed what were anciently used as defensive toxins to protect the snail from predators have evolved into a major class of excitatory peptides used by piscivorous cone snails to capture prey (Aman et al., 2015; Jin et al., 2015). The δ-conotoxins promote VGSC activity through inhibition of channel inactivation by binding to neurotoxin binding site 6 located at the S3/S4 linker region of domain-IV (ie, a voltage sensor). δ-Conotoxin binding prevents fast inactivation and can result in a more negative membrane potential threshold for generation of action potentials. The result is prolonged action potentials and persistent firing of the neuron (Leipold, Hansel, Olivera, Terlau, & Heinemann, 2005). Leipold and coworkers identified the binding site for the δ-conotoxins as a group of three hydrophobic residues (Y1433-F1434-V1435) in the voltage sensor through cysteine-scanning mutagenesis of the conserved residues in this region. Furthermore, it has been hypothesized that δ-conotoxins participate in hydrophobic interactions with the channel (Leipold et al., 2005). The most important residues for δ-conotoxin activity are three large hydrophobic amino acids conserved across members of this toxin family from fish-hunting *Conus* species, namely Phe9, Ile12, and Leu16 (Fig. 3) (Bulaj et al., 2001).

3.1.2 ι-Conotoxins

Less well understood, with respect to VGSC agonists, are the ι-conotoxins, which possess four disulfide bridges between eight cysteine residues connected in either a type III (CC-C-C-CC) or type XI (-C-C-CC-CC-C-C) framework. To date, only two members of this toxin class have been characterized (Fiedler et al., 2008; Wang et al., 2009) with several other potential members of this class being observed in the venoms of piscivorous *Conus* (Buczek et al., 2008; Jimenez et al., 2003). These peptides promote VGSC activation by shifting the voltage dependence of activation in the hyperpolarizing direction (Buczek et al., 2007). ι-Conotoxins, unlike δ-conotoxins, have no observable effect on the voltage dependence of steady-stage inactivation (Buczek et al., 2007). The precise site of interaction within site 4 has yet to be defined, but it has been proposed that the less-well-ordered C-terminal tail of these peptides may be important for interactions with the channel (Fiedler et al., 2008). Several members of the ι-conotoxin class possess a single post-translationally modified amino acid residue (ie, L- to D-isomerization) near the C-terminus of the mature peptide (Buczek et al., 2008; Buczek, Yoshikami, Bulaj, Jimenez, & Olivera, 2005) that has been demonstrated to be critical for biological activity (Buczek et al., 2007).

3.2 Na_V1 antagonists

In contrast to those that promote neuronal firing through increased VGSC activity, several classes of conotoxins act to decrease the flow of Na^+ across the membrane. Antagonists of VGSC activity can be broadly classified into two groups: those that block the channel by physical occlusion of the Na^+ conductance pore and those that bind to sites distinct from the pore region and modulate channel gating.

3.2.1 μ-Conotoxins

The μ-conotoxins are perhaps the most well characterized of all conotoxins that act upon VGSCs (Green, Bulaj, & Norton, 2014). These peptides appear unique to the venom repertoires of piscivorous *Conus* species and are typified by three disulfide bridges between six cysteine residues arranged in a type III (CC-C-C-CC) framework (Fig. 4). The μ-conotoxin binding

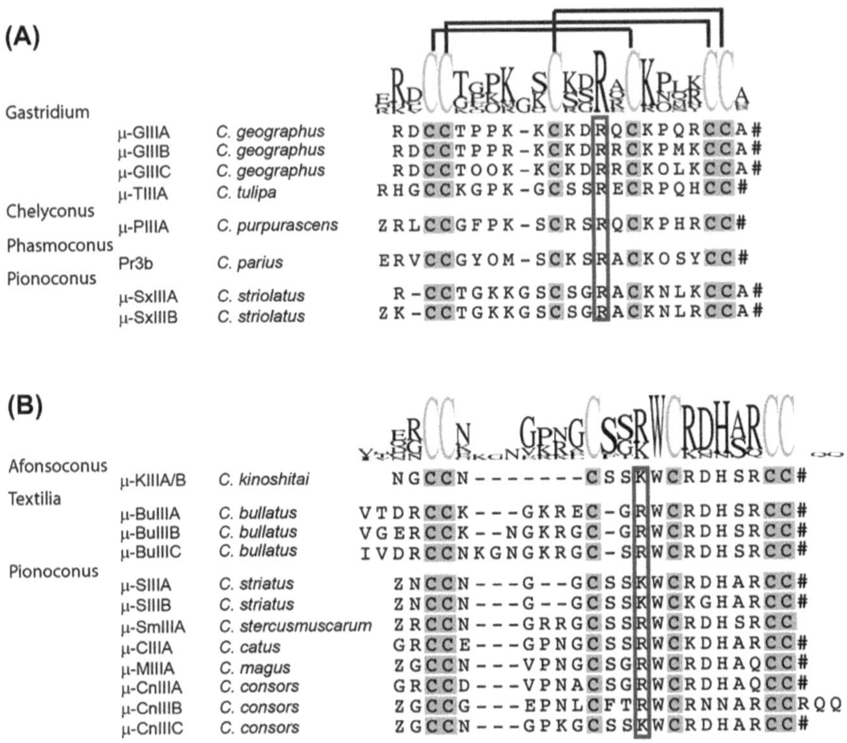

Figure 4 Sequences of M4 (Panel A) and M5 (Panel B) μ-conotoxins. Alignments were constrained by the cysteine framework. The basic residue important for voltage-gated sodium channel (VGSC) blockade (eg, Arg13 in μ-GIIIA) is indicated by the *red box* (light gray in print versions). # denotes C-terminal amidation of the peptide. Sequence alignment and positional conservation of amino acids were constructed using Geneious software.

site (site 1) overlaps with that of small alkaloid neurotoxins such as tetrodotoxin (TTX) and saxitoxin (STX) (Fozzard & Lipkind, 2010). However, given their larger molecular footprint, the μ-conotoxins exhibit greater VGSC subtype specificity compared to small molecule neurotoxins. Generally, the μ-conotoxins are highly basic in character, which is critical for their binding to acidic residues of the P-loop within the outer vestibule of the sodium-conducting pore. It has also been shown that a single basic residue within the second intercysteine loop of several μ-conotoxins (eg, Arg13 in μ-GIIIA) is critical for all-or-none block of single VGSCs (Fig. 4) (Chang, French, Lipkind, Fozzard, & Dudley, 1998; Hui, Lipkind, Fozzard, & French, 2002). Furthermore, numerous basic residues in the C-terminal half of these peptides are key for coordination of the peptide within the channel and were important in the determination of the tertiary structure of the channel, that being the arrangement of the four domains in a clockwise orientation about the pore, when viewed from an extracellular perspective (Choudhary, Aliste, Tieleman, French, & Dudley, 2007; Dudley et al., 2000; Li et al., 2001).

The first members of this conotoxin class were obtained from the venoms of *Conus geographus* (eg, μ-GIIIA, GIIIB, and GIIIC) and were found to possess exquisite specificity for sodium channels in muscle tissue over those from the brain (Cruz et al., 1985). In the 30 years since the initial discovery of μ-conotoxins GIIIA-GIIIC, over 20 members of this class of peptides have been identified. Several members of this conotoxin class possess altered selectivity profiles (Zhang et al., 2006). Most recently, a number of μ-conotoxins have been identified that preferentially block VGSC subtypes in the brain (eg, $Na_V1.2$) (Bulaj et al., 2005; Zhang et al., 2007). Structure–activity studies of μ-KIIIA from *Conus kinoshitai* showed that mutations made to residues on the C-terminal α-helical region of the peptide (specifically, alanine substitution of Lys7, Trp8, or Asp11) could increase discrimination between the neuronal and muscle subtypes in favor of $Na_V1.2$ (Zhang et al., 2007). Also, looking at the effects of mutations made in this region, particularly to Arg14, showed that subtype preference could be shifted in favor of sodium channels in the peripheral nervous system (eg, $Na_V1.7$) (McArthur et al., 2011) giving hope to the potential for engineering of these peptides in favor of pain-relevant VGSC subtypes. The ability of certain μ-conotoxins to discriminate between different VGSC subtypes has allowed for the pharmacological fractionation of VGSCs in certain neurons. The utility of μ-conotoxins as pharmacological tools was recently illustrated in work by Zhang, Wilson, Gajewiak, et al. (2013) where combinations of members of this class were used to determine the Na_V1-subunit composition of DRG neurons.

One issue facing the development of μ-conotoxins as viable therapeutic options is the high degree of sequence homology among VGSC subtypes, which are >50% identical with respect to amino acid sequence in their extracellular and transmembrane domains. As demonstrated by the structure–activity relationship studies of μ-KIIIA, these peptides are highly amenable to modification and certain amino acid replacements that can alter the selectivity profiles of these peptides. Also, given that there are at least 100 known piscivorous cone snails (Aman et al., 2015), with only a small number of these examined for the presence of μ-conotoxins, there remains an enormous potential to discover new members of this class with specificity for pain-relevant VGSC subtypes.

3.2.2 μO-contotoxins

Approximately a decade after the first report of VGSC pore-blocking conotoxins, a new family of peptides was discovered that bore no sequence similarity to the μ-conotoxins (McIntosh et al., 1995). Structurally, the μO-conotoxins are more closely related to the δ- and ω-conotoxins, in that the three disulfide bridges are topologically arranged in an ICK motif, which is hypothesized to contribute to the structural stability of these peptides (Fig. 5) (Daly et al., 2004). Of the μO-conotoxins (μO-MrVIA, MrVIB, and MfVIA) described to date, all are hydrophobic peptides that bind to site 4 on the extracellular loop of the domain II voltage sensor (S3/S4 loop), which is also the site of interaction for the scorpion β-toxins. μO-conotoxins effectively "trap" the channel in a resting state thereby inhibiting the movement of the voltage sensor and prevent the Na^+-conducting pore from opening. The μO-conotoxins are of particular interest due to their ability to inhibit both TTX-sensitive and TTX-resistant sodium currents in primary sensory neurons (Bulaj et al., 2006; Zorn et al., 2006). Furthermore, members of this class exhibit potent, and long-lasting, analgesic activity in multiple animal models of pain (Bulaj et al., 2006). It is thought that the analgesic activity of the μO-conotoxins arises from the ability to potently and selectively inhibit $Na_V1.8$, a subtype that is highly expressed in peripheral sensory and nociceptive neurons (Bulaj et al., 2006; Vetter et al., 2012).

3.2.3 μO§-conotoxins

As evidence of the continuing potential of cone snail venoms to yield new pharmacologically interesting compounds, the founding member of a new

Figure 5 Sequences of μO-conotoxins. Alignments were constrained by the 14 cysteine framework. Sequence alignment and positional conservation of amino acids were constructed using Geneious software.

class of VGSC-inhibiting peptides was identified in the venom of *C. geographus*, namely μO§-GVIIJ (Gajewiak et al., 2014). Unlike the other peptide classes that inhibit VGSC activity, members of this class possess a unique disulfide framework containing an odd number of cysteine residues (C-C-CC-C-C*-C) where the "extra" cysteine (C*) is posttranslationally modified through S-cysteinylation (Gajewiak et al., 2014). This unique feature is important to the mechanism of action of μO§-conotoxins, specifically that members of this class form a covalent association (eg, disulfide bridge) through disulfide exchange with a free cysteine residue on the surface of the channel and elicit nearly irreversible blockade of six of the eight VGSC subtypes tested (excluding $Na_V1.5$ and 1.8) (Gajewiak et al., 2014; Zhang et al., 2015). The importance of the "extra" cysteine residue was recently explored through isosteric replacement of Cys24 with serine, and it was demonstrated that the extra cysteine was critical for the persistent block of VGSCs, but not for the inhibition Na^+ current per se (Green et al., 2016). μO§-conotoxin GVIIJ was also important in identifying a new neurotoxin binding site (site 8) on the surface of the channel that was both separate and distinct from the binding sites of the μ- and μO-conotoxins (sites 1 and 4, respectively) (Gajewiak et al., 2014; Green et al., 2016). Beyond the importance of Cys24, recent structure–activity relationship studies identified additional amino acid residues that are important for the activity of μO§-GVIIJ (Green et al., 2016). These studies identified three residues in the sequence of μO§-GVIIJ (Lys12, Arg14, and Tyr16) that are crucial for activity. These residues reside on a structurally less well-defined loop region of the peptide, separate from the β-hairpin loop that presents Cys24. Given the relatively small size of this peptide, it is difficult to say that these regions are "topologically distinct." However, these studies demonstrated a "functionally bipartite" mode of action by this peptide in that the identity of the residue at position 24 is responsible for k_{off} (ie, off-rate) and the residues in loop 2 are critical for k_{on} (ie, on-rate). Interestingly, the ability of μO§-GVIIJ to inhibit VGSCs is also influenced by the presence of the "accessory" β-subunits (Gajewiak et al., 2014), and this will be discussed in greater detail in the following sections.

4. $Na_V\beta$ EFFECTS ON CONOTOXIN ACTIVITY

As the pore-bearing Na_V1-subunit is known to be the minimal requisite structure for a fully functional sodium channel, conotoxin activity has traditionally been assessed against the α-subunit alone in heterologous expression systems such as *Xenopus* oocytes. However, more recently the importance of the roles of $Na_V\beta$-subunits has come to be realized. Neurons in mammals express one or more Na_V1-subtypes, each associated with one or more $Na_V\beta$-subunits (Catterall et al., 2007). These associations are not well understood, and VGSC-blocking peptides have proven useful in shedding light on the in vivo composition of VGSC subtypes and the roles of these proteins with respect to toxin binding and action.

4.1 Effects of $Na_V\beta$ on μ-conotoxin inhibition

The effects of coexpression of various β-subunits with TTX-sensitive α-subunits (eg, $Na_V1.1$, 1.2, 1.6, and 1.7) on μ-conotoxin activity was recently assessed in channels heterologously expressed in oocytes (Zhang, Wilson, Azam, et al., 2013). The presence of $Na_V\beta$-subunits was shown to have little or no effect on VGSC block by site 1 small molecule inhibitors (ie, TTX and STX), but was shown to have significant effects on the affinities of the μ-conotoxins (Zhang, Wilson, Azam, et al., 2013). The effects of $Na_V\beta$-subunit coexpression were examined in the context of both binary (eg, 1 Na_V1:1 $Na_V\beta$ or αβ) and ternary (eg, 1 Na_V1:2 $Na_V\beta$ or αββ) VGSC complexes. Generally, coexpression of an Na_V1-subunit with either $Na_V\beta1$ or β3 led to modest increases in k_{on} and coexpression with $Na_V\beta2$ or β4 led to significant reduction in k_{on} (Zhang, Wilson, Azam, et al., 2013). Moreover, coexpression with either $Na_V\beta1$ or β3 had little effect on off-rate (k_{off}) and a pronounced decrease in off-rate was generally observed with coexpression of $Na_V\beta2$ or β4 (Zhang, Wilson, Azam, et al., 2013). In the ternary complexes, these results appeared consistent, albeit less pronounced. The differences between $Na_V\beta$ coexpression on small molecule and μ-conotoxin kinetics led the authors to propose that β-subunits are capable of encroaching upon the more superficial site occupied by μ-conotoxins (ie, site 1b), but not upon the site of action of the alkaloid neurotoxins which is deeper in the pore (ie, site 1a) (Zhang, Wilson, Azam, et al., 2013).

4.2 Effects of $Na_V\beta$ on gating-modifier conotoxins

The effects of $Na_V\beta$-subunit coexpression on the susceptibility of pain-relevant Na_V1-subtypes to gating-modifier toxins (eg, μO- and μO§-conotoxins) have also been explored (Wilson et al., 2011, 2015). Mentioned earlier, μO-conotoxins such as μO-MrVIB act by blocking channel activation. Coexpression of certain $Na_V\beta$-subunits altered the time course of fast inactivation, as well as the voltage sensitivity of activation and inactivation

(Wilson et al., 2011). Coexpression of any one of the four NaVβ-subunits increased k_{on} and decreased k_{off}, which translated to higher-affinity block by μO-MrVIB. A ternary complex consisting of NaV1.8 coexpressed with both NaVβ1 and β2was blocked by μO-MrVIB with on-rates that fell between those of either β-subunit in binary combination with the α-subunit (Wilson et al., 2011).

Most recently, the coexpression of each of the four NaVβ-subunits with NaV1-subtypes with respect to μO§-conotoxin activity was studied (Gajewiak et al., 2014; Wilson et al., 2015). In the initial pharmacological characterization, analogs of μO§-GVIIJ were screened against rat NaV1.7 coexpressed with NaVβ1-4 in *Xenopus* oocytes. When NaVβ1 or β3 were coexpressed, the toxin blocked in a manner similar to what was observed against the α-subunit alone, whereas coexpression with NaVβ2 or β4 made the channel completely resistant to block (Gajewiak et al., 2014). It is interesting to note that both NaVβ2 and β4 are covalently associated with the α-subunit via a disulfide bond (Gilchrist, Das, Van Petegem, & Bosmans, 2013). It has been recently shown that Cys910 in the NaV1.2 domain II pore loop, which is also the site for μO§-GVIIJ binding (site 8 (Gajewiak et al., 2014)), is the point of association between the NaV1- and NaVβ2/β4-subunits (Das, Gilchrist, Bosmans, & Van Petegem, 2016). The composition of NaVβ-subunits in rodent sciatic nerve was investigated by examining the ability of μO§-GVIIJ to block compound action potentials of large, myelinated fibers (A-CAPs) and those of small, unmyelinated fibers (C-CAPs). The peptide blocked neither A-CAPs nor C-CAPs suggesting that the NaV1s in these axons were associated with NaVβ2- or β4-subunits (Wilson et al., 2015). To test whether this was a result of an inability of μO§-GVIIJ to block endogenously expressed VGSCs, the peptide was also assessed for activity against dissociated rat DRG neurons. Partial block of the sodium current was observed in small neurons while sodium currents of large neurons were not blocked (Wilson et al., 2015). Together, these results demonstrate the potential for members of this newly described family of toxins to serve as tools to pharmacologically dissect the NaVβ-subunit composition of VGSCs of different neurons.

5. CONCLUSIONS AND FUTURE PROSPECTS

Much has been learned about the roles of various ion channels in electrical signaling since the early efforts of Hodgkin and Huxley (1952).

Discovery and design of compounds that modulate the activities of these channels has been greatly aided by crystal structures, allowing researchers to identify potential sites of interaction that may promote or inhibit channel activity. Despite recent successes in obtaining crystal structures of several bacterial sodium channels (Catterall, 2001; Payandeh, Scheuer, Zheng, & Catterall, 2011), the structure of the mammalian sodium channel has yet to be solved. This has left researchers almost completely reliant on either homology models, based on similarly structured ion channels (Korkosh, Zhorov, & Tikhonov, 2014), or on pharmacological information obtained from exposure of the vertebrate channels to compounds that influence VGSC activity. Conotoxins have been, and continue to be, valuable tools to understand VGSC structure and function. Alongside peptide neurotoxins from other venomous organisms (eg, spiders, scorpions, snakes, and others), conotoxins have been instrumental in identifying several distinct binding sites on the channel (Catterall et al., 2007).

Beyond their roles as tools to study VGSCs, members of the μ-, μO-, and μO§-conotoxin classes, in particular, have an enormous potential as therapeutics for severe pain and other neurological disorders. Such disorders commonly stem from dysfunction of the processes responsible for regulating neuronal firing. Spontaneous and rapid firing of neurons may contribute to the neuronal hyperexcitability related to neuropathic pain and a host of other neurological disorders including: epilepsy, multiple sclerosis, Huntington disease, and others. Several of the nine described VGSC subtypes have been demonstrated to play important roles in numerous neurological disorders, including neuropathic pain. Many of these subtypes exhibit increased activity, or their expression is significantly upregulated, following illness or injury. Given the intrinsic ability of conotoxins to potently and selectively target VGSCs, and their amenability to modification for enhanced subtype specificity, these peptides may serve as promising alternatives to currently available therapeutics that exhibit potentially dangerous off-target effects.

The venoms of *Conus* are a largely untapped resource for compounds that modulate VGSC activity with less than 0.3% of the estimated 50,000–100,000 neuroactive peptides having been characterized so far. Furthermore, these peptides have been obtained from the venoms of only 37 of the over 500 known species of cone snails (www.conoserver.org). In addition to *Conus*, there are an estimated 12,000 species of venomous mollusks, spread across different genera, most of these having remained largely unexplored for VGSC-targeting peptides. In the 35 years since the first report of VGSC-inhibiting μ-GIIIA/B/C, over 44 distinct conotoxins, distributed across five toxin families, have been discovered that interact specifically with sodium channels. The discovery of new venom-derived toxins is expected to increase significantly with the recent advances made in the field of transcriptomics. Using methods such as mass-spec matching, the crude venoms can be rapidly analyzed, and related peptides can be identified for pharmacological characterization and potential therapeutic development.

ACKNOWLEDGMENTS

The work of the authors has been supported by National Institute of General Medical Sciences Grant GM048677. We are grateful to Dr Doju Yoshikami for his review of this chapter and for the integral work performed by him and Dr Minmin Zhang.

REFERENCES

Aman, J. W., Imperial, J. S., Ueberheide, B., Zhang, M. M., Aguilar, M., Taylor, D., … Olivera, B. M. (2015). Insights into the origins of fish hunting in venomous cone snails from studies of *Conus tessulatus*. *Proceedings of the National Academy of Sciences of the United States of America, 112*(16), 5087–5092.

Buczek, O., Jimenez, E. C., Yoshikami, D., Imperial, J. S., Watkins, M., Morrison, A., & Olivera, B. M. (2008). I1-superfamilly conotoxins and prediction of single D-amino acid occurrence. *Toxicon: Official Journal of the International Society on Toxinology, 51*(2), 218–229.

Buczek, O., Wei, D., Babon, J. J., Yang, X., Fiedler, B., Chen, P., … Norton, R. S. (2007). Structure and sodium channel activity of and excitatory I1-superfamily conotoxin. *Biochemistry, 46*(35), 9929–9940.

Buczek, O., Yoshikami, D., Bulaj, G., Jimenez, E. C., & Olivera, B. M. (2005). Post-translational amino acid isomerization: a functionally important d-amino acid in an excitatory peptide. *Journal of Biological Chemistry, 280*(6), 4247–4253. http://dx.doi.org/10.1074/jbc.M405835200.

Bulaj, G., De La Cruz, R., Azimi-Zonooz, A., West, P., Watkins, M., Yoshikami, D., & Olivera, B. M. (2001). Delta-conotoxin structure/function through a cladistic analysis. *Biochemistry, 40*(44), 13201–13208.

Bulaj, G., West, P. J., Garrett, J. E., Watkins, M., Zhang, M. M., Norton, R. S., … Olivera, B. M. (2005). Novel conotoxins from *Conus striatus* and *Conus kinoshitai* selectively block TTX-resistant sodium channels. *Biochemistry, 44*(19), 7259–7265.

Bulaj, G., Zhang, M. M., Green, B. R., Fiedler, B., Layer, R. T., Wei, S., … Olivera, B. M. (2006). Synthetic μO-conotoxin MrVIB blocks TTX resistant sodium channel NaV1.8 and has long-lasting analgesic activity. *Biochemistry, 45*(23), 7404–7414.

Catterall, W. A. (1995). Structure and function of voltage-gated sodium channels. *Annual Review of Biochemistry, 64*, 493–531.

Catterall, W. A. (2001). Physiology. A one-domain voltage-gated sodium channel in bacteria. *Science, 294*(5550), 2306–2308.

Catterall, W. A. (2012). Voltage-gated sodium channels at 60: structure, function and pathophysiology. *Journal of Physiology, 590*(11), 2577–2589.

Catterall, W. A., Cestèle, S., Yarov-Yarovoy, V., Yu, F. H., Konoi, K., & Scheuer, T. (2007). Voltage-gated ion channels and gating modifier toxins. *Toxicon: Official Journal of the International Society on Toxinology, 49*(2), 124–141.

Chang, N. S., French, R. J., Lipkind, G. M., Fozzard, H. A., & Dudley, S. C. J. (1998). Predominant interactions between m-conotoxin Arg-13 and the skeletal muscle Na^+ channel localized by mutant cycle analysis. *Biochemistry, 37*(13), 4407–4419.

Choudhary, G., Aliste, M. P., Tieleman, D. P., French, R. J., & Dudley, S. C. J. (2007). Docking of μ-conotoxin GIIIA in the sodium channel outer vestibule. *Channels (Austin), 1*(5), 344–352.

Cruz, L. J., Gray, W. R., Olivera, B. M., Zeikus, R. D., Kerr, L., Yoshikami, D., & Moczylowski, E. (1985). *Conus geographus* toxins that discriminate between neuronal and muscle sodium channels. *Journal of Biological Chemistry, 260*(16), 9280–9288.

Daly, N. L., Ekberg, J. A., Thomas, L., Adams, D. J., Lewis, R. L., & Craik, D. J. (2004). Structures of μO-conotoxins from *Conus marmoreus*. Inhibitors of tetrodotoxin (TTX)-sensitive and TTX-resistant sodium channels in mammalian sensory neurons. *Journal of Biological Chemistry, 279*(24), 25774–25782.

Das, S., Gilchrist, J., Bosmans, F., & Van Petegem, F. (2016). Binary architecture of the NaV1.2-β2 signalling complex. *eLife, 5*, 1–21.

Dudley, S. C. J., Chang, N., Hall, J., Lipkind, G., Fozzard, H. A., & French, R. J. (2000). μ-Conotoxin GIIIA interactions with the voltage-gated Na^+ channel predict a clockwise arrangement of the domains. *Journal of General Physiology, 116*(5), 679–690.

Elliger, C. A., Richmond, T. A., Lebaric, Z. N., Pierce, N. T., Sweedler, J. V., & Gilly, W. F. (2011). Diversity of conotoxin types from *Conus californicus* reflects a diversity of prey types and a novel evolutionary history. *Toxicon: Official Journal of the International Society on Toxinology, 57*(2), 311–322. http://dx.doi.org/10.1016/j.toxicon.2010.12.008.

Fiedler, B., Zhang, M. M., Buczek, O., Azam, L., Bulaj, G., Norton, R. S., … Yoshikami, D. (2008). Specificity, affinity and efficacy of iota-conotoxin RXIA, an agonist of voltage-gated sodium channels NaV1.2, 1.6 and 1.7. *Biochemical Pharmacology, 75*(12), 2334–2344.

Fozzard, H. A., & Lipkind, G. (2010). The tetrodotoxin binding site is within the outer vestibule of the sodium channel. *Marine Drugs, 8*(2), 219–234.

Gajewiak, J., Azam, L., Imperial, J., Walewska, A., Green, B. R., Bandyopadhyay, P. K., … Zhang, M. M. (2014). A disulfide tether stabilizes the block of sodium channels by the conotoxin μO§-GVIIJ. *Proceedings of the National Academy of Sciences of the United States of America, 111*(7), 2758–2763.

Gilchrist, J., Das, S., Van Petegem, F., & Bosmans, F. (2013). Crystallographic insights into sodium-channel modulation by the β4 subunit. *Proceedings of the National Academy of Sciences of the United States of America, 110*(51), 5016–5024.

Gilly, W. F., Richmond, T. A., Duda, T. F., Elliger, C., Lebaric, Z., Schulz, J., … Sweedler, J. V. (2011). A diverse family of novel peptide toxins from an unusual cone snail, *Conus californicus*. *Journal of Experimental Biology, 214*(1), 147–161. http://dx.doi.org/10.1242/jeb.046086.

Gray, W. R., Olivera, B. M., & Cruz, L. J. (1988). Peptide toxins from venomous *Conus* snails. *Annual Review of Biochemistry, 57*, 665–700.

Green, B. R., Bulaj, G., & Norton, R. S. (2014). Structure and function of μ-conotoxins, peptide-based sodium channel blockers with analgesic activity. *Future Medicinal Chemistry, 6*(15), 1677–1698.

Green, B. R., Gajewiak, J., Chhabra, S., Skalicky, J. J., Zhang, M. M., Rivier, J. E., … Norton, R. S. (2016). Structural basis for the inhibition of voltage-gated sodium channels by conotoxin μO§-GVIIJ. *Journal of Biological Chemistry, 291*(13), 7205–7220.

Hasson, A., Shon, K., Olivera, B. M., & Spira, M. E. (1995). Alterations of voltage-activated Na current by a novel conotoxin from the venom of *Conus gloriamaris*. *Journal of Neurophysiology, 73*, 1295–1301.

Hodgkin, A. L., & Huxley, A. F. (1952). A quantitative description of membrane current and its application to conduction and excitation in nerve. *Journal of Physiology, 117*(4), 500–544.

Hui, K., Lipkind, G., Fozzard, H. A., & French, R. J. (2002). Electrostatic and steric contributions to block of the skeletal muscle sodium channel by μ-conotoxin. *Journal of General Physiology, 119*(1), 45–54. http://dx.doi.org/10.1085/jgp.119.1.45.

Isom, L. L. (2001). Sodium channel beta subunits: anything but auxiliary. *Neuroscientist, 7*(1), 42–54.

Jimenez, E. C., Shetty, R. P., Lirazan, M., Rivier, J., Walker, C., Abogadie, F. C., ... Olivera, B. M. (2003). Novel excitatory *Conus* peptides define a new conotoxin family. *Journal of Neurochemistry, 85*(3), 610–621.

Jin, A.-H., Israel, M. R., Inserra, M. C., Smith, J. J., Lewis, R. L., Alewood, P. F., ... Dutertre, S. (2015). δ-Conotoxin SuVIA suggests an evolutionary link between ancestral predator defence and the origin of fish-hunting behaviour in carnivorous cone snails. *Proceedings. Biological Sciences, 282*(1811), 20150817.

Korkosh, V. S., Zhorov, B. S., & Tikhonov, D. B. (2014). Folding similarity of the outer pore region in prokaryotic and eukaryotic sodium channels revealed by docking of conotoxins GIIIA, PIIIA and KIIIA in a NaVAb-based model of NaV1.4. *Journal of General Physiology, 144*(3), 231–244.

Leipold, E., Hansel, A., Olivera, B. M., Terlau, H., & Heinemann, S. H. (2005). Molecular interaction of delta-conotoxins with voltage-gated sodium channels. *FEBS Letters, 579*(18), 3881–3884.

Li, R. A., Ennis, I. L., French, R. J., Dudley, S. C. J., Tomaselli, G. F., & Marbán, E. (2001). Clockwise domain arrangement of the sodium channel revealed by μ-conotoxin (GIIIA) docking orientation. *Journal of Biological Chemistry, 276*(14), 11072–11077.

McArthur, J. R., Singh, G., McMaster, D., Winkfein, R., Tieleman, D. P., & French, R. J. (2011). Interactions of key charged residues contributing to selective block of neuronal sodium channels by μ-conotoxin KIIIA. *Molecular Pharmacology, 80*(4), 573–584.

McIntosh, J. M., Hasson, A., Spira, M. E., Gray, W. R., Li, W., Marsh, M., ... Olivera, B. M. (1995). A new family of conotoxins that blocks voltage-gated sodium channels. *Journal of Biological Chemistry, 270*(28), 16796–16802.

Payandeh, J., Scheuer, T., Zheng, N., & Catterall, W. A. (2011). The crystal structure of a voltage-gated sodium channel. *Nature, 475*(7356), 353–358.

Puillandre, N., Bouchet, P., Duda, T. F., Jr., Kauferstein, S., Kohn, A. J., Olivera, B. M., ... Meyer, C. (2014). Molecular phylogeny and evolution of the cone snails (Gastropoda, Conoidea). *Molecular Phylogenetics and Evolution, 78*, 290–303. http://dx.doi.org/10.1016/j.ympev.2014.05.023.

Shon, K.-J., Hasson, A., Spira, M. E., Cruz, L. J., Gray, W. R., & Olivera, B. M. (1994). δ-Conotoxin GmVIA, a novel peptide from the venom of *Conus gloriamaris*. *Biochemistry, 33*, 11420–11455.

Terlau, H., & Olivera, B. M. (2004). *Conus* venoms: a rich source of novel ion channel-targeted peptides. *Physiological Reviews, 84*(1), 41–68.

Terlau, H., Shon, K. J., Grilley, M., Stocker, M., Stühmer, W., & Olivera, B. M. (1996). Strategy for rapid immobilization of prey by a fish-hunting marine snail. *Nature, 381*(6578), 148–151

Vetter, I., Dekan, Z., Knapp, O., Adams, D. J., Alewood, P. F., & Lewis, R. L. (2012). Isolation, characterization and total regioselective synthesis of the novel μO-conotoxin MfVIA from *Conus magnificus* that targets voltage-gated sodium channels. *Biochemical Pharmacology, 84*(4), 540–548.

Wang, L., Liu, J., Pi, C., Zeng, M., Zhou, M., Jiang, X., … Xu, A. (2009). Identification of a novel M-superfamily conotoxin with the ability to enhance tetrodotoxin sensitive sodium currents. *Archives of Toxicology, 83*(10), 925–932.

Wilson, M. J., Zhang, M. M., Azam, L., Olivera, B. M., Bulaj, G., & Yoshikami, D. (2011). NaVβ subunits modulate the inhibition of NaV1.8 by the analgesic gating modifier μO-conotoxin MrVIB. *Journal of Pharmacology and Experimental Therapeutics, 338*(2), 687–693.

Wilson, M. J., Zhang, M. M., Gajewiak, J., Azam, L., Rivier, J., Olivera, B. M., & Yoshikami, D. (2015). α- and β-subunit composition of voltage-gated sodium channels investigated with μ-conotoxins and the recently discovered μO§-conotoxin GVIIJ. *Journal of Neurophysiology, 113*(7), 2289–2301.

Zhang, M. M., Fiedler, B., Green, B. R., Catlin, P., Watkins, M., Garrett, J. E., … Bulaj, G. (2006). Structural and functional diversities among μ-conotoxins targeting TTX-resistant sodium channels. *Biochemistry, 45*(11), 3723–3732.

Zhang, M. M., Green, B. R., Catlin, P., Fiedler, B., Azam, L., Chadwick, A., … Bulaj, G. (2007). Structure/function characterization of μ-conotoxin KIIIA, and analgesic, nearly irreversible blocker of mammalian neuronal sodium channels. *Journal of Biological Chemistry, 282*(42), 30699–30706.

Zhang, M. M., Wilson, M. J., Azam, L., Gajewiak, J., Rivier, J., Bulaj, G., … Yoshikami, D. (2013). Co-expression of NaVβ subunits alters the kinetics of inhibition of voltage-gated sodium channels by pore-blocking μ-conotoxins. *British Journal of Pharmacology, 168*(7), 1597–1610.

Zhang, M. M., Wilson, M. J., Gajewiak, J., Rivier, J. E., Bulaj, G., Olivera, B. M., & Yoshikami, D. (2013). Pharmacological fractionation of tetrodotoxin-sensitive currents in rat dorsal root ganglion neurons by μ-conotoxins. *British Journal of Pharmacology, 169*(1), 102–114.

Zhang, M. M., Gajewiak, J., Azam, L., Bulaj, G., Olivera, B. M., & Yoshikami, D. (2015). Probing the redox states of sodium channel cysteines at the binding site of μO§-conotoxin GVIIJ. *Biochemistry, 54*(25), 3911–3920.

Zorn, S., Leipold, E., Hansel, A., Bulaj, G., Olivera, B. M., Terlau, H., & Heinemann, S. H. (2006). The μO-conotoxin MrVIA inhibits voltage-gated sodium channels by association with domain-3. *FEBS Letters, 580*(5), 1360–1364.

CHAPTER FOUR

Convergent Evolution of Tetrodotoxin-Resistant Sodium Channels in Predators and Prey

G. Toledo*, C. Hanifin§, S. Geffeney§ and E.D. Brodie, III*,[1]
*University of Virginia, Charlottesville, VA, United States
§Utah State University Uintah Basin, Vernal, UT, United States
[1]Corresponding author: E-mail: bbrodie@virginia.edu

Contents

Abstract

Convergent evolution of similar adaptive traits may arise from either common or disparate molecular and physiological mechanisms. The forces that determine the degree of underlying mechanistic similarities across convergent phenotypes are highly debated and poorly understood. Some garter snakes are able to consume newts that possess the channel blocking compound tetrodotoxin (TTX). Despite belonging to unrelated lineages, both the predators and prey have independently evolved remarkably similar physiological mechanisms of resistance to TTX that involve chemical and structural changes in voltage-gated sodium channels (Na_V). The evolution of TTX resistance in this predator–prey pair constitutes a natural experiment that allows us to explore

Current Topics in Membranes, Volume 78
ISSN 1063-5823
http://dx.doi.org/10.1016/bs.ctm.2016.07.006

the causes of molecular convergence. Here, we review broad patterns of convergence at the level of amino acid changes in Na_V channels of animals that evolved TTX resistance and make comparisons to known TTX-resistant channels that did not evolve under the selective pressures imposed by TTX. We conclude that convergence likely stems from the interplay of the target specificity of TTX and functional constraints of Na_V that are shared among taxa. These and other factors can limit channel evolution to favor a few functionally permissible paths of adaptation, which can explain the observed predictability of changes to channel structure. By studying the functional causes of convergence in Na_V channels, we can further our understanding of the role of these important channel proteins at the center of the evolution of the nervous system.

1. INTRODUCTION

Convergent phenotypes arise when unrelated taxa evolve similar adaptive solutions to meet demands imposed by similar environmental pressures. Explorations of the mechanisms of adaptation from a molecular perspective have revealed that phenotypic convergence can arise from either common or disparate underlying developmental and physiological bases (Arendt & Reznick, 2008; Arnegard, Zwickl, Lu, & Zakon, 2010). In some cases, convergent phenotypes involve modifications of the same genes and even the same substitutions within genes. In other examples, different structural and molecular building blocks contribute to phenotypic similarities (Manceau, Domingues, Linnen, Rosenblum, & Hoekstra, 2010). The probability that a common gene will contribute to convergence in different taxa increases with phylogenetic relatedness, but convergence associated with a common gene can still occur in unrelated lineages (Conte, Arnegard, Peichel, & Schluter, 2012). The phenomena influencing the level of underlying mechanistic similarity in convergent phenotypes are not well understood, and resolving the role of functional constraint in this process is critical.

Predator—prey systems in which the fitness of each species is governed by the presence of, or resistance to, a toxin are ideal for testing hypotheses about convergent evolution. In these systems, the animals that feed on toxic organisms as well as the animals that use the toxin for defense must evolve resistance to the same toxin. Although the strength of selection will vary for prey and predators, both organisms must converge at some point along the adaptive path to resistance. Because predators and prey typically belong to evolutionarily divergent taxa, convergent phenotypes are less likely to arise through parallel adaptation at the molecular level because of differences in the physiological and functional requirements of these organisms. As a result, parallel changes associated with such predator–prey systems can provide important insight into the phenomena underlying mechanistic similarity in convergent phenotypes.

The broad taxonomic range of the neurotoxin tetrodotoxin (TTX) provides an exceptional case study of convergent evolution of toxin resistance. TTX has a very specific mode of action and has been extensively studied in the context of structure/function models of its protein target: voltage-gated sodium channels (Na_V) (Moczydlowski, 2013). Studies of the adaptive history and molecular basis of TTX resistance have been able to focus directly on the specific regions of Na_V channels that regulate TTX-binding (Choudhary, Yotsu-Yamashita, Shang, Yasumoto, & Dudley, 2003; Fozzard & Lipkind, 2010; Santarelli, Eastwood, Dougherty, Horn, & Ahern, 2007).

This toxin has been identified in a remarkable array of species that includes prokaryotes, multiple invertebrates including cephalopods, flatworms, gastropods, as well as many vertebrate taxa including fish and amphibians (Hanifin, 2010; Williams, Hanifin, Brodie, & Brodie, 2010). In animals, TTX serves a defensive role that protects prey species from predation. However, some predators, including snakes, do successfully prey on TTX-bearing taxa. Ultimately, this broad taxonomic distribution of a toxin that targets fundamental physiological processes allows examination of the mechanistic basis of phenotypic convergence across a broad spectrum of species and ecological roles.

Here we will focus on the mechanistic basis of TTX resistance in TTX-bearing salamanders and garter snakes that prey on those salamanders. The modern newts (family: Salamandridae) are a monophyletic clade of salamanders that include both New World and Old World genera. All genera in this group likely possess TTX, but the best studied of these is the genus *Taricha*, which includes four species known to possess high levels of TTX (Hanifin, Brodie, & Brodie, 2008). This genus is distributed along the western coast of North America and cooccurs with garter snakes in the genus *Thamnophis*. Some species and populations of *Thamnophis* are known to eat TTX-bearing *Taricha* and this predator–prey interaction has become one of the best-studied examples of arms race coevolution (Brodie, Ridenhour, & Brodie, 2002; Feldman, Brodie, Brodie, & Pfrender, 2012; Hanifin et al., 2008; Thompson, 2005). Both predator and prey lineages have evolved resistance to TTX through adaptive substitutions in their Na_V channels (Feldman et al., 2012; Geffeney, Fujimoto, Brodie, Brodie, & Ruben, 2005; Hanifin & Gilly, 2015).

2. TETRODOTOXIN AS AN AGENT OF SELECTION

The ability to initiate and propagate action potentials (APs) is a fundamental component of vertebrate nerve and muscle function. These signals are generated by the influx of sodium (Na^+) ions through Na_V channels, which are ubiquitous in vertebrates. TTX disrupts the initiation and propagation of APs in excitable tissues. The toxin selectively binds and blocks the outer pore of Na_V channels and prevents the influx of Na^+ ions that carry the current responsible for membrane depolarization during the initial phase of the AP (Hille, 2001; Narahashi, 2008). The detrimental consequences of blocking AP generation include the inability of nerve signals to reach targets and prevention of muscle fibers contraction. A primary symptom of TTX intoxication in animals is paralysis resulting from loss of muscle function (Clark, Williams, Nardt, & Manoguerra, 1999). This muscle paralysis can also interrupt a predator from ingesting a toxic prey animal, allowing the prey to survive the encounter. Ultimately, the effects of TTX kill the majority of nonresistant predators that might encounter the prey (Brodie, 1968; Williams et al., 2010). The rapid, lethal effects of TTX make it a powerful antipredator weapon for animals that can either produce or sequester high levels of the toxin (Hanifin & Gilly, 2015; Jost et al., 2008).

Because of the fundamental role of Na_V channels in neuromuscular function, the presence of endogenous TTX in TTX-bearing animals and exposure to TTX in predators of those animals exerts strong convergent selection on both predators and prey. Even exposure to nonlethal levels impairs locomotor performance and likely has significant negative effects on survivorship (Brodie & Brodie, 1999). Studies of coevolution between TTX-bearing newts and snake predators indicate that multiple populations of snakes have evolved elevated TTX resistance in response to evolution of elevated TTX levels in prey (Hanifin et al., 2008). Newts do not posses a means of sequestering the toxin, and the internal organs including heart, muscle, and lungs are exposed to high levels of TTX (Wakely, Fuhrman, Fuhrman, Fischer, & Mosher, 1966) generating strong selection favoring TTX resistance (Hanifin & Gilly, 2015).

2.1 The interaction of tetrodotoxin with Na_V

The alpha-subunit of Na_V channels, which forms the pore of the voltage-gated ion channel, is encoded by the *SCNA* gene family. In vertebrates, whole genome and tandem gene duplication events have given rise to multiple Na_V paralogs. The number of paralogs varies among lineages; amphibians have six, nonmammalian amniotes have nine, and placental mammals have 10 (Lopreato et al., 2001; Zakon, 2012; Zakon, Jost, & Lu, 2011; Zakon, Jost, Zwickl, Lu, & Hillis, 2009). Na_V channels are composed of four domains (DI—DIV), each of which includes six transmembrane segments (S1—S6). The extracellular linkers between S5 and S6 form the outer portion of the channel pore that controls ion selectivity and conductance of the channel (Chiamvimonvat, Perez-Garcia, Tomaselli, & Marban, 1996; Favre, Moczydlowski, & Schild, 1996; Heinemann, Terlau, Stühmer, Imoto, & Numa, 1992; Terlau et al., 1991; Yamagishi, Li, Hsu, Marbán, & Tomaselli, 2001). These linkers form a constriction in the ionic pathway, called the selectivity filter, that allows entry of Na^+ while inhibiting the influx of other cations such as calcium (Ca^{++}) and potassium (K^+) (Fig. 1; Schlief, Schönherr, Imoto, & Heinemann, 1996; Sun, Favre, Schild, & Moczydlowski, 1997). The selectivity filter consists of four amino acids, Asp-Glu-Lys-Ala (DEKA), one from each of the four domains. Three positions downstream from the selectivity filter in each domain is a ring of charged amino acids represented in mammals by the sequence Glu-Glu-Met-Asp (EEMD) (Penzotti, Fozzard, Lipkind, & Dudley, 1998; Terlau et al., 1991; Tikhonov & Zhorov, 2011, 2012). Negative charges in this outer ring create an electrostatic cloud that serves to attract Na^+ ions into the pore outer vestibule, and alterations in the amino acids of the selectivity filter or outer charged ring can severely reduce channel conductance of sodium ions (Mahdavi & Kuyucak, 2015; Terlau et al., 1991).

TTX blocks ion conduction through Na_V channels by binding to amino acids in the outer pore of the channel, thereby occluding the pore. The binding affinity of the channel for TTX can be dramatically altered by substitutions in a limited number of amino acids in the region that includes the selectivity filter, the downstream outer charged ring, and the resides that link these two rings (Choudhary et al., 2003; Noda, Suzuki, Numa, & Stühmer, 1989; Terlau et al., 1991). For example, substitution of a nonaromatic amino acid for an aromatic amino acid at a single position in the domain I outer pore decreases TTX-binding affinity of the channel by several orders of magnitude (Backx, Yue, Lawrence, Marban, & Tomaselli, 1992). Less dramatic shifts in channel binding affinity can be induced by substitutions outside this region in portions of the pore loops that affect the position of amino acids in the selectivity filter and outer charged ring (Yamagishi, Janecki, Marban, & Tomaselli, 1997). The specific role of many of the observed substitutions associated with TTX resistance in newts and snakes

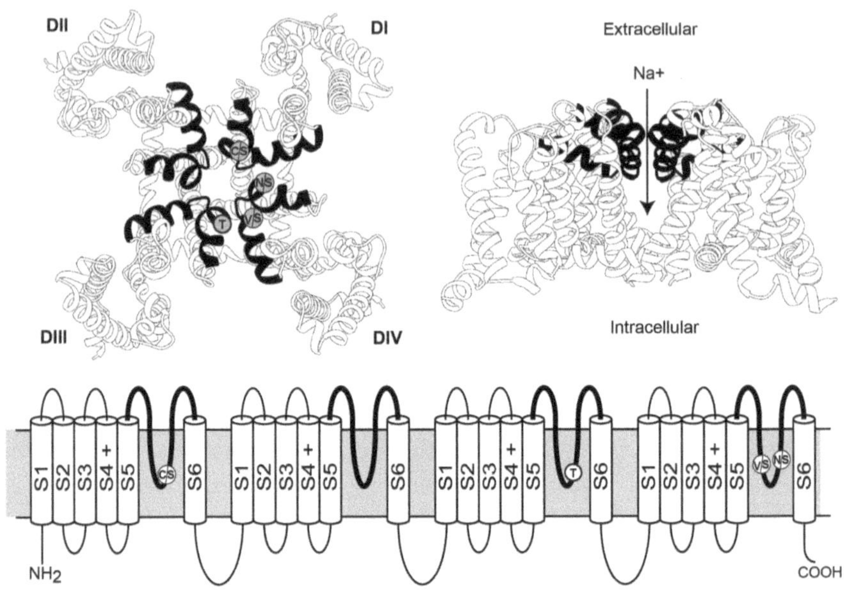

Figure 1 Positions of TTX-resistant amino acid substitutions in the pore loops of the voltage-gated sodium channel. The TTX-binding site/outer pore (in black) is identified in top-down (left) and lateral (right) views of the channel, with common TTX-resistant amino acid substitutions identified. At the bottom is a schematic view of the Na_V protein unwound, with outer pore regions in black and common amino acid substitutions identified. The Cys (C) in domain I is present in the mammalian cardiac muscle channel $Na_V1.5$, while a Ser (S) is found in the same position in mammalian neuronal channels $Na_V1.8$ and $Na_V1.9$. Nonaromatic residues (Cys or Ser) in this position decrease TTX-binding by several orders of magnitude compared to channels with an aromatic residue (Tyr or Phe). The Met—Thr substitution in domain III can reduce TTX-binding affinity in $Na_V1.4$ channels by approximately 15-fold. This substitution is found in the garter snake *Thamnophis couchii* (M1276T), in all TTX-bearing newts (M1116T) and in some pufferfish channels. In domain IV, the I1561V substitution common to all resistant *Thamnophis sirtalis* reduces TTX-binding affinity by approximately twofold; all toxic newts present a substitution in the same position I1424S. Also in domain IV, the Asp—Asn substitution is associated with high levels of organismal resistance in garter snakes (D1568N) and can increase channel TTX resistance 30- to 40-fold; toxic newts exhibit a Ser at the same position (D1431S). Channel protein model based on the crystal structure of the bacterial sodium channel and Tikhonov and Zhorov (2012). *Figure modified from Hanifin, C. T., & Gilly, W. F. (2015). Evolutionary history of a complex adaptation: tetrodotoxin resistance in salamanders.* Evolution: International Journal of Organic Evolution, *69(1), 232—244.*

is unclear. However, some substitutions appear to alter electrostatic interactions between TTX and the pore, whereas others seem to shift the position of electrostatically interacting residues in ways that reduce the binding affinity of TTX in the pore.

3. MOLECULAR PATTERNS OF CONVERGENCE IN TETRODOTOXIN-RESISTANT Na_V CHANNELS

Convergent molecular evolution associated with TTX resistance in predators and prey can be viewed at multiple levels: the genes/protein involved, the specific region of the protein where changes occur, the location of amino acid changes within these regions, and the identity of the substitutions themselves. In newts, snakes, and pufferfish, TTX resistance is not associated with a physiological means of sequestering TTX such as a TTX-binding protein or other form of detoxification. In all these taxa, resistance is associated with adaptive evolution in Na_V channels. Furthermore, adaptive TTX resistance in all these taxa does not appear to be associated with upregulation or other changes in expression patterns of Na_V channels. Instead, the evolution of TTX resistance in these taxa is associated with repeated patterns of amino acid substitutions across channel paralogs that also display remarkable patterns of convergence (but see Feldman et al., 2015).

The best-studied channel isoform in newts and snakes is $Na_V1.4$ (*SCN4a*). This channel is present in all tetrapod lineages. Although the channel is expressed at low levels in multiple tissues, it is the primary channel expressed in skeletal muscle of both newts and snakes and is responsible for propagating APs in this tissue in both lineages. $Na_V1.4$ is also present in pufferfish, but in puffers and other fish an independent whole genome duplication event has given rise to two copies of this channel (Novak et al., 2006). Examination of multiple populations and species of snakes and newts has painted a picture of dramatic convergence in this channel (Fig. 2).

In this ortholog, TTX-resistant garter snakes and newts exhibit amino acid substitutions in Na_V channels that are restricted to the pore loops of DIII and IV. Many of these changes occur at equivalent positions between newts and snakes, and in some instances the amino acid substitutions are identical.

3.1 Tetrodotoxin-resistant $Na_V1.4$ channels in predators

The original identification of TTX resistance in a predator resulted from the observation that some garter snakes (*Thamnophis sirtalis*) were capable of eating TTX-bearing *Taricha granulosa*. Early work showed that these snakes were not entirely immune to TTX, but that exposure to TTX resulted in a loss of righting reflex and a reduction in crawl speed (Brodie, 1968). Subsequent studies showed that skeletal muscle resistance to TTX was correlated

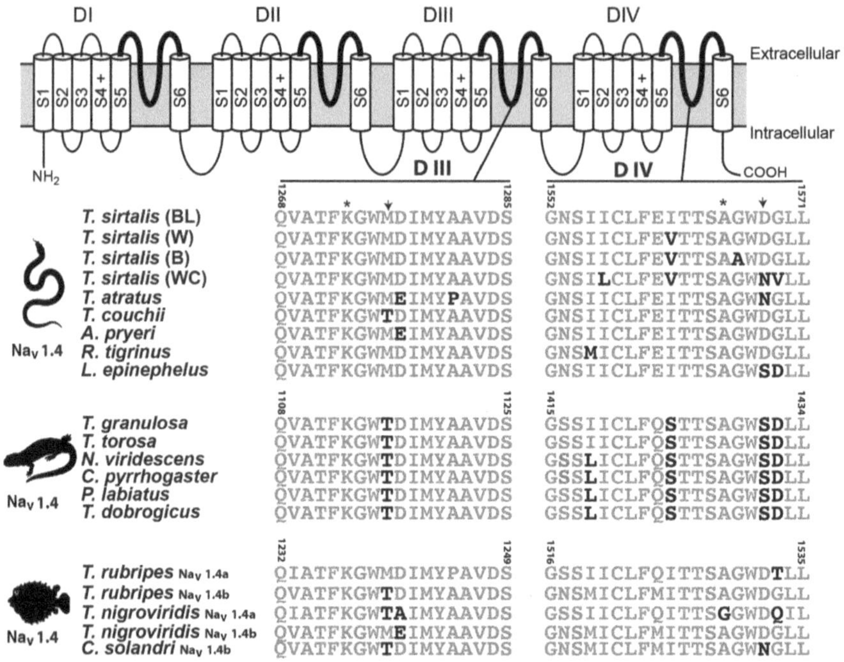

Figure 2 Sequence alignments of skeletal muscle voltage-gated sodium channel pore loops of domains III and IV showing pattern of convergent amino acid substitutions between TTX-resistant predators (snakes) and TTX-bearing prey (newts and pufferfish). The positions of several of these substitutions in the pore loops are shared among the different lineages and in some cases the changes are the same. The predators are represented by snakes known to consume TTX-bearing amphibians including *Thamnophis sirtalis, Thamnophis atratus, Thamnophis couchii, Amphiesma pryeri, Rhabdophis tigrinus,* and *Liophis epinephelus,* with the exception of *T. sirtalis* from Bear Lake, ID (BL) that does not eat toxic newts, and possesses a TTX-sensitive channel. Within *T. sirtalis,* pore loop substitutions have evolved independently in different populations: W, Warrenton, OR; B, Benton County, OR; WC, Willow Creek, CA. The salamanders are represented by *Taricha granulosa, Taricha torosa, Notophthalmus viridescens, Cynops pyrrhogaster, Pachytriton labiatus,* and *Triturus dobrogicus.* The pufferfish species are *Takifugu rubripes, Tetraodon nigroviridis,* and *Canthigaster solandri.* Amino acid substitutions are shown in black; the positions of selectivity filter residues for DIII (K, lysine) and DIV (A, alanine) are marked with an *asterisk,* and the positions of the outer charged ring residues for DIII (M, methionine) and DIV (D, aspartic acid) are marked with an *arrowhead.* Amino acid positions for snakes, newts, and pufferfish follow those used in Feldman et al. (2012), Hanifin and Gilly (2015), and Jost et al. (2008), respectively. Sequences from snakes, newts, and pufferfish were obtained from Geffeney et al. (2005), Feldman et al. (2012), Hanifin and Gilly (2015), and Jost et al. (2008), respectively.

with whole-animal resistance among populations of garter snakes (Geffeney, Brodie, Ruben & Brodie, 2002). Sequence analysis of the voltage-gated sodium channel from skeletal muscle ($Na_V1.4$) from these populations revealed four sites of amino acid substitution in the pore loops of DIV (Geffeney et al., 2005). To investigate whether these pore loop substitutions were responsible for the differences in resistance, Geffeney et al. (2005) created chimeric channels in which pore regions of mammalian $Na_V1.4$ contained snake sequence. These channel clones were used for electrophysiological recordings in frog oocytes to measure current transmission in the presence of TTX. The measured K_d50 (TTX concentration necessary to block 50% of Na^+ currents) mirrored the pattern of muscle resistance among populations of *T. sirtalis*. Channels with whole-snake sequence that were tested along with chimeras did not have significantly different K_d values, indicating that the substitutions in the DIV pore loops accounted for the differences in TTX sensitivity of the channels.

3.2 Interactions of amino acid replacements with tetrodotoxin

In garter snakes, a single mutation (I1561V) is shared by all resistant *T. sirtalis* populations that have been studied (Fig. 2). This substitution is associated with the lowest level of physiological resistance and alone reduces the binding affinity of the channel for TTX by approximately twofold (Geffeney et al., 2005). The I1561V substitution appears to have arisen at least twice independently in distinct lineages of *T. sirtalis*. Because this amino acid site is not known to interact directly with TTX, its effect on affinity may be the result of outer pore shape alterations that could hinder interactions between TTX and key residues in the outer vestibule (Tikhonov & Zhorov, 2011). A single substitution, D1568N found in snakes from the *T. sirtalis* Willow Creek population increases resistance 30- to 40-fold in rat $Na_V1.4$ (Choudhary et al., 2003; Penzotti et al., 1998). This substitution may eliminate or weaken the hydrogen bond between the carboxyl group in Asp and a hydroxyl group in TTX (Choudhary et al., 2003). Neither the individual effects of the remaining changes found in *T. sirtalis* populations nor the possible additive or epistatic effects of combinations of these changes on the binding affinity of TTX have been explored.

Other snake species that prey on TTX-bearing amphibians were subsequently investigated and revealed additional pore loop amino acid substitutions in DIII and DIV of $Na_V1.4$ (Fig. 2; Feldman et al., 2012). Character state reconstruction revealed that resistant mutations had evolved independently at least six times in colubroid snakes. Some of these substitutions occurred in sites already known to interact with TTX from previous functional studies of the channel. For example, the Asp—Asn substitution in DIV (D1568N) that confers high resistance in *T. sirtalis* from Willow Creek, CA was also found in another garter snake species, *Thamnophis atratus*. A change to an alternate residue in the same position, D1568S, was observed in *Liophis epinephelus*, a South American snake that consumes tetrodotoxic *Atelopus* toads. Changes at this position are often accompanied by substitutions of the adjacent Gly (1569), suggesting the possibility of combinatorial effects on resistance. Finally, a change in DIII (D1277E) evolved independently in the $Na_V1.4$ of *T. atratus* and also in the Asian snake *Amphiesma pryeri*, which consumes tetrodotoxic newts in the genus *Cynops* (Fig. 2).

3.3 Tetrodotoxin-resistant $Na_V1.4$ channels in prey

Recent investigations have revealed that skeletal muscle and its related sodium channel $Na_V1.4$ is also resistant to TTX in the TTX-bearing newts (Hanifin & Gilly, 2015). This group of "modern newts" evolved roughly 50 MYA and includes taxa from Europe, Asia, and the New World. Hanifin and Gilly (2015) showed that fibers from *Taricha* skeletal muscle could fire APs in high concentrations (30 μM) of TTX. A total of four amino acid substitutions in the DIII and DIV pore loops appeared to explain the resistance to TTX in newts. As in the garter snake *Thamnophis couchii*, a Met—Thr substitution occurs in the DIII pore loop (M1116T in the newt protein) in all toxic newts. This same substitution is present in pufferfish and has been shown to reduce TTX-binding affinity in $Na_V1.4$ channels by approximately 15-fold (Jost et al., 2008). In addition, toxic newts present the I1424S substitution located in the position equivalent to the Ile—Val (I1561V) in snakes (Fig. 2). Substitutions at positions equivalent to D1568 and G1569 in snakes were also identified in TTX-bearing newt species (positions D1431S and G1432D, respectively). This pair of substitutions is present in all genera of newts that possess TTX.

TTX-bearing salamanders are not the only lineage of TTX-bearing animals to evolve resistant Na_V channels. Jost et al. (2008) found resistant amino acid substitutions in the pore loops of all eight Na_V paralogs from pufferfish species. These paralogs are expressed in skeletal muscle, heart, and nervous system (Novak et al., 2006). Phylogenetic analyses showed that substitutions in pufferfish channels evolved independently in multiple genera, yet some changes were shared among paralogs and across species.

Some of the substitutions found in tetrodotoxic pufferfish Na_V channels are similar or identical to those observed in the skeletal muscle channels of resistant snakes and newts (Fig. 2).

4. CONVERGENCE ACROSS OTHER Na_V PARALOGS

Different Na_V proteins, encoded by unique but related paralogous genes, are expressed in different tissues across vertebrates. The specific number of the channels varies among tetrapod lineages and the orthology and evolutionary history of these genes are still being explored (Zakon et al., 2011). Ecologically relevant whole-animal resistance to TTX likely requires that TTX-resistant Na_V channels are expressed in most, if not all, tissues. Other channels are less well studied than $Na_V1.4$, but recent work has also shown strong patterns of convergence in these channel paralogs.

McGlothlin et al. (2014) investigated whether Na_V paralogs other than $Na_V1.4$ were resistant in garter snakes using genomic data to examine sequence from two channels expressed primarily in peripheral nervous system. They identified amino acid substitutions in sites associated with TTX resistance in the pore loops of DIII and DIV of both the $Na_V1.6$ and the $Na_V1.7$ ancestor channels in snakes (Fig. 3A). In mammals, these channels are identified as $Na_V1.6$ (*SCN8a*) and $Na_V1.7$ (*SCN9a*). Snakes and other reptiles do not possess a true $Na_V1.7$ (*SCN9a*) ortholog but instead express the ancestral copy that gave rise to $Na_V1.7$ (*SCN9a*) and Na_X (*SCN10a*) in mammals.

In garter snakes, $Na_V1.6$ exhibited the Ile–Val substitution in DIV (I1709V), that is common in $Na_V1.4$ of resistant *T. sirtalis* and that halves TTX-binding affinity (Figs. 2 and 3A). The $Na_V1.7$ ancestor presented four substitutions including the DIII Asp–Glu substitution (D1393E) also found in the $Na_V1.4$ of resistant snakes *T. atratus* and *A. pryeri*, the DIV Asp–Asn substitution (D1684N) that confers extreme resistance also found in $Na_V1.4$ of *T. atratus* and *T. sirtalis* from Willow Creek, and the DIV A1681G substitution that occurs at the selectivity filter also found in the skeletal muscle channel $Na_V1.4a$ of resistant pufferfish, which can reduce TTX-binding affinity by 1.5-fold (Figs. 2 and 3A; Jost et al., 2008).

Other Na_V paralogs in snakes do not have derived amino acid substitutions in the pore region that are expected to lower TTX-binding affinity (McGlothlin et al., 2014). In mammals, three channel types ($Na_V1.1$, $Na_V1.2$, $Na_V1.3$) are expressed in CNS tissue. If expression patterns in

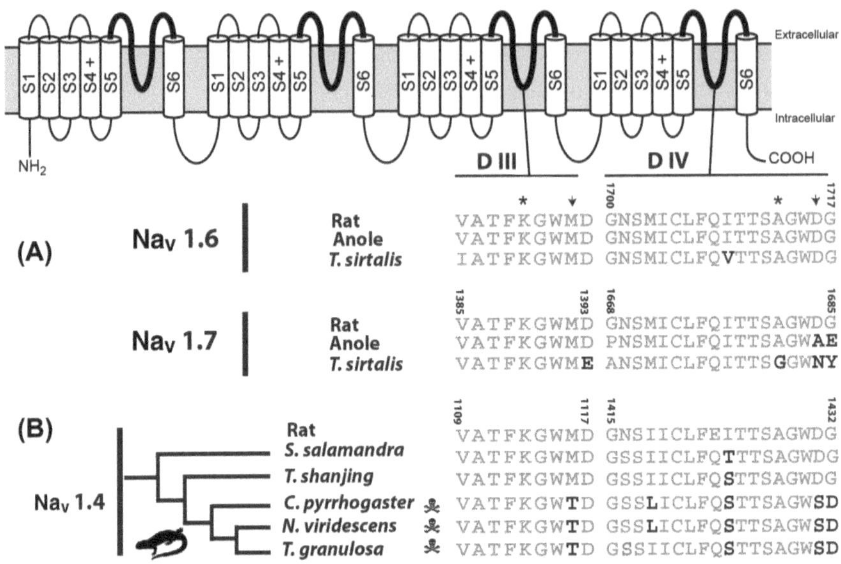

Figure 3 Sequence alignments of voltage-gated sodium channel pore loops of domains III and IV showing amino acid substitutions (black) in sites known to influence TTX-binding. (A) Garter snake peripheral nerve channels $Na_V1.6$ and $Na_V1.7$ present substitutions previously associated with increased TTX resistance. These substitutions are lacking in the sensitive mammalian channels, but the anole $Na_V1.7$ channel also presents substitutions suggesting channel changes in $Na_V1.7$ could have appeared ancestrally in reptiles. Snake sequences and amino acid positions taken from McGlothlin et al. (2014). Mammalian sequences obtained from GenBank: $Na_V1.6$ (rat AF049240.1, anole NC_014,777.1) $Na_V1.7$ (rat AF000368.1), anole H9G4U9 obtained from UniProt. (B) Phylogenetic distribution of amino acid substitutions in pore loops of salamander and newt skeletal muscle channel $Na_V1.4$ in domains III and IV. Substitutions M1116T, D1431S, and G1432D are present in TTX-bearing species (skull). These three substitutions are associated with extreme increases in the TTX resistance in modern newts while substitutions Thr and Ser at I1424 are associated with moderate increases in resistance in nontoxic species suggesting moderate resistance may be the ancestral state of this group possibly predating evolution of toxicity. Salamander and newt species are *Salamandra salamandra, Tylototriton shanjing, Cynops pyrrhogaster, Notophthalmus viridescens, and Taricha granulosa.* The TTX-sensitive rat sequence is presented for comparison. Salamander sequences, phylogenetic tree, and amino acid positions correspond to Hanifin and Gilly (2015). The positions of selectivity filter residues for DIII (K, lysine) and DIV (A, alanine) are marked with an *asterisk*; and the positions of the outer charged ring residues for DIII (M, methionine) and DIV (D, aspartic acid) are marked with an *arrowhead.*

reptiles are similar, these channel isoforms would be protected by the blood–brain barrier, which is thought to prevent exposure to ingested toxins like TTX. Consistent with the hypothesis that they are not under selection for TTX resistance, none of these three paralogs show evidence of derived amino acid substitutions in the pore region of Na_V in garter snakes.

Amphibians do not possess true orthologs of any of the CNS channel isoforms but instead possess the ancestor of $Na_V1.2$ and $Na_V1.3$ (Zakon et al., 2011). In TTX-bearing newts, this channel has evolved resistance to TTX and provides an exception to patterns observed in other channels. In the TTX-bearing newt *Cynops pyrrhogaster* TTX resistance is associated with an Ala substitution in the domain I pore that replaces an aromatic amino acid residue and mirrors substitutions identified in mammalian $Na_V1.5$, $Na_V1.8$, and $Na_V1.9$ channels. This change renders the channel highly resistant to TTX (Fig. 4; Hirota, Kaneko, Matsumoto, & Hanyu, 1999).

5. CONVERGENCE OF Na_V IN INVERTEBRATES

Invertebrates posses two channel isoforms: Na_V1 and Na_V2 (Moran, Barzılai, Liebeskind, & Zakon, 2015). The Na_V2 isoform is present only in invertebrates and is thought to represent an evolutionary intermediate between voltage-gated Ca^{++} channels and true Na_V channels. The Na_V1 isoform is thought to be orthologous to the entire family of Na_V channels present in vertebrates (Moran et al., 2015). TTX-insensitive currents have been reported in multiple invertebrate species including jellyfishes, flatworms, annelids, some insects, and arachnids (Du, Nomura, Liu, Huang, & Dong, 2009). Resistance in these channels mostly results from changes to Na_V1 outer ring EEMD in DIII and IV (Fig. 4). Some of these changes have been tested functionally to confirm they reduce TTX-binding affinity. Changes to single residues in the EEMD outer ring of the mite *Varroa destructor* channel ($VdNa_V1a$) to mimic the changes seen in other invertebrate channels resulted in TTX-resistant forms (Du et al., 2009). These invertebrate channels share pore loop amino acid substitutions with the resistant channels in snakes, newts, and pufferfish. Notably, the negatively charged Asp (position 1717 in invertebrate channel alignments, analogous to position 1431 in newts and position 1568 in snakes) is replaced by other residues.

6. IMPLICATIONS OF MOLECULAR CONVERGENCE IN Na_V

Convergent patterns of evolution have long intrigued biologists because convergence suggests fundamental rules that underlie adaptation

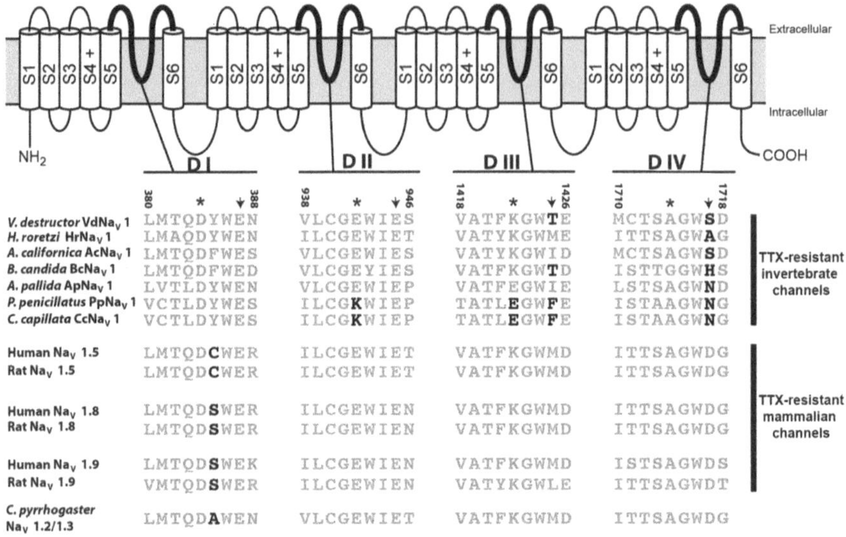

Figure 4 Sequence alignments of voltage-gated sodium channel pore loops of domains I–IV showing amino acid substitutions (black) in sites known to influence TTX-binding in invertebrate and mammalian channels. With the exception of flatworms (Salvitti, Wood, Taylor, McNabb, & Cary, 2015), there is no clear evidence that these species are exposed to TTX suggesting the acquisition of these substitutions may respond to selective pressures to change channel biophysical properties. Substitutions in invertebrate channels constitute deviations from the selectivity filter DEKA motif and the outer charged ring EEMD motif. Invertebrate sequences include the mite *Varroa destructor* (AY259834), the tunicate *Halocynthia roretzi* (D17311), sea hare *Aplysia californica* (U66915), flatworm *Bdelloura candida* (U93074), sea anemone *Aiptasia pallida* (AF041851), hydrozoan jellyfish *Polyorchis penicillatus* (AF047380), and scyphozoan jellyfish *Cyanea capillata* (L15445) with amino acid positions corresponding to rat $Na_V1.2$ (X03639) as reviewed in Du et al. (2009). TTX-resistant mammalian cardiac muscle channel $Na_V1.5$ (human BC051374.1, rat AF353637.1), neuronal channels $Na_V1.8$ (human AC116038, rat AJ623271.1), and $Na_V1.9$ (human AY686224.1, rat AF059030.2) show DI residues Cys and Ser in the same positions as the aromatic residue Tyr present in other paralogs. Nonaromatic residues (Cys or Ser) in this position decrease TTX-binding by several orders of magnitude. Interestingly, the amphibian neuronal channel thought to be the ancestor of $Na_V1.2$ and $Na_V1.3$ (*Cynops pyrrhogaster*, AF123593.1) presents a nonaromatic residue Ala in DI suggesting this channel is TTX resistant. The DEKA motif residues for all domains are marked with *asterisks*; and the EEMD motif residues are marked with *arrowheads*.

(Losos, 2011). Convergence at the molecular level suggests that protein evolution also follows predictable rules (Stern & Orgogozo, 2009; Weinreich, Delaney, DePristo, & Hartl, 2006). For Na_V channels, convergent evolution of TTX resistance across lineages can shed important light into the structure and function of these proteins.

In contrast to other voltage-gated ion channel gene families, the voltage-gated sodium channel gene family expanded during a time that tetrapod-vertebrates evolved greater forebrain complexity and novel sensory receptors (Zakon et al., 2011). This expansion appears to have allowed for greater specialization of sodium channel gene expression in tissues and subcellular locations, as well as enhanced processing of sensory input from the skin and body. Yet, because of their key role in signaling of nerve and muscle tissue, Na_V channels are under strong functional constraint and appear to have limited evolutionary lability. If adaptation has produced sequence level convergence in taxa as widely diverged as newts and snakes, it suggests that there are universal processes linked to Na_V channel structure and/or function that generate that convergence.

Repeated solutions to the TTX resistance "problem" might result from multiple nonmutually exclusive sources. Convergence could result from the molecular specificity of the mode of action of TTX. TTX interacts with only a small number of amino acids located in the outer pore of the channel. Changes that neutralize the negative charges in the pore that make contact with TTX have the largest effect on resistance (Choudhary et al., 2003; Penzotti et al., 1998). Convergence can also result from functional constraints on the channel itself. The pore of the Na_V channel is responsible for the rapid but regulated influx of Na^+ ions. TTX interacts with charged amino acids that form a region of the channel responsible for the selective influx of Na^+ and blocks the influx of other ions. For example, disruption of the selectivity filter can increase TTX resistance but simultaneously causes increased calcium ion (Ca^{2+}) permeability, which would disrupt the signaling properties of the channel (Heinemann et al., 1992). Similarly, substitution of pore residues can both decrease the TTX-binding affinity of the channel and severely disrupt the permeability of the channel to Na^+ and reduce Na^+ conductance (Terlau et al., 1991). As a result, convergence could result because only a small number of amino acid site changes render a channel resistant, and an even smaller number of these changes may be permissible due to functional constraints.

Although experimental data suggest that amino acid substitutions in DI can greatly reduce TTX-binding (Satin et al., 1992; Terlau et al., 1991), the molecular pattern of TTX-resistant amino acid changes that evolution has fixed in snakes and newts is restricted to DIII and IV (with the exception of the $Na_V1.2/1.3$ ancestor in newts). This pattern suggests that functional constraints may exist that limit the advantage of charge-changing substitutions in DI of the pore. Although this hypothesis has not been tested directly, there is evidence that pore substitutions in DI and DII, which increase TTX resistance more than a 1000-fold, can also reduce Na^+ conductance to less than 1% of that seen in wild-type channels (Terlau et al., 1991). In contrast, the DIV substitution Asp to Asn observed in garter snakes (D1568N) significantly increases resistance by removing the negative charge of Asp, but its effect on Na^+ conductance is less dramatic than the effects of similar substitutions in DI and DII (Terlau et al., 1991). Although not tested directly, the same effect could be true for the analogous D1431S substitution found in newts.

In both snakes and newts, negative charge-neutralizing substitutions are accompanied by other pore loop substitutions whose contribution to TTX resistance is not well understood. Electrophysiological data show that some of these substitutions alone can have a small to moderate effect in reducing TTX-binding (Du et al., 2009; Geffeney et al., 2005; Jost et al., 2008). This reduction could be due to alterations in the shape of the pore that indirectly hamper TTX-docking in other sites (Tikhonov & Zhorov, 2011). These small effect substitutions also exhibit a convergent pattern between snakes and newts, which suggests they may have an important role in the evolution of TTX resistance in these channels.

There is a dearth of studies looking at epistatic interactions between these substitutions, but it is possible that changes occur together because they interact nonadditively to increase the degree of resistance. For example, the mite *Varroa destructor* possesses a TTX-resistant channel $VdNa_V1$ that exhibits two pore loop substitutions in the outer charged ring in DIII and IV (positions 1425 and 1717, respectively) (Fig. 4). These substitutions change the arthropod EEID (Glu-Glu-Ile-Asp) outer charged ring motif into EETS (Glu-Glu-Thr-Ser). Du et al. (2009) introduced mutations in the resistant $VdNa_V1$ (containing EETS) to change this channel to the sensitive type by changing each site individually resulting in EEIS and EETD and a double mutation EEID. Each single change resulted in approximately a 10-fold increase in sensitivity, but the sensitive double mutant exhibited a remarkable 400-fold increase in TTX sensitivity. This work is a clear demonstration that single amino acid changes in different channel domains can interact nonadditively to affect TTX resistance.

6.1 Costs of tetrodotoxin resistance

A correlated amino acid substitution could be favored if it helped compensate for negative shifts in biophysical properties caused by large-effect TTX-resistant substitutions. Resistant substitutions, though adaptive, can have detrimental pleiotropic effects on other channel properties. As discussed earlier, ionic conductance and permeation can be affected by charge-changing or neutralizing amino acid substitution near the selectivity filter. However, proper channel function relies closely on gating properties as well. In addition to controlling ion flow, amino acid substitutions in the pore region of the channel can affect channel gating, including the voltage dependence of activation and slow inactivation (Hilber et al., 2005; Lee, Jones, Ahern, Sarhan, & Ruben, 2011; Vilin & Ruben, 2001; Xiong et al., 2006).

Lee et al. (2011) used electrophysiological recordings of the mammalian skeletal muscle channel $Na_V1.4$ expressed in frog oocytes to examine the possible functional consequences of pore TTX-resistant snake substitutions. They measured biophysical properties of $Na_V1.4$ containing DIV pore loop substitutions seen in resistant *T. sirtalis* from populations of low and high resistance. Their study showed that some of these substitutions could significantly shift the voltage dependence of slow inactivation. To explain these findings the authors postulated that changes in the pore might result in a channel structure that hinders voltage sensor movement and therefore shifts the voltage threshold at which the channel undergoes slow inactivation.

Disturbances of voltage dependence of slow inactivation generate channel hyperexcitability that can result in a slower rate of current decay or an increase in window-current amplitude (Bendahhou, Cummins, Kula, Fu, & Ptácek, 2002; Bennett & Woods, 2014, Cannon, 2000, 2015). Substitutions at I1561 in garter snakes result in a depolarized shift in the voltage dependence of slow inactivation of a magnitude comparable to shifts in the same biophysical property associated with human channelopathies (Jurkat-Rott, Holzherr, Fauler, & Lehmann-Horn, 2010; Vicart, Sternberg, Fontaine, & Meola, 2005). Newts also posses substitutions at this position S1424, but their effects have not been quantified. Although there is no direct evidence linking the TTX-resistant $Na_V1.4$ mutations that shift biophysical properties to a detrimental whole-body phenotype, there is some evidence that suggests the evolution of TTX resistance in garter snakes could involve physiological trade-offs. Brodie and Brodie (1999) compared crawl speeds of nonresistant and resistant garter snakes and found that resistant snakes crawled more slowly than nonresistant snakes. This pattern was true at the individual, family and population levels, suggesting that, as with human muscle channelopathies, changes in the snake $Na_V1.4$ channels could result in deleterious whole-body phenotypes.

In wild populations, mutations to $Na_V1.4$ channels that cause deleterious or suboptimal phenotypes would normally be selected out of the population. However, if these normally "deleterious" mutations confer an adaptive advantage in the ecological context of a predator—prey interaction, they could rise to fixation despite the negative effects. Whole-body effects of Na_V channels with altered biophysical properties could explain why resistant snakes crawl slower. Snakes that crawl more slowly may be at higher risk of predation and less apt to capture prey. However, the fitness benefits of resistance may outweigh the physiological disadvantages in snake populations under high selective pressures to survive encounters with highly toxic newts. The fact that pore loop mutations spread through the population despite possible cost to channel function indicates that functional tradeoff may be unavoidable and that there are very few changes that would decrease TTX-binding while minimally upsetting important channel biophysical properties.

6.2 Compensatory changes associated with tetrodotoxin resistance

As correlated negative effects may be an inescapable consequence of acquiring amino acid substitutions that lower TTX-binding, these may be more likely to cooccur with other changes that ameliorate or compensate for the negative shifts in biophysical properties. Although such compensatory substitutions are not necessarily restricted to the channel pore, they could arise in the pore loops as changes in this region of the channel have been shown to affect channel properties. The acquisition of compensatory amino acid substitutions after adaptive ones is a very common phenomenon in the evolution of drug resistance in microbes. Although harder to study in larger organisms, the fine-tuning of adaptations through the evolution of compensatory mechanism is believed to be as important in the evolution of phenotypes in eukaryotes (Pavlicev & Wagner, 2012).

The skeletal muscle channels of resistant newts exhibit the D1431S substitution, which, along with the analogous D1568N in snakes, removes a negative charge that interacts with TTX. Elimination of this charge decreases but does not eliminate ionic conductance (Terlau et al., 1991). In toxic newts, the residue immediately downstream from the S1431 is changed to an Asp from the ancestral Gly. The G1432D change could be compensatory as it replaces the lost negative charge at the upstream positions 1431, which could restore reduced conductance (Fig. 1 and 2). Interestingly, the same pair of substitutions evolved independently in the unrelated snake

L. epinephelus as well as multiple invertebrate taxa, suggesting this combination of substitutions may effectively confer TTX resistance while reducing loss of ionic conductance.

In the evolution of TTX-resistant Na_V channels, changes that ameliorate negative effects of highly resistant substitutions can either be favored after the acquisition of pleiotropic substitutions or they may constitute functional prerequisites predating the acquisition of highly resistant but functionally costly substitutions. Amino acid substitutions could predispose the acquisition of other (TTX-resistant) substitutions if these initial changes either confer limited resistance that allows the organism to survive exposure to low levels of TTX or reduce the level of functional/structural barriers constraining the acquisition of resistance substitutions.

6.3 Coincidental tetrodotoxin resistance

Substitutions in the pore region of Na_V proteins can have pleiotropic effects that alter both fundamental biophysical properties and TTX-binding. Some of the channel types that are functionally "TTX-resistant" appear to have evolved that resistance in the absence of, or prior to, exposure to TTX in the environment. Pore loop substitutions could be adaptive if they modify channel biophysical properties to modulate channel function in other adaptive ways. Na_V paralogs and orthologs differ in biophysical properties as a result of functional specialization in different tissues. The $Na_V1.5$ channel, for example, is expressed in the mammalian cardiac muscle. Its specific role in controlling excitability of cardiac myocytes suggests it has been selected for unique biophysical properties that enhance performance within the cardiac tissue. In TTX-sensitive channels, an aromatic residue (Tyr or Phe) at pore loop in DI forms important chemical interactions with TTX. This interaction is negated in $Na_V1.5$ due to the presence of a nonaromatic Cys at this position, therefore rendering the channel TTX resistant (Fig. 4; Satin et al., 1992). The presence of a Cys instead of an aromatic residue in mammalian $Na_V1.5$ DI pore loop probably serves a group-specific adaptive role. Because pore loop changes have the potential to make the channel TTX-insensitive by changing the charge or the availability of TTX contact sites, it is possible that TTX resistance was acquired as the corollary during the acquisition of a pore loop substitution in $Na_V1.5$ that was favored because it changed channel properties in a way that was adaptive within the cardiac muscle. A similar argument might hold for the mammalian nerve channels $Na_V1.8$ and $Na_V1.9$ that are TTX-resistant due the presence of a nonaromatic residue (Ser) in the same DI pore loop position (Fig. 4; Leffler, Herzog, Dib-Hajj, Waxman, & Cummins, 2005).

The presence of TTX-resistant channels in taxa that are not exposed to TTX presents an intriguing question. Comparisons with other vertebrate channels suggest that TTX-resistant forms of Na_V1.5, Na_V1.8, and Na_V1.9 evolved independently in mammals (and from each other). However, there is no evidence that ecological exposure to TTX played any role in this process, and it is generally assumed that TTX resistance in these channels is a by-product or secondary artifact associated with selection that favored substitutions associated with changes in other biophysical properties of these channels.

Furthermore, these channels show strong patterns of convergence; changes associated with resistance in these channels all occur in one specific amino acid position in the DI pore. Changes in DI pore loop that result in TTX-resistant channels may be the result of selection to modulate function, whereas pore loop changes in DIII and IV may be associated with evolution to withstand TTX. This pattern of protein evolution could reveal important details about the inherent functional constraints in the evolution of the Na_V protein family.

6.4 Evolutionary predisposition for tetrodotoxin resistance

The "accidental" acquisition of TTX resistance by means of Na_V amino acid substitutions that adaptively modulate channel biophysical properties could function as a preadaptation (or "exaptation") that allows organisms to exploit environments or resources where TTX is present. Mildly resistant organisms could then harbor TTX for defense or could exploit new TTX-containing dietary niches. In fact, it is possible that this type of exaptation was present in both snakes and newts that evolved TTX resistance. In garter snakes, two peripheral nerve channels (Na_V1.6 and Na_V1.7) possess pore loop mutations previously associated with TTX resistance. These substitutions are shared among TTX-resistant and nonresistant garter snakes, suggesting that peripheral nerve TTX resistance preceded ecological interactions with toxic *Taricha* (McGlothlin et al., 2014).

As in pufferfish, whole-body TTX resistance may be the product of possessing several resistant Na_V paralogs. Resistant peripheral nerves may be crucial to organismal resistance in snakes. The peripheral nerve Na_V1.6 is found in nodes of Ranvier in mammals, where it is involved in AP propagation (Caldwell, Schaller, Lasher, Peles, & Levinson, 2000). In snakes, which lack a diaphragm, axial muscles innervated by peripheral nerves regulate expansion of the lungs. Possible disruption of the signal to these muscles due to TTX blockage of $Na_V1.6$ could result in death by asphyxiation in these reptiles (McGlothlin et al., 2014). $Na_V1.7$ is expressed in sympathetic peripheral nerves that regulate vasodilation. One detrimental symptom of TTX intoxication is severe hypotension (Zimmer, 2010), possibly as a result of Na_V channel blockage in sympathetic nerves (Dib-Hajj, Yang, Black, & Waxman, 2013). Due to the severe detrimental effect of TTX in these tissues, resistance of $Na_V1.6$ and $Na_V1.7$ could be a requirement for exposed snakes.

Similarly, TTX resistance precedes the origin of the TTX-toxic clade of newts. Moderate skeletal muscle resistance was identified in basal salamandrid lineages that lack TTX. Sequence analysis of the channel protein showed that this moderate resistance results from substitutions in pore loops of DIV (Fig. 3B). In the nontoxic *Salamandra salamandra* the Ile1424 residue is a Thr and in nontoxic *Tylototriton shanjing*, which is evolutionarily intermediate to *S. salamandra* and the modern newts this position is a Ser, but these outgroup genera lack other substitutions in DIII or DIV. Skeletal muscle resistance in modern newts is an order of magnitude higher than in any snake measured (Hanifin & Gilly, 2015). It is not clear why ancestral salamanders possess pore loop substitutions in $Na_V1.4$, but it is possible they also constitute an exaptation resulting from group-specific adaptive modifications of channel properties.

TTX-resistant Na^+ currents have been identified in several invertebrate Na_V1 channels (Du et al., 2009). It is possible that some of these animals are exposed to TTX or STX (saxitoxin, a marine toxin chemically similar to TTX) since these toxins are prevalent in environments where some of these animals occur (Zimmer & Ferrer, 2007). However, in many cases there is no definitive evidence of exposure to these toxins. In these cases, the evolution of biophysical functions could have led to coincidentally TTX-resistant forms. For example, Na^+ currents in the motor neurons of jellyfishes differ from the currents in the excitable cell of higher metazoan in that they fire fast transient overshooting APs. This results in rapid contraction of the myothelium of the medusa wall, which is a crucial adaptation for fast escape swimming. In the jellyfishes *Polyorchis penicillatus*, *Cyanea capillata* and *Aglantha digitale*, voltage-clamping of these motor neurons showed that these fast currents, though dependent on Na^+, are resistant to TTX (Anderson, 1987; Meech & Mackie, 1993; Spafford, Grigoriev, & Spencer, 1996). It is possible that pore substitutions in areas of the jellyfish channel affecting

TTX-binding were favored as part of evolution to modify a swimming behavior and not because jellyfishes were evolving to the presence of TTX.

7. CONCLUSION

The deep and specific convergence of TTX-resistant Na_V channels points to a confluence of molecular evolutionary phenomena including protein sequence conservation, functional constraints, epistasis, and pleiotropy. This combination of factors has resulted in highly predictable molecular routes to adaptation across deeply divided lineages. In each of the resistant taxa, TTX resistance involves amino acid changes in the pore loops of Na_V that reduce the binding affinity of the toxin by either changing the charge of specific residues that interact with TTX or by altering the availability of toxin contact points in the channel pore. These mutations are remarkably repeatable and often occur in predictable combinations suggesting limits to the permissible structural variation at the pore region accompanied by the existence of sites that interact epistatically. This bias may limit the number of possible mutational paths to a few that are frequently seen in TTX-resistant Na_V channels across widely separated lineages.

Adaptive evolution has produced remarkable patterns of convergent evolution in the Na_V channels of predators and prey. Interpreting observed patterns of substitutions suggests that multiple overlapping mechanisms likely explain this convergence at the molecular level. TTX interacts with a very specific region of Na_V channels and, as a result, only a small number of substitutions in these channels can increase TTX resistance of the channel. Cost associated with function may also play a role in driving convergence in these lineages. Substitutions in the outer pore can negatively impact important biophysical properties of the channel and a subset of substitutions observed in newts and snakes seem to maximize increases in TTX resistance while simultaneously minimized this negative effect. However, there is an additional subset of convergent changes whose effects are unclear. These changes do not occur at positions that interact with TTX but appear to be correlated with substitutions at other sites that do interact with TTX. One possibility is that epistatic interactions between these residues and other substitutions may be driving convergent patterns. Finally, convergence between newts and snakes suggests the possibility that exaptation and pleiotropy may play an important role in the evolution of TTX resistance.

ACKNOWLEDGMENTS

We thank the editors of this volume for the opportunity to share our research on evolution with a new audience. We would like to thank our many collaborators who helped us understand the evolution and neurophysiology of sodium channels and TTX resistance and generated critical data discussed herein, including E. D. Brodie, Jr., C. Feldman, J. McGlothlin, and M. Pfrender. Support for this work was provided by NSF grants to E.D.B.III and E.D.B. Jr. and GT, USU internal funding to CTH and SLG, and an NIH F32 award to CTH.

REFERENCES

Anderson, P. A. (1987). Properties and pharmacology of a TTX-insensitive Na^+ current in neurones of the jellyfish *Cyanea capillata*. *Journal of Experimental Biology, 133*(1), 231–248.

Arendt, J., & Reznick, D. (2008). Convergence and parallelism reconsidered: what have we learned about the genetics of adaptation? *Trends in Ecology & Evolution, 23*(1), 26–32.

Arnegard, M. E., Zwickl, D. J., Lu, Y., & Zakon, H. H. (2010). Old gene duplication facilitates origin and diversification of an innovative communication system—twice. *Proceedings of the National Academy of Sciences of the United States of America, 107.51*(2010), 22172–22177.

Backx, P. H., Yue, D. T., Lawrence, J. H., Marban, E., & Tomaselli, G. F. (1992). Molecular localization of an ion-binding site within the pore of mammalian sodium channels. *Science, 257*(5067), 248–251.

Bendahhou, S., Cummins, T. R., Kula, R. W., Fu, Y.-H., & Ptácek, L. J. (2002). Impairment of slow inactivation as a common mechanism for periodic paralysis in DIIS4-S5. *Neurology, 58*(8), 1266–1272.

Bennett, D. L., & Woods, C. G. (2014). Painful and painless channelopathies. *Lancet Neurology, 13*(6), 587–599.

Brodie, E. D., III, & Brodie, E. D., Jr. (1999). Costs of exploiting poisonous prey: evolutionary trade-offs in a predator-prey arms race. *Evolution: International Journal of Organic Evolution*, 626–631.

Brodie, E. D., Jr. (1968). Investigations on the skin toxin of the adult rough-skinned newt, *Taricha granulosa*. *Copeia*, 307–313.

Brodie, E. D., Jr., Ridenhour, B. J., & Brodie, E. D., III (2002). The evolutionary response of predators to dangerous prey: hotspots and coldspots in the geographic mosaic of coevolution between garter snakes and newts. *Evolution: International Journal of Organic Evolution, 56*(10), 2067–2082.

Caldwell, J. H., Schaller, K. L., Lasher, R. S., Peles, E., & Levinson, S. R. (2000). Sodium channel Nav1.6 is localized at nodes of Ranvier, dendrites, and synapses. *Proceedings of the National Academy of Sciences of the United States of America, 97*(10), 5616–5620.

Cannon, S. C. (2000). Spectrum of sodium channel disturbances in the nondystrophic myotonias and periodic paralyses. *Kidney International, 57*(3), 772–779.

Cannon, S. C. (2015). Channelopathies of skeletal muscle excitability. *Comprehensive Physiology*, 761–790.

Chiamvimonvat, N., Perez-Garcia, M. T., Tomaselli, G. F., & Marban, E. (1996). Control of ion flux and selectivity by negatively charged residues in the outer mouth of rat sodium channels. *Journal of Physiology, 491*(1), 51–59

Choudhary, G., Yotsu-Yamashita, M., Shang, L., Yasumoto, T., & Dudley, S. C. (2003). Interactions of the C-11 hydroxyl of tetrodotoxin with the sodium channel outer vestibule. *Biophysical Journal, 84*(1), 287–294.

Clark, R. F., Williams, S. R., Nardt, S. P., & Manoguerra, A. S. (1999). A review of selected seafood poisonings. *Undersea & Hyperbaric Medicine, 26*(3), 175.

Conte, G. L., Arnegard, M. E., Peichel, C. L., & Schluter, D. (2012). The probability of genetic parallelism and convergence in natural populations. *Proceedings of the Royal Society of London B: Biological Sciences, 279*(1749), 5039–5047.
Dib-Hajj, S. D., Yang, Y., Black, J. A., & Waxman, S. G. (2013). The Na(V)1.7 sodium channel: from molecule to man. *Nature Reviews. Neuroscience, 14*(1), 49–62.
Du, Y., Nomura, Y., Liu, Z., Huang, Z. Y., & Dong, K. (2009). Functional expression of an arachnid sodium channel reveals residues responsible for tetrodotoxin resistance in invertebrate sodium channels. *Journal of Biological Chemistry, 284*(49), 33869–33875.
Favre, I., Moczydlowski, E., & Schild, L. (1996). On the structural basis for ionic selectivity among Na^+, K^+, and Ca^{2+} in the voltage-gated sodium channel. *Biophysical Journal, 71*(6), 3110.
Feldman, C. R., Brodie, E. D., Jr., Brodie, E. D., III, & Pfrender, M. E. (2012). Constraint shapes convergence in tetrodotoxin-resistant sodium channels of snakes. *Proceedings of the National Academy of Sciences of the United States of America, 109*(12), 4556–4561.
Feldman, C. R., Durso, A. M., Hanifin, C. T., Pfrender, M. E., Ducey, P. K., Stokes, A. N., … Brodie, E. D., Jr. (2015). Is there more than one way to skin a newt? Convergent toxin resistance in snakes is not due to a common genetic mechanism. *Heredity, 116*(1), 84–91.
Fozzard, H. A., & Lipkind, G. M. (2010). The tetrodotoxin binding site is within the outer vestibule of the sodium channel. *Marine Drugs, 8*(2), 219–234.
Geffeney, S., Brodie, E. D., Jr., Ruben, P. C., & Brodie, E. D., III (2002). Mechanisms of adaptation in a predator-prey arms race: TTX-resistant sodium channels. *Science, 297*(5585), 1336–1339.
Geffeney, S. L., Fujimoto, E., Brodie, E. D., Jr., Brodie, E. D., III, & Ruben, P. C. (2005). Evolutionary diversification of TTX-resistant sodium channels in a predator–prey interaction. *Nature, 434*(7034), 759–763.
Hanifin, C. T. (2010). The chemical and evolutionary ecology of tetrodotoxin (TTX) toxicity in terrestrial vertebrates. *Marine Drugs, 8*(3), 577–593.
Hanifin, C. T., Brodie, E. D., Jr., & Brodie, E. D., III (2008). Phenotypic mismatches reveal escape from arms-race coevolution. *PLoS Biology, 6*(3), e60.
Hanifin, C. T., & Gilly, W. F. (2015). Evolutionary history of a complex adaptation: tetrodotoxin resistance in salamanders. *Evolution: International Journal of Organic Evolution, 69*(1), 232–244.
Heinemann, S. H., Terlau, H., Stühmer, W., Imoto, K., & Numa, S. (1992). Calcium channel characteristics conferred on the sodium channel by single mutations. *Nature, 356*(6368), 441–443.
Hilber, K., Sandtner, W., Zarrabi, T., Zebedin, E., Kudlacek, O., Fozzard, H. A., & Todt, H. (2005). Selectivity filter residues contribute unequally to pore stabilization in voltage-gated sodium channels. *Biochemistry, 44*(42), 13874–13882.
Hille, B. (2001). *Ion channels of excitable membranes* (Vol. 507). Sunderland, MA: Sinauer.
Hirota, K., Kaneko, Y., Matsumoto, G., & Hanyu, Y. (1999). Cloning and distribution of a putative tetrodotoxin-resistant Na^+ channel in newt retina. *Zoological Science, 16*(4), 587–594.
Jost, M. C., Hillis, D. M., Lu, Y., Kyle, J. W., Fozzard, H. A., & Zakon, H. H. (2008). Toxin-resistant sodium channels: parallel adaptive evolution across a complete gene family. *Molecular Biology and Evolution, 25*(6), 1016–1024.
Jurkat-Rott, K., Holzherr, B., Fauler, M., & Lehmann-Horn, F. (2010). Sodium channelopathies of skeletal muscle result from gain or loss of function. *Pflügers Archiv – European Journal of Physiology, 460*(2), 239–248.

Lee, C. H., Jones, D. K., Ahern, C., Sarhan, M. F., & Ruben, P. C. (2011). Biophysical costs associated with tetrodotoxin resistance in the sodium channel pore of the garter snake, *Thamnophis sirtalis*. *Journal of Comparative Physiology A: Sensory, Neural, and Behavioral Physiology, 197*(1), 33–43.

Leffler, A., Herzog, R. I., Dib-Hajj, S. D., Waxman, S. G., & Cummins, T. R. (2005). Pharmacological properties of neuronal TTX-resistant sodium channels and the role of a critical serine pore residue. *Pflügers Archiv – European Journal of Physiology, 451*(3), 454–463.

Lopreato, G. F., Lu, Y., Southwell, A., Atkinson, N. S., Hillis, D. M., Wilcox, T. P., & Zakon, H. H. (2001). Evolution and divergence of sodium channel genes in vertebrates. *Proceedings of the National Academy of Sciences of the United States of America, 98*(13), 7588–7592.

Losos, J. B. (2011). Convergence, adaptation, and constraint. *Evolution: International Journal of Organic Evolution, 65*(7), 1827–1840.

Mahdavi, S., & Kuyucak, S. (2015). Mechanism of ion permeation in mammalian voltage-gated sodium channels. *PLoS One, 10*(8), e0133000.

Manceau, M., Domingues, V. S., Linnen, C. R., Rosenblum, E. B., & Hoekstra, H. E. (2010). Convergence in pigmentation at multiple levels: mutations, genes and function. *Philosophical Transactions of the Royal Society of London. Series B, Biological Sciences, 365*(1552), 2439–2450.

McGlothlin, J. W., Chuckalovcak, J. P., Janes, D. E., Edwards, S. V., Feldman, C. R., Brodie, E. D., Jr., … Brodie, E. D., III (2014). Parallel evolution of tetrodotoxin resistance in three voltage-gated sodium channel genes in the garter snake *Thamnophis sirtalis*. *Molecular Biology and Evolution, 31*(11), 2836–2846.

Meech, R. W., & Mackie, G. O. (1993). Ionic currents in giant motor axons of the jellyfish, *Aglantha digitale*. *Journal of Neurophysiology, 69*(3), 884–893.

Moczydlowski, E. G. (2013). The molecular mystique of tetrodotoxin. *Toxicon: Official Journal of the International Society on Toxinology, 63*(C), 165–183.

Moran, Y., Barzilai, M. G., Liebeskind, B. J., & Zakon, H. H. (2015). Evolution of voltage-gated ion channels at the emergence of Metazoa. *Journal of Experimental Biology, 218*(4), 515–525.

Narahashi, T. (2008). Tetrodotoxin: a brief history. *Proceedings of the Japan Academy. Series B, Physical and Biological Sciences, 84*(5), 147–154.

Noda, M., Suzuki, H., Numa, S., & Stühmer, W. (1989). A single point mutation confers tetrodotoxin and saxitoxin insensitivity on the sodium channel II. *FEBS Letters, 259*(1), 213–216.

Novak, A. E., Jost, M. C., Lu, Y., Taylor, A. D., Zakon, H. H., & Ribera, A. B. (2006). Gene duplications and evolution of vertebrate voltage-gated sodium channels. *Journal of Molecular Evolution, 63*(2), 208–221.

Pavlicev, M., & Wagner, G. P. (2012). A model of developmental evolution: selection, pleiotropy and compensation. *Trends in Ecology & Evolution, 27*(6), 316–322.

Penzotti, J. L., Fozzard, H. A., Lipkind, G. M., & Dudley, S. C. (1998). Differences in saxitoxin and tetrodotoxin binding revealed by mutagenesis of the Na^+ channel outer vestibule. *Biophysical Journal, 75*(6), 2647–2657.

Salvitti, L., Wood, S. A., Taylor, D. I., McNabb, P., & Cary, S. C. (2015). First identification of tetrodotoxin (TTX) in the flatworm *Stylochoplana* sp.; a source of TTX for the sea slug *Pleurobranchaea maculata*. *Toxicon: Official Journal of the International Society on Toxinology, 95*, 23–29.

Santarelli, V. P., Eastwood, A. L., Dougherty, D. A., Horn, R., & Ahern, C. A. (2007). A cation-π interaction discriminates among sodium channels that are either sensitive or resistant to tetrodotoxin block. *Journal of Biological Chemistry, 282*(11), 8044–8051.

Satin, J., Kyle, J. W., Chen, M., Bell, P., Cribbs, L. L., Fozzard, H. A., & Rogart, R. B. (1992). A mutant of TTX-resistant cardiac sodium channels with TTX-sensitive properties. *Science, 256*(5060), 1202–1205.

Schlief, T., Schönherr, R., Imoto, K., & Heinemann, S. H. (1996). Pore properties of rat brain II sodium channels mutated in the selectivity filter domain. *European Biophysics Journal, 25*(2), 75–91.

Spafford, J., Grigoriev, N., & Spencer, A. (1996). Pharmacological properties of voltage-gated Na^+ currents in motor neurones from a hydrozoan jellyfish *Polyorchis penicillatus. Journal of Experimental Biology, 199*(Pt 4), 941–948.

Stern, D. L., & Orgogozo, V. (2009). Is genetic evolution predictable? *Science, 323*(5915), 746–751.

Sun, Y. M., Favre, I., Schild, L., & Moczydlowski, E. (1997). On the structural basis for size-selective permeation of organic cations through the voltage-gated sodium channel effect of alanine mutations at the DEKA locus on selectivity, inhibition by Ca^{2+} and H^+, and molecular sieving. *Journal of General Physiology, 110*(6), 693–715.

Terlau, H., Heinemann, S. H., Stühmer, W., Pusch, M., Conti, F., Imoto, K., & Numa, S. (1991). Mapping the site of block by tetrodotoxin and saxitoxin of sodium channel II. *FEBS Letters, 293*(1–2), 93–96.

Thompson, J. N. (2005). *The geographic mosaic of coevolution*. University of Chicago Press.

Tikhonov, D. B., & Zhorov, B. S. (2011). Possible roles of exceptionally conserved residues around the selectivity filters of sodium and calcium channels. *Journal of Biological Chemistry, 286*(4), 2998–3006.

Tikhonov, D. B., & Zhorov, B. S. (2012). Architecture and pore block of eukaryotic voltage-gated sodium channels in view of NavAb bacterial sodium channel structure. *Molecular Pharmacology, 82*(1), 97–104.

Vicart, S., Sternberg, D., Fontaine, B., & Meola, G. (2005). Human skeletal muscle sodium channelopathies. *Neurological Sciences: Official Journal of the Italian Neurological Society and of the Italian Society of Clinical Neurophysiology, 26*(4), 194–202.

Vilin, Y. Y., & Ruben, P. C. (2001). Slow inactivation in voltage-gated sodium channels: molecular substrates and contributions to channelopathies. *Cell Biochemistry and Biophysics, 35*(2), 171–190.

Wakely, J. F., Fuhrman, G. J., Fuhrman, F. A., Fischer, H. G., & Mosher, H. S. (1966). The occurrence of tetrodotoxin (tarichatoxin) in Amphibia and the distribution of the toxin in the organs of newts (*Taricha*). *Toxicon: Official Journal of the International Society on Toxinology, 3*(3), 195–203.

Weinreich, D. M., Delaney, N. F., DePristo, M. A., & Hartl, D. L. (2006). Darwinian evolution can follow only very few mutational paths to fitter proteins. *Science, 312*(5770), 111–114.

Williams, B. L., Hanifin, C. T., Brodie, E. D., Jr., & Brodie, E. D., III (2010). Tetrodotoxin affects survival probability of rough-skinned newts (*Taricha granulosa*) faced with TTX-resistant garter snake predators (*Thamnophis sirtalis*). *Chemoecology, 20*(4), 285–290.

Xiong, W., Farukhi, Y. Z., Tian, Y., DiSilvestre, D., Li, R. A., & Tomaselli, G. F. (2006). A conserved ring of charge in mammalian Na^+ channels: a molecular regulator of the outer pore conformation during slow inactivation. *Journal of Physiology, 576*(3), 739–754.

Yamagishi, T., Janecki, M., Marban, E., & Tomaselli, G. F. (1997). Topology of the P segments in the sodium channel pore revealed by cysteine mutagenesis. *Biophysical Journal, 73*(1), 195.

Yamagishi, T., Li, R. A., Hsu, K., Marbán, E., & Tomaselli, G. F. (2001). Molecular architecture of the voltage-dependent Na channel functional evidence for α helices in the pore. *Journal of General Physiology, 118*(2), 171–182.

Zakon, H. H. (2012). Adaptive evolution of voltage-gated sodium channels: the first 800 million years. *Proceedings of the National Academy of Sciences of the United States of America, 109*(Suppl. 1), 10619–10625.
Zakon, H. H., Jost, M. C., & Lu, Y. (2011). Expansion of voltage-dependent Na^+ channel gene family in early tetrapods coincided with the emergence of terrestriality and increased brain complexity. *Molecular Biology and Evolution, 28*(4), 1415–1424.
Zakon, H. H., Jost, M. C., Zwickl, D. J., Lu, Y., & Hillis, D. M. (2009). Molecular evolution of Na^+ channels in teleost fishes. *Integrative Zoology, 4*(1), 64–74.
Zimmer, R. K., & Ferrer, R. P. (2007). Neuroecology, chemical defense, and the keystone species concept. *Biological Bulletin, 213*(3), 208–225.
Zimmer, T. (2010). Effects of tetrodotoxin on the mammalian cardiovascular system. *Marine Drugs, 8*(3), 741–762.

SECTION THREE

Computational Studies and Modeling

CHAPTER FIVE

Computational Structural Pharmacology and Toxicology of Voltage-Gated Sodium Channels

B.S. Zhorov*,§,1 and D.B. Tikhonov§
*McMaster University, Hamilton, ON, Canada
§Russian Academy of Sciences, St. Petersburg, Russian Federation
[1]Corresponding author: E-mail: zhorov@mcmaster.ca

Contents

Current Topics in Membranes, Volume 78
ISSN 1063-5823
http://dx.doi.org/10.1016/bs.ctm.2015.12.001

Abstract

Voltage-gated sodium channels are targets for many toxins and medically important drugs. Despite decades of intensive studies in industry and academia, atomic mechanisms of action are still not completely understood. The major cause is a lack of high-resolution structures of eukaryotic channels and their complexes with ligands. In these circumstances a useful approach is homology modeling that employs as templates X-ray structures of potassium channels and prokaryotic sodium channels. On one hand, due to inherent limitations of this approach, results should be treated with caution. In particular, models should be tested against relevant experimental data. On the other hand, docking of drugs and toxins in homology models provides a unique possibility to integrate diverse experimental data provided by mutational analysis, electrophysiology, and studies of structure—activity relations. Here we describe how homology modeling advanced our understanding of mechanisms of several classes of ligands. These include tetrodotoxins and mu-conotoxins that block the outer pore, local anesthetics that block of the inner pore, batrachotoxin that binds in the inner pore but, paradoxically, activates the channel, pyrethroid insecticides that activate the channel by binding at lipid-exposed repeat interfaces, and scorpion alpha and beta-toxins, which bind between the pore and voltage-sensing domains and modify the channel gating. We emphasize importance of experimental data for elaborating the models.

Abbreviations

DDT 1,1,1-trichloro-2,2-bis(*p*-chlorophenyl) ethane
EM Electron microscopy
LAs Local anesthetics
MC Monte Carlo
MCM MC minimization
MD Molecular dynamics
STX Saxitoxin
TTX Tetrodotoxin

1. INTRODUCTION

Voltage-gated sodium channels play key roles in physiology of excitable cells. They are targeted by many classes of toxins and drugs. Among naturally occurring toxins are commonly known tetrodotoxin (TTX) and batrachotoxin (BTX), and large families of toxins produced by marine cones, scorpions, spiders, plants, and other species. Pain and various health disorders are treated with sodium channel drugs. Well-known examples include local anesthetics and medications for treatment of angina, arrhythmias, schizophrenia, and epilepsies. Another important application of sodium channel ligands are insecticides: pyrethroid agonists, DDT, and sodium channel blockers. A large number of original papers and reviews are published in this field (eg, Catterall, 2014; Catterall & Swanson, 2015; Stevens, Peigneur, & Tytgat, 2011).

Despite long history of pharmacological and toxicological studies and broad range of practical applications, binding sites and atomic mechanisms of action of sodium channel ligands are still subjects of intensive research in academia and industry. Examples include local anesthetics (Yarov-Yarovoy et al., 2002), antiarrhythmics (Wang, Russell, & Wang, 2013), anticonvulsants (Catterall, 2014), conotoxins (CTX) (McArthur, Singh, et al., 2011), insecticides (Du et al., 2013), and scorpion and spider toxins (Bosmans & Swartz, 2010; Wang et al., 2011). A cause of so long activity in the field is a huge diversity of the ligands that bind to distinct sites in sodium channels and act by different and complex mechanisms. In particular, studies of various aspects of use-dependent action, which includes state-dependent access to the binding sites, and different affinity of drugs to the open, closed, and inactivated states of sodium channels are very important for understanding the drugs action in physiological and pathological conditions.

Another problem is a lack of high-resolution structures of eukaryotic sodium channels. Currently, homology modeling is the only approach to gain insights in atomic details of ligand interaction with these channels. Below we describe some achievements and problems of the modeling approach, which is used to understand atomic mechanisms of action of drugs and toxins on the voltage-gated sodium channels.

2. STRUCTURE OF SODIUM CHANNELS

Eukaryotic voltage-gated sodium channels contain the pore domain and four voltage-sensing domains within a single polypeptide chain that has four homologous repeats. Each repeat has six transmembrane segments (S1–S6) and extra- and intracellular loops, including membrane-reentering P-loop (P). Segments S1–S4 from each repeat form four voltage-sensing domains. Segments S5 (the outer helices), P, and S6 (the pore-lining inner helices) from the four repeats jointly contribute to the pore domain (Fig. 1). P-loops contain membrane-descending helices (P1) and ascending limbs with residues that contribute to the selectivity filter. In eukaryotic sodium channels the selectivity filter is formed by the so-called DEKA ring, ie, D, E, K, and A residues in homologous positions of the four ascending limbs (Fig. 1D). The narrow DEKA ring divides the ion-conducting pathway in

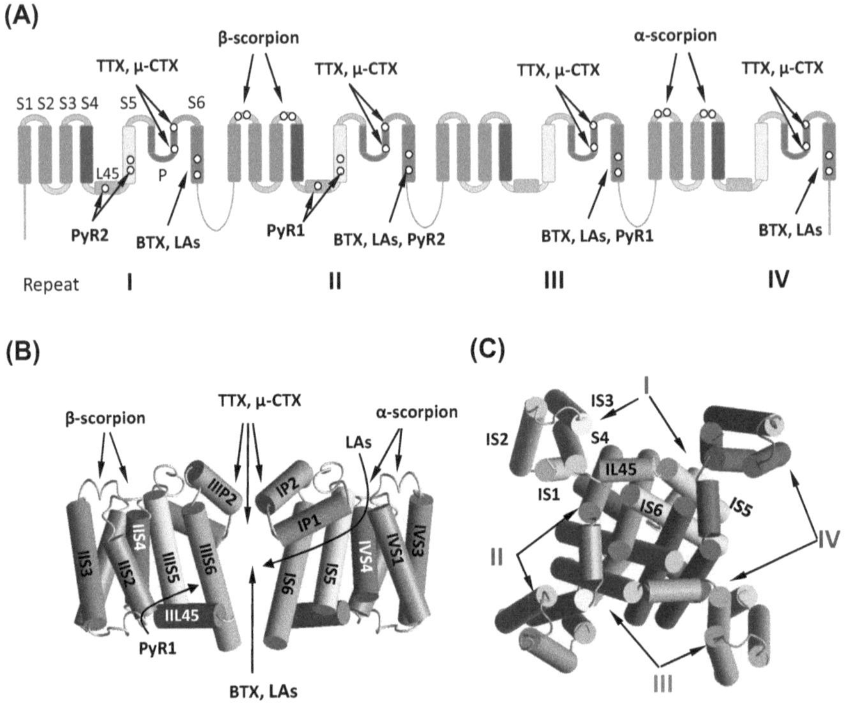

(D)

Kv1.2	L45-S5	SKGLQILGQTLK ASMRELGLLI FFLFIGVILF SSAVYFAE
NavAb	L45-S5	VPQMRKIVSALI SVIPGMLSVI ALMTLFFYIF AIMATQLF
Na$_v$1.4	I L45-S5	IPGLKTIVGALI QSVKKLSDVM ILTVFCLSVF ALVGLQLF
	II L45-S5	WPTLNMLIKIIG NSVGALGNLT LVLAIIVFIF AVVGMQLF
	III L45-S5	FEGMRVVVNALL GAIPSIMNVL LVCLIFWLIF SIMGVNLF
	IV L45-S5	AKGIRTLLFALM MSLPALFNIG LLLFLVMFIY SIFGMSNF
Kv1.2	P	FPSIPDAF WWAVVSMTTV GYGDMVPTT
NavAb	P	FGTLGESF YTLFQVMTLE SWSMGIVRP
Na$_v$1.4	IP	YDTFSWAF LALFRLMTQD YW•ENLFQL
	IIP	MHDFFHSF LIVFRILCGE •WIETMWDC
	IIIP	YDNVGLGY LSLLQVATFK GW•MDIMYA
	IVP	FETFGNSI ICLFEITTSA GW•DGLLNP
Kv1.2	S6	IGGKIVGSLC AIAGVLTIAL PVPVIVSNFN YFY
NavAb	S6	PYAWVFFIPF IFVVTFVMIN LVVAICVDAM AIL
Na$_v$1.4	IS6	KTYMIFFVVI IFLGSFYLIN LILAVVAMAY AEQ
	IIS6	AMCLTVFLMV MVIGNLVVLN LFLALLLSSF SAD
	IIIS6	LYMYLYFVIF IIFGSFFTLN LFIGVIIDNF NQQ
	IVS6	SIGICFFCSY IIISFLIVVN MYIAIILENF NVA

Figure 1 Ligand-binding sites in sodium channels. (A) Transmembrane topology of eukaryotic channels. (B) Side view of the NavAb X-ray structure. Only two repeats are shown for clarity. Segments are colored as in (A). (C) Intracellular view of the NavAb X-ray structure with P-loops removed for clarity. Individual subunits are shown with different colors. Roman numerals indicate repeats in heterotetrameric eukaryotic channels. (D) Aligned sequences of helical segments in the pore-forming domain of potassium and sodium channels. Segment labels are colored as in (A) and (B).

two parts: the inner pore, which is exposed to the cytoplasm in the open channel, and the outer pore, which is exposed to the extracellular space. The activation gate is formed by cytoplasmic parts of the inner helices, which converge to a tight bundle in the closed state and diverge in the open state to create rather wide inner vestibule. Activation by membrane voltage is controlled by the voltage-sensing domains. The S4 helices contain positively charged residues, which move as the voltage changes. This induces movements of the L45 linkers that in turn result in the activation gate opening.

Pioneering crystallographic studies of potassium channels by Roderick MacKinnon et al. opened a possibility to analyze structural details of other channels that are believed to have a similar folding, including sodium channels. The crystal structure of a pH-gated potassium channel KcsA (Doyle et al., 1998) demonstrated general features of the pore domain in the closed state. Next, crystal structure of a calcium-activated potassium channel MthK (Jiang et al., 2002) revealed the principal structure of the open pore and a mechanism of the activation gating. Structures of voltage-gated prokaryotic potassium channel KvAP (Jiang et al., 2003) and eukaryotic potassium channel Kv1.2 (Long, Campbell, & Mackinnon, 2005) revealed mutual disposition of the pore and voltage-sensing domains. Recent crystal structures of several bacterial voltage-gated sodium channels, such as NavAb (Payandeh, Scheuer, Zheng, & Catterall, 2011), NavRh (X. Zhang et al., 2012), and NavMs (McCusker et al., 2012), are particularly important for studies of eukaryotic sodium channels.

Comparison of available X-ray structures reveals both conserved and different details of folding. For example, membrane-descending P1 helices are conserved, whereas ascending parts of P-loops, which include selectivity filter residues, are different (Fig. 2A). In potassium channels the selectivity filter is formed by several rings of backbone carbonyls from the VGYG motifs, whereas in bacterial sodium channels the filter is a ring of serine or glutamate side chains. Another important difference is presence of ascending P2 helices in sodium channels and their absence in P-loops of potassium channels. The general disposition of the S5 and S6 helices is rather similar in all the channels. However, the middle parts of neighboring S6s are close to each other in potassium channels, whereas in sodium channels, they line wide intersubunit fenestrations (Fig. 2B). Comparison of the open and closed channels shows that P1 helices and extracellular parts of S5 and S6 helices are structurally conserved, whereas the intracellular parts of S6s diverge in the open channels to form a wide access pathway to the inner pore

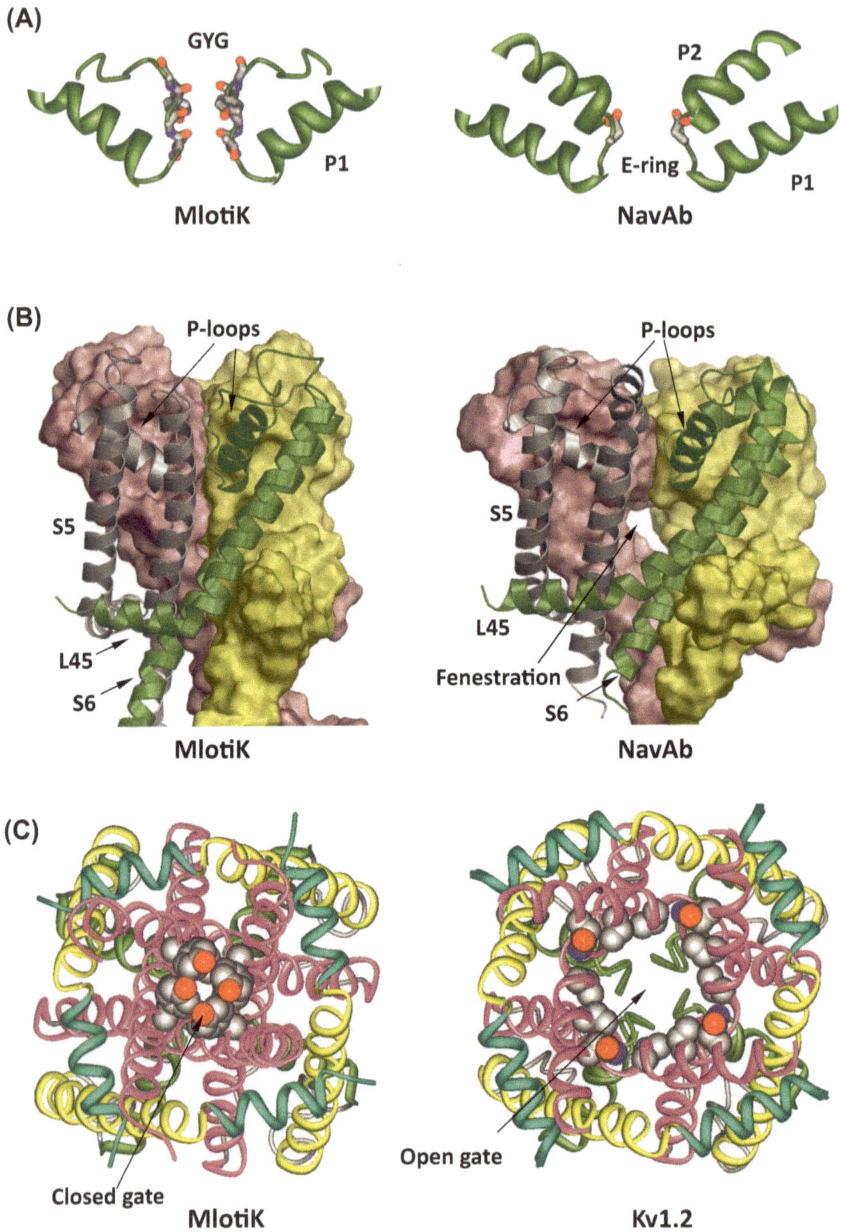

Figure 2 Structural templates used for homology modeling of sodium channels. (A) P-loops and selectivity filters in 4x6TM potassium (MlotiK) and sodium (NavAb) channels. (B) Subunit interfaces viewed from lipids. Front subunits are shown as helices and rear subunits as surfaces. A wide fenestration is seen in the sodium channel. (C) Cytoplasmic views at the activation gates in the closed (MlotiK) and open (Kv1.2) channels. Individual segments are colored as in Fig. 1A. Residues that form the gate constriction in the closed state are space-filled. In Kv1.2 these are Val, Ile, and Asn residues, which are located, respectively, 8, 12, and 16 positions downstream from the gating-hinge glycine in helix S6 of Kv1.2 (Fig. 1D). In MlotiK only the Tyr residue 16 positions downstream from the gating-hinge glycine is seen.

(Fig. 2C). Furthermore, the degree of S6 divergence varies between the open channels KvAP, MthK, and Kv1.2. Additional insights are provided by the open-state model of a prokaryotic sodium channel NavCt that is based on the low-resolution EM structure of the pore domain (Tsai et al., 2013).

3. HOMOLOGY MODELING

Despite the impressive progress in crystallographic studies of ion channels, X-ray structures of eukaryotic sodium channels are still unavailable. Therefore, understanding of atomic details of ligand action on these channels is currently possible only through homology modeling. This approach is based on the assumption that, despite differences in their sequences, homologous proteins have similar general folding patterns. The assumption allows one to use available high-resolution data (usually X-ray structures) as templates to model proteins, whose 3D structures are unknown. The homology modeling method is widely used in theoretical studies of various proteins and their complexes with drugs (Petrey et al., 2015; Schmidt, Bergner, & Schwede, 2014).

3.1 Computational procedures

Any computational modeling involves calculations of atomic interactions using a set of semiempirical functions (force fields). Structure optimization algorithms involve sampling of thousands to millions different conformations. There are two methods of sampling: molecular dynamics (MD) and Monte Carlo energy minimizations (MCM). Both approaches can use the same force fields, but have different strategies of calculation. MD provides numerical solutions of the Newton's equations of motion, thus simulating realistic trajectories of movement of all atoms. An important advantage of MD is that all atoms move simultaneously, making MD indispensible for simulation of large systems, like ion channels in lipid and water environments. The major problem is a very small (femtosecond) time step and therefore, huge computational resources required to simulate microsecond trajectories (Jensen et al., 2012). The MCM approach involves "jumps" over energy barriers, which speed up the search for low-energy conformations. The major problem is that MCM samples one geometrical parameter at a time so that exploring large macromolecules requires huge computational resources. However, when the number of geometrical parameters to be sampled is limited, eg, docking a ligand in a certain region of a protein with implicit solvent, MCM efficiently solves the multiminima problem. MCM is used in various programs for ligand docking, eg, AutoDock, Glide, Gold, ICM, and ZMM (see Garden & Zhorov, 2010). Some studies use MCM to find energetically optimal positions of ligands and then MD to simulate behavior of the complexes (Marzian et al., 2013).

The predictive power of ligand-docking programs is limited due to imprecise force fields and huge conformational hyperspace, where finding the global minimum of free energy is not guaranteed. Comparison of X-ray structures of ligand—protein complexes with results of ligand docking in respective binding sites shows that the docking success rate does not exceed ~75%. Thus, results of hands-free ligand docking are not absolutely reliable.

The major cause of the limited reliability of homology modeling is not the above mentioned limitations of computational procedures, but imprecision of models per se. Attempts to model sodium channels encountered serious problems due to their poor sequence similarity with potassium channels. Therefore, the key initial step in homology modeling of sodium channels, the alignment of sequences, is a nontrivial task, which cannot be solved by standard alignment methods. Different alignments result in vastly different properties of the models, especially in helical segments where orientation of $C^{\alpha}-C^{\beta}$ bonds of adjacent residues differs by ~104 degree. As a result, just a single-position shift in the sequence alignment completely changes interactions of the respective helix with the rest of the system.

Another problem is that a strong 3D similarity between a template and a query protein is questionable when their sequence similarity is poor. A similar general folding of voltage-gated channels is an underlying assumption of homology modeling, but exact positions and orientations of individual segments may be different (see above). The degree of similarity between 3D structures of symmetric homotetrameric prokaryotic sodium channels and asymmetric pseudoheteromeric eukaryotic channels is unclear. Therefore, building homology models of mammalian sodium channels remains a problem for which standard computational protocols are hardly applicable.

3.2 Using experimental data in homology modeling

The only and best way to increase precision of homology modeling of sodium channels is to employ experimental data as constraints. The most important information, which allows one to both simplify the docking procedure and increase its reliability, comes from studies using site-directed mutagenesis. It is tempting to assume that if a residue substitution affects action of a ligand, the residue contributes to the ligand binding site. However, such a straightforward data interpretation can be misleading because the substitution can affect the drug action indirectly, by at least two possible mechanisms. First, a mutation beyond the binding site may change the channel conformation. Second, a mutation may affect activation and/or inactivation gating. A mutation that, eg, shifts the equilibrium between the open and closed channel states would influence the use-dependent drug action. Analysis of the latter possibility requires careful study of the mutant channel properties.

Very important information is provided by experimental studies of molecular mechanisms of drug action. Indeed, a modeling study should not only predict a drug-binding site, but also explain (at least qualitatively) such peculiarities as use, voltage, and ion dependence; cooperativity; and ingress and egress pathways. Especially interesting are experimental studies that explore series of ligands. Analysis of structure–function relationships helps understand structural determinants of the ligand action and determine some features of the binding site organization.

These experimental data may help tune homology models and focus ligand docking to certain binding regions. Major problems are possible ambiguity of structural interpretations of experimental data, existence of various low-energy structures that may be consistent with experimental data, imprecise calculations of free energy, especially its entropic and electrostatic components, and possible involvement of permeant ions in ligand–channel interactions, which is difficult to predict without employing quantum chemistry computations or polarizable force fields. In view of these problems, the merit of a model is related to its ability to integrate diverse experimental data, which otherwise seem scattered, and especially to provide experimentally testable predictions. Thus, modeling of sodium channels with ligands involves finding a consensus between the model energy (which must be low, but not necessarily the lowest one) and the model consistency with experimental data. Below we describe several modeling studies, which contributed to the field.

4. PROGRESS IN MODELING SODIUM CHANNEL WITH LIGANDS

4.1 Outer pore blockers

4.1.1 Overview

Classical ligands acting on the outer pore are TTX, saxitoxin (STX) (and analogs) as well as peptide μ-CTXs. Intensive studies with site-directed mutagenesis revealed that the binding of these toxins is controlled by the selectivity filter residues and several residues downstream. Among these the key role belongs to the ring of outer carboxylates, ie, acidic residues three to four positions downstream from the DEKA residues (Fig. 1D). Detailed analysis revealed many pairwise contacts between certain residues of the channels and particular chemical groups of TTX and STX (Terlau et al., 1991) or certain residues of mu-CTXs (Chang, French, Lipkind, Fozzard, & Dudley, 1998; Xue, Ennis, Sato, French, & Li, 2003). For some contacts, the free energy of interaction was estimated using mutant cycle analysis (Choudhary, Aliste, Tieleman, French, & Dudley, 2007; Li et al., 2001). Although the patterns of residues interacting with different toxins are not identical, their binding sites essentially overlap. Particularly the outer carboxylates, all of which are proposed to contribute to a composite-binding site for permeant sodium ions (Khan, Romantseva, Lam, Lipkind, & Fozzard, 2002), are critically important for the action of TTX, STX, and μ-CTXs. There is an intriguing difference between TTX and STX, which completely abolishes the conductance, and some native and mutant μ-CTXs, which can reduce the conductance to certain degree.

4.1.2 Models

The pioneering structural model of the sodium channel outer pore with TTX and STX was proposed by Lipkind and Fozzard (1994) before any crystal structure of P-loop channels was available. Of particular importance for the modeling were the data that the TTX and STX action is dramatically diminished by mutations of the selectivity filter aspartate and glutamate in repeats I and II, respectively, and outer carboxylates in repeats I, II, and IV (Terlau et al., 1991). The model visualized the cationic guanidinium group of TTX bound to the selectivity filter and repeat II outer carboxylate forming H-bonds with TTX hydroxyl groups. The model also proposed that repeats I, II, III, and IV have a clockwise arrangement at the extracellular view, and this important prediction was confirmed by the mutant cycle analysis of the channel interaction with μ-CTX GIIIA (Dudley et al., 2000).

When the KcsA X-ray structure become available, Lipkind and Fozzard (2000) elaborated a KcsA-based homology model of Nav1.4 and employed the experimental data on the toxin—channel interactions to dock TTX and STX. Since the KcsA outer pore is too narrow to accommodate the bulky semirigid toxins, the authors placed the selectivity filter residues at the C-ends of P-helices and suggested that the P-helices and ascending limbs are more distant from the pore axis as compared to potassium channels.

In this model, the DEKA ring is placed at the border between the central cavity and the outer pore. To provide a structural rationale for numerous experimentally defined pairwise contacts between big peptide μ-CTXs and sodium channels, it was necessary to further modify the KcsA-based model of the sodium channel by shifting the P-helices farther from the pore axis and increasing their slope (Fig. 3A) (Choudhary et al., 2007).

Alternative KcsA-based homology models of Nav1.4 with TTX and STX were elaborated assuming a stronger 3D similarity between potassium and sodium channels (Tikhonov & Zhorov, 2005a, 2011). These models were built to explain the same experimental data on toxin–channel contacts, but the selectivity filter DEKA residues were placed in the middle of the ascending limbs rather at the turns of P-loops. As a result, it was possible to dock the TTX and STX in the outer pore without displacing P-helices from the positions, which are seen in potassium channels (Fig. 3B).

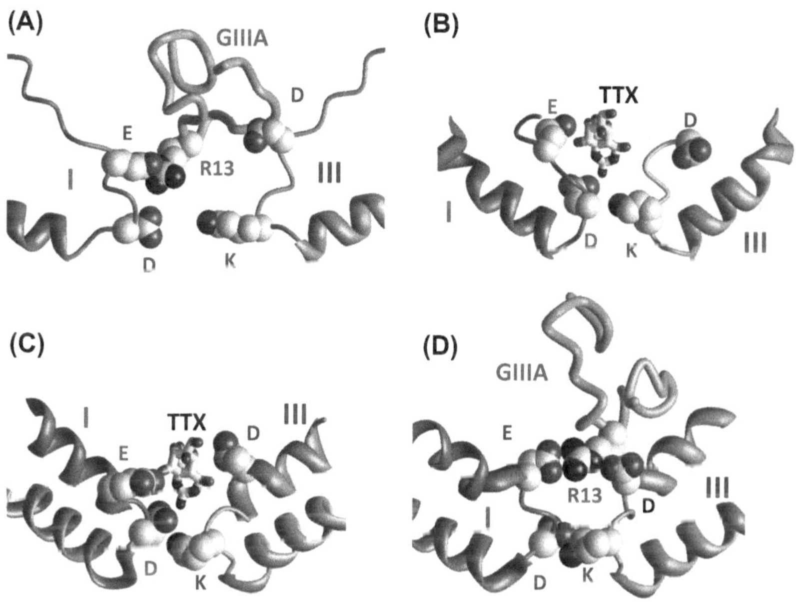

Figure 3 Structural models of toxin-binding sites in the sodium channel outer pore region. Only repeats I and III are shown for clarity. (A) μ-Conotoxin GIIIA in the sodium channel model (Choudhary et al., 2007), which was modified from the original KcsA-based model (Lipkind & Fozzard, 2000). (B) TTX-binding in the KcsA-based model (Tikhonov & Zhorov, 2005a). P-helices are located exactly as in KcsA. (C) TTX-binding in the NavAb-based model (Tikhonov & Zhorov, 2012). (D) GIIIA binding in the NavAb-based model (Korkosh et al., 2014).

The X-ray structures of bacterial sodium channels (Payandeh et al., 2011; X. Zhang et al., 2012) provided evidence in favor of the latter models (Tikhonov & Zhorov, 2005a, 2011). The selectivity filter glutamates are in the middle of the ascending limbs and positions of P-helices in NavAb are very close to those in potassium channels (Fig. 2A). However, both models (Lipkind & Fozzard, 2000; Tikhonov & Zhorov, 2005a) failed to predict folding of the C-terminal halves of P-loops. The NavAb structure demonstrates that unlike in potassium channels, the C-terminal halves of P-loops contain P2 helices that diverge in the extracellular direction, making the outer pore wide enough to accommodate large-size toxins.

It remains unclear, however, to which extent the X-ray structures of the homotetrameric prokaryotic sodium channels are close to the structures of eukaryotic channels. In particular, MD simulations of TTX interactions with the NavAb-based model of Nav1.4, which was built using a straightforward sequence alignment of P-loops, predicts that the TTX long axis is perpendicular to the pore axis and TTX forms H-bonds with the outer carboxylates in domains I, II, and III (Chen & Chung, 2014). This result is not in a good agreement with the experimental data on TTX interactions with the sodium channel. In particular, the model does not explain a strong contribution of the outer carboxylate in repeat IV and a weak contribution of the outer carboxylate in repeat III in the action of TTX.

In a NavAb-based model of Nav1.4, which was built using the straightforward sequence alignment of P-loops (Tikhonov & Zhorov, 2012), the outer carboxylates are also unable to interact with TTX as effectively as implied by mutational data. A possible solution for the problem was found by using an adjusted sequence alignment between bacterial and eukaryotic sodium channels (Fig. 1D). Proposed deletions in the sequence near the DEKA ring (Tikhonov & Zhorov, 2012) provide orientation of the outer carboxylates in agreement with the experimental data, which were summarized by Lipkind and Fozzard.

The Nav1.4 model (Fig. 3C), which is based on the adjusted sequence alignment with NavAb (Tikhonov & Zhorov, 2012), was further used to dock peptide toxins GIIIA, PIIIA, and KIIIA (Korkosh, Zhorov, & Tikhonov, 2014). Docking of the best-studied toxin, GIIIA, was facilitated due to employment of distance constraints between the toxin and the channel residues, which represent pairwise interactions obtained from the mutant cycle analysis (Choudhary et al., 2007). The distance constraints have been satisfied without backbone deformations in the channel and toxin (Fig. 3D). Furthermore, computed energies of specific interactions correlated with the experimental estimations (Choudhary et al., 2007), thus supporting both the experimental data and the model.

Robert French et al. further explored the voltage dependence of action of PIIIA and its mutants, and determined the depths of residues in the membrane electric field (McArthur, Singh, et al., 2011). These data were not used to dock PIIIA in the sodium channels, but the model predicts the depths of the charged residues penetration into the outer pore in good agreement with the experimental data (Korkosh et al., 2014).

The model of KIIIA-bound channel (Korkosh et al., 2014) rationalized intriguing observations that TTX and KIIIA can simultaneously bind to the sodium channel (Khoo et al., 2012; Stevens et al., 2012; Van Der Haegen, Peigneur, & Tytgat, 2011; Wilson et al., 2011; Zhang et al., 2009). In this model, TTX can pass between Nav1.4 and the channel-bound KIIIA to reach its binding site at the selectivity filter.

For the first time, the model (Korkosh et al., 2014) explained interesting phenomena of incomplete block, which is produced by some native and mutant CTXs (Hui, Lipkind, Fozzard, & French, 2002; McArthur, Ostroumov, Al-Sabi, McMaster, & French, 2011; Wilson et al., 2011). Unlike TTX, which enters the narrow part of the outer pore and completely plugs it, large peptide toxins bind above the outer pore and cover it. The toxin's charged residues form salt bridges with the outer carboxylates thus blocking the ion permeation. In the absence of some of these basic residues, at least one carboxylate does not form bridges with the toxin, thus allowing permeant ion to pass, albeit with smaller rate than when the toxin is not present.

Thus, employing the modeling approach has allowed to proceed from schematic views on the outer pore ligands toward detailed pictures, which explain different phenomena of action. However, because of the lack of high-resolution structures of toxin—channel complexes, it remains unclear how close are the models to reality?

4.2 Inner pore blockers

4.2.1 Overview

The inner pore region is targeted by local anesthetics (LAs) and related drugs, which block the channel conductance. Despite the simple structure of these ligands and common principal mechanism of the pore block, phenomenology of their action is complex. LAs demonstrate pronounced use dependence of action. Rather weak "tonic block" is observed under infrequent channel activations. Activation by trains of depolarization and by long depolarizations enhances the blocking action, effects usually called "use-dependent block." To explain these effects, Bertil Hille (1977) proposed a "modulated receptor hypothesis". It suggests a weak affinity of an LA molecule to the closed channel and significant increase of the affinity as the channel adopts the open or inactivated states. This hypothesis also proposes two access pathways of LA molecules to the binding site: a hydrophilic pathway through the open activation gate and a hydrophobic pathway through the membrane. Mutagenesis studies revealed residues in the pore-lining S6 segments of repeats I, III, and IV, which control action of LAs (Ragsdale, McPhee, Scheuer, & Catterall, 1994; Wang, Quan, & Wang, 1998; Yarov-Yarovoy et al., 2001, 2002). This strongly suggests binding of LAs in the inner pore region. Interestingly, influence of particular substitutions on the tonic block, frequency-dependent block, and hydrophobic access are unequal (Li, Galue, Meadows, & Ragsdale, 1999).

Quaternary membrane-impermeant compounds like QX-222 and QX-314, which are not LAs, block the open channels when applied from inside the cell (Hille, 1977). They also block cardiac sodium channels when applied extracellularly (Alpert, Fozzard, Hanck, & Makielski, 1989; Qu, Rogers, Tanada, Scheuer, & Catterall, 1995). Mutational analysis revealed critical residues controlling the hydrophobic access pathway (Qu et al., 1995; Ragsdale et al., 1994; Sunami, Glaaser, & Fozzard, 2001; Wang et al., 1998).

4.2.2 Structural models rationalize use-dependent block

The major challenge of modeling the sodium channel with LAs and related drugs is to provide structural rationale for the complex mechanism of action, which is addressed by the "modulated receptor hypothesis" (Hille, 1977). The hydrophilic access pathway obviously involves the open activation gate, which is located more intracellularly from the LA-binding site. However, localization of the hydrophobic pathway, which provides access of LAs to the closed channel, remained unclear. Residues, which control this pathway were found in the middle of S6 and in the P-loop (Qu et al., 1995; Sunami, Dudley, & Fozzard, 1997; Sunami, Glaaser, & Fozzard, 2000; Wang et al., 1998).

LAs have a larger affinity to the open/inactivated state than to the closed state. Lipkind and Fozzard (2005) employed the first published structure of an open potassium channel, MthK, as a template to model the open sodium channel. Because the degree of structural similarity between MthK and sodium channels was unclear, the authors did not attempt to find the energetically optimal binding mode, which is critically sensitive to structural details, but docked a representative series of compounds to rationalize experimental data on mutations and structure—activity relationships. Particularly, the authors assumed that alkylamine and aromatic moieties of LAs are located, respectively, near LA-sensing phenylalanine and tyrosine residues in IVS6, which were previously discovered by William Catterall et al. (Ragsdale et al., (1994). Energetically appropriate binding modes for different LAs, which are consistent with mutational data, were obtained (Fig. 4A). An important conclusion was that LAs do not completely plug the inner pore and block the channel by primarily electrostatic mechanism, ie, they create a positive electrostatic barrier for cation permeation (Lipkind & Fozzard, 2005). The authors further concluded that lidocaine-like molecules are too big to fit into the KcsA-based closed-channel model in the same way as they fit into the MthK-based open-channel model.

In the subsequent KvAP-based model of the open Nav1.4 (Tikhonov & Zhorov, 2007), the optimal binding mode of lidocaine is similar to that in the MthK-based model (Lipkind & Fozzard, 2005). In the closed-state model, big LAs like tetracaine cannot fit the central cavity and their aromatic rings protrude into the repeat interface between helices IIIP, IIIS6, and IVS6 (Bruhova, Tikhonov, & Zhorov, 2008). It was noted that some residues controlling the hydrophobic access line the interface. The model inspired a proposition that the III/IV repeat interface serves as a hydrophobic access pathway ("sidewalk") for LAs into the closed sodium channel (Bruhova et al., 2008; Tikhonov & Zhorov, 2005a, 2007).

The sodium channel models, which are based on potassium channel templates, have two weak points. First, the subunit interfaces in potassium channels are narrow and the possibility of an LA molecule to pass through respective interfaces in sodium channel models is questionable. Second, some residues, which affect drug access into the closed channel, face the selectivity filter region rather than the repeat interface (Sunami et al., 1997, 2000).

Recent X-ray structures of sodium channels support the model-based propositions on the hydrophobic access pathway location. Subunit interfaces in these channels are much wider than in potassium channels (Fig. 2B) and could provide access pathways for big LAs from the membrane into the inner pore. Of special interest is the structure a bacterial sodium channel NavMs with a bromophenyl-containing LA-like blocker (Bagneris et al., 2014). Because of a limited resolution, only the bromine atom of the ligand is seen, but its position in the subunit interface is close to that of the

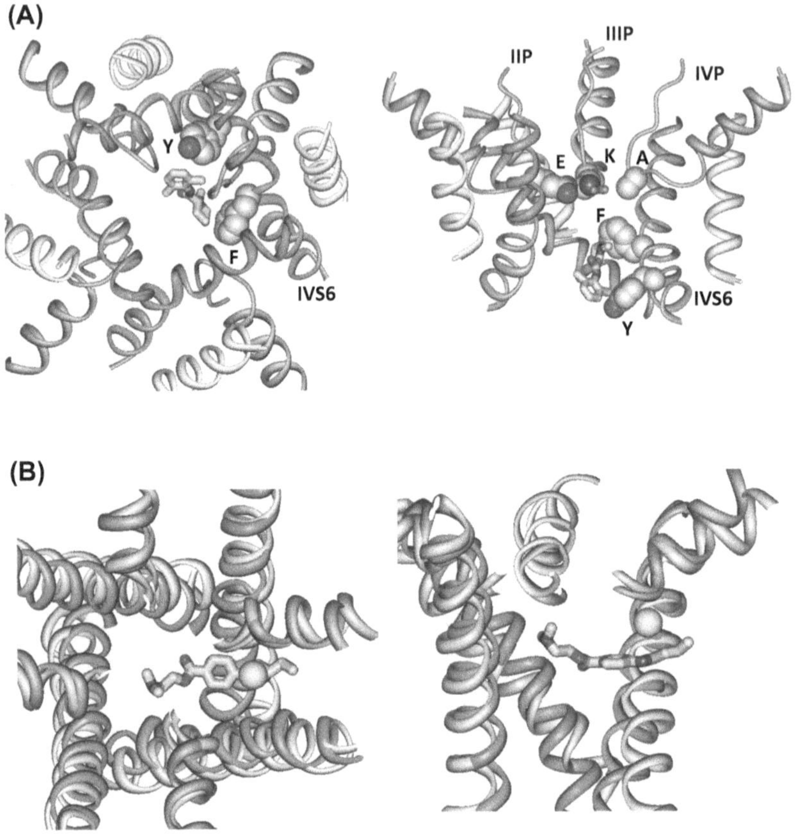

Figure 4 Structural models of the sodium channel with local anesthetics in the inner pore region. (A) Cytoplasmic and side views at the MthK-based open-state model with mepivacaine bound to space-filled phenylalanine and tyrosine residues in helix IVS6 (Lipkind & Fozzard, 2005). Helices S5, S6, and P are colored as in Fig. 1A and B. At the side view, repeat I is removed for clarity and the selectivity-filter residues are shown. (B) Extracellular and side views at the superposition of the KcsA-based model of Nav1.5 with tetracaine (Bruhova et al., 2008) and the X-ray structure of NavMs (gray ribbons) with the bromine atom of LA-like ligand shown as a *yellow sphere* (light gray in print versions) (Bagneris et al., 2014). The selectivity filter region is not shown for clarity.

tetracaine Ph-N nitrogen in the closed Nav1 model (Bruhova et al., 2008) (Fig. 4B). Moreover, in the presence of the blocker, the ionic occupation of the NavMs selectivity filter is reduced indicating displacement of sodium ion(s) in the presence of the drug. This agrees with a hypothesis on the coupled movement of organic and inorganic cations, according to which an LA molecule enters the closed channel through the repeat interface and displaces the resident ion from the central cavity, which can leave the channel only through the selectivity filter (Bruhova et al., 2008). The latter model explains intriguing observations that the outer pore mutations or the outer pore blocks by TTX prevent the closed-channel block by LAs (Bruhova et al., 2008).

Thus, the models of sodium channels with the inner pore-targeting blockers explain effects of mutations on the ligand action. The models allowed us to make important predictions on the ligands orientation in the open and closed channels and on the access pathways (Bruhova et al., 2008). Recent X-ray structures agree with these predictions.

4.3 Steroidal sodium channel activators

4.3.1 Overview

Naturally occurring BTX, aconitine, veratridine, and grayanotoxins are known as steroidal sodium channel activators (agonists). Location of the binding sites and mechanisms of action of these toxins are perhaps the most intriguing puzzle regarding sodium channel ligands. In the presence of such compounds the activation threshold is shifted to more negative values so that the channels open near the resting membrane potential. Additionally, the inactivation is abolished and the channels remain persistently open. The complex effects of these lipid-soluble molecules on the gating and permeation properties initially inspired a hypothesis on the allosteric mechanism of action of the ligands, which were believed to bind at the lipid–channel interface (Catterall, 1979, Hille, 2001). However, subsequent mutational data were surprising. BTX-sensing residues were found in S6 segments of all four repeats of the channel. Moreover, patterns of LA-sensing and BTX-sensing residues suggested that receptors of BTX and LAs may overlap. Some use dependence characteristics of activators and blockers are also similar (Wang & Wang, 2003). For instance, large-size blockers prevent the channel closure (Yamamoto, 1986; Yeh & Narahashi, 1977).

4.3.2 Models predict binding region and mechanism of sodium channel activators

In concept-breaking models, BTX, veratridine, and aconitine are proposed to bind to the inner pore without blocking the ion permeation (Tikhonov & Zhorov, 2005b). In these models the channel-bound activators do not completely occlude the ion permeation pathway. Rather, they exhibit cation-attractive groups (the so-called oxygen triad) to a pore-facing hydrophilic residue of the channel, thus allowing the ion permeation between the agonist and the inner pore wall. Furthermore, binding of these relatively big toxins prevents channel closure. This model and its subsequent development (Du, Garden, Wang, Zhorov, & Dong, 2011) suggest that BTX acts in the ion channel like a surgical stent that acts in a blood vessel (Fig. 5A).

The models inspired mutational studies, which identified several new BTX-sensing residues in agreement with modeling predictions (Du et al., 2011; Wang, Mitchell, Tikhonov, Zhorov, & Wang, 2006; Wang, Tikhonov, Zhorov, Mitchell, & Wang, 2007). Of special interest is a lysine substitution of the pore-facing asparagine in the middle of II6, which converts the BTX action from the channel activation to the irreversible channel block (Wang, Tikhonov, Mitchell, Zhorov, & Wang, 2007). This demonstrates a general similarity in the binding sites and binding modes of the

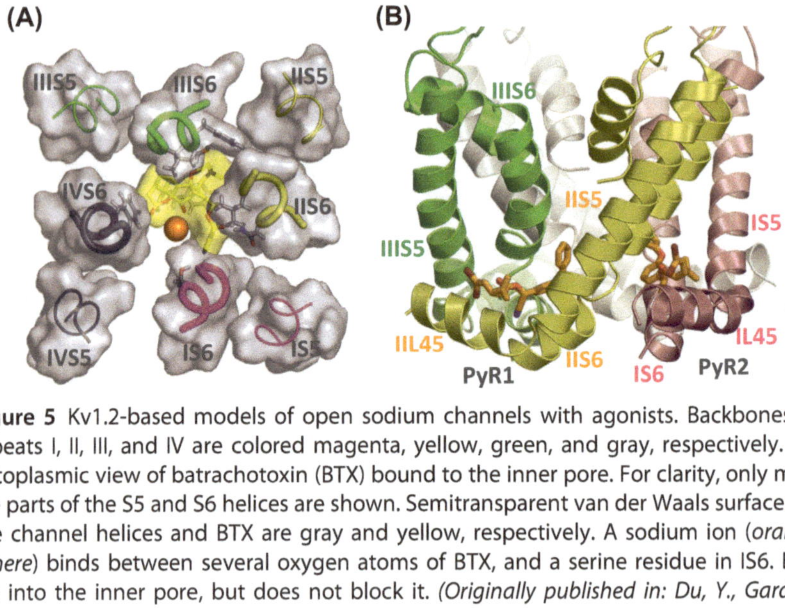

Figure 5 Kv1.2-based models of open sodium channels with agonists. Backbones in repeats I, II, III, and IV are colored magenta, yellow, green, and gray, respectively. (A) Cytoplasmic view of batrachotoxin (BTX) bound to the inner pore. For clarity, only middle parts of the S5 and S6 helices are shown. Semitransparent van der Waals surfaces of the channel helices and BTX are gray and yellow, respectively. A sodium ion (*orange sphere*) binds between several oxygen atoms of BTX, and a serine residue in IS6. BTX fits into the inner pore, but does not block it. *(Originally published in: Du, Y., Garden, D.P., Wang, L., Zhorov, B.S., & Dong, K. (2011). Identification of new batrachotoxin-sensing residues in segment IIIS6 of the sodium channel. Journal of Biological Chemistry, 286(15), 13151–13160. © The American Society for Biochemistry and Molecular Biology.)* (B) Side view at the pore domain with two deltamethrin molecules docked into pyrethroid receptors PyR1 and PyR2. Each ligands binds between four helices (L45, S5, and two S6s). Helix IIS6 contributes to both PyR1 and PyR2. *(Reproduced with permission from Du, Y., Nomura, Y., Zhorov, B.S., & Dong, K. (2015). Rotational symmetry of two pyrethroid receptor sites in the mosquito sodium channel. Molecular Pharmacology, 88(2), 273–280.)*

pore-targeting steroidal molecules, which, depending on their structures and amino acid composition of the pore-lining inner helices, can either block or activate sodium channels.

4.4 Pyrethroid insecticides

4.4.1 Overview

Another class of sodium channels activators includes pyrethroids, a large class of synthetic analogs of natural pyrethrins from the flowers of the *Pyrethrum daisy* (Dong et al., 2014). Pyrethroids, which have relatively low mammalian toxicity and relatively favorable environmental properties, are widely used as one of the most effective control measures against agriculturally deleterious arthropod pests and mosquito-borne diseases, including malaria and dengue. Another well-known sodium channel agonist is DDT, an early insecticide, which is believed to bind at the site that overlaps with the pyrethroid receptor. Intensive use of DDT and then pyrethroids over several decades has led to the development of resistance in many insect populations. A major mechanism of resistance, known as knockdown resistance (*kdr*), has been documented globally in almost all medically and agriculturally significant arthropod pests, and currently more than 50 sodium channel mutations are associated with pyrethroid resistance in diverse arthropod pests (Silver et al., 2014). This problem motivates mapping of pyrethroid-binding sites with the goal to develop new insecticides.

4.4.2 Models

A pioneering model integrated mutational data, including kdr mutations, and predicted binding of pyrethroids in the lipid-exposed II/III repeat interface (O'Reilly et al., 2006). Analysis of additional experimental data motivated modeling of the second pyrethroid receptor in the I/II repeat interface, and model-driven mutagenesis unveiled new pyrethroid-sensing residues (Du et al., 2013). The latter includes residues, which are different between insect and mammalian sodium channels, thus explaining the molecular basis of the selective toxicity of pyrethroids. Recently, a mosquito sodium channel model was elaborated with two pyrethroid receptors, PyR1 and PyR2 (Fig. 5B), exhibiting rotational quasisymmetry around the pore axis (Du, Nomura, Zhorov, & Dong, 2015). The model-driven mutagenesis has led to discovery of seven new pyrethroid-sensing residues, indicating that each pyrethroid receptor is located between four helices (L45, S5, and two S6s), which belong to neighboring repeats, with helix IIS6 contributing four residues to PyR1 and another four residues to PyR2 (Du et al., 2015).

It is proposed that binding of ligands to both PyR1 and PyR2 is necessary to activate the channel.

4.5 Gating modifiers

4.5.1 Overview

In the X-ray structures of homotetrameric voltage-gated channels, the extracellular S1—S2 and S3—S4 loops in a given subunit are close to the S5—S6 loop of the adjacent subunit (Fig. 1B). A peptide toxin bound to these loops between the pore domain and a voltage-sensing domain can trap the latter in a specific state and thus modify the channel gating. Such toxins are produced by scorpions (Wang et al., 2011; Zhang et al., 2011; J.Z. Zhang et al., 2012), spiders (Bosmans & Swartz, 2010; Minassian et al., 2014), and sea anemones (Xiao, Blumenthal, & Cummins, 2014). Alpha and beta scorpion toxins bind to different sites and cause different effects. Alpha toxins slow down inactivation, while β-toxins cause a negative shift of activation by trapping the voltage sensor of repeat II in the activated state. The extracellular S1—S1 and S3—S4 loops contribute to the receptors for α- and β-scorpion toxins, but the binding regions are located at the opposite sides of the channel and involve repeat IV for α-toxins and repeat II for β-toxins (Cestele et al., 2006; Rogers, Qu, Tanada, Scheuer, & Catterall, 1996; Thomsen & Catterall, 1989) (Fig. 1). Surprisingly, β-toxin-sensing residues are found not only in the IIS1—S2 and IIS3—S4 loops, but also in the IIIP2—S6 loop (J.Z. Zhang et al., 2012).

4.5.2 Models

Combined mutational and molecular-modeling studies have lead to the emerging atomic-scale picture of the scorpion toxin action (Catterall et al., 2007). A particular problem was involvement of flexible channel-specific extracellular loops, whose structure is unknown. To address the problem, the authors generated several energetically possible models and selected the model in which key toxin-sensing residues are clustered (J.Z. Zhang et al., 2012). The wedge-shaped body of a β-toxin, CssIV, is proposed to bind deeply in a slot between loops IIS1—S2 and IIS3—S4. Docking of the toxin in this model (Fig. 6A and B) reproduced most of experimentally known specific interactions. However, an important experimentally detected residue, D1445 (J.Z. Zhang et al., 2012), is beyond the proposed binding site. This may be evidence of a limited precision of the model or result from an indirect effect of the residue substitution on the toxin action.

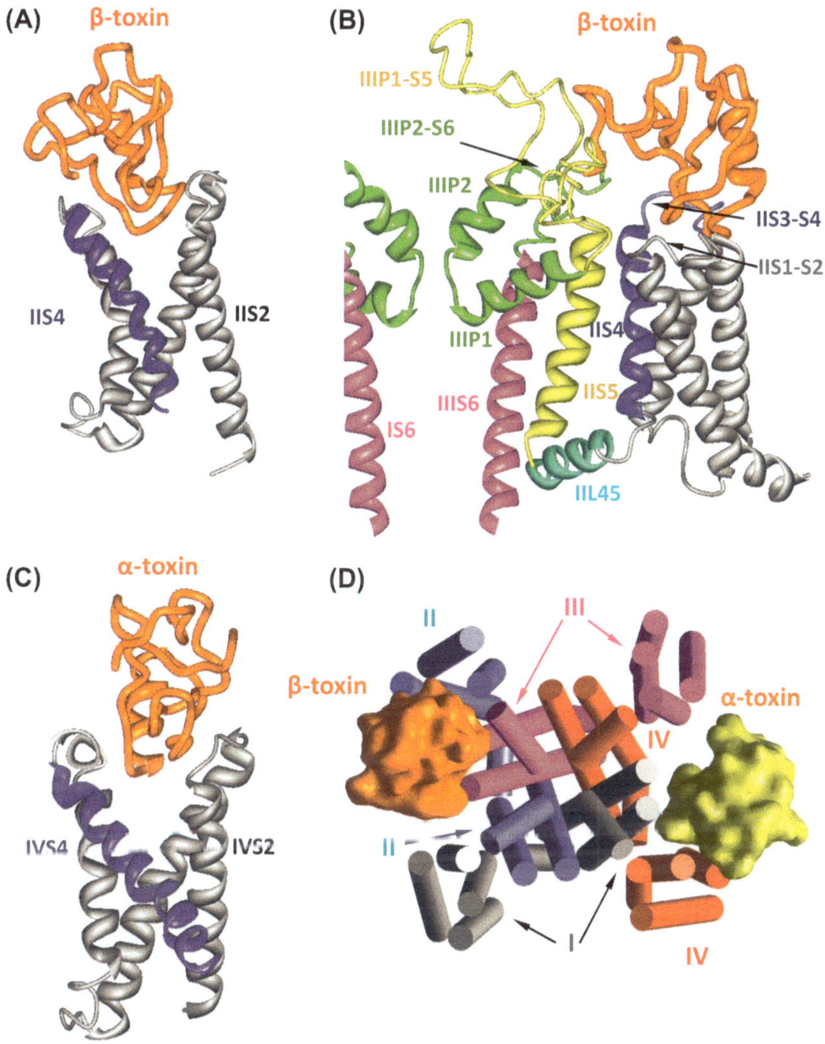

Figure 6 Sodium channel models with peptide scorpion toxins. (A) and (B) β-scorpion toxin binds in the slot between helices IIS4 and IIS2 of the voltage-sensing domain and interacts with the extracellular linkers IIIP1–S5 and IIIP2–S5 of the pore domain (J.Z. Zhang et al., 2012). (C) α-scorpion toxin binds in the slot between helices IVS4 and IVS2 of the voltage-sensing domain (Wang et al., 2011). (D) Schematic view on the quasisymmetrical binding of α- and β-toxins. *The images are generated using coordinates provided by Dr. Yarov-Yarovoy.*

Similar analysis has been performed for α-toxins, which appear to bind quasisymmetrically relative to the β-toxins (Wang et al., 2011). The modeling suggests that they fit a tapered slot between loops IVS1–S2 and

IVS3–S4 (Fig. 6C). The difference between the main effects of α-toxins (inhibiting fast inactivation) and β-toxins (enhancing activation) may be due to the structural and functional asymmetry of the channel repeats. Particularly, it is known that the voltage sensors in repeats I and II rapidly move in response to the membrane depolarization, whereas the voltage sensor in repeat IV moves slowly, and its activation time course is similar to the fast inactivation (Chanda & Bezanilla, 2002).

Ligand binding to the same region can cause unequal and sometimes opposite effects. For example, both LAs and BTX target the inner pore region, but LAs block the channel whereas BTX activates it. Other examples may involve toxins, which bind to the outer loops. Spider toxins like huwentoxin-IV from *Ornithoctonus huwena* and some hainantoxins were at first considered to bind in the outer pore like TTX because they inhibit the sodium conductance (Li et al., 2003). However, huwentoxin-sensing residues and residues affecting β-scorpion toxin are located in the same region (Leipold et al., 2007). A model with huwentoxin-IV bound in the cleft between IIS1–S2 and IIS3–S4 (Minassian et al., 2014) is remarkably similar with the α-scorpion toxin-binding model. Apparently, the main difference is that β-scorpion toxins trap the voltage sensor in the activated state, whereas huwentoxin-IV and related toxins trap the voltage sensor in the resting state. Details of these specific interactions require further analysis.

5. CONCLUSIONS

Here we described some contributions of structural modeling to the field of molecular pharmacology and toxicology of voltage-gated sodium channels. Structural models provide a possibility to integrate data from different experiments including site-directed mutagenesis, electrophysiology, and studies of structure-function relations of ligands. Such models can provide mechanistic explanations for complex effects and help design new experiments. Examples include model-based rationale for complete and incomplete channel block by μ-CTXs, state-dependent action and access pathways of LAs, paradoxical action of BTX (a bulky toxin that binds in the pore without blocking the current), existence of multiple pyrethroid-sensing residues in different repeats, and complex binding sites and different effects of spider and scorpion toxins.

It should be emphasized that a ligand docking in a homology model does not necessarily provide an unambiguous solution. Modern computers are powerful, force fields are reliable, and sampling protocols are efficient. However, precision of homology models is still limited. Besides inherent uncertainties of homology models (especially when the sequence similarity between the template and query proteins is poor) there are problems for which standard computational protocols are inefficient or lacking. Particularly difficult is accounting for entropy, ionization of residues, interactions with permeant ions, membrane voltage, and influence of auxiliary subunits. Therefore, a sensitive touchstone of a model's validity is its ability to reproduce a representative "training set" of experimental data. But the key criterion of a model's strength is its capability to rationalize the "examining set" of experimental data, which were not used at the stage of the model building.

X-ray structures of eukaryotic sodium channels with ligands are still lacking. However, new X-ray structures of bacterial sodium channels provide additional templates, which facilitate development of new homology models. An open-state model of a bacterial sodium channel NavCt, which is based on the low-resolution EM data (Tsai et al., 2013), is a new template to model open eukaryotic sodium channels.

Future X-ray structures of eukaryotic sodium channels are unlikely to diminish importance of molecular modeling. Indeed, the X-ray structure of Kv1.2 is available since 2005, but there are no X-ray structures of medically more important potassium channels Kv1.3 and Kv1.5. Computational studies produce sometimes controversial models of ligand binding in these channels despite each model is consistent with certain set of experimental data. A major problem is that a mutation may affect action of a ligand either directly or indirectly. Elaboration of computational approaches that would help discriminate between these effects seems an important direction in computational structural pharmacology and toxicology.

ACKNOWLEDGMENTS

We thank Dr. Yarov-Yarovoy for coordinates of the sodium channel with α- and β-scorpion toxins. This work was supported by grant RGPIN-2014-04894 to B.S.Z. from the Natural Sciences and Engineering Council of Canada and grant to D.B.T. from the RAS program "Molecular and Cell Biology."

REFERENCES

Alpert, L. A., Fozzard, H. A., Hanck, D. A., & Makielski, J. C. (1989). Is there a second external lidocaine binding site on mammalian cardiac cells? *American Journal of Physiology, 257*(1 Pt 2), H79–H84.

Bagneris, C., DeCaen, P. G., Naylor, C. E., Pryde, D. C., Nobeli, I., Clapham, D. E., & Wallace, B. A. (2014). Prokaryotic NavMs channel as a structural and functional model for eukaryotic sodium channel antagonism. *Proceedings of the National Academy of Sciences of the United States of America, 111*(23), 8428–8433.

Bosmans, F., & Swartz, K. J. (2010). Targeting voltage sensors in sodium channels with spider toxins. *Trends in Pharmacological Sciences, 31*(4), 175–182.

Bruhova, I., Tikhonov, D. B., & Zhorov, B. S. (2008). Access and binding of local anesthetics in the closed sodium channel. *Molecular Pharmacology, 74*(4), 1033–1045.

Catterall, W. A. (1979). Neurotoxins as allosteric modifiers of voltage-sensitive sodium channels. *Advances in Cytopharmacology, 3*, 305–316.

Catterall, W. A. (2014). Sodium channels, inherited epilepsy, and antiepileptic drugs. *Annual Review of Pharmacology and Toxicology, 54*, 317–338.

Catterall, W. A., Cestele, S., Yarov-Yarovoy, V., Yu, F. H., Konoki, K., & Scheuer, T. (2007). Voltage-gated ion channels and gating modifier toxins. *Toxicon, 49*(2), 124–141. http://dx.doi.org/10.1016/j.toxicon.2006.09.022.

Catterall, W. A., & Swanson, T. M. (2015). Structural basis for pharmacology of voltage-gated sodium and calcium channels. *Molecular Pharmacology, 88*.

Cestele, S., Yarov-Yarovoy, V., Qu, Y., Sampieri, F., Scheuer, T., & Catterall, W. A. (2006). Structure and function of the voltage sensor of sodium channels probed by a beta-scorpion toxin. *Journal of Biological Chemistry, 281*(30), 21332–21344. http://dx.doi.org/10.1074/jbc.M603814200.

Chanda, B., & Bezanilla, F. (2002). Tracking voltage-dependent conformational changes in skeletal muscle sodium channel during activation. *Journal of General Physiology, 120*(5), 629–645.

Chang, N. S., French, R. J., Lipkind, G. M., Fozzard, H. A., & Dudley, S., Jr. (1998). Predominant interactions between mu-conotoxin Arg-13 and the skeletal muscle Na^+ channel localized by mutant cycle analysis. *Biochemistry, 37*(13), 4407–4419. http://dx.doi.org/10.1021/bi9724927.

Chen, R., & Chung, S. H. (2014). Mechanism of tetrodotoxin block and resistance in sodium channels. *Biochemical and Biophysical Research Communications, 446*(1), 370–374.

Choudhary, G., Aliste, M. P., Tieleman, D. P., French, R. J., & Dudley, S. C., Jr. (2007). Docking of mu-conotoxin GIIIA in the sodium channel outer vestibule. *Channels (Austin), 1*(5), 344–352.

Dong, K., Du, Y., Rinkevich, F., Nomura, Y., Xu, P., Wang, L.… Zhorov, B. S. (2014). Molecular biology of insect sodium channels and pyrethroid resistance. *Insect Biochemistry and Molecular Biology, 50*, 1–17. http://dx.doi.org/10.1016/j.ibmb.2014.03.012.

Doyle, D. A., Morais Cabral, J., Pfuetzner, R. A., Kuo, A., Gulbis, J. M., Cohen, S. L.… MacKinnon, R. (1998). The structure of the potassium channel: molecular basis of K^+ conduction and selectivity. *Science, 280*(5360), 69–77.

Du, Y., Garden, D. P., Wang, L., Zhorov, B. S., & Dong, K. (2011). Identification of new batrachotoxin-sensing residues in segment IIIS6 of the sodium channel. *Journal of Biological Chemistry, 286*(15), 13151–13160.

Du, Y., Nomura, Y., Satar, G., Hu, Z., Nauen, R., He, S. Y.… Dong, K. (2013). Molecular evidence for dual pyrethroid-receptor sites on a mosquito sodium channel. *Proceedings of the National Academy of Sciences of the United States of America, 110*(29), 11785–11790.

Du, Y., Nomura, Y., Zhorov, B. S., & Dong, K. (2015). Rotational symmetry of two pyrethroid receptor sites in the mosquito sodium channel. *Molecular Pharmacology, 88*(2), 273–280.

Dudley, S. C., Jr., Chang, N., Hall, J., Lipkind, G., Fozzard, H. A., & French, R. J. (2000). mu-conotoxin GIIIA interactions with the voltage-gated Na(+) channel predict a clockwise arrangement of the domains. *Journal of General Physiology, 116*(5), 679–690.

Garden, D. P., & Zhorov, B. S. (2010). Docking flexible ligands in proteins with a solvent exposure- and distance-dependent dielectric function. *Journal of Computer Aided Molecular Design, 24*(2), 91–105.

Hille, B. (1977). Local anesthetics: hydrophilic and hydrophobic pathways for the drug-receptor reaction. *Journal of General Physiology, 69*(4), 497–515.

Hille, B. (2001). *Ion channels of excitable membranes* (3rd ed.). Sunderland, MA, USA: Sinauer Associates, Inc.

Hui, K., Lipkind, G., Fozzard, H. A., & French, R. J. (2002). Electrostatic and steric contributions to block of the skeletal muscle sodium channel by mu-conotoxin. *Journal of General Physiology, 119*(1), 45–54.

Jensen, M. O., Jogini, V., Borhani, D. W., Leffler, A. E., Dror, R. O., & Shaw, D. E. (2012). Mechanism of voltage gating in potassium channels. *Science, 336*(6078), 229–233.

Jiang, Y., Lee, A., Chen, J., Cadene, M., Chait, B. T., & MacKinnon, R. (2002). Crystal structure and mechanism of a calcium-gated potassium channel. *Nature, 417*(6888), 515–522. http://dx.doi.org/10.1038/417515a.

Jiang, Y., Lee, A., Chen, J., Ruta, V., Cadene, M., Chait, B. T., & MacKinnon, R. (2003). X-ray structure of a voltage-dependent K^+ channel. *Nature, 423*(6935), 33–41. http://dx.doi.org/10.1038/nature01580.

Khan, A., Romantseva, L., Lam, A., Lipkind, G., & Fozzard, H. A. (2002). Role of outer ring carboxylates of the rat skeletal muscle sodium channel pore in proton block. *Journal of Physiology, 543*(Pt 1), 71–84.

Khoo, K. K., Gupta, K., Green, B. R., Zhang, M. M., Watkins, M., Olivera, B. M.… Norton, R. S. (2012). Distinct disulfide isomers of mu-conotoxins KIIIA and KIIIB block voltage-gated sodium channels. *Biochemistry, 51*(49), 9826–9835. http://dx.doi.org/10.1021/bi301256s.

Korkosh, V. S., Zhorov, B. S., & Tikhonov, D. B. (2014). Folding similarity of the outer pore region in prokaryotic and eukaryotic sodium channels revealed by docking of conotoxins GIIIA, PIIIA, and KIIIA in a NavAb-based model of Nav1.4. *Journal of General Physiology, 144*(3), 231–244.

Leipold, E., DeBie, H., Zorn, S., Borges, A., Olivera, B. M., Terlau, H., & Heinemann, S. H. (2007). muO conotoxins inhibit NaV channels by interfering with their voltage sensors in domain-2. *Channels (Austin), 1*(4), 253–262.

Li, D., Xiao, Y., Hu, W., Xie, J., Bosmans, F., Tytgat, J., & Liang, S. (2003). Function and solution structure of hainantoxin-I, a novel insect sodium channel inhibitor from the Chinese bird spider *Selenocosmia hainana*. *FEBS Letters, 555*(3), 616–622.

Li, H. L., Galue, A., Meadows, L., & Ragsdale, D. S. (1999). A molecular basis for the different local anesthetic affinities of resting versus open and inactivated states of the sodium channel. *Molecular Pharmacology, 55*(1), 134–141.

Li, R. A., Ennis, I. L., French, R. J., Dudley, S. C., Jr., Tomaselli, G. F., & Marban, E. (2001). Clockwise domain arrangement of the sodium channel revealed by (mu)-conotoxin (GIIIA) docking orientation. *Journal of Biological Chemistry, 276*(14), 11072–11077.

Lipkind, G. M., & Fozzard, H. A. (1994). A structural model of the tetrodotoxin and saxitoxin binding site of the Na^+ channel. *Biophysical Journal, 66*(1), 1–13.

Lipkind, G. M., & Fozzard, H. A. (2000). KcsA crystal structure as framework for a molecular model of the Na(+) channel pore. *Biochemistry, 39*(28), 8161–8170.

Lipkind, G. M., & Fozzard, H. A. (2005). Molecular modeling of local anesthetic drug binding by voltage-gated sodium channels. *Molecular Pharmacology, 68*(6), 1611–1622. http://dx.doi.org/10.1124/mol.105.014803.

Long, S. B., Campbell, E. B., & Mackinnon, R. (2005). Crystal structure of a mammalian voltage-dependent Shaker family K^+ channel. *Science, 309*(5736), 897–903.

Marzian, S., Stansfeld, P. J., Rapedius, M., Rinne, S., Nematian-Ardestani, E., Abbruzzese, J. L.… Decher, N. (2013). Side pockets provide the basis for a new mechanism of Kv channel-specific inhibition. *Nature Chemical Biology, 9*(8), 507–513. http://dx.doi.org/10.1038/nchembio.1271.

McArthur, J. R., Ostroumov, V., Al-Sabi, A., McMaster, D., & French, R. J. (2011). Multiple, distributed interactions of mu-conotoxin PIIIA associated with broad targeting among voltage-gated sodium channels. *Biochemistry, 50*(1), 116–124. http://dx.doi.org/10.1021/bi101316y.

McArthur, J. R., Singh, G., O'Mara, M. L., McMaster, D., Ostroumov, V., Tieleman, D. P., & French, R. J. (2011). Orientation of mu-conotoxin PIIIA in a sodium channel vestibule, based on voltage dependence of its binding. *Molecular Pharmacology, 80*(2), 219–227.

McCusker, E. C., Bagneris, C., Naylor, C. E., Cole, A. R., D'Avanzo, N., Nichols, C. G., & Wallace, B. A. (2012). Structure of a bacterial voltage-gated sodium channel pore reveals mechanisms of opening and closing. *Nature Communications, 3*, 1102.

Minassian, N. A., Gibbs, A., Shih, A. Y., Liu, Y., Neff, R. A., Sutton, S. W.… Wickenden, A. D. (2014). Analysis of the structural and molecular basis of voltage-sensitive sodium channel inhibition by the spider toxin huwentoxin-IV (mu-TRTX-Hh2a). *Journal of Biological Chemistry, 288*(31), 22707–22720.

O'Reilly, A. O., Khambay, B. P., Williamson, M. S., Field, L. M., Wallace, B. A., & Davies, T. G. (2006). Modelling insecticide-binding sites in the voltage-gated sodium channel. *Biochemical Journal, 396*(2), 255–263.

Payandeh, J., Scheuer, T., Zheng, N., & Catterall, W. A. (2011). The crystal structure of a voltage-gated sodium channel. *Nature, 475*(7356), 353–358.

Petrey, D., Chen, T. S., Deng, L., Garzon, J. I., Hwang, H., Lasso, G.… Honig, B. (2015). Template-based prediction of protein function. *Current Opinion in Structural Biology, 32C*, 33–38.

Qu, Y., Rogers, J., Tanada, T., Scheuer, T., & Catterall, W. A. (1995). Molecular determinants of drug access to the receptor site for antiarrhythmic drugs in the cardiac Na^+ channel. *Proceedings of the National Academy of Sciences of the United States of America, 92*(25), 11839–11843.

Ragsdale, D. S., McPhee, J. C., Scheuer, T., & Catterall, W. A. (1994). Molecular determinants of state-dependent block of Na^+ channels by local anesthetics. *Science, 265*(5179), 1724–1728.

Rogers, J. C., Qu, Y., Tanada, T. N., Scheuer, T., & Catterall, W. A. (1996). Molecular determinants of high affinity binding of alpha-scorpion toxin and sea anemone toxin in the S3-S4 extracellular loop in domain IV of the Na^+ channel alpha subunit. *Journal of Biological Chemistry, 271*(27), 15950–15962.

Schmidt, T., Bergner, A., & Schwede, T. (2014). Modelling three-dimensional protein structures for applications in drug design. *Drug Discovery Today, 19*(7), 890–897.

Silver, K. S., Du, Y., Nomura, Y., Oliveira, E. E., Salgado, V. L., Zhorov, B. S., & Dong, K. (2014). Voltage-gated sodium channels as insecticide targets. In E. Cohen (Ed.), *Advances in insect physiology* (Vol. 46). Oxford, UK: Academic Press.

Stevens, M., Peigneur, S., Dyubankova, N., Lescrinier, E., Herdewijn, P., & Tytgat, J. (2012). Design of bioactive peptides from naturally occurring mu-conotoxin structures. *Journal of Biological Chemistry, 287*(37), 31382–31392. http://dx.doi.org/10.1074/jbc.M112.375733.

Stevens, M., Peigneur, S., & Tytgat, J. (2011). Neurotoxins and their binding areas on voltage-gated sodium channels. *Frontiers in Pharmacology, 2*, 71.

Sunami, A., Dudley, S. C., Jr., & Fozzard, H. A. (1997). Sodium channel selectivity filter regulates antiarrhythmic drug binding. *Proceedings of the National Academy of Sciences of the United States of America, 94*(25), 14126–14131.

Sunami, A., Glaaser, I. W., & Fozzard, H. A. (2000). A critical residue for isoform difference in tetrodotoxin affinity is a molecular determinant of the external access path for local anesthetics in the cardiac sodium channel. *Proceedings of the National Academy of Sciences of the United States of America, 97*(5), 2326–2331.

Sunami, A., Glaaser, I. W., & Fozzard, H. A. (2001). Structural and gating changes of the sodium channel induced by mutation of a residue in the upper third of IVS6, creating an external access path for local anesthetics. *Molecular Pharmacology, 59*(4), 684−691.

Terlau, H., Heinemann, S. H., Stuhmer, W., Pusch, M., Conti, F., Imoto, K., & Numa, S. (1991). Mapping the site of block by tetrodotoxin and saxitoxin of sodium channel II. *FEBS Letters, 293*(1−2), 93−96.

Thomsen, W. J., & Catterall, W. A. (1989). Localization of the receptor site for alpha-scorpion toxins by antibody mapping: implications for sodium channel topology. *Proceedings of the National Academy of Sciences of the United States of America, 86*(24), 10161−10165.

Tikhonov, D. B., & Zhorov, B. S. (2005a). Modeling P-loops domain of sodium channel: homology with potassium channels and interaction with ligands. *Biophysical Journal, 88*(1), 184−197.

Tikhonov, D. B., & Zhorov, B. S. (2005b). Sodium channel activators: model of binding inside the pore and a possible mechanism of action. *FEBS Letters, 579*(20), 4207−4212.

Tikhonov, D. B., & Zhorov, B. S. (2007). Sodium channels: ionic model of slow inactivation and state-dependent drug binding. *Biophysical Journal, 93*(5), 1557−1570.

Tikhonov, D. B., & Zhorov, B. S. (2011). Possible roles of exceptionally conserved residues around the selectivity filters of sodium and calcium channels. *Journal of Biological Chemistry, 286*(4), 2998−3006.

Tikhonov, D. B., & Zhorov, B. S. (2012). Architecture and pore block of eukaryotic voltage-gated sodium channels in view of NavAb bacterial sodium channel structure. *Molecular Pharmacology, 82*(1), 97−104.

Tsai, C. J., Tani, K., Irie, K., Hiroaki, Y., Shimomura, T., McMillan, D. G. … Li, X. D. (2013). Two alternative conformations of a voltage-gated sodium channel. *Journal of Molecular Biology, 425*(22), 4074−4088. http://dx.doi.org/10.1016/j.jmb.2013.06.036.

Van Der Haegen, A., Peigneur, S., & Tytgat, J. (2011). Importance of position 8 in mu-conotoxin KIIIA for voltage-gated sodium channel selectivity. *FEBS Journal, 278*(18), 3408−3418. http://dx.doi.org/10.1111/j.1742-4658.2011.08264.x.

Wang, G. K., Quan, C., & Wang, S. (1998). A common local anesthetic receptor for benzocaine and etidocaine in voltage-gated mu1 Na^+ channels. *Pflugers Archiv: European Journal of Physiology, 435*(2), 293−302.

Wang, G. K., Russell, G., & Wang, S. Y. (2013). Persistent human cardiac Na^+ currents in stably transfected mammalian cells: robust expression and distinct open-channel selectivity among Class 1 antiarrhythmics. *Channels (Austin)*, 7(4), 263−274.

Wang, J., Yarov-Yarovoy, V., Kahn, R., Gordon, D., Gurevitz, M., Scheuer, T., & Catterall, W. A. (2011). Mapping the receptor site for alpha-scorpion toxins on a Na^+ channel voltage sensor. *Proceedings of the National Academy of Sciences of the United States of America, 108*(37), 15426−15431.

Wang, S. Y., Mitchell, J., Tikhonov, D. B., Zhorov, B. S., & Wang, G. K. (2006). How batrachotoxin modifies the sodium channel permeation pathway: computer modeling and site-directed mutagenesis. *Molecular Pharmacology, 69*(3), 788−795.

Wang, S. Y., Tikhonov, D. B., Mitchell, J., Zhorov, B. S., & Wang, G. K. (2007). Irreversible block of cardiac mutant Na^+ channels by batrachotoxin. *Channels (Austin), 1*(3), 179−188.

Wang, S. Y., Tikhonov, D. B., Zhorov, B. S., Mitchell, J., & Wang, G. K. (2007). Serine-401 as a batrachotoxin- and local anesthetic-sensing residue in the human cardiac Na^+ channel. *Pflugers Archiv: European Journal of Physiology, 454*(2), 277−287.

Wang, S. Y., & Wang, G. K. (2003). Voltage-gated sodium channels as primary targets of diverse lipid soluble neurotoxins. *Cellular Signalling, 15*(2), 151−159.

Wilson, M. J., Yoshikami, D., Azam, L., Gajewiak, J., Olivera, B. M., Bulaj, G., & Zhang, M. M. (2011). mu-Conotoxins that differentially block sodium channels NaV1.1 through 1.8 identify those responsible for action potentials in sciatic nerve.

Proceedings of the National Academy of Sciences of the United States of America, 108(25), 10302–10307. http://dx.doi.org/10.1073/pnas.1107027108.

Xiao, Y., Blumenthal, K., & Cummins, T. R. (2014). Gating-pore currents demonstrate selective and specific modulation of individual sodium channel voltage-sensors by biological toxins. *Molecular Pharmacology, 86*(2), 159–167. http://dx.doi.org/10.1124/mol.114.092338.

Xue, T., Ennis, I. L., Sato, K., French, R. J., & Li, R. A. (2003). Novel interactions identified between micro-Conotoxin and the Na^+ channel domain I P-loop: implications for toxin-pore binding geometry. *Biophysical Journal, 85*(4), 2299–2310.

Yamamoto, D. (1986). Dynamics of strychnine block of single sodium channels in bovine chromaffin cells. *Journal of Physiology, 370*, 395–407.

Yarov-Yarovoy, V., Brown, J., Sharp, E. M., Clare, J. J., Scheuer, T., & Catterall, W. A. (2001). Molecular determinants of voltage-dependent gating and binding of pore-blocking drugs in transmembrane segment IIIS6 of the Na(+) channel alpha subunit. *Journal of Biological Chemistry, 276*(1), 20–27.

Yarov-Yarovoy, V., McPhee, J. C., Idsvoog, D., Pate, C., Scheuer, T., & Catterall, W. A. (2002). Role of amino acid residues in transmembrane segments IS6 and IIS6 of the Na^+ channel alpha subunit in voltage-dependent gating and drug block. *Journal of Biological Chemistry, 277*(38), 35393–35401.

Yeh, J. Z., & Narahashi, T. (1977). Kinetic analysis of pancuronium interaction with sodium channels in squid axon membranes. *Journal of General Physiology, 69*(3), 293–323.

Zhang, J. Z., Yarov-Yarovoy, V., Scheuer, T., Karbat, I., Cohen, L., Gordon, D.… Catterall, W. A. (2011). Structure-function map of the receptor site for beta-scorpion toxins in domain II of voltage-gated sodium channels. *Journal of Biological Chemistry, 286*(38), 33641–33651.

Zhang, J. Z., Yarov-Yarovoy, V., Scheuer, T., Karbat, I., Cohen, L., Gordon, D.… Catterall, W. A. (2012). Mapping the interaction site for a beta-scorpion toxin in the pore module of domain III of voltage-gated Na(+) channels. *Journal of Biological Chemistry, 287*(36), 30719–30728.

Zhang, M. M., McArthur, J. R., Azam, L., Bulaj, G., Olivera, B. M., French, R. J., & Yoshikami, D. (2009). Synergistic and antagonistic interactions between tetrodotoxin and mu-conotoxin in blocking voltage-gated sodium channels. *Channels (Austin), 3*(1), 32–38.

Zhang, X., Ren, W., DeCaen, P., Yan, C., Tao, X., Tang, L.… Yan, N. (2012). Crystal structure of an orthologue of the NaChBac voltage-gated sodium channel. *Nature, 486*(7401), 130–134.

CHAPTER SIX

Understanding Sodium Channel Function and Modulation Using Atomistic Simulations of Bacterial Channel Structures

C. Boiteux* and T.W. Allen*,§,1
*RMIT University, Melbourne, VIC, Australia
§University of California Davis, Davis, CA, United States
1Corresponding author: E-mail: toby.allen@rmit.edu.au

Contents

Abstract

Sodium channels are chief proteins involved in electrical signaling in the nervous system, enabling critical functions like heartbeat and brain activity. New high-resolution X-ray structures for bacterial sodium channels have created an opportunity to see how these proteins operate at the molecular level. An important challenge to overcome is establishing relationships between the structures and functions of mammalian and

Current Topics in Membranes, Volume 78
ISSN 1063-5823
http://dx.doi.org/10.1016/bs.ctm.2016.07.002

bacterial channels. Bacterial sodium channels are known to exhibit the main structural features of their mammalian counterparts, as well as several key functional characteristics, including selective ion conduction, voltage-dependent gating, pore-based inactivation and modulation by local anesthetic, antiarrhythmic and antiepileptic drugs. Simulations have begun to shed light on each of these features in the past few years. Despite deviations in selectivity signatures for bacterial and mammalian channels, simulations have uncovered the nature of the multiion conduction mechanism associated with Na^+ binding to a high-field strength site established by charged glutamate side chains. Simulations demonstrated a surprising level of flexibility of the protein, showing that these side chains are active participants in the permeation process. They have also uncovered changes in protein structure, leading to asymmetrical collapses of the activation gate that have been proposed to correspond to inactivated structures. These observations offer the potential to examine the mechanisms of state-dependent drug activity, focusing on pore-blocking and pore-based slow inactivation in bacterial channels, without the complexities of inactivation on multiple timescales seen in eukaryotic channels. Simulations have provided molecular views of the interactions of drugs, consistent with sites predicted in mammalian channels, as well as a wealth of other sites as potential new drug targets. In this chapter, we survey the new insights into sodium channel function that have emerged from studies of simpler bacterial channels, which provide an excellent learning platform, and promising avenues for mechanistic discovery and pharmacological development.

1. INTRODUCTION

Since first identified as the main agent for the transmission of electric signals in nervous cells by Hodgkin and Huxley (1952a), voltage-gated sodium (Na_V) channels have been the center of interest of a rapidly increasing numbers of physiological, biophysical, and pharmacological studies. Na_V channels are one of the fundamental molecular components of signaling in our nervous systems. They act to depolarize excitable cells through selective transport of Na^+ ions across cell membranes in response to changes in membrane electrical potential. Na_V channel isoforms can be found in almost all electrically active tissues in the body, from neurons to cardiac and skeletal muscles cells (Goldin, 1999), and are important in physiological processes such as regulation of heartbeat, pain sensation, transmission of nervous signals, and enablement of cognitive processes (Hille, 2001). Ion channels are a focus in the development of therapeutic agents (Ashcroft, 2000; Clare, 2010), representing 13% of all known drugs (Overington, Al-Lazikani, & Hopkins, 2006), with particular interest in Na_V channels, being targets for epilepsy, cardiac arrhythmias, chronic and neuropathic pain, migraine, spasticity, tremor, neurodegenerative disease, bipolar disorder, anxiety, obsessive–compulsive disorder, myotonia, dystonia, schizophrenia, restless legs syndrome, posttraumatic stress syndrome, alcohol dependence, and withdrawal (Kyle & Ilyin, 2007; Zuliani et al., 2009).

Research into Na_V channels entered a new era at the beginning of the 2000s with the discovery of the first bacterial Na_V channel, NaChBac (Ren et al., 2001). Unlike its mammalian counterparts, where four similar domains are linked to form a long complex single unit, NaChBac presents a simpler structure, reminiscent of potassium channels, with four identical and independent subunits (Fig. 1) arranged symmetrically to shape a pore across the membrane (Fig. 2). The discovery of this first channel was soon followed by many other close relatives (Ito et al., 2004; Koishi et al., 2004), all displaying the same structural features (Payandeh & Minor, 2015). Their simple structure, along with the added practicality of working with prokaryotic material, allowed for rapid progresses in crystallization and determination of their high-resolution X-ray structures, still unavailable for mammalian channels (whose only structure available to date was obtained by

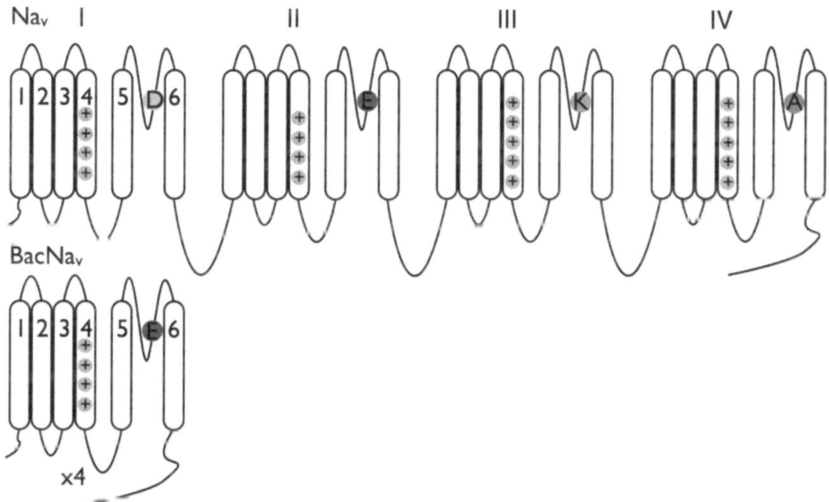

Figure 1 Comparing Na_V and $BacNa_V$ channels. Arrangement of the transmembrane α-subunit of a mammalian Na_V channel showing the four-domain (I to IV) structure with six α-helical transmembrane segments in each domain (top). Segments S1–S4 form the voltage-sensing domain. The positively charged residues of gating segments S4 are highlighted in purple. The S5 and S6 segments and the pore loop region between them form the pore domain. The residues forming the selectivity locus DEKA are highlighted as *colored circles*, in green, red, blue, and orange, respectively. Arrangement of the transmembrane α-subunit of a bacterial Na_V channel, showing the similarity to one domain of the Na_V channel α-subunit (bottom).

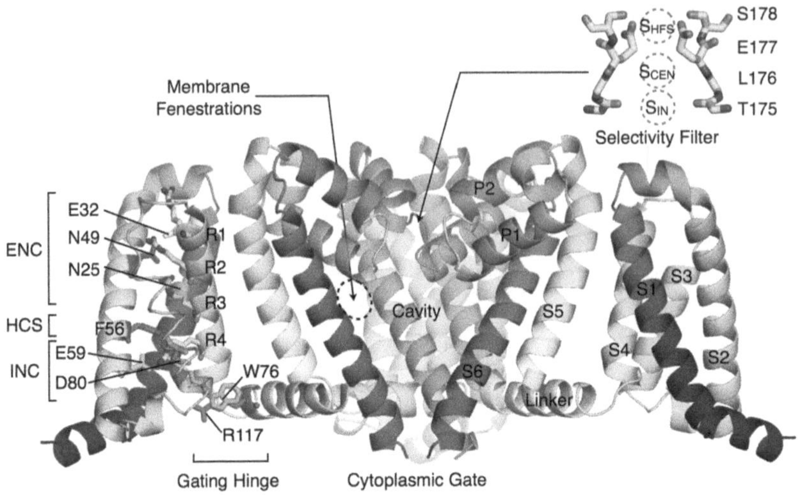

Figure 2 The structure of channel Na_VAb. Main structural elements of channel Na_VAb showing the arrangement of segments S1–S6 forming the pore domain and the voltage-sensing domains. For clarity, the segments forming the pore domain of one subunit and the voltage-sensing domain of another are omitted. A more detailed representation of the selectivity filter is added as an insert (*colored sticks*, C in cyan (light gray in print versions), N in blue (dark gray in print versions), and O in red (gray in print versions), two subunits shown), displaying its specific sequence of residues TLES and the three binding sites for Na^+: S_{HFS}, S_{CEN}, and S_{IN} as *dashed circles*. Within the voltage-sensing domain, the key residues playing a role in gating are represented as *colored sticks* with labels. Additionally, the residues forming the gating hinge, the intracellular negatively charged cluster (INC), the extracellular negatively charged cluster (ENC), and the hydrophobic constriction site (HCS) are also indicated. One of the openings between the membrane and the pore ("Membrane Fenestrations") formed by adjacent S6 segments is circled.

low-resolution cryoelectron microscopy (Sato et al., 2001), showing a central pore surrounded by four peripheral regions and giving limited insight into their structure/function relationship).

Since the publication of the first X-ray structure of the bacterial Na_VAb channel in 2011 (Payandeh et al., 2011), several other channels have been solved (Bagneris et al., 2014; McCusker et al., 2012; Payandeh et al., 2012; Shaya et al., 2014; Zhang et al., 2012). This bacterial family of channels, $BacNa_V$, is both structurally and functionally simpler than the human family, while sharing essential features of selective ion conduction, voltage activation, inactivation, and drug modulation (Bagneris et al., 2014; Finol-Urdaneta et al., 2013; Lee, Goodchild, & Ahern, 2012; McArthur et al., 2011; Vilin & Ruben, 2001). This makes them very attractive to molecular level studies and explains the recent emergence of several computational studies (Amaral et al., 2012; Boiteux, Vorobyov, & Allen, 2014; Chakrabarti et al., 2013; Corry & Thomas, 2012; Finol-Urdaneta et al., 2014; Furini & Domene, 2012; Ulmschneider et al., 2013; Xia et al., 2013; Zhang et al., 2013). In this chapter, we explore recent computational investigations that have used bacterial channel structures to begin to explore function and modulation of sodium channels.

The ability of an ion channel to choose which ions it conducts is essential to the generation of electrical signals. Yet the way in which Na_V channels selectively conduct Na^+ over physiologically abundant K^+, Ca^{2+}, and other ions remains an open question. Na_V channels possess an extracellular selectivity filter that is known to control ion discrimination (Fig. 2). Na_V channels present glutamate (Glu/E) or aspartate (Asp/D) side chains, thought to create a high-field strength site (S_{HFS}) that preferentially stabilizes the smaller Na^+ over K^+ (Payandeh et al., 2011). However, selectivity is not as simple as a thermodynamic preference at one site and emerges from subtle differences in multiple-ion conduction mechanisms (Kim & Allen, 2011; Thompson et al., 2009). These mechanisms are not well defined, and we have not yet pinpointed what is the critical chemical or physical property that makes a protein pore Na^+ selective. The $BacNa_V$ selectivity signature consists of a tetrameric Glu ring, unlike human channels that exhibit a DEKA (Asp-Glu-Lys-Ala in domains I-IV) or related locus of residues (Charalambous & Wallace, 2011) (Figs. 1 and 3). While this may suggest different strategies exist to enforce Na^+ selectivity (Finol-Urdaneta et al., 2014), investigations are beginning to comprehend the underlying processes responsible for selective conduction across the phylogenetic tree.

Na_V channels are responsible for the rising phase of action potentials. When the neuronal membrane reaches a threshold potential, these channels undergo a conformational change within a few 100 μs, causing channel opening (gating). In this process, charged S4 segments in the voltage-sensing domain (residues R1-4 in Fig. 2) move due to the voltage change, pulling on the S4–S5 linker that opens a gate in the pore domain (Yarov-Yarovoy et al., 2012). After some time (milliseconds or longer), ion flow terminates due to other changes that lead to a nonconducting/inactive structure. Na_V channels rapidly cycle through closed, open, and inactivated states, allowing neurons to fire high-frequency trains of action potentials for brain function (Rogawski & Loscher, 2004). Drugs may act via blocking the pore directly or by stabilizing a specific state of the channel. As many drugs target the inactivated states of the channel (Kaczorowski

NaChBac	FQVVTLESWASG
NavAb	FQVMTLESWSMG
NavMs	FQVMTLESWSMG
NavRh	FQVLTLSSWEM-
Nav1.2 DI	FRLMTQDFWEN-
Nav1.2 DII	FRVLCGE-WIE-
Nav1.2 DIII	LQVATFKGWMD-
Nav1.2 DIV	FQITTSAGWDG-
Cav1.2 DI	FQCITMEGWTDV
Cav1.2 DII	FQILTGEDWNSV
Cav1.2 DIII	FTVSTFEGWPEL
Cav1.2 DIV	FRCARGEAWQDI

Figure 3 Sequence alignment for the selectivity filters of bacterial NaChBac, Na_VAb, Na_VMs, Na_VRh and mammalian $Na_V1.2$ and $Ca_V1.2$ channels. Negatively charged residues are colored in light gray and positively charged residues in dark gray.

et al., 2008), understanding the molecular events associated with inactivation is important for future pharmacological developments. Functionally, inactivation occurs on two timescales, with fast inactivation resulting from channel occlusion by a domain linker (Goldin, 2003) on a timescale comparable to channel opening, and slow inactivation, due to rearrangements in the pore domain (Capes et al., 2012; Pavlov et al., 2005) on a time course of milliseconds and longer (Rogawski & Loscher, 2004). Slow inactivation plays an important role in drug activity (Fozzard, Lee, & Lipkind, 2005; Zeng & French, 2011a) with successful new antiepileptic drug (AED) molecules acting via that mechanism (Errington et al., 2008). Bacterial channels lack the linker responsible for fast inactivation, allowing one to focus on slow inactivation, as done with K channels (Cuello et al., 2010).

Na_V inhibitor drugs have the remarkable ability to target specific cells to control sensory or motor functions without threatening vital processes. This selectivity allows for uses as local anesthetic (LA), AEDs, and analgesic and anti-arrhythmic agents. Understanding drug modulation has been hampered by conflicting models that have postulated different active sites, pathways, and target states. In the 1970s, Hille proposed separate lipophilic and aqueous pathways for drug binding (Hille, 1977a) with the more hydrophobic drugs acting via the former, and charged or hydrophilic drugs, such as titratable amines and quaternary ammonium derivatives, via the latter. Charged drugs are thought to block the open channel, whereas neutral drugs bind and dissociate even if the pore is nonconducting, apparently via lipid-bound fenestrations (Barber et al., 2014; Boiteux, Vorobyov, French et al., 2014; Hille, 2001; Martin & Corry, 2014; Payandeh et al., 2011). Evidence suggests that therapeutic drugs inhibit via a modulated receptor mechanism, binding more strongly to inactivated channels (Li et al., 1999), thereby reducing high-frequency firing. Despite historical focus on fast inactivation (Kuo & Bean, 1994), it is known that slow inactivation is affected by drugs (Chen et al., 2000; Fozzard et al., 2005) and that mutation of key residues for drug binding impairs slow inactivation (Carboni et al., 2005). Structurally defined Na_VBac channels (which only slow inactivate) exhibit drug inhibition like human channels (Bagneris et al., 2014; Lee et al., 2012; Pavlov et al., 2003), making them ideal targets for drug discovery.

Molecular dynamics (MD) simulation is an essential tool for studying biological systems at the molecular level (Karplus & McCammon, 2002). In the last decade, dramatic increases in computer power have led to bigger and longer simulations than ever before (eg, Freddolino et al., 2006; Jensen et al., 2012), allowing studies of an ever-widening array of biological phenomena. The availability of Na_VBac X-ray structures has enabled the use of atomic level simulations to further understand function. In addition, modern computational architectures have enabled simulations of realistic models of membrane ion channel systems on the multi-μs timescales, revealing ion movements and protein conformational changes relevant to activity. In this chapter, we will explore progress in understanding the function of the Na_V family, as well as their drug modulation, using MD simulations of bacterial channels.

2. SODIUM CHANNEL ARCHITECTURE

2.1 Mammalian Na_V channel structure

The mammalian Na_V channel family is composed of nine known members, $Na_V1.1$ to $Na_V1.9$, presenting a high level of homology and sharing a set of common features. In the absence of high-resolution structural data for this family of channels, our understanding of their structure relies on the analysis of their sequence and the interpretation of data obtained through electrophysiology and mutagenesis. Their pore-forming subunit is composed of four homologous but not identical domains linked together to form a single polypeptide chain. Each domain (DI to DIV) presents a high level of sequence identity throughout the family and is made up of six transmembrane helices (S1−S6), where segments S1−S4 form the voltage-sensing domain, while S5 and S6 form the pore domain connected together via a cytoplasmic helix (S4−S5 linker). Within the pore domain, segments S5 and S6 are linked together by two short helices P1 and P2, and a coil-like structure forming the selectivity filter (Fig. 1).

The roles of each segment and their relations to function have been extensively investigated, highlighting the particular importance of segment S4 in voltage sensing (Bezanilla, 2000; Hille, 2001) and segment S6 in gating and drug binding (Ragsdale et al., 1996; Yarov-Yarovoy et al., 2001; Yarov-Yarovoy et al., 2002). Similarly, the extra- and intracellular flexible loops linking domains and segments have also been the subject of investigation and play specific key roles in the binding of therapeutic agents (Cestele & Catterall, 2000) or for self-regulation (Vassilev, Scheuer, & Catterall, 1988). A pattern of conserved residues have been identified in each domain as responsible for sodium selectivity, in particular the DEKA locus (Chiamvimonvat et al., 1996; Favre, Moczydlowski, & Schild, 1996; Schlief et al., 1996; Tsushima, Li, & Backx, 1997), and systematic mutagenesis has shown that these residues need to be in close proximity to maintain the tetrodotoxin binding property characteristic of many Na_V channels (Penzotti et al., 1998; Terlau et al., 1991), which is consistent with the idea of a narrow selective filter.

While this accumulation of functional data has started to piece together an adequate overall design of the sodium channel, striking many similarities with other voltage-gated ion channels, it was the solution of the X-ray structure for bacterial channel Na_VAb in 2011 (Payandeh et al., 2011) that finally offered a complete visualization of channel architecture.

2.2 Bacterial Na_V channel structure

Bacterial Na_V channels present a much simpler arrangement than their mammalian counterparts; assembling into homotetramers where each subunit is made up of six segments. The make-up of each subunit closely resembles what is seen in mammalian Na_V channels with segments S1−S4 forming the voltage-sensing domain and segments S5 and S6 making up the pore domain (Fig. 1). Since the publication of the first X-ray structure of Na_VAb in 2011 (Payandeh et al., 2011), subsequent structures, including for Na_VMs (Bagneris et al., 2014; McCusker et al., 2012), Na_VRh (Zhang et al., 2012), Na_VAe (Shaya et al., 2014), and other structures for Na_VAb (Payandeh et al., 2012), have confirmed this overall architecture. Taken together, these bacterial channels offer compelling insight into the structure and function relationships for the $BacNa_V$ family.

2.2.1 Overview of the pore domain

Published $BacNa_V$ X-ray structures have been proposed to represent the pore domain in its closed (Payandeh et al., 2011; Shaya et al., 2014), open (Bagneris et al., 2014; McCusker et al., 2012), and inactivated (Payandeh et al., 2012; Zhang et al., 2012) conformations. Despite the proposed disparity of conformations, all structures display remarkably similar pore domains. The $BacNa_V$ pore domain is reminiscent of that of K channels with a comparable type of folding showing segments S5 and S6 arranged symmetrically around a central axis to form the ion-conducting pore (Fig. 2). The transmembrane segments are linked together by two pore helices, P1 and P2, and a coil forming the selectivity filter, on the extracellular side of the pore, constituting the P-loop. The existence of a second pore helix, P2, unexpected as it was not seen in K channels, raises questions as to its purpose; though it has been suggested to play a role in coordinating channel activity between the extracellular and intracellular regions of the pore domain (Boiteux, Vorobyov, French et al., 2014). On the cytoplasmic side, helices S6 entangle to form the activation gate, beginning with a wider cavity below the selectivity filter (near the P-loop region), and narrowing toward the intracellular side to form a hydrophobic constriction (Fig. 2). This bundle crossing thus creates a long hydrophobic channel (Fig. 2), which is largely conserved in diverse ion channels, creating a preferentially stable dewetted pore that can be regulated by external stimuli, such as voltage sensor activation (eg, Jensen et al., 2010), or agonist binding in ligand-gated channels (eg, Zhu & Hummer, 2012).

A most interesting feature of the $BacNa_V$ pore domain is the presence of lateral openings (fenestrations) between the pore and the lipid bilayer (Payandeh et al., 2011). These are formed by adjacent S6 segments at the level of the cavity (Fig. 2) and have been shown to allow the passage of lipid tails in both X-ray structures (Payandeh et al., 2011, 2012) and computational studies (Bagneris et al., 2014; Boiteux, Vorobyov, & Allen, 2014). They are also expected to permit the passage of small lipophilic drugs into the pore, blocking the channel; consistent with the proposed mechanisms by Hille in the 1970s (Hille, 1977a). Early studies showed that small hydrophobic molecules can enter and leave the pore even when the cytoplasmic gate is closed (Hille, 1977a, 1977b; Hondeghem & Katzung, 1977), suggesting the existence of a pathway via the membrane. The recent discovery that $BacNa_V$ channels share a similar pharmacology with Na_V channels (Bagneris et al., 2014; Lee et al., 2012) opens the way to promising perspectives in drug design and understanding of channel modulation by drugs, using atomistic simulations to unveil the binding modes and mechanisms of channel modulation.

2.2.2 Ion selectivity filter

While the structure of the pore domain is apparently well conserved between mammalian and bacterial channels, the selectivity filter of $BacNa_V$ is the most striking departure from its mammalian counterpart. Instead of the expected DEKA locus, essential for the selectivity of eukaryotic channels, prokaryotic Na_V channels (with the exception of Na_VRh (Zhang et al., 2012)) display a ring of glutamates EEEE, more commonly associated with Ca_V channels in eukaryotic channels (Tyson & Snutch, 2013) (Fig. 3). Despite this discrepancy, $BacNa_V$s exhibit selective permeabilities similar to eukaryotic Na_Vs (Finol-Urdaneta et al., 2014; Ren et al., 2001; Yue et al., 2002).

The selectivity filter of $BacNa_V$ channels is very different to the well-studied K channels (Andersen, 2011). Where the K channels present a narrow, relatively rigid backbone carbonyl-lined filter, allowing only for a single file of ions and water molecules, the filter of $BacNa_V$ appears wider and more flexible, suggesting a different selectivity mechanism. Its most distinctive characteristic is the presence of glutamate side chains inside the pore, lining the filter with mobile and (presumed) negatively charged carboxylates (Fig. 2). It is also wide enough to allow the permeation of partially hydrated Na^+ ions (Furini & Domene, 2012; Qiu, Shen, & Guo, 2012). This specific structure creates at least three binding sites for the ions: a "high-field strength" site, S_{HFS}, formed by the side chains of E177; a central site, S_{CEN}, at the level of L176 carbonyls; and an inner site, S_{IN}, created by the carbonyls and hydroxyls of T175 (Fig. 2).

It is currently not known if Na^+ selectivity occurs by different mechanisms in $BacNa_V$ and eukaryotic channels, owing to the distinct EEEE and DEKA signatures, although this has been suggested (Finol-Urdaneta et al., 2014). This is because the selectivity filter sequence contains other residues that contribute to selective permeation. Mammalian channels also display an additional section of negatively charged D/E residues surrounding the outer mouth of the filter (Fig. 3). These additional charged residues help discern Na^+ (Tang et al., 2014; Yue et al., 2002), such that understanding mammalian Na^+ selectivity, and how it relates to bacterial selectivity, requires consideration of the entire selectivity filter sequence, not just the DEKA signature, as previously attempted (Xia et al., 2013). It remains plausible that these additional residues could cause an asymmetric arrangement of charged side chains that results in a similar high field strength Na^+-selective site to that observed in $BacNa_V$.

2.2.3 Voltage-sensing domains

The voltage-sensing domains are key to the gating process in voltage-gated ion channels, transducing transmembrane electric fields into protein conformational changes that trigger the opening or closing of the ion-conducting pore, essential for the generation and propagation of cellular signals (Bezanilla, 2000; Li et al., 2014). Voltage-sensing domains share a common architecture throughout all families of voltage-gated ion channels (K_V, Ca_V, and Na_V channels), with four transmembrane segments, S1–S4, linked by extra- and intracellular short helices and loops of various sizes. These domains carry out signal transduction through movements of positive charges in the S4 helices (Papazian et al., 1991; Stuhmer et al., 1989). Segment S4 carries four to eight positively charged residues, generally arginine but sometimes lysine, separated by two hydrophobic residues in a repeated RxxR motif. The domains are connected to the pore domain via a helix-shaped linker located at the cytoplasmic interface between S4 and S5, which transmits the up and down movements of S4 during activation to the gate to cause a lateral opening motion (Fig. 2).

Analysis of $BacNa_V$ voltage-sensor sequences and structures (Payandeh et al., 2011, 2012; Zhang et al., 2012) reveals similarities with K_V channels (Li et al., 2014), but is missing long extra- and intracellular loops. The $BacNa_V$ S4 segment carries four highly conserved charged residues R1–R4 (R99, R102, R105 and R108 in Na_VAb). These positively charged residues can interact with negatively charged clusters of residues carried by segments S1–S3. Two clusters can be identified, the extracellular negatively charged cluster (ENC, N25, E32 and N49 in Na_VAb) and the intracellular negatively charged cluster (INC, E59 and D80 in Na_VAb), sitting on both sides of the hydrophobic constriction site (HCS, F56 in Na_VAb), which blocks the passage of ions and protons and confers it its hourglass shape (Fig. 2).

In the activated conformation, charged residues R1−R4 interact with the ENC, while in the resting state they are bound to the INC.

In their respective X-ray structures, Na_VAb appears to be almost completely activated (Payandeh et al., 2011, 2012), while Na_VRh is in a completely activated (or "up") state (Zhang et al., 2012). Despite the states of their voltage-sensing domains, the cytoplasmic gates are closed or collapsed in both channels. This creates a mismatch of linked voltage-sensing and pore domains that may impact on the details of those structures. At the bottom of the voltage-sensing domains, a hinge formed by cation−π stacking of a charged residue on the linker (R117 in Na_VAb) and a highly conserved aromatic residue of S3 (W76) seems to play a crucial role in the stabilization of the open/closed states of the gate. This gating hinge contributes to the closed conformation of Na_VAb in particular. The absence of a channel structure with a "down" or resting state configuration of the voltage-sensing domain makes it difficult to examine gating motions, though we may learn from other channels (Guo et al., 2016; Takeshita et al., 2014; Vargas, Bezanilla, & Roux, 2011), as we discuss further.

3. PRINCIPLES OF SODIUM CONDUCTION AND SELECTIVITY

$BacNa_V$ channels, with their wide selectivity filters, are expected to conduct ions via a different process to the single-file knock-on mechanism now commonly accepted in K channels (Berneche & Roux, 2001; Hodgkin & Keynes, 1955). Added to this, the signature glutamate side chains that line the upper part of the filter are highly flexible, demonstrated by recent simulations to play active roles in facilitating ion translocation, enabled by this flexibility (Boiteux, Vorobyov, & Allen, 2014; Chakrabarti et al., 2013).

The $BacNa_V$ X-ray structures reveal three ion binding sites: S_{HFS}, S_{CEN} and S_{IN} (Payandeh et al., 2011; Zhang et al., 2012), and an additional site located in the vestibule of the channel S_{OUT} (Shaya et al., 2014), allowing up to four ions in the upper part of the pore. Simulations of Na_VAb have shown that the filter typically binds one to four ions, but with a net preference for the two- and three-ion occupancy states (Boiteux, Vorobyov, & Allen, 2014; Chakrabarti et al., 2013). Free energy surfaces describing these two preferred occupancy states reveal a conduction mechanism involving transitions between two- and three-ion occupancy (Fig. 4), reminiscent of

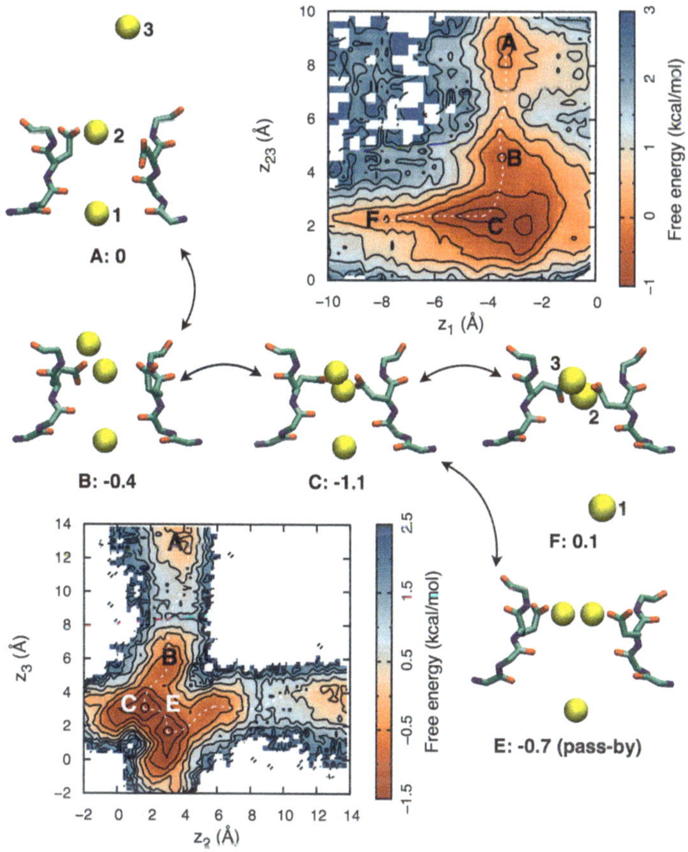

Figure 4 The conduction process of Na^+ ions in Na_vAb relies on knock-on and pass-by processes in the presence of three ions. 2-D free energy projections for Na_vAb, when three ions occupy the selectivity filter, with z_1, z_2, and z_3 corresponding to the positions of the bottom (1), middle (2), and top (3) ions, and z_{23} the center of mass of the upper ions (2 and 3). Contouring is at 0.5 kcal/mol, with lowest free energy pathways as *white dashed curves*. The permeation mechanism is shown, with free energies relative to state A in kcal/mol (~±0.5 kcal/mol). *Adapted from Boiteux, C., Vorobyov, I., & Allen, T. W. (2014). Ion conduction and conformational flexibility of a bacterial voltage-gated sodium channel.* Proceedings of the National Academy of Sciences of the United States of America, 111*(9), 3454–3459.*

knock-on conduction in K channels (Allen & Chung, 2001; Berneche & Roux, 2001; Hodgkin & Keynes, 1955; Jensen et al., 2010).

In the presence of two ions in the selectivity filter, the ions move freely between binding sites S_{HFS}, S_{IN} and S_{CEN}, but seem trapped in the filter in the absence of a third ion (not shown). The entrance of that third ion in the filter reveals an interesting competition of conduction mechanisms, as shown in Fig. 4. One possibility involves the upper ion entering the filter from above, joining the lower ion at the ring of glutamates (EEEE), and then knocking the bottom ion downward, while the alternative is for one ion to pass the other, coexisting at the EEEE ring before pushing the bottom ion into the cavity. These direct knock-on and "pass-by" mechanisms encounter similar low (<1 kcal/mol) free energy barriers and would both contribute to conduction. Interestingly, the binding of a third ion always involves the rupture of the H-bonds seen between E177 and S178 in the crystal structure (Payandeh et al., 2011), leading to a rearrangement of the side chains of E177, provided sufficient sampling time in simulations.

Ion permeation in Na_VAb can therefore be described as a three-ion process (Fig. 4). During the entire conduction process, the largest free energy barrier encountered is ~1.2 kcal/mol (~$2k_BT$), allowing for rapid conduction. This three-ion mechanism is supported by other recent computational investigations (Chakrabarti et al., 2013) and by previous studies of simplified Glu-lined pores (Corry et al., 2001), but not in studies that especially focused on a two-ion process (Corry & Thomas, 2012; Furini & Domene, 2012). Other studies have focused on generating ion conduction through a constrained-open pore (McCusker et al., 2012; Ulmschneider et al., 2013) using large membrane potentials, in contrast to the aforementioned studies that focused on equilibrium sampling, appropriate to physiological voltages, despite using a closed Na_VAb structure.

The specific make-up of the filter and its flexibility are at the center of this dynamic conduction mechanism. The interplay between rotameric changes of the side chains of residues E177 and the movements of permeating Na^+ ions creates an ensemble of distinct bound states, and the rapid exchange between those states is an essential part of the conduction process, allowing for near-diffusion permeation rates in $BacNa_V$ channels (ie, the flexibility of the side chains, and their ability to undergo structural isomerizations, leads to a continuum of low energy configurations connecting distinct free energy minima, facilitating conduction by lowering activation barriers (Boiteux, Vorobyov, & Allen, 2014; Chakrabarti et al., 2013)). Such intimate involvement in ion conduction is only achieved when simulating on the timescales of conduction itself and is thus only adequately sampled through hundreds of nanoseconds or more of simulation, not achieved in early studies (Corry & Thomas, 2012; Furini & Domene, 2012) but adequately in subsequent simulations (Boiteux, Vorobyov, & Allen, 2014; Chakrabarti et al., 2013).

In contrast to the narrow, ion-sized selectivity filter of the K channel, the $BacNa_V$ filter is wider and lined not only by protein backbone (residues 175–178) but also by the side chains of a ring of four glutamates (E177) (Fig. 2). Curiously, this ring is conserved within the $BacNa_V$ family but is indicative of a Ca^{2+}-selective (Tyson & Snutch, 2013) and not of a Na^+-selective mammalian channel, which has a conserved DEKA locus (Charalambous & Wallace, 2011) (Fig. 3). Despite the EEEE-lined selectivity filter, bacterial channels nominally favor Na^+ over Ca^{2+} or K^+ with relative permeabilities $5 < P_{Na}/P_K < 171$ and $7 < P_{Na}/P_{Ca} < 70$ in NaChBac (Finol-Urdaneta et al., 2014; Ren et al., 2001; Yue et al., 2002). This Na^+ selectivity emerging from a Glu-lined pore leads us to question the mechanism by which these channels optimally conduct Na^+ ions.

Simple physical models have proposed a charge/space competition theory to explain the variable Na^+/Ca^{2+} selectivity of Glu-lined selectivity filters, where the volume and electrostatic environment both play a role in the number of ions that the channels can bind and efficiently conduct (Boda et al., 2009; Corry et al., 2001; Yue et al., 2002). It has been suggested (Boda et al., 2007) that confinement of charged ligands in a smaller volume will favor the binding of divalent ions over monovalent ions. Another factor is the extent of additional negative charges in the vicinity of the selectivity filter, with increased number expectantly increasing propensity for attracting Ca^{2+} (Yue et al., 2002). Clearly molecular simulations are needed to better explore valence selectivity in the sodium channel. A major hindrance has been the availability of accurate models of Ca^{2+} ions to capture the interactions with water, protein carboxylates, and carbonyl groups (Yoo, Wilson, & Aksimentiev, 2016) with recent attempts observing significant overbinding of the divalent ions (Boiteux, Vorobyov, & Allen, 2014).

To understand competition between the monovalent cations Na^+ and K^+ in the sodium channel, we can learn much by returning to the case of K^+ selectivity in a K channel. A long-running hypothesis is that the K channel thermodynamically chooses to bind K^+ over Na^+ ions with greater affinity (Neyton & Miller, 1988) (although this apparent fact has been recently challenged (Kim & Allen, 2011)). While it has been argued that size alone might achieve such difference in thermodynamic stability (Doyle et al., 1998), this has been disputed simply because proteins are highly flexible with even their backbones fluctuating on the Å-scale. These fluctuations determine the flexible response of the coordinating ligands that control ion binding (Allen, Andersen, & Roux, 2004; Noskov, Berneche, & Roux, 2004), deeming sub-Å differences in ion size of little energetic consequence. While there have been suggestions of such discrimination based on small differences in the coordinating environments of Na^+ and K^+ within the Na_V filter (Corry & Thomas, 2012; Furini & Domene, 2012), flexible response is an even more important contributor in the sodium channel, due to its dynamic coordinating glutamate side chains, leading not only to strong Na^+ binding but also a highly efficient Na^+ conduction mechanism.

We may also learn by comparing the nature of the coordinating ligands between Na and K channels. Selectivity between the monovalent cations Na^+ and K^+ may be interpreted in terms of ligand field strength (Eisenman, 1962), where amide carbonyls provide weaker fields that favor larger ions, like K^+, whereas carboxylates create a high-field strength site favorable to Na^+. Yet in both the sodium and potassium channels, selectivity is an inherently dynamic process (Thompson et al., 2009), rather than governed strictly by the geometry and chemical nature of the ligands encountered by the ions. Furthermore, despite the likely effects of ligand chemistry and dynamics on the binding of ions, we must acknowledge that selective permeation of the channels is ultimately a kinetic process (Bezanilla & Armstrong, 1972). In a K channel, selectivity has been shown to arise from a well-optimized multiple-ion conduction mechanism, where K^+ ions encounter low free energy barriers to permeation. This mechanism breaks down in the presence of Na^+, as those ions bind to different (though adjacent) sites, disrupting the low free energy pathways (Thompson et al., 2009). This kinetic view of selective exclusion of ions suggests that one can only truly observe and understand selectivity by examining the free energy landscape governing the movements of all of the ions in the pore.

For the sodium channel, we have also observed that the multiion scheme is a highly optimized knock-on mechanism for Na^+ ions (Boiteux, Vorobyov, & Allen, 2014; Chakrabarti et al., 2013). In this process, we have observed how essential E177 rotamer changes and the formation of ionic clusters with their carboxylate groups are to efficient permeation. We must then turn to examine the behavior of K^+ ions in the sodium channel to begin to learn about its selectivity process. Preliminary results obtained during MD simulations of the permeation of K^+ ions in Na_VAb show that this ion is unable to form the same kind of interactions with E177 (Boiteux, Vorobyov, & Allen, 2014), and that reduced clustering of K^+ with those side chains leads to an overall reduction in ion occupancy (Boiteux, Vorobyov, & Allen, 2014). Yet, simulations in the presence of K^+ ions indicate that the free energy surface for a reduced one- to two-ion knock-on mechanism may still permit conduction of K^+ ions, albeit at a reduced rate (Boiteux et al., In preparation). Key to the difference thus appears to be the ability of Na^+ to form a three-ion occupancy state of the filter due to stronger interactions with charged carboxylates.

What we have learned so far from simulations of bacterial channels can aid us in describing the likely situation in the mammalian channel, with its distinct DEKA locus in the selectivity filter. The DEKA signature does indeed contain two carboxylates that may bind sodium ions in a similar fashion to the bacterial channel. We again note that the selectivity filter contains outer rings of charged residues (Fig. 3) that may contribute to the binding of multiple ions (as suggested by their mutagenesis (Yue et al., 2002)). What then is the role of the lysine residue? It is known experimentally that its mutation affects selectivity in mammalian channels (Favre et al., 1996; Sun et al., 1997). It has been suggested theoretically that this Lys may act to repel Na^+ ions toward the opposing D and E residues (Xia et al., 2013). However, it is also possible that simulations on the timescales of permeation may reveal that the asymmetric channel and flexible D and E (and potentially outer ring carboxylates) provide plenty of binding partners for a multiion mechanism involving Na^+, as well as for the charged (or neutralized) Lys within the pore. Computational work on mammalian models, based on BacNa$_V$ templates, should be helpful in explaining Na^+ selectivity in mammalian channels.

4. STRUCTURAL BASIS FOR ACTIVATION AND INACTIVATION

4.1 Activation/gating

There has been much progress in understanding voltage-sensing movements in the family of voltage-gated ion channels in general, allowing us a reasonable understanding about this process for the sodium channels (Yarov-Yarovoy et al., 2012), with changes in the voltage-sensing and pore domains described earlier. At this point in time, published X-ray structures have been proposed to represent the pore domain in its closed (Payandeh et al., 2011; Shaya et al., 2014), open (Bagneris et al., 2014; McCusker et al., 2012) and inactivated (Payandeh et al., 2012, Zhang et al., 2012) conformations; seemingly offering structural definitions of the important states in the function of the channel.

Gating motions in the pore domain of bacterial Na$_V$ channels have been a matter of some contention. The initial unavailability of an open state

$BacNa_V$ meant that studies of ion conduction were forced to simulate a closed pore based on Na_VAb. The proposal of an open state form of $BacNa_V$ (Bagneris et al., 2014; McCusker et al., 2012) did not uncover the degree of conformational change and pore opening anticipated. Comparison of pore radii in Fig. 5 shows very little difference at the level of the cytoplasmic gate, suggesting a strong bias toward a closed pore, even if the S5 and S6 arrange themselves slightly differently in these structures. In fact, Fig. 5 suggests that the first published Na_VMs (proposed open) structure (McCusker et al., 2012) is just as closed at the previous structures. In the most recent Na_VMs structure, we now see a small deflection at the very bottom in Fig. 5, corresponding only to deflection of the end of S6 in PDB:4CBC (Bagneris et al., 2014).

The marginal change observed when comparing closed and proposed open channels has led to the proposition of a subtle gating mechanism involving a slight bend at Thr209 on S6 (McCusker et al., 2012). Whether this bend is sufficient to constitute a distinct conformation on the free energy surface has yet to be determined. Moreover, simulations have indicated that this may merely correspond to a natural fluctuation of the pore within the one state (Boiteux, Vorobyov, & Allen, 2014) (although

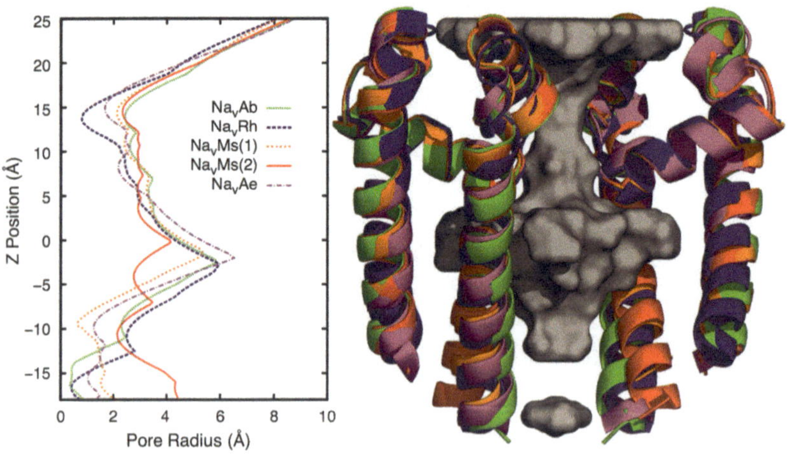

Figure 5 Gating in $BacNa_V$ channel pore domain. Comparison of the pore domain of published X-ray structures of bacterial Na_V channels. Plot of pore radius versus position along the *z*-axis for Na_VAb (pdb 3rvy, green short dashed line), Na_VRh (pdb 4dwx, blue dashed line), $Na_VMs(1)$ (pdb 4f4l, orange dotted line), $Na_VMs(2)$ (pdb 4cbc, red solid line), and Na_VAe (pdb 4lto, violet dash dot line) and backbone superimposition of the same channels using ribbons with the same color code, with Na_VAb pore volume represented as a gray mesh. For clarity, only two subunits of the pore domain are represented. Pore volume and radii were calculated with HOLE (Smart, Neduvelil, Wang, Wallace, & Sansom, 1996).

also appears linked to changes linked to inactivation—see below). Nevertheless, unless heavily constrained to stay open (Ulmschneider et al., 2013), simulations of this pore domain-only structure, either isolated or used as a template to build more complete open models of Na_VAb or Na_VRh involving the voltage-sensing domains, have been unsuccessful in maintaining an open pore (Boiteux and Allen, Unpublished). A significant part of the problem lies with the absence of voltage-sensing domains in the Na_VMs structure, and the issues in merging with the complete structures of Na_VAb or Na_VRh, which have been solved in "preactive" states with fully or partly activated voltage-sensing domains and closed pores. Such a mismatch means the interactions of the protein and its S4–S5 linker with S6 may be unfavorable. Solution of a complete channel (pore and voltage sensor domains) in an open state remains an important challenge in the field.

4.2 Inactivation

Na_V channels cycle through resting, open, and inactivated states during their function. They typically open on depolarization and close either rapidly upon repolarization or more slowly on sustained depolarization. The latter mechanism is called inactivation and causes a refractory period during which channels remain unresponsive to further depolarization. Inactivation was observed as early as 1924 (Woronzow, 1924), where a long, sustained pulse had to be applied to restore excitability. This phenomenon was later on associated with specific characteristic of Na^+ channels, when Hodgkin and Huxley (1952a, 1952b) first described neuron excitability in terms of sodium and potassium conductance and identified an inactive state for the channels.

Inactivation occurs on different timescales (Adelman & Palti, 1969; Narahashi, 1964), leading to at least two different classes: fast and slow inactivation, also called N-type and C-type; terms initially coined to describe inactivation in the Shaker K channel (Hoshi, Zagotta, & Aldrich, 1991). Inactivation is an essential physiological process, protecting cells from overexcitability, which has been associated with periodic paralysis (Cannon, Brown, & Corey, 1991; Cannon & Strittmatter, 1993), arrhythmia (Grant, 2001), long QT syndrome (Wang et al., 1995), or epilepsy (Lossin et al., 2003; Rhodes et al., 2004) among others issues.

4.2.1 Fast inactivation

Fast inactivation in eukaryotic Na_V channels typically occurs during sustained depolarization on a timescale of the order of a millisecond (varying among channel isoforms (O'Leary, 1998)) via a "ball and chain" or "hinged lid" mechanism during which a cytoplasmic part of the protein (the ball) obstructs the pore by binding to a docking site. The inactivating particle has been identified as the cytoplasmic loop linking DIII and DIV (Benz et al., 1997; Vassilev et al., 1988), in particular three highly conserved residues forming the IFM motif (Kellenberger, Scheuer, & Catterall, 1996; Rohl et al., 1999; West et al., 1992). The docking site is made up of numerous cytoplasmic regions surrounding the gate, including the linker S4–S5 of domains III and IV (McPhee et al., 1998) and the cytoplasmic terminal of segment S6 in DIV (Hilber et al., 2002). Fast inactivation plays an important role for action potential repolarization, in particular within the nodes of Ranvier (Chiu et al., 1979).

In the absence of cytoplasmic linker between subunits (Payandeh et al., 2011) prokaryotic Na_V channel structures offer limited additional insight into fast inactivation in mammalian channels. However, MD studies of homology models have been used to examine flexible linker interactions with both the pore and voltage sensors to provide insight into the fast inactivation process (Lam, Wu, & Schulten, 2016). While both fast and slow inactivation play roles in function and modulation (eg, Fozzard et al., 2005; Zeng & French, 2011b), the simplicity of the bacterial channels allows for experimental and computational studies to easily focus on the one process of pore-based slow inactivation, as done with K channels (Cuello et al., 2010).

4.2.2 Slow inactivation

Slow inactivation is distinct from fast inactivation and does not involve the cytoplasmic linker specific to eukaryotic channels. It is a less mechanically defined process determined to involve the region of the pore. Slow inactivation has no clearly defined timescale, and can span multiple orders of magnitude, from milliseconds to minutes (Ellerkmann et al., 2001; Ulbricht, 2005).

Structural investigations have shown that different pore regions can be associated with slow inactivation (Fig. 6). In particular, mutations around the selectivity filter (rNa_V1.4 DI-selectivity filter W402C (Balser et al, 1996a; Tomaselli et al., 1995) and W402A (Kambouris et al., 1998), equivalent to W179 in Na_VAb, DI-selectivity filter E403C (Zhang et al., 2003), aligned with S180 in Na_VAb, and DIII-selectivity filter F1236C (Ong, Tomaselli, & Balser, 2000), aligned with L176 in Na_VAb) and the segments

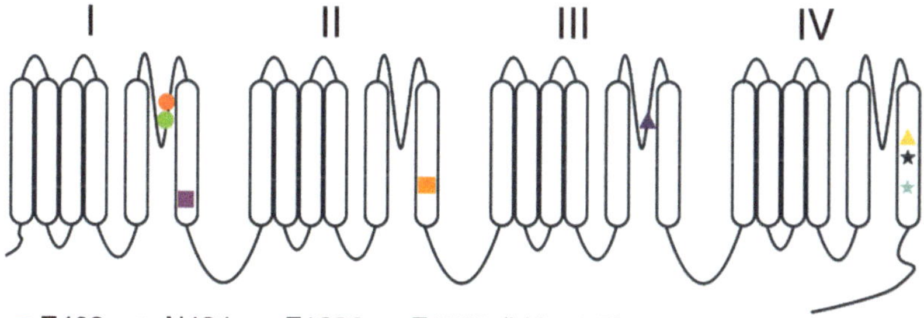

Figure 6 Slow inactivation in mammalian Na_V channels. Schematic representation of a mammalian Na_V channel and the residues known to modulate slow inactivation: DI-selectivity filter W402 (Balser et al, 1996a; Kambouris et al., 1998; Tomaselli et al., 1995) (*green dot*), DI-selectivity filter E403 (Zhang et al., 2003) (*red dot*), DIII-selectivity filter F1236 (Ong et al., 2000) (*blue triangle*), DI-S6 N434 (Wang & Wang, 1997) (*purple square*), DII-S6 V787 (O'Reilly et al., 2001) (*orange square*), DIV-S6 F1764 (Carboni et al., 2005) (*black star*), and DIV-S6 I1575 (*yellow triangle*) and DIV-S6 V1583 (*light blue star*) (Vedantham & Cannon, 2000). The numbering corresponds to the rat $Na_V1.4$ channel sequence unless indicated otherwise.

S6 forming the cytoplasmic gate (r$Na_V1.4$ DI-S6 N434A (Wang & Wang, 1997), equivalent to N211 in Na_VAb, DII-S6 V787C (O'Reilly, Wang, & Wang, 2001), aligned with Na_VAb M209, DIV-S6 V1583C (Vedantham & Cannon, 2000), DIV-S6 I1575A (Zarrabi et al., 2010), equivalent to I202 in Na_VAb, and h$Na_V1.5$ DIV-S6 F1760A (Carboni et al., 2005)) can modulate slow inactivation rates. The distribution of the residues potentially involved, or located near conformational changes associated with slow inactivation, signify a complex and poorly understood process. The regions may suggest linking of the selectivity filter and segment S6 in a collapse throughout the entire pore domain.

Prokaryotic Na_V channels display some form of inactivation (necessarily pore-based), spanning milliseconds to seconds (Payandeh et al., 2012; Ren et al., 2001; Zhang et al., 2012), compatible with the timescales described for slow inactivation in mammalian Na_V channels. Additionally, structures of potentially inactivated or partially inactivated states of Na_VAb (Payandeh et al., 2012) and Na_VRh (Zhang et al., 2012) show an asymmetrically collapsed gate, involving bonding between L123 (linker S4–S5) and S6 N211 (equivalent to N434 involved in slow inactivation in $Na_V1.4$ (Wang & Wang, 1997)) in Na_VAb (Payandeh et al., 2012). Another common feature of these inactivated structures is a narrower selectivity filter, described as closed in Na_VRh (Zhang et al., 2012). However this closed selectivity filter was shown to spontaneously open during MD with only a slight reorientation of the side chains of the residues lining it but remained poorly conductive (Zhang et al., 2013). The structural data collected in eukaryotic Na_V channels are consistent with the conformational changes observed in $BacNa_V$ channels, suggesting that they share a common mechanism for slow inactivation, further suggesting bacterial channels are good models for understanding the process.

While equilibrium with a true slow inactivated state is not likely to be achieved on accessible timescales using MD simulation, recurrent behavior relevant to inactivation can be sampled. As described, Na_VAb's selectivity filter is very flexible, and this appears to be key to conduction and selectivity (Boiteux, Vorobyov, & Allen, 2014; Chakrabarti et al., 2013). Interestingly, studies on long timescales have shown that conformational changes within the filter are correlated with significant distortion along segments S6 (Boiteux, Vorobyov, & Allen, 2014), coincident with pore collapses that are much akin to the proposed inactivated forms of the channel (Payandeh et al., 2012; Zhang et al., 2012) (Fig. 7). Analysis of the backbone fluctuations in the segments S6 of Na_VAb shows some instability between residues V196 and P200, in the top part of the helices, and around residues I202 and F203, located in the middle of S6, at the level of the fenestrations. The

Figure 7 Superimposed bottom views of the proposed inactivated asymmetric conformation described by Payandeh et al. (2012). (PDB:4EKW segments AB, *yellow* (light gray in print versions) *and orange cylinders* (gray in print versions)) and the conformation obtained after a 2.5 μs simulation of Na_VAb (*blue* (dark gray in print versions) and *purple cylinders* (black in print versions)), showing the displacement of the S6 helices. The *black dashed line* indicates the asymmetrically collapsed gate, being similar in both structures.

structure of S6 depends strongly on the state of two concurrent I202–T206 interactions: the backbone-to-backbone helix forming H-bond and the interfering backbone to side chain H-bond, potentially helix weakening. While serine is more notably described as a helix-breaker (Ballesteros et al., 2000; McGregor, Islam, & Sternberg, 1987), threonine displays a similar mechanism of interfering with α-helical structure and is also known to modulate the bending commonly induced by proline in α-helices (Deupi et al., 2004). In Na_VAb, the methyl group of residue T206 is surrounded by a large cluster of nonpolar side chains from neighboring residues that helps direct the hydroxyl group toward the carbonyl oxygen of I202, leading to weakening of the α-helix forming H-bond. This helix-weakening effect around I202 and T206 is reinforced by the significant deformation introduced by residue P200 between F198 and I202, leading to deep bending of segment S6. In the most extreme cases, this bending is associated with a tilt of P2 that pulls on the selectivity filter from above and contributes to large-scale selectivity filter distortion (Fig. 8).

Taken together, these conformational changes result in protein structures that resemble the proposed inactivated structures of Na_VAb and Na_VRh (Payandeh et al., 2012; Zhang et al., 2012) (Fig. 7). These structures present asymmetrically collapsed S6 activation gates (asymmetry of 5 Å at the level of residues 215–218 for Na_VAb but only ~1 Å for Na_VRh), as well as an associated narrower selectivity filter and a reshaped cavity. During simulations, similar gate asymmetry has been observed, reaching as high as 12 Å, and a narrower to collapsed selectivity filter (Boiteux, Vorobyov, & Allen, 2014).

This bending of segment S6 appears to be a natural feature of the channel. Significantly, it can be linked to a cluster of key residues located at the level of the cavity fenestrations and associated with slow inactivation in both mammalian and bacterial channels (Fig. 8). Mutation of the helix-weakening T206, described as the hinge point in open Na_VMs (McCusker et al., 2012), has been shown to prevent inactivation in NaChBac (T220) (Zhao, Scheuer, & Catterall, 2004), while mutation of the highly conserved equivalent to I202 dramatically alters slow inactivation in $Na_V1.4$ (Zarrabi et al., 2010). Moreover, mutation of a phenylalanine at that same level (not equivalent but in similar position and orientation to F203 in Na_VAb) has been associated with a strong inhibition of slow inactivation in mammalian channels (Carboni et al., 2005). While these rapid conformational changes would need to be stabilized on much longer timescales (milliseconds to seconds) to yield a truly slow inactivated channel, they give some

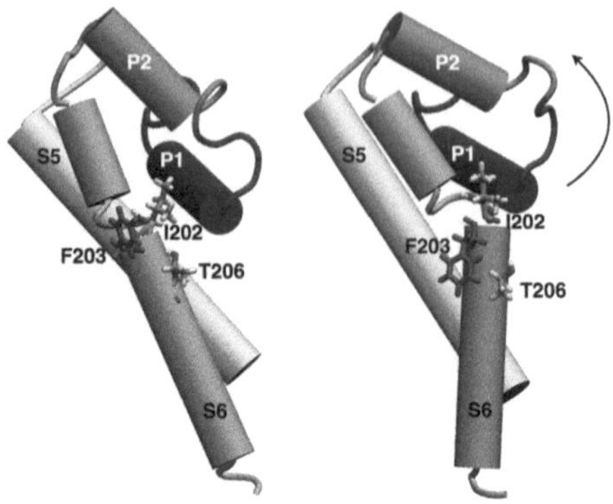

Figure 8 Conformational changes relevant to slow inactivation in the pore domain of Na_VAb. Segment S6 (*yellow cylinder* (light gray in print versions)) bends near residue F203 (*red sticks* (dark gray in print versions)), breaking the bond between residues I202 and T206 (*colored sticks*), and induces a tilt of the top pore helix, P2 (*light blue cylinder* (gray in print versions)), and an upward displacement of a selectivity filter subunit, leading to a collapse of the filter.

significant insights into the structural changes occurring during the transition from a conducting to an inactivated state. It is also of much interest that changes in S6 helices are correlated with conformational changes in the filter, a phenomenon described during C-type (slow) inactivation in K channels (Cuello et al., 2010), and suggested by NaChBac studies.

5. DRUG BINDING AND MODULATION

Na_V channels are involved in a vast range of pathologies, including neurological, muscular, and cardiac disorders (Ashcroft, 2000; Catterall, 2010; England & de Groot, 2009; George, 2005; Kyle & Ilyin, 2007; Tarnawa, Bolcskei, & Kocsis, 2007; Zuliani et al., 2009). As such, they are the primary targets for many drugs, such as LAs, general anesthetics, AEDs and antiarrhythmics (Kyle & Ilyin, 2007; Mike & Lukacs, 2010; Tarnawa et al., 2007). The modulation of Na_V channels by pharmacological agents is a complex phenomenon, involving voltage- and state-dependent block of current, specific interaction with various binding sites in different activated and inactivated states, and modification of the recovery time from these inactivated states (Hille, 2001). It has long been postulated that the fast-inactivated state was the conformational target for state-specific binding (Bean, Cohen, & Tsien, 1983; Hille, 1977a; Wang et al., 1987), and LAs, for example, can see their potency reduced in fast inactivation defective channels (Balser et al, 1996b; Bennett et al., 1995; Cahalan, 1978). However, the focus has recently begun to include slow inactivation, as both LAs and AEDs have been shown to modulate slow inactivated channels (Balser et al, 1996c; Errington et al., 2008; Hebeisen et al., 2015; Sheets et al., 2008; Sheets, Jarecki, & Cummins, 2011), and mutation of key residues forming the receptor site for LAs affects slow inactivation (Carboni et al., 2005).

Functional studies have successfully identified two major binding sites for LA and AEDs: a high-affinity site, that appears to be state and voltage dependent, and a low-affinity site apparently involving closed channels (Fozzard, Sheets, & Hanck, 2011; Sheets et al., 2010). The high-affinity site has been the subject of extensive investigation (McNulty et al., 2007; Mike & Lukacs, 2010; Qu et al., 1995; Wang, Quan, & Wang, 1998) since its first identification in 1994 (Ragsdale et al., 1994) and appears to be common to many drugs (Ragsdale et al., 1996; Wang et al., 1998). This site has been located on segment S6 of domain DIV and involves a phenylalanine located in the middle of the helix (rNa_V1.2 F1764 (Ragsdale et al., 1994) and hNa_V1.5 F1760 (Carboni et al., 2005; McNulty et al., 2007)) and a tyrosine belonging to the lower half of the helix (rNa_V1.2 Y1771 (Ragsdale et al., 1994) and hNa_V1.5 Y1766 (McNulty et al., 2007)). The low-affinity site is not as well characterized and has been described as resulting from a "combination of hydrophobic interactions" occurring when the channel is maintained at hyperpolarized voltages (Fozzard et al., 2011).

Despite this recent accumulation of knowledge about the binding of drugs, there is still little known about the mechanism by which these drugs alter the functions of sodium channels. The development of new therapeutic agents for specific disorders, while limiting or avoiding altogether possible side effects, will rely on our ability to understand the complex interactions controlling selective drug binding to Na_V channels. Bacterial Na_V channels can be considered models for understanding drug binding as they display sensitivity to common Na_V inhibitors (Bagneris et al., 2014, Barber et al., 2014; Lee et al., 2012; Raju et al., 2013), and key binding regions are well conserved between mammalian and bacterial channels (Bagneris et al., 2014; Payandeh et al., 2011). In silico studies of BacNa_V channels in the presence of LAs and AEDs have been successfully used to shed light on binding sites and inhibition mechanisms (Bagneris et al., 2014; Boiteux, Vorobyov, French et al., 2014; Martin & Corry, 2014). Additionally, crystal complexes of brominated drug-like compounds bound to the

Na_VMs channel have shown binding in the cavity near what appears to be the high-affinity site (Bagneris et al., 2014).

5.1 Access and binding to the central cavity and high-affinity site

MD simulations of the LA benzocaine (Boiteux, Vorobyov, French et al., 2014) (as well as other LA and AEDs; Boiteux et al., In preparation) show extensive binding in the hydrophobic cavity, and more specifically to residue F203, consistent with its role as the high-affinity binding site in Na_VAb. The simulations reveal two different access pathways, via the lipid membrane through the fenestrations, and, unexpectedly, directly through the cytoplasmic gate despite it being closed (benzocaine) (Fig. 9). The coexistence of these two access routes is in excellent agreement with the hydrophobic (fenestration) and hydrophilic (gate) pathways predicted by Hille (1977a) but never before demonstrated at the molecular level.

Na_VAb may offer unique similarity to mammalian Na_V channels for the investigation of drug binding within the high-affinity pore site because of the presence of the pore-facing F203. Long unbiased MD simulations

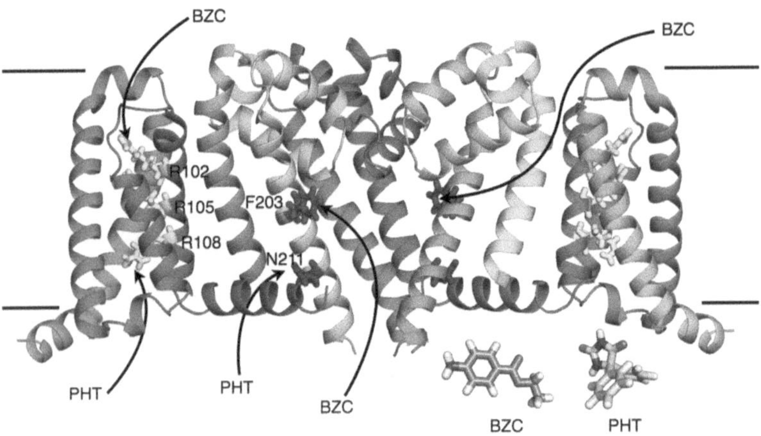

Figure 9 Binding sites and access pathways revealed by MD simulation for benzocaine and phenytoin in Na_VAb. For clarity only two subunits are shown for the voltage-sensing domain and three for the pore domain. Residues important for binding are shown as colored sticks (F203 in red, N211 in dark blue, and S4 R102, R105, and R108 in yellow; seen as different shades of grey in print versions). *Horizontal black lines* mark the limit of the lipid bilayer while black arrows show simplified access pathways to specific binding sites for the drugs specified. Benzocaine (green) and phenytoin (blue) are shown as sticks with labels.

(Boiteux, Vorobyov, French et al., 2014) and metadynamics simulations (Martin & Corry, 2014) of LA benzocaine and AED phenytoin have suggested binding in the cavity of Na_VAb, having preferential interactions with F203, following passage through fenestrations to this high-affinity site. Other bacterial Na_V channels only possess Phe residues on S6 that are directed away from the pore (Boiteux, Vorobyov, French et al., 2014). Interestingly, however, binding may still occur in the absence of this residue, as demonstrated for brominated drug-like compounds to more distant F214 on S6 in Na_VMs (Bagneris et al., 2014). In this case, docking of known Na_V antagonists showed preferential binding in the top half of the cavity and around the selectivity filter, rather than through the fenestrations (Bagneris et al., 2014). A study of a model of NaChBac in the presence of the general anesthetic isoflurane has also shown access to the cavity via the fenestrations, though suggests binding to a phenylalanine that sits low on the S6 helix (F227, equivalent Y1771) (Raju et al., 2013), despite the suggestion that no Phe side chain on NaChBac is capable of facing the pore (Boiteux, Vorobyov, French et al., 2014). The shared binding in the pores of different bacterial (and mammalian) channels suggests a dominant binding in a hydrophobic pore region, influenced by available aromatic and polar groups in each protein.

5.2 Binding around the cytoplasmic gate and the linker

The AED phenytoin (Boiteux, Vorobyov, French et al., 2014) binds to residue N211 at the level of the cytoplasmic gate between segments S6 and the linkers S4–S5 (Fig. 9). In the case of phenytoin, the drug interacts via H-bonding with N211 and D219 of adjacent subunits, effectively keeping the gate in its closed state. For both drugs, this binding site forms a wedge between the outside of the gate and the linker, and would be more accessible when the voltage-sensing domain is in an activated state, lifting the linker up, and when the gate is closed or collapsed; a conformation compatible with a slow-inactivated state. Binding at this site would be expected to stabilize this conformation. Similar results were obtained during the simulation of isoflurane with NaChBac, showing binding in the same location with N425 (equivalent to N211 in Na_VAb) (Raju et al., 2013). Mutation of the equivalent of residue N211 in channel $rNa_V1.4$ (DI-S6 N434A (Wang & Wang, 1997)) has been shown to modulate slow inactivation, reinforcing the significance of binding at this site for AEDs that target and enhance slow inactivation (Errington et al., 2008; Hebeisen et al., 2015; Kuo & Bean, 1994; Quandt, 1988).

5.3 Interactions within voltage-sensing domains

While direct interaction with the pore domain seems the most intuitive inhibition process, other modulating mechanisms should not be discounted. Channel inhibitors are known to alter the voltage sensor to gate coupling through various processes, either by preventing the recovery from an inactivated state or by stabilizing or actively inducing a nonconductive state (Mike & Lukacs, 2010). LAs in particular have been shown to affect the voltage sensors in domains III and IV of Na_V channels, disrupting their coupling with the pore domain and rendering the channel insensitive to voltage (Muroi & Chanda, 2009). During simulations, all investigated drugs display some level of interaction with the voltage-sensing domains and in particular with the charged residues of S4. Benzocaine is seen interacting via cation–π stacking with R2, phenytoin forms H-bonds with R4 (Fig. 9), while other drugs have been seen to interact with both R2 and R4, as well as R3 (not shown; Boiteux et al., In preparation).

LAs and AEDs are known to modulate the response of Na_V channels to voltage (Muroi & Chanda, 2009), but this effect is usually considered as an indirect consequence of the binding of the drugs around S6 and subsequent effect on the gate and permeation of ions. While some toxin-binding sites have been clearly identified at the top of some of the voltage-sensing domains (Cestele & Catterall, 2000), the direct interaction of drugs with specific residues of Na_V is not well documented. However, a study has recently shown that the AED ztz240 interacts directly with the charged residues of S4 and the conserved Phe of the HCS (S2) of $K_V7.2$ to modulate its activation (Li et al., 2013). No such results can be found for Na_V channels but the high level of conservation of both the charged residues of S4 and the Phe forming the HCS suggests that the same kind of interaction can be expected.

The effects of such interactions remain difficult to interpret and would be highly state dependent. In the activated conformation, residues R2 and R3 would be more exposed and likely to become attractive binding sites, the up conformation would then be stabilized and the channel rendered less sensitive to hyperpolarization and subsequent depolarizations. Such a phenomenon is consistent with the effects of a drug targeting and enhancing slow inactivation. We shall be better positioned to use MD simulation to understand voltage sensor contributions to drug modulation, once an X-ray structure of a downstate voltage sensor has been solved.

6. DISCUSSION AND CONCLUSIONS

While experimental techniques can probe the structures and activities of ion channels, they remain limited by experimental length and timescales, and their abilities to observe the detailed motions and interactions of the atoms that give rise to channel function. In the past few years, breakthroughs in several atomic structures of Na_V channels have provided the necessary springboard from which we can begin to explore function using atomistic simulations. In this chapter, we have described recent developments in the use of structurally defined bacterial channels, in conjunction with modern atomistic MD simulations that push toward physiological timescales. These studies have, for the first time, uncovered molecular-level mechanisms underlying neuronal signaling, and how they may be modulated by different molecules.

The solution of atomic resolution structures for Na^+-selective ion channels has enabled computer simulations to shed light on Na^+ conduction mechanisms and the ability of the protein to discriminate against other physiologically abundant cations. Although the bacterial channels exhibit a ring of glutamates that is distinct from the mammalian signature sequence (involving glutamate, aspartate, lysine, and alanine), those simulations revealed how carboxylate groups preferentially attract Na^+ and facilitate low-barrier ion translocation across a membrane channel. Simulations demonstrated a different mechanism to that we had become familiar within K channels, where a regular single-file movement of multiple K^+ ions passes through a conformationally rigid selectivity filter. In the sodium channel, the filter itself is an active participant, with flexible Glu side chains forming complexes with multiple ions, and essentially carrying the ions across the narrow constriction. This conduction is either via a familiar knock-on mechanism or via a pass-by mechanism where the high affinity for the Glu site enables concurrent binding of two Na^+ ions. Simulations have suggested that this does not occur for the larger K^+, with lower affinity for the high-field strength site, and reduced overall occupancy. The challenge now remains to create more accurate empirical models of the also abundant Ca^{2+} ion, to resolve the mystery of why a Glu-lined filter is Na^+ selective in one organism and Ca^{2+} selective in another. The further challenge remains to extrapolate these findings to understand mammalian selectivity mechanisms, through carefully constructed homology models or through breakthroughs in eukaryotic channel crystal structures.

Atomic simulations have also shed light on the possible conformational changes in the sodium channel, and how they may relate to activation and inactivation mechanisms. Simulations have reported significant protein conformational flexibility on the multi-μs timescale, influencing the conducting state of the pore, as well as its interplay with the lipid membrane. Simulations have observed pore collapse, involving bending of helix S6, leading to an asymmetric arrangement that resembles the inactivated state proposed experimentally. These changes have been shown to involve residues that are critical to slow inactivation in bacterial and mammalian channels. They also reveal communication throughout the protein, through S6 and the P2 helix of the P-loop, and steric interactions with the selectivity filter itself. They also are associated with the opening and closing of membrane-bound fenestrations, giving clues as to the processes that underscore inactivation and drug inhibition of mammalian Na_V channels.

It is known experimentally that drugs influence both slow and fast inactivation in mammalian sodium channels, and that these drugs inhibit bacterial channels in a similar fashion, targeting pore-based slow inactivation. This motivates the use of newly solved crystal structures to explore the binding of LA, AED, and general anesthetic molecules to bacterial channels, as simplified models of human sodium channels. Simulations of neutral, hydrophobic drugs revealed high-affinity pore-blocking, approached via paths that were either lipid-bound fenestrations or through a closed intracellular gate. In Na_VAb, the high-affinity site involves a Phe side chain that mimics the side chain that controls binding in mammalian channels, while in other bacterial channels, binding is still achieved inside the pore of the channel with similar affinity. A range of lower affinity sites have been identified, which are expected to modulate channel function by altering the stability of different states of the gate (with different size and asymmetry), communication of change between voltage sensors and the gate, or the voltage sensor movements themselves. Evidence suggests that these binding sites are dependent on the chemical nature of the drug, suggesting that such extensive sampling of binding pathways and sites is required for each drug candidate.

These recent studies represent the first stage in understanding the molecular origins of sodium channel inhibition. More complete investigations must focus on the state dependence of binding and ultimately capture the modulation of the receptor by the shifting of the equilibrium of conformational states, presenting a significant challenge to atomistic MD simulation. These studies have, however, demonstrated that simpler bacterial channels offer a wealth of mechanistic information that can be used to learn modulation processes in humans and provide new targets for therapeutically selective sodium channel drugs.

ACKNOWLEDGMENTS

This work was supported by National Institutes of Health U01-HL126273, DE Shaw Anton (PSCA00061P; NRBSC, through NIH RC2GM093307), VLSCI (VR0200), and NCI (dd7).

REFERENCES

Adelman, W. J., Jr., & Palti, Y. (1969). The effects of external potassium and long duration voltage conditioning on the amplitude of sodium currents in the giant axon of the squid, *Loligo pealei*. *Journal of General Physiology, 54*(5), 589–606.

Allen, T. W., Andersen, O. S., & Roux, B. (2004). On the importance of atomic fluctuations, protein flexibility, and solvent in ion permeation. *Journal of General Physiology, 124*(6), 679–690.

Allen, T. W., & Chung, S. H. (2001). Brownian dynamics study of an open-state KcsA potassium channel. *Biochimica et Biophysica Acta, 1515*(2), 83–91.

Amaral, C., et al. (2012). Exploring conformational states of the bacterial voltage-gated sodium channel Na_VAb via molecular dynamics simulations. *Proceedings of the National Academy of Sciences of the United States of America, 109*(52), 21336–21341.

Andersen, O. S. (2011). Perspectives on: ion selectivity. *Journal of General Physiology, 137*(5), 393–395.

Ashcroft, F. (2000). *Ion channels and disease*. UK: University of Oxford (Elsiever).

Bagneris, C., et al. (2014). Prokaryotic Na_VMs channel as a structural and functional model for eukaryotic sodium channel antagonism. *Proceedings of the National Academy of Sciences of the United States of America, 111*(23), 8428–8433.

Ballesteros, J. A., et al. (2000). Serine and threonine residues bend α-helices in the $\chi_1 = g$ conformation. *Biophysical Journal, 79*(5), 2754–2760.

Balser, J. R., et al. (1996a). External pore residue mediates slow inactivation in μ1 rat skeletal muscle sodium channels. *Journal of Physiology, 494*(Pt 2), 431–442.

Balser, J. R., et al. (1996b). Local anesthetics as effectors of allosteric gating. Lidocaine effects on inactivation-deficient rat skeletal muscle Na channels. *Journal of Clinical Investigation, 98*(12), 2874–2886.

Balser, J. R., et al. (1996c). Functional consequences of lidocaine binding to slow-inactivated sodium channels. *Journal of General Physiology, 107*(5), 643–658.

Barber, A. F., et al. (2014). Modulation of a voltage-gated Na^+ channel by sevoflurane involves multiple sites and distinct mechanisms. *Proceedings of the National Academy of Sciences of the United States of America, 111*(18), 6726–6731.

Bean, B. P., Cohen, C. J., & Tsien, R. W. (1983). Lidocaine block of cardiac sodium channels. *Journal of General Physiology, 81*(5), 613–642.

Bennett, P. B., et al. (1995). On the molecular nature of the lidocaine receptor of cardiac Na^+ channels. Modification of block by alterations in the α-subunit III-IV interdomain. *Circulation Research, 77*(3), 584–592.

Benz, I., et al. (1997). Two types of modified cardiac Na^+ channels after cytosolic interventions at the α-subunit capable of removing Na^+ inactivation. *European Biophysics Journal: EBJ, 25*(3), 189–200.

Berneche, S., & Roux, B. (2001). Energetics of ion conduction through the K^+ channel. *Nature, 414*, 73–77.

Bezanilla, F. (2000). The voltage sensor in voltage-dependent ion channels. *Physiological Reviews, 80*(2), 555–592.

Bezanilla, F., & Armstrong, C. M. (1972). Negative conductance caused by entry of sodium and cesium ions into the potassium channels of squid axons. *Journal of General Physiology, 60*(5), 588–608.

Boda, D., et al. (2007). Combined effect of pore radius and protein dielectric coefficient on the selectivity of a calcium channel. *Physical Review Letters, 98*(16), 168102.

Boda, D., et al. (2009). Ionic selectivity in L-type calcium channels by electrostatics and hard-core repulsion. *Journal of General Physiology, 133*(5), 497–509.

Boiteux, C., Vorobyov, I., French, R. J., French, C., Yarov-Yarovoy, V., & Allen, T. W. (2014). Local anesthetic and antiepileptic drug access and binding to a bacterial voltage-gated sodium channel. *Proceedings of the National Academy of Sciences of the United States of America, 111*(36), 13057–13062.

Boiteux, C., Vorobyov, I., & Allen, T. W. (2014). Ion conduction and conformational flexibility of a bacterial voltage-gated sodium channel. *Proceedings of the National Academy of Sciences of the United States of America, 111*(9), 3454–3459.

Cahalan, M. D. (1978). Local anesthetic block of sodium channels in normal and pronase-treated squid giant axons. *Biophysical Journal, 23*(2), 285–311.

Cannon, S. C., Brown, R. H., Jr., & Corey, D. P. (1991). A sodium channel defect in hyperkalemic periodic paralysis: potassium-induced failure of inactivation. *Neuron, 6*(4), 619–626.

Cannon, S. C., & Strittmatter, S. M. (1993). Functional expression of sodium channel mutations identified in families with periodic paralysis. *Neuron, 10*(2), 317–326.

Capes, D. L., et al. (2012). Gating transitions in the selectivity filter region of a sodium channel are coupled to the domain IV voltage sensor. *Proceedings of the National Academy of Sciences of the United States of America, 109*(7), 2648–2653.

Carboni, M., et al. (2005). Slow sodium channel inactivation and use-dependent block modulated by the same domain IV S6 residue. *Journal of Membrane Biology, 207*(2), 107–117.

Catterall, W. A. (2010). Ion channel voltage sensors: structure, function, and pathophysiology. *Neuron, 67*(6), 915–928.

Cestele, S., & Catterall, W. A. (2000). Molecular mechanisms of neurotoxin action on voltage-gated sodium channels. *Biochimie, 82*(9–10), 883–892.

Chakrabarti, N., et al. (2013). Catalysis of Na^+ permeation in the bacterial sodium channel Na_VAb. *Proceedings of the National Academy of Sciences of the United States of America, 110*(28), 11331–11336.

Charalambous, K., & Wallace, B. A. (2011). NaChBac: the long lost sodium channel ancestor. *Biochemistry, 50*(32), 6742–6752.

Chen, Z., et al. (2000). Lidocaine induces a slow inactivated state in rat skeletal muscle sodium channels. *Journal of Physiology, 524*(Pt 1), 37–49.

Chiamvimonvat, N., et al. (1996). Depth asymmetries of the pore-lining segments of the Na^+ channel revealed by cysteine mutagenesis. *Neuron, 16*(5), 1037–1047.

Chiu, S. Y., et al. (1979). A quantitative description of membrane currents in rabbit myelinated nerve. *Journal of Physiology, 292*, 149–166.

Clare, J. J. (2010). Targeting voltage-gated sodium channels for pain therapy. *Expert Opinion on Investigational Drugs, 19*(1), 45–62.

Corry, B., et al. (2001). Mechanisms of permeation and selectivity in calcium channels. *Biophysical Journal, 80*(1), 195–214.

Corry, B., & Thomas, M. (2012). Mechanism of ion permeation and selectivity in a voltage gated sodium channel. *Journal of the American Chemical Society, 134*(3), 1840–1846.

Cuello, L. G., et al. (2010). Structural mechanism of C-type inactivation in K^+ channels. *Nature, 466*(7303), 203–208.

Deupi, X., et al. (2004). Ser and Thr residues modulate the conformation of Pro-kinked transmembrane α-helices. *Biophysical Journal, 86*(1 Pt 1), 105–115.

Doyle, D. A., et al. (1998). The structure of the potassium channel: molecular basis of K^+ conduction and selectivity. *Science, 280*(5360), 69–77.

Eisenman, G. (1962). Cation selective glass electrodes and their mode of operation. *Biophysical Journal, 2*(2 Pt 2), 259–323.

Ellerkmann, R. K., et al. (2001). Slow recovery from inactivation regulates the availability of voltage-dependent Na^+ channels in hippocampal granule cells, hilar neurons and basket cells. *Journal of Physiology, 532*(2), 385–397.

England, S., & de Groot, M. J. (2009). Subtype-selective targeting of voltage-gated sodium channels. *British Journal of Pharmacology, 158*(6), 1413–1425.

Errington, A. C., et al. (2008). The investigational anticonvulsant lacosamide selectively enhances slow inactivation of voltage-gated sodium channels. *Molecular Pharmacology, 73*(1), 157–169.

Favre, I., Moczydlowski, E., & Schild, L. (1996). On the structural basis for ionic selectivity among Na^+, K^+, and Ca^{2+} in the voltage-gated sodium channel. *Biophysical Journal, 71*(6), 3110–3125.

Finol-Urdaneta, R. K., et al. (2013). Polymodal, high affinity actions of mu-conotoxins on a bacterial voltage-gated sodium channel. *Biophysical Journal, 104*(2), 136a–137a.

Finol-Urdaneta, R. K., et al. (2014). Sodium channel selectivity and conduction: prokaryotes have devised their own molecular strategy. *Journal of General Physiology, 143*(2), 157–171.

Fozzard, H. A., Lee, P. J., & Lipkind, G. M. (2005). Mechanism of local anesthetic drug action on voltage-gated sodium channels. *Current Pharmaceutical Design, 11*(21), 2671–2686.

Fozzard, H. A., Sheets, M. F., & Hanck, D. A. (2011). The sodium channel as a target for local anesthetic drugs. *Frontiers in Pharmacology, 2*, 68.

Freddolino, P. L., et al. (2006). Molecular dynamics simulations of the complete satellite tobacco mosaic virus. *Structure, 14*(3), 437–449.

Furini, S., & Domene, C. (2012). On conduction in a bacterial sodium channel. *PLOS Computational Biology, 8*(4), e1002476.

George, A. L., Jr. (2005). Inherited disorders of voltage-gated sodium channels. *Journal of Clinical Investigation, 115*(8), 1990–1999.

Goldin, A. L. (1999). Diversity of mammalian voltage-gated sodium channels. *Annals of the New York Academy of Sciences, 868*, 38–50.

Goldin, A. L. (2003). Mechanisms of sodium channel inactivation. *Current Opinion in Neurobiology, 13*(3), 284–290.

Grant, A. O. (2001). Molecular biology of sodium channels and their role in cardiac arrhythmias. *American Journal of Medicine, 110*(4), 296–305.

Guo, J., et al. (2016). Structure of the voltage-gated two-pore channel TPC1 from *Arabidopsis thaliana*. *Nature, 531*(7593), 196–201.

Hebeisen, S., et al. (2015). Eslicarbazepine and the enhancement of slow inactivation of voltage-gated sodium channels: a comparison with carbamazepine, oxcarbazepine and lacosamide. *Neuropharmacology, 89*, 122–135.

Hilber, K., et al. (2002). Interaction between fast and ultra-slow inactivation in the voltage-gated sodium channel. Does the inactivation gate stabilize the channel structure? *Journal of Biological Chemistry, 277*(40), 37105–37115.

Hille, B. (1977a). Local anesthetics: hydrophilic and hydrophobic pathways for the drug-receptor reaction. *Journal of General Physiology, 69*(4), 497–515.

Hille, B. (1977b). The pH-dependent rate of action of local anesthetics on the node of Ranvier. *Journal of General Physiology, 69*(4), 475–496.

Hille, B. (2001). *Ionic channels of excitable membranes* (3rd ed.). Sunderland: Sinauer.

Hodgkin, A. L., & Huxley, A. F. (1952a). A quantitative description of membrane current and its application to conduction and excitation in nerve. *Journal of Physiology, 117*(4), 500–544.

Hodgkin, A. L., & Huxley, A. F. (1952b). The dual effect of membrane potential on sodium conductance in the giant axon of *Loligo*. *Journal of Physiology, 116*(4), 497–506.

Hodgkin, A. L., & Keynes, R. D. (1955). The potassium permeability of a giant nerve fibre. *Journal of Physiology, 128*(1), 61–88.

Hondeghem, L. M., & Katzung, B. G. (1977). Time- and voltage-dependent interactions of antiarrhythmic drugs with cardiac sodium channels. *Biochimica et Biophysica Acta (BBA) — Reviews on Biomembranes, 472*(3–4), 373–398.

Hoshi, T., Zagotta, W. N., & Aldrich, R. W. (1991). Two types of inactivation in Shaker K^+ channels: effects of alterations in the carboxy-terminal region. *Neuron,* 7(4), 547–556.

Ito, M., et al. (2004). The voltage-gated Na^+ channel Na_vBP has a role in motility, chemotaxis, and pH homeostasis of an alkaliphilic *Bacillus*. *Proceedings of the National Academy of Sciences of the United States of America, 101*(29), 10566–10571.

Jensen, M. O., et al. (2010). Principles of conduction and hydrophobic gating in K^+ channels. *Proceedings of the National Academy of Sciences of the United States of America, 107*(13), 5833–5838.

Jensen, M. O., et al. (2012). Mechanism of voltage gating in potassium channels. *Science, 336*(6078), 229–233.

Kaczorowski, G. J., et al. (2008). Ion channels as drug targets: the next GPCRs. *Journal of General Physiology, 131*(5), 399–405.

Kambouris, N. G., et al. (1998). Mechanistic link between lidocaine block and inactivation probed by outer pore mutations in the rat micro1 skeletal muscle sodium channel. *Journal of Physiology, 512*(Pt 3), 693–705.

Karplus, M., & McCammon, J. A. (2002). Molecular dynamics simulations of biomolecules. *Nature Structural Biology, 9*(9), 646–652.

Kellenberger, S., Scheuer, T., & Catterall, W. A. (1996). Movement of the Na^+ channel inactivation gate during inactivation. *Journal of Biological Chemistry, 271*(48), 30971–30979.

Kim, I., & Allen, T. W. (2011). On the selective ion binding hypothesis for potassium channels. *Proceedings of the National Academy of Sciences of the United States of America, 108*(44), 17963–17968.

Koishi, R., et al. (2004). A superfamily of voltage-gated sodium channels in bacteria. *Journal of Biological Chemistry, 279*(10), 9532–9538.

Kuo, C. C., & Bean, B. P. (1994). Slow binding of phenytoin to inactivated sodium channels in rat hippocampal neurons. *Molecular Pharmacology, 46*(4), 716–725.

Kyle, D. J., & Ilyin, V. I. (2007). Sodium channel blockers. *Journal of Medicinal Chemistry, 50*(11), 2583–2588.

Lam, K., Wu, Z., & Schulten, K. (2016). Computational modeling of sodium channel inactivation. *Biophysical Journal, 110*(3), 108a.

Lee, S., Goodchild, S. J., & Ahern, C. A. (2012). Local anesthetic inhibition of a bacterial sodium channel. *Journal of General Physiology, 139*(6), 507–516.

Li, H. L., et al. (1999). A molecular basis for the different local anesthetic affinities of resting versus open and inactivated states of the sodium channel. *Molecular Pharmacology, 55*(1), 134–141.

Li, P., et al. (2013). The gating charge pathway of an epilepsy-associated potassium channel accommodates chemical ligands. *Cell Research, 23*(9), 1106–1118.

Li, Q., et al. (2014). Structural mechanism of voltage-dependent gating in an isolated voltage-sensing domain. *Nature Structural & Molecular Biology, 21*(3), 244–252.

Lossin, C., et al. (2003). Epilepsy-associated dysfunction in the voltage-gated neuronal sodium channel SCN1A. *Journal of Neuroscience: The Official Journal of the Society for Neuroscience, 23*(36), 11289–11295.

Martin, L. J., & Corry, B. (2014). Locating the route of entry and binding sites of benzocaine and phenytoin in a bacterial voltage gated sodium channel. *PLOS Computational Biology, 10*(7), e1003688.

McArthur, J. R., et al. (2011). Interactions of key charged residues contributing to selective block of neuronal sodium channels by mu-conotoxin KIIIA. *Molecular Pharmacology, 80*(4), 573–584.

McCusker, E. C., et al. (2012). Structure of a bacterial voltage-gated sodium channel pore reveals mechanisms of opening and closing. *Nature Communications, 3*, 1102.

McGregor, M. J., Islam, S. A., & Sternberg, M. J. (1987). Analysis of the relationship between side-chain conformation and secondary structure in globular proteins. *Journal of Molecular Biology, 198*(2), 295–310.

McNulty, M. M., et al. (2007). Charge at the lidocaine binding site residue Phe-1759 affects permeation in human cardiac voltage-gated sodium channels. *Journal of Physiology, 581*(Pt 2), 741–755.

McPhee, J. C., et al. (1998). A critical role for the S4-S5 intracellular loop in domain IV of the sodium channel α-subunit in fast inactivation. *Journal of Biological Chemistry, 273*(2), 1121–1129.

Mike, A., & Lukacs, P. (2010). The enigmatic drug binding site for sodium channel inhibitors. *Current Molecular Pharmacology, 3*(3), 129–144.

Muroi, Y., & Chanda, B. (2009). Local anesthetics disrupt energetic coupling between the voltage-sensing segments of a sodium channel. *Journal of General Physiology, 133*(1), 1–15.

Narahashi, T. (1964). Restoration of action potential by anodal polarization in lobster giant axons. *Journal of Cellular Physiology, 64*(1), 73–96.

Neyton, J., & Miller, C. (1988). Discrete Ba^{2+} block as a probe of ion occupancy and pore structure in the high-conductance Ca^{2+}-activated K^{+} channel. *Journal of General Physiology, 92*(5), 569–586.

Noskov, S. Y., Berneche, S., & Roux, B. (2004). Control of ion selectivity in potassium channels by electrostatic and dynamic properties of carbonyl ligands. *Nature, 431*(7010), 830–834.

O'Leary, M. E. (1998). Characterization of the isoform-specific differences in the gating of neuronal and muscle sodium channels. *Canadian Journal of Physiology and Pharmacology, 76*(10–11), 1041–1050.

Ong, B.-H., Tomaselli, G. F., & Balser, J. R. (2000). A structural rearrangement in the sodium channel pore linked to slow inactivation and use dependence. *Journal of General Physiology, 116*(5), 653–662.

O'Reilly, J. P., Wang, S. Y., & Wang, G. K. (2001). Residue-specific effects on slow inactivation at V787 in D2-S6 of $Na_V1.4$ sodium channels. *Biophysical Journal, 81*(4), 2100–2111.

Overington, J. P., Al-Lazikani, B., & Hopkins, A. L. (2006). How many drug targets are there? *Nature Reviews. Drug Discovery, 5*(12), 993–996.

Papazian, D. M., et al. (1991). Alteration of voltage-dependence of Shaker potassium channel by mutations in the S4 sequence. *Nature, 349*(6307), 305–310.

Pavlov, E., et al. (2003). Block of a prokaryotic voltage-gated sodium channel (NaChBac) by phenytoin and lidocaine. *Biophysical Journal, 84*.

Pavlov, E., et al. (2005). The pore, not cytoplasmic domains, underlies inactivation in a prokaryotic sodium channel. *Biophysical Journal, 89*(1), 232–242.

Payandeh, J., et al. (2011). The crystal structure of a voltage-gated sodium channel. *Nature, 475*(7356), 353–358.

Payandeh, J., et al. (2012). Crystal structure of a voltage-gated sodium channel in two potentially inactivated states. *Nature, 486*(7401), 135–139.

Payandeh, J., & Minor, D. L., Jr. (2015). Bacterial voltage-gated sodium channels (BacNas) from the soil, sea, and salt lakes enlighten molecular mechanisms of electrical signaling and pharmacology in the brain and heart. *Journal of Molecular Biology, 427*(1), 3–30.

Penzotti, J. L., et al. (1998). Differences in saxitoxin and tetrodotoxin binding revealed by mutagenesis of the Na^+ channel outer vestibule. *Biophysical Journal, 75*(6), 2647–2657.
Qiu, H., Shen, R., & Guo, W. (2012). Ion solvation and structural stability in a sodium channel investigated by molecular dynamics calculations. *Biochimica et Biophysica Acta, 1818*(11), 2529–2535.
Qu, Y., et al. (1995). Molecular determinants of drug access to the receptor site for antiarrhythmic drugs in the cardiac Na^+ channel. *Proceedings of the National Academy of Sciences of the United States of America, 92*(25), 11839–11843.
Quandt, F. N. (1988). Modification of slow inactivation of single sodium channels by phenytoin in neuroblastoma cells. *Molecular Pharmacology, 34*(4), 557–565.
Ragsdale, D. S., et al. (1994). Molecular determinants of state-dependent block of Na^+ channels by local anesthetics. *Science, 265*(5179), 1724–1728.
Ragsdale, D. S., et al. (1996). Common molecular determinants of local anesthetic, antiarrhythmic, and anticonvulsant block of voltage-gated Na^+ channels. *Proceedings of the National Academy of Sciences of the United States of America, 93*(17), 9270–9275.
Raju, S. G., et al. (2013). Exploring volatile general anesthetic binding to a closed membrane-bound bacterial voltage-gated sodium channel via computation. *PLoS Computational Biology, 9*(6), e1003090.
Ren, D., et al. (2001). A prokaryotic voltage-gated sodium channel. *Science, 294*(5550), 2372–2375.
Rhodes, T. H., et al. (2004). Noninactivating voltage-gated sodium channels in severe myoclonic epilepsy of infancy. *Proceedings of the National Academy of Sciences of the United States of America, 101*(30), 11147–11152.
Rogawski, M. A., & Loscher, W. (2004). The neurobiology of antiepileptic drugs for the treatment of nonepileptic conditions. *Nature Medicine, 10*(7), 685–692.
Rohl, C. A., et al. (1999). Solution structure of the sodium channel inactivation gate. *Biochemistry, 38*(3), 855–861.
Sato, C., et al. (2001). The voltage-sensitive sodium channel is a bell-shaped molecule with several cavities. *Nature, 409*(6823), 1047–1051.
Schlief, T., et al. (1996). Pore properties of rat brain II sodium channels mutated in the selectivity filter domain. *European Biophysics Journal: EBJ, 25*(2), 75–91.
Shaya, D., et al. (2014). Structure of a prokaryotic sodium channel pore reveals essential gating elements and an outer ion binding site common to eukaryotic channels. *Journal of Molecular Biology, 426*(2), 467–483.
Sheets, M. F., et al. (2010). Sodium channel molecular conformations and antiarrhythmic drug affinity. *Trends in Cardiovascular Medicine, 20*(1), 16–21.
Sheets, P. L., et al. (2008). Differential block of sensory neuronal voltage-gated sodium channels by lacosamide [(2R)-2-(acetylamino)-N-benzyl-3-methoxypropanamide], lidocaine, and carbamazepine. *Journal of Pharmacology and Experimental Therapeutics, 326*(1), 89–99.
Sheets, P. L., Jarecki, B. W., & Cummins, T. R. (2011). Lidocaine reduces the transition to slow inactivation in $Na_V1.7$ voltage-gated sodium channels. *British Journal of Pharmacology, 164*(2b), 719–730.
Smart, O. S., Neduvelil, J. G., Wang, X., Wallace, B. A., & Sansom, M. S. P. (1996). HOLE: a program for the analysis of the pore dimensions of ion channel structural models. *Journal of Molecular Graphics, 14*(6), 354–360.
Stuhmer, W., et al. (1989). Structural parts involved in activation and inactivation of the sodium channel. *Nature, 339*(6226), 597–603.
Sun, Y. M., et al. (1997). On the structural basis for size-selective permeation of organic cations through the voltage-gated sodium channel. Effect of alanine mutations at the DEKA locus on selectivity, inhibition by Ca^{2+} and H^+, and molecular sieving. *Journal of General Physiology, 110*(6), 693–715.

Takeshita, K., et al. (2014). X-ray crystal structure of voltage-gated proton channel. *Nature Structural & Molecular Biology, 21*(4), 352–357.

Tang, L., et al. (2014). Structural basis for Ca^{2+} selectivity of a voltage-gated calcium channel. *Nature, 505*(7481), 56–61.

Tarnawa, I., Bolcskei, H., & Kocsis, P. (2007). Blockers of voltage-gated sodium channels for the treatment of central nervous system diseases. *Recent Patents on CNS Drug Discovery, 2*(1), 57–78.

Terlau, H., et al. (1991). Mapping the site of block by tetrodotoxin and saxitoxin of sodium channel II. *FEBS Letters, 293*(1–2), 93–96.

Thompson, A. N., et al. (2009). Mechanism of potassium-channel selectivity revealed by Na^+ and Li^+ binding sites within the KcsA pore. *Nature Structural & Molecular Biology, 16*(12), 1317–1324.

Tomaselli, G. F., et al. (1995). A mutation in the pore of the sodium channel alters gating. *Biophysical Journal, 68*(5), 1814–1827.

Tsushima, R. G., Li, R. A., & Backx, P. H. (1997). Altered ionic selectivity of the sodium channel revealed by cysteine mutations within the pore. *Journal of General Physiology, 109*(4), 463–475.

Tyson, J. R., & Snutch, T. P. (2013). Molecular nature of voltage-gated calcium channels: structure and species comparison. *Wiley Interdisciplinary Reviews: Membrane Transport and Signaling, 2*(5), 181–206.

Ulbricht, W. (2005). Sodium channel inactivation: molecular determinants and modulation. *Physiological Reviews, 85*(4), 1271–1301.

Ulmschneider, M. B., et al. (2013). Molecular dynamics of ion transport through the open conformation of a bacterial voltage-gated sodium channel. *Proceedings of the National Academy of Sciences of the United States of America, 110*(16), 6364–6369.

Vargas, E., Bezanilla, F., & Roux, B. (2011). In search of a consensus model of the resting state of a voltage-sensing domain. *Neuron, 72*(5), 713–720.

Vassilev, P. M., Scheuer, T., & Catterall, W. A. (1988). Identification of an intracellular peptide segment involved in sodium channel inactivation. *Science, 241*(4873), 1658–1661.

Vedantham, V., & Cannon, S. C. (2000). Rapid and slow voltage-dependent conformational changes in segment IVS6 of voltage-gated Na^+ channels. *Biophysical Journal, 78*(6), 2943–2958.

Vilin, Y. Y., & Ruben, P. C. (2001). Slow inactivation in voltage-gated sodium channels: molecular substrates and contributions to channelopathies. *Cell Biochemistry and Biophysics, 35*(2), 171–190.

Wang, G. K., et al. (1987). Inhibition of sodium currents by local anesthetics in chloramine-T-treated squid axons. The role of channel activation. *Journal of General Physiology, 89*(4), 645–667.

Wang, G. K., Quan, C., & Wang, S. (1998). A common local anesthetic receptor for benzocaine and etidocaine in voltage-gated mu1 Na^+ channels. *Pflugers Archiv, 435*(2), 293–302.

Wang, Q., et al. (1995). Cardiac sodium channel mutations in patients with long QT syndrome, an inherited cardiac arrhythmia. *Human Molecular Genetics, 4*(9), 1603–1607.

Wang, S. Y., & Wang, G. K. (1997). A mutation in segment I-S6 alters slow inactivation of sodium channels. *Biophysical Journal, 72*(4), 1633–1640.

West, J. W., et al. (1992). A cluster of hydrophobic amino acid residues required for fast Na^+-channel inactivation. *Proceedings of the National Academy of Sciences of the United States of America, 89*(22), 10910–10914.

Woronzow. (1924). Über die Einwirkung des konstanten Stromes auf den mit Wasser, Zuckerlosung, Alkali-und Erdalkalichloridlosungen behandelten Nerven. *Pftug. Arch. ges. Physiol, 203*, 300–318.

Xia, M., et al. (2013). The mechanism of Na^+/K^+ selectivity in mammalian voltage-gated sodium channels based on molecular dynamics simulation. *Biophysical Journal, 104*(11), 2401–2409.

Yarov-Yarovoy, V., et al. (2001). Molecular determinants of voltage-dependent gating and binding of pore-blocking drugs in transmembrane segment IIIS6 of the Na^+ channel α subunit. *Journal of Biological Chemistry, 276*(1), 20–27.

Yarov-Yarovoy, V., et al. (2002). Role of amino acid residues in transmembrane segments IS6 and IIS6 of the Na^+ channel α subunit in voltage-dependent gating and drug block. *Journal of Biological Chemistry, 277*(38), 35393–35401.

Yarov-Yarovoy, V., et al. (2012). Structural basis for gating charge movement in the voltage sensor of a sodium channel. *Proceedings of the National Academy of Sciences of the United States of America, 109*(2), E93–E102.

Yoo, J., Wilson, J., & Aksimentiev, A. (2016). Improved model of hydrated calcium ion for molecular dynamics simulations using classical biomolecular force fields. *Biopolymers.* http://dx.doi.org/10.1002/bip.22868 [Epub ahead of print].

Yue, L., et al. (2002). The cation selectivity filter of the bacterial sodium channel, NaChBac. *Journal of General Physiology, 120*(6), 845–853.

Zarrabi, T., et al. (2010). A molecular switch between the outer and the inner vestibules of the voltage-gated Na^+ channel. *Journal of Biological Chemistry, 285*(50), 39458–39470.

Zeng, Z., & French, C. (2011a). The modulation of sodium conductance by phenytoin is mediated by slower rather than fast inactivation processes, and is unaffected by fast inactivation removal. In *Australian neurosciences society* (Auckland).

Zeng, Z., & French, C. (2011b). The effects of anti-epileptic drug phenytoin on slower processes of Na_V channels. In *Australian neurosciences society* (Auckland).

Zhang, X., et al. (2012). Crystal structure of an orthologue of the NaChBac voltage-gated sodium channel. *Nature, 486*(7401), 130–134.

Zhang, X., et al. (2013). Analysis of the selectivity filter of the voltage-gated sodium channel Na_VRh. *Cell Research, 23*(3), 409–422.

Zhang, Z., et al. (2003). A negatively charged residue in the outer mouth of rat sodium channel determines the gating kinetics of the channel. *American Journal of Physiology. Cell Physiology, 284*(5), C1247–C1254.

Zhao, Y., Scheuer, T., & Catterall, W. A. (2004). Reversed voltage-dependent gating of a bacterial sodium channel with proline substitutions in the S6 transmembrane segment. *Proceedings of the National Academy of Sciences of the United States of America, 101*(51), 17873–17878.

Zhu, F., & Hummer, G. (2012). Drying transition in the hydrophobic gate of the GLIC channel blocks ion conduction. *Biophysical Journal, 103*(2), 219–227.

Zuliani, V., et al. (2009). Recent advances in the medicinal chemistry of sodium channel blockers and their therapeutic potential. *Current Topics in Medicinal Chemistry, 9*(4), 396–415.

CHAPTER SEVEN

Voltage-Gated Sodium Channels: Mechanistic Insights From Atomistic Molecular Dynamics Simulations

V. Oakes*, S. Furini[§] and C. Domene*,[¶,1]

*King's College London, London, United Kingdom
[§]University of Siena, Siena, Italy
[¶]University of Oxford, Oxford, United Kingdom
[1]Corresponding author: E-mail: carmen.domene@kcl.ac.uk

Contents

Abstract

The permeation of ions and other molecules across biological membranes is an inherent requirement of all cellular organisms. Ion channels, in particular, are responsible for the conduction of charged species, hence modulating the propagation of electrical signals. Despite the universal physiological implications of this property, the molecular functioning of ion channels remains ambiguous. The combination of atomistic structural data with computational methodologies, such as molecular dynamics (MD) simulations, is now considered routine to investigate structure—function relationships in biological systems. A fuller understanding of conduction, selectivity,

Current Topics in Membranes, Volume 78
ISSN 1063-5823
http://dx.doi.org/10.1016/bs.ctm.2015.12.002

and gating, therefore, is steadily emerging due to the applicability of these techniques to ion channels. However, because their structure is known at atomic resolution, studies have consistently been biased toward K^+ channels, thus the molecular determinants of ionic selectivity, activation, and drug blockage in Na^+ channels are often overlooked. The recent increase of available crystallographic data has eminently encouraged the investigation of voltage-gated sodium (Na_V) channels via computational methods. Here, we present an overview of simulation studies that have contributed to our understanding of key principles that underlie ionic conduction and selectivity in Na^+ channels, in comparison to the K^+ channel analogs.

List of Abbreviations

PD Pore domain
SF Selectivity filter
TRP Transient receptor potential
VGIC Voltage-gated ion channel
VSD Voltage sensor domain

1. INTRODUCTION

Ion channels are an important family of integral membrane proteins that are responsible for passive transport across biological membranes. Gene sequencing has identified over 400 alleged ion channels, yet only a fragment of these have been cloned, functionally tested, and crystallized (Bagal et al., 2013). Therefore, despite their ubiquitous distribution in tissues throughout all living organisms and their decisive roles in fundamental processes such as cell proliferation, nerve transmission, muscle contraction, sensory transduction and blood pressure regulation ion channels are considerably underexploited by targeted drug therapies (Bagal et al., 2013). Furthermore, conglomerates of ailments, known as "channelopathies," are the consequence of ion channel mutations and subsequent functional alterations (Zaydman, Silva, & Cui, 2012).

The widespread physiological implications of ion channels emanate from the modulation of electrical signals across the cellular membrane. This occurs by facilitating the passive diffusion of ionic species down their electrochemical gradient between periplasmic and cytoplasmic compartments. The permeation pathway is ordinarily "gated" on the application of external stimuli; for example, adjustment of transmembrane potentials (voltage-gated ion channels (VGIC)), ligand binding (ligand-gated ion channels), and mechanical stretch (mechanosensitive channels) can induce transformations between open and closed states. A constricted region of the pore, known as the selectivity filter, imposes further restraints on conduction, by discrimination between ionic species. Selectivity, as this property is known, is responsible for K^+/Na^+ permeability ratios of 1000:1 and 1:8 in K^+ and Na^+ channels, respectively (Hille, 1992).

The understanding of such channel properties at the atomic level using computational techniques has traditionally been focused toward K^+ channels due to the availability of crystallographic data, while the principles determining conduction in Na^+ channels remain largely unsolved. The publication of the crystal structure of the Na_VAb α-subunit (Payandeh, Scheuer, Zheng, & Catterall, 2011), a mutated voltage-gated Na^+ channel from *Arcobacter butzleri* (Fig. 1A), and subsequent analogous structures from bacterial sources (Bagnéris et al., 2013; McCusker et al., 2012; Payandeh, Gamal El-Din, Scheuer, Zheng, & Catterall, 2012; Shaya et al., 2014) in the last 5 years has invigorated interest into the workings of Na^+ channels, resulting in numerous exploratory computational studies. In this chapter, we provide a comprehensive survey of the available X-ray crystal data and accompanying simulation studies that have been undertaken to discern the fundamental features regulating permeation and selectivity in Na^+ channels, with particular focus on voltage-gated sodium (Na_V) channels. The atomic structures of associated β-subunits have recently been determined (Gilchrist, Das, Van Petegem, & Bosmans, 2013; Namadurai et al., 2014), yet in this review, the focus is on the principal subunits (designated α or α1) which are responsible for the voltage gated ion conductance, the defining characteristic of these proteins.

2. VOLTAGE-GATED SODIUM CHANNELS

Na_V channels form part of the VGIC family, which contains a variety of potassium, sodium, calcium, chloride, and transient receptor potential (TRP) ion channels, of which Na_V was the first to be cloned (Noda et al., 1984). Currently, based on common structural and functional properties, with distinct expression, regulatory, and pharmacological profiles, nine members of the Na_V family have been identified (Na_V1.1–Na_V1.9).

Na_V channels are critical for the generation of action potentials in neurons and other excitable cells. Most biological membranes experience a high-Na^+/low-K^+ concentration on the extracellular surface and a low-Na^+/high-K^+ concentration on the intracellular surface, generating

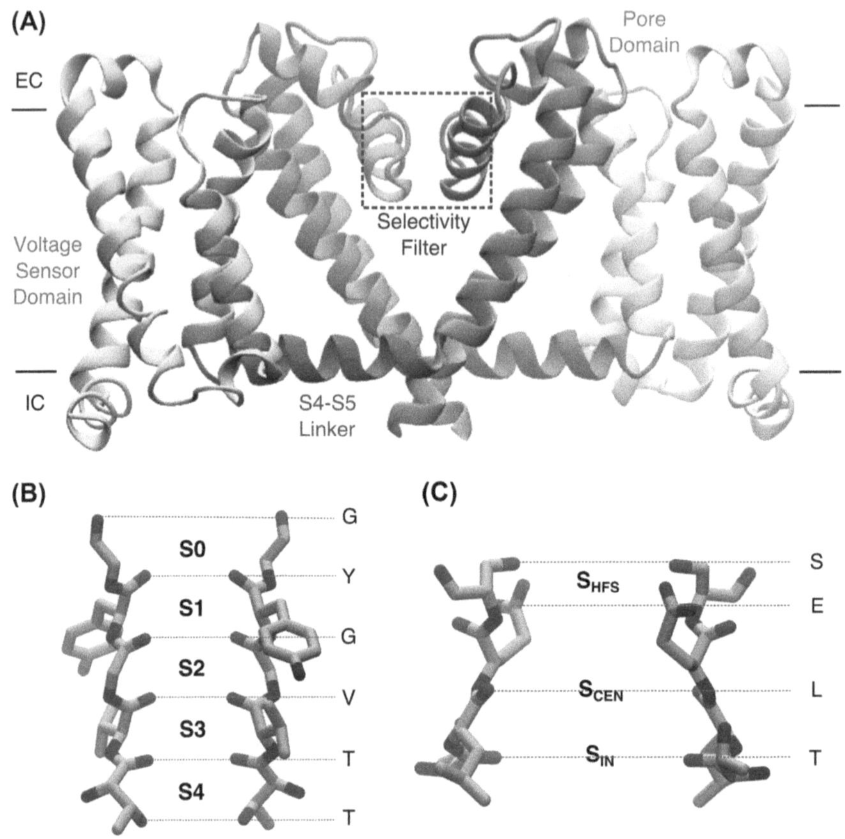

Figure 1 (A) Overall architecture of a voltage-gated ion channel as represented by the crystal structure of Na_VAb (PDB ID: 3RVY) in a closed conformation. Only two chains are shown for clarity. The voltage sensor domain (VSD) comprises transmembrane helices S1–S4, with S5 and S6 constituting the pore domain (PD) in cyan (light gray in print versions); the S4–S5 linker (in orange (gray in print versions)) is responsible for conferring conformational changes between the VSD and the PD on voltage activation. The selectivity filter is an integral part of the PD at the extracellular entrance to the pore. (B) Structure of a K^+ selective selectivity filter from the prototypical K^+ channel, KcsA (PDB ID: 1K4C). Binding sites are clearly defined by surrounding carbonyl oxygen atoms directed toward the pore axis. (C) Structure of a prototypical Na^+ selective selectivity filter from Na_VAb (PDB ID: 3RVY), with proposed binding sites labeled.

apt conditions for passive Na^+ influx and K^+ efflux in their respective channels. At typical negative resting potential (ie, −60 mV), both K_V and Na_V activation gates are closed. Under depolarizing conditions Na_V channels open, allowing diffusion of Na^+ ions through the central pore; this causes further depolarization inducing a large inward current. K_V channels respond slower than Na_V channels to depolarization; therefore, delayed K^+ efflux is observed and an overall positive membrane potential results. At the peak of depolarization, repolarization is initiated by the inactivation of Na_V and the opening of K_V channels. By the generation of a nascent electrical signal in this way, VGICs initiate action potentials in excitable cells in various tissues in the heart, brain, and nervous system.

The expression on Na_V channels is not limited to excitable cells; the role of Na_V channels in astrocytes, NG2 cells, microglia, macrophages, and cancer cells is currently under investigation (Black & Waxman, 2013). The dysfunction of Na_V channels has been affiliated with a myriad of disorders such as pain, migraine, impaired movement, epilepsy, and cardiac arrhythmias (Mantegazza, Curia, Biagini, Ragsdale, & Avoli, 2010). Currently, treatment of these disorders is serviced by the binding of small organic molecules, such as local anesthetics, to block the channels (Bagal et al., 2013). However, due to the high structural and sequence homology (Fig. 2), subtype targeting can rarely be achieved, increasing the risk of inimical side effects (England & de Groot, 2009). Therefore, understanding of the molecular determinants of conduction, gating, and drug binding in ion channels is required to develop the next generation of subtype specific pharmaceuticals.

3. STRUCTURAL BASIS OF K^+ VERSUS Na^+ CONDUCTION

Hodgkin and Keynes provided the first insights into conduction in K^+ channels, proposing a "knock-on" mechanism of ionic conduction, associated with concerted transitions of individual ions in a single-file manner through a narrow pore (Hodgkin & Keynes, 1955). In addition, various theories for selectivity were proposed concerning the hydration free energy (Eisenman, 1962; Eisenman & Horn, 1983), coordination number, and the size difference of K^+ and Na^+ ions (Bezanilla & Armstrong, 1972). Furthermore, sequence analysis and complementary mutagenesis studies of K^+ channels identified a conserved sequence, TVGYG, which confers ionic selectivity, hence coined the selectivity filter (Heginbotham, Lu, Abramson, & MacKinnon, 1994; Smart, Goodfellow, & Wallace, 1993). The X-ray crystal structure of KcsA, from *Streptomyces lividans*, provided the first atomistic structural information, revealing key insights into the conduction/selectivity mechanisms of K^+ channels (Doyle et al., 1998). Developments concerning expression, purification, and crystallization of ion

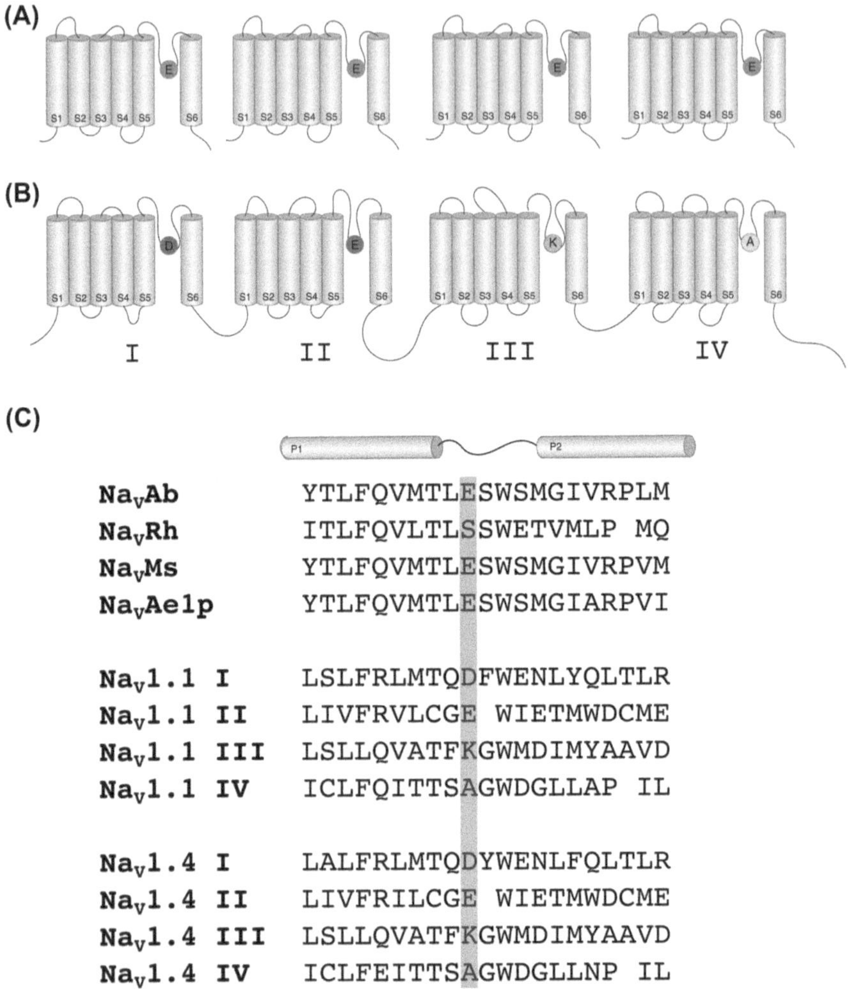

Figure 2 Topology and sequence data of voltage-gated sodium channels from prokaryotic and eukaryotic origins. (A) General representation of Na_V channel architecture from bacterial channels: four homologous subunits comprising voltage-sensor (S1–S4) and pore-forming (S5–S6) domains, containing the signature EEEE sequence, form the channel. (B) Eukaryotic channels share a similar tetrameric architecture, yet in a continuous polypeptide chain, with nonconserved sequences in individual domains. In this case, DEKA constitutes the characteristic SF ring, as illustrated in (C). Sequence alignments of the selectivity filter and surrounding pore domain in select bacterial (Na_VAb, Na_VRh, Na_VMs, NaVAe1p) and mammalian (Na_V1.1, Na_V1.4) channels; the signature SF sequence is highlighted in orange (light gray in print versions) (Payandeh & Minor, 2015).

channels have now lead to a wealth of available K^+ channel crystal structures (Brohawn, del Marmol, & MacKinnon, 2012; Doyle et al., 1998; Jiang et al., 2003; Kuo et al., 2003; Long, Campbell, & Mackinnon, 2005; Miller & Long, 2012). K^+ channels can exist in numerous activation states, most simply defined as open, closed, and inactivated. A range of these is sampled throughout these structures, providing additional clues into the gating mechanism of K^+ channels.

Throughout K^+ channels, a fundamental pore structure is conserved. The channel is formed from a symmetrical arrangement of four identical subunits composed of inner and outer transmembrane α-helices connected by a pore loop. The latter contains a pore helix and the signature selectivity sequence, TVGYG. A water-filled cavity, directly connected to the cytoplasm, is present underneath the selectivity filter to overcome large energetic barriers accompanying penetration of the hydrophobic core of the surrounding membrane by ionic species (Parsegian, 1969). In the open state of the channel, passage of ions is unrestricted from the cavity to the cytoplasm. In the closed state, the inner helices coalesce, via bending at a conserved glycine residue, preventing ion translocation and intra- and extracellular water exchange. A wide range of external stimuli modulates cytoplasmic gating of this sort, interceded by supplementary structural domains in some cases.

In K_V channels, an additional four transmembrane segments in each subunit comprise the voltage sensor domain (VSD), which regulates channel opening. The molecular basis of voltage sensing and gating was elucidated by the X-ray crystal structure of K_VAP from *Aeropyrum pernix* at 3.2 Å resolution with the isolated VSD at 1.9 Å (Jiang et al., 2003). The VSD segments, denoted S1−S4, arrange in an antiparallel α-helical bundle, attached to the "traditional" pore helices, S5−S6. Between four and seven repeat motifs containing a positively charged residue, ordinarily arginine, succeeded by two hydrophobic residues, constitutes the conserved S4 VSD segment that is deemed the activation gate (Stuhmer et al., 1989) The assemblages of charge clusters in this region stabilize S4 within the membrane and facilitate channel gating by interactions with S1−S3 (Long, Tao, Campbell, & MacKinnon, 2007). When stimulated by a change in the transmembrane voltage, movement of S4 prompts rearrangement of the pore-lining helices, via the S4−S5 linker, thus controlling the passage of ions.

Ion permeation, of ionic species other than K^+, is blocked at the selectivity filter. The KcsA structure revealed a selectivity filter optimized for favorable binding and efficient conduction of K^+ ions. In physiological K^+

concentrations, the selectivity filter residues arrange to form rings of carbonyl oxygen atoms directed toward the center of the pore axis (Fig. 1B). With the added contribution of threonine side-chain oxygen atoms, four adjoining sites are formed, namely S1—S4, which are able to bind dehydrated ions in a cage-like structure. Further binding sites for "partially" hydrated K^+ have been identified at the extracellular entrance (S0, S_{EXT}) and within the central cavity (S_C). Stabilization of ions in S_C is a consequence of the helix-dipole effect from the above pore helices (Aqvist, Luecke, Quiocho, & Warshel, 1991; Furini, Zerbetto, & Cavalcanti, 2007; Jogini & Roux, 2005). Specific salt bridges and hydrogen bonding with proximal helices and solvent generally stabilize the selectivity filter (Cordero-Morales, Jogini, Chakrapani, & Perozo, 2011). However, a further KcsA structure, in low K^+/high Na^+ concentrations, displayed significant distortions in the selectivity filter rendering the channel impermeable to K^+, Na^+, and even water (Zhou, Morais-Cabral, Kaufman, & MacKinnon, 2001). This structure has been associated with an inactivated state, reminiscent of C-type inactivation, a mechanism of gating characterized by conformational changes in the proximity of the selectivity filter which prevent conduction (Cuello, Jogini, Cortes, & Perozo, 2010).

In contrast to K_V channels, the molecular structure of eukaryotic Na_V channels remains uncharacterized. Crystallographic data of bacterial Na_V channels, of which the first structure, (mutant) Na_VAb, was published in 2011 (Payandeh et al., 2011) with numerous full length (Na_VRh, wild-type Na_VAb) (Bagnéris et al., 2013; Payandeh et al., 2012; Zhang et al., 2012) and pore-only structures to follow (Na_VMs, Na_VAe1p) (McCusker et al., 2012; Shaya et al., 2014), have provided, to date, the greatest molecular insights of Na_V channels at the atomic level.

Bacterial Na_V channels display a conserved structure similar to that of K^+ channels (Fig. 1A). As noted previously, four identical subunits arranged around a large central cavity that can accommodate and stabilize hydrated ions. The P-loop region contains a similar α-helix, with the addition of a unique second pore helix in an antiparallel configuration (Fig. 1A). The available structures display remarkable similarities with Cα superposition of monomer subunits comfortably under 1 Å in most cases (Payandeh & Minor, 2015), and only subtle differences in the symmetry of the pore helices with respect to the pore axis (Furini & Domene, 2013).

A novel finding in Na_VAb pore architecture is the presence of lateral openings, termed fenestrations, directly connecting the central pore to the surrounding membrane. In Na_VAb, membrane phospholipids penetrate these cavities, providing a further physical block to conduction in the pore

(Payandeh et al., 2011). Furthermore, these fenestrations may provide an additional entry route for small neutral or hydrophobic molecules and will therefore have implications for targeted channel blockage. Analogous fenestrations have recently been identified in multiple members of the two-pore domain K^+ channel (K_{2P}) family (Brohawn et al., 2012; Miller & Long, 2012).

The Na_V VSDs are analogous to those found in K_v channels. In Na_VAb, interactions of S4 with conserved intracellular and extracellular charge clusters suggest that the VSD is in an activated conformation (Payandeh et al., 2011). Therefore, Na_VAb is in a supposed "preopen" state in which the VSD is in an active state yet the cytoplasmic entrance is closed. Subtle differences in the conformational, and hence activation, state are observed throughout the available Na_V VSD structures; perspectives on voltage-sensing mechanisms as a consequence of these observations are outlined in a recent review by Payandeh and Minor (2015).

The resolved selectivity filter structure has also provided significant contributions to the understanding of Na^+ conduction. The sequence and structure is clearly divergent from K^+ channels, being notably wider and shorter, and comprised of four amino acids whose side chains partially line the pore (Fig. 1C). An arrangement of four glutamates, the so-called EEEE ring, forms the mouth of the selectivity filter in Na_VAb, consistent with experimental studies that implicated an anionic region of high-field strength (Hille, 1971, 1972) composed of Glu side chains (Heinemann, Terlau, Stuhmer, Imoto, & Numa, 1992, Sun, Favre, Schild, & Moczydlowski, 1997) in the selective conduction of Na^+ over K^+. The equivalent sequence in eukaryotic sodium channels, the DEKA motif, has also been affiliated with channel selectivity by mutagenesis studies (Favre, Moczydlowski, & Schild, 1996; Sun et al., 1997). Sequence alignments of the SF and surrounding PD in select bacterial (Na_VAb, Na_VRh, Na_VMs, Na_VAc1p) and mammalian (Na_V1.1, Na_V1.4) channels are shown in Fig. 2.

Na_VAb (with mutations I217C and M221C) and Na_VMs displayed symmetrical selectivity filter structures as above and are therefore considered conductive states. However, Na_VRh (Zhang et al., 2012) contains a dissimilar selectivity filter sequence, TLSSWF substituting TLESWS; therefore, the outer ring is comprised of serine residues, which adopt nonuniform conformations resulting in a 1–2 Å reduction in selectivity filter radii. Notable asymmetry is also observed in wild-type Na_VAb. Therefore, both are thought to correspond to inactivated states.

Despite the inability to resolve the location of Na^+ ions within the selectivity filter, on the basis of structural data three binding sites were proposed in Na_VAb. Firstly, the side chain of Glu177 (S_{HFS}) provides a negatively charged region at the extracellular entrance that can feasibly attract positively charged ions. Deeper sites formed of rings of carbonyl atoms from Leu76 (S_{CEN}) and Thr175 (S_{IN}) are proposed to form the remainder of the permeation pathway, with the latter sites potentially binding Na^+ ions via hydrogen bonding with a square array of water molecules forming the ions solvation shell. All binding sites are, therefore, amenable to hydrated ions, suggesting that in stark contrast to K^+ channels, sodium ions can traverse the entire length of the channel without full desolvation.

It is worth noting that all published structures to date are of prokaryotic origin; eukaryotic Na_V channels are comprised of a single polypeptide chain containing over ~2000 amino acids, arranged into four homologous transmembrane domains. A schematic representation of the basic topology of Na_V channels from prokaryotic and eukaryotic origins is presented in Fig. 2. The general representation of Na_V channel architecture from bacterial channels shows four homologous subunits comprising voltage-sensor (S1–S4) and pore-forming (S5–S6) domains, containing the signature EEEE sequence (Fig. 2). Further differences may be observed in eukaryotic species where the equivalent selectivity filter ring sequence is DEKA in contrast with EEEE (Na_VAb) and SSSS (Na_VRh). Note that eukaryotic channels share a similar tetrameric architecture, yet in a continuous polypeptide chain, with nonconserved sequences in individual domains. Therefore, crystallization of eukaryotic channels will markedly aid the advancement of this field.

4. OVERVIEW OF COMPUTATIONAL STUDIES

Crystallographic data essentially only provides static structural information, with limited indications of the operations in a biological setting. Therefore, computational studies utilizing molecular dynamics (MD) simulations are employed to supplement experimental data and gain time-dependent information on biological systems.

In MD simulations, integration of Newton's equations of motion, via an initial potential energy function, yields a trajectory containing the time evolution of the system. Accordingly, dynamical information can be gained at the atomistic level. The analytic expression for the energy, known as the *force field*, is composed of numerous functions approximating inter- and intramolecular energetic contributions. Various forms of this expression have been developed for simulations of biological molecules, including, but not limited to, the CHARMM (Chemistry at Harvard Molecular Mechanics) (Brooks et al., 1983), AMBER (Assisted Model Building with Energy Requirement) (Cornell et al., 1995), GROMACS (Groningen Machine for Chemical Simulation) (Berendsen, van der Spoel, & van Drunen, 1995) and OPLS (Optimized Potentials for Liquid Simulations) (Jorgensen & Tirado-Rives, 1988) force fields. The availability of force field parameters for proteins, lipids, ions, and water, alongside defined protocols for the parameterization of novel drug molecules, equips the user with appropriate tools to study the behavior of ion channels embedded in model membranes and the identification of potential drug binding sites, for example.

The availability of accurate force fields is the first necessary condition for the analysis of ion conduction by atomistic simulations, the second one being the possibility to obtain statistically robust estimates of the conduction properties by sampling the ion configurations with MD trajectories. The most direct method to sample ion configurations in an MD trajectory is certainly to simulate the ion channel in the presence of an electrochemical driving force and to observe ion conduction as it happens. In simulations with periodic boundary conditions, a straightforward strategy for simulating the presence of a driving force is to apply a constant electric field in the direction orthogonal to the membrane (Gumbart, Khalili-Araghi, Sotomayor, & Roux, 2012). An alternative strategy is to include in the simulation domain two parallel lipid membranes, which divide the space in two separate compartments. In this way, it is possible to impose an electrochemical gradient between the two sides of each membrane by controlling the ion concentrations in the two compartments (Kutzner, Grubmüller, de Groot, & Zachariae, 2011). In order to analyze ion conduction by these strategies it is necessary to simulate trajectories sufficiently long to sample numerous conduction events. For a typical ion channel (conductance 10–50 pS) at physiological condition (membrane potential in the range ±100 mV) it is necessary to simulate tens of microseconds in order to obtain a statistically robust estimate of the conductance. A different approach for the analysis of the conduction properties of an ion channel is to estimate the potential of mean force (PMF) along a restricted set of reaction coordinates. Several algorithms exist to accelerate sampling along a predefined set of reaction coordinates. "Umbrella sampling" which uses a biased potential to transform a system between thermodynamic states has been widely used in the field of ion channels, but examples of applications of other approaches, such as metadynamics (Stock, Delemotte, Carnevale, Treptow, & Klein, 2013), adaptive biasing force (Zuniga, Marquez, Gonzalez-Nilo, Chipot, Cid, Sepulveda, & Niemeyer, 2011), or steered-MD (Ngo, Stefanovski, Haas, & Farley, 2014) were also reported in the literature. Compared to the direct sampling of ion conduction in simulations with a driving force, the computational cost of PMF calculations might be significantly lower. However, these methods do not provide a direct estimate of the channel conductance, and the choice of the reaction coordinates might be critical.

MD simulations, and accompanying techniques, have become established methods to study K^+ channels in atomistic detail. We direct the reader to accomplished reviews (Furini & Domene, 2013; Roux, 2005) for an extensive overview of this field, yet we highlight a notable success in determining the conduction mechanism: the predominant selectivity filter conformations (Morais-Cabral, Zhou, & MacKinnon, 2001), w-S2-w-S4 and S1-w-S3-w, were identified in multiple MD simulations (Bernèche & Roux, 2000; Domene, Klein, Branduardi, Gervasio, & Parrinello, 2008; Domene & Sansom, 2003) and confirmed by free energy perturbation (FEP) techniques (Aqvist & Luzhkov, 2000) leading to the recognition of a "knock-on" mechanism, driven by the approach of K^+ from the central cavity and subsequent ion—ion repulsion, summarized in Fig. 3 (Morais-Cabral et al., 2001). Free energy calculations later showed that cation pairs moving through the selectivity filter without intervening water molecules are energetically feasible (Furini & Domene, 2009). In this mechanism, in addition to direct K^+—K^+ contacts, vacancies are also involved during the permeation event. A schematic representation of both mechanisms is illustrated in Fig. 3. FEP methods were also used to demonstrate a significant energetic preference for K^+ over Na^+ in the selectivity filter of K^+ channels (Berneche & Roux, 2001; Noskov & Roux, 2006). Specifically, S2 was identified as a highly selective, S1 and S3 displayed moderate selectivity whereas S4 provided a nonselective site, giving insights into the molecular determinants of selectivity in K^+ channels (Luzhkov & Åqvist, 2001).

Technical progressions concerning the development of parallel computing algorithms have resulted in an expansion of the phenomena that can be sampled throughout molecular simulations. D.E. Shaw and colleagues have been instrumental in enabling the performance of long-scale simulations of ion channels by the construction of the supercomputer Anton, which performs over an order of magnitude faster than any other machine utilizing an MD code. Simulations on an extended timescale under the influence of different membrane potentials have allowed the first direct observation of an ion channel gating transition (Jensen et al., 2012).

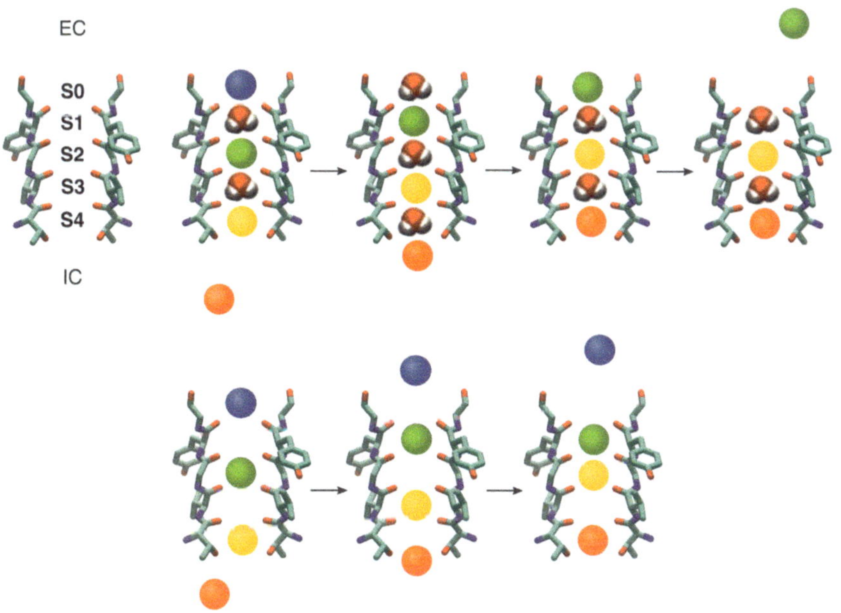

Figure 3 Proposed mechanisms of ion conduction in K^+ channels: (upper panel) chains of alternating K^+ ions and water molecules cross the selectivity filter, and/or (lower panel) cation move through the selectivity filter without intervening water molecules. Colored spheres represent individual K^+ ions. Approach of an ion from the central cavity prompts simultaneous ion movement throughout the selectivity filter, resulting in the exit of one ion that rapidly diffuses into the extracellular solution. The initial/final configuration of the selectivity filter was confirmed to be the most energetically favorable state using FEP methods (Aqvist & Luzhkov, 2000; Furini & Domene, 2009).

It is also worth noting that the Lennard–Jones interaction parameters in nonpolarizable *force fields* are derived from the standard Lorentz–Berthelot combination rule, optimized to emulate the free energies of ions in bulk water. As a consequence, in heterogeneous environments, such as the cell membrane, this method has inherent limitations. Therefore, the continued improvement of polarizable force fields will enhance the accuracy to which ion permeation can be simulated (Illingworth & Domene, 2009; Vanommeslaeghe & MacKerell, 2015; Xu et al., 2015).

4.1 Application to Na_V channels

Following the publication of the Na_VAb structure, numerous studies began to emerge in an attempt to identify the molecular determinants of Na^+ conduction mechanism, employing the vast range of available methods. In what follows, an exhaustive review of the MD studies that have been published to date is provided, placing particular emphasis on the atomistic structure of the central pore that has key implications for conduction and selectivity.

The first published MD simulations demonstrated that the original Na_VAb structure was stable in the presence and absence of Na^+ ions (Carnevale, Treptow, & Klein, 2011). The cytoplasmic gate remained closed for the duration of the simulations; therefore, access to the central pore was limited to entrance at the periplasmic side via the selectivity filter. Observed ion dynamics confirmed earlier binding site predictions (Fig. 4). In Na^+ channels, binding sites for hydrated or partially hydrated sodium ions have been identified at the extracellular entrance (S_{HFS}) in the core region of the selectivity filer (S_{IN} and S_{CEN}) and within the central cavity (S_C). A considerable occupation time was observed at S_{HFS} (~ 10 ns), comprised of both symmetric and asymmetric conformations, with the former predominant. Electrostatic interactions with two opposite Glu177 residues, H-bonding between coordinating water molecules and remaining Glu177 residues, and further interactions with surrounding water stabilized ions in the on-axis binding mode. In contrast, the ion preferentially contacts one Glu177 and adjoined Ser178 in the off-axis mode at this site. The overall octahedral geometry mimics that observed in sodium hydrate crystals. Ions were observed to traverse deeper into the selectivity filter, mediated by surrounding water molecules to bind at S_{IN} via S_{CEN}, as postulated from the crystal structure. It is apparent that favorable binding sites are established on the basis of retaining a full hydration shell, with contributions from both water molecules and protein atoms in a manner consistent with K^+ conduction through Kv channels. In contrast, ions are partially hydrated throughout and interact in an asymmetric fashion to lining residues. Finally, water permeation is apparently uncoupled to the presence of bound species and therefore ion permeation. The nature of water transport in K^+ channels remains controversial in the

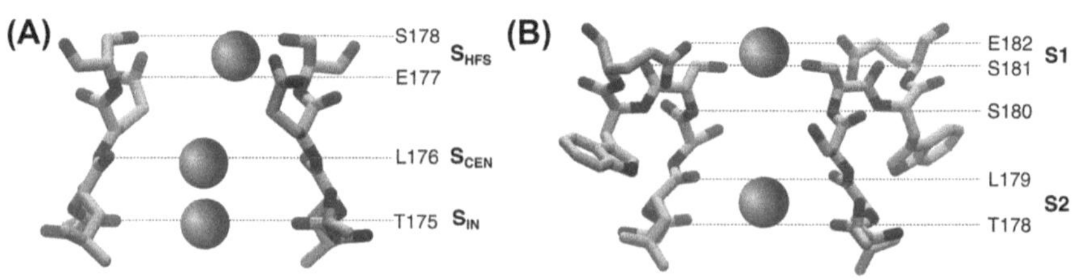

Figure 4 Binding sites in the selectivity filter of a Na^+ channel confirmed by MD simulations in (A) Na_VAb and (B) Na_VRh. Analogous binding sites as Na_VAb were also identified in Na_VMs.

literature, robust water flux has been observed in a K^+ depleted channel (Furini, Beckstein, & Domene, 2009), contrary to various accounts suggesting ions and water molecules are transported on a 1:1 ratio.

The Na_VAb binding sites of channels were also corroborated by the presence of defined energy minima in single and multiion PMF conduction profiles (Corry & Thomas, 2012; Furini & Domene, 2012) and various normal and long-scale MD simulations. A successive study provided insights into the stability of these binding sites; the overall selectivity filter structure remained remarkably similar in numerous ionic configurations, albeit a constriction of the pore in the presence of ions that can be attributed to favorable electrostatic interactions (Qiu, Shen, & Guo, 2012). This renders evidence the selectivity filter is stable independent of ionic occupancy, contrasting with K^+ channels that exhibit significant distortions in the absence of K^+ ions (Domene et al., 2008; Furini & Domene, 2009; Zhou et al., 2001).

Simulations of Na_V orthologs confirmed conserved selectivity filter binding sites across the bacterial Na_V family. The location and habitation of analogous binding sites observed in MD simulations of Na_VMs (Ke, Timin, & Stary-Weinzinger, 2014) is consistent with electron density detected in the Na_VMs crystal structure (McCusker et al., 2012). In Na_VRh, due to the nonconserved selectivity filter sequence, the characteristic inner "EEEE ring" and outer "SSSS ring" at the extracellular mouth are in reversed positions. Despite the supposed collapsed conformation of these residues, reorientation at the side chain of the inner ring (Ser181) leads to spontaneous opening of the selectivity filter in the first stages of the simulation (Zhang et al., 2013). Thereafter, occupation of binding sites at the vestibule (Ser181/Glu183 side chains: S1) and the inner of the selectivity filter (Thr178/Leu179 backbone carbonyls: S2) is observed.

To provide a feasible conduction pathway, an understanding of energetic barriers accompanying ionic movement, in addition to characterization of favorable binding sites is required. The application of free energy techniques to K^+ channels proved integral in elucidating the mechanism of K^+ conduction, therefore similar studies for Na_V emerged. Furthermore, long-scale MD simulations allow for the production of statistically significant conduction rates that can be validated against experimental single channel conductance measurements.

4.2 Single versus multiion conduction

The question of pore ion occupancy has ensued debate in the literature for many years, with flux measurements (Benos, Hyde, & Latorre, 1983) and reversal potentials (Begenisich & Cahalan, 1980) suggesting independent and multiion movement respectively, with the latter supported by Brownian dynamics (Vora, Corry, & Chung, 2008). Therefore, single and multiion profiles of Na^+ conduction, utilizing the Na_VAb structure, emerged from numerous research groups (Corry & Thomas, 2012; Furini & Domene, 2012).

1D PMF profiles exploring movement of a single ion as a function of the pore axis follow the expected permeation pathway, subsequently occupying S_{HFS}, S_{CEN}, and S_{IN} until release into the central cavity. Due to the attractive interactions with the "EEEE ring", a deep energy well (between −5 kcal/mol (Corry, 2013) and −8 kcal/mol (Furini & Domene, 2012) relative to the extracellular solution) is observed at S_{HFS} corresponding to the off-axis binding mode described previously. Despite maintaining a full hydration shell throughout, penetration further into the selectivity filter is accompanied by an energetic barrier of ~4 kcal/mol (Corry, 2013; Furini & Domene, 2012).

Multiion profiles demonstrated distinct differences; the presence of a second ion in close proximity to a bound ion in S_{HFS} significantly lowers the barrier for selectivity filter entrance. A further reduced energy well is also observed when both S_{CEN} and S_{HFS} are occupied. Therefore, the minimum energy pathway between initial (ions bound in the selectivity filter and external medium, denoted extracellular) and final states (ions bound in the selectivity filter and the central cavity, denoted intracellular) follows the multiion configurations: extracellular/S_{HFS}, S_{HFS}/S_{HFS}, S_{HFS}/S_{CEN}, S_{HFS}/S_{IN}, S_{HFS}/intracellular (Fig. 5). Considering this route, permeation can occur from the extracellular region into the central cavity with an energetic cost less than 3 kcal/mol. This value is consistent with the magnitude of energetic barriers previously determined for K^+ channels and the production of high flux rates (Aqvist & Luzhkov, 2000; Berneche & Roux, 2001).

Further efforts to characterize permeation through the Na_VAb selectivity filter were undertaken using metadynamics simulations (Stock, Delemotte, et al., 2013). Binding as a function of the radial distribution around the pore axis, as well as the distance along the pore axis, was taken into consideration to analyze previously unexplored two-ion configurations, and hence alternative conduction pathways. In particular, this allowed inspection of doubly occupied binding sites, a possible consequence of the wide selectivity filter.

Five local minima were recognized; potential wells, relative to the uppermost sampled state, of ~−12 kcal/mol corresponded to occupancy of S_{HFS}/S_{CEN} and S_{CEN}/S_{CEN} whereas shallower regions between −7

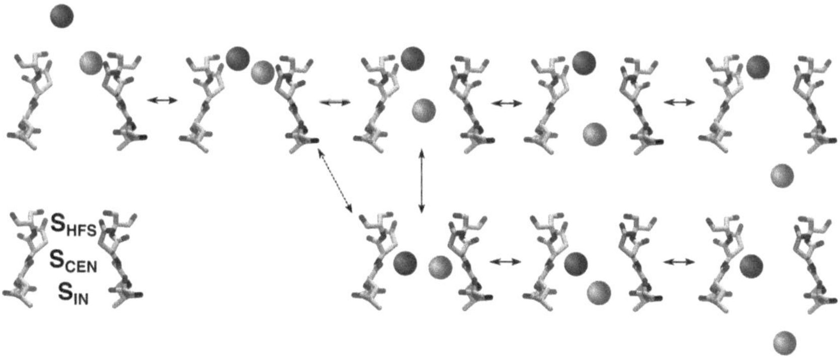

Figure 5 Proposed conduction pathways in the presence of two Na^+ ions in the selectivity filter. Selectivity filter residues are represented in licorice and Na^+ ions are orange (dark gray in print versions) and green (gray in print versions) spheres. 2D PMF profiles (Corry & Thomas, 2012; Furini & Domene, 2012) considering the position of individual ions along the pore axis, identified the minimum energy pathway following the sequence extracellular/S_{HFS}, S_{HFS}/S_{HFS}, S_{HFS}/S_{CEN}, S_{HFS}/S_{IN}, S_{HFS}/intracellular, represented in the upper array. Stock, Delemotte, et al. (2013) proposed an alternative mechanism with ions sequentially occupying extracellular/S_{HFS}, S_{HFS}/S_{HFS}, S_{HFS}/S_{CEN}, S_{CEN}/S_{CEN}, S_{CEN}/S_{IN}, and S_{HFS}/intracellular. The latter proved predominant in accompanying voltage-driven MD simulations.

and −9 kcal/mol coincided with S_{HFS}/S_{HFS}, S_{HSF}/S_{IN}, and S_{CEN}/S_{IN} configurations. These states form a complex network leading between the initial and final states of inward conduction, of which four are sampled in the minimum energy pathway (S_{HFS}/S_{HFS}, S_{HFS}/S_{CEN}, S_{CEN}/S_{CEN}, and S_{CEN}/S_{IN}). A direct S_{HFS}/S_{HFS} to S_{HFS}/S_{HFS} transition is also viable, in this case, illustrating transitions can be independent of the preceding ions, and conduction can occur via "knock-on" or "drive-by" mechanisms through the selectivity filter. These findings provide additional feasible routes to two-ion conduction, summarized in Fig. 5.

Overall, it is apparent that the conduction of a single Na^+ ion is promoted by the presence of additional ions. Weak coupling such as this can be described as a moderate version of the "knock-on" mechanism. In dications of a conserved mechanism throughout bacterial Na_V channels have been acknowledged. For example, in Na_VRh, FEP methods were used to determine if entrance to the outer state would destabilize ion binding at the inner site. Binding affinity was found to diminish from 3.7 kcal/mol to 2.6 kcal/mol, hence promoting onward conduction to the central cavity.

Throughout the examples described previously, further characterization of the permeation pathway succeeding the central cavity is restricted due to the closed state of the Na_VAb structure. MD simulations in the same Stock, Delemotte, et al. (2013) study utilized an open-state structure of Na_VAb conferred from comparative studies of the available Na_VAb structure with activated-open and resting-closed states of $K_V1.2$ (Amaral, Carnevale, Klein, & Treptow, 2012) allowing for analysis of whole channel permeation. The onset of a hyperpolarized potential (−600 mV) to promote inward permeation events identified an average selectivity filter occupancy of ~1.8 ions throughout, providing further evidence that in the presence of a negative potential simultaneous binding of two ions is part of the normal functioning of the channel. Furthermore, ions were found to predominantly follow the minimum energy pathway identified in the previously described metadynamics calculations.

Similarly, simulations based on pore-only channel structure from Na_VMs (McCusker et al., 2012), containing an open activation gate, also identified an average selectivity filter occupancy of ~1.8 ions (Ulmschneider et al., 2013). Both single and multiion-binding events were observed, with the latter predominant; configurations with ions occupying an outer (extracellular site or S_{HFS}) and inner site (S_{CEN}, S_{IN} or an intermediate region 3 Å below S_{HFS}) monopolize approximately 70% of the simulation. The use of microsecond simulations, in combination with a constant electric field along the channel axis as in this study, allowed accurate estimation of conductance rate (34 ± 6 pS), in comparison to single channel conductance of transfected HEK cells (~33 pS) (Ulmschneider et al., 2013). This accordance with experimental data suggests that both conduction mechanisms are physiologically relevant, albeit to different extents. Furthermore, the rate of water transport was quantified of $\sim 3 \times 10^9\ s^{-1}$ in both directions, with a dwell time of two orders of magnitude less than Na^+ (<1 ns). In contrast to strongly coupled transport in K^+ channels (Alcayaga, Cecchi, Alvarez, & Latorre, 1989; Ando, Kuno, Shimizu, Muramatsu, & Oiki, 2005), this confirms earlier proposals that the permeation of water was uncoupled to that of ions in Na^+ channels.

Further large-scale MD simulations of the Na_VAb structure, totaling ~21.6 μs, in equilibrium conditions (under 0 mV), provided unique insights into selectivity filter occupancy and its relation to the conformational state of the selectivity filter, specifically the EEEE motif (Chakrabarti et al., 2013). These simulations displayed direct observations of "knock-on" and "knock-off" conduction events. Due to the closed nature of the channel, it is likely the occupation of the central cavity is higher than expected in an open-state structure in the presence of an electrochemical gradient.

Analysis of the selectivity filter alone confirmed an approximate 70% population of doubly occupied configurations; however, the identity of the remaining states digressed, with a triply occupied arrangement constituting 23%, increasing the average selectivity filter occupancy to 2.09 ± 0.05. The observed translocation kinetics has been attributed to the presence of highly degenerate ionic binding modes concerning the conformational isomerization of the EEEE ring. Glu177 side chains can either point out toward the extracellular vestibule or point in toward the selectivity filter pore. Inward facing conformations, referred to as "dunked" in this study, are shown to be favored in the presence of Na^+ and consequently promote Na^+ binding in the selectivity filter. The amalgamation of Na^+ ions and multiple carboxyl conformers leads to an extensive plethora of favorable states. Dynamic transitions between these ionic clusters occur on a <100 ps timescale and resembling that of a highly disordered liquid. The lack of preference for a particular state, and the separation by low-lying barriers, therefore promotes diffusion in the selectivity filter and high throughput Na^+ permeation. The authors suggest that voltage-driven nonequilibrium MD simulations may bias the dynamic behavior of the EEEE motif and discourage the occupation of degenerate conformational states in earlier studies, thereby overlooking this phenomenon.

Subsequently, Boiteux and coworkers identified a double potential well at S_{HFS} in PMF profiles, corresponding to available conformers of Glu177, reducing the barrier between selectivity filter sites (Boiteux, Vorobyov, & Allen, 2014). An average occupancy of 2.3 ions is reported in the selectivity filter, lending further evidence to a three-ion conduction mechanism in which the selectivity filter switches between one, two, and three-ion configurations facilitated by the isomerization of Glu177. In this case, three-ion binding events are promoted after the cleavage of H-bonding between Glu177 carbonyls and Ser178 hydroxyl groups, which restrict Glu177 movement during the first microsecond of the simulation. This remark illustrates that short scale MD simulations are also unlikely to sample this phenomena.

The coexistence of conduction mechanisms with two or three ions inside the selectivity filter was observed also in bias-exchange metadynamics simulations (Domene, Barbini & Furini, 2015). The significance of triply occupied states requires further investigation with regards to the possibility of contrasting mechanisms of Na^+ of inward and outward conduction. Comparative studies of Na_VAb (Stock, Delemotte, et al., 2013) and Na_VMs (Ke et al., 2014) channels under depolarized and hyperpolarized membrane potentials, to emulate influx and efflux respectively, have therefore been undertaken to elucidate this proposal.

Both studies displayed an increased maximum barrier to conduction between S_{CEN} and S_{HFS}, (4.6 kcal/mol versus 0.4 kcal/mol and 2.1 kcal/mol versus 2.3 kcal/mol, respectively) with considerably higher dwell times in the selectivity filter (13.5 ± 0.6 ns versus 20.1 ± 1.1 ns) and hence lower flux rates (15 ± 3 pS versus $\sim 27 \pm 3$ pS) in Na_VMs. Furthermore, an increased average occupancy of ~ 2.3 ions during efflux in comparison to ~ 1.8 ions during influx was observed in Na_VAb, thereby suggesting independent conduction mechanisms with the presence of a third ion more predominant in the outward direction (Stock, Delemotte, et al., 2013).

An additional "nudging" ion is observed to initiate the process in both systems. Therefore, to overcome the heightened barrier, this is considered to be an essential requirement. Moreover, in a Na_VMs study, reversible rotations of Glu177 residues play a pivotal role in efflux (Ke et al., 2014). A significantly larger proportion of flipped states is observed resulting in appreciable differences in the distribution of the ions throughout the selectivity filter ($32 \pm 5.9\%$ versus $2.7 \pm 0.5\%$ for influx). Binding at S_{HFS} was found to be generally on-axis. Furthermore, S_{CEN} and an intermediate region below S_{HFS} were more densely populated. Additional simulations with a restrained "one-flip" ring conformation displayed increased efflux rates (relative to "no-flip" simulations), reiterating these proposals, yet appeared uncoupled to the Na^+ influx rate, disagreeing with Chakrabarti's earlier conclusions that this phenomenon was essential for the catalysis of Na^+ permeation regardless of the directionality (Chakrabarti et al., 2013). The discrepancies between previous studies are attributed to use of an open-state structure, as opposed to closed Na_VAb, which will induce repulsive effects when the central cavity is inhabited, in conjunction with the use of different force fields (Cordomí, Caltabiano, & Pardo, 2012).

4.3 Protonation state of Glu177

The conformational states of the EEEE signature sequence are further complicated by the prospect that one or more glutamate residues may be protonated. With potential implications for the electrostatics at the extracellular vestibule, and hence the ability to attract cations, and the overall structural stability of the selectivity filter and its enclosed binding sites, the dependency of the conduction mechanism on the protonation state requires thorough investigation.

Corry and Thomas (2012) calculated the 1D PMF profiles for Na_VAb with one and two Glu177 (in opposite side chains) residues protonated; the former exhibited a shallower, yet similar, profile with lower occupation of S_{HFS}, expected from the reduced negative potential of the selectivity filter. In contrast, when two protonated Glu177 residues are present, the ability of S_{HFS} to beckon Na^+ ions out of extracellular solution is diminished, in addition to an unexpectedly high energy at S_{CEN}, stunting Na^+ permeation.

Similar conclusions were reached from long-scale MD simulations (Boiteux et al., 2014), suggesting an essentially barrierless (~1.5 kcal/mol maximum barrier to conduction) multiion "knock-on" or "pass-by" mechanism is more likely in the fully charged state, promoted by repulsive interactions between charged side chains. The presence of one protonated glutamate, similarly, leads to a reduced charge density at the mouth of the selectivity filter. Rotational movements of Glu177 are integral in smoothing the energetic landscape and deeming conduction energetically feasible in this state, yet discourage three-ion events in the selectivity filter suggesting permeation occurs to a less extent. When two or more residues are protonated, hydrogen bonds between protonated and deprotonated glutamates occlude the permeation pathway causing significant barriers in the plane of the EEEE motif. Intrapore proton block such as this has been identified experimentally in $Na_V1.4$ (Khan, Kyle, Hanck, Lipkind, & Fozzard, 2006) and selected L-type Ca^{2+} channels (Chen, Bezprozvanny, & Tsien, 1996).

Overall, it is apparent that systems containing multiple protonated glutamates are unfeasible states for ion conduction, a fully charged EEEE ring is optimal for conduction; yet, the structural and functional state of Na_VAb containing a single uncharged Glu177 is presently dubious. An additional study by Furini, Barbini and Domene (2014) identified a noticeable increase in flexibility of the protonated glutamate relative to its deprotonated counterparts, residing in novel side chain conformations pointing toward S_{CEN} and laying at the back of the selectivity filter. As a consequence, structural fluctuations throughout the entire selectivity filter were experienced, with a substantial impact in both 1D and 2D PMF's, the energetics of which are likely to markedly impede ion flow. Recognized differences with prior microsecond simulations are attributed to allowed transitions to alternative metastable states that are unlikely to be sampled during typically short umbrella sampling trajectories, employed in this study. The appropriateness of these methods to the question in hand remains ambiguous; therefore, the physiological relevance of identified structural and permeation properties of a singly protonated Na_V selectivity filter cannot be delineated at present.

An interesting observation worth noting, however, is the extent to which structural changes at the selectivity filter are propagated throughout the entire pore domain. Restriction of the selectivity filter, in a singly protonated system, seemingly adjusts the cavity shape and the orientation of S6 (Boiteux et al., 2014). The observed bent conformation in S6, on the level of P200–T206, is parallel to that of an inactivated structure suggested by Payandeh et al. (2012). Such rapid conformational flux appears to be an inherent property of Na_V behavior on a microsecond timescale, potentially related to slow inactivation (millisecond to second).

Structural rearrangements in this region also have implications for fenestrations between the central core and the surrounding membrane. The size of these openings is directly related to the activation state of S6; specifically, the rotation of F203 of phenyl side chains reversibly blocks the pathway, consistent with experimental reports this residue is essential for inactivation and drug binding (Ahern, Eastwood, Dougherty, & Horn, 2008; Carboni, Zhang, Neplioueva, Starmer, & Grant, 2005; Li, Galue, Meadows, & Ragsdale, 1999; Ragsdale, McPhee, Scheuer, & Catterall, 1994). The open fenestrations have a maximum ~4 Å width, feasible for the entry of lipophilic drugs (ie, local anesthetics), suggesting binding at this site may influence slow channel inactivation propagated through this region from the selectivity filter (Carboni et al., 2005; Chen et al., 2000; Li et al., 1999).

4.4 Selectivity of Na^+ over other monovalent and divalent ions

To elucidate the molecular determinants of selectivity in Na^+ channels, comparison of the energy profiles of Na^+ and K^+ was undertaken (Corry & Thomas, 2012; Furini & Domene, 2012). A similar landscape was observed, with elevated barriers at S_{HFS} and S_{CEN} of the order of 5–6 kcal/mol, independent of the presence of multiple ions.

Further analysis of the PMF displayed a similar binding affinity of both ions at S_{HFS}, due to flexibility and availability of water molecules, with a notable barrier to conduction arising in the plane of the Glu177 side chains. Corry et al. proposed a mechanism for this concerning the orientation of the proximal residues. The ideal geometry of S_{HFS} during a binding event involves a straight ion-water-carbonyl complex. However, the increased distance of direct coordination between K^+ and Glu177 means this conformation is improbable unless reorganization of the binding site occurs. Only slight differences in the binding site size are observed, between K^+ and Na^+, insufficient for this purpose. Therefore, even though K^+ can fit through the pore with a full shell it is not in an optimum geometry.

This hypothesis was tested further by analysis of the relationship between selectivity and pore radius. The free energies of K^+, relative to Na^+, to travel through Glu177 ring of a model pore illustrated a lack of selectivity until bridging water molecules can no longer fit in plane of the pore, and a peak of selectivity at 6.3 Å, optimum for Na^+ and in plane binding of water molecules. Similar free energies are observed over a wide range of values, suggesting favorable conduction does not require specific alignment with the pore axis, as in K^+ channels.

In both studies (Corry & Thomas, 2012; Furini & Domene, 2012), the calculated energetic penalty of K^+ conduction through a Na_V channel was 3 kcal/mol, as opposed to 6 kcal/mol in the reverse situation of Na^+ conduction in K^+ channels. Therefore, the feasibility of K^+ conduction via this mechanism is consistent with the lower selectivity identified in Na_V channels compared to K^+ channels; K^+ channels select for K^+ ions at a rate of approximately 1000 to 1 whereas in Na^+ channels this is reduced to only 8 to 1.

The close relationship between Na^+ and Ca^{2+} selective channels complicates the situation of determining the origins of channel selectivity. A similar tetrameric architecture with each subunit containing six transmembrane helices and selectivity filter structure, in addition to interchangeable selectivity in both channel families on the mutation of selected selectivity filter residues, suggests only subtle differences in protein structure are responsible for the observed conduction ratios (Heinemann et al., 1992; Shaya et al., 2011; Sun et al., 1997; Yue, Navarro, Ren, Ramos, & Clapham, 2002). The similar ionic radii lead to proposals that selectivity is based on the inability of Na^+ channels to accommodate the increased charge density of Ca^{2+} (Corry, 2013).

Similarly to comparative simulations of Na^+ and K^+ described previously, off-axis binding modes between Glu177 and Ser178 are observed in the external site with Ca^{2+}, with an energetic barrier while penetrating the "EEEE ring" (Corry, 2013). A ~3 kcal/mol energetic penalty (1 kcal/mol for Na^+) is observed accompanying movement from the selectivity filter to central cavity, emphasizing an increased binding affinity for Ca^{2+} in this site. Furthermore, an additional barrier was identified deeper into the pore before reaching the central/internal sites, in contrast to Na^+ and K^+. An average lower coordination number observed at this position, and corresponding energies of dehydration of about 8 kcal/mol to reduce the coordination shell from seven to six, suggests the removal of water molecules is a feasible explanation of the energy barrier. The lack of ion–protein interactions, from the selectivity filter lining residues, to compensate for the reduced ion–water interactions further substantiates this claim. Increased affinity for water binding in Ca^{2+} is also observed in the second solvation shell, likely propagating this effect further. The presence of a second ion lowers the maximum barrier to conduction between S_{HFS} and S_{CEN} to ~4.5 kcal/mol; therefore, an analogous weakly coupled "knock-on" mechanism is expected here. In a mixed ionic solution, barriers to Ca^{2+} conduction are slightly lower than those observed in the single-ion conduction. Na^{+} permeation, in the presence of bound Ca^{2+}, appears to be the most likely scenario in this case, however, despite an increased barrier in comparison to Na^{+} alone, suggesting the presence of Ca^{2+} diminishes Na^{+} conduction. This is consistent with experimental reports of Ca^{2+} obstruction (French, Worley, Wonderlin, Kularatna, & Krueger, 1994; Ravindran, Schild, & Moczydlowski, 1991; Yamamoto, Yeh, & Narahashi, 1984) albeit to a considerably lesser extent than predicted computationally (Ren et al., 2001). The disparity is likely due to the uniform protonation states of Glu177 residues and limitations regarding the use of nonpolarizable force fields for divalent cations (Bakó, Hutter, & Pálinkás, 2002; Bucher & Kuyucak, 2008).

Simulations undertaken simultaneously by Ke, Zangerl, and Stary-Weinzinger (2013) proposed contrasting conclusions. After a ~40 ns occupation of S_{HFS}, the ion entered an intermediate site below the EEEE ring, termed $S_{S178/L176}$ in this study, formed from the backbone nitrogen's of Ser178 and Leu176, for a further 55 ns. The single ion entered S_{CEN} thereafter, which is populated for the remainder of the simulation, excluding the entrance of further ions. The general landscape of the PMF energy profile is in general agreement with those calculated by the Corry group for single-ion conduction. An elevated barrier is observed in the plane of the EEEE ring, suggesting discrimination against Ca^{2+} may occur in this region due to size constraints in a similar mechanism proposed for K^{+} selectivity (Corry & Thomas, 2012). In addition, a significant barrier is observed after S_{CEN}, suggesting diffusion into the central cavity may be the rate-limiting step in the case of inward conduction of Ca^{2+}. The high binding affinity for S_{CEN} is consistent with crystallographic data of the closely related Na_VRh channel.

The presence of a potential well at $S_{S178/L176}$ requires further investigation; in this site, two rotated Glu177 residues bind Ca^{2+} to optimize the available electrostatic network. When the ion vacates this site, one glutamate moved back to its original orientation, with the second revolving further to accommodate ion translocation to S_{CEN} demonstrating the flexibility of this residue in this case. Similar potential wells are observed in PMF profiles of Na^+ (Ke et al., 2013), Na^+ and K^+ (Furini & Domene, 2012), and Ca^{2+} (Boiteux et al., 2014) yet absent in analogous calculations of Na^+, K^+, and Ca^{2+} permeation (Corry & Thomas, 2012) highlighting further contentions in the literature concerning the behavior of Glu177.

In Na_VRh, all simulations with variable concentrations of Ca^{2+} show persistent binding at S1; strong binding determined by electrostatic interactions with Glu183, as well as interactions with Ser180, results in an insurmountable energy barrier (~11.38 kcal/mol), hindering progression deeper in the selectivity filter (Zhang et al., 2013). Therefore, the authors suggest this site confers selectivity in Na_VRh. The binding mode of Na^+ at S1 is reflective of preferred interactions with Ser181; this conformation is unreachable when Ca^{2+} is bound, consistent with extracellular millimolar blockage of Na^+ current by Ca^{2+}. Simulations with ions beginning in the central cavity were used to examine binding in S2. Here, Ca^{2+} exists preferentially in the center, approximately equidistant from the eight surrounding carbonyls. In contrast, Na^+ binds off-axis, in close proximity to one or two of the carbonyls, occasionally jumping between positions, subsequently persuading asymmetric selectivity filter conformations. Elements of asymmetry are observed to some extent throughout the entire Na^+ translocation process, despite the homologous sequence present in all subunits. Relative to other bacterial Na_V channels, the selectivity filter sequence and asymmetric binding features of Na_VRh could be more applicable to mammalian Na_V channels, which contain nonconserved selectivity filter sequences; therefore, it is likely more studies utilizing this structure will emerge in the coming years.

5. CONCLUSIONS

In this review, we have primarily focused on the mechanisms of conduction and selectivity in Na^+ channels and a comparison with K^+ channels has been established. Aspects of voltage-gating and drug blockage have likewise been studied by computational methods to which numerous reviews are available (Bagnéris, Naylor, McCusker, & Wallace, 2015; Stock, Souza, & Treptow, 2013; Vargas et al., 2012). There is a general consensus about the main characteristics of ion conduction in bacterial Na^+ channels emerging from atomistic stimulations. Conduction requires the cooperative movement of at least two Na^+ ions across the selectivity filter of the channel. In sharp contrast to K^+ channels, the movements of the ions inside the selectivity filter of Na^+ channels are only loosely correlated, and the passage of water molecules is uncoupled from ion conduction. The central region of the selectivity filter can host a hydrated Na^+ ion but it cannot host a hydrated K^+ ion, which might explain the low selectivity of bacterial Na^+ channels for Na^+ ions over K^+ ions. In spite of this overall picture, there are still some controversial issues about the conformation of Glu177 and the number of ions involved in conduction events. Extensive atomistic simulations proved that the side chain of Glu177 is highly mobile, conferring a liquidlike nature to the selectivity filter of bacterial Na^+ channels. Instead, in other simulations of ion conduction, the side chain of Glu177 did not show significant deviations from the crystallographic structure. Therefore, the importance of the conformational states of Glu177 on ion conduction is still disputed. The movements of Glu177 can also modify the number of Na^+ ions inside the selectivity filter and consequently the number of ions involved in conduction events. Conduction mechanisms with two and three ions have been observed, and it has been proposed that inward and outward conduction might follow different routes, with the third ion possessing a more critical role in outward conduction. It is possible that several conduction mechanisms coexist for bacterial Na^+ channels, as it has been suggested for K^+ channels (Furini & Domene, 2009), and that the distribution of the conductive trajectories among these different mechanisms is influenced by the membrane potential and by the concentration of ions in the intracellular and extracellular compartments. These open questions can be investigated by atomistic simulations, and they will provide a description of ion conduction in bacterial Na^+ channels at an unprecedented level of detail.

ACKNOWLEDGMENTS

Work in the Domene lab is supported by the Hartree Center and the Engineering and Physical Sciences Research Council via the UK National Service for Computational Chemistry Software (NSCCS) and Archer. V.O is supported by a CASE studentship from the Biotechnology and Biological Sciences Research Council and Pfizer Neusentis.

REFERENCES

Ahern, C. A., Eastwood, A. L., Dougherty, D. A., & Horn, R. (2008). Electrostatic contributions of aromatic residues in the local anesthetic receptor of voltage-gated sodium channels. *Circulation Research, 102*(1), 86–94.

Alcayaga, C., Cecchi, X., Alvarez, O., & Latorre, R. (1989). Streaming potential measurements in Ca^{2+}-activated K^+ channels from skeletal and smooth muscle. Coupling of ion and water fluxes. *Biophysical Journal, 55*(2), 367–371.

Amaral, C., Carnevale, V., Klein, M. L., & Treptow, W. (2012). Exploring conformational states of the bacterial voltage-gated sodium channel NavAb via molecular dynamics simulations. *Proceedings of the National Academy of Sciences of the United States of America, 109*(52), 21336–21341.

Ando, H., Kuno, M., Shimizu, H., Muramatsu, I., & Oiki, S. (2005). Coupled K^+-water flux through the HERG potassium channel measured by an osmotic pulse method. *Journal of General Physiology, 126*(5), 529–538.

Aqvist, J., Luecke, H., Quiocho, F. A., & Warshel, A. (1991). Dipoles localized at helix termini of proteins stabilize charges. *Proceedings of the National Academy of Sciences of the United States of America, 88*(5), 2026–2030.

Aqvist, J., & Luzhkov, V. (2000). Ion permeation mechanism of the potassium channel. *Nature, 404*(6780), 881–884.

Bagal, S. K., Brown, A. D., Cox, P. J., Omoto, K., Owen, R. M., Pryde, D. C., & Swain, N. A. (2013). Ion channels as therapeutic targets: a drug discovery perspective. *Journal of Medicinal Chemistry, 56*(3), 593–624.

Bagnéris, C., DeCaen, P. G., Hall, B. A., Naylor, C. E., Clapham, D. E., Kay, C. W. M., & Wallace, B. A. (2013). Role of the C-terminal domain in the structure and function of tetrameric sodium channels. *Nature Communications, 4*.

Bagnéris, C., Naylor, C. E., McCusker, E. C., & Wallace, B. A. (2015). Structural model of the open–closed–inactivated cycle of prokaryotic voltage-gated sodium channels. *Journal of General Physiology, 145*(1), 5–16.

Bakó, I., Hutter, J., & Pálinkás, G. (2002). Car–Parrinello molecular dynamics simulation of the hydrated calcium ion. *Journal of Chemical Physics, 117*(21), 9838–9843.

Begenisich, T. B., & Cahalan, M. D. (1980). Sodium channel permeation in squid axons. I: reversal potential experiments. *Journal of Physiology, 307*, 217–242.

Benos, D. J., Hyde, B. A., & Latorre, R. (1983). Sodium flux ratio through the amiloride-sensitive entry pathway in frog skin. *Journal of General Physiology, 81*(5), 667–685.

Berendsen, H. J. C., van der Spoel, D., & van Drunen, R. (1995). GROMACS: a message-passing parallel molecular dynamics implementation. *Computer Physics Communications, 91*(1–3), 43–56.

Bernèche, S., & Roux, B. (2000). Molecular dynamics of the KcsA K^+ channel in a bilayer membrane. *Biophysical Journal, 78*(6), 2900–2917.

Berneche, S., & Roux, B. (2001). Energetics of ion conduction through the K^+ channel. *Nature, 414*(6859), 73–77.

Bezanilla, F., & Armstrong, C. M. (1972). Negative conductance caused by entry of sodium and cesium ions into the potassium channels of squid axons. *Journal of General Physiology, 60*(5), 588–608.

Black, J. A., & Waxman, S. G. (2013). Noncanonical roles of voltage-gated sodium channels. *Neuron, 80*(2), 280–291.

Boiteux, C., Vorobyov, I., & Allen, T. W. (2014). Ion conduction and conformational flexibility of a bacterial voltage-gated sodium channel. *Proceedings of the National Academy of Sciences of the United States of America, 111*(9), 3454–3459.

Brohawn, del Marmol, J., & MacKinnon, R. (2012). Crystal structure of the human K2P TRAAK, a lipid- and mechano-sensitive K^+ ion channel. *Science, 335*, 436–441.

Brooks, B. R., Bruccoleri, R. E., Olafson, B. D., States, D. J., Swaminathan, S., & Karplus, M. (1983). CHARMM: a program for macromolecular energy, minimization, and dynamics calculations. *Journal of Computational Chemistry, 4*(2), 187–217.

Bucher, D., & Kuyucak, S. (2008). Polarization of water in the first hydration shell of K^+ and Ca^{2+} ions. *Journal of Physical Chemistry B, 112*(35), 10786–10790.

Carboni, M., Zhang, Z. S., Neplioueva, V., Starmer, C. F., & Grant, A. O. (2005). Slow sodium channel inactivation and use-dependent block modulated by the same domain IV S6 residue. *Journal of Membrane Biology, 207*(2), 107–117.

Carnevale, V., Treptow, W., & Klein, M. L. (2011). Sodium ion binding sites and hydration in the lumen of a bacterial ion channel from molecular dynamics simulations. *Journal of Physical Chemistry Letters, 2*(19), 2504–2508.

Chakrabarti, N., Ing, C., Payandeh, J., Zheng, N., Catterall, W. A., & Pomès, R. (2013). Catalysis of Na^+ permeation in the bacterial sodium channel NaVAb. *Proceedings of the National Academy of Sciences of the United States of America, 110*(28), 11331–11336.

Chen, X. H., Bezprozvanny, I., & Tsien, R. W. (1996). Molecular basis of proton block of L-type Ca^{2+} channels. *Journal of General Physiology, 108*(5), 363–374.

Chen, Z., Ong, B.-H., Kambouris, N. G., Marbán, E., Tomaselli, G. F., & Balser, J. R. (2000). Lidocaine induces a slow inactivated state in rat skeletal muscle sodium channels. *Journal of Physiology, 524*, 37–49.

Cordero-Morales, J. F., Jogini, V., Chakrapani, S., & Perozo, E. (2011). A multipoint hydrogen-bond network underlying KcsA C-type inactivation. *Biophysical Journal, 100*(10), 2387–2393.

Cordomí, A., Caltabiano, G., & Pardo, L. (2012). Membrane protein simulations using AMBER force field and Berger lipid parameters. *Journal of Chemical Theory and Computation, 8*(3), 948–958.

Cornell, W. D., Cieplak, P., Bayly, C. I., Gould, I. R., Merz, K. M., Ferguson, D. M. … Kollman, P. A. (1995). A second generation force field for the simulation of proteins, nucleic acids, and organic molecules. *Journal of the American Chemical Society, 117*(19), 5179–5197.

Corry, B. (2013). Na^+/Ca^{2+} selectivity in the bacterial voltage-gated sodium channel NavAb. *PeerJ, 1*, e16.

Corry, B., & Thomas, M. (2012). Mechanism of ion permeation and selectivity in a voltage gated sodium channel. *Journal of the American Chemical Society, 134*(3), 1840–1846.

Cuello, L. G., Jogini, V., Cortes, D. M., & Perozo, E. (2010). Structural mechanism of C-type inactivation in K^+ channels. *Nature, 466*(7303), 203–208.

Domene, C., Barbini, P., & Furini, S. (2015). Bias-exchange metadynamics simulations: An efficient strategy for the analysis of conduction and selectivity in ion channels. *Journal of Chemical Theory and Computation, 11*(4), 1896–1906. http://dx.doi.org/10.1021/ct501053x.

Domene, C., Klein, M. L., Branduardi, D., Gervasio, F. L., & Parrinello, M. (2008). Conformational changes and gating at the selectivity filter of potassium channels. *Journal of the American Chemical Society, 130*(29), 9474–9480.

Domene, C., & Sansom, M. S. (2003). Potassium channel, ions, and water: simulation studies based on the high resolution X-ray structure of KcsA. *Biophysical Journal, 85*(5), 2787–2800.

Doyle, D. A., Morais Cabral, J., Pfuetzner, R. A., Kuo, A., Gulbis, J. M., Cohen, S. L. … MacKinnon, R. (1998). The structure of the potassium channel: molecular basis of K^+ conduction and selectivity. *Science, 280*(5360), 69–77.

Eisenman, G. (1962). Cation selective glass electrodes and their mode of operation. *Biophysical Journal, 2*, 259–323.

Eisenman, G., & Horn, R. (1983). Ionic selectivity revisited: the role of kinetic and equilibrium processes in ion permeation through channels. *Journal of Membrane Biology, 76*(3), 197–225.

England, S., & de Groot, M. J. (2009). Subtype-selective targeting of voltage-gated sodium channels. *British Journal of Pharmacology, 158*(6), 1413–1425.

Favre, I., Moczydlowski, E., & Schild, L. (1996). On the structural basis for ionic selectivity among Na^+, K^+, and Ca^{2+} in the voltage-gated sodium channel. *Biophysical Journal, 71*, 3110–3125.

French, R. J., Worley, J. F., Wonderlin, W. F., Kularatna, A. S., & Krueger, B. K. (1994). Ion permeation, divalent ion block, and chemical modification of single sodium channels. Description by single- and double-occupancy rate-theory models. *Journal of General Physiology, 103*, 447–470.
Furini, S., Barbini, P., & Domene, C. (2014). Effects of the protonation state of the EEEE motif of a bacterial Na(+)-channel on conduction and pore structure. *Biophysical Journal, 106*(10), 2175–2183. http://dx.doi.org/10.1016/j.bpj.2014.04.005.
Furini, S., Beckstein, O., & Domene, C. (2009). Permeation of water through the KcsA K^+ channel. *Proteins, 74*(2), 437–448.
Furini, S., & Domene, C. (2009). Atypical mechanism of conduction in potassium channels. *Proceedings of the National Academy of Sciences of the United States of America, 106*(38), 16074–16077.
Furini, S., & Domene, C. (2012). On conduction in a bacterial sodium channel. *PLoS Computational Biology, 8*(4), e1002476.
Furini, S., & Domene, C. (2013). K^+ and Na^+ conduction in selective and nonselective ion channels via molecular dynamics simulations. *Biophysical Journal, 105*(8), 1737–1745.
Furini, S., Zerbetto, F., & Cavalcanti, S. (2007). Role of the intracellular cavity in potassium channel conductivity. *Journal of Physical Chemistry B, 111*(50), 13993–14000.
Gilchrist, J., Das, S., Van Petegem, F., & Bosmans, F. (2013). Crystallographic insights into sodium-channel modulation by the β4 subunit. *Proceedings of the National Academy of Sciences of the United States of America, 110*(51), E5016–E5024.
Gumbart, J., Khalili-Araghi, F., Sotomayor, M., & Roux, B. (2012). Constant electric field simulations of the membrane potential illustrated with simple systems. *Biochim Biophys Acta, 1818*(2), 294–302.
Heginbotham, L., Lu, Z., Abramson, T., & MacKinnon, R. (1994). Mutations in the K^+ channel signature sequence. *Biophysical Journal, 66*(4), 1061–1067.
Heinemann, S. H., Terlau, H., Stuhmer, W., Imoto, K., & Numa, S. (1992). Calcium channel characteristics conferred on the sodium channel by single mutations. *Nature, 356*, 441–443.
Hille, B. (1971). The permeability of the sodium channel to organic cations in myelinated nerve. *Journal of General Physiology, 58*(6), 599–619.
Hille, B. (1972). The permeability of the sodium channel to metal cations in myelinated nerve. *Journal of General Physiology, 59*(6), 637–658.
Hille, B. (1992). *Ionic channels of excitable membranes* (2nd ed.). Sunderland, MA: Sinauer Associates Inc.
Hodgkin, A. L., & Keynes, R. D. (1955). The potassium permeability of a giant nerve fibre. *Journal of Physiology, 128*(1), 61–88.
Illingworth, C. J., & Domene, C. (2009). Many-body effects and simulations of potassium channels. *Proceedings of the Royal Society of London, Series A, 465*(2106), 1701–1716.
Jensen, M.Ø., Jogini, V., Borhani, D. W., Leffler, A. E., Dror, R. O., & Shaw, D. E. (2012). Mechanism of voltage gating in potassium channels. *Science, 336*(6078), 229–233.
Jiang, Y., Lee, A., Chen, J., Ruta, V., Cadene, M., Chait, B. T., & MacKinnon, R. (2003). X-ray structure of a voltage-dependent K^+ channel. *Nature, 423*(6935), 33–41.
Jogini, V., & Roux, B. (2005). Electrostatics of the intracellular vestibule of K^+ channels. *Journal of Molecular Biology, 354*(2), 272–288.
Jorgensen, W. L., & Tirado-Rives, J. (1988). The OPLS [optimized potentials for liquid simulations] potential functions for proteins, energy minimizations for crystals of cyclic peptides and crambin. *Journal of the American Chemical Society, 110*(6), 1657–1666.
Ke, S., Timin, E. N., & Stary-Weinzinger, A. (2014). Different inward and outward conduction mechanisms in NaVMs suggested by molecular dynamics simulations. *PLoS Computational Biology, 10*(7), e1003746.

Ke, S., Zangerl, E.-M., & Stary-Weinzinger, A. (2013). Distinct interactions of Na^+ and Ca^{2+} ions with the selectivity filter of the bacterial sodium channel NaVAb. *Biochemical and Biophysical Research Communications, 430*(4), 1272–1276.

Khan, A., Kyle, J. W., Hanck, D. A., Lipkind, G. M., & Fozzard, H. A. (2006). Isoform-dependent interaction of voltage-gated sodium channels with protons. *Journal of Physiology, 576*, 493–501.

Kuo, A., Gulbis, J. M., Antcliff, J. F., Rahman, T., Lowe, E. D., Zimmer, J., & Doyle, D. A. (2003). Crystal structure of the potassium channel KirBac1.1 in the closed state. *Science, 300*(5627), 1922–1926.

Kutzner, C., Grubmüller, H., de Groot, B. L., & Zachariae, U. (2011). Computational electrophysiology: The molecular dynamics of ion channel permeation and selectivity in atomistic detail. *Biophysical Journal, 101*(4), 809–817. http://dx.doi.org/10.1016/j.bpj.2011.06.010.

Li, H. L., Galue, A., Meadows, L., & Ragsdale, D. S. (1999). A molecular basis for the different local anesthetic affinities of resting versus open and inactivated states of the sodium channel. *Molecular Pharmacology, 55*(1), 134–141.

Long, S. B., Campbell, E. B., & Mackinnon, R. (2005). Crystal structure of a mammalian voltage-dependent Shaker family K^+ channel. *Science, 309*(5736), 897–903.

Long, S. B., Tao, X., Campbell, E. B., & MacKinnon, R. (2007). Atomic structure of a voltage-dependent K^+ channel in a lipid membrane-like environment. *Nature, 450*(7168), 376–382.

Luzhkov, V. B., & Åqvist, J. (2001). K^+/Na^+ selectivity of the KcsA potassium channel from microscopic free energy perturbation calculations. *BBA – Protein Structure and Molecular Enzymology, 1548*(2), 194–202.

Mantegazza, M., Curia, G., Biagini, G., Ragsdale, D. S., & Avoli, M. (2010). Voltage-gated sodium channels as therapeutic targets in epilepsy and other neurological disorders. *Lancet Neurology, 9*(4), 413–424.

McCusker, E. C., Bagneris, C., Naylor, C. E., Cole, A. R., D'Avanzo, N., Nichols, C. G., & Wallace, B. A. (2012). Structure of a bacterial voltage-gated sodium channel pore reveals mechanisms of opening and closing. *Nature Communications, 3*, 1102.

Miller, A. N., & Long, S. B. (2012). Crystal structure of the human Two–Pore domain potassium channel K2P1. *Science, 335*(6067), 432–436.

Morais-Cabral, J. H., Zhou, Y., & MacKinnon, R. (2001). Energetic optimization of ion conduction rate by the K^+ selectivity filter. *Nature, 414*(6859), 37–42.

Namadurai, S., Balasuriya, D., Rajappa, R., Wiemhöfer, M., Stott, K., Klingauf, J., & Jackson, A. P. (2014). Crystal structure and molecular imaging of the Nav channel β3 subunit indicates a trimeric assembly. *Journal of Biological Chemistry, 289*(15), 10797–10811.

Ngo, V., Stefanovski, D., Haas, S., & Farley, R. A. (2014). Non-equilibrium dynamics contribute to ion selectivity in the KcsA channel. *PLoS One, 9*(1), e86079. http://dx.doi.org/10.1371/journal.pone.0086079.

Noda, M., Shimizu, S., Tanabe, T., Takai, T., Kayano, T., Ikeda, T.… Numa, S. (1984). Primary structure of electrophorus electricus sodium channel deduced from cDNA sequence. *Nature, 312*(5990), 121–127.

Noskov, S. Y., & Roux, B. (2006). Ion selectivity in potassium channels. *Biophysical Chemistry, 124*(3), 279–291.

Parsegian, A. (1969). Energy of an ion crossing a low dielectric membrane: solutions to four relevant electrostatic problems. *Nature, 221*(5183), 844–846.

Payandeh, J., Gamal El-Din, T. M., Scheuer, T., Zheng, N., & Catterall, W. A. (2012). Crystal structure of a voltage-gated sodium channel in two potentially inactivated states. *Nature, 486*(7401), 135–139.

Payandeh, J., & Minor, D. L., Jr. (2015). Bacterial voltage-gated sodium channels (BacNaVs) from the soil, sea, and salt lakes enlighten molecular mechanisms of electrical signaling and pharmacology in the brain and heart. *Journal of Molecular Biology, 427*(1), 3–30.

Payandeh, J., Scheuer, T., Zheng, N., & Catterall, W. A. (2011). The crystal structure of a voltage-gated sodium channel. *Nature, 475*(7356), 353–358.

Qiu, H., Shen, R., & Guo, W. (2012). Ion solvation and structural stability in a sodium channel investigated by molecular dynamics calculations. *Biochimica et Biophysica Acta, 1818*(11), 2529–2535.

Ragsdale, D. S., McPhee, J. C., Scheuer, T., & Catterall, W. A. (1994). Molecular determinants of state-dependent block of Na^+ channels by local anesthetics. *Science, 265*(5179), 1724–1728.

Ravindran, A., Schild, L., & Moczydlowski, E. (1991). Divalent cation selectivity for external block of voltage-dependent Na^+ channels prolonged by batrachotoxin. Zn^{2+} induces discrete substates in cardiac Na^+ channels. *Journal of General Physiology, 97*(1), 89–115.

Ren, D., Navarro, B., Xu, H., Yue, L., Shi, Q., & Clapham, D. E. (2001). A prokaryotic voltage-gated sodium channel. *Science, 294*(5550), 2372–2375.

Roux, B. (2005). Ion conduction and selectivity in K^+ channels. *Annual Review of Biophysics and Biomolecular Structure, 34*(1), 153–171.

Shaya, D., Findeisen, F., Abderemane-Ali, F., Arrigoni, C., Wong, S., Nurva, S. R., & Minor, D. L., Jr. (2014). Structure of a prokaryotic sodium channel pore reveals essential gating elements and an outer ion binding site common to eukaryotic channels. *Journal of Molecular Biology, 426*(2), 467–483.

Shaya, D., Kreir, M., Robbins, R. A., Wong, S., Hammon, J., Brüggemann, A., & Minor, D. L. (2011). Voltage-gated sodium channel (NaV) protein dissection creates a set of functional pore-only proteins. *Proceedings of the National Academy of Sciences of the United States of America, 108*(30), 12313–12318.

Smart, O. S., Goodfellow, J. M., & Wallace, B. A. (1993). The pore dimensions of gramicidin A. *Biophysical Journal, 65*(6), 2455–2460.

Stock, L., Delemotte, L., Carnevale, V., Treptow, W., & Klein, M. L. (2013). Conduction in a biological sodium selective channel. *Journal of Physical Chemistry B, 117*(14), 3782–3789.

Stock, L., Souza, C., & Treptow, W. (2013). Structural basis for activation of voltage gated cation channels. *Biochemistry, 52*(9), 1501–1513.

Stuhmer, W., Conti, F., Suzuki, H., Wang, X., Noda, M., Yahagi, N., & Numa, S. (1989). Structural parts involved in activation and inactivation of the sodium channel. *Nature, 339*(6226), 597–603.

Sun, Y. M., Favre, I., Schild, L., & Moczydlowski, E. (1997). On the structural basis for size-selective permeation of organic cations through the voltage-gated sodium channel. Effect of alanine mutations at the DEKA locus on selectivity, inhibition by Ca^{2+} and H^+, and molecular sieving. *Journal of General Physiology, 110*(6), 693–715.

Ulmschneider, M. B., Bagnéris, C., McCusker, E. C., DeCaen, P. G., Delling, M., Clapham, D. E. … Wallace, B. A. (2013). Molecular dynamics of ion transport through the open conformation of a bacterial voltage-gated sodium channel. *Proceedings of the National Academy of Sciences of the United States of America, 110*(16), 6364–6369.

Vanommeslaeghe, K., & MacKerell, A. D., Jr. (2015). CHARMM additive and polarizable force fields for biophysics and computer-aided drug design. *BBA, 1850*(5), 861–871.

Vargas, E., Yarov-Yarovoy, V., Khalili-Araghi, F., Catterall, W. A., Klein, M. L., Tarek, M., & Roux, B. (2012). An emerging consensus on voltage-dependent gating from computational modeling and molecular dynamics simulations. *Journal of General Physiology, 140*(6), 587–594.

Vora, T., Corry, B., & Chung, S. H. (2008). Brownian dynamics study of flux ratios in sodium channels. *European Biophysics Journal, 38*(1), 45–52.

Xu, P., Wang, J., Xu, Y., Chu, H., Liu, J., Zhao, M. … Li, G. (2015). Advancement of polarizable force field and its use for molecular modeling and design. In D. Wei, Q. Xu, T. Zhao, & H. Dai (Eds.), *Advance in structural bioinformatics* (Vol. 827, pp. 19–32). Netherlands: Springer.

Yamamoto, D., Yeh, J. Z., & Narahashi, T. (1984). Voltage-dependent calcium block of normal and tetramethrin-modified single sodium channels. *Biophysical Journal, 45*(1), 337–344.

Yue, L., Navarro, B., Ren, D., Ramos, A., & Clapham, D. E. (2002). The cation selectivity filter of the bacterial sodium channel, NaChBac. *Journal of General Physiology, 120*, 845–853.

Zaydman, M. A., Silva, J. R., & Cui, J. (2012). Ion channel associated diseases: overview of molecular mechanisms. *Chemical Reviews, 112*(12), 6319–6333.

Zhang, X., Ren, W., DeCaen, P., Yan, C., Tao, X., Tang, L., & Yan, N. (2012). Crystal structure of an orthologue of the NaChBac voltage-gated sodium channel. *Nature, 486*(7401), 130–134.

Zhang, X., Xia, M., Li, Y., Liu, H., Jiang, X., Ren, W., & Gong, H. (2013). Analysis of the selectivity filter of the voltage-gated sodium channel NavRh. *Cell Research, 23*(3), 409–422.

Zhou, Morais-Cabral, Kaufman, & MacKinnon. (2001). Chemistry of ion coordination and hydration revealed by a K^+ channel-Fab complex at 2.0 angstrom resolution. *Nature, 414*, 43–48.

Zuniga, L., Marquez, V., Gonzalez-Nilo, F. D., Chipot, C., Cid, L. P., Sepulveda, F. V., & Niemeyer, M. I. (Jan 25, 2011). Gating of a pH-sensitive K-2P potassium channel by an electrostatic effect of basic sensor residues on the selectivity filter. *PLoS One, 6*(1), e16141. http://dx.doi.org/10.1371/journal.pone.0016141.

CHAPTER EIGHT

Simulation Studies of Ion Permeation and Selectivity in Voltage-Gated Sodium Channels

C. Ing[*,§] **and R. Pomès**[*,§,1]
[*]Hospital for Sick Children, Toronto, ON, Canada
[§]University of Toronto, Toronto, ON, Canada
[1]Corresponding author: E-mail: pomes@sickkids.ca

Contents

Current Topics in Membranes, Volume 78
ISSN 1063-5823
http://dx.doi.org/10.1016/bs.ctm.2016.07.005

Abstract

Voltage-gated ion channels are responsible for the generation and propagation of action potentials in electrically excitable cells. Molecular dynamics simulations have become a useful tool to study the molecular basis of ion transport in atomistic models of voltage-gated ion channels. The elucidation of several three-dimensional structures of bacterial voltage-gated sodium channels (Nav) in 2011 and 2012 opened the way to detailed computational investigations of this important class of membrane proteins. Here we review the numerous simulation studies of Na^+ permeation and selectivity in bacterial Nav channels published in the past 5 years. These studies use a variety of simulation methodologies differing in force field parameters, molecular models, sampling algorithms, and simulation times. Although results disagree on the details of ion permeation mechanisms, they concur in the presence of two primary Na^+ binding sites in the selectivity filter and support a loosely coupled knock-on mechanism of Na^+ permeation. Comparative studies of Na^+, K^+, and Ca^{2+} permeation reveal sites within Nav channels that are Na^+ selective, yet a consensus model of selectivity has not been established. We discuss the agreement between simulation and experimental results and propose strategies that may be used to resolve discrepancies between simulation studies in order to improve future computational studies of permeation and selectivity in ion channels.

1. INTRODUCTION

In both eukaryotic and prokaryotic organisms, ion channels are a class of transmembrane proteins that facilitate critical physiological processes such as nerve impulses, muscle contraction, and cell signaling (Hille, 2001; Zheng & Trudeau, 2015). In excitable cells, two types of voltage-gated ion channels called Nav and Kv channels specialize in Na^+ and K^+ permeation, respectively. These channels open and close in response to transmembrane voltage and allow for one or the other of these cations to flow according to its electrochemical gradient. In mammalian skeletal muscle, the extracellular (EC) ion concentrations of Na^+ and K^+ are 145 mM and 4 mM, while intracellular (IC) ion concentrations are 12 mM and 155 mM, respectively (Ashcroft, 2000; Hille, 2001). This concentration asymmetry is exploited in a process involving selective ion permeation, with Na^+ influx through Nav channels and K^+ efflux through Kv channels giving rise to the rising and decay phases of an action potential, respectively. To understand the function of wild-type ion channels as well as disease-causing mutations (Ashcroft, 2000), it is essential to elucidate the molecular basis for selective ion permeation.

The selectivity of ion channels can be measured experimentally by several methods: from the ratio of ionic conductance, the reversal potential for currents under biionic conditions (Alam & Jiang, 2009; Hille, 2001; Liu & Lockless, 2013; Sauer et al., 2013), crystallographic titration (Alam & Jiang, 2009; Liu & Lockless, 2013; Neyton & Miller, 1988a, 1988b; Nimigean, 2002; Sauer et al., 2013), or through channel block experiments (Im & Roux, 2002; Kutzner et al., 2011; Neyton & Miller, 1988a, 1988b; Nimigean, 2002; Sotomayor et al., 2007). The selectivity of ion channels can also be estimated computationally from the ratio of ion conductance (Im & Roux, 2002; Kim & Allen, 2011; Kutzner et al., 2011; Noskov, Bernèche, & Roux, 2004; Sotomayor et al., 2007), by free-energy perturbation (Kim & Allen, 2011; Nimigean & Allen, 2011; Noskov et al., 2004; Roux et al., 2011), or by comparing the potential of mean force (PMF) for the permeation of different ion species (Catterall, 2012; Hille, 2001; Nimigean & Allen, 2011; Roux et al., 2011). In a PMF or free energy profile of ion permeation, ion binding affinity is reflected by free-energy well depth, and permeation rates are correlated with the height of free energy barriers, which together may be used to compute selectivity.

The aim of this review is to present a critical analysis and discussion of the major findings of the computational studies published to date on ion permeation and selectivity in voltage-gated Nav channels. First, we summarize the known structures of voltage-gated sodium channels (Section 2), followed by a brief discussion of permeation and selectivity in potassium channels (Section 3). We then present an overview of published computational studies of sodium channels, discussing methodological considerations such as force field, channel model, and sampling technique (Section 4). We systematically review the Na^+ binding sites and permeation mechanisms observed in simulation studies of Nav channels (Section 5). We then review computational studies of the selectivity of Nav channel for Na^+ over K^+, as well as Na^+ over Ca^{2+} (Section 6). Finally, we discuss the agreement and disagreement of these research findings (Section 7) and make recommendations for future computational studies of voltage-gated sodium channels (Section 8).

2. VOLTAGE-GATED SODIUM CHANNELS

Voltage-gated sodium channels (Nav channels) are responsible for initiating electrical impulses in neurons and other excitable cells (Catterall, 2012; Hille, 2001). Mutations in the genes encoding for human Na^+ channels result in devastating diseases, including periodic paralysis, long QT syndrome, febrile convulsions, and epilepsy (Ashcroft, 2000). Nav channels were first characterized by Hodgkin and Huxley in a series of landmark papers published in 1952. Using the voltage-clamp technique on the membrane of the squid giant axon, Hodgkin and Huxley were able to describe complex phenomena in sodium channels such as voltage-dependent activation, fast inactivation, and Na^+ selectivity over other ions (Hille, 1971; Hodgkin & Huxley 1952a, 1952b, 1952c, 1952d). In 1971, before any ion channel sequence or structure was determined, Hille (1971, 1972) was able to construct a structural model of the narrowest constriction in Nav channels, the selectivity filter or SF, by measuring the permeability of a range of organic cations. Hille predicted a rectangular SF of dimensions 3 × 5 Å that was lined with oxygen atoms and predicted that permeating Na^+ could be partially hydrated and make direct contacts with up to three oxygen ligands of the SF. In a follow-up study, Hille measured the permeability of a range of metal cations through sodium channels and found a characteristic ordering of selectivity ($Na^+ \approx Li^+ > Tl^+ > K^+ > Ca^{2+}$) (Eisenman & Horn, 1983; Hille, 1972). Based on the "field strength" theory of Eisenman (Eisenman & Horn, 1983; Noda & Numa, 1987), strong negatively charged ligands, such as carboxyl oxygen atoms, could create "high field" binding sites for small cations preferentially, and Hille found that his results were consistent with this model.

Sequencing of the mammalian Nav channel (Noda & Numa, 1987; Noda et al., 1989; Terlau et al., 1991), along with the identification of toxin binding sites within the SF (Favre, Moczydlowski, & Schild, 1996; Noda et al., 1989; Terlau et al., 1991), led to the identification of a ring of four amino acids of sequence "DEKA" responsible maintaining Na^+ selectivity. As predicted in the Eisenman model, a mutation of the glutamic acid in "DEKA" resulted in a loss of Na^+ selectivity over K^+ (Cibulsky & Sather, 2000; Ellinor et al., 1995; Favre et al., 1996; Heinemann et al., 1992). Similarly, site-directed mutagenesis studies were utilized to identify a ring of four amino acids of sequence "EEEE" responsible for maintaining Ca^{2+} selectivity in the SF of mammalian voltage-gated calcium channels (Cibulsky & Sather, 2000; Ellinor et al., 1995; Heinemann et al., 1992; Hille, 2001). Nav and Cav channels are expected to be very similar in domain architecture and detail of the SF (Catterall & Zheng, 2015; Hille, 2001; Yu & Catterall, 2004) and likely share the prokaryotic sodium channel as a common ancestor (Catterall & Zheng, 2015; Koishi et al., 2004; Ren et al., 2001; Yu & Catterall, 2004).

The bacterial sodium channel NaChBac and several prokaryotic relatives were found to possess the same "EEEE" motif as observed in SF of Cav channels. However, bacterial sodium channels form homotetrameric assemblies of a single monomer sequence, whereas the mammalian Nav and Cav channels consist of a single chain containing four domains with four unique sequences (Heinemann et al., 1992; Koishi et al., 2004; Ren et al., 2001; Shaya et al., 2011; Sun, Favre, & Schild, 1997; Tang et al., 2014; Yue, 2002). Mutagenesis studies reveal that the SF of both eukaryotic and prokaryotic Nav channels can be modified to impart Ca^{2+} selectivity (Cibulsky & Sather, 2000; Ellinor et al., 1995; Heinemann et al., 1992; Shaya et al., 2011; Sun et al., 1997; Tang et al., 2014; Yue, 2002). Reciprocally, mutations in the SF of Cav channels can be introduced to reduce Ca^{2+} selectivity (Catterall, 2012; Cibulsky & Sather, 2000; Ellinor et al., 1995; Payandeh et al., 2012). Bacterial voltage-gated sodium channels offer an ideal model for gaining an understanding of the principles of Na^{+} permeation and selectivity in a large class of voltage-gated ion channels.

Sixty years after the seminal work of Hodgkin and Huxley, the first three-dimensional structure of a bacterial voltage-gated sodium channel, NavAb, was solved by X-ray crystallography (Catterall, 2012; Payandeh et al., 2012). High-resolution structures are essential for elucidating the relationship between structure and function of Nav channels, providing a foundation for understanding the molecular basis for disease-causing mutations and opening the way toward structure-based drug design for this class of proteins. A molecular rendering of a simulation cell containing a lipid-embedded voltage-gated Na^{+} channel NavAb is depicted in Fig. 1A. Subsequently, structures of Nav channels were obtained in closed [NavAb (Payandeh et al., 2012) and NavAe (Shaya et al., 2014)], inactivated [NavAb (Payandeh et al., 2013) and NavRh (Zhang, Ren, et al., 2013)], and open [NavMs (Bagnéris et al., 2013; McCusker et al., 2012; Naylor et al., 2016)] states. These states refer to the conformations of the voltage-sensing domains and the pore domain as the channel moves through a closed-open-inactive state cycle (Amaral et al., 2012; Bagneris et al., 2014). Although both NavMs and NavAe structures lack voltage-sensing domains, experimental studies suggest that pore-only Nav channels still undergo opening and closing transitions along with selective ion flux (McCusker et al., 2011; Shaya et al., 2011).

The pore domain of bacterial Nav channels consists of four identical subunits comprised of two transmembrane helices (S5–S6). As ions enter the pore through a region called the EC funnel, they must first pass through the SF. The SF and surrounding features of the EC funnel are the primary structural determinants of Na^{+} selectivity. As ions proceed beyond the SF,

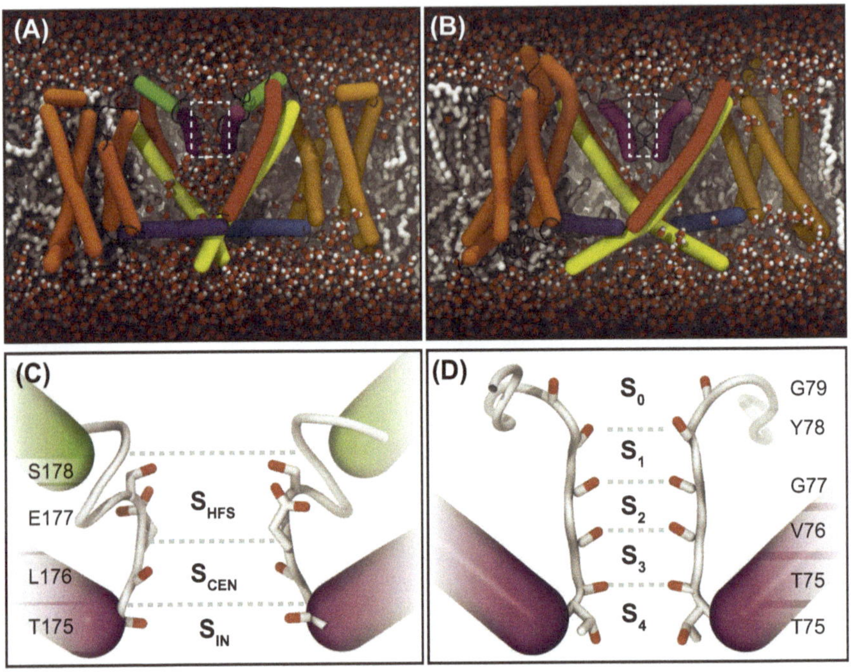

Figure 1 Molecular rendering of voltage-gated ion channels and selectivity filter structure of Na^+ and K^+ selective channels. Molecular snapshots of the closed states of (A) sodium channel NavAb (Payandeh et al., 2012) and (B) a potassium channel chimera, Kv1.2/Kv2.1 (Long et al., 2007), after orientation in a lipid bilayer using coarse-grained molecular dynamics (Stansfeld et al., 2015). Only two of four subunits are depicted to assist in visualizing the protein interior. Lipid molecules are shown in white and water molecules in red and white. Cylinders depict the helices forming the voltage-sensing domain (helices S1–S4, orange), the S4–S5 linker (blue), pore-domain helix S5 (red), the pore helix (purple), and pore-domain helix S6 (yellow). NavAb contains an additional "P2" helix near the selectivity filter (green). A *white dashed box* indicates the position of the selectivity filter. Molecular renderings of two opposite subunits of the selectivity filter in (C) NavAb (Payandeh et al., 2012) and (D) KcsA (Zhou et al., 2001). Residue IDs indicate the sequence of channel ligands in the selectivity filter. Inside the SF, each label indicates an Na^+ or K^+ binding site observed in molecular simulations of NavAb and KcsA, respectively. All structures were rendered using visual molecular dynamics (VMD) with the Bendix algorithm (Dahl, Chavent, & Sansom, 2012; Humphrey, Dalke, & Schulten, 1996).

they enter a water-filled expansion within the channel, named the central cavity (CC). At the base of the CC is a constriction called the IC gate that is formed by close packing of S6 helices when the pore domain is in a closed state. Each subunit of the central pore domain is flanked by four transmembrane helices (S1–S4) that form a voltage-sensing domain found in both Nav and Kv channels (Fig. 1A and B). The voltage-sensing domain is connected to the pore domain through the S4—S5 linker. Upon membrane depolarization, it is thought that the S4 transmembrane helix shifts in the direction of the EC surface, and this motion is translated through the linker to open the IC gate of the pore domain through rearrangement of S6 helix packing (Catterall & Zheng, 2015; Payandeh et al., 2012). Much of what is known about the structural basis for voltage-sensitive gating comes from studies of Kv channels (Jensen et al., 2012; Vargas et al., 2012). Similar to Nav channels, each subunit of voltage-gated potassium channels is comprised of six transmembrane helices. However, important structural differences between these two classes of channels are found in the SF.

The sequence of each of the four loops making up the SF in bacterial channels NavAb (Payandeh et al., 2012), NavMs (McCusker et al., 2012), and NavAe (Shaya et al., 2014) is TLES (Fig. 1C). The Nav channel SF is shorter than that of the prototypical K^+ channel SF (Fig. 1D) (Doyle et al., 1998; Hille, 1971) and is wide enough to accommodate partially or even fully hydrated Na^+ ions, in agreement with Hille's prediction (Hille, 1971; Payandeh et al., 2012). The relative Na^+ and K^+ currents through NavAb suggest that this channel is only weakly Na^+ selective by a ratio of 10:1 (McCusker et al., 2012; Payandeh et al., 2012). No bound cation was observed in the NavAb structure. Putative ligands of permeating cations in the lumen of the SF include the backbone carbonyl groups of residues T175 and L176 (in the residue numbering of NavAb), as well as the carboxylate and hydroxyl groups of the side chains of residues E177 and S178, respectively.

Based on the available structures of Nav channel SFs, whose minimum van der Waals diameter is ~5 Å (Eisenman & Horn, 1983; McCusker et al., 2012; Payandeh et al., 2012), partially hydrated Na^+ may readily pass through the SF and make direct interactions with channel ligands. In accord with the high-field-strength model of Eisenman (Eisenman & Horn, 1983; Payandeh et al., 2012), it was posited that Na^+ ions would exchange some of the water molecules in their first solvation shell with carboxylate oxygen atoms of E177 at a site referred to as S_{HFS} (Payandeh et al., 2012). Upon deeper entry of Na^+ into the SF, the ion was suggested to become fully hydrated at a site centrally coordinated by backbone carbonyls of L176 at a site referred to as S_{CEN}. Finally, ions may also become coordinated to T175 at a site referred to as S_{IN} before entering the CC.

Several mammalian voltage-gated Ca^{2+} channels possess the same characteristic EEEE ring that was shown to impart Na^+ selectivity in the SF of bacterial Nav channels. A 4.2-Å resolution cryo-EM structure of the rabbit Cav1.1 channel was recently solved, but its resolution was not high enough to accurately resolve the side chain orientations within the SF (Wu et al., 2015). Interestingly, Tang et al. were able to create a Ca^{2+} selective mutant of NavAb by introducing DDDD rings into the SF of NavAb (Catterall & Zheng, 2015; Tang et al., 2014). The high-resolution crystallographic structure of these mutants revealed that there were only minor changes in backbone geometry of the SF compared to NavAb. Together, these studies suggest that selectivity is governed predominately by side chain interactions in this channel rather than the backbone geometry of the SF per se. Based on electron density within the SF of crystallographic structures obtained with Ca^{2+} in the crystallization medium, three binding sites were identified on the central axis of the channel, implying that Ca^{2+} is bound in a hydrated form. While this work supports a mechanism for Ca^{2+} conduction based on structural evidence, it does not isolate the energetic factors which govern Ca^{2+} selectivity, or provide an explanation for Na^+ selectivity over Ca^{2+} in NavAb.

In K^+ channel structures, electron density attributable to ions was observed directly within the SF and used to infer a permeation mechanism (Zhou et al., 2001). Although no electron density was observed at the expected SF binding sites in the NavAb structure (Payandeh et al., 2012), three highly occupied Na^+ binding sites were observed in the SF of a 2.7-Å NavMs structure (Naylor et al., 2016). All three binding sites were identified on the central axis of the channel, with positions that correspond to S_{HFS} site, a site slightly above S_{CEN}, and a site slightly above S_{IN}, respectively. From these observed binding sites, the authors infer that Na^+ remains fully hydrated as it permeates the SF. In addition, on-axis Ca^{2+} binding sites were identified, likely fully hydrated, above S_{HFS} and at S_{CEN}/S_{IN} in NavAe (Shaya et al., 2014) and NavRh structures (Zhang, Ren, et al., 2013), respectively. While electron density provides experimental support for ion binding, ion permeation per se is a dynamic process that may not be captured in static crystallographic structures.

3. MECHANISM OF ION PERMEATION AND SELECTIVITY IN K^+ CHANNELS

The first crystallographic structure of a bacterial K^+ channel, KcsA (Doyle et al., 1998), was a landmark in the study of the molecular basis of biological ion transport. The pore domain of this channel is homologous to that of a large class of channels including eukaryotic and prokaryotic Kv channels. Multiple structures of KcsA have been solved, as well as the structures of voltage-gated potassium channels [KvAP (Jiang et al., 2003), Kv1.2 (Long, Campbell, & Mackinnon, 2005), the Kv1.2/Kv2.1 chimera (Fig. 1B) (Long et al., 2007), and KvLm (Santos et al., 2012)], which reveal structural information about voltage regulation. Thanks to these advances, detailed structural analysis can now be performed on a wide class of K^+ channels, including members of the inward-rectifying, two-pore-domain, and ligand-gated channel families (Kuang, Purhonen, & Hebert, 2015; Tian et al., 2013).

The SF of K^+ channels contains four sequential K^+ binding sites (S_1–S_4) formed by the oxygen atoms of protein backbone carbonyl groups, with one additional site (S_0) lying outside the SF (Fig. 1D) (Zhou et al., 2001). As ions move from the bulk solution to the channel, the small radius of the SF requires that K^+ enter the SF in a dehydrated state. The energy penalty of removing water molecules from the first hydration shell of the ion is compensated by direct contacts to as many as eight oxygen atom ligands in the core binding sites of the SF. Although this process is thought to confer high selectivity of K^+ channels against Na^+ (Nimigean & Allen, 2011), the molecular basis for K^+ selectivity is still discussed in the literature (Andersen, 2011).

Selective permeation in ion channels reflects the balance between a multitude of pairwise interactions involving ions, protein groups, and water molecules. As ions enter the confined geometry of a pore, the energetic cost of ion and pore desolvation must be compensated by favorable ion–protein interactions which must be weak enough to prevent bound ions from blocking permeation (Alam & Jiang, 2011; Roux et al., 2011). A long-standing mechanism of K^+ permeation called "knock-on" was hypothesized by Hodgkin and Keynes (1955) and later observed in computer simulations (Bernèche & Roux, 2001, 2003). In this mechanism, the entry of K^+ into an SF occupied by multiple single-file K^+ ions causes the simultaneous exit of a K^+ on the opposite side of the SF through ion–ion repulsion. K^+ permeation is now widely accepted to occur through a knock-on mechanism (Åqvist & Luzhkov, 2000; Domene & Sansom, 2003; Jensen et al., 2010; Roux, 2005), although the details of water–ion cotranslation in this process have recently been called into question in a study emphasizing the destabilizing role of direct contacts between permeating cations in achieving a rate of K^+ permeation approaching the diffusion limit (Kopfer et al., 2014).

Experimental and computational studies have led to several proposed permeation and selectivity mechanisms for K^+ channels (Andersen, 2011; Bernèche & Roux, 2001; Domene & Sansom, 2003; Furini & Domene, 2013; Kopfer et al., 2014; Zhou et al., 2001). The "snug-fit" model of selectivity suggests that the SF distinguishes ions based on ionic radius (Na^+, 1.16 Å; K^+, 1.52 Å; Ca^{2+}, 1.14 Å). In this model, the distance from an ion to "cages" of channel ligands largely determines ion selectivity, and conduction occurs through electrostatic repulsion between ions (Bezanilla & Armstrong, 1972; Mullins, 1960). However, this model was unable to reconcile the measurement that thermal fluctuations of the SF were larger than the difference between the ionic radius of Na^+ and K^+ (Noskov et al., 2004), supporting a model of ion selectivity emphasizing the interplay of attractive ion–ligand interactions and repulsive ligand–ligand interactions in a dynamic pore interior (Noskov et al., 2004; Noskov & Roux, 2006). Subsequent simulation-based models of K^+ channel selectivity rely on both thermodynamic and kinetic factors (Bostick & Brooks, 2007; Derebe et al., 2011; Sauer et al., 2011; Thompson et al., 2009; Varma & Rempe, 2007). This perspective is supported by recent studies of selective and nonselective mutants of NaK channels, which suggest that both thermodynamic and kinetic factors must be considered to arrive at an accurate model of K^+ selectivity (Liu & Lockless, 2013; Sauer et al., 2013).

While these studies have provided models for the molecular basis of K^+ selectivity, there is no consensus model described in the literature. Although the applicability of these models to the mechanism of Na^+ selectivity in Nav channels is also unknown, the availability of multiple voltage-gated sodium channel structures makes it possible to refine mechanistic hypotheses about ion permeation and selectivity which were originally proposed for K^+ channels.

4. MOLECULAR SIMULATIONS OF SODIUM CHANNELS

Computer simulations of ion channels may be used to identify the molecular driving forces for ion binding, permeation, and selectivity. To study these concepts, a simulation methodology must be selected that establishes a channel model, a force field, a sampling algorithm, and the extent of sampling. The accuracy of simulation results depends heavily on these methodological choices due to a combination of both systematic and statistical errors. Statistical errors refer to limitations due to insufficient data set sizes, whereas systematic errors reflect biases resulting not only from inherent limitations of the force field and of the channel model, but also from the simulation protocol itself. Sampling errors include both statistical and systematic errors. Systematic sampling errors arise from imperfect initial conditions, which incur a need for equilibration of the system and from rare conformational transitions due to "hidden" or unforeseen energy barriers, which can make equilibration exceed the time of the simulations (Neale et al., 2013). Before discussing simulation results, in this section we provide an overview of the methods utilized in computational studies of Nav channels (see Tables 1 and 2) and comment briefly on their relationship to simulation error.

4.1 Channel model and force field

Known structures of sodium channels differ not only in the amino acid sequence of the pore domain, but also in the presence of voltage-sensing or IC domains and in the openness of the IC gate (as discussed in Section 2). The latter two elements may contribute to systematic simulation errors that cannot be reduced by long simulation timescales. Most studies of ion permeation and selectivity were conducted in a closed NavAb channel under the assumption that the state of the IC gate is decoupled from the SF—an assumption supported by the high degree of structural similarity of the SF in open and closed channel structures (Bagneris et al., 2014)—and that permeation and selectivity are therefore unaffected. In simulations conducted in the open NavMs channel, strong structural restraints on transmembrane helices were needed to prevent the closure of the IC gate. Although the NavMs channel is functional without voltage-sensing domains, the effect of structural restraints on the structure and dynamic fluctuations of the pore domain has not been quantified and presents a possible source of systematic force field error. Simulations utilizing structural restraints during production simulations are noted in Tables 1 and 2.

An accurate model to study ion conduction and selectivity is one in which the channel samples conformations as close as possible to the physiological or experimental environment. Most ion channel models obtained from crystallographic structures are inherently out of equilibrium due to the crystal packing, which may not be representative of the native environment of the lipid bilayer, and underdetermined side chain orientations due to limited structural resolution. As such, these models must be allowed to equilibrate at physiological temperature, pressure, and ion concentration in a lipid bilayer so as to adopt a conformation that is less biased by these factors (Kandt, Ash, & Peter Tieleman, 2007). All simulations were performed at physiological temperature (300–320 K) and pressure (1 atm), but higher-than-physiological ion concentrations were often used (300–500 mM). A high ion concentration is frequently utilized in unbiased simulations to increase sampling of ion binding and permeation events. The lipid composition of each bilayer varied considerably across studies, including DMPC, DPPC, DOPC, and POPC with various bilayer patch sizes (Tables 1 and 2). However, the rationale for selecting a lipid species is not strongly justified in any of the studies reviewed herein. A model lipid was picked either to reproduce experimental conditions [NavAb was crystallized in a DMPC-based bicelle (Payandeh et al., 2012)], to match the size of a hydrophobic belt on the protein–lipid interface of the channel, or because simulations using a specific lipid are frequently performed in the literature. All patch clamp studies of bacterial sodium channels (NavAb, NavMs, and NaChBac) have been performed in mammalian cells (Finol-Urdaneta et al., 2014; Payandeh et al., 2012; Ulmschneider et al., 2013), and it is extremely challenging to reproduce these complex experimental conditions with molecular models. The indirect effects of the lipid bilayer composition and ionic concentration on the conformational fluctuations of sodium channels are thus an additional source of systematic error.

Simulations of Nav channels have been performed with a number of protein and ion force fields (Tables 1 and 2). The majority of these simulations have utilized CHARMM parameters (MacKerell et al., 1998, 2004), with additional studies conducted using OPLS (Jorgensen, Maxwell, & Tirado-Rives, 1996; Kaminski et al., 2001) and AMBER parameters (Hornak et al., 2006). Ions were modeled using parameters of Åqvist (1990), Joung and Cheatham, (2008), Beglov and Roux (1994), and Noskov and Roux (2008). Unless otherwise stated, studies are assumed to use the default ion parameters packaged with protein force field. Some studies using the CHARMM force field included additional nonbonded interaction adjustments for ions and carbonyl oxygen atoms (Noskov et al., 2004). While force fields are parameterized to reproduce similar experimental and quantum mechanical measurements for folded proteins, they can result in remarkably different results (Lindorff-Larsen et al., 2012; Rauscher et al., 2015). As such, force fields are a primary source of systematic error that must be carefully considered when comparing simulation studies.

Given that a diverse combination of channel models and force fields have been used to study Nav channels, it is tempting to attribute any possible discrepancies to these factors. However, to assess the effect of these methodological choices possible, statistical and systematic sampling errors must first be reduced through appropriate sampling.

4.2 Sampling technique

Various molecular dynamics sampling techniques have been utilized to study ion permeation and selectivity in ion channels. These techniques can be broadly classified as biased (Table 1) and unbiased (Table 2) sampling, where bias refers to additional energy terms added on top of the potential energy function or force field. The accuracy of both of these approaches relies on the convergence of sampling data sets, such that time-averaged measurements can approach true statistical ensemble average measurements in the limit of long sampling times. In practice, different simulation observables converge on different timescales, requiring careful examination of observables as a function of simulation time. Such convergence plots, however, are rarely presented in molecular simulation studies of sodium channels. Sampling is further complicated by long timescale processes inherent in biological systems, whereby the existence of rare (or unanticipated) transitions in "hidden" degrees of freedom can impose an unintended bias on conformational sampling even when statistical convergence of an observable appears to have been achieved (Neale et al., 2011). Here we review both biased and unbiased simulation methodologies used to study sodium channels and describe the advantages and disadvantages of each approach.

4.2.1 Biased sampling

In biased-sampling studies of ion channels, the movement of permeants is forced to follow a predefined pathway on a reaction coordinate. Biased-sampling algorithms comprise umbrella sampling (Torrie & Valleau, 1974), metadynamics (Laio & Gervasio, 2008; Laio & Parrinello, 2002), and nonequilibrium pulling approaches based on Jarzynski's (1997) equality. Standard umbrella sampling methodology in the context of ion channels consists of multiple simulations which include biasing potentials (usually harmonic) on the position of one or more permeating ions. The outcome of these simulations can be rigorously debiased and assembled a posteriori to produce a PMF profile that represents the reversible thermodynamic work or free energy for ion permeation along the predefined reaction coordinate. This reaction coordinate can be mono- or bidimensional, leading to 1D- or 2D-PMFs. In metadynamics, the PMF is determined by introducing a history-dependent potential energy term that biases the motion of ions to escape free-energy minima. Pulling experiments are used to study nonequilibrium processes (such as the movement of ions along an electrochemical gradient) by gradually moving the center of a harmonic biasing force along a reaction coordinate. The primary advantage of these techniques is that they can efficiently sample the energetics of rare events that may exist along the predefined reaction coordinate. As such, the effectiveness of biased sampling largely depends on the definition of a reaction coordinate that can capture the physiological conduction mechanism, including but not limited to the number of permeating ions, their relative arrangement, and the range of their motion within the pore. Umbrella sampling studies typically utilize aggregate sampling of tens to hundreds of nanoseconds to converge the potential energy landscape of ion permeation in ion channels (Allen, Andersen, & Roux, 2006b; Fowler et al., 2013). However, even though biased approaches facilitate sampling along the reaction coordinate, longer timescale simulations may still be required to converge the PMF when other slowly relaxing degrees of freedom are present. With this caveat in mind, biased sampling can be used to efficiently compute estimates of ionic binding affinity and free energy barriers, and conduction rates can be approximated from the PMF using transition state theory and stochastic simulations (Bernèche & Roux, 2003; Crouzy, Woolf, & Roux, 1994).

4.2.2 Unbiased sampling

In unbiased-sampling approaches, the movement of ions is not restrained and simulations can be performed either at equilibrium or nonequilibrium conditions (such as in the presence of an ionic gradient or an external applied voltage). In this approach, ions diffuse in and out of the channel as governed by the force field and molecular dynamics algorithm. The main advantage of this approach is that ion conduction can proceed without requiring assumptions about a reaction coordinate (such as ionic degrees of freedom or ionic occupancy of the pore). Consequently, conformational states and transitions can be observed that may have otherwise been precluded or discouraged in biased simulations as a result of a suboptimal choice of reaction coordinate. Variable ionic occupancy of the pore may promote the relaxation of orthogonal degrees of freedom within the channel, such as side chain rearrangement within the SF. Inversely, biased sampling can effectively reduce the number of accessible pathways between states, leading to artificially slow transitions. Unbiased simulations allow for ionic currents to be computed directly by counting permeation events, rather than by building a kinetic model based on the barrier heights and well depths in a PMF. Comparing nonequilibrium and equilibrium sampling requires careful consideration due to possible systematic errors introduced by applying an external voltage. Nevertheless, an advantage of nonequilibrium over equilibrium sampling is that applying an external voltage can accelerate the number of conduction events in open-state models of the channel, improving estimates of ion conduction. In general, unbiased-sampling approaches are well suited to studies of Nav channels provided sufficient computational resources are available, given that permeation events occur spontaneously over the course of microsecond simulations and that optimal reaction coordinates for ion conduction have yet to be established.

5. ION BINDING AND PERMEATION IN Nav CHANNELS

Here we review studies aimed at identifying Na^+ binding sites and proposed permeation mechanisms in Nav channels. Methodological choices pertaining to force field, channel model, and sampling methods can have a significant impact on computed ion binding and permeation properties. These factors contribute to systematic and statistical errors that prevent direct comparisons between many studies. We note that simulation analysis and aggregate data set sizes are similar for studies with a common sampling approach. As such, we present binding and permeation studies grouped by sampling approach; biased (Section 5.1), unbiased equilibrium (without external applied voltage, Section 5.2), and unbiased nonequilibrium sampling (with external applied voltage, Section 5.3). Unless otherwise specified, all glutamic acid side chains in the "EEEE" motif of the Nav channel SF were deprotonated, although several studies considered the effect of protonating one or more carboxylic groups in the SF on permeation properties (Section 5.4).

5.1 Biased simulations

Biased simulations were performed for NavAb in multiple studies, utilizing aggregate sampling data sets from 100 to 1000 ns (Table 1) (Corry, 2013; Corry & Thomas, 2012; Domene, Barbini, & Furini, 2015; Finol-Urdaneta et al., 2014; Furini, Barbini, & Domene, 2014; Furini & Domene, 2012; Mahdavi & Kuyucak, 2015; Ngo et al., 2016; Stock et al., 2013). These studies are largely in agreement with respect to the presence of a loosely coupled knock-on mechanism involving two ions and do not report extensive conformational fluctuations within the SF.

By using umbrella sampling with aggregate simulation times of 0.2–1 μs (Table 1), Corry et al. identified energetic minima at the binding sites S_{HFS}, S_{CEN}, and S_{IN} (Corry, 2013; Corry & Thomas, 2012), in agreement with the sites suggested by Payandeh et al. (2012). These single-ion PMFs feature a 3–4 kcal/mol barrier for Na^+ to move from S_{HFS} to S_{CEN}, and 1 kcal/mol to traverse a S_{CEN}-to-S_{IN} barrier to enter the CC. A 2D-PMF revealed that this S_{HFS} to S_{CEN} barrier was lowered to 2.5 kcal/mol when the SF was occupied by two ions. Corry et al. inferred that the mechanism of permeation involves a "loosely coupled knock-on" process whereby an incoming Na^+ ion displaces a bound Na^+ at S_{HFS} and causes an ion at S_{CEN} or S_{IN} to enter the CC. Unlike knock-on permeation in K^+ channels, this mechanism is "loosely coupled" because the motion of ions are not necessarily concerted.

A short unbiased simulation of NavAb (140 ns) was performed by Carnevale, Treptow, and Klein (2011) leading to qualitative observations of Na^+ binding at sites S_{HFS} and S_{IN}. Ions were found to bind with a complete first hydration shell on the central axis of the SF and to occasionally coordinate channel ligands of both glutamic acid and leucine directly. These observations were validated quantitatively by metadynamics studies of two-ion Na^+ conduction (Stock et al., 2013). A multiion 1D-PMF indicated that the higher-affinity binding sites S_{HFS} and S_{CEN} were equally populated and separated by a 2.5 kcal/mol barrier. 2D-PMFs indicated that the minimum-energy pathway for ion conduction occurs when there is double Na^+ occupancy of S_{HFS}, followed by S_{HFS} and S_{CEN} occupancy, double occupancy of S_{CEN}, and finally occupancy of S_{CEN} and S_{IN}. This work suggests a loosely coupled knock-on permeation mechanism, whereby an incoming ion may pass another at both dual-occupancy sites.

Umbrella sampling calculations found a similar knock-on conduction mechanism involving Na^+ binding sites at both S_{HFS} and S_{CEN} (Furini et al., 2014; Furini & Domene, 2012). Single-ion PMF calculations revealed a barrier between S_{HFS} and S_{CEN} of 3.5 and 1.5 kcal/mol in outward and inward permeation directions, respectively, with the S_{CEN} site favored by 1.5 kcal/mol (Furini et al., 2014). When channel occupancy is restricted to two Na^+, 2D-PMFs indicate that conduction proceeds through a single minimum-energy pathway. Subsequently, a sampling algorithm called bias-exchange metadynamics was used to study ion conduction in a TM-helix truncated model of the NavAb SF (Domene et al., 2015). This algorithm utilizes exchanges between system conformations within an ensemble of simulation replicas where single ions are subject to a history-dependent potential to accelerate sampling of unexplored regions of the reaction coordinate. A 1D-PMF profile computed from multiion simulations indicated that there are four distinct binding sites in the SF separated by 3 kcal/mol barriers (two binding sites near S_{HFS} and two near S_{CEN}). A 2D-PMF surface computed with metadynamics identified the same minimum free energy pathway as described in Furini and Domene (2012), suggesting a loosely coupled conduction mechanism. However, this approach also identified multiple low-barrier conduction pathways involving two to three ions in the SF, including a state with a global free energy minimum with two Na^+ in S_{CEN} and one Na^+ in S_{HFS}.

Nonequilibrium pulling simulations were used by Ngo et al. (2016) to study the permeation of three Na^+ ions through the pore domain. The ionic occupancy of the SF was fixed based on unbiased simulations that suggest the conductive state of NavAb may involve as many as three ions (Boiteux, Vorobyov, & Allen, 2014; Chakrabarti et al., 2013). These simulations utilized a novel stepwise pulling methodology that includes multiple relaxation periods (0.5 ns or 3 ns) as all three ions were simultaneously pulled along the channel axis (Ngo, 2012). Nonequilibrium pulling simulations were used to compute the energetics of inward and outward conduction under conditions similar to applied voltage. Ion biasing force constants were selected to reproduce applied forces similar to biological membrane potentials (−100 to 100 mV). These simulations support a loosely coupled knock-on mechanism in single-file arrangement whereby ions line up parallel to the channel axis. During inward pulling simulations, all three Na^+ were at global free-energy minima when the group was centered at and below the E177 side chains (between S_{HFS} and S_{CEN}), with a secondary energy minimum when ions were interacting with S178 and E177 (S_{HFS}). During outward pulling experiments, a single free energy minimum was observed when ions were positioned about T175 (S_{IN}). This result suggests that Na^+ conduction asymmetry exists for Na^+ at low applied voltages, in agreement with two other nonequilibrium studies (Ke, Timin, & Stary-Weinzinger, 2014; Stock et al., 2013).

Since no eukaryotic structure is available as yet, homology models of mammalian channel Nav1.4 have been constructed using NavAb as a template (Chen, Robinson, & Chung, 2014; Gosselin-Badaroudine et al., 2012; Mahdavi & Kuyucak, 2014; Tikhonov & Zhorov, 2012). A single-ion PMF with aggregate sampling of 270 ns suggests that Na^+ permeation occurs through a simple knock-on mechanism with single-ion occupancy within the SF and single-ion occupancy of a site in the EC vestibule (Mahdavi & Kuyucak, 2015).

Table 1 Summary of biased simulations of Nav channels

Author	Protein model	Force field/restraints	System preparation	Data	Ion concentration
Corry and Thomas (2012)	Closed pore w/VSDs NavAb I217C (3RVY)[1]	CHARMM22 + CMAP protein,[a] CHARMM27 POPC lipids,[b] Joung and Cheatham ions.[j] 0.01 kcal/mol/A^2 restraints on TM helices.	dt = 1 fs, NPT T = 298 K Umbrella sampling Three E177 protonation states	1D-PMF 71 × 1 ns 2D-PMF 2000 × 0.5 ns	300 mM NaCl
Furini and Domene (2012)	Closed pore, no VSDs NavAb I217C (3RVY)[1]	CHARMM22 + CMAP protein,[a] CHARMM27 DOPC lipids,[b] CHARMM22 ions.[c]	dt = 2 fs, NPT T = 300 K Umbrella sampling	1D-PMF 20 × 0.5 ns 2D-PMF 270 × 0.5 ns	~66 mM NaCl or ~66 mM KCl
Corry (2013)	Closed pore (Q115 to M221), no VSDs NavAb I217C (3RVY)[1]	CHARMM22 + CMAP protein,[a] CHARMM36 POPC lipids,[c] Joung and Cheatham ions.[j] 0.01 kcal/mol/A^2 restraints on TM helices.	dt = 1 fs, NPT T = 298 K Umbrella sampling	1D-PMF 101 × 2.0 ns 2D-PMF 695 × 2.0 ns	250 mM NaCl or 250 mM $CaCl_2$
Stock et al. (2013)	Open-Activated Model w/VSDs (Amaral et al., 2012) NavAb wild-type (3RVY)[1]	CHARMM22 + CMAP protein,[a] CHARMM27 POPC lipids,[b] CHARMM22 ions.[d]	dt = 2 fs, NVT T = 300 K Metadynamics	1 × 0.18 μs	100 mM NaCl
Furini et al. (2014)	Closed pore, no VSDs NavAb I217C (3RVY)[1]	CHARMM22 + CMAP protein,[a] CHARMM27 DOPC lipids,[b] CHARMM22 ions.[d]	dt = 2 fs, NPT T = 300 K Umbrella sampling All E177 protonation states	1D-PMF 22 × ~2.4 ns 2D-PMF 175 × ~2.3 ns	150 mM NaCl

Finol-Urdaneta et al. (2014)	WT/E177D. Closed pore w/VSDs NavAb (3RVY)[1]	CHARMM22+CMAP protein [a] CHARMM27 DMPC lipids,[b] CHARMM36 ions with carbonyl NBFIX.[d,e]	dt = 1 fs, NPT T = 315 K Umbrella sampling	1D-PMF 50 × 3 ns 2D-PMF 1045 × 0.5 ns	150 mM NaCl
Domene et al. (2015)	SF Model (E154 to Y193), no S5–S6, no VSDs NavAb I217C (3RVY)[1]	CHARMM22 + CMAP protein,[a] CHARMM22 ions.[d] 1.19 kcal/mol/Å^2 restraints on TM helices.	dt = 2 fs, NPT T = 300 K Bias-exchange metadynamics Umbrella sampling	2D-PMF 3 × 100 ns 2D-PMF 534 × 2.0 ns	0 mM NaCl (+counterions)
Mahdavi and Kuyucak (2015)	Nav1.4 Model, closed pore, no VSDs NavAb I217C (3RVY),[1] NavAb WT (4EKW)[5]	CHARMM36 + CMAP protein,[a] CHARMM36 POPC lipids,[c] CHARMM22 ions with carbonyl NBFIX.[d,e]	dt = 2 fs, NPT T = 300 K	1D-PMF (no ions in SF) 45 × 6 ns 1D-PMF (1 ion in SF) 45 × 6 ns	100 mM NaCl
Ngo et al. (2016)	Closed pore w/VSDs NavAb I217C (3RVY)[1]	CHARMM22 + CMAP protein,[a] CHARMM36 POPC,[c] CHARMM36 ions with carbonyl NBFIX.[d,e]	dt = 2 fs, NPT T = 300 K	Nonequilibrium pulling in incremental steps 18 × 3 ns (inward for 3 Na^+, 3 K^+, 2 Na^+ 1 K^+, outward for 3 Na^+, 3 K^+)	~150 mM NaCl

All constant pressure simulations were conducted at 1 bar, and all water models were TIP3P. Restraints were applied to C-alpha atoms in all studies where restraints were used.

Structures/models: [1](Payandeh et al., 2012), [2](Zhang, Ren, et al., 2013), [3](McCusker et al., 2012), [4](Amaral et al., 2012), [5](Zhang, Ren, et al., 2013; Payandeh et al., 2013).

Force fields: [a](MacKerell et al., 1998, 2004), [b](Feller & MacKerell, 2000), [c](Klauda, Venable, & Freites, 2010), [d](Beglov & Roux, 1994), [e](Beglov & Roux, 1994; Noskov & Roux, 2008), [f](Hornak et al., 2006), [g](Siu et al., 2008), [h](Berger, Edholm, & Jähnig, 1997), [i](Cordomí, Caltabiano, & Pardo, 2012), [j](Joung & Cheatham, 2008), [k](Jorgensen et al., 1996; Kaminski et al., 2001), [l](Åqvist, 1990), [m](Best et al., 2012).

Table 2 Summary of unbiased simulations of Nav channels

Author	Protein model	Force field/restraints	System preparation	Data	Ion concentration
Carnevale et al. (2011)	Closed pore w/VSDs NavAb wild-type (3RVY)[1]	CHARMM22 + CMAP protein,[a] CHARMM27 POPC lipids,[b] CHARMM22 ions.[c]	dt = 2 fs, NPT T = 300 K	1 × 0.14 μs	70 mM NaCl
Zhang, Xia, et al. (2013)	WT/DEKA/DEAA. Closed pore (E118 to L228), no VSDs NavRh, (4DXW)[2]	CHARMM22 + CMAP protein,[a] CHARMM27 POPC lipids,[b] CHARMM22 ions.[c]	dt = 2 fs, NPT T = 310 K	2 × 0.05 μs 2 × 0.05 μs 2 × 0.05 μs …	70 mM NaCl 70 mM $CaCl_2$ 100 mM NaCl +33 mM $CaCl_2$ More
Stock et al. (2013)	Open-Activated Model w/VSDs[4] NavAb wild-type (3RVY)[1]	CHARMM22 + CMAP protein,[a] CHARMM27 POPC lipids,[b] CHARMM22 ions.[c]	dt = 2 fs, NVT T = 300 K Charge imbalance (−600 mV)	1 × 0.5 μs (each voltage)	100 mM NaCl
Ke, Zangerl, and Stary-Weinzinger (2013)	Closed pore w/VSDs NavAb wild-type (3RVY)[1]	AMBER99sb protein,[f] AMBER GAFF DOPC lipids,[g] Aqvist ions.[l]	dt = 2 fs, NPT T = 310 K	1 × 0.1 μs 1 × 0.1 μs	100 mM NaCl 100 mM $CaCl_2$
Ulmschneider et al. (2013)	Open pore, no VSDs NavMs (4F4L)[3]	CHARMM22 + CMAP protein,[a] CHARMM36 POPC lipids,[c] Joung and Cheatham ions.[j] 1 kcal/mol/A^2 restraints on TM helices.	dt = 2 fs, NPT T = 310 K External applied field (0,… , -600 mV)	1 × 1 μs (each voltage) 1 × 1 μs (each voltage)	500 mM NaCl 500 mM KCl
Chakrabarti et al. (2013)	Closed pore w/VSDs NavAb I217C (3RVY)[1]	OPLS-AA/L protein,[k] Berger DMPC lipids,[h] Aqvist ions.[l]	dt = 2 fs, NPT T = 300 K	47 × 0.5 μs 5 × 0.2 μs	150 mM NaCl 0 mM NaCl
Xia et al. (2013)	WT/DEKA/DEAA. Closed pore (E118 to L228), no VSDs NavRh, (4DXW)[2]	CHARMM22 + CMAP protein,[a] CHARMM27 POPC lipids,[b] CHARMM22 ions with carbonyl NBFIX.[d,e] 5 kcal/mol/A^2 restraints on residue 118−123.	dt = 2 fs, NPT T = 310 K	1 × 0.05 μs 1 × 0.05 μs 1 × 0.05 μs	70 mM NaCl 70 mM KCl 70 mM NaCl +70 mM KCl

Boiteux et al. (2014)	Closed pore w/VSDs NavAb wild-type (3RVY)[1]	CHARMM22 + CMAP protein,[a] CHARMM36 DPPC lipids,[c] CHARMM22 ions with carbonyl NBFIX.[d,e]	dt = 2 fs, NVT T = 323 K All E177 protonation states	1 × 2.5 μs 1 × 0.5 μs 1 × 0.6 μs	150 mM NaCl 300 mM NaCl/ KCl
Ke et al. (2014)	Open pore, no VSDs NavMs symmetrized (4F4L)[3]	AMBER99SB protein,[f] Berger POPC lipids,[i] oung and Cheatham ions.[j] 1 kcal/mol/A[2] restraints on TM helices.	dt = 4 fs, NPT T = 300 K Double membrane (±565 mV)	4 × 0.5 μs 3 × 0.5 μs	500 mM NaCl 0 mM NaCl
		Above, and glutamic acid χ_2 restraints in up-facing or down-facing states.		4 × 0.3 μs 4 × 0.3 μs	500 mM NaCl
Mahdavi and Kuyucak (2015)	Nav1.4 Model, closed pore, no VSDs. NavAb I217C (3RVY),[1] NavAb WT (4EKW)[5]	CHARMM36 + CMAP protein,[a,m] CHARMM36 POPC lipids,[c] CHARMM22 ions with carbonyl NBFIX.[d,e]	dt = 2 fs, NPT T = 300K	1 × 0.05 μs	100 mM NaCl

All constant pressure simulations were conducted at 1 bar, and all water models were TIP3P. Restraints were applied to C-alpha atoms in all studies where restraints were used.

Structures/models. [1](Payandeh et al., 2012), [2](Zhang, Ren, et al., 2013), [3](McCusker et al., 2012), [4](Amaral et al., 2012), [5](Zhang, Ren, et al., 2013; Payandeh et al., 2013).

Force fields: [a](MacKerell et al., 1998, 2004), [b](Feller & MacKerell, 2000), [c](Klauda et al., 2010), [d](Beglov & Roux, 1994), [e](Beglov & Roux, 1994; Noskov & Roux, 2008), [f](Hornak et al., 2006), [g](Siu et al., 2008), [h](Berger et al., 1997), [i](Cordomí et al., 2012), [j](Joung & Cheatham, 2008), [k](Jorgensen et al., 1996; Kaminski et al., 2001), [l](Åqvist, 1990), [m](Best et al., 2012).

5.2 Unbiased equilibrium simulations

Unbiased equilibrium simulations of NavAb were performed in two studies, each reporting aggregate sampling times of the order of 1–10 μs (Table 1) (Boiteux et al., 2014; Chakrabarti et al., 2013). Both studies report spontaneous and reversible diffusion of Na^+ through the SF between the EC medium and the CC, with ion movement coupled to conformational fluctuations of glutamic acid side chains. Many of these findings were the result of long timescale simulations.

Dozens of repeats of 450–500 ns simulations of NavAb were generated by Chakrabarti et al. (2013). Although all crystallographic structures of Nav channels solved to date feature glutamic acid side chains of the "EEEE" ring in an up-facing state (Fig. 1C), these simulations demonstrate rapid and reversible conformational isomerization of glutamic acid (E177) side chains to a lumen-facing "dunked state" in the presence of ions. This process resulted in a highly degenerate, liquidlike energy landscape with many ion binding modes exchanging rapidly with one another. In this mechanism, the conformational isomerization of E177 side chains promotes the rapid diffusion of Na^+ through two main binding sites characterized by coordination to either E177-only or E177 and L176 channel ligands. Although these simulations were performed in a closed channel, a current of ~1 pA was computed based on SF permeation events, consistent with expectations from electrophysiological studies. The average Na^+ occupancy of the SF was 2.09 ± 0.05. Consistent with a loosely coupled knock-on mechanism, permeation events in the SF occurred via one- or three-ion intermediate states.

Ion permeation in NavAb in the absence of applied voltage was also studied by Boiteux et al. (2014) using unrestrained simulations. This study reports an average SF occupancy of 2.3 ions. A multiion 1D-PMF indicates that Na^+ primarily binds at S_{HFS} and S_{IN}, with a 1.5 kcal/mol barrier between these states at S_{CEN}. Conformational isomerization of glutamic acid side chains was observed, frequently resulting in dual occupancy of the S_{HFS} site. A multiion 2D-PMF supports both direct knock-on and pass-by permeation mechanisms, which differ primarily in ion interactions at S_{HFS}. When Na^+ is bound at both S_{HFS} and S_{CEN}, an incoming Na^+ cation either binds at S_{HFS}, causing a knock-on event, or passes by the ion bound at S_{HFS}. Both events result in the lower ion being forced out of the SF into the CC. Both mechanisms involve free energy barriers lower than 1 kcal/mol and concerted movement of ion and glutamate side chains.

5.3 Unbiased nonequilibrium simulations

An open-state model of NavAb was obtained by forcing the backbone of the pore domain and voltage sensors to adopt the activated open-state conformation of an equilibrated Kv1.2 structure (Amaral et al., 2012; Treptow & Tarek, 2006). Starting from the lowest-energy two-ion state of the SF (with ions at S_{HFS} and S_{CEN}) obtained from metadynamics simulations, hyperpolarizing and depolarizing voltages were applied to study inward and outward conduction (Stock et al., 2013). At both voltages, permeation proceeded along a 2D minimum-energy pathway similar to that determined from 0 mV metadynamics simulations, yet average SF occupancy was ~1.8 during inward conduction and ~2.3 for outward conduction. The analysis of conditional probabilities of ionic occupancy suggested an asymmetric permeation mechanism whereby inward conduction proceeds through a two-ion mechanism, whereas a third ion at S_{CEN} or S_{IN} is required for outward conduction to occur.

The open-state NavMs pore-domain structure was studied under a range of inward applied voltages by Ulmschneider et al. (2013). The magnitude of inward voltage applied in these studies varies from 0 to 665 mV (beyond the physiological range). Pseudo-free-energy profiles for ion permeation computed under these nonequilibrium conditions indicate that Na^+ preferentially binds at S_{HFS} and at a shared binding site formed by both S_{CEN} and S_{IN}, matching the location of electron density at S_{CEN} and S_{IN} in the crystallographic structure of NavMs. Minor populations were observed at a site beyond the contact distance of carboxylate oxygen atoms of E177 in an up-facing state, where electron density was observed (McCusker et al., 2012; Ulmschneider et al., 2013). The average total occupancy of the channel was 1.8 ions, irrespective of the magnitude of applied voltage. All binding sites were occupied by fully hydrated Na^+ ions with the exception of S_{HFS}, where ions exchanged first hydration shell water molecules with carboxylate oxygen atoms. In this model, Na^+ permeates the channel in a single-ion, multisite process in which each of the primary channel binding sites is singly occupied, with ions unable to pass each other. The mechanism of Na^+ translocation described by Ulmschneider et al. (and shown in animations) involves one to three ions, diffusing largely in single-file, and does not involve a knock-on mechanism per se. Ulmschneider et al. constructed an I-V curve and computed Na^+ conductance (~34 pS) in agreement with single-channel conductance measurements (~33 pS). Contrary to the studies by Chakrabarti et al. (2013) and Boiteux et al. (2014), no conformational isomerization of E side chains was reported and at zero voltage no ions permeated the channel.

Ion permeation in NavMs was also studied by Ke et al. (2014) in the presence of an ionic gradient using the computational electrophysiology simulation protocol. This protocol consists of stacking two simulation cells and maintaining an ionic charge imbalance between the two aqueous compartments as ions permeate channels in both inward and outward directions (Kutzner et al., 2011). A multiion 1D-PMF revealed that the primary binding sites were S_{HFS} and $S_{CEN} + S_{IN}$ for inward conduction, along with S_{CEN} and S_{HFS} for outward conduction, in agreement with the binding sites described by Ulmschneider et al. (2013). While the largest energy barriers for inward and outward conduction were comparable (2.1 kcal/mol and 2.3 kcal/mol, respectively), the energy landscape is more rugged below S_{HFS} during outward conduction, resulting in reduced conduction rates. Ke et al. observed inward Na^+ conduction via a loosely coupled knock-on mechanism. Outward conduction rates were found to be half of inward rates, by means of a tightly coupled knock-off mechanism facilitated by a glutamic acid side chain directly coordinated to Na^+. In unrestrained simulations, conformational isomerization of Glu side chains was observed only when Na^+ was inside the SF during efflux. Repeating simulations with Glu side chains forced to adopt either lumen-facing or up-facing states suggested that influx conduction rates are independent of glutamic acid conformation, whereas efflux conduction rates increase when a carboxylate group faces the lumen.

5.4 Protonation studies

Experimental studies of mammalian Nav channels have shown that lowering the external pH can alter ion conductance (Khan et al., 2006; Baker & Bostock, 1999). This effect may be due to protonation of carboxylate side chains of E177, reducing ion binding and preventing conduction. Whereas the protonation and deprotonation of titratable groups occurs dynamically in physiological conditions, protonation states are fixed in conventional MD simulations. Accordingly, several simulation studies have examined the effect of fixed carboxylate protonation on binding and permeation in Nav channels.

Corry and Thomas (2012) used umbrella sampling simulations to compute the PMF for Na^+ permeation through the SF of NavAb with one or two opposite glutamic acids protonated. Although Na^+ binding sites found in deprotonated simulations were also populated in singly protonated

PMFs, binding was less favorable throughout the SF by approximately 1 kcal/mol. No ion binding was observed in the SF with two adjacent Glu protonated.

Furini et al. (2014) used biased-sampling simulations to compute 1D-PMFs for all possible protonation states of the EEEE ring, and 2D-PMFs for deprotonated, singly protonated, and doubly protonated states. Upon protonation of a single or two opposite E side chains, the preference of Na^+ for S_{CEN} over S_{HFS} increased from 1.5 to 8.2 kcal/mol. Protonation of two adjacent, three, or four side chains reduced this binding asymmetry to 4–5 kcal/mol, but also reduced the overall binding affinity of Na^+ cations to the SF with respect to the bulk. The conformational isomerization of protonated E177 side chains resulted in Na^+ coordination to carboxylic acid groups as far into the SF as the S_{CEN} site. When Na^+ was located between S_{CEN} and S_{HFS}, conformational isomerization of E177 side chains was observed in 51% and 63% of frames with one or two opposite E side chains protonated, respectively. A higher propensity of E177 adopting a lumen-facing state was correlated with higher occupancy of the S_{CEN} site (with respect to the S_{HFS} site).

Unrestrained simulations of NavAb on the microsecond timescale were performed with the EEEE ring in all possible protonation states (Boiteux et al., 2014). Protonating one glutamic acid side chain reduced average Na^+ occupancy from 2.3 to 2.0, but still supported a conduction mechanism involving two to three ions in S_{HFS} and S_{IN} binding sites. Unlike previous free energy studies, protonation of a single E side chain was found to eliminate the barrier between S_{HFS} and S_{IN} sites, making it well suited for Na^+ conduction. Protonation of one E side chain also induced a loss of stability in the SF that induced helix breaking in the S6 helix of the pore domain on the timescale of hundreds of nanoseconds. Finally, Na^+ binding at S_{HFS} was unfavorable when two or more glutamic acid side chains were protonated, as hydrogen bonds formed between protonated and deprotonated glutamic acids, blocking the channel.

6. IONIC SELECTIVITY OF Nav CHANNELS

A striking feature of the Nav channel structure is that the diameter of the SF is significantly larger than that of the K^+ channel (Fig. 1). Given that the crystallographic radius of Na^+ is smaller than K^+, one may have expected a narrow SF in which fully dehydrated Na^+ ions directly coordinate channel ligands or a larger SF in which Na^+ remains fully hydrated, both of which could support a "snug-fit" hypothesis for selectivity (Mullins, 1960; Bezanilla & Armstrong, 1972). However, the diameter of the SF in the crystallographic structure lies between these two limits; accordingly, Na^+ remains only partially hydrated as it moves through the SF. Simulation studies have revealed dynamic fluctuations of glutamic acid side chains that effectively modulate the diameter of the SF as ions bind and diffuse through it (Chakrabarti et al., 2013; Boiteux et al., 2014). These findings raise the question of how the wider, flexible SF controls the weak (10:1) selectivity of the channel for Na^+ over K^+ (Payandeh et al., 2012), which contrasts with the strongly selective K^+ channels (1000:1) (LeMasurier, Heginbotham, & Miller, 2001). In this section, we review studies of Na^+ selectivity over K^+ and Ca^{2+} separately. These studies are further divided into studies that used biased and unbiased sampling. As discussed in Section 4, differences in systematic and statistical errors may result from this choice and make comparisons between these studies problematic.

6.1 Studies of selectivity over K^+ using biased sampling

Single-ion PMFs computed by Corry and Thomas (2012) indicated that the barrier between S_{HFS} and S_{CEN} is 3 kcal/mol higher for K^+ than for Na^+, despite a minimal difference in binding affinity at S_{HFS}. These authors suggest that selectivity emerges due to the radius of the Nav channel pore at S_{HFS}. Solvated Na^+ is proposed to possess ideal geometry for off-center binding at S_{HFS}, with two solvation shell water molecules interacting favorably with carboxylate groups, while the same cannot be achieved for K^+ at this pore radius.

Furini and Domene (2012) presented single- and double-ion PMFs for K^+ permeation featuring a 4 kcal/mol destabilization of the S_{HFS} binding site relative to that of Na^+. This difference contributes to barriers of 5.0 and 2.3 kcal/mol between S_{CEN} and S_{HFS} compared to barriers of 3.5 and 2.4 for Na^+, for outward and inward permeation directions, respectively. The authors proposed that selectivity for Na^+ over K^+ is due to the existence of a binding site for hydrated Na^+ between S_{HFS} and S_{CEN}, which was not observed for K^+. This hypothesis was subsequently tested by using metadynamics to compute mixed-cation PMF profiles, with one or two K^+ in the pore in the presence of two additional Na^+ (Domene et al., 2015). This study indicated that one or two K^+ had no influence on the free energy surface of Na^+ and that K^+ was strongly excluded from binding at the Na^+-selective site between S_{CEN} and S_{HFS}, in agreement with their previous calculations (Furini & Domene, 2012).

In another study of the NavAb closed state, mixed-ion 2D-PMFs showed similar binding sites for Na^+ and K^+, with S_{IN} being the most energetically stable for both ions (Finol-Urdaneta et al., 2014). The authors presented conductance and selectivity measurements of voltage-sensitive bacterial channel NaChBac that revealed concentration-dependent changes in selectivity. With nearly symmetric concentrations, the authors computed a Na^+:K^+ selectivity ratio of 50:1 when the IC ion was K^+, but only 5:1 when the IC ion was Na^+. Moreover, the ratio of double-Na^+ and double-K^+ equilibrium dissociation constants from 2D-PMF calculations led to an estimated selectivity ratio of only ~2.3:1. Experimental measurements showed that mutating the EEEE ring of this channel to DDDD resulted in a reduction of Na^+:K^+ selectivity to ~1.7:1. PMF calculations showed that the E177D mutation eliminated multiple binding sites within the SF, notably destabilizing S_{HFS} for both Na^+ and K^+. The selectivity ratio was not computed in the E177D mutant. However, this mutation led carboxylate coordination to increase by 60% for K^+ and to decrease by 35% for Na^+ near their respective free-energy minima, resulting in a relative advantage for K^+ within the SF. Based on molecular renderings and coordination plots, conformational isomerization of E177 (and potentially D177) side chains occurred during the simulations, suggesting that dunking may be linked to selectivity. To address the asymmetry of selectivity in reversal potential measurements, a study of Na^+ selectivity over K^+ using nonequilibrium pulling experiments was performed (Ngo et al., 2016). In this work, pulling 3 Na^+ ions through the SF was favored over 3 K^+ by approximately 1.5 kcal/mol. Furthermore, the presence of K^+ within the SF was sufficient to block K^+ permeation during inward conduction but not during outward K^+ conduction, potentially explaining why NavAb is much less permeable to the inward flux of K^+ ions than to outward K^+ current.

6.2 Studies of selectivity over K^+ using unbiased sampling

The conductances of Na^+ and K^+ were compared in the open-state of NavMs using nonequilibrium sampling (inward external applied voltage) in the microsecond time range. The calculated Na^+:K^+ selectivity ratios based on the ratio of ionic currents below 166 mV were in the range of 11–18:1, which is commensurate with the experimental selectivity ratio for this channel (24:1) (Ulmschneider et al., 2013). However, it should be noted that the ratio of Na^+ to K^+ conductance at low applied voltage has poor statistics due to few K^+ conductance events (only three K^+ permeation events were observed in 2 μs of simulation at 100 mV). At voltages greater than 410 mV (i.e., outside the physiological voltage range), K^+ conductance was roughly half that of Na^+ (~40 pS and ~17 pS, for Na^+ and K^+, not including a scaling factor to correct from the simulation temperature to the experimental temperature) and Na^+:K^+ selectivity dropped to approximately 2–4:1, which does not match experimental measurements. Na^+ was found to bind at multiple sequential sites in the SF, yet K^+ binding was largely restricted to S_{HFS}. Reduced K^+ conduction rates in NavMs were a result of a ~5 kcal/mol barrier between S_{HFS} and S_{CEN}, with two ions at S_{HFS} required to initiate a conduction event.

Competitive binding studies of Na^+ and K^+ in the closed state of NavAb led to a rough estimate of ion binding affinity for each ion at S_{HFS} and S_{CEN} (Boiteux et al., 2014). 1D-PMF profiles obtained from these equilibrium simulations revealed that K^+ binds at all the same Na^+ binding sites, yet with ~1 kcal/mol lower affinity. Because complete sampling of Na^+ and K^+ along the channel axis was not part of the design of this study, a measure of selectivity was not calculated.

To study the mechanism of selective permeation of Na^+ in eukaryotic Nav channels, Xia, Liu, Li, Yan, and Gong, (2013) mutated the ring of S180 residues to the "DEKA" sequence in the SF of the asymmetric channel NavRh (Zhang, Ren, et al., 2013). Unrestrained simulations with 70 mM NaCl, 70 mM KCl, or both were used to study conduction and competitive binding. The lysine of the DEKA motif was suggested to act as a switch that stays open by forming a salt bridge with an acidic residue outside the SF, a potentially voltage-regulated process. In this conformation, cation permeation proceeds through coordination with Glu and Asp of the DEKA locus, a binding site that is Na^+-selective over K^+. Furthermore, mutation of this locus to DEAA was shown to eliminate Na^+ selectivity in experimental studies (Favre et al., 1996) in agreement with simulations reported in this study.

6.3 Studies of selectivity over Ca^{2+} using unbiased and biased sampling

In a biased-sampling study with aggregate simulation times of 1–3 μs, Corry (2013) compared free energy profiles for Na^+ and Ca^{2+} permeation in NavAb. Single-ion PMF profiles feature a global energy minimum at S_{HFS} for both Na^+ and Ca^{2+}, with another two distinct binding sites near

S_{CEN} strongly favoring Na^+. An 8-kcal/mol barrier opposing movement of Ca^{2+} between the latter sites, but not observed for Na^+, was ascribed to the partial desolvation of Ca^{2+}. Although this barrier was reduced to 4.5 kcal/mol when two Ca^{2+} ions occupied the channel in a 2D-PMF, it could still account for a selectivity ratio on the order of 10:1 for Na^+ over Ca^{2+}. This study suggests that Ca^{2+} permeation occurs through a knock-on mechanism. In a study of Na^+ and Ca^{2+} binding in NavAb using a different force field, Ke et al. (2013) produced a single-ion PMF with a 4.3 kcal/mol barrier for Ca^{2+} between S_{HFS} and S_{CEN} suggested to result from a less-favorable coordination geometry than Na^+. These studies are in agreement with respect to selective permeation of Na^+ over Ca^{2+}.

Both biased and unbiased simulations were utilized to study Na^+ selectivity over Ca^{2+} in the asymmetric bacterial channel NavRh (Zhang, Ren, et al., 2013; Zhang, Xia, et al., 2013), which differs from NavAb and NavMs in its SF sequence **TLSS**WET (Zhang, Ren, et al., 2013). In this work, Zhang et al. report that Na^+ permeated the SF via two binding sites, one involving S and E side chains and the other involving backbone carbonyl groups of T and L. In mixed-cation simulations, Ca^{2+} was found to block the SF at both of these sites, preventing Na^+ flux. Similarly, competitive binding studies from unbiased equilibrium simulations with various arrangements of Na^+ and Ca^{2+} in the SF of NavAb by Boiteux et al. (2014) revealed a Na^+-blocking site for Ca^{2+} characterized by a high coordination number to E177 carboxylate oxygen atoms (Boiteux et al., 2014).

6.4 Studies of selectivity over Ca^{2+} using quantum calculations

High-accuracy quantum mechanical calculations have been used on simplified models of channel SFs to study ion binding and selectivity (Dudev & Lim, 2014). Dudev and Lim (2012) optimized the geometry of Na^+ and Ca^{2+} bound to a crown–ether complex designed to mimic the EEEE locus of Nav channels. Using these structures, they computed the free energy of transferring Na^+ or Ca^{2+} from the gas-phase to the complex while systematically varying glutamic acid protonation state, pore size, pore rigidity, and solvent accessibility. Results suggest that the EEEE ring is selective for Na^+ when solvation effects dominate over Na^+-ligand electrostatic interactions. Conversely, the EEEE ring was selective for Ca^{2+} when attractive Ca^{2+}-ligand electrostatic interactions prevail over unfavorable solvation effects, and when pore size is decreased.

7. CRITICAL DISCUSSION

Several of the reported findings were consistent across multiple studies, implying that some general properties of ion conduction are robust to the choice of force field, channel model, and sampling methodology. However, specific properties relating to the average coordination of ions and the free energy surface of ion conduction often differed between simulation studies. In this section, we outline the consensus and divergent findings from computer simulations of Nav channels and comment on the agreement of some of these measurements to experimental data. Where discrepancies were found, we discuss possible sources of error in light of methodological considerations such as force field, channel model, and sampling approach.

7.1 Consensus findings

In both biased and unbiased simulations, Na^+ conduction proceeds through two primary SF binding sites, separated by free energy barriers in the range of 1.5–3 kcal/mol, with a total average occupancy of ~2 Na^+ cations. Both multidimensional free energy surfaces and the analysis of SF-bound states with highest probability support a loosely coupled knock-on Na^+ conduction mechanism, with the exception of one study that described a single-ion multisite permeation mechanism (Ulmschneider et al., 2013). In all studies, one or more water molecules in the first solvation shell of Na^+ were exchanged for channel ligands and Na^+ ions were found to be capable of passing one another somewhere within the SF.

Single-channel conductance measurements suggest that most ion channels can mediate at least 1 pA of current, corresponding to ~10^6 ions per second. Ion currents are modulated by voltage, and the ratio of current by voltage is referred to as the conductance. Although single-channel conductance has not been measured for NavAb, the Na^+ conductance of homologous bacterial Nav channels were on the order of 10–40 pS (Shaya et al., 2011; Ulmschneider et al., 2013; Ren et al., 2001). Sodium currents estimated from SF conduction events with no external applied voltage were on the expected order of magnitude for an Nav channel (Chakrabarti et al., 2013). Simulations of NavMs subject to a range of external applied voltage resulted in an estimated Na^+ and K^+ conductance in agreement with experimental values, particularly at low applied voltage (Ulmschneider et al., 2013). These authors also estimated ionic selectivity from the ratio of conductance for Na^+ and K^+, which was in agreement with experimental values. Ngo et al. (2016) were also able to use nonequilibrium pulling simulations to reproduce the magnitude of inward K^+ currents necessary to overcome K^+ block under biionic conditions measured in a homologous channel (Finol-Urdaneta et al., 2014).

7.2 Divergent findings

Simulation studies of Nav channels disagree on predicted ionic binding site location and occupancy. First, they differ in the location of the lower binding site, at S_{CEN} (Corry & Thomas, 2012; Stock et al., 2013; Domene et al., 2015; Furini et al., 2014; Furini & Domene, 2012; Ke et al., 2013, 2014), S_{IN} (Boiteux et al., 2014; Finol-Urdaneta et al., 2014), S_{CEN}/S_{IN} (Ulmschneider et al., 2013; Carnevale et al., 2011), or a site between S_{HFS} and S_{CEN} (Ke et al., 2014; Chakrabarti et al., 2013). Also note that Ke et al. (2014) identified a different lower binding site for inward and outward conduction. Similarly, while some studies describe two separate binding sites in the lower portion of the SF (S_{IN} and S_{CEN}) (Corry & Thomas, 2012; Boiteux et al., 2014; Stock et al., 2013; Ke et al., 2013, 2014), other studies describe only one (Domene et al., 2015; Furini & Domene, 2012; Ulmschneider et al., 2013; Furini et al., 2014; Chakrabarti et al., 2013). Although in all studies the average SF occupancy was ~2, Na^+ occupancies of the upper and lower SF binding sites were not consistent across studies, resulting in discrepancies on the molecular details of the ion permeation mechanism. Finally, binding of two Na^+ cations was observed either only in the upper site (Boiteux et al., 2014; Corry & Thomas, 2012; Furini & Domene, 2012), only in the lower site (Chakrabarti et al., 2013; Carnevale et al., 2011; Finol-Urdaneta et al., 2014), in both sites (Stock et al., 2013; Domene et al., 2015), or in neither sites (Ulmschneider et al., 2013).

All studies reported partial dehydration of Na^+, but average coordination to water in the SF was in disagreement. Where coordination plots were provided, at S_{HFS}, Na^+ was reported to lose an average of one (Ke et al., 2014; Ulmschneider et al., 2013; Corry & Thomas, 2012; Furini & Domene, 2012), two (Finol-Urdaneta et al., 2014), or three water molecules (Chakrabarti et al., 2013) in favor of E177 carboxylate oxygen atoms. Moreover, some studies report binding to an average of one water molecule to the S178 hydroxyl oxygen (Furini & Domene, 2012; Corry & Thomas, 2012; Finol-Urdaneta et al., 2014). At the S_{CEN} site, Na^+ was found to lose an average of one (Corry & Thomas, 2012; Finol-Urdaneta et al., 2014;

Boiteux et al., 2014; Furini & Domene, 2012), or one to two water molecules (Ulmschneider et al., 2013; Chakrabarti et al., 2013) in favor of L176 or T175 carbonyl oxygen atoms.

Conformational isomerization of E177 side chains ("dunking") was observed in simulation studies using unbiased equilibrium (Boiteux et al., 2014; Chakrabarti et al., 2013), unbiased nonequilibrium (Ke et al., 2014), and biased sampling (Noskov & Roux, 2008; Domene et al., 2015; Ngo et al., 2016; Finol-Urdaneta et al., 2014). Two additional studies reported conformational isomerization when glutamic acid side chains were protonated (Boiteux et al., 2014; Furini et al., 2014). Corry (2013) and Ke et al. (2013) also noted E177 side chain conformational isomerization upon Ca^{2+} binding but not for Na^{+}. In a nonequilibrium study by Ke et al. E177 dunking only occurred when outward external voltage was applied, suggesting that the conformation of E177 side chains may be voltage dependent. Some studies provide cursory evidence of dunking through descriptions and molecular renderings (Ngo et al., 2016; Finol-Urdaneta et al., 2014), but otherwise, dunking was either not observed in simulations or not explicitly analyzed in all other studies. The χ_2 torsional angle of the E177 side chain occupies basins corresponding to undunked and dunked conformations, centered at approximately (−60 degrees, 50 degrees) (Domene et al., 2015), (−60 degrees, 60 degrees) (Chakrabarti et al., 2013), or (−70 degrees, 60 degrees) (Ke et al., 2014; Boiteux et al., 2014), with a broad agreement across studies (here, the angles were shifted to the −180 degrees to 180 degrees range). In an unrelated simulation study of the nicotinic acetylcholine receptor, the Glu side chain conformation was also found to be correlated to Na^{+} conduction, suggesting that this process is not specific to Nav channels (Harpole & Grosman, 2014).

There is little consensus on the effect of Glu protonation on ion binding. In simulations with one protonated E177 side chain, Corry and Thomas (2012) describe a moderate overall loss of Na^{+} binding affinity in the SF (~1 kcal/mol) and Boiteux et al. (2014) observe sub-kcal/mol differences with deprotonated-Glu simulations at major binding sites and the elimination of the S_{HFS}-to-S_{IN} barrier. By contrast, Furini et al. (2014) describe large destabilization of S_{HFS} not observed in either aforementioned studies.

In studies of ionic selectivity, Na^{+} and K^{+} were found to bind at sites S_{HFS}, S_{CEN}, and S_{IN} in the SF of Nav channels, but the relative stability of each site varied considerably depending on simulation methodology. The primary site of Na^{+} selectivity over K^{+} was identified either as the S_{HFS} binding site (Corry & Thomas, 2012; Ngo et al., 2016), the S_{IN}

binding site (Finol-Urdaneta et al., 2014), or the S_{HFS}-S_{CEN} barrier (Furini & Domene, 2012; Domene et al., 2015; Ulmschneider et al., 2013). The primary site of Na^{+} selectivity over Ca^{2+} was the S_{HFS}-S_{CEN} barrier (Ke et al., 2013; Corry, 2013). Quantitative estimates of Na^{+}:Ca^{2+} selectivity did not match experimental values.

7.3 Sources of discrepancy

The aggregate sampling times used in published studies of Nav channels vary by three orders of magnitude, from ~100 ns to ~20 μs (Tables 1 and 2). As is commonly performed in biased sampling approaches such as umbrella sampling, multiple permeation studies generate a collection of short simulations [0.5–6 ns per umbrella sampling "window" (Furini et al., 2014; Corry & Thomas, 2012; Furini & Domene, 2012; Finol-Urdaneta et al., 2014; Ngo et al., 2016)] during which the simulations are assumed to be long enough that all the degrees of freedom orthogonal to the reaction coordinate undergo equilibration. Long-timescale unbiased simulations suggest that conformational isomerization of one or more E177 side chains can alter the ion conduction pathway and that this process occurs on the timescale of 0.1–10 ns (Boiteux et al., 2014; Chakrabarti et al., 2013). Accordingly, in the majority of short-timescale biased simulations, E177 dunking was likely either missing or undersampled, resulting in both systematic and statistical sampling errors (Corry & Thomas, 2012; Furini & Domene, 2012; Finol-Urdaneta et al., 2014; Ngo et al., 2016). To the extent that Glu isomerization modulates ion binding and dynamics, as suggested by Chakrabarti et al. (2013), these sampling errors may have significant effects on both thermodynamic and kinetic properties. Limited sampling times may also result in a source of discrepancy when comparing results of biased (Furini et al., 2014; Corry & Thomas, 2012) and unbiased equilibrium PMFs (Boiteux et al., 2014) from studies of Glu protonation.

Discrepancies between simulations of Nav channels may also result from the choice of protein, solvent, and ionic force fields. However, force field comparisons can only be performed in the long-timescale sampling limit, once both statistical and systematic sampling errors are significantly reduced, and the studies utilize a common sampling methodology. For these reasons, we restrain this part of our analysis to two sets of comparable studies. Chakrabarti et al. (2013) and Boiteux et al. (2014) both used long-timescale unbiased equilibrium sampling with a closed channel model, permitting the comparison of OPLS-AA/L (Jorgensen et al., 1996; Kaminski et al., 2001) and CHARMM22 (MacKerell et al., 1998, 2004) protein force fields [with different ion and lipid parameters but the same water model, TIP3P (Jorgensen et al., 1983), in both studies], respectively. Both of these studies describe conformational isomerization of E177 and a similar conduction mechanism involving two-to-three-ion occupancy within the SF. However, these force fields support double Na^+ binding in the SF at different sites—namely, E177 coordination for CHARMM22 and joint E177 and L176 coordination for OPLS-AA/L. Na^+ binding at the lower-SF site defined by coordination to backbone carbonyl oxygen atoms of L176 and T175 is greatly favored in CHARMM22 over OPLS-AA/L. This comparison suggests that interactions with channel oxygen atoms may be weaker in OPLS-AA/L than CHARMM22 and may partly explain the highly degenerate energy landscape of ion conduction described by Chakrabarti et al. (2013). Because the agreement of these two studies to experimental conduction or selectivity measurements is not provided, it is not possible to determine which of these force fields leads to more accurate results for this particular system.

Second, Ulmschneider et al. (2013) and Ke et al. (2014) both used unbiased nonequilibrium sampling in the microsecond timescale with a restrained open channel model, permitting the comparison of CHARMM22 (MacKerell et al., 1998, 2004) and AMBER99SB (Hornak et al., 2006) force fields (with identical ion parameters), respectively. Although these studies differed in their method for applying voltage, both computed comparable Na^+ conduction rates and similar ionic occupancies of the SF. Conformational isomerization was not observed in either study during inward conduction. Similar Na^+ binding sites were observed in both studies, but S_{CEN} and S_{IN} sites were joined and had higher binding affinity in the simulations of Ulmschneider et al. (2013). Although CHARMM22 and AMBER force fields with ionic parameters from Joung and Cheatham (2008) produce similar results for Na^+ conduction, K^+ was not simulated by Ke et al. (2014) and as such, the mechanism of selectivity in these force fields cannot be compared. Although methodological differences abound between these two studies and those of Chakrabarti et al. (2013) and Boiteux et al. (2014), it is possible that the presence of inward applied voltage disfavors conformational isomerization.

In a number of studies, several modifications to the standard membrane protein simulation protocol (Kandt et al., 2007) were utilized without validation. To accelerate the number of ion conduction events, several models utilized higher-than-physiological ion concentrations (300–500 mM NaCl) (Corry & Thomas, 2012; Ke et al., 2014; Ulmschneider et al., 2013) and external applied voltages (300–600 mV) (Stock et al., 2013; Ke et al., 2014; Ulmschneider et al., 2013). Similarly, in simulations of the open-state of NavMs, strong structural restraints (1 kcal/mol/A^2) were imposed on the transmembrane helices of the channel pore domain to prevent spontaneous closure of the pore (Ulmschneider et al., 2013; Ke et al., 2014). Although nonequilibrium simulations were effective at reproducing experimental conductance and selectivity in NavAb (Ulmschneider et al., 2013; Ke et al., 2014), microsecond-timescale simulations in the presence of external applied voltage using with the same force field were unable to reproduce the experimentally measured K^+ conductance of Kv1.2/2.1 and gramicidin A (Jensen et al., 2013). The reason for this discrepancy is unclear, especially given that Jensen et al. show that permeation is limited by infrequent recruitment of ions into the pore lumen, particularly at high voltage, an effect that is not specific to K^+ channels. Specific to Nav channels, simulations by Ke et al. (2014) suggest voltage dependence of glutamic acid side chain conformations. Modifications to the channel model with transmembrane restraints along with the application of voltage may not necessarily introduce significant bias in the study of conduction. However, they may explain the limited, or potentially unreported, E177 conformational isomerization along with a predominately single-file conduction mechanism observed in both microsecond-timescale, unbiased, nonequilibrium studies published to date (Ulmschneider et al., 2013; Ke et al., 2014).

While estimates of mean quantities and distributions are improved by the use of long aggregate sampling times, using multiple repeats of moderate length (hundreds of nanoseconds) with different initial conditions can greatly assist in sampling uncorrelated ion positions to improve convergence. Combining multiple simulation repeats with different initial conditions helps mitigate systematic sampling errors (Neale et al., 2014). In unbiased equilibrium studies by Chakrabarti et al. (2013), Na^+ cations were found to reversibly enter and exit the SF with concomitant rearrangements of sodium ions occurring on timescales greater than 100 ns. The time series of ion movement in the channel lumen shown in Fig. 2 illustrate the importance of repeating simulations with differing initial conditions. In Fig. 2A, reversible influx (190 ns) and efflux (370 ns) of a blue ion from the EC compartment results in the movement of the red ion in and out of the CC, in a loosely coupled manner. Fig. 2B features long-lived three-ion occupancy of the SF, where after 125 ns, no large-scale movements of ions in and out of the SF are observed on the length of our simulation. By contrast, in Fig. 2C, it takes almost 400 ns for a blue ion to enter the SF, resulting in

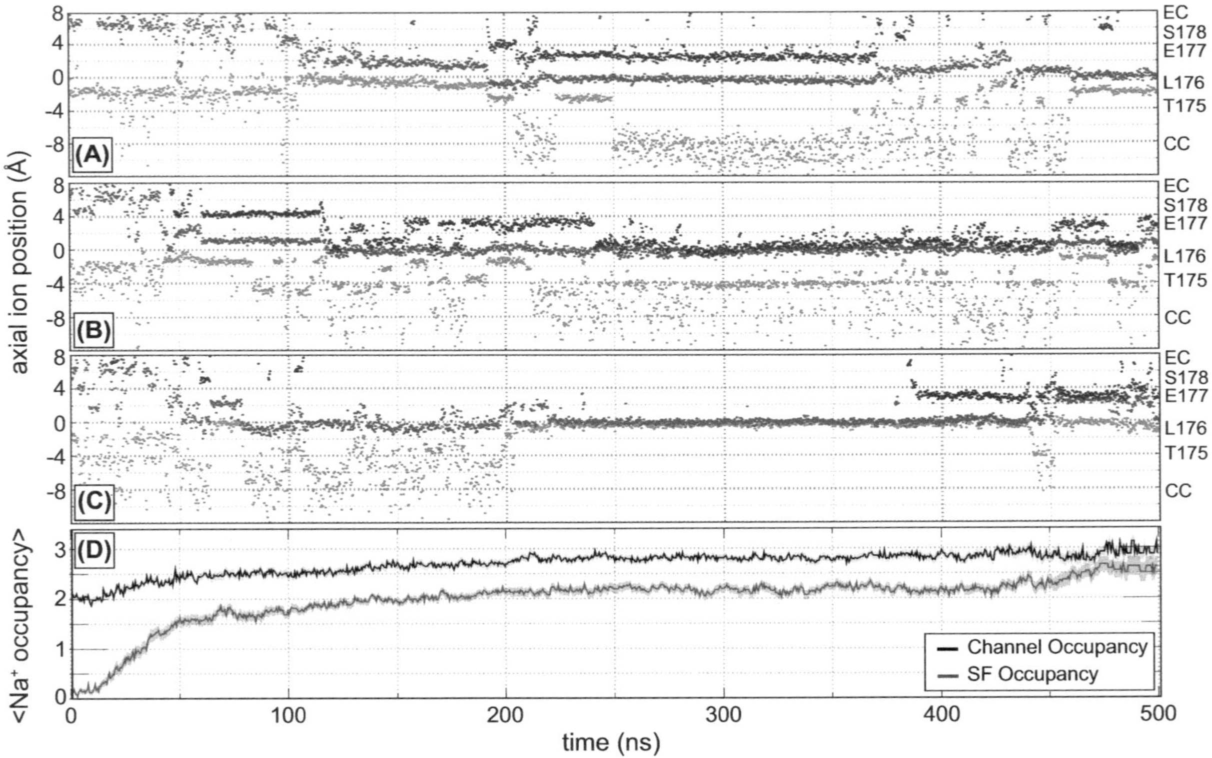

Figure 2 Time series of axial Na^+ positions and average occupancy from unbiased equilibrium simulations of NavAb (Chakrabarti et al., 2013). (A–C) Three out of forty-seven simulation repeats, where Na^+ reversibly diffuse in and out of the SF. Na^+ cations are colored by depth rank within the SF from intracellular to extracellular [ordered red (light gray in print versions), green (gray in print versions), blue (dark gray in print versions) in color]. The channel was initially devoid of cations and simulations differed in initial random velocities. (D) Time dependence of average Na^+ occupancy across all 47 simulation repeats for the channel (black, defined as any Na^+ within axial range 8 Å to −12 Å) and SF [red (light gray in print versions), defined as any Na^+ with first-shell coordination to channel ligands E177 and L176]. Error bars are computed using the standard error of mean across all simulation repeats. The simulation length of individual repeats varied from 450 to 500 ns, resulting in a smaller data set and increased noise in the last 50 ns of this time series.

a significantly different ionic arrangement within the SF. These three time series differ considerably as a whole and in 100 ns blocks of simulation time, and each would lead to different conclusions on the ion transport mechanism if considered separately. The diversity of ion binding modes is due in large part to conformational fluctuations of E177 in the SF. Performing multiple simulation repeats makes it possible to compute average properties across an ensemble and to establish the simulation timescale required for the convergence of equilibrium properties. This is reflected in the average sodium ion occupancy within the channel and SF, which takes ~150 ns to approach a stable value (Fig. 2D).

Although biased-sampling methods have been used to quantify the favorable binding sites of Na^+ and K^+ in the Nav channel SF, extensive unbiased equilibrium simulations have not been performed in the study of ionic selectivity. In the field-strength model of K^+ selectivity by Noskov et al. it was suggested that the number and the type of ligands could directly influence selectivity (Noskov & Roux, 2006; Noskov et al., 2004). This model is interesting when viewed in the context of Nav channels, since conformational isomerization of E177 side chains may result in a variable number of coordinating ligands. Chakrabarti et al. (2013) observe an average SF occupancy of 2.09 ± 0.05 Na^+ cations coordinated by an average of 2.23 ± 0.04 carboxylate groups, whereas the average SF occupancy of K^+ and its average coordinate to carboxylate side chains is yet to be determined. Simulations by Ngo et al. (2016) suggest that carboxylate coordination contributes to K^+ blockage during inward conduction, but they did not explicitly study the effect of conformational isomerization with respect to selectivity of Na^+. Using biomimetic graphene pores containing carboxylate or carbonyl groups, He et al. (2013) studied the voltage dependence of Na^+ and K^+ currents as well as selectivity. A graphene pore with three carboxylate groups was found to be Na^+-selective at low voltage and K^+-selective at high voltage in mixed-cation solutions. Voltage was also found to induce conformational changes in carboxylate group orientation with respect to the direction of the electric field, with an out-facing state associated with K^+ selectivity and lumen-facing state associated with Na^+ selectivity. While there are many differences between such simplified models and the complex environment of the SF of Nav channels, this study supports a putative mechanism of selectivity involving the conformational isomerization of E177 side chains that remains to be examined in Nav channels.

Multi-microsecond simulations have yet to be used to study the selectivity of Na^+ over Ca^{2+} in Nav channels. In addition to long simulation timescales, polarizable force fields may be required to reach quantitative estimates of Na^+:Ca^{2+} conductance ratios that match experimental values. Because the ionic radii of Na^+ and Ca^{2+} are very similar (1.16 Å and 1.14 Å, respectively), an accurate treatment of Ca^{2+} requires the use force fields that account for electronic polarization in the environment surrounding the ion, since the valence charge is greater for Ca^{2+} (Huang et al., 2014). Nonpolarizable models of Ca^{2+} fail to reproduce protein-ion binding energies obtained from quantum mechanical calculations over a diverse range of cation binding sites (Ngo et al., 2015). It has been shown that electronic polarization results in significant ion-induced dipole shifts on the carbonyl groups of the SF in KcsA (Allen, Andersen, & Roux, 2006a; Bucher et al., 2009), but this effect is unknown in Nav channels. As such, quantitative estimates from Nav channel simulations of Ca^{2+} using nonpolarizable force fields must be considered carefully.

8. PERSPECTIVES

Computational studies of bacterial Nav channels offer a unique perspective on molecular mechanisms of ionic permeation in a weakly selective channel. The consensus mechanism of Na^+ permeation consists of a weakly coupled knock-on involving approximately two ions on average. Multiple studies have described the lowest-energy, two-ion occupancy state of the SF as one in which ions are partially dehydrated and bind the SF in off-axis fashion in an upper site (S_{HFS}) and a lower site (S_{CEN} and/or S_{IN}). Permeation occurs via these two sites with low (~1–2 kcal/mol) intervening free energy barriers and the cations are able to pass one another, with occasional double occupancy of one or more sites in the SF. Although no consensus model of Na^+ selectivity has yet been established, several proposed mechanisms attribute selectivity to high permeation barriers between upper and lower binding sites for K^+ and Ca^{2+}.

In qualitative agreement with experiment, most of the simulation studies of cation permeation in Nav channels published to this date support a rapid conduction process. Equilibrium simulations of Na^+ in NavAb yielded estimates of permeation rates with the correct order of magnitude for a sodium channel (Chakrabarti et al., 2013). Estimates of Na^+ conductance computed from two nonequilibrium studies were in good agreement with experimental values (Ke et al., 2014; Ulmschneider et al., 2013). One of these studies also reported a Na^+:K^+ conductance ratio commensurate with experimental measurements, while all other studies of Na^+ selectivity over K^+ and Ca^{2+} were qualitatively correct. Future studies of ionic selectivity in Nav channels will benefit from more comprehensive sampling of different ionic species to assess the importance of conformational isomerization of Glu side chains in this process. In an effort to reproduce experimental selectivity ratios, it would be useful to conduct simulations of Nav channels in biionic conditions in the presence of applied voltage to compute reversal potentials.

To obtain simulation results that are minimally biased by systematic and statistical error, we suggest performing multiple repeats of unbiased equilibrium simulations as the optimal approach. Individual simulations should be long enough to observe SF conduction events (requiring computational resources accessible on nonspecialized supercomputers), with improved sampling due to randomized initial velocities and positions that may sample more uncorrelated ion binding modes than single long-time simulations. Equilibrium simulations (with 0 mV applied voltage) are a good approximation to the low physiological voltages—associated ion conduction (Hille, 2001), but nonequilibrium simulations may also be used in the limit of low applied voltage in the physiological regime. When biased sampling approaches are utilized, accelerated sampling algorithms such as bias-exchange metadynamics offer some benefits over umbrella sampling (Domene et al., 2015). However, even these approaches would benefit from multiple simulation repeats to ensure that infrequent structural transitions involving other degrees of freedom, such as conformational isomerization of E177 side chains, are appropriately sampled—failing which, reaction coordinates including channel degrees of freedom could be considered.

One important issue that must be considered in evaluating simulations is that they have been performed using different simulation methodologies (biased, unbiased equilibrium, and unbiased nonequilibrium sampling), different force fields (CHARMM, AMBER, OPLS), different channel structures (closed and open conformations), and different simulation conditions (such as temperature, ionic concentration, and applied voltage) (Tables 1 and 2). For Na^+ channels, these differences result in discrepancies in estimates of ion binding, permeation, and selectivity. Data set sizes varied between 0.1 and 10 μs of aggregate sampling between studies, and as such, it is not realistic to expect a consensus on the details of permeation and selectivity. In general, continued progress in computer simulation studies of ion channels will greatly benefit from the quantification of both systematic and statistical errors.

ACKNOWLEDGMENTS

This work was supported by Canadian Institutes of Health Research grant MOP-130461. Calculations were performed using supercomputer resources provided by SciNet, WestGrid, and Calcul Québec, managed by Compute Canada Calcul Canada.

REFERENCES

Alam, A., & Jiang, Y. (2009). Structural analysis of ion selectivity in the NaK channel. *Nature Structural & Molecular Biology, 16*(1), 35–41.

Alam, A., & Jiang, Y. (2011). Structural studies of ion selectivity in tetrameric cation channels. *Journal of General Physiology, 137*(5), 397–403.

Allen, T. W., Andersen, O. S., & Roux, B. (2006a). Ion permeation through a narrow channel: using gramicidin to ascertain all-atom molecular dynamics potential of mean force methodology and biomolecular force fields. *Biophysical Journal, 90*(10), 3447–3468.

Allen, T. W., Andersen, O. S., & Roux, B. (2006b). Molecular dynamics — potential of mean force calculations as a tool for understanding ion permeation and selectivity in narrow channels. *Biophysical Chemistry, 124*(3), 251–267.

Amaral, C., et al. (2012). Exploring conformational states of the bacterial voltage-gated sodium channel NavAb via molecular dynamics simulations. *Proceedings of the National Academy of Sciences of the United States of America, 109*(52), 21336–21341.

Andersen, O. S. (2011). Perspectives on: ion selectivity. *Journal of General Physiology, 137*, 393–395.

Åqvist, J. (1990). Ion-water interaction potentials derived from free energy perturbation simulations. *Journal of Physical Chemistry, 94*(21), 8021–8024.

Åqvist, J., & Luzhkov, V. (2000). Ion permeation mechanism of the potassium channel. *Nature, 404*(6780), 881–884.

Ashcroft, F. M. (2000). *Ion channel diseases*. Academic Press.

Bagnéris, C., et al. (2013). Role of the C-terminal domain in the structure and function of tetrameric sodium channels. *Nature Communications, 4*, 1–10.

Bagneris, C., et al. (2014). Structural model of the open-closed-inactivated cycle of prokaryotic voltage-gated sodium channels. *Journal of General Physiology, 145*(1), 5–16.

Baker, M. D., & Bostock, H. (1999). The pH dependence of late sodium current in large sensory neurons. *Neuroscience, 92*(3), 1119–1130.

Beglov, D., & Roux, B. (1994). Finite representation of an infinite bulk system: solvent boundary potential for computer simulations. *Journal of Chemical Physics, 100*(12), 9050.

Berger, O., Edholm, O., & Jähnig, F. (1997). Molecular dynamics simulations of a fluid bilayer of dipalmitoylphosphatidylcholine at full hydration, constant pressure, and constant temperature. *Biophysical Journal, 72*(5), 2002–2013.

Bernèche, S., & Roux, B. (2001). Energetics of ion conduction through the K^+ channel. *Nature, 414*(6859), 73–77.

Bernèche, S., & Roux, B. (2003). A microscopic view of ion conduction through the K^+ channel. *Proceedings of the National Academy of Sciences of the United States of America, 100*(15), 8644–8648.

Best, R. B., et al. (2012). Optimization of the additive CHARMM all-atom protein force field targeting improved sampling of the backbone ϕ, ψ and side-chain $\chi 1$ and $\chi 2$ dihedral angles. *Journal of Chemical Theory and Computation, 8*(9), 3257–3273.

Bezanilla, F., & Armstrong, C. M. (1972). Negative conductance caused by entry of sodium and cesium ions into the potassium channels of squid axons. *Journal of General Physiology, 60*(5), 588–608.

Boiteux, C., Vorobyov, I., & Allen, T. W. (2014). Ion conduction and conformational flexibility of a bacterial voltage-gated sodium channel. *Proceedings of the National Academy of Sciences of the United States of America, 111*(9), 3454–3459.
Bostick, D. L., & Brooks, C. L. (2007). Selectivity in K^+ channels is due to topological control of the permeant ion's coordinated state. *Proceedings of the National Academy of Sciences of the United States of America, 104*(22), 9260–9265.
Bucher, D., et al. (2009). Developing improved charge sets for the modeling of the KcsA K^+ channel using QM/MM electrostatic potentials. *Journal of Chemical Theory and Computation, 5*(8), 2173–2179.
Carnevale, V., Treptow, W., & Klein, M. L. (2011). Sodium ion binding sites and hydration in the lumen of a bacterial ion channel from molecular dynamics simulations. *Journal of Physical Chemistry Letters, 2*(19), 2504–2508.
Catterall, W. A. (2012). Voltage-gated sodium channels at 60: structure, function and pathophysiology. *Journal of Physiology, 590*(11), 2577–2589.
Catterall, W. A., & Zheng, N. (2015). Deciphering voltage-gated Na^+ and Ca^{2+} channels by studying prokaryotic ancestors. *Trends in Biochemical Sciences, 40*(9), 526–534.
Chakrabarti, N., et al. (2013). Catalysis of Na^+ permeation in the bacterial sodium channel NaVAb. *Proceedings of the National Academy of Sciences of the United States of America, 110*(28), 11331–11336.
Chen, R., Robinson, A., & Chung, S. H. (2014). Mechanism of μ-Conotoxin PIIIA binding to the voltage-gated Na channel NaV1.4. *PLoS One, 9*(3), e93267.
Cibulsky, S. M., & Sather, W. A. (2000). The EEEE locus is the sole high-affinity Ca^{2+} binding structure in the pore of a voltage-gated Ca^{2+} channel: block by Ca^{2+} entering from the intracellular pore entrance. *Journal of General Physiology, 116*(3), 349–362.
Cordomí, A., Caltabiano, G., & Pardo, L. (2012). Membrane protein simulations using AMBER force field and berger lipid parameters. *Journal of Chemical Theory and Computation, 8*(3), 948–958.
Corry, B. (2013). Na^+/Ca^{2+} selectivity in the bacterial voltage-gated sodium channel NavAb. *PeerJ, 1*(6), e16.
Corry, B., & Thomas, M. (2012). Mechanism of ion permeation and selectivity in a voltage gated sodium channel. *Journal of the American Chemical Society, 134*(3), 1840–1846.
Crouzy, S., Woolf, T. B., & Roux, B. (1994). A molecular dynamics study of gating in dioxolane-linked gramicidin A channels. *Biophysical Journal, 67*(4), 1370–1386.
Dahl, A. C. E., Chavent, M., & Sansom, M. S. P. (2012). Bendix: intuitive helix geometry analysis and abstraction. *Bioinformatics, 28*(16), 2193–2194.
Derebe, M. G., et al. (2011). Tuning the ion selectivity of tetrameric cation channels by changing the number of ion binding sites. *Proceedings of the National Academy of Sciences of the United States of America, 108*(2), 598–602.
Domene, C., Barbini, P., & Furini, S. (2015). Bias exchange metadynamics simulations: an efficient strategy for the analysis of conduction and selectivity in ion channels. *Journal of Chemical Theory and Computation, 11*(4), 1896–1906.
Domene, C., & Sansom, M. S. P. (2003). potassium channel, ions, and water: simulation studies based on the high resolution X-ray structure of KcsA. *Biophysical Journal, 85*(5), 2787–2800.
Doyle, D. A., et al. (1998). The structure of the potassium channel: molecular basis of K^+ conduction and selectivity. *Science, 280*(5360), 69–77.
Dudev, T., & Lim, C. (2012). Why voltage-gated Ca^{2+} and bacterial Na^+ channels with the same EEEE motif in their selectivity filters confer opposite metal selectivity. *Physical Chemistry Chemical Physics, 14*(36), 12451–12456.
Dudev, T., & Lim, C. (2014). Competition among metal ions for protein binding sites: determinants of metal ion selectivity in proteins. *Chemical Reviews, 114*, 538–556.

Eisenman, G., & Horn, R. (1983). Ionic selectivity revisited: the role of kinetic and equilibrium processes in ion permeation through channels. *Journal of Membrane Biology, 76*(3), 197–225.

Ellinor, P. T., et al. (1995). Ca^{2+} channel selectivity at a single locus for high-affinity Ca^{2+} interactions. *Neuron, 15*(5), 1121–1132.

Favre, I., Moczydlowski, E., & Schild, L. (1996). On the structural basis for ionic selectivity among Na^+, K^+, and Ca^{2+} in the voltage-gated sodium channel. *Biophysical Journal, 71*(6), 3110–3125.

Feller, S. E., & MacKerell, A. D. (2000). An improved empirical potential energy function for molecular simulations of phospholipids. *Journal of Physical Chemistry B, 104*(31), 7510–7515.

Finol-Urdaneta, R. K., et al. (2014). Sodium channel selectivity and conduction: prokaryotes have devised their own molecular strategy. *Journal of General Physiology, 143*(2), 157–171.

Fowler, P. W., et al. (2013). Energetics of multi-ion conduction pathways in potassium ion channels. *Journal of Chemical Theory and Computation, 9*(11), 5176–5189.

Furini, S., Barbini, P., & Domene, C. (2014). Effects of the protonation state of the EEEE motif of a Bacterial Na. *Biophysical Journal, 106*(10), 2175–2183.

Furini, S., & Domene, C. (2012)R. Livesay (Ed.), On Conduction in a Bacterial Sodium Channel D. *PLoS Computational Biology, 8*(4), e1002476.

Furini, S., & Domene, C. (2013). K^+ and Na^+ conduction in selective and nonselective ion channels via molecular dynamics simulations. *Biophysical Journal, 105*(8), 1737–1745.

Gosselin-Badaroudine, P., et al. (2012). Gating pore currents and the resting state of Nav1.4 voltage sensor domains. *Proceedings of the National Academy of Sciences of the United States of America, 109*(47), 19250–19255.

Harpole, T. J., & Grosman, C. (2014). Side-chain conformation at the selectivity filter shapes the permeation free-energy landscape of an ion channel. *Proceedings of the National Academy of Sciences of the United States of America, 111*(31), E3196–E3205.

He, Z., et al. (2013). Bioinspired graphene nanopores with voltage-tunable ion selectivity for Na^+ and K. *ACS Nano,* 7(11), 10148–10157.

Heinemann, S. H., et al. (1992). Calcium channel characteristics conferred on the sodium channel by single mutations. *Nature, 356*(6368), 441–443.

Hille, B. (1971). The permeability of the sodium channel to organic cations in myelinated nerve. *Journal of General Physiology, 58*(6), 599–619.

Hille, B. (1972). The permeability of the sodium channel to metal cations in myelinated nerve. *Journal of General Physiology, 59*(6), 637–658.

Hille, B. (2001). *Ion channels of excitable membranes* (3rd ed.). Sunderland, MA: Sinauer.

Hodgkin, A. L., & Huxley, A. F. (1952a). A quantitative description of membrane current and its application to conduction and excitation in nerve. *Journal of Physiology, 117*(4), 500–544.

Hodgkin, A. L., & Huxley, A. F. (1952b). Currents carried by sodium and potassium ions through the membrane of the giant axon of Loligo. *Journal of Physiology, 116*(4), 449–472.

Hodgkin, A. L., & Huxley, A. F. (1952c). The components of membrane conductance in the giant axon of Loligo. *Journal of Physiology, 116*(4), 473–496.

Hodgkin, A. L., & Huxley, A. F. (1952d). The dual effect of membrane potential on sodium conductance in the giant axon of Loligo. *Journal of Physiology, 116*(4), 497–506.

Hodgkin, A. L., & Keynes, R. D. (1955). The potassium permeability of a giant nerve fibre. *Journal of Physiology, 128*(1), 61–88.

Hornak, V., et al. (2006). Comparison of multiple amber force fields and development of improved protein backbone parameters. *Proteins: Structure, Function, and Bioinformatics, 65*(3), 712–725.

Huang, J., et al. (2014). Recent advances in polarizable force fields for macromolecules: microsecond simulations of proteins using the classical drude oscillator model. *Journal of Physical Chemistry Letters, 5*(18), 3144–3150.
Humphrey, W., Dalke, A., & Schulten, K. (1996). VMD: visual molecular dynamics. *Journal of Molecular Graphics, 14*(1), 33–38.
Im, W., & Roux, B. (2002). Ion permeation and selectivity of OmpF porin: a theoretical study based on molecular dynamics, Brownian dynamics, and continuum electrodiffusion theory. *Journal of Molecular Biology, 322*(4), 851–870.
Jarzynski, C. (1997). Nonequilibrium equality for free energy differences. *Physical Review Letters, 78*(14), 2690–2693.
Jensen, M. O., et al. (2010). Principles of conduction and hydrophobic gating in K^+ channels. *Proceedings of the National Academy of Sciences of the United States of America, 107*(13), 5833–5838.
Jensen, M. O., et al. (2012). Mechanism of voltage gating in potassium channels. *Science, 336*(6078), 229–233.
Jensen, M. O., et al. (2013). Atomic-level simulation of current-voltage relationships in single-file ion channels. *Journal of General Physiology, 90*(10), 3447.
Jiang, Y., et al. (2003). X-ray structure of a voltage-dependent K^+ channel. *Nature, 423*(6935), 33–41.
Jorgensen, W., et al. (1983). Comparison of simple potential functions for simulating liquid water. *Journal of Chemical Physics, 79*(2), 926–935.
Jorgensen, W. L., Maxwell, D. S., & Tirado-Rives, J. (1996). Development and testing of the OPLS all-atom force field on conformational energetics and properties of organic liquids. *Journal of the American Chemical Society, 118*(45), 11225–11236.
Joung, I. S., & Cheatham, T. E. (2008). Determination of alkali and halide monovalent ion parameters for use in explicitly solvated biomolecular simulations. *Journal of Physical Chemistry B, 112*(30), 9020–9041.
Kaminski, G. A., et al. (2001). Evaluation and reparametrization of the OPLS-AA force field for proteins via comparison with accurate quantum chemical calculations on peptides. *Journal of Physical Chemistry B, 105*(28), 6474–6487.
Kandt, C., Ash, W. L., & Peter Tieleman, D. (2007). Setting up and running molecular dynamics simulations of membrane proteins. *Methods: A Companion to Methods in Enzymology, 41*(4), 475–488.
Ke, S., Timin, E. N., & Stary-Weinzinger, A. (2014). Different inward and outward conduction mechanisms in Na_VMs suggested by molecular dynamics simulations. *PLoS Computational Biology, 10*(7), e1003746.
Ke, S., Zangerl, E.-M., & Stary-Weinzinger, A. (2013). Distinct interactions of Na^+ and Ca^{2+} ions with the selectivity filter of the bacterial sodium channel Na_VAb. *Biochemical and Biophysical Research Communications, 430*, 1272–1276.
Khan, A., et al. (2006). Isoform-dependent interaction of voltage-gated sodium channels with protons. *Journal of Physiology, 576*(2), 493–501.
Kim, I., & Allen, T. W. (2011). On the selective ion binding hypothesis for potassium channels. *Proceedings of the National Academy of Sciences of the United States of America, 108*(44), 17963–17968.
Klauda, J. B., Venable, R. M., & Freites, J. A. (2010). Update of the CHARMM all-atom additive force field for lipids: validation on six lipid types. *Journal of Physical Chemistry B, 114*, 7830–7843.
Koishi, R., et al. (2004). A superfamily of voltage-gated sodium channels in bacteria. *Journal of Biological Chemistry, 279*(10), 9532–9538.
Kopfer, D. A., et al. (2014). Ion permeation in K^+ channels occurs by direct Coulomb knock-on. *Science, 346*(6207), 352–355.

Kuang, Q., Purhonen, P., & Hebert, H. (2015). Structure of potassium channels. *Cellular and Molecular Life Sciences, 72*(19), 3677–3693.
Kutzner, C., et al. (2011). Computational electrophysiology: the molecular dynamics of ion channel permeation and selectivity in atomistic detail. *Biophysical Journal, 101*(4), 809–817.
Laio, A., & Gervasio, F. L. (2008). Metadynamics: a method to simulate rare events and reconstruct the free energy in biophysics, chemistry and material science. *Reports on Progress in Physics, 71*(12), 126601.
Laio, A., & Parrinello, M. (2002). Escaping free-energy minima. *Proceedings of the National Academy of Sciences of the United States of America, 99*(20), 12562–12566.
LeMasurier, M., Heginbotham, L., & Miller, C. (2001). Kcsa it's a potassium channel. *Journal of General Physiology, 118*(3), 303–314.
Lindorff-Larsen, K., et al. (2012)D. J. Muller (Ed.), Systematic Validation of Protein Force Fields Against Experimental Data. *PLoS One, 7*(2), e32131.
Liu, S., & Lockless, S. W. (2013). Equilibrium selectivity alone does not create K^+-selective ion conduction in K^+ channels. *Nature Communications, 4*(2746), 1–7.
Long, S. B., Campbell, E. B., & Mackinnon, R. (2005). Crystal structure of a mammalian voltage-dependent shaker family K^+ channel. *Science, 309*(5736), 897–903.
Long, S. B., et al. (2007). Atomic structure of a voltage-dependent K^+ channel in a lipid membrane-like environment. *Nature, 450*(7168), 376–382.
MacKerell, A. D., et al. (1998). All-atom empirical potential for molecular modeling and dynamics studies of proteins. *Journal of Physical Chemistry B, 102*(18), 3586–3616.
MacKerell, A. D., Jr., Feig, M., & Brooks, C. L., III (2004). Extending the treatment of backbone energetics in protein force fields: limitations of gas-phase quantum mechanics in reproducing protein conformational distributions in molecular dynamics simulations. *Journal of Computational Chemistry, 25*(11), 1400–1415.
Mahdavi, S., & Kuyucak, S. (2014)S. Barnes (Ed.), Molecular Dynamics Study of Binding of μ-Conotoxin GIIIA to the Voltage-Gated Sodium Channel Nav1.4. *PLoS One, 9*(8), e105300.
Mahdavi, S., & Kuyucak, S. (2015)S. Ye (Ed.), Mechanism of Ion Permeation in Mammalian Voltage-Gated Sodium Channels. *PLoS One, 10*(8), e0133000.
McCusker, E. C., et al. (2011). Simplified bacterial "pore" channel provides insight into the assembly, stability, and structure of sodium channels. *Journal of Biological Chemistry, 286*(18), 16386–16391.
McCusker, E. C., et al. (2012). Structure of a bacterial voltage-gated sodium channel pore reveals mechanisms of opening and closing. *Nature Communications, 3*(1102), 1–8.
Mullins, L. J. (1960). An analysis of pore size in excitable membranes. *Journal of General Physiology, 43*(5), 105–117.
Naylor, C. E., et al. (2016). Molecular basis of ion permeability in a voltage-gated sodium channel. *EMBO Journal*, e201593285.
Neale, C., et al. (2011). Statistical convergence of equilibrium properties in simulations of molecular solutes embedded in lipid bilayers. *Journal of Chemical Theory and Computation, 7*(12), 4175–4188.
Neale, C., et al. (2013). Accelerating convergence in molecular dynamics simulations of solutes in lipid membranes by conducting a random walk along the bilayer normal. *Journal of Chemical Theory and Computation, 9*(8), 3686–3703.
Neale, C., et al. (2014). Indolicidin binding induces thinning of a lipid bilayer. *Biophysical Journal, 106*(8), L29–L31.
Neyton, J., & Miller, C. (1988a). Discrete Ba^{2+} block as a probe of ion occupancy and pore structure in the high-conductance Ca^{2+}-activated K^+ channel. *Journal of General Physiology, 92*(5), 569–586.
Neyton, J., & Miller, C. (1988b). Potassium blocks barium permeation through a calcium-activated potassium channel. *Journal of General Physiology, 92*(5), 549–567.

Ngo, V., et al. (2015). Quantum effects in cation interactions with first and second coordination shell ligands in metalloproteins. *Journal of Chemical Theory and Computation, 11*(10), 4992–5001.
Ngo, V., et al. (2016)H. Weinstein (Ed.), K^+ Block Is the Mechanism of Functional Asymmetry in Bacterial Nav Channels. *PLoS Computational Biology, 12*(1), e1004482.
Ngo, V. A. (2012). Parallel-pulling protocol for free-energy evaluation. *Physical Review. E, Statistical, Nonlinear, and Soft Matter Physics, 85*(3), 036702.
Nimigean, C. M. (2002). Na^+ block and permeation in a K^+ channel of known structure. *Journal of General Physiology, 120*(3), 323–335.
Nimigean, C. M., & Allen, T. W. (2011). Origins of ion selectivity in potassium channels from the perspective of channel block. *Journal of General Physiology, 137*(5), 405–413.
Noda, M., & Numa, S. (1987). Structure and function of sodium channel. *Journal of Receptor and Signal Transduction Research,* 7(1–4), 467–497.
Noda, M., et al. (1989). A single point mutation confers tetrodotoxin and saxitoxin insensitivity on the sodium channel II. *FEBS Letters, 259*(1), 213–216.
Noskov, S. Y., Bernèche, S., & Roux, B. (2004). Control of ion selectivity in potassium channels by electrostatic and dynamic properties of carbonyl ligands. *Nature, 431*(7010), 830–834.
Noskov, S. Y., & Roux, B. (2006). Ion selectivity in potassium channels. *Biophysical Chemistry, 124*(3), 279–291.
Noskov, S. Y., & Roux, B. (2008). Control of ion selectivity in LeuT: two Na^+ binding sites with two different mechanisms. *Journal of Molecular Biology, 377*(3), 804–818.
Payandeh, J., et al. (2012). The crystal structure of a voltage-gated sodium channel. *Nature, 475*(7356), 353–358.
Payandeh, J., et al. (2013). Crystal structure of a voltage-gated sodium channel in two potentially inactivated states. *Nature, 486*(7401), 135–139.
Rauscher, S., et al. (2015). Structural ensembles of intrinsically disordered proteins depend strongly on force field: a comparison to experiment. *Journal of Chemical Theory and Computation, 11*(11), 5513–5524.
Ren, D., et al. (2001). A prokaryotic voltage-gated sodium channel. *Science, 294*(5550), 2372–2375.
Roux, B. (2005). Ion conduction and selectivity in K^+ channels. *Annual Review of Biophysics and Biomolecular Structure, 34*(1), 153–171.
Roux, B., et al. (2011). Perspectives on: ion selectivity: ion selectivity in channels and transporters. *Journal of General Physiology, 137*(5), 415–426.
Santos, J. S., et al. (2012). Crystal structure of a voltage-gated K^+ channel pore module in a closed state in lipid membranes. *Journal of Biological Chemistry, 287*(51), 43063–43070.
Sauer, D. B., et al. (2011). Protein interactions central to stabilizing the K^+ channel selectivity filter in a four-sited configuration for selective K^+ permeation. *Proceedings of the National Academy of Sciences of the United States of America, 108*(40), 16634–16639.
Sauer, D. B., et al. (2013). Sodium and potassium competition in potassium selective and non selective channels. *Nature Communications, 4*(2721), 1–9.
Shaya, D., et al. (2011). Voltage-gated sodium channel (NaV) protein dissection creates a set of functional pore-only proteins. *Proceedings of the National Academy of Sciences of the United States of America, 108*(30), 12313–12318.
Shaya, D., et al. (2014). Structure of a prokaryotic sodium channel pore reveals essential gating elements and an outer ion binding site common to eukaryotic channels. *Journal of Molecular Biology, 426*(2), 467–483.
Siu, S. W. I., et al. (2008). Biomolecular simulations of membranes: physical properties from different force fields. *Journal of Chemical Physics, 128*(12), 125103.
Sotomayor, M., et al. (2007). Ion conduction through MscS as determined by electrophysiology and simulation. *Biophysical Journal, 92*(3), 886–902.

Stansfeld, P. J., et al. (2015). MemProtMD: automated insertion of membrane protein structures into explicit lipid membranes. *Structure, 23*(7), 1350–1361.
Stock, L., et al. (2013). Conduction in a biological sodium selective channel. *Journal of Physical Chemistry B, 117*(14), 3782–3789.
Sun, Y. M., Favre, I., & Schild, L. (1997). On the structural basis for size-selective permeation of organic cations through the voltage-gated sodium channel. *Journal of General Physiology, 110*(6), 693–715.
Tang, L., et al. (2014). Structural basis for Ca^{2+} selectivity of a voltage-gated calcium channel. *Nature, 505*(7481), 56–61.
Terlau, H., et al. (1991). Mapping the site of block by tetrodotoxin and saxitoxin of sodium channel II. *FEBS Letters, 293*(1–2), 93–96.
Thompson, A. N., et al. (2009). Mechanism of potassium-channel selectivity revealed by Na^+ and Li^+ binding sites within the KcsA pore. *Nature Structural & Molecular Biology, 16*(12), 1317–1324.
Tian, C., et al. (2013). Potassium channels: structures, diseases, and modulators. *Chemical Biology & Drug Design, 83*(1), 1–26.
Tikhonov, D. B., & Zhorov, B. S. (2012). Architecture and pore block of eukaryotic voltage-gated sodium channels in view of NavAb bacterial sodium channel structure. *Molecular Pharmacology, 82*(1), 97–104.
Torrie, G. M., & Valleau, J. P. (1974). Monte Carlo free energy estimates using non-Boltzmann sampling: application to the sub-critical Lennard-Jones fluid. *Chemical Physics Letters, 28*(4), 578–581.
Treptow, W., & Tarek, M. (2006). Environment of the gating charges in the Kv1.2 Shaker potassium channel. *Biophysical Journal, 90*(9), L64–L66.
Ulmschneider, M. B., et al. (2013). Molecular dynamics of ion transport through the open conformation of a bacterial voltage-gated sodium channel. *Proceedings of the National Academy of Sciences of the United States of America, 110*(16), 6364–6369.
Vargas, E., et al. (2012). An emerging consensus on voltage-dependent gating from computational modeling and molecular dynamics simulations. *Journal of General Physiology, 140*(6), 587–594.
Varma, S., & Rempe, S. B. (2007). Tuning ion coordination architectures to enable selective partitioning. *Biophysical Journal, 93*(4), 1093–1099.
Wu, J., et al. (2015). Structure of the voltage-gated calcium channel Cav1.1 complex. *Science, 350*(6267), 1–10.
Xia, M., Liu, H., Li, Y., Yan, N., & Gong, H. (2013). The Mechanism of Na^+/K^+Selectivity in Mammalian Voltage-Gated Sodium Channels Based on Molecular Dynamics Simulation. *Biophysical Journal, 104*(11), 2401–2409. http://dx.doi.org/10.1016/j.bpj.2013.04.035.
Yu, F. H., & Catterall, W. A. (2004). The VGL-chanome: a protein superfamily specialized for electrical signaling and ionic homeostasis. *Science Signaling, 2004*(253), re15.
Yue, L. (2002). The cation selectivity filter of the bacterial sodium channel, NaChBac. *Journal of General Physiology, 120*(6), 845–853.
Zhang, X., Ren, W., et al. (2013). Crystal structure of an orthologue of the NaChBac voltage-gated sodium channel. *Nature, 486*(7401), 130–134.
Zhang, X., Xia, M., et al. (2013). Analysis of the selectivity filter of the voltage-gated sodium channel NavRh. *Cell Research, 23*(3), 409–422.
Zheng, J., & Trudeau, M. C. (2015). *Handbook of ion channels*. CRC Press.
Zhou, Y., et al. (2001). Chemistry of ion coordination and hydration revealed by a K^+ channel–Fab complex at 2.0 Å resolution. *Nature, 414*(6859), 43–48.

CHAPTER NINE

Voltage-Gated Sodium Channels: Evolutionary History and Distinctive Sequence Features

M.A. Kasimova, D. Granata and V. Carnevale[1]
Temple University, Philadelphia, PA, United States
[1]Corresponding author: E-mail: vincenzo.carnevale@temple.edu

Contents

Abstract

Voltage-gated sodium channels (Na_V) are responsible for the rising phase of the action potential. Their role in electrical signal transmission is so relevant that their emergence is believed to be one of the crucial factors enabling development of nervous system. The presence of voltage-gated sodium-selective channels in bacteria ($BacNa_V$) has raised questions concerning the evolutionary history of the ones in animals. Here we review some of the milestones in the field of Na_V phylogenetic analysis and discuss some of the most important sequence features that distinguish these channels from voltage-gated potassium channels and transient receptor potential channels.

Current Topics in Membranes, Volume 78
ISSN 1063-5823
http://dx.doi.org/10.1016/bs.ctm.2016.05.002

1. INTRODUCTION

1.1 Functional role of Na_V

To respond to changes in the external environment, cells propagate electrical signals generated by transient, highly controlled transmembrane ionic currents. Responsible for this process are ion channels, ubiquitous proteins that reside in membranes of excitable cells and convert chemical and electrical stimuli into influxes or effluxes of charged chemical species (Hille, 2001). The genomes of complex organisms encode for a vast array of ion channels (Harmar et al., 2009). Of particular relevance, from a biological perspective, are those channels that are vital for animals (metazoa). Indeed, the emergence of motility in multicellular organisms coincided with the development of the nervous system, a specialized electric network required to coordinate the response of distinct cells over long-range distances (Liebeskind, 2011; Liebeskind, Hillis, Zakon, & Hofmann, 2016; Marlow & Arendt, 2014).

Crucial for functioning of the nervous system is the ability of neurons (one of the components of the nervous system) to sustain the propagation of an electric signal with a well-defined or standardized waveform (Kandel & Schwartz, 2013). This requires the coordinated action of several types of ion channels each enabling, at a specific time, the passage of selected ionic species (Hille, 2001). In brief, a perturbation of the resting transmembrane potential in a given section of a neuron axon is propagated to the proximal one by the consecutive activation of channels that enhance or counteract such perturbation; the corresponding rising and falling phases of the electrical signal are the hallmarks of the propagating wave known as action potential (Elmslie, 2001).

Activation of voltage-gated sodium channels (Na_V) depolarizes the membrane and, as such, determines the rising phase of the action potential (Catterall, 2000; Guy & Seetharamulu, 1986). This role as *initiators* of the electrical signal makes them a critical component of the nervous system. Accordingly, mutations in Na_V have been found to be the causative agents of various diseases characterized by neuronal hypo- or hyperexcitability (George, 2005; Mantegazza, Curia, Biagini, Ragsdale, & Avoli, 2010). Due to this functional specialization, Na_V have been long considered a distinctive feature of metazoa. However, the discovery of an entire superfamily of sodium selective channels in bacteria ($BacNa_V$) has challenged this view (Durell & Guy, 2001; Ito et al., 2004; Koishi et al., 2004;

Payandeh & Minor, 2015; Payandeh, Scheuer, Zheng, & Catterall, 2011; Ren et al., 2001; Yue, Navarro, Ren, Ramos, & Clapham, 2002). Moreover, the identification of Na_V homologous genes in protists (that lack nervous system) has sparked a new interest on careful phylogenetic analyses that trace evolutionary history of these channels and put it in relation with the emergence of specific biological functions (Cai, 2012; Liebeskind, Hillis, & Zakon, 2011; Moran, Barzilai, Liebeskind, & Zakon, 2015).

1.2 Transmembrane domain architecture of Na_V

Na_V channels are integral membrane proteins with 24 membrane-spanning helical segments (Catterall, 2000; Noda et al., 1986, 1984). They are constituted by four homologous, nonidentical domains, referred to as DI through DIV, each containing six transmembrane helices (S1 through S6). The primary structure of each domain unambiguously indicates relatedness to the six-transmembrane-helix domain present in voltage-gated potassium channels (K_V) and known as 6TM domain. Since the functional assembly of K_V is a homotetramer, this homology suggests that Na_V and K_V ought to share the same overall architecture, with Na_V being pseudotetramers (Yu & Catterall, 2004). The determination of the structure of a member of the family of voltage-gated calcium channels (Ca_V) (Wu et al., 2015), pseudotetrameric channels closely related to Na_V, has provided a completely unambiguous experimental verification of this prediction. Moreover, this relatedness suggests that one of the ancestral Na_V genes was a single domain channel similar to K_V that underwent two rounds of duplications (Strong, Chandy, & Gutman, 1993). This hypothesis is indirectly reinforced by the high sequence similarity between Na_V and BacNa_V, a family of single domain channels that assemble in homotetramers.

As in all other voltage-gated ion channels, conduction of ions in Na_V occurs through the hydrophilic pathway lined by the four S6 helices (one from each domain). This pathway contains two crucial structural elements: (1) a gate, or hydrophobic constriction that hinders diffusion of ions when the channel is closed and (2) a selectivity filter, which is permeant to Na^+ ions. In Na_V, a quadruplet of amino acids, one from each domain, forms the distinctive DEKA motif lining the selectivity filter (Heinemann, Terlau, Stühmer, Imoto, & Numa, 1992).

1.3 Mechanism of gating of Na_V

Na_V contain voltage-dependent ion pores whose gating is governed by the polarization state of the embedding membrane. Their activation mechanism is in general the same as all the members of the voltage-gated ion channels superfamily, such as, for instance, K_V. While the gating occurs at the level of the pore domain, formed by the tetrameric assembly of the last two transmembrane helices S5 and S6, the domain responsible for voltage sensitivity is formed by the assembly of the first four transmembrane helices (S1−S4) and is called the voltage sensor domain (Fig. 1A) (McCusker et al., 2012; Payandeh et al., 2011; Shaya et al., 2014; Zhang et al., 2012). Thus, during activation, an allosteric signal is transmitted from the voltage sensor domain to the pore domain. Specifically, S4 contains positively charged residues (usually arginines) present with a periodicity of three residues. When the membrane polarization state is altered, S4 responds by translating vertically (along the normal to the membrane) in a stepwise helical screw motion, which in turn causes opening and closing of the S6 gate (Fig. 1B) (Amaral, Carnevale, Klein, & Treptow, 2012; DeCaen, Yarov-Yarovoy, Sharp, Scheuer, & Catterall, 2009; Delemotte, Tarek, Klein, Amaral, & Treptow, 2011; Freites, Schow, White, & Tobias, 2012; Henrion et al., 2012; Jensen et al., 2012; Khalili-Araghi et al., 2010; Vargas et al., 2012).

Voltage-gated potassium channel (K_V) crystal structures have revealed that the pore domain and the voltage sensor domain regions are coupled through an interfacial amphipathic helix, called the S4−S5 linker (Long, Tao, Campbell, & MacKinnon, 2007). Mutagenesis experiments have shown that mutation of the S4−S5 linker or the S6 C-term alone disrupts the coupling; however, when mutating both regions simultaneously, coupling can be restored, indicating that shape and chemical complementarity between these two regions is a requirement for proper electromechanical coupling (Labro et al., 2008; Lu, Klem, & Ramu, 2002; Tristani-Firouzi, Chen, & Sanguinetti, 2002; Kasimova, Zaydman, Cui, & Tarek, 2015). The mechanism proposed then involves conversion of the electrical signal into a mechanical signal in the form of S4 vertical translation, which is transmitted to the S6 C-term through an indirect interaction with the S4−S5 linker acting as a mediator (Blunck & Batulan, 2012).

While the molecular mechanism of activation is likely to be shared among all the members of the voltage-gated channels superfamily, inactivation shows significant differences and is peculiar of each family. In Na_V, inactivation plays a crucial role by preventing a runaway of the transmembrane potential during the action potential: when all the Na_V channels have been activated by a depolarizing potential, influx of Na^+ ions further increases the transmembrane potential, which, in turn, increases the sodium current, a positive feedback that would lead to an instability without a proper offset.

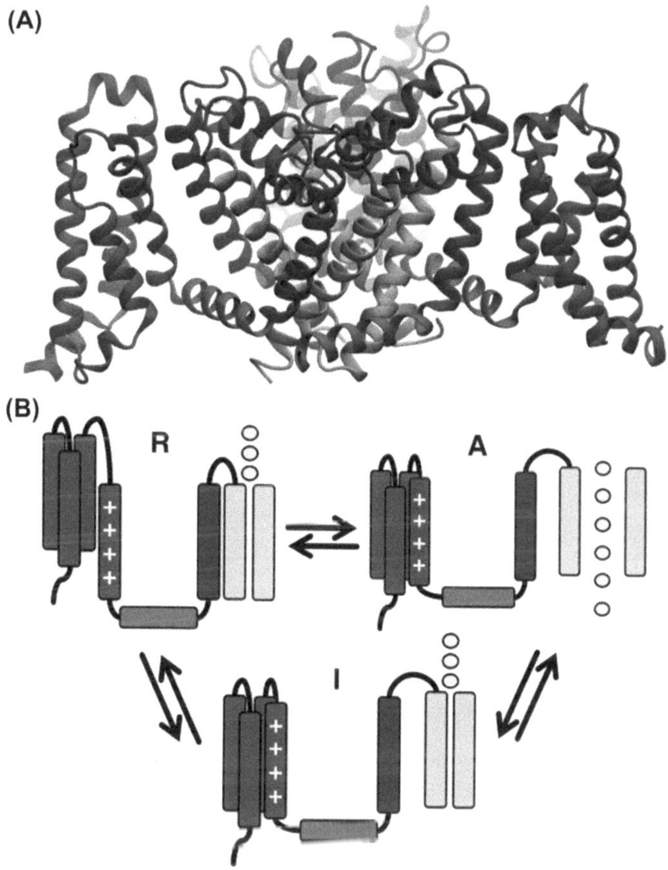

Figure 1 Na_V channels: structure and function. (A) Structure of the bacterial voltage-gated sodium channel Na_VAb (Payandeh et al., 2011). Unlike eukaryotic voltage-gated sodium channels, which consist of a single polypeptide chain containing 24 transmembrane (TM) segments, bacterial sodium channels consist of four identical subunits each containing six TM segments; however, the overall architecture of the TM part of eukaryotic and bacterial voltage-gated sodium channels is similar. The four TM helices from S1 to S4 of one subunit (or one domain in case of eukaryotic channels) constitute the voltage sensor (shown in violet (black in print versions)). The two TM helices, S5 (blue (gray in print versions)) and S6 (lime (light gray in print versions)), together with the two pore helices and the selectivity filter (shown in blue (gray in print versions)) from the four distinct subunits constitute the pore domain. The linker (shown in red (dark gray in print versions)) connects the voltage sensor to the pore domain. (B) Typical gating cycle of a Na_V channel: under depolarization, the resting state (R, the voltage sensors are in the resting conformation, the pore is closed) switches to the activated one (A, the voltage sensors are in the activated conformation, the pore is open); the activated state transitions quickly to the inactivated state (I, the voltage sensors are in the activated state, the pore is closed); finally, under repolarization, the channel transitions back from the inactivated to the resting state. The color code is identical to A. Sodium ions are shown in yellow (white in print versions).

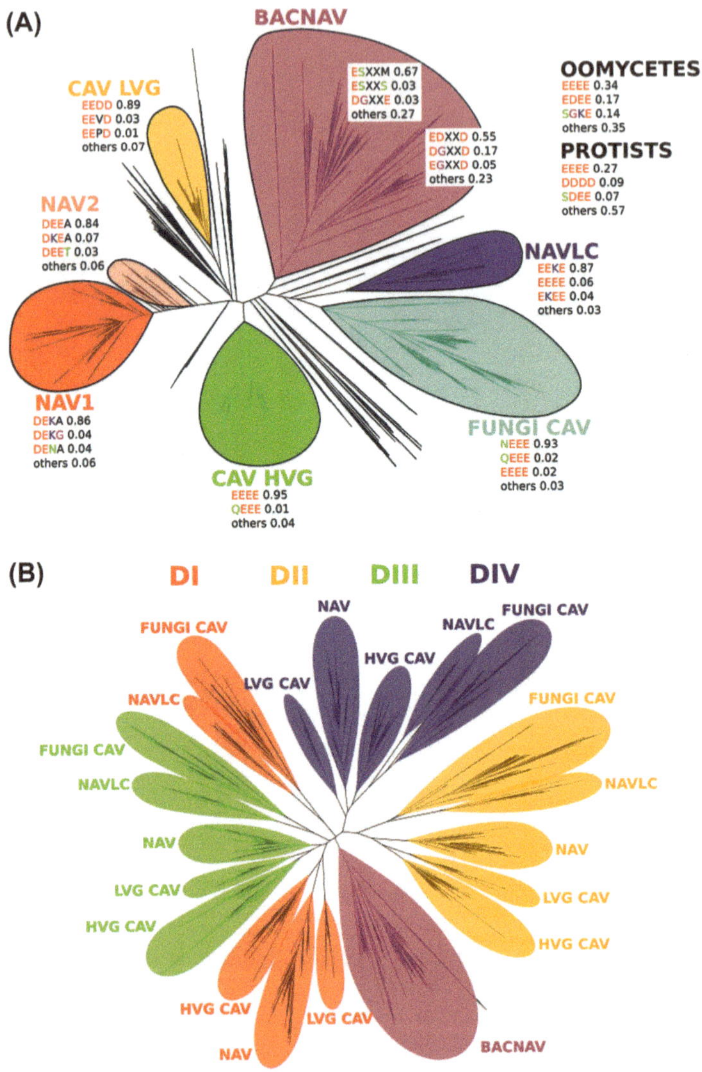

Figure 2 Phylogenetic trees of pseudotetrameric channels (A) and of individual domains of pseudotetrameric channels (B), considering the transmembrane part. The branch of bacterial voltage-gated sodium channels has been added as out-group to estimate the position of the root (in (A) these sequences were replicated four times in order to be consistent with the pseudotetrameric arrangement of Na_v and Ca_v channels). The trees are built using FastTree (Price, Dehal, & Arkin, 2010); the branches with the posterior probability higher than 0.7 are shown in color. The multiple sequence alignment (MSA) was obtained using hmmer (Finn, Clements, & Eddy, 2011). Redundant sequences were removed from the MSA so to have 95% as the maximum pairwise identity in the dataset. In total, ~3.100 sequences were used to build the trees. (A) The branches are colored according to the family. The unannotated braches correspond to pseudotetrameric channels found in oomycetes and protists. For each family

From a functional point of view, the inactivated state is a stable nonconductive state to which the channel transitions, usually from the activated state (Fig. 1B) (Ahern, 2013; Armstrong & Bezanilla, 1977; Armstrong, Bezanilla, & Rojas, 1973; Bezanilla & Armstrong, 1977).

The molecular mechanism of Na_V inactivation is thought to rely on a "hinged-lid" mechanism whereby a section of the linker connecting DIII to DIV interacts transiently with the S6 bundle occluding the channel and preventing ion permeation (Eaholtz, Scheuer, & Catterall, 1994; Rohl et al., 1999; Stühmer et al., 1989; Ulbricht, 2005; Vassilev, Scheuer, & Catterall, 1988; West et al., 1992). The sequence motif on the linker responsible for this interaction has been identified in the IFM conserved residues; however, no firm conclusion has been so far drawn about the residues of the pore interacting with such motif. Specific mutations in DI and DIV S6 have been shown to affect significantly the process of inactivation (McPhee, Ragsdale, Scheuer, & Catterall, 1994; Wang, Bonner, Russell, & Wang, 2003); however, since mutations often affect both activation and inactivation, it has not been possible to disentangle direct effects (binding of the IFM motif) from indirect ones resulting from altered activation kinetics. Irrespective of the specific microscopic details of the "hinged-lid" mechanism, it appears that inactivation relies on fine-tuned molecular interactions within the pore domain; it is reasonable to expect that this functional requirement resulted in an evolutionary pressure distinct from that exerted on the K_V pore domain and thus, ultimately, in different pore domain sequences.

1.4 Na_V are closely related to Ca_V, NALCN, and fungi-Ca_V

Na_V channels are not the only channels characterized by the pseudotetrameric architecture. In addition to the already-mentioned Ca_V, also fungi-Ca_V and sodium leak channels, nonselective (NALCN) (Fig. 2A) show a primary structure similar to that of Na_V. Fungi-Ca_V and NALCN are two closely related evolutionary families of channels that are not voltage gated (Lee, Cribbs, & Perez-Reyes, 1999; Liebeskind, Hillis, & Zakon, 2012).

(also for channels of oomycetes and protists), the sequences of the selectivity filter are shown (four residues each taken from one domain in case of pseudotetrameric channels, five residues from one subunit in case of $BacNa_V$ channels). (B) The branches are colored according to the domain: red for domain I (DI), orange for domain II (DII), green for domain III (DIII) and blue for domain IV (DIV).

While fungi-Ca_V are selective for Ca^{2+} ions, NALCN are nonselective. Owing to their phyletic distribution, this clade of pseudotetrameric groups of ion channels shares a common ancestor dating back more than one billion years. Accordingly, functional divergence resulted in channels responsive to different stimuli and with different selectivity profiles. However, and notwithstanding this functional diversity, these channels show strong sequence similarity in the pore domain (except for the selectivity filter).

These observations suggest that, in spite of the earlier-mentioned heterogeneity in ion selectivity and activating stimuli, specific sequence features have been selected and maintained along evolution, arguably because they underlie a gating mechanism common to all these channels (Irie et al., 2010; Koishi et al., 2004; Oelstrom, Goldschen-Ohm, Holmgren, & Chanda, 2014; Pavlov et al., 2005; Ren et al., 2001). This hypothesis is supported by the fact that these channels often share the pharmacological profile (Bagnéris et al., 2014; Barber, Carnevale, Klein, Eckenhoff, & Covarrubias, 2014; Boiteux et al., 2014; Bruhova, Tikhonov, & Zhorov, 2008; Corry, Lee, & Ahern, 2014; Covarrubias, Barber, Carnevale, Treptow, & Eckenhoff, 2015; Lee, Goodchild, & Ahern, 2012; Ouyang, Jih, Zhang, Correa, & Hemmings, 2007; Raju, Barber, LeBard, Klein, & Carnevale, 2013; Tikhonov & Zhorov, 2012). Discovery of such features can take advantage of a comparative analysis based on thousands of sequences of pseudotetrameric channels. These sequences are available, thanks to the many sequencing efforts that resulted in an almost exhaustive coverage of organisms from all clades and kingdoms. When a large collection of homologous channel sequences can be identified and aligned, statistical features in the resulting multiple sequence alignment (MSA), such as conservation or covariation, could be indicative of selective pressures common to all these channels (Halabi, Rivoire, Leibler, & Ranganathan, 2009; Palovcak, Delemotte, Klein, & Carnevale, 2014, 2015).

Using this approach, we summarize here sequence–function relationships common to the pseudotetrameric channel clade and previously reported in separate studies. To put these observations in an evolutionary context, we explored large repositories of protein sequences and collected the largest set of pseudotetrameric channel sequences thus far available. We then proceeded with a phylogenetic analysis of these sequences to align the 6TM region with the maximum possible accuracy. We used these data to first reconstruct the evolutionary history of Na_V and then to characterize their distinctive sequence features. To this end, we analyzed the MSAs using probabilistic generative models and statistics inspired by information theoretic concepts (Cover & Thomas, 1991). To characterize the evolutionary pressure exerted on specific segments of the transmembrane region and to discuss the functional relevance of networks of coevolving residues, we compare them to the already analyzed cases of TRP and K_V channels.

2. EVOLUTIONARY HISTORY OF Na_V

2.1 Na_V are part of a clade comprising Ca_V, NALCN, and fungi-Ca_V

Pseudotetrameric channels can be grouped into several families, including sodium-selective (Na_V), calcium-selective (Ca_V) and nonselective (NALCN) channels. These channels were shown to be present in animals, fungi, oomycetes, and protists (including algae) but never in green plants, which suggests that their common ancestor emerged at the level of protists but then was lost in green plants during evolution. On the phylogenetic tree, shown in Fig. 2A, there are five major branches of pseudotetrameric channels corresponding to Na_V (red), low- (orange) and high-voltage gated (green) Ca_V from animal, fungi-Ca_V (cyan), and NALCN (blue). We included the set of bacterial voltage-gated sodium (BacNa_V, magenta) channels sequences as an out-group to estimate the position of the root of the tree. Based on the position of the BacNa_V branch on the tree, the common ancestor of pseudotetrameric channels had a Ca_V-like sequence rather than Na_V-like. Moreover, this ancestor is closer to fungi-Ca_V and NALCN channels compared to animal Ca_V. Na_V channels are the most distant from the Bac-Na_V branch, which suggests that these are the most recent pseudotetramers. When BacNa_V were found in prokaryotes, they were suggested to be direct ancestors of eukaryotic Na_V (Charalambous & Wallace, 2011); however, our phylogenetic analysis places them closer to the Ca_V channels. Moreover, Liebeskind, Hillis, and Zakon (2013) found that CatSper is the closest group to BacNa_V. Several nonannotated branches shown on the tree correspond to pseudotetrameric channels found in oomycetes and protists. These channels are mostly clustered either with the branches of animal Ca_V or with the branches of fungi Ca_V and NALCN, suggesting that the splitting in these functional groups has happened early, at the level of protists.

2.2 Each family of pseudotetrameric channels has a distinctive sequence motif at the selectivity filter

Interestingly, while pseudotetrameric channels are grouped based on the similarity of their transmembrane parts (including the voltage sensor and the pore domains), each family has a distinctive sequence motif at the selectivity filter: DEKA for Na_V, EEEE or EEDD for animal Ca_V, EEKE for NALCN, and NEEE for fungi-Ca_V. This indicates a strong statistical coupling between the variation of the four residues of the selectivity filter and the entire transmembrane region (~1200 residues). Since the transmembrane part plays a crucial role in gating, this observation suggests that the nature of the permeant ion might affect gating properties such as modulation by voltage or cofactors and kinetics of gating. An interesting exception is the Na_V2.1-Na_V2.4 subfamilies, which group with the Na_V branch yet have been shown to be calcium selective (Zhang, Liu, Song, Du, & Dong, 2011; Zhou, Chung, Liu, Goldin, & Dong, 2004). Note that the Na_V2.5 subfamily is only selective for sodium (Gur Barzilai et al., 2012). The selectivity filter of oomycetes and protists channels, on the other hand, shows higher level of variability. However, even in this case, one can observe a dominant pattern (~30% in both oomycetes and protists), which is similar in these two groups and is EEEE. This sequence of the selectivity filter resembles the one of the calcium-selective channels (low- and high-voltage gated Ca_V from animals and fungi-Ca_V) rather than the one of the sodium-selective and nonselective channels (Na_V and NALCN). This suggests that the common ancestor emerged at the level of protists was calcium selective. Finally, most BacNa_V channels have the EEEE sequence playing the role of selectivity filter. Though this sequence is similar to that of Ca_V, BacNa_V channels were shown to be selective for sodium (Ren et al., 2001; Yue et al., 2002).

2.3 Sodium selectivity emerged independently several times during evolution

The evolution of ion selectivity in pseudotetrameric channels remains a topic of debate. There are two possible scenarios accounting for the evolution of ion selectivity (Liebeskind et al., 2013). One of the models implies that the common ancestor of pseudotetrameric and BacNa_V was calcium selective, and the selectivity to sodium was acquired independently in BacNa_V and Na_V (including Na_V1 and Na_V2.5 subfamilies). The other scenario, instead, posits that the common ancestor was sodium selective, and then calcium selectivity was acquired in Ca_V channels (animal and fungi-Ca_V) and in Na_V2.1-Na_V2.4. While both of the models show elements of credibility, the first one has been shown to be the only viable one on the basis of a rigorous phylogenetic analysis of the pseudotetrameric channels clade and of an ancestor sequence reconstruction of the selectivity filter of $BacNa_V$ (Liebeskind et al., 2012, 2013). We already mentioned that according to the tree shown in Fig. 2A the common ancestor of $BacNa_V$ and pseudotetrameric channels is Ca_V-like rather than Na_V-like, suggesting that this ancestor was selective to calcium. The ancestral state reconstruction performed by Liebeskind et al. also supports the first model and indicates that the sodium selectivity in $BacNa_V$ and Na_V arose independently (Liebeskind et al., 2013). Another piece of evidence comes from the CatSper channels, which share a more recent common ancestor with pseudotetrameric channels than $BacNa_V$. CatSper channels are calcium selective, and as discussed by Liebeskind et al. (2013) a late acquisition of calcium selectivity is unlikely in this family. Additionally, several computational and electrophysiological studies have shown that $BacNa_V$ have a conduction mechanism distinct from that of Na_V (Boiteux, Vorobyov, & Allen, 2014; Carnevale, Treptow, & Klein, 2011; Chakrabarti et al., 2013; Corry & Thomas, 2012; Finol-Urdaneta et al., 2014; Furini & Domene, 2012; Ke, Zangerl, & Stary-Weinzinger, 2013; Stock, Delemotte, Carnevale, Treptow, & Klein, 2013; Ulmschneider et al., 2013). Finally, the tree in Fig. 2A shows that the pseudotetrameric channels from protists are more similar to Ca_V-like than to Na_V-like channels, providing additional support to the hypothesis that sodium selectivity appeared independently in Na_V, and that the common ancestor of pseudotetrameric channels was likely calcium selective.

Interestingly, the $BacNa_V$ branch clusters in two smaller branches, each with a distinct sequence motif at the selectivity filter: ESXXM and EDXXD (here the sequence refers to each monomer as opposed to the case of pseudotetrameric channels, in which each residue is from a different domain) (Fig. 2A). While channels showing the first motif at the selectivity filter have been extensively characterized (Koishi et al., 2004; Yue et al., 2002), the other group of channels is largely unexplored. Furthermore, mutation of neutral residues into negatively charged one in the ESXXM selectivity filter of $BacNa_V$ channels has been shown to switch selectivity, rendering the channel calcium selective (Shaya et al., 2011; Tang et al., 2014; Yue et al., 2002). Thus, it is tempting to hypothesize that these two branches might correspond to sodium- and calcium-selective channels. If confirmed, this, together with the relative position of $BacNa_V$ and pseudotetrameric channels in the tree, would provide additional elements of support to the hypothesis that ancestral $BacNa_V$ were calcium selective and thus sodium selectivity was acquired independently in animals and bacteria.

2.4 Pseudotetramers result from two rounds of gene duplication

According to a widely accepted point of view, pseudotetrameric channels have a single origin, ie, an ancestral single-domain BacNa$_V$-like channel underwent two consecutive events of gene duplication to give rise to the common ancestor of all pseudotetrameric channels (Strong et al., 1993). Strong evidence in support of this view comes from the reconstruction of the phylogenetic tree based on single-domain alignments. Similar to the procedure of Strong et al. (1993), we split the sequences of pseudotetrameric channels in single domains and build the phylogenetic tree based on this alignment (Fig. 2B). On this tree, it is clear that sequences group according to the domain identity rather than channel family (ie, Na$_V$, Ca$_V$, NALCN). A notable exception is that of domains DI and DIII from fungi-Ca$_V$ and NALCN, which do not group with the corresponding sequences of Na$_V$ and animal Ca$_V$. The sequences are further grouped in two bigger branches containing the domains DI and DIII, and the domains DII and DIV, a further element supporting the hypothesis that the two rounds of gene duplication occurred only once.

3. DISTINCTIVE SEQUENCE FEATURES OF PSEUDOTETRAMERIC CHANNELS

3.1 Sequence features from relative entropy and pairwise couplings

Thanks to a structural alignment involving a prototypical K$_V$ structure (Long et al., 2007), the pseudotetrameric channel Ca$_V$1.1 (Wu et al., 2015) and a member of the TRP family (Cao, Liao, Cheng, & Julius, 2013; Liao, Cao, Julius, & Cheng, 2013), we are able to analyze the distributions of amino acids observed at each position of the pseudotetrameric channel in a comparative context. Structurally equivalent positions in the transmembrane region of each of these three families might or might not show similar distributions depending on the specific evolutionary pressure acting on the member of each family. Indeed, in large and diverse protein families, the peculiar deviation of the distribution of amino acids from the naturally occurring frequencies is often informative about the physicochemical properties required at each position to preserve protein functionality. Therefore, visual inspection of in-register sequence logos usually highlights the distinctive features of each family. In the case of K$_V$, TRP, and pseudotetrameric channels, despite the substantial structural similarity, only few positions share similar amino acid distributions in all the three channel families. To make this observation more quantitative, we show also the relative entropy for equivalent positions in the three MSAs. The relative entropy (or Kullback—Leibler divergence) is a measure of the difference between two probability distributions (Cover & Thomas, 1991). Used to compare pairs of MSAs, it quantifies the difference between the distributions observed at structurally equivalent position in the two families.

The second sets of properties that we discuss and compare are the networks of coevolving residues. In general, each given functional requirement, such as the responsiveness to voltage or ligand binding, leaves a distinct evolutionary imprint on the distribution of sequences encoding for the specific protein family. Large databases of sequences allow one to highlight this imprint in a way that goes beyond the measure of conservation at each site of the alignment. Especially interesting, in these respects, are joint frequency distributions involving pairs of positions in the MSA. In the native folded state, each residue engages in residue—residue interactions. The specific mutations allowed at a given position are then highly dependent on the chemical identity of neighboring residues. As a result, frequency distributions of amino acids at different positions are expected to be statistically dependent on each other. Detection of these correlations can potentially unveil the physical interactions sustaining the propagation of allosteric signals in ion channels and thus highlight differences between two functionally distinct families. To reconstruct the network of coevolving residues, one has to solve an inverse problem: finding the set of residue—residue pairwise interactions (in a statistical sense) that best reproduce the distributions of sequences observed in the MSA (Feinauer, Skwark, Pagnani & Aurell, 2014; Weigt, White, Szurmant, Hoch, & Hwa, 2009). Then the pairs of nodes of the network with the largest (statistically significant) interaction strength are identified with pairs of coevolving residues.

3.2 Sequence similarity highlights the modular nature of pseudotetrameric channels sequences

From the MSA of the pseudotetrameric channels in which each domain is treated as a separate unit, it is possible to estimate the entropy of the distribution of amino acid for each position in the transmembrane region. This allows one to get a convenient estimate of sequence variability at each position (Fig. 3). For comparison, here the same analysis is reported also for BacNa$_V$, TRP, and K$_V$ channels. Similar to K$_V$ channels, the voltage sensor domain of pseudotetrameric and BacNa$_V$ channels demonstrates periodic

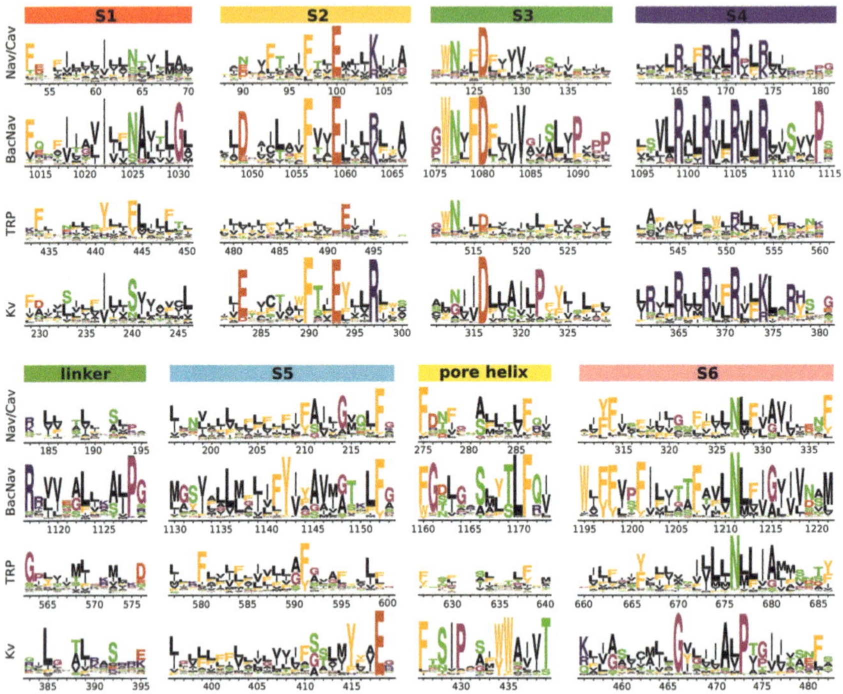

Figure 3 Aligned sequence logos for pseudotetrameric (Na_V/Ca_V), $BacNa_V$, TRP, and K_V channels in the voltage sensor and pore domains. The numbering corresponds to Ca_V1.1 (DI), Na_VAB, TRPV1, and K_V1.2.

pattern of residues with low entropies and thus low variability. These include the gating charges of S4, the negatively charged residues of S2 and S3 and several hydrophobic residues of S1, S2, and S3, which face the interior of the voltage sensor and create a so-called hydrophobic plug. Note that this pattern is not present in TRP channels, which are not (primarily) responsive to voltage. Interestingly, the S3 helices of pseudotetrameric channels and $BacNa_V$ share similar features with that of TRP channels. The linker of pseudotetrameric channels also resembles that of K_V rather than TRP. For the pore domain, however, the situation is the opposite: the sequence patterns of pseudotetrameric channels, $BacNa_V$, and TRP are similar to each other and distinct from those of K_V channels. This is especially evident for the S6 helix: pseudotetrameric channels, $BacNa_V$, and TRP have an extremely well-conserved asparagine in the middle of S6 surrounded by several bulky hydrophobic residues (Bagnéris, Naylor, McCusker, & Wallace, 2015; Barber et al., 2012; Shafrir, Durell, & Guy, 2008); in K_V this part shows two conserved residues (proline and glycine), which act as structural hinges allowing a kinked conformation of S6 during gating (Long et al., 2007). These data demonstrate that, as expected, the voltage sensor domain and the S4–S5 linker domain of pseudotetrameric channels show amino acid distributions extremely similar to those of K_V channels; less discussed so far in literature is, instead, the striking similarity between the distribution of amino acids in the pore domain of pseudotetrameric channels and TRP. To visually convey these patterns of similarity, we reported the relative entropy for the two pairs (pseudotetrameric/K_V and pseudotetrameric/TRP) and projected this quantity on the structure of one subunit of the channel (Fig. 4).

3.3 Networks of coevolving residues show similarity between pseudotetrameric channels and TRP

The similarity in the pore domain between pseudotetrameric channels and TRP is not limited to the amino acid distributions. Analysis of the network

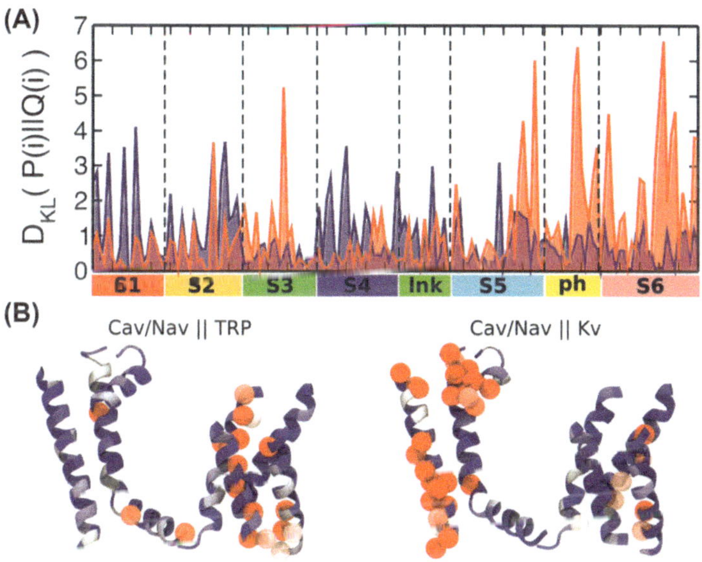

Figure 4 Relative entropy analysis. (A) Symmetrized relative entropy between Na_V/Ca_V and TRP channels (blue) and Na_V/Ca_V and K_V channels (red). Low/high values correspond to positions with similar/dissimilar distributions of amino acids. The major structural elements of the channel are indicated on the x-axis: transmembrane helices S1–S6, linker (lnk), and the pore helix (the first pore helix in case of Na_V/Ca_V). (B) Symmetrized relative entropy plotted on the structure of $Ca_V1.1$. Red, white, and blue correspond to high, intermediate, and low values respectively. The positions with relative entropy higher than 2.5 nats are highlighted with *red spheres*.

of coevolving residues indicates that, not only the chemical nature of several amino acids is preserved on passing from one family to the other, but also several key residue–residue interactions. It has been shown (Palovcak, Delemotte, Klein, & Carnevale, 2015) that both TRP and K_V are characterized by networks of strongly coevolving residues at the extracellular side of the pore domain, running across the top section of S5 and S6 and at the interface between S5 and S6 on one side and the pore helix on the other. However, the two families were shown to be significantly different in the number and the strength of coevolving pairs detected between the N-terminal part of S5 (right after the linker) and the C-terminal part of S6. Specifically, while TRP showed an extensive network of coevolving pairs, suggestive of a fine-tuned interaction between the two helices, K_V was characterized by few sparse coupled pairs. Here we show that pseudotetrameric channels show a network of coevolving residues more similar to TRP than K_V. These data are shown in Fig. 5, in which the statistical significance of each coupling has been evaluated via resampling techniques; the difference between pairs of channel families (pseudotetrameric channels, TRP, and K_V) is reported in terms of size of the effect (Fig. 5A); note how the differences between pseudotetrameric channels and TRP are hardly noticeable while those pseudotetrameric channels and K_V are much more significant. A similar conclusion can be drawn by inspecting the couplings for each family separately (Fig. 5B): although weaker, the statistical couplings are arranged in a pattern that bears similarity with that of TRP. Finally, the projection of the statistical couplings on the structure (Fig. 5B, right column) highlights how the couplings between the end of the linker and S6 present in the pseudotetrameric channels and in TRP are absent or weaker in K_V.

4. SUMMARY AND CONCLUSION: WHAT CAN WE LEARN FROM SEQUENCES ABOUT Na_V GATING?

The increased complexity of the body plan at the emergence of metazoa required the development of an apt system to propagate information in a fast and accurate way over long distances. The newly developed nervous system and its acquired functional needs promoted the evolution of a new class of voltage-gated ion channels, namely the sodium selective Na_V (Liebeskind, Hillis, & Zakon, 2015). These channels appeared thanks to a

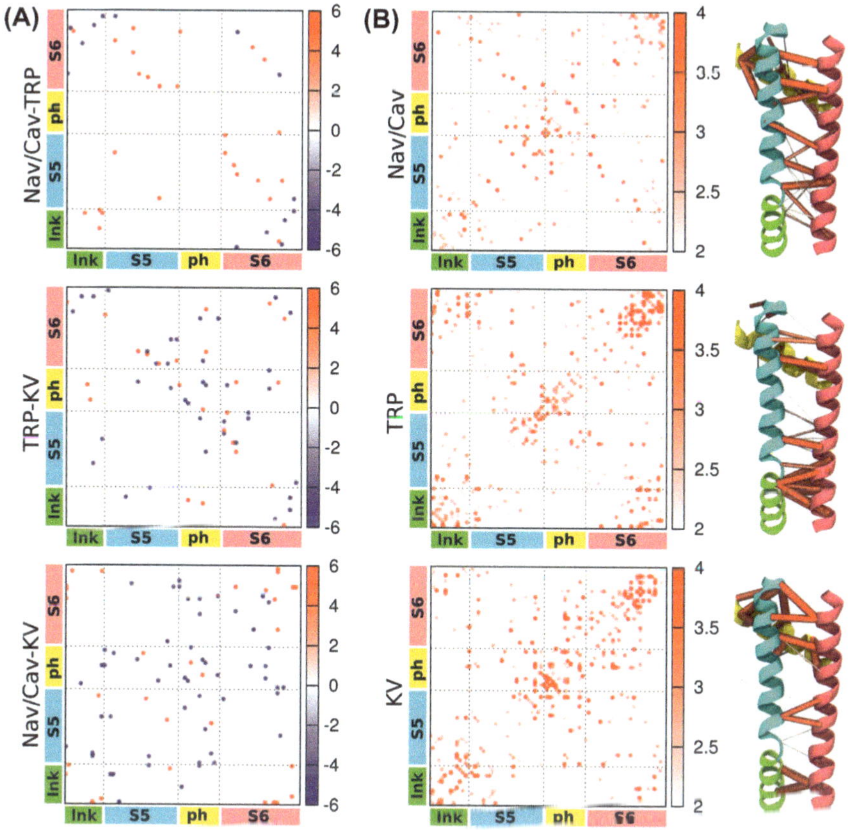

Figure 5 Coevolution analysis of the Na_V/Ca_V, K_V, and TRP families. Evolutionary couplings were calculated using the direct coupling analysis approach as discussed in (Feinauer et al., 2014). To characterize the statistical significance of each pairwise coupling we used the same approach as in Palovcak et al. (2015): in brief, we used a resampling technique to generate 1000 distinct multiple sequence alignments; from these, the mean and the variance of each pairwise coupling was estimated. (A) The difference in coupling strength is estimated for each pair of residues across distinct families and is reported in terms of effect size (difference between the two means divided by the pooled standard deviation). Note that these differences have the largest magnitude in the case of the Na_V/Ca_V-K_V comparison. (B) EC-scores (as defined in Palovcak et al. (2015)) are shown for the three families separately. Na_V/Ca_V shows couplings much weaker than in the case of K_V and TRP, possibly because in the homotetrameric cases statistical signals are expected to be stronger (a mutation in the polypeptide results in four residues changed on the structure). On the right hand side the couplings are projected onto the structure of Ca_V1.1. Note how in TRP and Na_V/Ca_V, the region between the linker and S5 shows a more extensive network of couplings than in K_V.

gene duplication event followed by a functional divergence of one of the two resulting paralogs (Strong et al., 1993). It is interesting to note that ancestral Na_V were variations on the structural theme of Ca_V rather than K_V. In fact, given the extremely well-coordinated action of Na_V and K_V in the nervous system, in an easy to envision scenario these two classes of channels might have arisen from subfunctionalization of an ancestral nonselective channel.

There are two possible reasons that might justify the observed lineage. The first possible explanation is that the first nervous systems that emerged in multicellular organisms were Ca^{2+} based rather than Na^+ based (Hille, 2001; Moran et al., 2015). In this scenario, an increased complexity of the neuronal network required a neat separation between electrical signal transmission and intracellular signaling. It is then conceivable that a change in selectivity might have produced an immediate increase of fitness and thus an evolutionary advantage for the hosting species. The second reason concerns the physicochemical promiscuity of the sodium and calcium-selective selectivity filters. For both $BacNa_V$ and Na_V, single residue mutations in the selectivity filter have been shown to change the selectivity profile so dramatically to render the mutated channels calcium selective. These experiments clearly delineate a possible evolutionary pathway to link Ca_V to Na_V. In fact, at least three independent changes of selectivity from Ca^{2+} to Na^+ are known (we discussed them earlier): the most important one is, of course, the passage from Ca_V to Na_V; the second is the change of selectivity within the Na_V2 family with the emergence of $Na_V2.5$, the former being calcium- and the latter sodium-selective; and the third one is the acquired sodium selectivity of $BacNa_V$ highlighted by the presence of putatively calcium-selective members within the family and by ancestral sequence reconstruction (Liebeskind et al., 2013). Notably, in spite of the wide phyletic distribution of K_V, no example is known of a K_V-like exquisitely sodium-selective voltage-gated ion channel, suggesting that sodium selectivity emerges preferentially from Ca_V-like channels.

This peculiar evolutionary history of Na_V results in sequence profiles that are substantially different from those of K_V. It is then noteworthy that, in spite of the low sequence identity, the structures of K_V and pseudotetrameric channels are strikingly similar, as evidenced by the recently solved structure of $Ca_V1.1$ (Wu et al., 2015). However, sharing the same overall fold does not imply that all the subtle structural rearrangements accompanying the conformational transitions (ie, from resting to activated or from activated to inactivated) are coincident. In other words, while both the channels respond to a change in the transmembrane potential, the molecular mechanism by which this stimulus is transmitted to the pore domain to affect its conformation might be substantially different.

It is instructive then to analyze Na_V in a wide evolutionary perspective that includes an additional class of closely related channels, namely TRP, that appears to be characterized by a very similar structure yet respond to a different set of stimuli. Accurate sequence analysis highlights a modular nature of Na_V sequences, whereby the voltage sensor and the linker domains are K_V-like, while the pore domain appears to be extremely similar to that of TRP (Fig. 6). Given the fact that they are both voltage sensitive, the similarity between K_V and Na_V in the voltage sensor and linker domains is unsurprising. Less expected from the functional point of view, however, is the similarity between Na_V and TRP. We discussed earlier that not only the chemical identity, but also the residue–residue interactions are conserved between the two families. The latter have been shown to be informative about the overall channel mechanical properties (Palovcak et al., 2014, 2015).

All these observations raise intriguing questions: Is there a common gating mechanism shared by TRP and Na_V? Moreover: Is there any relation between the two absorbing nonconductive states at the end of the activation cycle (inactivated for Na_V, desensitized for TRP)? These very speculative questions rest on a strong assumption: that a complex conformational transition relying on the concerted motion of several structural elements might be more conserved along evolution than aspects such as selectivity or voltage sensitivity. Whether or not this is true is the subject of future analyses and experiments.

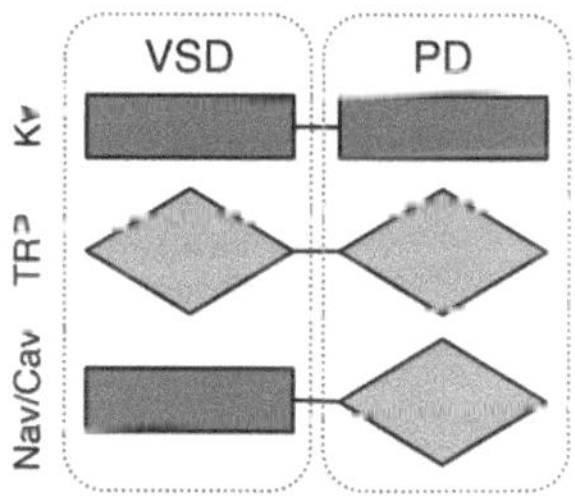

Figure 6 Modularity of pseudotetrameric channel sequences. Schematic representation of one subunit/domain of K_V, TRP, and pseudotetrameric (Na_V/Ca_V) channels. While the voltage sensor of Na_V/Ca_V is similar to that of K_V (*blue rectangular* (gray in print versions)), the pore domain resembles that of TRP (*orange diamond* (light dark gray in print versions)).

REFERENCES

Ahern, C. A. (2013). What activates inactivation? *Journal of General Physiology, 142*, 97–100. http://dx.doi.org/10.1085/jgp.201311046.

Amaral, C., Carnevale, V., Klein, M. L., & Treptow, W. (2012). Exploring conformational states of the bacterial voltage-gated sodium channel Na_VAb via molecular dynamics simulations. *Proceedings of the National Academy of Sciences of the United States of America, 109*, 21336–21341. http://dx.doi.org/10.1073/pnas.1218087109.

Armstrong, C. M., & Bezanilla, F. (1977). Inactivation of the sodium channel. II. Gating current experiments. *Journal of General Physiology, 70*, 567–590.

Armstrong, C. M., Bezanilla, F., & Rojas, E. (1973). Destruction of sodium conductance inactivation in squid axons perfused with pronase. *Journal of General Physiology, 62*, 375–391.

Bagnéris, C., DeCaen, P. G., Naylor, C. E., Pryde, D. C., Nobeli, I., Clapham, D. E., & Wallace, B. A. (2014). Prokaryotic Na_VMs channel as a structural and functional model for eukaryotic sodium channel antagonism. *Proceedings of the National Academy of Sciences of the United States of America, 111*, 8428–8433. http://dx.doi.org/10.1073/pnas.1406855111.

Bagnéris, C., Naylor, C. E., McCusker, E. C., & Wallace, B. A. (2015). Structural model of the open–closed–inactivated cycle of prokaryotic voltage-gated sodium channels. *Journal of General Physiology, 145*, 5–16. http://dx.doi.org/10.1085/jgp.201411242.

Barber, A. F., Carnevale, V., Klein, M. L., Eckenhoff, R. G., & Covarrubias, M. (2014). Modulation of a voltage-gated Na^+ channel by sevoflurane involves multiple sites and distinct mechanisms. *Proceedings of the National Academy of Sciences of the United States of America, 111*, 6726–6731. http://dx.doi.org/10.1073/pnas.1405768111.

Barber, A. F., Carnevale, V., Raju, S. G., Amaral, C., Treptow, W., & Klein, M. L. (2012). Hinge-bending motions in the pore domain of a bacterial voltage-gated sodium channel. *Biochimica et Biophysica Acta BBA – Biomembranes, 1818*, 2120–2125. http://dx.doi.org/10.1016/j.bbamem.2012.05.002.

Bezanilla, F., & Armstrong, C. M. (1977). Inactivation of the sodium channel. I. Sodium current experiments. *Journal of General Physiology, 70*, 549–566.

Blunck, R., & Batulan, Z. (2012). Mechanism of electromechanical coupling in voltage-gated potassium channels. *Frontiers in Pharmacology, 3*, 166. http://dx.doi.org/10.3389/fphar.2012.00166.

Boiteux, C., Vorobyov, I., & Allen, T. W. (2014). Ion conduction and conformational flexibility of a bacterial voltage-gated sodium channel. *Proceedings of the National Academy of Sciences of the United States of America, 111*, 3454–3459. http://dx.doi.org/10.1073/pnas.1320907111.

Boiteux, C., Vorobyov, I., French, R. J., French, C., Yarov-Yarovoy, V., & Allen, T. W. (2014). Local anesthetic and antiepileptic drug access and binding to a bacterial voltage-gated sodium channel. *Proceedings of the National Academy of Sciences of the United States of America, 111*, 13057–13062. http://dx.doi.org/10.1073/pnas.1408710111.

Bruhova, I., Tikhonov, D. B., & Zhorov, B. S. (2008). Access and binding of local anesthetics in the closed sodium channel. *Molecular Pharmacology, 74*, 1033–1045. http://dx.doi.org/10.1124/mol.108.049759.

Cai, X. (2012). Ancient origin of four-domain voltage-gated Na^+ channels predates the divergence of animals and fungi. *Journal of Membrane Biology, 245*, 117–123. http://dx.doi.org/10.1007/s00232-012-9415-9.

Cao, E., Liao, M., Cheng, Y., & Julius, D. (2013). TRPV1 structures in distinct conformations reveal activation mechanisms. *Nature, 504*, 113–118. http://dx.doi.org/10.1038/nature12823.

Carnevale, V., Treptow, W., & Klein, M. L. (2011). Sodium ion binding sites and hydration in the lumen of a bacterial ion channel from molecular dynamics simulations. *Journal of Physical Chemistry Letters, 2*, 2504–2508.

Catterall, W. A. (2000). From ionic currents to molecular mechanisms: the structure and function of voltage-gated sodium channels. *Neuron, 26*, 13–25.

Chakrabarti, N., Ing, C., Payandeh, J., Zheng, N., Catterall, W. A., & Pomès, R. (2013). Catalysis of Na^+ permeation in the bacterial sodium channel Na_VAb. *Proceedings of the National Academy of Sciences of the United States of America, 110*, 11331–11336. http://dx.doi.org/10.1073/pnas.1309452110.

Charalambous, K., & Wallace, B. A. (2011). NaChBac: the long lost sodium channel ancestor. *Biochemistry (Moscow), 50*, 6742–6752. http://dx.doi.org/10.1021/bi200942y.

Corry, B., Lee, S., & Ahern, C. A. (2014). Pharmacological insights and quirks of bacterial sodium channels. In P. C. Ruben (Ed.), *Voltage gated sodium channels* (pp. 251–267). Springer Berlin Heidelberg.

Corry, B., & Thomas, M. (2012). Mechanism of ion permeation and selectivity in a voltage gated sodium channel. *Journal of the American Chemical Society, 134*, 1840–1846. http://dx.doi.org/10.1021/ja210020h.

Covarrubias, M., Barber, A. F., Carnevale, V., Treptow, W., & Eckenhoff, R. G. (2015). Mechanistic insights into the modulation of voltage-gated ion channels by inhaled anesthetics. *Biophysical Journal, 109*, 2003–2011. http://dx.doi.org/10.1016/j.bpj.2015.09.032.

Cover, T. M., & Thomas, J. A. (1991). *Elements of information theory*. New York, NY, USA: Wiley-Interscience.

DeCaen, P. G., Yarov-Yarovoy, V., Sharp, E. M., Scheuer, T., & Catterall, W. A. (2009). Sequential formation of ion pairs during activation of a sodium channel voltage sensor. *Proceedings of the National Academy of Sciences of the United States of America, 106*, 22498–22503. http://dx.doi.org/10.1073/pnas.0912307106.

Delemotte, L., Tarek, M., Klein, M. L., Amaral, C., & Treptow, W. (2011). Intermediate states of the Kv1.2 voltage sensor from atomistic molecular dynamics simulations. *Proceedings of the National Academy of Sciences of the United States of America, 108*, 6109–6114. http://dx.doi.org/10.1073/pnas.1102724108.

Durell, S. R., & Guy, H. R. (2001). A putative prokaryote voltage-gated Ca^{2+} channel with only one 6TM motif per subunit. *Biochemical and Biophysical Research Communications, 281*, 741–746. http://dx.doi.org/10.1006/bbrc.2001.4408.

Eaholtz, G., Scheuer, T., & Catterall, W. A. (1994). Restoration of inactivation and block of open sodium channels by an inactivation gate peptide. *Neuron, 12*, 1041–1048.

Elmslie, K. S. (2001). Action potential: ionic mechanisms. In *eLS*. John Wiley & Sons, Ltd.

Feinauer, C., Skwark, M. J., Pagnani, A., & Aurell, E. (2014). Improving Contact Prediction along Three Dimensions. *PLoS Comput Biol, 10*(10), e1003847. http://dx.doi.org/10.1371/journal.pcbi.1003847.

Finn, R. D., Clements, J., & Eddy, S. R. (2011). HMMER web server: interactive sequence similarity searching. *Nucleic Acids Research, 39*, W29–W37. http://dx.doi.org/10.1093/nar/gkr367.

Finol-Urdaneta, R. K., Wang, Y., Al-Sabi, A., Zhao, C., Noskov, S. Y., & French, R. J. (2014). Sodium channel selectivity and conduction: prokaryotes have devised their own molecular strategy. *Journal of General Physiology, 143*, 157–171. http://dx.doi.org/10.1085/jgp.201311037.

Freites, J. A., Schow, E. V., White, S. H., & Tobias, D. J. (2012). Microscopic origin of gating current fluctuations in a potassium channel voltage sensor. *Biophysical Journal, 102*, L44–L46. http://dx.doi.org/10.1016/j.bpj.2012.04.021.

Furini, S., & Domene, C. (2012). On conduction in a bacterial sodium channel. *PLoS Computational Biology, 8*, e1002476. http://dx.doi.org/10.1371/journal.pcbi.1002476.

George, A. L. (2005). Inherited disorders of voltage-gated sodium channels. *Journal of Clinical Investigation, 115*, 1990–1999. http://dx.doi.org/10.1172/JCI25505.

Gur Barzilai, M., Reitzel, A. M., Kraus, J. E. M., Gordon, D., Technau, U., Gurevitz, M., & Moran, Y. (2012). Convergent evolution of sodium ion selectivity in metazoan neuronal signaling. *Cell Reports [Electronic Resource], 2*, 242–248. http://dx.doi.org/10.1016/j.celrep.2012.06.016.

Guy, H. R., & Seetharamulu, P. (1986). Molecular model of the action potential sodium channel. *Proceedings of the National Academy of Sciences of the United States of America, 83*, 508–512.

Halabi, N., Rivoire, O., Leibler, S., & Ranganathan, R. (2009). Protein sectors: evolutionary units of three-dimensional structure. *Cell, 138*, 774–786. http://dx.doi.org/10.1016/j.cell.2009.07.038.

Harmar, A. J., Hills, R. A., Rosser, E. M., Jones, M., Buneman, O. P., Dunbar, D. R., … Spedding, M. (2009). IUPHAR-DB: the IUPHAR database of G protein-coupled receptors and ion channels. *Nucleic Acids Research, 37*, D680–D685. http://dx.doi.org/10.1093/nar/gkn728.

Heinemann, S. H., Terlau, H., Stühmer, W., Imoto, K., & Numa, S. (1992). Calcium channel characteristics conferred on the sodium channel by single mutations. *Nature, 356*, 441–443. http://dx.doi.org/10.1038/356441a0.

Henrion, U., Renhorn, J., Börjesson, S. I., Nelson, E. M., Schwaiger, C. S., Bjelkmar, P., … Elinder, F. (2012). Tracking a complete voltage-sensor cycle with metal-ion bridges. *Proceedings of the National Academy of Sciences of the United States of America*. http://dx.doi.org/10.1073/pnas.1116938109.

Hille, B. (2001). *Ion channels of excitable membranes* (3rd ed.). Sunderland, MA: Sinauer Associates, 814 pp.

Irie, K., Kitagawa, K., Nagura, H., Imai, T., Shimomura, T., & Fujiyoshi, Y. (2010). Comparative study of the gating motif and C-type inactivation in prokaryotic voltage-gated sodium channels. *Journal of Biological Chemistry, 285*, 3685–3694. http://dx.doi.org/10.1074/jbc.M109.057455.

Ito, M., Xu, H., Guffanti, A. A., Wei, Y., Zvi, L., Clapham, D. E., & Krulwich, T. A. (2004). The voltage-gated Na^+ channel Na_VBP has a role in motility, chemotaxis, and pH homeostasis of an alkaliphilic *Bacillus*. *Proceedings of the National Academy of Sciences of the United States of America, 101*, 10566–10571. http://dx.doi.org/10.1073/pnas.0402692101.

Jensen, M.Ø., Jogini, V., Borhani, D. W., Leffler, A. E., Dror, R. O., & Shaw, D. E. (2012). Mechanism of voltage gating in potassium channels. *Science, 336*, 229–233. http://dx.doi.org/10.1126/science.1216533.

Kandel, E., & Schwartz, J. (2013). *Principles of neural science* (5th ed.). McGraw Hill Professional, 1761 pp.

Kasimova, M. A., Zaydman, M. A., Cui, J., & Tarek, M. (2015). PIP_2-dependent coupling is prominent in $K_v7.1$ due to weakened interactions between S4-S5 and S6. *Scientific reports, 5*, 7474. http://dx.doi.org/10.1038/srep07474.

Ke, S., Zangerl, E.-M., & Stary-Weinzinger, A. (2013). Distinct interactions of Na^+ and Ca^{2+} ions with the selectivity filter of the bacterial sodium channel Na_VAb. *Biochemical and Biophysical Research Communications, 430*, 1272–1276. http://dx.doi.org/10.1016/j.bbrc.2012.12.055.

Khalili-Araghi, F., Jogini, V., Yarov-Yarovoy, V., Tajkhorshid, E., Roux, B., & Schulten, K. (2010). Calculation of the gating charge for the $K_v1.2$ voltage-activated potassium channel. *Biophysical Journal, 98*, 2189–2198. http://dx.doi.org/10.1016/j.bpj.2010.02.056.

Koishi, R., Xu, H., Ren, D., Navarro, B., Spiller, B. W., Shi, Q., & Clapham, D. E. (2004). A superfamily of voltage-gated sodium channels in bacteria. *Journal of Biological Chemistry, 279*, 9532–9538. http://dx.doi.org/10.1074/jbc.M313100200.

Labro, A. J., Raes, A. L., Grottesi, A., Hoorick, D. V., Sansom, M. S. P., & Snyders, D. J. (2008). Kv channel gating requires a compatible S4-S5 linker and bottom part of S6, constrained by non-interacting residues. *Journal of General Physiology, 132*, 667–680. http://dx.doi.org/10.1085/jgp.200810048.

Lee, J. H., Cribbs, L. L., & Perez-Reyes, E. (1999). Cloning of a novel four repeat protein related to voltage-gated sodium and calcium channels. *FEBS Letters, 445*, 231–236.

Lee, S., Goodchild, S. J., & Ahern, C. A. (2012). Local anesthetic inhibition of a bacterial sodium channel. *Journal of General Physiology, 139*, 507–516. http://dx.doi.org/10.1085/jgp.201210779.

Liao, M., Cao, E., Julius, D., & Cheng, Y. (2013). Structure of the TRPV1 ion channel determined by electron cryo-microscopy. *Nature, 504*, 107–112. http://dx.doi.org/10.1038/nature12822.

Liebeskind, B. J. (2011). Evolution of sodium channels and the new view of early nervous system evolution. *Communicative & Integrative Biology, 4*, 679–683.

Liebeskind, B. J., Hillis, D. M., & Zakon, H. H. (2011). Evolution of sodium channels predates the origin of nervous systems in animals. *Proceedings of the National Academy of Sciences of the United States of America, 108*, 9154–9159. http://dx.doi.org/10.1073/pnas.1106363108.

Liebeskind, B. J., Hillis, D. M., & Zakon, H. H. (2012). Phylogeny unites animal sodium leak channels with fungal calcium channels in an ancient, voltage-insensitive clade. *Molecular Biology and Evolution, 29*, 3613–3616. http://dx.doi.org/10.1093/molbev/mss182.

Liebeskind, B. J., Hillis, D. M., & Zakon, H. H. (2013). Independent acquisition of sodium selectivity in bacterial and animal sodium channels. *Current Biology, 23*, R948–R949. http://dx.doi.org/10.1016/j.cub.2013.09.025.

Liebeskind, B. J., Hillis, D. M., & Zakon, H. H. (2015). Convergence of ion channel genome content in early animal evolution. *Proceedings of the National Academy of Sciences of the United States of America, 112*, E846–E851. http://dx.doi.org/10.1073/pnas.1501195112.

Liebeskind, B. J., Hillis, D. M., Zakon, H. H., & Hofmann, H. A. (2016). Complex homology and the evolution of nervous systems. *Trends in Ecology & Evolution, 31*, 127–135. http://dx.doi.org/10.1016/j.tree.2015.12.005.

Long, S. B., Tao, X., Campbell, E. B., & MacKinnon, R. (2007). Atomic structure of a voltage-dependent K^+ channel in a lipid membrane-like environment. *Nature, 450*, 376–382. http://dx.doi.org/10.1038/nature06265.

Lu, Z., Klem, A. M., & Ramu, Y. (2002). Coupling between voltage sensors and activation gate in voltage-gated K^+ channels. *Journal of General Physiology, 120*, 663–676.

Mantegazza, M., Curia, G., Biagini, G., Ragsdale, D. S., & Avoli, M. (2010). Voltage-gated sodium channels as therapeutic targets in epilepsy and other neurological disorders. *Lancet Neurology, 9*, 413–424. http://dx.doi.org/10.1016/S1474-4422(10)70059-4.

Marlow, H., & Arendt, D. (2014). Evolution: ctenophore genomes and the origin of neurons. *Current Biology, 24*, R757–R761. http://dx.doi.org/10.1016/j.cub.2014.06.057.

McCusker, E. C., Bagnéris, C., Naylor, C. E., Cole, A. R., D'Avanzo, N., Nichols, C. G., & Wallace, B. A. (2012). Structure of a bacterial voltage-gated sodium channel pore reveals mechanisms of opening and closing. *Nature Communications, 3*, 1102. http://dx.doi.org/10.1038/ncomms2077.

McPhee, J. C., Ragsdale, D. S., Scheuer, T., & Catterall, W. A. (1994). A mutation in segment IVS6 disrupts fast inactivation of sodium channels. *Proceedings of the National Academy of Sciences of the United States of America, 91*, 12346–12350.

Moran, Y., Barzilai, M. G., Liebeskind, B. J., & Zakon, H. H. (2015). Evolution of voltage-gated ion channels at the emergence of metazoa. *Journal of Experimental Biology, 218*, 515–525. http://dx.doi.org/10.1242/jeb.110270.

Noda, M., Ikeda, T., Suzuki, H., Takeshima, H., Takahashi, T., Kuno, M., & Numa, S. (1986). Expression of functional sodium channels from cloned cDNA. *Nature, 322*, 826–828. http://dx.doi.org/10.1038/322826a0.

Noda, M., Shimizu, S., Tanabe, T., Takai, T., Kayano, T., Ikeda, T., … Numa, S. (1984). Primary structure of *Electrophorus electricus* sodium channel deduced from cDNA sequence. *Nature, 312*, 121–127. http://dx.doi.org/10.1038/312121a0.

Oelstrom, K., Goldschen-Ohm, M. P., Holmgren, M., & Chanda, B. (2014). Evolutionarily conserved intracellular gate of voltage-dependent sodium channels. *Nature Communications, 5*, 3420. http://dx.doi.org/10.1038/ncomms4420.

Ouyang, W., Jih, T.-Y., Zhang, T.-T., Correa, A. M., & Hemmings, H. C. (2007). Isoflurane inhibits NaChBac, a prokaryotic voltage-gated sodium channel. *Journal of Pharmacology and Experimental Therapeutics, 322*, 1076–1083. http://dx.doi.org/10.1124/jpet.107.122929.

Palovcak, E., Delemotte, L., Klein, M. L., & Carnevale, V. (2014). Evolutionary imprint of activation: the design principles of VSDs. *Journal of General Physiology, 143*, 145–156. http://dx.doi.org/10.1085/jgp.201311103.

Palovcak, E., Delemotte, L., Klein, M. L., & Carnevale, V. (2015). Comparative sequence analysis suggests a conserved gating mechanism for TRP channels. *Journal of General Physiology, 146*, 37–50. http://dx.doi.org/10.1085/jgp.201411329.

Pavlov, E., Bladen, C., Winkfein, R., Diao, C., Dhaliwal, P., & French, R. J. (2005). The pore, not cytoplasmic domains, underlies inactivation in a prokaryotic sodium channel. *Biophysical Journal, 89*, 232–242. http://dx.doi.org/10.1529/biophysj.104.056994.

Payandeh, J., & Minor, D. L., Jr. (2015). Bacterial voltage-gated sodium channels (BacNa$_V$s) from the soil, sea, and salt lakes enlighten molecular mechanisms of electrical signaling and pharmacology in the brain and heart. *Journal of Molecular Biology, 427*, 3–30. http://dx.doi.org/10.1016/j.jmb.2014.08.010.

Payandeh, J., Scheuer, T., Zheng, N., & Catterall, W. A. (2011). The crystal structure of a voltage-gated sodium channel. *Nature, 475*, 353–358. http://dx.doi.org/10.1038/nature10238.

Price, M. N., Dehal, P. S., & Arkin, A. P. (2010). FastTree 2 – approximately maximum-likelihood trees for large alignments. *PLoS One, 5*, e9490. http://dx.doi.org/10.1371/journal.pone.0009490.

Raju, S. G., Barber, A. F., LeBard, D. N., Klein, M. L., & Carnevale, V. (2013). Exploring volatile general anesthetic binding to a closed membrane-bound bacterial voltage-gated sodium channel via computation. *PLoS Computational Biology, 9*, e1003090. http://dx.doi.org/10.1371/journal.pcbi.1003090.

Ren, D., Navarro, B., Xu, H., Yue, L., Shi, Q., & Clapham, D. E. (2001). A prokaryotic voltage-gated sodium channel. *Science, 294*, 2372–2375. http://dx.doi.org/10.1126/science.1065635.

Rohl, C. A., Boeckman, F. A., Baker, C., Scheuer, T., Catterall, W. A., & Klevit, R. E. (1999). Solution structure of the sodium channel inactivation gate. *Biochemistry (Moscow), 38*, 855–861. http://dx.doi.org/10.1021/bi9823380.

Shafrir, Y., Durell, S. R., & Guy, H. R. (2008). Models of voltage-dependent conformational changes in NaChBac channels. *Biophysical Journal, 95*, 3663–3676. http://dx.doi.org/10.1529/biophysj.108.135335.

Shaya, D., Findeisen, F., Abderemane-Ali, F., Arrigoni, C., Wong, S., Nurva, S. R., … Minor, D. L., Jr. (2014). Structure of a prokaryotic sodium channel pore reveals essential gating elements and an outer ion binding site common to eukaryotic channels. *Journal of Molecular Biology, 426*, 467–483. http://dx.doi.org/10.1016/j.jmb.2013.10.010.

Shaya, D., Kreir, M., Robbins, R. A., Wong, S., Hammon, J., Brüggemann, A., & Minor, D. L. (2011). Voltage-gated sodium channel (Na_V) protein dissection creates a set of functional pore-only proteins. *Proceedings of the National Academy of Sciences of the United States of America, 108*, 12313–12318. http://dx.doi.org/10.1073/pnas.1106811108.

Stock, L., Delemotte, L., Carnevale, V., Treptow, W., & Klein, M. L. (2013). Conduction in a biological sodium selective channel. *Journal of Physical Chemistry B, 117*, 3782–3789.

Strong, M., Chandy, K. G., & Gutman, G. A. (1993). Molecular evolution of voltage-sensitive ion channel genes: on the origins of electrical excitability. *Molecular Biology and Evolution, 10*, 221–242.

Stühmer, W., Conti, F., Suzuki, H., Wang, X., Noda, M., Yahagi, N., … Numa, S. (1989). Structural parts involved in activation and inactivation of the sodium channel. *Nature, 339*, 597–603. http://dx.doi.org/10.1038/339597a0.

Tang, L., Gamal El-Din, T. M., Payandeh, J., Martinez, G. Q., Heard, T. M., Scheuer, T., … Catterall, W. A. (2014). Structural basis for Ca^{2+} selectivity of a voltage-gated calcium channel. *Nature, 505*, 56–61. http://dx.doi.org/10.1038/nature12775.

Tikhonov, D. B., & Zhorov, B. S. (2012). Architecture and pore block of eukaryotic voltage-gated sodium channels in view of Na_VAb bacterial sodium channel structure. *Molecular Pharmacology, 82*, 97–104. http://dx.doi.org/10.1124/mol.112.078212.

Tristani-Firouzi, M., Chen, J., & Sanguinetti, M. C. (2002). Interactions between S4-S5 linker and S6 transmembrane domain modulate gating of HERG K^+ channels. *Journal of Biological Chemistry, 277*, 18994–19000. http://dx.doi.org/10.1074/jbc.M200410200.

Ulbricht, W. (2005). sodium channel inactivation: molecular determinants and modulation. *Physiological Reviews, 85*, 1271–1301. http://dx.doi.org/10.1152/physrev.00024.2004.

Ulmschneider, M. B., Bagnéris, C., McCusker, E. C., DeCaen, P. G., Delling, M., Clapham, D. E., … Wallace, B. A. (2013). Molecular dynamics of ion transport through the open conformation of a bacterial voltage-gated sodium channel. *Proceedings of the National Academy of Sciences of the United States of America, 110*, 6364–6369. http://dx.doi.org/10.1073/pnas.1214667110.

Vargas, E., Yarov-Yarovoy, V., Khalili-Araghi, F., Catterall, W. A., Klein, M. L., Tarek, M., … Roux, B. (2012). An emerging consensus on voltage-dependent gating from computational modeling and molecular dynamics simulations. *Journal of General Physiology, 140*, 587–594. http://dx.doi.org/10.1085/jgp.201210873.

Vassilev, P. M., Scheuer, T., & Catterall, W. A. (1988). Identification of an intracellular peptide segment involved in sodium channel inactivation. *Science, 241*, 1658–1661. http://dx.doi.org/10.1126/science.2458625.

Wang, S.-Y., Bonner, K., Russell, C., & Wang, G. K. (2003). Tryptophan scanning of D1S6 and D4S6 C-Termini in voltage-gated sodium channels. *Biophysical Journal, 85*, 911–920.

Weigt, M., White, R. A., Szurmant, H., Hoch, J. A., & Hwa, T. (2009). Identification of direct residue contacts in protein–protein interaction by message passing. *Proceedings of the National Academy of Sciences of the United States of America, 106*, 67–72. http://dx.doi.org/10.1073/pnas.0805923106.

West, J. W., Patton, D. E., Scheuer, T., Wang, Y., Goldin, A. L., & Catterall, W. A. (1992). A cluster of hydrophobic amino acid residues required for fast Na(+)-channel inactivation. *Proceedings of the National Academy of Sciences of the United States of America, 89*, 10910–10914.

Wu, J., Yan, Z., Li, Z., Yan, C., Lu, S., Dong, M., & Yan, N. (2015). Structure of the voltage-gated calcium channel $Ca_V1.1$ complex. *Science, 350*, aad2395. http://dx.doi.org/10.1126/science.aad2395.

Yu, F. H., & Catterall, W. A. (2004). The VGL-Chanome: a protein superfamily specialized for electrical signaling and ionic homeostasis. *Science Signaling, 2004*, re15. http://dx.doi.org/10.1126/stke.2532004re15.

Yue, L., Navarro, B., Ren, D., Ramos, A., & Clapham, D. E. (2002). The cation selectivity filter of the bacterial sodium channel, NaChBac. *Journal of General Physiology, 120*, 845–853.

Zhang, T., Liu, Z., Song, W., Du, Y., & Dong, K. (2011). Molecular characterization and functional expression of the DSC1 channel. *Insect Biochemistry and Molecular Biology, 41*, 451–458. http://dx.doi.org/10.1016/j.ibmb.2011.04.010.

Zhang, X., Ren, W. L., DeCaen, P., Yan, C. Y., Tao, X., Tang, L., … Yan, N. (2012). Crystal structure of an orthologue of the NaChBac voltage-gated sodium channel. *Nature, 486*, 130–134. http://dx.doi.org/10.1038/nature11054.

Zhou, W., Chung, I., Liu, Z., Goldin, A. L., & Dong, K. (2004). A voltage-gated calcium-selective channel encoded by a sodium channel-like gene. *Neuron, 42*, 101–112.

CHAPTER TEN

Cardiac Na Channels: Structure to Function

K.R. DeMarco and C.E. Clancy[1]
University of California, Davis, Davis, CA, United States
[1]Corresponding author: E-mail: ceclancy@ucdavis.edu

Contents

Abstract

Heart rhythms arise from electrical activity generated by precisely timed opening and closing of ion channels in individual cardiac myocytes. Opening of the primary cardiac voltage-gated sodium ($Na_V1.5$) channel initiates cellular depolarization and the propagation of an electrical action potential that promotes coordinated contraction of the heart. The regularity of these contractile waves is critically important since it drives

Current Topics in Membranes, Volume 78
ISSN 1063-5823
http://dx.doi.org/10.1016/bs.ctm.2016.05.001

the primary function of the heart: to act as a pump that delivers blood to the brain and vital organs. When electrical activity goes awry during a cardiac arrhythmia, the pump does not function, the brain does not receive oxygenated blood, and death ensues. Perturbations to $Na_V1.5$ may alter the structure, and hence the function, of the ion channel and are associated downstream with a wide variety of cardiac conduction pathologies, such as arrhythmias.

1. Na^+ CHANNELS: I_{Na} THE FAST VOLTAGE-GATED Na^+ CHANNEL

Overview: Voltage-gated sodium (Na_V) channels are transmembrane proteins of approximately 260 kDa that underlie rapid depolarization to initiate action potentials in electrically excitable cells, such as neurons and cardiac myocytes. They are topologically organized into primary and auxiliary subunits, α and β (Fig. 1). The primary α subunit contains pore-forming

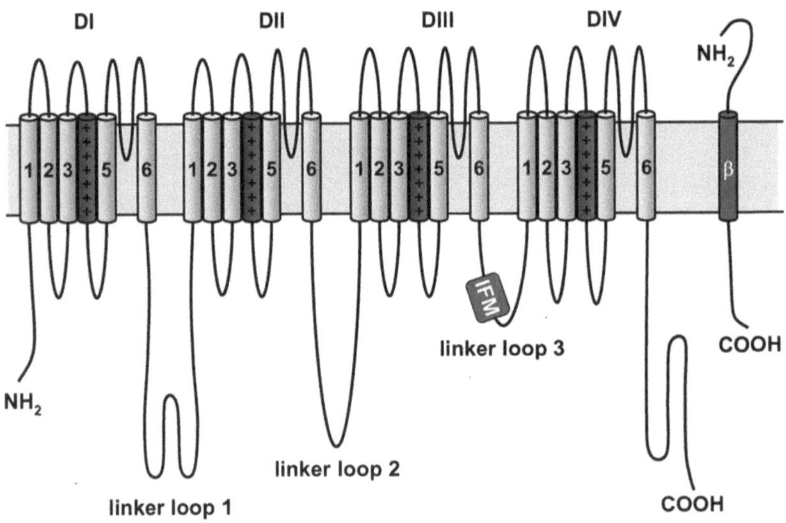

Figure 1 ***Membrane topology of $Na_V1.5$.*** Illustration of the four domains (DI–DIV) of the α subunit along with an associated β subunit. The transmembrane S1–S4 helices that comprise the voltage sensing domains are depicted with the charged S4 helix highlighted in green (dark gray in print versions). The S5 and S6 helices form the central pore domain. The DIII–DIV inactivation loop is highlighted with the critical hydrophobic isoleucine, phenylalanine, and methionine (IFM) sequence. The N-terminal and C-terminal domains of each subunit also labeled. *Copyright Hugues Abriel, Oxford University Press, 2007. http://dx.doi.org/10.1016/j.cardiores.2007.07.019*

and voltage-sensing domains (VSDs) that control the gating and ion selection functions of the channel. The auxiliary β subunits modulate gating, regulate expression, and also play a role in the oligomerization of individual channels (Isom, 2001; Namadurai et al., 2015; Yu & Catterall, 2003).

Na_V channels can adopt at least three structurally distinct states: open, closed, and inactivated. In the absence of an electrical stimulus, at the cell membrane resting potential, Na_V channels reside in a functionally deactivated (closed) state. Upon membrane depolarization, they activate (open) but rapidly become inactivated (nonconducting) again (Yellen, 1998). The functional distinction between deactivated and inactivated states is that channels deactivate during cellular hyperpolarization, while an inactivated channel occurs in response to cellular depolarization, rendering the channel unavailable to open and conduct ions. The relative residency in the closed and inactivated states is critical because the upstroke of the action potential depends on the availability of the channels to open (Kleber & Rudy, 2004). Even a subtle depolarization from the cell resting potential may promote channel inactivation that may reduce cellular excitability. Recovery from inactivation (RFI) occurs only after membrane repolarization, during which the channel recovers from the inactivated state and in a separate transition, deactivates to a closed available state. The time and voltage dependence of Na_V channel transitions between discrete channel states depends on a complex interplay of structural, chemical, genetic, and electrical factors that determine channel function (Chen-Izu et al., 2015, Herren, Bers, & Grandi, 2013; Marionneau & Abriel, 2015; Rook et al., 2012).

The alpha (α) subunit: There are 10 genes (*SCN1A–SCN10A*) that encode mammalian Na_V α subunit isoforms, each of which exhibits distinct structures related to their distinct physiological function. *SCN5A* is the gene that encodes the α subunit of the primary cardiac isoform ($Na_V1.5$) of the voltage-gated sodium channel. $Na_V1.5$ consists of a continuous chain of four heterologous domains (DI–DIV), each containing six transmembrane spanning segments (S1–S6) organized around a central, ion-selective pore (Fig. 2). A ring of DEKA residues between segments S5 and S6 of each domain forms the ion selectivity filter (SF) (Sun et al., 1997; Yamagishi et al., 2001). Positively charged residues are clustered in the S4 segments that comprise the voltage sensor (Kontis & Goldin, 1997; Stuhmer et al., 1989). In some Na_V isoforms, the intracellular linker between domains three and four, DIII/DIV, includes a hydrophobic isoleucine-phenylalanine-methionine (IFM) motif, which acts as a gating mechanism to confer rapid

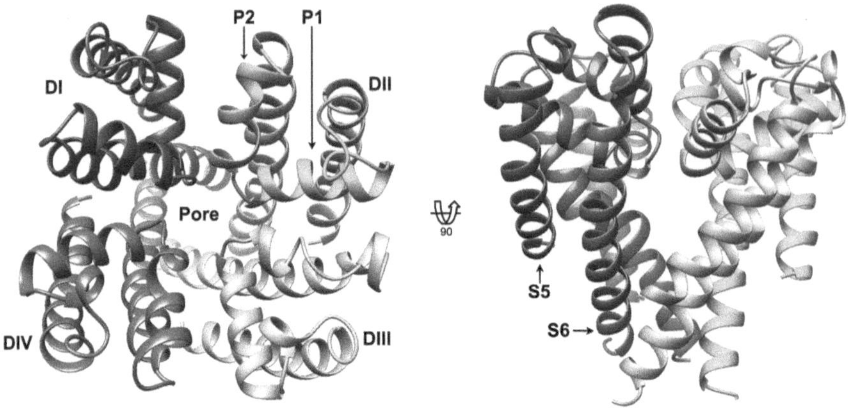

Figure 2 ***Homology model of the $Na_V1.5$ channel pore domain.*** Top view of the extracellular face (left), and transmembrane view (right), of the pore domain of $Na_V1.5$, based on the bacterial template Na_VRh (PDB ID code 4DXW). Structure is shown in ribbon representation with each domain colored from DI (blue (dark gray in print versions)) to DIV (red (light gray in print versions)). Pore helices P1 and P2 are labeled in DII, and transmembrane helices S5 and S6 are labeled in DIV for representative purposes. *Kevin R. DeMarco.*

occlusion of the channel pore, resulting in fast inactivation subsequent to channel opening (Stuhmer et al., 1989; West et al., 1992). Other studies also suggest a role for the intracellular DIV carboxy-terminus (C-terminus) in channel inactivation in both the brain and cardiac isoforms ($Na_V1.1$ and $Na_V1.5$, respectively) (Cormier et al., 2002; Mantegazza et al., 2001; Motoike et al., 2004).

Additionally, the resolved structure (using cryoelectron microscopy) of the rabbit voltage-gated skeletal calcium channel, $Ca_V1.1$, provides new insight into the mammalian architecture of voltage-gated ion channels, as sodium and calcium channels are both four-domain channels closely related among the voltage-gated ion channel superfamily (Wu et al., 2015; Yu et al., 2005). This Ca_V structure has the potential to serve as a template to inform more accurate homology models of the human Na_V channels for use in computational studies (Fig. 3).

Subunit β: There are only four genes (*SCN1B–SCN4B*) that encode the mammalian Na_V β subunits, all of which are expressed in cardiac tissue (Maier et al., 2004). They are membrane proteins consisting of a single transmembrane segment connecting an extracellular Immunoglobin (Ig) domain to an intracellular C-terminus (Brackenbury & Isom, 2011; Cusdin, Clare, & Jackson, 2008). Studies have shown that subunits β1

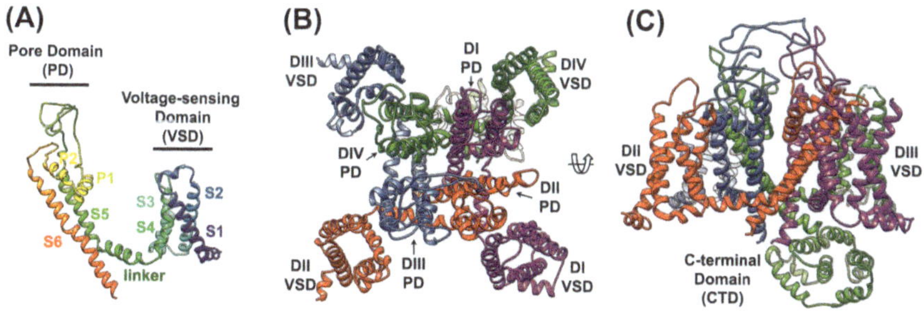

Figure 3 ***Cryoelectron microscopy reveals structural elements of Ca_V1.1.*** (A) Transmembrane view of a single subunit of Ca_V1.1 (PDB accession 3JBR) to illustrate the architecture of the pore domain (P1, P2, S5, and S6), voltage-sensing domain (S1–S4), and linker regions common to voltage-gated sodium and calcium channels. The N-terminus is colored blue, and the C-terminus is colored red in ribbon representation. (B) Top view of Ca_V1.1 colored to highlight the heterotetrameric arrangement of the voltage-sensing and pore domains, with the ion-conducting pore in the center. (C) Transmembrane view of Ca_V1.1 highlighting the C-terminal domain important for ion channel regulation, as well as a transmembrane fenestration important for local anesthetic ingress. *Kevin R. DeMarco.*

and β3 have the highest sequence homology out of the four and that they are noncovalently bound to α subunits, while subunits β2 and β4 bind to α subunits through a disulfide bond (Chen et al., 2012; Namadurai et al., 2015). Recently, the X-ray crystal structures of the human β3 and β4 subunits have been resolved (Gilchrist et al., 2013; Namadurai et al., 2014), and evidence suggests that the presence of membrane-embedded glutamic acid residues in the helical region of β3 may promote oligomerization (into dimers and trimers) which may play a role in channel localization (Morgan et al., 2000; Namadurai et al., 2015). β4 subunits, on the other hand, lack these residues and have been implicated in "resurgent current" in hippocampal neurons (Lewis & Raman, 2011; Namadurai et al., 2015). Resurgent current arises from Na channel RFI and reopening during repolarization, which results in a transient inward current during the action potential. The proposed mechanism posits that the β4 subunit competes with the inactivation particle upon depolarization, forming a blocked state of the channel that is distinct from the fast inactivated state. Repolarization favors removal of the β4 particle, resulting in a channel that briefly conducts before inactivating or deactivating, depending on the membrane potential (Lewis & Raman, 2014). Cardiac cells have also been shown to exhibit resurgent currents, termed nonequilibrium gating, although the specific molecular mechanism is not clear, β4 subunits are expressed in cardiac myocytes (Clancy et al., 2003; George, 2009; Moreno & Clancy, 2012).

2. ACTIVATION OF THE CARDIAC Na^+ CHANNEL Na_V1.5

Activation of the Na_V family of ion channels depends fundamentally on voltage sensing, whereby changes in the membrane potential promote changes to channel structure. Under resting conditions Na_V channels are closed, but open in response to local membrane depolarization caused by a capacitive current stimulus. If a sufficient number of channels open, then permeation of sodium ions occurs through the channel pore down the electrochemical gradient into the cell, depolarizing it and thus beginning a positive feedback loop in which more sodium channels open in response to the changing membrane potential. This promotes even further depolarization, which amplifies the electrical signal and drives the cellular action potential (Catterall, 2010; Huxley & Kendrew, 1952).

2.1 Structure and mechanism of the Na_V voltage-sensing domains

The S1—S4 helices of each asymmetric domain (DI—DIV) of the α subunit comprise the VSD of the Na_V. The amphiphilic S4 helix contains an alternating motif of a positively charged residue followed by two hydrophobic residues. This well-conserved "3_{10}-helical" arrangement serves as the voltage-sensing element of activation in each homologous domain (Schwaiger et al., 2011; Yarov-Yarovoy et al., 2012). The charged residues on S4 are hypothesized to cause an outward motion of the helix, toward the extracellular side of the membrane, which in turn induces conformational changes within the pore (Armstrong, 1981; Capes et al., 2013; Payandeh et al., 2011; Yarov-Yarovoy, Baker, & Catterall, 2006). This is the "helical screw" model, which is supported both by experimental evidence as well as extensive studies involving structural modeling of the Na_V VSD (Catterall, 1986, 2000, 2010; Yarov-Yarovoy et al., 2006, 2012). In this model, the S4 voltage sensor motion is coupled to pore opening and closing via the S4—S5 linker (Long, Campbell, & Mackinnon, 2005). The combination of outward and rotating movement of the S4 segment has been predicted to impose a gating movement of the S4—S5 linker that is roughly orthogonal to the transmembrane orientation of the pore-forming helices in the bacterial NaChBac VSD. Moreover, investigation into charge-neutralized mutants of the mammalian $Na_V1.4$ sodium channel has found that S4, in all four domains, are involved in pore gating, albeit to different extents (Chanda & Bezanilla, 2002; Yarov-Yarovoy et al., 2012). In particular, charges on DIV S4 seem to have a larger effect on inactivation than activation, whereas the S4 helices in DI-DIII seem to influence activation (Chahine et al., 1994; Chanda & Bezanilla, 2002; Kontis, Rounaghi, & Goldin, 1997; Stuhmer et al., 1989). This may imply that the DIV voltage sensor is tightly coupled to fast inactivation (Chanda & Bezanilla, 2002).

An integral component in the helical screw model of Na_V activation is the existence of negative countercharges that stabilize and shield the positively charged intramembrane S4 voltage-sensing helix as it responds to membrane depolarization. To overcome the energetically unfavorable translocation of charges over a hydrophobic barrier, the S4 helix of the VSD has been predicted, in molecular modeling studies, to favor one or more intermediate steps (Chanda & Bezanilla, 2002). This has been verified for the active states of the bacterial NachBac channel in which disulfide cross-linking and mutant cycle studies have identified molecular interactions between the S4 helix and surrounding S1–S3 helices that facilitate activation (DeCaen et al., 2008, 2009, 2011; Paldi & Gurevitz, 2010; Yarov-Yarovoy et al., 2012). Structurally, the role of countercharges in mammalian sodium channels have been studied using a combination of molecular modeling and molecular dynamics simulations and are predicted to involve three intermediate states of S4 translocation (Gosselin-Badaroudine et al., 2012). Functional investigation into mammalian VSDs using disulfide cross-linking has not been possible because these channels contain native cysteine residues that eliminate the possibility of intuiting disulfide bonding for any specific pair. However, mutagenesis studies of the VSDs of $Na_V1.4$ demonstrate that reversing the charge of the S1–S3 countercharges had domain-specific outcomes including inhibition of activation in DI–DIII but destabilization of fast inactivation in DIV (Groome & Winston, 2013). While these studies thus far do not include the specific structural aspects of the $Na_V1.5$ VSDs, they provide valuable insight into the intricacies of activation of Na_V channels.

2.2 Naturally occurring mutations can cause delayed $Na_V1.5$ activation

Isolated cardiac conduction disease (ICCD) is observed on the ECG in a widening of the QRS complex that indicates delays in ventricular excitation (Grant et al., 2002; Tan et al., 2001). Mutations in $Na_V1.5$ have been shown to cause ICCD and typically result from a depolarizing shift of the Na^+ channel activation curve (Grant et al., 2002; Tan et al., 2001). This shift most likely results from a reduction in the rate of channel activation or decreased channel sensitivity to the voltage required for activation. Mutant channels may require a longer time to reach depolarized membrane potentials at which the maximum I_{Na} occurs. The lag in activation of ICCD mutant Na^+ channels results in a reduction of the action potential upstroke velocity, which is a primary determinant of conduction velocity, leading to an increase in the amount of stimulus current required to excite coupled tissue (Rohr, Kucera, & Kleber, 1998; Shaw & Rudy, 1997; Tan et al., 2001). Computational modeling supports the hypothesis that a depolarizing shift in the $Na_V1.5$ activation and corresponding reduced I_{Na} produces right ventricular conduction delay and a transmural activation gradient, which is a mechanism that is consistent with the findings of a study investigating the G514C mutation linked to ICCD (Tan et al., 2001).

3. INACTIVATION

Inactivation of $Na_V1.5$ is critically important because the completeness of the inactivation process influences both action potential duration and action potential frequency (restitution) in the myocardium. There are at least three distinct forms of inactivation that have been identified in $Na_V1.5$ channels: fast, intermediate, and slow (Goldin, 2003; Wang et al., 2000).

4. FAST INACTIVATION

Fast inactivation in $Na_V1.5$ is a time- and voltage-dependent process that occurs via a "hinged lid" mechanism in which the cytoplasmic linking region between DIII and DIV binds into the hydrophobic pore, thereby occluding ion conduction (Kellenberger et al., 1997; McPhee et al., 1998; Smith & Goldin, 1997). An NMR solution structure of this linking region, or inactivation particle, has been resolved as a 53 amino acid peptide, consisting of a single alpha-helix structure flanked by two disordered tail regions (Rohl et al., 1999). Several early mutagenesis studies have implicated the importance of a hydrophobic sequence of isoleucine, phenylalanine, and methionine (IFM) within the DIII–DIV linker as critical for sodium channel inactivation: mutation to glutamine (Q) at one or more of the IFM

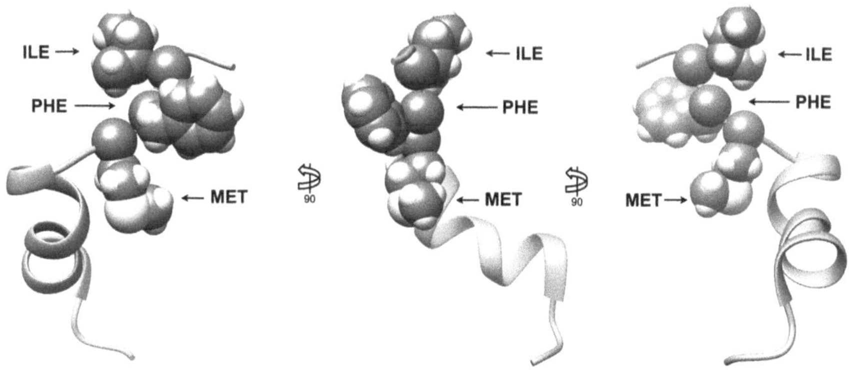

Figure 4 ***Solution NMR structure of the DIII–DIV inactivation loop.*** Three views of the putative sodium channel inactivation loop segment (PDB ID code 1BYY). The hydrophobic sequence of residues (IFM), demonstrated to be critical for $Na_V1.5$ inactivation, are highlighted in space-filling representation. The remainder of the segment is shown in ribbon representation. *Kevin R. DeMarco.*

residues either diminished or abolished inactivation (Eaholtz et al., 1999; Eaholtz, Scheuer, & Catterall, 1994; Eaholtz, Zagotta, & Catterall, 1998; Goldin, 2003; Patton et al., 1992; West et al., 1992) Fig. 4.

Experimental studies have implicated an adjacent threonine (T) residue as being important for the binding of the inactivation particle within the mouth of the pore, since its substation with a methionine (M) residue increased by 24% sustained noninactivating currents (Miyamoto et al., 2001). In $Na_V1.5$, cysteine scanning of the IFM sequence has shown it to be critical for stable binding of the inactivation particle to the pore (Deschenes, Trottier, & Chahine, 1999). Additionally, a recent study has shown that a shortened peptide containing the IFM sequence (KIFMK) can not only block inactivation-deficient open $Na_V1.5$ channels (L409C, A410W) but must be cleared from the pore before it can close, as evidenced by tail currents recorded upon repolarization. The same authors also found that an F1760K mutation will eliminate fast inactivation altogether, suggesting that the inactivation particle itself is important for stabilizing the inactivated state of the channel (Wang & Wang, 2005). Hence, there is ample evidence that fast inactivation occurs via a hydrophobic packing of the DIII–DIV linker within the intracellular mouth of the pore. Additionally, site-directed glutamine mutations in the DIII S4–S5 linker have been shown to lead to a positive shift in the voltage dependence of fast inactivation, suggesting that the abolition of hydrophobic residues in that region functionally corresponds to the loss of interaction with the hydrophobic

IFM triad (Smith & Goldin, 1997). There is limited consensus on the mechanism of the "hinged lid" hypothesis of fast inactivation, as initial investigation suggested that the glycine and proline residues in the linker (G1484, G1485, and P1512) form the hinge (Kellenberger et al., 1997), but theoretical models in which the inactivation particle forms a hairpin motif centered around G1505 only implicate P1512 in functioning of the hinge (Sirota, Pascutti, & Anteneodo, 2002).

4.1 Structural and functional aspects of fast inactivation gating

A recent study has revealed that fast inactivation in $Na_V1.4$ strictly depends on the motion of the DIV voltage sensor. The authors hypothesize that the opening of the sodium channel actually occurs in two steps: one in which the DI, DII, and DIII voltage sensors open the pore and allow ion permeation and another in which activation of the DIV voltage sensor renders the channel in an intermediate state (S2) and induces a conformational change in the pore itself, which serves as a necessary precursor to the binding of the inactivation particle (Goldschen-Ohm et al., 2013). Using an inactivation-deficient mutant (L435W/L437C/A438W, DI-S6), they argue that the frequent observation of subconductance states in electrophysiological recordings of the inactivation-deficient $Na_V1.4$ channel may indicate a discrete open state that occurs following DIV voltage sensor movement albeit before the fast inactivation particle enters the pore. These secondary conductance states are normally hidden by rapid channel inactivation. The implications for this gating paradigm are significant in that fast inactivation may be a mechanism in which a two-step structural process is translated, functionally, into a single step. In addition, this finding may help to explain conflicting data supporting high-affinity block of both open and inactivated states by local anesthetics (Bennett et al., 1995a; Goldschen-Ohm et al., 2013; Wang et al., 2004). While this study was conducted using WT and mutant forms of $Na_V1.4$, the proposed existence of the preinactivated S2 state should be applicable to $Na_V1.5$ given the high level of homology between both isoforms.

4.2 A role for the carboxy-terminus and allosteric proteins in inactivation

The intracellular C-terminus of the α subunit of $Na_V1.5$ plays an important role in modulating fast inactivation (An et al., 1998; Cormier et al., 2002; Mantegazza et al., 2001; Tateyama et al., 2004). In this channel an uncoupling of the C-terminus from the DIII–DIV inactivation loop has been shown to increase inward current during repolarization leading to delays in cellular repolarization. These studies have suggested that the C-terminus is involved in the structural stabilization of the fast-inactivated state (Motoike et al., 2004). An S1885stop deletion experiment further revealed that the highly charged portion of the C-terminal S6 is likely involved in electrostatic interactions with the inactivation particle (Cormier et al., 2002). One possibility is that the C-terminal stabilization of the DIII–DIV inactivation loop constitutes a mode of intermediate inactivation (Cormier et al., 2002; Kass, 2006; Mantegazza et al., 2001), although there is lack of consensus regarding the precise structural determinants of such an intermediate inactivated state.

Fast inactivation in $Na_V1.5$ can also be modulated by accessory proteins such as fibroblast growth factor homologous factors (FHFs), which are a family of intracellular signaling proteins that interact with the C-terminus. It has been suggested that a tethered FHF-derived N-terminal particle competes for the induction of the intrinsic fast inactivation particle of the Na_V at depolarized transitions near the open state (Dover et al., 2010). One study has found that FGF12, an FHF that is the most abundant in human heart, has diminished interaction with a C-terminal H1849R mutant of $Na_V1.5$, resulting most notably in a depolarizing (~−9 mV) shift in fast (steady state) inactivation, which is a gain-of-function phenotype associated with arrhythmia and long QT syndrome (Musa et al., 2015). Another study has identified the acidic-rich C terminal region (1773–1832) of $Na_V1.5$ as critical for FGF12 binding, and that the presence of FGF12 modulates the voltage dependence of inactivation, with no effect on activation of the channel (Liu et al., 2003). Congenital long QT syndrome (LQT3) is a disorder characterized by a delay in cardiac cellular repolarization, a condition that can lead to cardiac arrhythmias and sudden death. Other mutations in this C-terminal region, such as E1784K, 1795insD, and Y1795C interfere with fast inactivation and evoke small, sustained currents associated with LQT3 as well (An et al., 1998; Rivolta et al., 2001; Veldkamp et al., 2000).

4.3 Channelopathies arising from altered fast inactivation

The first naturally occurring mutation that was identified in $Na_V1.5$ is ΔKPQ: a 3 amino acid deletion (lysine, proline, glutamine at positions 1505–1507) in the DIII–DIV linker that was shown to affect fast inactivation and was causally linked to LQT3 (Bennett et al., 1995b). This motif is known to be critical for fast inactivation of the Na^+ channel, and the ΔKPQ mutation results in a small (<1% of peak current amplitude) persistent non-inactivating current ("bursting") during normal cellular repolarization. This mutation was also the first to be modeled in a study of an in silico mutation that linked genotype to phenotype (Clancy & Rudy, 1999). Model simulations showed that delayed ventricular repolarization promotes a substrate for triggered activity via early afterdepolarizations (EADs), which result from reactivation of L-type Ca^{2+} channels. Since that first study, numerous modeling and simulation studies have been used to predict effects of arrhythmia-linked mutations in genes underlying ion channels and accessory and regulatory proteins (Noble & Rudy, 2001).

4.4 Modulation of fast inactivation by β subunits

Fast inactivation can also be modulated by associated β subunits. In particular, the β3 subunit, which is normally expressed in ventricular and Purkinje cardiac cells, shifts steady state (fast) inactivation leftward, indicating reduced channel availability. When a ΔECD mutation is made in β3, abolition of this gating shift in $Na_V1.5$ and a persistent current is observed, implicating its interference with the DIII–DIV inactivation particle, as has previously been suggested for the β1 subunit (McCormick et al., 1998; Yu et al., 2005).

5. SLOW INACTIVATION

Slow inactivation of $Na_V1.5$ is a process spanning seconds or tens of seconds that can occur during prolonged or repetitive depolarization (Kass, 2004). Slow inactivation is independent of the DIII–DIV linker associated with fast inactivation and is more likely a result of the structural deformation of the pore, such as with C-type inactivation in potassium channels, although the exact mechanism is not yet known (Liu, Jurman, & Yellen, 1996; Vassilev, Scheuer, & Catterall, 1989). Early experiments involving the transfer of all four $Na_V1.5$ P-loop regions into the $Na_V1.4$ transmembrane backbone conferred the cardiac type of slow inactivation (Russ et al., 1996; Vilin et al., 1999). These regions include the DEKA SF, as well as the highly conserved P-loop region that forms the extracellular funnel of the pore domain. For example, a single residue difference between $Na_V1.4$ (V754) and $Na_V1.5$ (I891) in the DII P-loop region caused a higher probability of slow inactivation in the cardiac isoform (Vilin & Ruben, 2001), and a tryptophan mutation (W53C) in the DI P-loop region has been shown to diminish single channel conductance (Tomaselli et al., 1995). Furthermore, a mutation in the DEKA SF itself has been reported to increase the probability of Na_V channels entering a slow inactivated state (Hilber et al., 2001). The many regions of the Na_V channel that have been implicated in slow inactivation may contribute to an as yet poorly understood co-ordinated gating process. A study has correlated the onset of slow inactivation in $Na_V1.4$ with immobilization of the VSDs of DI and DII (Silva & Goldstein, 2013), suggesting that slow inactivation may involve additional protein regions and not be solely be due to deformation of the pore domain.

6. RECOVERY FROM INACTIVATION

Repeated regular firing of cardiac action potentials depends crucially on the process of RFI of $Na_V1.5$ channels. RFI is an l process that indicates the fraction of Na_V channels that are available for activation. The recovery process occurs on a millisecond timescale, with exponential kinetics that are accelerated with increased hyperpolarization (Bezanilla & Armstrong, 1977; Huxley & Kendrew, 1952). Both antiarrhythmic drugs and naturally occurring mutations have been shown to target and modulate RFI. In a physiological setting, RFI participates in a tightly orchestrated relationship with deactivation, ensuring minimization of leak current during channel repriming (Kuo & Bean, 1994).

6.1 DIV voltage sensor movement is rate limiting

The kinetics of recovery from fast inactivation has been studied in both wild-type $Na_V1.5$ sodium channels and in mutants in which the positive S4 arginine gating charges of each domain of the VSD have been neutralized by mutation to glutamine. Results demonstrate that compared to the wild type, while all $Na_V1.5$ S4 mutants slowly recover, the charge-neutralized DIV mutant exhibited the largest effect, with a pronounced recovery lag of almost 0.5 ms; suggesting that the motion of the DIV VSD may be rate limiting for RFI (Capes et al., 2013). Such findings are consistent with the fact that site-3 toxins, small peptide toxins derived from venoms that bind with high specificity to the extracellular surface of Na_V channels inhibit the outward motion of the DIV S4 voltage sensor, consequently inducing a faster RFI (Hanck & Sheets, 2007). During RFI, the voltage sensors of DI–DIII are likely required to deactivate first, closing the channel and preventing conduction, after which the slower inward motion of DIV S4 acts to drive the inactivation particle from its vestibule (Armstrong, 2006; Kuo & Bean, 1994). Furthermore, an accelerated RFI and slower onset of fast inactivation has been found to result from mutations in DIII S1−S2 linker (R1232W) and the DIV S3−S4 linker (T1620M) resulting in idiopathic ventricular fibrillation (Vilin, Fujimoto, & Ruben, 2001), suggesting again an important role for DIV VSD in RFI.

6.2 Modulation of recovery from inactivation by β subunits

RFI of $Na_V1.5$ channels is modulated by at least two of β subunits. One study has shown that the transmembrane helix, plus either the intracellular or extracellular region of β1 is necessary to accelerate its RFI (Fahmi et al., 2001). β1 is known to accelerate RFI, not only in $Na_V1.5$, but also in the skeletal muscle and neuronal isoforms of the channel (Zimmer & Benndorf, 2002). β3 also accelerates RFI of $Na_V1.5$ in oocytes, which may be physiologically significant because β3 is expressed in cardiac ventricular cells and Purkinje fibers (Fahmi et al., 2001).

6.3 Recovery from inactivation channelopathy and disease phenotypes

Examples of LQT-3 mutations that do not result in a classical gain of function as measured during square wave voltage clamp experiments, include D1790G, E1295K, and I1768V. Interestingly, these mutations are not localized to a single region of the channel, rather they are found in varied regions including the C-terminal domain and in the DIII S3−S4 linker region. Studies have shown that these mutations rather result from an acceleration of RFI. Experiments and mathematical simulations have revealed that under nonequilibrium conditions during repolarization, channel reopening results from faster RFI at membrane potentials that facilitate the activation transition (Clancy et al., 2003). Mathematical modeling and simulation approaches predicted that the additional inward current through the defective channels was sufficient to cause prolonged repolarization, consistent with the long QT syndrome phenotype (Clancy et al., 2003).

Additionally, reduction in $Na_V1.5$ have been attributed to changes in RFI that can lead to Brugada syndrome (diagnosed by ST segment elevation on the ECG) by several distinct alterations to ion channel gating, including reduced rates of RFI as well as faster inactivation subsequent to channel opening and protein trafficking defects (Clancy & Rudy, 2002). For example, the 1795insD mutant channel showed a rightward shift in the recovery curve, indicating a delayed RFI in $Na_V1.5$ of 100 ms or more. The resulting loss of current was shown to be sufficient to explain the Brugada phenotype.

7. REGULATION OF CARDIAC $Na_V1.5$

There has been work focused on the regulation of $Na_V1.5$ by posttranslational processes (Marionneau & Abriel, 2015). There is also a growing literature on how posttranslational modifications of $Na_V1.5$ are facilitated through the presence of macromolecular complexes that allow selective and specific protein modification to confer functional effects (Abriel, Rougier, & Jalife, 2015). Multiple phosphorylation mechanisms through a variety of kinase pathways as well as glycosylation, arginine methylation, S-nitrosylation, ubiquitylation, and ion homeostatic mechanisms are all implicated as functional regulators of $Na_V1.5$ (Marionneau & Abriel, 2015).

7.1 Phosphorylation

One of the key functional modulators of $Na_V1.5$ is phosphorylation by Ca^{2+}/calmodulin-dependent protein kinase II (CaMKII) that results in changes to macroscopic $Na_V1.5$ gating including a reduction in channel availability, increased intermediate inactivation, and slowed RFI (Wagner et al., 2006). Multiple phosphorylation sites in the intracellular loop linking $Na_V1.5$ domains I and II have been identified (S516, S571, and T594) and implicated in the CaMKII-induced changes to I_{Na} gating (Ashpole et al., 2012; Hund et al., 2010), although, how each of the three sites confers functional effects is still not clear.

While all of the effects of CaMKII on macroscopic $Na_V1.5$ current result in peak current reduction, CaMKII activation also results in substantial current increases in the late or persistent component of current that can cause $Na_V1.5$ gain of function associated proarrhythmia (Anderson, Brown, & Bers, 2011; Bers & Grandi, 2009; Erickson & Anderson, 2008; Grandi et al., 2007; Horvath et al., 2013; Hund et al., 2008; Ma et al., 2012; Maier, 2011; Wagner et al., 2011). During AP-clamp experiments in canine ventricular myocytes, inhibition of CaMKII has been shown to reduce $Na_V1.5$ late current (Horvath et al., 2013; Maltsev et al., 2008). In addition to phosphorylation by CaMKII, increased in intracellular Ca^{2+} modifies and promotes $Na_V1.5$ late current (Mori et al., 2000; Wingo et al., 2004) through a variety of mechanisms including activation of CaMKII (Wagner et al., 2006), increased mitochondrial oxidative phosphorylation (Belardinelli, Shryock, & Fraser, 2006), and consequent increased reactive oxygen species (ROS) (Kohlhaas et al., 2010; Liu et al., 2010). Interestingly, an increase $Na_V1.5$ late current can also activate CaMKII, thereby resulting in a positive feedback cycle between $Na_V1.5$ late current and CaMKII (Yao et al., 2011). The effects conferred by CaMKII on both peak and late $Na_V1.5$ current has been suggested to occur through a complex with βIV-spectrin that positions CaMKII to phosphorylate the channel (Hund et al., 2010).

In addition to CaMKII, in rodent heart cells cAMP-dependent protein kinase (PKA) has been shown to phosphorylate serine residues at positions 525 and 528 and confer functional effects such as reducing channel availability, promoting $Na_V1.5$ trafficking, and increasing $Na_V1.5$ macroscopic current (Hallaq et al., 2006; Marionneau & Abriel, 2015; Matsuda, Lee, & Shibata, 1992; Murphy et al., 1996; Tateyama et al., 2003; Zhou et al., 2002). These effects are similar to those observed in expressed $Na_V1.5$ channels following PKC-dependent phosphorylation of S1503, which additionally promotes the $Na_V1.5$ late current (Qu et al., 1994; West et al., 1991).

Other kinases have also been implicated in posttranslational modification of $Na_V1.5$ current. Included are the phosphatidylinositol kinase (PI3K), whose inhibition promotes $Na_V1.5$ late current (Lu et al., 2012), while Fyn and Adenosine Monophosphate-activated Protein Kinases (AMPK) promote $Na_V1.5$ current by increasing channel availability and promoting RFI (Ahern et al., 2005; Light, Wallace, & Dyck, 2003).

7.2 Glycosylation

In additional to kinase modulation of $Na_V1.5$ via phosphorylation, glycosylation of $Na_V1.5$ is an additional presumed mode of posttranslational modification. Multiple putative asparagines have been identified in $Na_V1.5$ as potential N-linked glycosylation sites (Ahmad et al., 2007). However, thus far only asparagines in the ancillary $Na_V\beta1$ subunit have been demonstrated to functionally modify the cardiac Na^+ current, by causing leftward shifts of the inactivation and activation curves, promoting fast inactivation, and slowing RFI (Ednie et al., 2013; Stocker & Bennett, 2006; Ufret-Vincenty et al., 2001).

7.3 Methylation and ubiquitylation

A set of arginine residues in $Na_V1.5$ located in the DI—DII linker has been shown to undergo arginine methylation (Beltran-Alvarez et al., 2014).

The effect of methylation of these residues is to increase the $Na_V1.5$ surface expression and hence current density (Beltran-Alvarez et al., 2013). Exactly the opposite effects on channel function have been shown to result from ubiquitylation, which results in $Na_V1.5$ internalization and reduction in current (van Bemmelen et al., 2004; Laedermann, Decosterd, & Abriel, 2014). The ubiquitylation sites are presently unknown.

7.4 Reactive oxygen species regulation

There are multiple modes of redox regulation of cardiac $Na_V1.5$ through various signaling mechanism, both direct and indirect. A prominent indirect mechanism is by ROS activation of CaMKII, which confers the gain-of-function effects on $Na_V1.5$ late current described earlier (Marionneau & Abriel, 2015). Nitric oxide (generally derived from endothelial nitric oxide synthase (eNOS)), also potentiates $Na_V1.5$ late current, which can occur via direct S-nitrosylation of the channel (Ahern et al., 2000; Cheng et al., 2013). Changes in the NAD(H) balance has been shown to promote mitochondrial ROS production, which leads to a reduction in $Na_V1.5$ peak current, which may provide a link between altered metabolism, excitability, conduction block, and arrhythmic risk (Liu et al., 2013, 2010). Indeed, naturally occurring disease-linked mutations that have been shown to prevent the protein kinase A potentiation of $Na_V1.5$ during oxidative stress, leading to loss of Na channel function and manifestations of Brugada syndrome (Aiba et al., 2014).

8. SUMMARY

Understanding how $Na_V1.5$ structure underlies channel function is continually improving due the development of improved experimental and computational approaches that allow more concrete links between structure and function. Connecting ion channel function to emergent behavior in the heart to cause disease is a critical area of research. The ubiquity of disruption of $Na_V1.5$ channels in cardiac disorders of excitability emphasizes the importance of fundamental understanding of the associated mechanisms and disease processes to ultimately reveal new targets for human therapy.

ACKNOWLEDGMENTS

This work was supported by The American Heart Association Pre-doctoral Fellowship (16PRE27260295), Western States Affiliate (KRD) and The American Heart Association

(GIAs (15GRNT25700136), Western States Affiliate), and National Institutes of Health NHLBI 1U01HL126273 (CEC).

REFERENCES

Abriel, H., Rougier, J. S., & Jalife, J. (2015). Ion channel macromolecular complexes in cardiomyocytes: roles in sudden cardiac death. *Circulation Research, 116*(12), 1971–1988.

Ahern, C. A., et al. (2005). Modulation of the cardiac sodium channel $Na_V1.5$ by Fyn, a Src family tyrosine kinase. *Circulation Research, 96*(9), 991–998.

Ahern, G. P., et al. (2000). Induction of persistent sodium current by exogenous and endogenous nitric oxide. *Journal of Biological Chemistry, 275*(37), 28810–28815.

Ahmad, I., et al. (2007). In silico determination of intracellular glycosylation and phosphorylation sites in human selectins: implications for biological function. *Journal of Cellular Biochemistry, 100*(6), 1558–1572.

Aiba, T., et al. (2014). A mutation causing Brugada syndrome identifies a mechanism for altered autonomic and oxidant regulation of cardiac sodium currents. *Circulation Cardiovascular Genetics,* 7(3), 249–256.

An, R. H., et al. (1998). Novel LQT-3 mutation affects Na^+ channel activity through interactions between alpha- and beta1-subunits. *Circulation Research, 83*(2), 141–146.

Anderson, M. E., Brown, J. H., & Bers, D. M. (2011). CaMKII in myocardial hypertrophy and heart failure. *Journal of Molecular and Cellular Cardiology, 51*(4), 468–473.

Armstrong, C. M. (1981). Sodium channels and gating currents. *Physiological Reviews, 61*(3), 644–683.

Armstrong, C. M. (2006). Na channel inactivation from open and closed states. *Proceedings of the National Academy of Sciences of the United States of America, 103*(47), 17991–17996.

Ashpole, N. M., et al. (2012). Ca^{2+}/calmodulin-dependent protein kinase II (CaMKII) regulates cardiac sodium channel $Na_V1.5$ gating by multiple phosphorylation sites. *Journal of Biological Chemistry, 287*(24), 19856–19869.

Belardinelli, L., Shryock, J. C., & Fraser, H. (2006). Inhibition of the late sodium current as a potential cardioprotective principle: effects of the late sodium current inhibitor ranolazine. *Heart, 92*(Suppl. 4), iv6–iv14.

Beltran-Alvarez, P., et al. (2013). Protein arginine methyl transferases-3 and -5 increase cell surface expression of cardiac sodium channel. *FEBS Letters, 587*(19), 3159–3165.

Beltran-Alvarez, P., et al. (2014). Identification of N-terminal protein acetylation and arginine methylation of the voltage-gated sodium channel in end-stage heart failure human heart. *Journal of Molecular and Cellular Cardiology, 76*, 126–129.

van Bemmelen, M. X., et al. (2004). Cardiac voltage-gated sodium channel $Na_v1.5$ is regulated by Nedd4-2 mediated ubiquitination. *Circulation Research, 95*(3), 284–291.

Bennett, P. B., et al. (1995a). On the molecular nature of the lidocaine receptor of cardiac Na^+ channels. Modification of block by alterations in the alpha-subunit III-IV interdomain. *Circulation Research,* 77(3), 584–592.

Bennett, P. B., et al. (1995b). Molecular mechanism for an inherited cardiac arrhythmia. *Nature, 376*(6542), 683–685.

Bers, D. M., & Grandi, E. (2009). Calcium/calmodulin-dependent kinase II regulation of cardiac ion channels. *Journal of Cardiovascular Pharmacology, 54*(3), 180–187.

Bezanilla, F., & Armstrong, C. M. (1977). Inactivation of the sodium channel. I. Sodium current experiments. *Journal of General Physiology, 70*(5), 549–566.

Brackenbury, W. J., & Isom, L. L. (2011). Na channel beta subunits: overachievers of the ion channel family. *Frontiers in Pharmacology, 2*, 53.

Capes, D. L., et al. (2013). Domain IV voltage-sensor movement is both sufficient and rate limiting for fast inactivation in sodium channels. *Journal of General Physiology, 142*(2), 101–112.

Catterall, W. A. (1986). Molecular properties of voltage-sensitive sodium channels. *Annual Review of Biochemistry, 55*, 953–985.
Catterall, W. A. (2000). From ionic currents to molecular mechanisms: the structure and function of voltage-gated sodium channels. *Neuron, 26*(1), 13–25.
Catterall, W. A. (2010). Ion channel voltage sensors: structure, function, and pathophysiology. *Neuron, 67*(6), 915–928.
Chahine, M., et al. (1994). Sodium channel mutations in paramyotonia congenita uncouple inactivation from activation. *Neuron, 12*(2), 281–294.
Chanda, B., & Bezanilla, F. (2002). Tracking voltage-dependent conformational changes in skeletal muscle sodium channel during activation. *Journal of General Physiology, 120*(5), 629–645.
Chen, C., et al. (2012). Identification of the cysteine residue responsible for disulfide linkage of Na^+ channel alpha and beta2 subunits. *Journal of Biological Chemistry, 287*(46), 39061–39069.
Cheng, J., et al. (2013). Caveolin-3 suppresses late sodium current by inhibiting nNOS-dependent S-nitrosylation of SCN5A. *Journal of Molecular and Cellular Cardiology, 61*, 102–110.
Chen-Izu, Y., et al. (2015). Na^+ channel function, regulation, structure, trafficking and sequestration. *Journal of Physiology, 593*(6), 1347–1360.
Clancy, C. E., & Rudy, Y. (1999). Linking a genetic defect to its cellular phenotype in a cardiac arrhythmia. *Nature, 400*(6741), 566–569.
Clancy, C. E., & Rudy, Y. (2002). Na^+ channel mutation that causes both Brugada and long-QT syndrome phenotypes – a simulation study of mechanism. *Circulation, 105*(10), 1208–1213.
Clancy, C. E., et al. (2003). Non-equilibrium gating in cardiac Na^+ channels: an original mechanism of arrhythmia. *Circulation, 107*(17), 2233–2237.
Cormier, J. W., et al. (2002). Secondary structure of the human cardiac Na^+ channel C terminus – evidence for a role of helical structures in modulation of channel inactivation. *Journal of Biological Chemistry, 277*(11), 9233–9241.
Cusdin, F. S., Clare, J. J., & Jackson, A. P. (2008). Trafficking and cellular distribution of voltage-gated sodium channels. *Traffic, 9*(1), 17–26.
DeCaen, P. G., et al. (2008). Disulfide locking a sodium channel voltage sensor reveals ion pair formation during activation. *Proceedings of the National Academy of Sciences of the United States of America, 105*(39), 15142–15147.
DeCaen, P. G., et al. (2009). Sequential formation of ion pairs during activation of a sodium channel voltage sensor. *Proceedings of the National Academy of Sciences of the United States of America, 106*(52), 22498–22503.
DeCaen, P. G., et al. (2011). Gating charge interactions with the S1 segment during activation of a Na^+ channel voltage sensor. *Proceedings of the National Academy of Sciences of the United States of America, 108*(46), 18825–18830.
Deschenes, I., Trottier, E., & Chahine, M. (1999). Cysteine scanning analysis of the IFM cluster in the inactivation gate of a human heart sodium channel. *Cardiovascular Research, 42*(2), 521–529.
Dover, K., et al. (2010). Long-term inactivation particle for voltage-gated sodium channels. *Journal of Physiology, 588*(Pt 19), 3695–3711.
Eaholtz, G., et al. (1999). Block of brain sodium channels by peptide mimetics of the isoleucine, phenylalanine, and methionine (IFM) motif from the inactivation gate. *Journal of General Physiology, 113*(2), 279–294.
Eaholtz, G., Scheuer, T., & Catterall, W. A. (1994). Restoration of inactivation and block of open sodium channels by an inactivation gate peptide. *Neuron, 12*(5), 1041–1048.
Eaholtz, G., Zagotta, W. N., & Catterall, W. A. (1998). Kinetic analysis of block of open sodium channels by a peptide containing the isoleucine, phenylalanine, and methionine (IFM) motif from the inactivation gate. *Journal of General Physiology, 111*(1), 75–82.

Ednie, A. R., et al. (2013). Expression of the sialyltransferase, ST3Gal4, impacts cardiac voltage-gated sodium channel activity, refractory period and ventricular conduction. *Journal of Molecular and Cellular Cardiology, 59*, 117–127.

Erickson, J. R., & Anderson, M. E. (2008). CaMKII and its role in cardiac arrhythmia. *Journal of Cardiovascular Electrophysiology, 19*(12), 1332–1336.

Fahmi, A. I., et al. (2001). The sodium channel beta-subunit SCN3b modulates the kinetics of SCN5a and is expressed heterogeneously in sheep heart. *Journal of Physiology, 537*(Pt 3), 693–700.

George, A. L., Jr. (2009). Genetic modulation of impaired cardiac conduction: sodium channel beta4 subunit missing in action. *Circulation Research, 104*(11), 1238–1239.

Gilchrist, J., et al. (2013). Crystallographic insights into sodium-channel modulation by the beta4 subunit. *Proceedings of the National Academy of Sciences of the United States of America, 110*(51), E5016–E5024.

Goldin, A. L. (2003). Mechanisms of sodium channel inactivation. *Current Opinion in Neurobiology, 13*(3), 284–290.

Goldschen-Ohm, M. P., et al. (2013). Multiple pore conformations driven by asynchronous movements of voltage sensors in a eukaryotic sodium channel. *Nature Communications, 4*, 1350.

Gosselin-Badaroudine, P., et al. (2012). Gating pore currents and the resting state of $Na_V1.4$ voltage sensor domains. *Proceedings of the National Academy of Sciences of the United States of America, 109*(47), 19250–19255.

Grandi, E., et al. (2007). Simulation of Ca-calmodulin-dependent protein kinase II on rabbit ventricular myocyte ion currents and action potentials. *Biophysical Journal, 93*(11), 3835–3847.

Grant, A. O., et al. (2002). Long QT syndrome, Brugada syndrome, and conduction system disease are linked to a single sodium channel mutation. *Journal of Clinical Investigation, 110*(8), 1201–1209.

Groome, J. R., & Winston, V. (2013). S1-S3 counter charges in the voltage sensor module of a mammalian sodium channel regulate fast inactivation. *Journal of General Physiology, 141*(5), 601–618.

Hallaq, H., et al. (2006). Quantitation of protein kinase A-mediated trafficking of cardiac sodium channels in living cells. *Cardiovascular Research, 72*(2), 250–261.

Hanck, D. A., & Sheets, M. F. (2007). Site-3 toxins and cardiac sodium channels. *Toxicon, 49*(2), 181–193.

Herren, A. W., Bers, D. M., & Grandi, E. (2013). Post-translational modifications of the cardiac Na channel: contribution of CaMKII-dependent phosphorylation to acquired arrhythmias. *American Journal of Physiology. Heart and Circulatory Physiology, 305*(4), H431–H445.

Hilber, K., et al. (2001). The selectivity filter of the voltage-gated sodium channel is involved in channel activation. *Journal of Biological Chemistry, 276*(30), 27831–27839.

Horvath, B., et al. (2013). Dynamics of the late Na^+ current during cardiac action potential and its contribution to after depolarizations. *Journal of Molecular and Cellular Cardiology, 64*(0), 59–68.

Hund, T. J., et al. (2008). Role of activated CaMKII in abnormal calcium homeostasis and I(Na) remodeling after myocardial infarction: insights from mathematical modeling. *Journal of Molecular and Cellular Cardiology, 45*(3), 420–428.

Hund, T. J., et al. (2010). A beta(IV)-spectrin/CaMKII signaling complex is essential for membrane excitability in mice. *Journal of Clinical Investigation, 120*(10), 3508–3519.

Huxley, H. E., & Kendrew, J. C. (1952). Extractability of the Lotmar-Picken material from dried muscle. *Nature, 170*(4334), 882.

Isom, L. L. (2001). Sodium channel beta subunits: anything but auxiliary. *Neuroscientist, 7*(1), 42–54.

Kass, R. S. (2004). Sodium channel inactivation goes with the flow. *Journal of General Physiology, 124*(1), 7–8.
Kass, R. S. (2006). Sodium channel inactivation in heart: a novel role of the carboxy-terminal domain. *Journal of Cardiovascular Electrophysiology, 17*(Suppl. 1), S21–S25.
Kellenberger, S., et al. (1997). Molecular analysis of potential hinge residues in the inactivation gate of brain type IIA Na^+ channels. *Journal of General Physiology, 109*(5), 607–617.
Kleber, A. G., & Rudy, Y. (2004). Basic mechanisms of cardiac impulse propagation and associated arrhythmias. *Physiological Reviews, 84*(2), 431–488.
Kohlhaas, M., et al. (2010). Elevated cytosolic Na^+ increases mitochondrial formation of reactive oxygen species in failing cardiac myocytes. *Circulation, 121*(14), 1606–1613.
Kontis, K. J., & Goldin, A. L. (1997). Sodium channel inactivation is altered by substitution of voltage sensor positive charges. *Journal of General Physiology, 110*(4), 403–413.
Kontis, K. J., Rounaghi, A., & Goldin, A. L. (1997). Sodium channel activation gating is affected by substitutions of voltage sensor positive charges in all four domains. *Journal of General Physiology, 110*(4), 391–401.
Kuo, C. C., & Bean, B. P. (1994). Na^+ channels must deactivate to recover from inactivation. *Neuron, 12*(4), 819–829.
Laedermann, C. J., Decosterd, I., & Abriel, H. (2014). Ubiquitylation of voltage-gated sodium channels. *Handbook of Experimental Pharmacology, 221*, 231–250.
Lewis, A. H., & Raman, I. M. (2011). Cross-species conservation of open-channel block by Na channel beta4 peptides reveals structural features required for resurgent Na current. *Journal of Neuroscience: The Official Journal of the Society for Neuroscience, 31*(32), 11527–11536.
Lewis, A. H., & Raman, I. M. (2014). Resurgent current of voltage-gated Na(+) channels. *Journal of Physiology, 592*(Pt 22), 4825–4838.
Light, P. E., Wallace, C. H., & Dyck, J. R. (2003). Constitutively active adenosine monophosphate-activated protein kinase regulates voltage-gated sodium channels in ventricular myocytes. *Circulation, 107*(15), 1962–1965.
Liu, C. J., et al. (2003). Modulation of the cardiac sodium channel $Na_V1.5$ by fibroblast growth factor homologous factor 1B. *Journal of Biological Chemistry, 278*(2), 1029–1036.
Liu, M., et al. (2013). Mitochondrial dysfunction causing cardiac sodium channel downregulation in cardiomyopathy. *Journal of Molecular and Cellular Cardiology, 54*, 25–34.
Liu, M., Liu, H., & Dudley, S. C., Jr. (2010). Reactive oxygen species originating from mitochondria regulate the cardiac sodium channel. *Circulation Research, 107*(8), 967–974.
Liu, T., Brown, D. A., & O'Rourke, B. (2010). Role of mitochondrial dysfunction in cardiac glycoside toxicity. *Journal of Molecular and Cellular Cardiology, 49*(5), 728–736.
Liu, Y., Jurman, M. E., & Yellen, G. (1996). Dynamic rearrangement of the outer mouth of a K^+ channel during gating. *Neuron, 16*(4), 859–867.
Long, S. B., Campbell, E. B., & Mackinnon, R. (2005). Voltage sensor of Kv1.2: structural basis of electromechanical coupling. *Science, 309*(5736), 903–908.
Lu, Z., et al. (2012). Suppression of phosphoinositide 3-kinase signaling and alteration of multiple ion currents in drug-induced long QT syndrome. *Science Translational Medicine, 4*(131), 131ra50.
Ma, J., et al. (2012). Calmodulin kinase II and protein kinase C mediate the effect of increased intracellular calcium to augment late sodium current in rabbit ventricular myocytes. *American Journal of Physiology. Cell Physiology, 302*(8), C1141–C1151.
Maier, L. S. (2011). CaMKII regulation of voltage-gated sodium channels and cell excitability. *Heart Rhythm, 8*(3), 474–477.
Maier, S. K., et al. (2004). Distinct subcellular localization of different sodium channel alpha and beta subunits in single ventricular myocytes from mouse heart. *Circulation, 109*(11), 1421–1427.

Maltsev, V. A., et al. (2008). Modulation of late sodium current by Ca^{2+}, calmodulin, and CaMKII in normal and failing dog cardiomyocytes: similarities and differences. *American Journal of Physiology. Heart and Circulatory Physiology, 294*(4), H1597–H1608.

Mantegazza, M., et al. (2001). Role of the C-terminal domain in inactivation of brain and cardiac sodium channels. *Proceedings of the National Academy of Sciences of the United States of America, 98*(26), 15348–15353.

Marionneau, C., & Abriel, H. (2015). Regulation of the cardiac Na^+ channel $Na_V1.5$ by post-translational modifications. *Journal of Molecular and Cellular Cardiology, 82*, 36–47.

Matsuda, J. J., Lee, H., & Shibata, E. F. (1992). Enhancement of rabbit cardiac sodium channels by beta-adrenergic stimulation. *Circulation Research, 70*(1), 199–207.

McCormick, K. A., et al. (1998). Molecular determinants of Na^+ channel function in the extracellular domain of the beta1 subunit. *Journal of Biological Chemistry, 273*(7), 3954–3962.

McPhee, J. C., et al. (1998). A critical role for the S4-S5 intracellular loop in domain IV of the sodium channel -subunit in fast inactivation. *Journal of Biological Chemistry, 273*(2), 1121–1129.

Miyamoto, K., et al. (2001). Solution structures of the inactivation gate particle peptides of rat brain type-IIA and human heart sodium channels in SDS micelles. *Journal of Peptide Research, 57*(3), 203–214.

Moreno, J. D., & Clancy, C. E. (2012). Pathophysiology of the cardiac late Na current and its potential as a drug target. *Journal of Molecular and Cellular Cardiology, 52*(3), 608–619.

Morgan, K., et al. (2000). Beta 3: an additional auxiliary subunit of the voltage-sensitive sodium channel that modulates channel gating with distinct kinetics. *Proceedings of the National Academy of Sciences of the United States of America, 97*(5), 2308–2313.

Mori, M., et al. (2000). Novel interaction of the voltage-dependent sodium channel (VDSC) with calmodulin: does VDSC acquire calmodulin-mediated Ca^{2+}-sensitivity? *Biochemistry, 39*(6), 1316–1323.

Motoike, H. K., et al. (2004). The Na^+ channel inactivation gate is a molecular complex: a novel role of the COOH-terminal domain. *Journal of General Physiology, 123*(2), 155–165.

Murphy, B. J., et al. (1996). cAMP-dependent phosphorylation of two sites in the alpha subunit of the cardiac sodium channel. *Journal of Biological Chemistry, 271*(46), 28837–28843.

Musa, H., et al. (2015). SCN5A variant that blocks fibroblast growth factor homologous factor regulation causes human arrhythmia. *Proceedings of the National Academy of Sciences of the United States of America, 112*(40), 12528–12533.

Namadurai, S., et al. (2014). Crystal structure and molecular imaging of the Na_v channel beta3 subunit indicates a trimeric assembly. *Journal of Biological Chemistry, 289*(15), 10797–10811.

Namadurai, S., et al. (2015). A new look at sodium channel beta subunits. *Open Biology, 5*(1), 140192.

Noble, D., & Rudy, Y. (2001). Models of cardiac ventricular action potentials: iterative interaction between experiment and simulation. *Philosophical Transactions of the Royal Society of London Series A-Mathematical Physical and Engineering Sciences, 359*(1783), 1127–1142.

Paldi, T., & Gurevitz, M. (2010). Coupling between residues on S4 and S1 defines the voltage-sensor resting conformation in NaChBac. *Biophysical Journal, 99*(2), 456–463.

Patton, D. E., et al. (1992). Amino acid residues required for fast Na(+)-channel inactivation: charge neutralizations and deletions in the III-IV linker. *Proceedings of the National Academy of Sciences of the United States of America, 89*(22), 10905–10909.

Payandeh, J., et al. (2011). The crystal structure of a voltage-gated sodium channel. *Nature, 475*(7356), 353–358.

Qu, Y., et al. (1994). Modulation of cardiac Na^+ channels expressed in a mammalian cell line and in ventricular myocytes by protein kinase C. *Proceedings of the National Academy of Sciences of the United States of America, 91*(8), 3289–3293.

Rivolta, I., et al. (2001). Inherited Brugada and long QT-3 syndrome mutations of a single residue of the cardiac sodium channel confer distinct channel and clinical phenotypes. *Journal of Biological Chemistry, 276*(33), 30623–30630.

Rohl, C. A., et al. (1999). Solution structure of the sodium channel inactivation gate. *Biochemistry, 38*(3), 855–861.

Rohr, S., Kucera, J. P., & Kleber, A. G. (1998). Slow conduction in cardiac tissue, I – effects of a reduction of excitability versus a reduction of electrical coupling on microconduction. *Circulation Research, 83*(8), 781–794.

Rook, M. B., et al. (2012). Biology of cardiac sodium channel $Na_V1.5$ expression. *Cardiovascular Research, 93*(1), 12–23.

Russ, U., et al. (1996). Effects of the Na^+/H^+-exchange inhibitor Hoe 642 on intracellular pH, calcium and sodium in isolated rat ventricular myocytes. *Pflugers Archiv, 433*(1–2), 26–34.

Schwaiger, C. S., et al. (2011). 3(1)(0)-helix conformation facilitates the transition of a voltage sensor S4 segment toward the down state. *Biophysical Journal, 100*(6), 1446–1454.

Shaw, R. M., & Rudy, Y. (1997). Ionic mechanisms of propagation in cardiac tissue – roles of the sodium and L-type calcium currents during reduced excitability and decreased gap junction coupling. *Circulation Research, 81*(5), 727–741.

Silva, J. R., & Goldstein, S. A. (2013). Voltage-sensor movements describe slow inactivation of voltage-gated sodium channels I: wild-type skeletal muscle Na(V)1.4. *Journal of General Physiology, 141*(3), 309–321.

Sirota, F. L., Pascutti, P. G., & Anteneodo, C. (2002). Molecular modeling and dynamics of the sodium channel inactivation gate. *Biophysical Journal, 82*(3), 1207–1215.

Smith, M. R., & Goldin, A. L. (1997). Interaction between the sodium channel inactivation linker and domain III S4-S5. *Biophysical Journal, 73*(4), 1885–1895.

Stocker, P. J., & Bennett, E. S. (2006). Differential sialylation modulates voltage-gated Na^+ channel gating throughout the developing myocardium. *Journal of General Physiology, 127*(3), 253–265.

Stuhmer, W., et al. (1989). Structural parts involved in activation and inactivation of the sodium channel. *Nature, 339*(6226), 597–603.

Sun, Y. M., et al. (1997). On the structural basis for size-selective permeation of organic cations through the voltage-gated sodium channel – effect of alanine mutations at the DEKA locus on selectivity, inhibition by Ca^{2+} and H^+, and molecular sieving. *Journal of General Physiology, 110*(6), 693–715.

Tan, H. L., et al. (2001). A sodium-channel mutation causes isolated cardiac conduction disease. *Nature, 409*(6823), 1043–1047.

Tateyama, M., et al. (2003). Modulation of cardiac sodium channel gating by protein kinase A can be altered by disease-linked mutation. *Journal of Biological Chemistry, 278*(47), 46718–46726.

Tateyama, M., et al. (2004). Structural effects of an LQT-3 mutation on heart Na^+ channel gating. *Biophysical Journal, 86*(3), 1843–1851.

Tomaselli, G. F., et al. (1995). A mutation in the pore of the sodium channel alters gating. *Biophysical Journal, 68*(5), 1814–1827.

Ufret Vincenty, C. A., et al. (2001). Role of sodium channel deglycosylation in the genesis of cardiac arrhythmias in heart failure. *Journal of Biological Chemistry, 276*(30), 28197–28203.

Vassilev, P., Scheuer, T., & Catterall, W. A. (1989). Inhibition of inactivation of single sodium channels by a site-directed antibody. *Proceedings of the National Academy of Sciences of the United States of America, 86*(20), 8147–8151.

Veldkamp, M. W., et al. (2000). Two distinct congenital arrhythmias evoked by a multidysfunctional Na^+ channel. *Circulation Research, 86*(9), E91–E97.

Vilin, Y. Y., Fujimoto, E., & Ruben, P. C. (2001). A novel mechanism associated with idiopathic ventricular fibrillation (IVF) mutations R1232W and T1620M in human cardiac sodium channels. *Pflugers Archiv, 442*(2), 204–211.

Vilin, Y. Y., et al. (1999). Structural determinants of slow inactivation in human cardiac and skeletal muscle sodium channels. *Biophysical Journal,* 77(3), 1384–1393.

Vilin, Y. Y., & Ruben, P. C. (2001). Slow inactivation in voltage-gated sodium channels: molecular substrates and contributions to channelopathies. *Cell Biochemistry and Biophysics, 35*(2), 171–190.

Wagner, S., et al. (2006). Ca^{2+}/calmodulin-dependent protein kinase II regulates cardiac Na^+ channels. *Journal of Clinical Investigation, 116*(12), 3127–3138.

Wagner, S., et al. (2011). Reactive oxygen species-activated Ca/calmodulin kinase IIdelta is required for late I(Na) augmentation leading to cellular Na and Ca overload. *Circulation Research, 108*(5), 555–565.

Wang, D. W., et al. (2000). Enhanced Na(+) channel intermediate inactivation in Brugada syndrome. *Circulation Research, 87*(8), E37–E43.

Wang, S. Y., et al. (2004). Block of inactivation-deficient Na^+ channels by local anesthetics in stably transfected mammalian cells: evidence for drug binding along the activation pathway. *Journal of General Physiology, 124*(6), 691–701.

Wang, S. Y., & Wang, G. K. (2005). Block of inactivation-deficient cardiac Na(+) channels by acetyl-KIFMK-amide. *Biochemical and Biophysical Research Communications, 329*(2), 780–788.

West, J. W., et al. (1991). A phosphorylation site in the Na^+ channel required for modulation by protein kinase C. *Science, 254*(5033), 866–868.

West, J. W., et al. (1992). A cluster of hydrophobic amino acid residues required for fast Na(+)-channel inactivation. *Proceedings of the National Academy of Sciences of the United States of America, 89*(22), 10910–10914.

Wingo, T. L., et al. (2004). An EF-hand in the sodium channel couples intracellular calcium to cardiac excitability. *Nature Structural & Molecular Biology, 11*(3), 219–225.

Wu, J., et al. (2015). Structure of the voltage-gated calcium channel Cav1.1 complex. *Science, 350*(6267), aad2395.

Yamagishi, T., et al. (2001). Molecular architecture of the volta re-dependent Na channel: functional evidence for at helices in the pore. *Journal of General Physiology, 118*(2), 171–181.

Yao, L., et al. (2011). $Na_v1.5$-dependent persistent Na^+ influx activates CaMKII in rat ventricular myocytes and N1325S mice. *American Journal of Physiology. Cell Physiology, 301*(3), C577–C586.

Yarov-Yarovoy, V., Baker, D., & Catterall, W. A. (2006). Voltage sensor conformations in the open and closed states in ROSETTA structural models of K(+) channels. *Proceedings of the National Academy of Sciences of the United States of America, 103*(19), 7292–7297.

Yarov-Yarovoy, V., et al. (2012). Structural basis for gating charge movement in the voltage sensor of a sodium channel. *Proceedings of the National Academy of Sciences of the United States of America, 109*(2), E93–E102.

Yellen, G. (1998). The moving parts of voltage-gated ion channels. *Quarterly Reviews of Biophysics, 31*(3), 239–295.

Yu, E. J., et al. (2005). Distinct domains of the sodium channel beta3-subunit modulate channel-gating kinetics and subcellular location. *Biochemical Journal, 392*(Pt 3), 519–526.

Yu, F. H., & Catterall, W. A. (2003). Overview of the voltage-gated sodium channel family. *Genome Biology, 4*(3), 207.

Yu, F. H., et al. (2005). Overview of molecular relationships in the voltage-gated ion channel superfamily. *Pharmacological Reviews, 57*(4), 387–395.

Zhou, J., et al. (2002). Phosphorylation and putative ER retention signals are required for protein kinase A-mediated potentiation of cardiac sodium current. *Circulation Research, 91*(6), 540–546.

Zimmer, T., & Benndorf, K. (2002). The human heart and rat brain IIA Na^+ channels interact with different molecular regions of the beta1 subunit. *Journal of General Physiology, 120*(6), 887–895.

SECTION FOUR

Functional Studies – Modulation by Other Molecules and Roles of Inactivation and Prolonged Currents

CHAPTER ELEVEN

Developmental and Regulatory Functions of Na^+ Channel Non–pore-forming β Subunits

J.J. Winters* and L.L. Isom*,§,1

*University of Michigan Neuroscience Program, Ann Arbor, MI, United States
§University of Michigan Medical School, Ann Arbor, MI, United States
[1]Corresponding author: E-mail: lisom@umich.edu

Contents

Abstract

Voltage-gated Na^+ channels (VGSCs) isolated from mammalian neurons are heterotrimeric complexes containing one pore-forming α subunit and two non–pore-forming β subunits. In excitable cells, VGSCs are responsible for the initiation of action potentials.

Current Topics in Membranes, Volume 78
ISSN 1063-5823
http://dx.doi.org/10.1016/bs.ctm.2016.07.003

VGSC β subunits are type I topology glycoproteins, containing an extracellular amino-terminal immunoglobulin (Ig) domain with homology to many neural cell adhesion molecules (CAMs), a single transmembrane segment, and an intracellular carboxyl-terminal domain. VGSC β subunits are encoded by a gene family that is distinct from the α subunits. While α subunits are expressed in prokaryotes, β subunit orthologs did not arise until after the emergence of vertebrates. β subunits regulate the cell surface expression, subcellular localization, and gating properties of their associated α subunits. In addition, like many other Ig-CAMs, β subunits are involved in cell migration, neurite outgrowth, and axon pathfinding and may function in these roles in the absence of associated α subunits. In sum, these multifunctional proteins are critical for both channel regulation and central nervous system development.

1. INTRODUCTION TO THE β SUBUNITS

Voltage-gated Na^+ channels (VGSCs), anticipated by the classic voltage clamp experiments of Hodgkin and Huxley, were first discovered and biochemically isolated from rat brain over 30 years ago. The purified constituents included one large (~260 kD) α subunit and two much smaller (~33–36 kD) β subunit proteins (Beneski & Catterall, 1980; Hartshorne & Catterall, 1984; Hartshorne, Messner, Coppersmith, & Catterall, 1982; Hodgkin & Huxley, 1952). Although Na^+ currents can be measured by expression of an α subunit alone, eg, in *Xenopus* oocytes (Goldin et al., 1986; Noda et al., 1986) in order to approximate physiological kinetics and gating properties, β subunits need to be co-expressed (Isom et al., 1992; Isom, Ragsdale, et al., 1995). Mammalian VGSCs are heterotrimeric complexes (Messner & Catterall, 1985). At the center of each channel is a pore-forming α subunit, flanked by one noncovalently associated β subunit (β1 or β3) and one covalently associated β subunit (β2 or β4) (Hartshorne & Catterall, 1981; Isom et al., 1992; Isom, Ragsdale, et al., 1995; Morgan et al., 2000; Yu et al., 2003). As components of the VGSC complex, β subunits function in ion conduction by regulating channel expression at the cell surface, anchoring the complex to intracellular proteins, and by directly modulating the gating properties of the α subunit (Aman et al., 2009; Bennett, Makita, & George, 1993; Brackenbury et al., 2010; Chen et al., 2004; Isom et al., 1992; Isom, Ragsdale, et al., 1995; Isom, Scheuer, et al., 1995; Kazarinova-Noyes et al., 2001; Kazen-Gillespie et al., 2000; Ko, Lenkowski, Lee, Mounsey, & Patel, 2005; Lopez-Santiago, Brackenbury, Chen, & Isom, 2011; Lopez-Santiago et al., 2007; McEwen, Meadows, Chen, Thyagarajan, & Isom, 2004; Medeiros-Domingo et al., 2007; Patino et al., 2009; Watanabe et al., 2008;

Yu et al., 2003; Zhou, Potts, Trimmer, Agnew, & Sigworth, 1991). In addition, β1 subunits have now been shown to modulate currents carried by voltage-gated K^+ channels (VGKCs), expanding the known β subunit functions outside of the realm of Na^+ channel physiology (Brackenbury, Yuan, O'Malley, Parent, & Isom, 2013; Deschenes & Tomaselli., 2002; Marionneau et al., 2012; Nguyen et al., 2012). Beyond that, it is now clear that in the central nervous system (CNS) β subunit proteins are involved in a wide array of developmental functions that may be independent of the VGSC complex (Brackenbury & Isom, 2011; Brackenbury et al., 2008, 2013; Davis, Chen, & Isom, 2004; Fein, Wright, Slat, Ribera, & Isom, 2008; Patino et al., 2011). β subunits each contain an extracellular immunoglobulin (Ig) domain, common to many cell adhesion molecules (CAMs), with important roles in brain development (Isom, Ragsdale, et al., 1995; Maness & Schachner, 2007). This feature makes them unique among ion channel subunits and places them in the Ig-superfamily of CAMs. Binding partners include other VGSC β subunits, CAMs such as contactin and neurosfascin-186 (NF186), and extracellular matrix proteins such as tenascin (Malhotra, Thyagarajan, Chen, & Isom, 2004; McEwen & Isom, 2004; Ratcliffe, Westenbroek, Curtis, & Catterall, 2001; Srinivasan, Schachner, & Catterall, 1998; Xiao et al., 1999). Finally, β1 interacts with ankyrins. This interaction may regulate VGSC subcellular localization and brain development (Malhotra, Kazen-Gillespie, Hortsch, & Isom, 2000; Malhotra et al., 2002, 2004). In sum, VGSC β subunits are promiscuous and multifunctional. This chapter will review the known regulatory and developmental functions of the VGSC β subunits, β1, β2, β3, β4, and β1B.

2. β SUBUNIT EVOLUTION, GENES, AND STRUCTURE

2.1 Genes and evolution

Ancestral orthologs to the mammalian VGSC β subunits have accompanied α subunits in the genome since the evolution of early vertebrate species (Chopra, Watanabe, Zhong, & Roden, 2007; Vilella et al., 2009). Mammalian β subunits are glycoproteins encoded by a gene family distinct from that of the α subunits, with the genes *SCN1B* through *SCN4B* encoding β1 through β4, respectively (Isom et al., 1992; Isom, Ragsdale, et al., 1995; Morgan et al., 2000; Qin et al., 2003; Yu et al., 2003). The gene *SCN1B* also encodes the secreted splice variant β1B, originally called β1A (Kazen-Gillespie et al., 2000; Patino et al., 2011). While VGSC α subunit orthologs are expressed in prokaryotes, the β subunits evolved much later. In prokaryotes, VGSC α subunit genes are closely related to those of voltage-gated Ca^{2+} channels. Bacterial Na^{+} channels and Ca^{2+} channels are each composed of six transmembrane repeats that form homotetramers to construct the ion-conducting pore. They have a high degree of sequence similarity, especially in the pore domains, exhibit similar pharmacological profiles, and ultimately share common descent from VGKCs (Catterall, 2000; Goldin, 2001; Ren et al., 2001). By contrast, the VGSC β subunits share sequence homology and common ancestry with other proteins that have V-type Ig domains, including myelin P_0 (McCormick et al., 1998). β subunits are present in mammals, birds, and fish but are absent from invertebrates (Chopra et al., 2007; Isom et al., 1992; Patino et al., 2009; Ren et al., 2001). β subunits were notably absent from purified VGSCs of the electric organ and brain of *Electrophorus electricus* (electric eel), shown to consist of α subunits alone (Miller, Agnew, & Levinson, 1983; Sutkowski & Catterall, 1990). The recently reported *E. electricus* genome provided the opportunity to search for sequences homologous to the human proteins (Gallant et al., 2014). Our protein Basic Local Alignment Search Tool (pBLAST) analysis reveals orthologs to all four human β subunits. Perhaps in *E. electricus* these proteins serve functions separate from the VGSC complex. A β1 ortholog is expressed in the electric organ of the weakly electric fish *Sternopygus macrurus*. This ortholog modulates the properties of human Nav1.4 and the orthologous α subunit smNav1.4b, accelerating the rate of inactivation (Liu, Ming—Ming, & Zakon, 2007). The genome of another teleost, *Danio rerio* (zebrafish), contains six β subunit genes, homologous to β2, β3, and two duplicates each of β1 and β4, respectively (Chopra et al., 2007; Fein et al., 2008). Remarkably, important structural features of mammalian β subunits, including the extracellular Ig domain, an intracellular tyrosine phosphorylation site in β1, and potential sites for N-linked glycosylation, are conserved in the zebrafish and electric eel orthologs (Chopra et al., 2007; Gallant et al., 2014). Zebrafish *scn1b* and *scn2b* have several splice variants and, similar to mammalian *SCN1B* encoding β1B, the zebrafish *scn1b* splice variants have variable carboxyl-termini (Chopra et al., 2007; Fein, Meadows, Chen, Slat, & Isom, 2007). Analysis of the zebrafish genome suggested that β2/β4 and β1/β3 subunits arose from the duplication of two distinct precursor genes early in vertebrate evolution (Chopra et al., 2007). In agreement with this estimate, no genes homologous to the sequences of zebrafish and human β subunits exist in the genomes of the invertebrate filter-feeding tunicates *Ciona intestinalis* and

Ciona savignyi (sea squirts) (Azumi et al., 2003; Chôpra et al., 2007; Du Pasquier, Zucchetti, & De Santis, 2004). The *Drosophila melanogaster* VGSC is composed of the pore-forming α subunit *paralytic* and the associated subunit *tipE*, but this protein is not homologous to vertebrate β subunits (Li, Waterhouse, & Zdobnov, 2011). Vertebrate genomes including jawless fishes, cartilaginous fishes, ray-finned fishes, and lobe-finned fishes all have predicted β subunit orthologs (Amemiya et al., 2013; Shin et al., 2014; Smith et al., 2013; Venkatesh et al., 2014; Vilella et al., 2009). The genomes of *Gallus gallus* (chicken) and other birds reveal that they have only three β subunit genes, with β1 being absent from all available sequences (Hillier, Miller, Birney, Warren, & Hardison, 2004; Vilella et al., 2009). Fig. 1 shows a phylogenetic tree of selected vertebrate and invertebrate species, revealing homologous genes to human *SCN1B-SCN4B* found in the sequenced genomes of those organisms. Note that the vertebrate jawless fish, *Petromyzon marinus* (sea lamprey) is the most evolutionarily distant species from humans that possesses a predicted β subunit ortholog. This finding suggests that these proteins arose in early fishes at least 500 million years ago. Notably, orthologs to *MPZ*, the human gene which encodes myelin P_0, occur in ray-finned fish genomes but not in more distantly related jawless fishes or in invertebrates (Vilella et al., 2009). This corresponds with the evolutionary origin of myelin, the insulating sheath that is a necessary feature of saltatory axonal conduction, which characterizes the nervous systems of hinged-jawed fishes and their descendants but is absent in jawless fishes such as lampreys and hagfish (Zalc, Goujet, & Colman, 2008). Myelinated axons have specialized structures, the nodes of Ranvier, where VGSC complexes, including β subunits, are highly concentrated (Buffington & Rasband, 2013; Chen et al., 2002, 2004; Kaplan, Cho, Ullian, & Isom, 2001; Patino et al., 2009). This could explain why genes encoding these proteins emerged alongside myelin proteins. Myelin P_0, one of the most abundant proteins in the peripheral myelin sheath, shares a highly homologous Ig domain with VGSC β subunits and is predicted to have a common ancestral gene (Vilella et al., 2009).

2.2 Structure

The structural features of VGSC β subunits predict diverse intracellular and extracellular interactions with ion channels, adhesion molecules, and cytoskeletal elements. Fig. 2 diagrams the structures of β1 and β2, illustrating many of these important features. Mammalian β1, β2, β3, and β4 are type I topology transmembrane proteins, each having an extracellular aminoterminus containing an Ig domain, a single transmembrane segment, and an

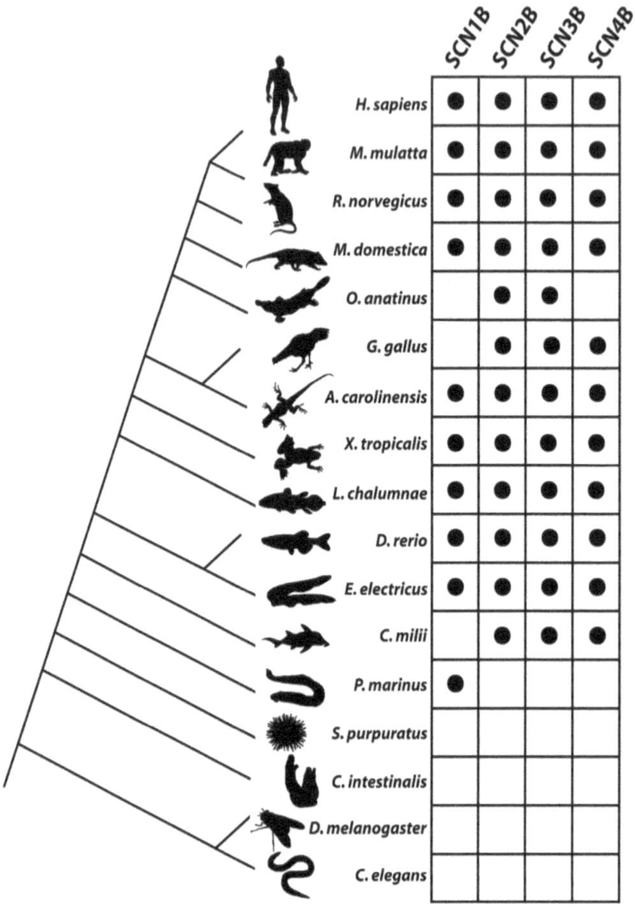

Figure 1 ***Phylogenetic tree of selected vertebrate and invertebrate species showing genes homologous to human* SCN1B-SCN4B.** *Macaca mulatta* = rhesus macaque (Gibbs et al., 2007); *Rattus norvegicus* = Norway rat; *Monodelphis domestica* = opossum (Mikkelsen et al., 2007); *Ornithorhynchus anatinus* = duck-billed platypus (Warren et al., 2008); *Gallus gallus* = chicken (Hillier et al., 2004); *Anolis carolinensis* = anole lizard (Alföldi et al., 2011); *Xenopus tropicalis* = western clawed frog (Hellsten et al., 2010); *Latimeria chalumnae* = coelacanth (Amemiya et al., 2013); *Danio rerio* = zebrafish (Chopra et al., 2007; Fein et al., 2008); *Electrophorus electricus* = electric eel (Gallant et al., 2014); *Callorhinchus milii* = elephant shark (Venkatesh et al., 2014); *Petromyzon marinus* = sea lamprey (Smith et al., 2013); *Strongylocentrotus purpuratus* = sea urchin (Sodergren et al., 2006); *Ciona intestinalis* = sea squirt (Azumi et al., 2003); *Drosophila melanogaster* = fruit fly (Adams et al., 2000); *Caenorhabditis elegans* = nematode (The *C. elegans* Sequencing Consortium, 1998). Homologous genes and evolutionary history were analyzed using Ensembl (Vilella et al., 2009).

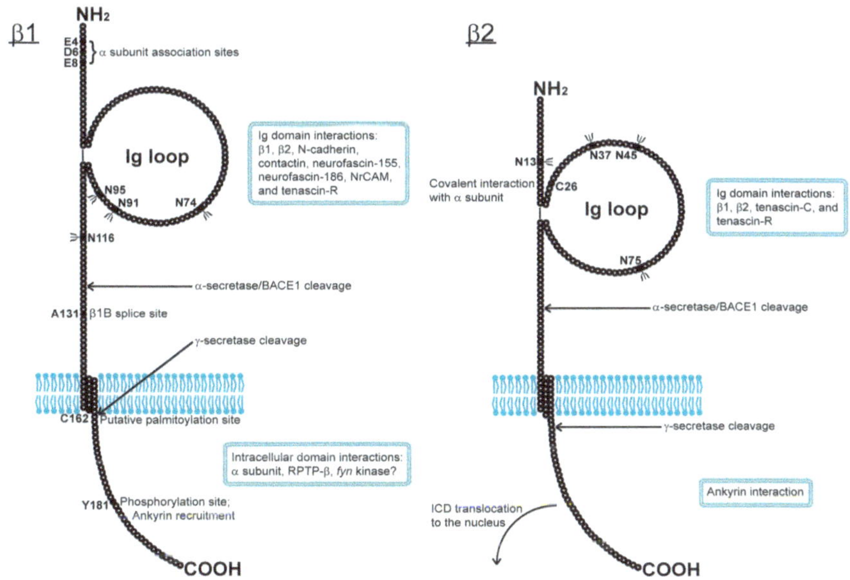

Figure 2 ***Structural features of β1 and β2.*** Features shown for β1 include extracellular sites for α subunit interaction (McCormick et al., 1998), N-linked glycosylation (Ψ) sites (McCormick et al., 1998), the splice site for β1B indicating the first residue (A131) that is not shared by the splice variant (Kazen-Gillespie et al., 2000; Patino et al., 2011; Qin et al., 2003), a putative palmitoylation site (McEwen & Isom, 2004), secretase cleavage sites (Wong et al., 2005), and a tyrosine phosphorylation site (Y181) which interacts with ankyrin (Malhotra et al., 2002, 2004). *Boxes* indicate the known Ig domain homophilic interaction (Malhotra et al., 2000) and heterophilic interactions with β2, N-cadherin, contactin, neurofascin-155 and -186, NrCAM (McEwen & Isom, 2004; Ratcliffe et al., 2001), and tenascin-R (Xiao et al., 1999) and ICD interactions with the α subunit (Spampanato et al., 2004), RPTP-β (Ratcliffe et al., 2000), and, potentially, *fyn* kinase (Brackenbury et al., 2008; Malhotra et al., 2002, 2004). Features shown for β2 include N-linked glycosylation (Ψ) sites (Isom, Ragsdale, et al., 1995), the cysteine residue (C26) that interacts covalently with the associated α subunit (Chen et al., 2012), and secretase cleavage sites (Wong et al., 2005). Also noted is the translocation of the cleaved β2 ICD to the nucleus (Kim et al., 2007). *Boxes* indicate known Ig-domain homophilic interactions (Malhotra et al., 2000) and heterophilic interactions with β1 (McEwen & Isom, 2004), tenascin-C and tenascin-R (Srinivasan et al., 1998; Xiao et al., 1999) as well as the intracellular ankyrin interaction (Malhotra et al., 2000).

intracellular carboxyl-terminus (Isom, 2001). By contrast, β1B, formed by in-frame extension of the *SCN1B* third exon into intron 3, shares the extracellular amino-terminal domain of β1 but lacks a transmembrane domain and has instead an alternate carboxyl-terminus (Kazen-Gillespie et al., 2000; Patino et al., 2011). This results in a soluble protein that is secreted from the cell. Two subunits, β1 and β3, which share 57% sequence homology, interact exclusively in a noncovalent manner, via their amino- and carboxyl-termini, with the associated α subunit (McCormick et al., 1998; Meadows, Malhotra, Stetzer, Isom, & Ragsdale, 2001; Morgan et al., 2000; Spampanato et al., 2004). β2 and β4, which share 35% sequence identity, each interact covalently with the α subunit by means of a single extracellular disulfide bond between cysteine residue 26 of β2 or cysteine residue 28 of β4 (residue 58 from the initiator methionine) and one of the cysteine residues in the α subunit S5–S6 loops (Buffington & Rasband, 2013; Chen et al., 2012; Gilchrist, Das, Van Petegem, & Bosmans, 2013; Hartshorne & Catterall, 1984; Hartshorne et al., 1982; Messner & Catterall, 1985; Yu et al., 2003). The extracellular Ig domain, common to all five β subunits, is homologous to the V-type Ig loop motifs of members of the Ig-superfamily of CAMs (Ig-CAMs) (Isom & Catterall, 1996; Kazen-Gillespie et al., 2000; McCormick et al., 1998; Morgan et al., 2000; Yu et al., 2003). Fig. 3 compares the structures of the V-type Ig-CAMs neural cell adhesion molecule (N-CAM), L1, neurofascin, neuronal cell adhesion molecule (NrCAM), contactin, and myelin P_0 with the VGSC β subunits. Molecules of the Ig-superfamily are characterized by one or more tandem Ig domain repeats with homology to the motif first identified in Ig proteins of the immune system. Neural Ig-CAMs, including N-CAM, L1, and contactin, bind homophilically and heterophilically and mediate important developmental processes such as neurite outgrowth, axon fasciculation, and cell migration (Crossin & Krushel, 2000; Vaughn & Bjorkman, 1996; Williams & Barclay, 1988). Sequence analysis of the β2 Ig loop shows remarkable homology to the third Ig repeat of the CAM contactin (Isom, Ragsdale, et al., 1995). The β1 and β3 Ig loops show homology to another Ig-CAM, myelin P_0 (Isom, Ragsdale, et al., 1995; McCormick et al., 1998; Morgan et al., 2000) (Fig. 3). The recently reported β3 and β4 extracellular domain crystal structures revealed that the Ig loop of each β subunit, containing multiple β-sheets, is formed by a disulfide bridge between two cysteine residues (Gilchrist et al., 2013; Namadurai et al., 2014). This structure is consistent with V-type Ig domains (Vaughn & Bjorkman, 1996). An additional intramolecular disulfide bridge, unique to β3, results in further protein stability. Interestingly, the β3 crystal structure showed the formation of homodimers and homotrimers. Since Nav1.5 contains multiple sites for interaction with β3, it follows that it may form oligomeric complexes (Namadurai et al., 2014).

β subunits contain recognition sequences that subject them to sequential proteolytic cleavage by BACE1 β-secretase (BACE1) and γ-secretases, resulting in the release of amino-terminal and carboxyl-terminal

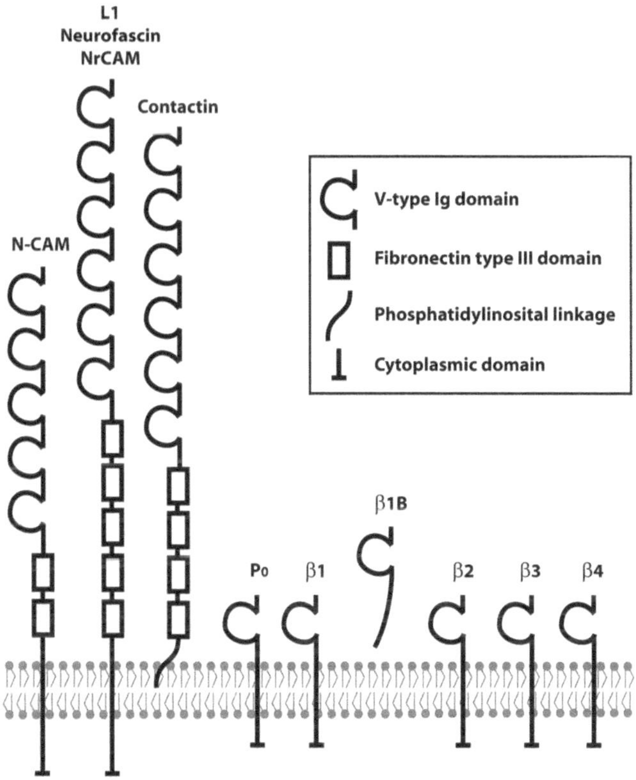

Figure 3 ***Comparison of VGSC β subunits with other V-type members of the Ig superfamily of CAMs.***

fragments (Wong et al., 2005). This processing has important consequences. In particular, β1-mediated neurite outgrowth requires γ-secretase activity (Brackenbury & Isom, 2011). β2-cleavage by γ-secretase leads to translocation of the intracellular domain (ICD) to the nucleus and subsequent changes in VGSC gene expression (Kim et al., 2007). Finally, the carboxyl-termini of β1 and β2 interact with the cytoskeleton-linked proteins ankyrinG and ankyrinB (Malhotra et al., 2000, 2002). This interaction, which may be critical to localizing channel proteins to specific cellular domains, is disrupted in β1 by phosphorylation of an intracellular tyrosine residue (Malhotra et al., 2002, 2004). Together, the extracellular and intracellular features of β subunit structures equip them for involvement in a diverse range of cellular functions.

3. β SUBUNIT EXPRESSION, LOCALIZATION, AND POSTTRANSLATIONAL MODIFICATION

3.1 Expression

Specific combinations of VGSC α and β subunits (and other proteins of the channelome complexes in which they reside) contribute to the particular physiological characteristics of divergent cell types. A model of such a complex is represented in Fig. 4, which shows an α subunit associated with β1 and β2 to form a heterotrimer in addition to potential interacting partners that make up the channelome. Along with other molecular components of these channel complexes, VGSC β subunits are highly expressed in excitable cells of the CNS, peripheral nervous system (PNS), heart and skeletal muscle (Brackenbury & Isom, 2011; Isom et al., 1992; Isom, Ragsdale, et al., 1995; Lopez-Santiago et al., 2011, 2006; Maier et al., 2004; Morgan et al., 2000; Yu et al., 2003). VGSC β subunits are also expressed in nonexcitable cells, including astrocytes and radial glia, vascular endothelial cells, and cancer cells (Andrikopoulos et al., 2011; Aronica et al., 2003; Chioni, Brackenbury, Calhoun, Isom, & Djamgoz, 2009; Davis et al., 2004; Diss et al., 2008; Fein et al., 2008; Oh & Waxman, 1995). In a mouse model of breast cancer, β1 is postulated to enhance adhesion and increase tumor growth and invasion (Nelson, Millican-Slater, Forrest, & Brackenbury, 2014; Patel & Brackenbury, 2015). Reactive astrocytes in a rat temporal lobe epilepsy (TLE) model and those associated with human pathologies show increased β1 protein expression, which might affect their ability to migrate toward an area of CNS injury (Aronica et al., 2003; Gorter, van Vliet, da Silva, Isom, & Aronica, 2002). In zebrafish; however, the *SCN1B* orthologs *scn1ba* and *scn1bb*, are differentially expressed in cells of the CNS and PNS, with *scn1ba* expression being greatest in neurons and skeletal muscle cells and *scn1bb* being expressed in Schwann cells, supporting cells of the olfactory pit and inner ear, and CNS myelinating glia (Fein et al., 2007, 2008). In addition, β subunit expression is developmentally regulated, with certain subunits generally decreasing during development and others generally increasing (Sutkowski & Catterall, 1990). In rodents, β1B and β3 are highly expressed prenatally with decreasing abundance after birth (Kazen-Gillespie et al., 2000; Patino et al., 2011; Shah, Stevens, Pinnock, Dixon, & Lee, 2001). By contrast, β1 and β2 expression increases during postnatal development and remains high in adulthood (Isom, Ragsdale, et al., 1995; Kazen-Gillespie et al., 2000; Patino et al., 2011). The

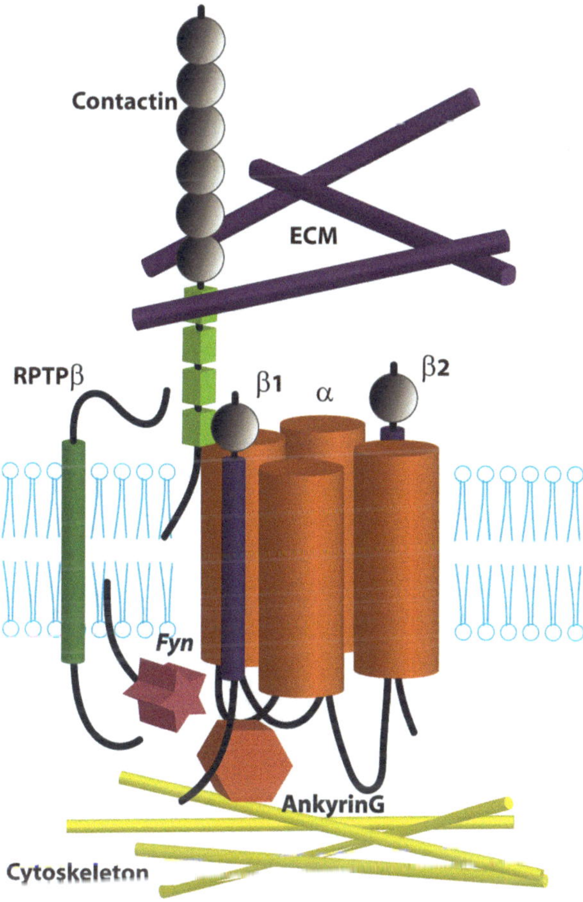

Figure 4 ***Model of the VGSC complex along with potential neuronal β subunit-interacting proteins.*** At the center is the pore-forming α subunit, composed of four contiguous domains (*orange cylinders*). β1 and β2 (*blue cylinders*) are represented flanking the α subunit, each with one extracellular Ig domain (*gray sphere*). Contactin, which has four fibronectin type-III domains (*green rectangles*) and six tandem Ig domains (*gray spheres*), is shown anchored to the membrane adjacent to β1. Extracellular matrix (ECM) proteins (*purple tubes*), such as tenascin-C and tenascin-R, which are known partners for β1 and β2, are positioned above the complex. AnkyrinG (*red hexagon*), which binds to the intracellular domains of α and β1, is shown associated with the cytoskeleton (*yellow filaments*). Finally, the kinase *fyn* (*pink star*) and RTPβ (*green cylinder*) appear to the left of the complex near their sites of interaction with the β1 carboxyl terminus.

developmental time course of β4 expression is unknown, but it is expressed postnatally in rat (Yu et al., 2003). Notable exceptions to these generalities exist. For example, β1B expression is maintained in adult heart, and β3 continues to be expressed in adult dorsal root ganglion (DRG) neurons (Kazen-Gillespie et al., 2000; Takahashi et al., 2003).

3.2 Subcellular localization in neurons

As components of VGSCs, β subunits localize to subcellular compartments where they engage in specialized tasks. In neurons, these compartments include the axon initial segment (AIS) and the nodes of Ranvier. β subunits may also be present in the dendrites of certain neurons, such as subicular neurons, which have reduced dendritic arborization in a *Scn1b* mutant mouse (Reid et al., 2014). Viewed ultrastructurally, Nav1.6 localizes to presynaptic and postsynaptic membranes in the cortex and cerebellum (Caldwell, Schaller, Lasher, Peles, & Levinson, 2000). It is likely, although it has not been shown, that β subunits associate with the Nav1.6 channel complex at these synapses. The AIS, a specialized neuronal structure at the base of an axon near the soma, is the site of action potential initiation. Within this structure neuronal subtypes have diverse molecular compositions (Yoshimura & Rasband, 2014). In cortical neurons, Nav1.6 is enriched in the distal AIS along with VGKCs, whereas Nav1.2 is localized to the proximal AIS (Hu et al., 2009; Yoshimura & Rasband, 2014). Depending on the subtype, specific VGSC α and β subunits localize at high density to the neuron's AIS along with the cytoskeletal adaptor protein ankyrinG, and CAMs such as β1 binding partners NrCAM and NF186. β2 and β4 localize to the AIS in some neuronal subtypes (Buffington & Rasband, 2013; Chen et al., 2012). Nodes of Ranvier are axonal structures found in the gaps between myelinated segments. They are necessary for rapid, long-distance electrical signaling by means of saltatory conduction. A high density of ion channels and CAMs, including VGSC α and β subunits, are localized to nodes of Ranvier in CNS and PNS. In the CNS, β1 and β2 are localized to optic nerve nodes of Ranvier, with a subset also expressing β4 (Buffington & Rasband, 2013; Kaplan et al., 2001; O'Malley, Shreiner, Chen, Huffnagle, & Isom, 2009; Patino et al., 2009). In the PNS, β1 and β2 localize to nodes of Ranvier in the sciatic nerve, whereas β4 localizes to most nodes of Ranvier in the dorsal and ventral spinal cord roots (Buffington & Rasband, 2013; Chen et al., 2002, 2004). Of note, an important binding partner for β1, the Ig-CAM contactin, is transiently colocalized with VGSCs at newly formed PNS nodes of Ranvier and persistently colocalized at CNS nodes of Ranvier from development to adulthood (Kazarinova-Noyes et al., 2001). During neuronal development, another subcellular structure, the axon growth cone, has been shown to express CAMs, including β1 (Brackenbury et al., 2010). The axon growth cone is the dynamic, growing tip which senses its environment by1 means of cell adhesion and intracellular signaling to accurately direct axon pathfinding (Lowery & Van Vactor, 2009).

3.3 Subcellular localization in cardiomyocytes

VGSC β subunits are differentially localized to specialized subcellular domains of cardiomyocytes. Intercalated disks, formed at junctions between cardiomyocytes, are necessary for mechanical and electrical coupling. Transverse (T) tubules are invaginations of the plasma membrane, linking it to the sarcoplasmic reticulum to allow for efficient Ca^{2+}-induced Ca^{2+} release upon stimulation (Brette & Orchard, 2003). β1 has been shown to localize to both intercalated disks and T-tubules in ventricular cardiomyocytes and to intercalated discs in atrial cardiomyocytes (Kaufmann et al., 2013; Maier et al., 2004; Malhotra et al., 2001, 2004). β2 has been alternately reported to localize to intercalated disks or to T-tubules in ventricular cardiomyocytes (Maier et al., 2004; Malhotra et al., 2001). In the atrium, β2 is localized to T-tubules (Kaufmann et al., 2013). β3 localizes to T-tubules in ventricular cardiomyocytes and to intercalated disks in the atrium (Kaufmann et al., 2013; Maier et al., 2004). β4, by contrast, localizes to intercalated disks in ventricular cardiomyocytes (Maier et al., 2004). β1B is also expressed in atrial and ventricular cardiomyocytes and may be particularly important in junctions between cells (Kazen-Gillespie et al., 2000). Nav1.5 is highly expressed at intercalated disks, and β1B has been shown to regulate these α subunits in heterologous cells, increasing peak Na$^+$ current and negatively shifting the voltage dependence of activation and inactivation (Watanabe et al., 2008).

3.4 Glycosylation

N-linked glycosylation sites of VGSC β subunits are conserved across highly divergent species (Chopra et al., 2007). Approximately one-third of the total β subunit molecular weight is glycosylation. The Ig domain of each of these proteins contains three or four N-linked glycosylation sites (Fig. 2; Isom et al., 1992; Messner & Catterall, 1985). The potential importance of glycosylation in β subunit function has been demonstrated in heterologous systems. In Chinese hamster ovary (CHO) cells lacking the capacity for sialylation, a prominent type of glycosylation in VGSC proteins, β1 is unable to modulate the gating of Nav1.2, Nav1.5, or Nav1.7 (Johnson, Montpetit, Stocker, & Bennett, 2004). In the same type of assay, β2 modulates Nav1.5 in a sialic acid–dependent manner but modulates Nav1.2 even without sialic acid (Johnson & Bennett, 2006). These in vitro studies suggest that glycosylation is likely to be important for the modulation and increased cell surface expression of certain α subunits by their partner β subunits. Notably, myelin P_0, which has homology to β subunit Ig domains, requires glycosylation for its homophilic interactions (Filbin & Tennekoon, 1993). This suggests that β subunit CAM functions may also depend critically on carbohydrate components.

3.5 Phosphorylation

β1 is regulated by phosphorylation of its carboxyl-terminal domain. Basal tyrosine phosphorylation of β1 can be detected in rat brain membranes, and a tyrosine residue at position 181 is phosphorylated in vitro (Fig. 2; Malhotra et al., 2002). In *Drosophila* S2 cells expressing a β1 mutant construct (β1Y181E) which mimics phosphorylation, CAM functions are intact but interaction with ankyrinG is disrupted (Malhotra et al., 2002). β1Y181E increases cell surface expression of α in transfected Chinese hamster lung (CHL) fibroblasts but does not modulate Na^+ current as β1 does (McEwen et al., 2004). In cardiomyocytes, tyrosine-phosphorylated β1 colocalizes with tetrodotoxin (TTX)-resistant Nav1.5 at intercalated disks, whereas nonphosphorylated β1 localizes at T-tubules, where the TTX-sensitive channels, Nav1.1, Nav1.3, and Nav1.6 are known to reside (Malhotra et al., 2004). This suggests that tyrosine phosphorylation may regulate the subcellular localization of β1 and its interactions with other proteins within the cell. A screen for phosphorylated proteins in mouse synaptosomes revealed that β1 is also phosphorylated on threonine residues, but the functional impact of this modification remains unknown (Trinidad et al., 2012). Finally, both α and β1, but not β2, interact with receptor protein tyrosine phosphatase β (RPTPβ), in a developmentally regulated manner resulting in altered VGSC function (Ratcliffe et al., 2000). Data suggest that RPTPβ is a member of the VGSC signaling complex along with β subunits and other Ig-CAMs (Fig. 4; Ratcliffe et al., 2000).

3.6 Proteolytic regulation

The transmembrane VGSC β subunits are targets for sequential proteolytic cleavage by BACE1 and γ-secretase (Fig. 2). First, BACE1 cleaves and releases the Ig-loop—containing extracellular domain, which may function as a soluble CAM (Wong et al., 2005). The β1 construct β1Fc and the secreted β1B subunit have been proposed to function in this way (Davis et al., 2004; McEwen & Isom, 2004; Patino et al., 2011). Following that, γ-secretase cleavage releases the ICD into the cytoplasm (Wong et al., 2005). Pharmacological inhibition of γ-secretase prevents neurite outgrowth mediated by homophilic β1-β1 interactions, suggesting a role for the β1 ICD in signal transduction (Brackenbury & Isom, 2011). The cleaved β2 ICD has been shown to translocate to the nucleus, leading to increased *SCN1A* messenger RNA (mRNA) expression and Nav1.1 protein levels in vitro (Fig. 2; Kim et al., 2007). *Scn1b*-null mice have increased levels of *Scn3a* and *Scn5a* mRNA in the heart (Lin et al., 2015; Lopez-Santiago et al., 2007). This may result from the loss of β1 ICD regulation of *Scn3a* and *Scn5a* transcription in the nucleus. No function is yet known for the β3 and β4 ICDs.

4. β SUBUNITS MODULATE α SUBUNIT LOCALIZATION AND FUNCTION

4.1 Modulation in heterologous cells

Expression of VGSC α and β subunits in heterologous cells has provided much of what we know about the functions and interactions of these proteins. Results in these systems are not always consistent with what is seen in the physiological context of the neuron or cardiomyocyte but have, nevertheless, yielded valuable insights. Early VGSC studies expressing Nav1.2 in *Xenopus* oocytes revealed Na^+ currents that activated and inactivated much more slowly than those recorded in neurons (Auld et al., 1988; Goldin et al., 1986; Joho et al., 1990). The first suggestion that modulatory proteins were necessary to achieve physiological currents came from co-injection of α subunit mRNA with low-molecular weight mRNA isolated from rat brain into oocytes (Auld et al., 1988, Krafte et al., 1990). Recordings from these cells showed increased Na^+ current density, more rapid inactivation, and altered voltage dependence of inactivation, mimicking Na^+ currents recorded from neurons. Because VGSC α subunits purified from rat brain to theoretical homogeneity contained β subunits, these were predicted to be the modulatory proteins encoded in the low-molecular weight mRNA (Hartshorne & Catterall, 1984; Hartshorne et al., 1982). The complementary DNAs (cDNAs) for β1 and β2 were subsequently cloned, and mRNAs encoding these proteins were co-expressed in oocytes along with Nav1.2 (Isom et al., 1992; Isom, Ragsdale, et al., 1995; Isom, Scheuer, et al., 1995). Co-expression of Nav1.2 with β1 revealed increased peak Na^+ current density (2.5-fold), negatively shifted voltage dependence of inactivation (~19 mV) and accelerated inactivation compared with Nav1.2 alone (Isom et al., 1992). Co-expression of Nav1.2 with β2 also led to increased Na^+ current density (1.8- to 2.9-fold) and accelerated channel inactivation. In addition, β2 expression in the absence of α increased the oocyte surface area, likely by stimulating the fusion of intracellular vesicles with the plasma membrane. The largest Na^+ current levels (3.7-fold) were achieved when all three subunits were co-expressed. Most importantly, co-expression of Nav1.2 with both β1 and β2 resulted in rapidly inactivating currents similar to those recorded from neurons (Isom, Ragsdale, et al., 1995; Isom, Scheuer, et al., 1995). β1 also modulates K^+ currents carried by the K^+ channels Kv1.1, Kv1.2, Kv1.3, Kv1.6, Kv4.2, and Kv7.2 when co-expressed in oocytes (Deschenes & Tomaselli., 2002; Nguyen et al., 2012). In this regard, the Kv4.2 channel complex immunoprecipitated from mouse brain contains VGSC β1. Furthermore, knockdown of β1 in isolated cortical neurons resulted in reduced A-type K^+ currents and prolonged action potential waveforms (Marionneau et al., 2012). These experiments show that β subunits are capable of promiscuous alteration of ion channel function.

Mammalian cell lines have been used extensively to study VGSC physiology. These heterologous cells provide a more physiological context than *Xenopus* oocytes, but nevertheless do no fully recapitulate the VGSC properties seen in neurons and cardiomyocytes. Co-expression of Nav1.2 with β1 in CHL cells recapitulates the results seen in oocytes with regard to increased peak current density. β1 also causes negative shifts in the voltage dependence of activation (2−11 mV) and inactivation (10−12 mV) although these effects are smaller than those measured in oocytes (Isom, Ragsdale, et al., 1995; Patino et al., 2009). The modulatory and localization effects exerted by β1 on Nav1.2 in mammalian cells are further amplified by interaction with other Ig-superfamily CAMs (Kazarinova-Noyes et al., 2001; McEwen et al., 2004). For example, when contactin is co-expressed with Nav1.2 and β1 in CHL cells, α subunit surface expression and Na^+ current density are both increased dramatically (~4-fold) over that seen with just Nav1.2 and β1 (Kazarinova-Noyes et al., 2001). Co-expression of NF186 with Nav1.2 and β1 also increases the expression of α subunits on the cell surface beyond that seen with Nav1.2 and β1 alone (McEwen et al., 2004). Co-expression of Nav1.2 with β2 in CHL cells results in Na^+ currents that are reduced or unchanged compared to Nav1.2 expressed alone (Kazarinova-Noyes et al., 2001; McEwen et al., 2004). This result is in disagreement with experiments conducted in oocytes. However, when β1 and β2 are both co-expressed with Nav1.2, β2 increases cell surface expression of α subunits above that seen with Nav1.2 and β1 alone. Thus, at least in this heterologous system, β2 requires expression of β1 to exert its effects (Kazarinova-Noyes et al., 2001). When co-expressed with Nav1.3 in CHO cells, either β1 or β3 induced a negative shift (~10 mV) in the voltage of inactivation without affecting the rate of inactivation. β2 co-expression had no effect on Nav1.3 in these cells (Meadows, Chen, Powell, Clare, & Ragsdale, 2002). In rodent development β3 and Nav1.3 are expressed at their highest levels early, so the interaction of these two subunits could be relevant physiologically (Shah et al., 2001). β3 co-expressed with Nav1.5 in CHO cells, results in a negative shift in the voltage dependence of inactivation and slows the rate of inactivation (Ko et al., 2005). In human embryonic kidney (HEK) cells, co-expression of β4 with either Nav1.2 or Nav1.4 leads to a negative shift (6–7 mV) in the voltage dependence of activation (Yu et al., 2003). Co-expression of β4 with Nav1.1 in HEK cells has the same effect but also increases the amount of noninactivating current (Aman et al., 2009). Co-expression of Nav1.5 and β4 in HEK cells induces a negative shift in the voltage of inactivation (Medeiros-Domingo et al., 2007). Resurgent current, the influx of Na^+ through VGSCs during membrane repolarization in neurons, has been proposed to involve the ICD of β4 as a blocking particle of the open channel (Bean, 2005; Grieco, Malhotra, Chen, Isom, & Raman, 2005). Indeed, in HEK cells a β4 peptide was able to mediate resurgent current when coexpressed with Nav1.7, but not Nav1.6 (Chen et al., 2008; Theile & Cummins, 2011). Co-expression of β1B with Nav1.2 in CHL cells, reveals a strong increase in α subunit surface expression and Na^+ current density with only a subtle effect on channel activation and inactivation (Kazen-Gillespie et al., 2000). By contrast, when co-expressed in CHO cells, β1B has no effect on cell surface localization of Nav1.3 or its Na^+ current density, despite both proteins being preferentially expressed in prenatal brain (Patino et al., 2011). β1B, which is also expressed in heart, interacts in an interesting way with Nav1.5, remaining at the cell surface along with the α subunit and increasing its expression there (Patino et al., 2011; Watanabe et al., 2008). Thus, β1B may function as a soluble CAM in some contexts and, in others, as an α subunit modulator.

4.2 Modulation in native cells

The best experimental representation of VGSC physiology is found by investigating the native channels in neurons and cardiomyocytes. In acutely isolated primary cells and tissue slices, β subunit regulatory effects on α subunit function and surface expression are more subtle than in heterologous systems and differ between cell types. *Scn1b*-null mice lack both β1 and β1B subunits. These mice exhibit spontaneous seizures, ataxic gait, and premature death (Chen et al., 2004). Recordings from P10-18 *Scn1b*-null acutely isolated hippocampal pyramidal (excitatory) neurons and bipolar (inhibitory) neurons showed no differences in Na^+ current properties compared to neurons from wild type animals (Chen et al., 2004; Patino et al., 2009). By contrast, slice recordings from similarly aged *Scn1b*-null brain reveal hyperexcitability in hippocampal CA3 neurons (Patino et al., 2009). Because brain slices preserve intact dendrites and axons, β1 and β1B in these neuronal processes, not just the soma, may be important in regulating the cell's activity. Immunofluorescence experiments show a reduction in Nav1.1 and an increase in Nav1.3 in *Scn1b*-null CA3 neurons relative to wild type neurons (Chen et al., 2004). P16 *Scn1b*-null brain slices display epileptiform activities in cortical and hippocampal extracellular field potential recordings (Brackenbury et al., 2013). These effects may require the presence of neuronal processes since they are not apparent in acutely isolated cells. In slice recordings, cerebellar granule neurons (CGNs) from P12-13 *Scn1b*-null mice have reduced resurgent current compared to wild type controls, but normal transient current. Thus, loss of β1 and β1B results in reduced CGN excitability. These cells also have reduced Nav1.6 at the AIS with concomitant increase in Nav1.1 (Brackenbury et al., 2010). Acutely dissociated *Scn1b*-null DRG neurons, by contrast, are hyperexcitable (Lopez-Santiago et al., 2011). *Scn1b*-null ventricular myocytes have increased Na^+ current, and *Scn1b*-null mice exhibit prolonged QT and RR intervals. These cardiac cells also have increased *Scn5a* mRNA, Nav1.5 protein, and [^{3}H]-saxitoxin (STX) binding to cell surface channels compared to wild type (Lopez-Santiago et al., 2007). Moreover, deletion of *Scn1b* in the heart, results in increased TTX-sensitive Na^+ currents and higher expression of *Scn3a*, encoding Nav1.3 (Lin et al., 2015). Together, these data clearly demonstrate that β1 and β1B are important regulators of α subunit function and subcellular localization, in vivo. Importantly, short hairpin RNA (shRNA)-mediated knockdown of *Scn1b* results in the reduction of K^+ current carried by Kv4 and β1 co-immunoprecipitates with

Kv4.2 from mouse brain (Marionneau et al., 2012). Furthermore, prolonged action potentials and increased repetitive firing measured in *Scn1b*-null cortical pyramidal neurons supports a link between the β1 and β1B subunits and K^+ current (Brackenbury et al., 2013; Marionneau et al., 2012).

Unlike *Scn1b*-null mice, *Scn2b*-null mice do not have seizures or die early but rather have increased susceptibility to pharmacologically induced seizures. Nevertheless, compared to wild type, acutely dissociated hippocampal neurons from *Scn2b*-null mice exhibit a negative shift in the voltage dependence of inactivation and significantly decreased Na^+ current density. Measurement of cell surface [^{3}H]-STX binding in *Scn2b*-null brain neurons is decreased, whereas [^{3}H]-STX binding in *Scn2b*-null brain membranes is unchanged, showing the importance for this β subunit as an α subunit chaperone (Chen et al., 2002). In the PNS, a subset of *Scn2b*-null DRG neurons have reduced Nav1.7 Na^+ current density and slowed rates of activation and inactivation (Lopez-Santiago et al., 2006). Thus, like β1 and β1B, β2 has different roles depending on the cell type in which it is expressed.

Scn3b-null mice have defects in cardiac excitability. Specifically, ventricular and atrial tachycardia and fibrillation occur in stimulated Langendorff-perfused *Scn3b*-null, but rarely wild type, hearts (Hakim et al., 2010, 2008).

Transfection experiments carried out in cultured hippocampal neurons and CGNs suggest that β4 promotes resurgent Na^+ current (Aman et al., 2009; Bant & Raman, 2010; Bean, 2005; Grieco et al., 2005). This is an important adaptation for high-frequency firing exhibited by certain types of neurons, such as Purkinje cells (Burgess et al., 1995, Raman, Sprunger, Meisler, & Bean, 1997). In acutely dissociated hippocampal CA3 neurons, which lack endogenous β4, addition of a peptide consisting of the carboxyl-terminal portion of β4 generates resurgent current (Grieco et al., 2005). Knockdown of β4 expression in CGNs by short interfering RNA (siRNA) reduces resurgent current and decreases repetitive action potential firing (Bant & Raman, 2010). These data implicate the ICD of β4 as an open-channel blocking particle that allows for resurgent current in neurons (Grieco et al., 2005). Evidence suggests that Nav1.6 contributes strongly to resurgent current in Purkinje neurons and DRG neurons (Cummins, Dib-Hajj, Herzog, & Waxman, 2005; Raman & Bean, 1997). Despite this, in HEK cells, co-expression of Nav1.6 and β4 is not sufficient to produce resurgent current (Chen et al., 2008). This suggests that other proteins or cellular processes found in neurons are required for physiological responses. These data support the emerging understanding that β subunits impact neuronal and cardiac physiology in a manner that is dependent on the particular cell type.

5. β SUBUNITS ARE CAMS THAT HAVE ROLES IN BRAIN DEVELOPMENT

5.1 CAM functions

Members of the Ig superfamily are neural recognition molecules involved in developmental functions ranging from axon growth and guidance to synaptogenenesis (Maness & Schachner, 2007). VGSC β subunits each have an extracellular Ig domain with homology to the V-set of Ig-CAMs (Fig. 3; Isom, Ragsdale, et al., 1995). These subunits, especially β1, have been shown experimentally to function as CAMs. *Drosophila* S2 cells transfected to express β1 or β2 form aggregates, showing that these molecules adhere to one another homophilically. Visualized by immunofluorescence, *Drosophila* ankyrin is recruited to sites of contact between these cells. Ankyrin recruitment occurs in the presence of full-length β subunits, but not when truncated β1 or β2 constructs lacking the intracellular carboxyl-terminal domain are expressed (Malhotra et al., 2000). Homophilic binding is disrupted in the epilepsy-linked mutant β1C121W, which lacks a cysteine residue that normally forms a critical disulfide bridge for Ig loop formation (Meadows, Malhotra, et al., 2002; Wallace et al., 1998). This suggests that in human patients, loss of cell adhesive functions may contribute to *SCN1B*-linked pathologies. In vitro experiments have shown that β1 also interacts with other neuronal and glial CAMs (Fig. 2). These include contactin, N-cadherin, neurofascin-155 (NF155), neurofascin-186 (NF186), NrCAM, and VGSC β2 (Malhotra et al., 2004; McEwen & Isom, 2004; Ratcliffe et al., 2001). β1 also interacts with the extracellular matrix protein tenascin-R (Xiao et al., 1999). β2 participates in homophilic as well as heterophilic binding, in vitro, with partners including β1 and the extracellular matrix proteins tenascin-R and tenascin-C (Fig. 2; Malhotra et al., 2000; McEwen & Isom, 2004; Srinivasan et al., 1998; Xiao et al., 1999). Both β1 and β2 regulate migration of transfected CHL cells away from tenascin-R (Xiao et al., 1999). These data were the first evidence that β1 and β2 could function in cell migration. While β3 homophilic binding leading to cellular aggregation was not observed in transfected *Drosophila* S2 experiments, a more recent study reported *trans*-homophilic interactions between β3 subunits expressed in HEK cells (McEwen, Chen, Meadows,

Lopez-Santiago, & Isom, 2009; Yereddi et al., 2013). In contrast to β1, β3 does not associate with either contactin or ankyrinG in transfected CHL cells (McEwen et al., 2009). With the exception of VGSC α subunit proteins, no known binding partners for β4 have been reported. In agreement, β1 and β2, but not β4, influence neurite outgrowth (Davis et al., 2004). β subunit involvement in this process will be described in detail in the following section. Taken together, these data indicate that β subunits are functional adhesion molecules similar to other Ig-CAMs like NCAM and L1.

5.2 Neurite outgrowth

Extensive experimental evidence has been collected showing a role for β1 in neurite outgrowth, at least in CGNs. CGNs extend longer neurites when grown on monolayers of CHL fibroblasts expressing β1 than on mock-transfected CHL cells. Importantly, *Scn1b*-null CGNs do not produce longer neurites in this assay, suggesting that *trans*-homophilic binding is required (Davis et al., 2004). β1-mediated neurite outgrowth also does not occur in CGNs isolated from contactin-null mice. These mice have reduced β1 protein expression, further implying a functional connection between contactin and β1. Interestingly, RPTPβ, which associates with β1, has previously been demonstrated to promote neurite outgrowth through interactions with contactin and NrCAM (Peles et al., 1995; Sakurai et al., 1997). Epidermal growth factor (EGF) receptor and fibroblast growth factor (FGF) receptor signaling have been implicated in L1-mediated adhesion and NCAM-mediated neurite outgrowth, respectively (Islam, Kristiansen, Romani, García-Alonso, & Hortsch, 2004; Niethammer et al., 2002). Investigation of the intracellular signaling cascade responsible for β1-mediated neurite outgrowth shows that EGF and FGF receptor kinase inhibitors fail to reduce the effect, so a separate pathway is likely involved. When grown on β1-expressing monolayers CGNs isolated from *fyn*-null mice do not extend longer neurites (Brackenbury et al., 2008). In detergent resistant membrane fractions solubilized from mouse brain *fyn*, a member of the src family of tyrosine kinases, associates with β1 subunit polypeptides (Brackenbury et al., 2008). Contactin, a glycosylphosphatidylinositol linked CAM, also associates with this membrane fraction (Krämer et al., 1999). These data suggest a model in which β1-β1 *trans*-homophilic binding, combined with β1-contactin heterophilic interactions in *cis*, within lipid rafts results in transmembrane signaling to induce neurite extension via a mechanism involving *fyn* kinase (Brackenbury & Isom, 2011). The model of the VGSC complex shown in Fig. 4 includes these and other important interacting partners. β1-mediated neurite outgrowth is abrogated in the absence of Nav1.6 or following TTX treatment. Thus for neurite outgrowth, neuronal activity seems to be a requirement. Significantly, TTX treatment has no inhibitory effect on FGF-mediated neurite outgrowth, further implicating a separate signaling cascade (Brackenbury et al., 2010). Proteolytic processing of β1 is also important, as pharmacological inhibition of γ-secretase activity prevents β1-mediated neurite outgrowth (Brackenbury & Isom, 2011). In cultured breast cancer cells that express β1, *trans*-homophilic interaction with β1-expressing CHL cells promotes the outgrowth of neurite-like extensions (Nelson et al., 2014). Thus, β1 may have developmental functions outside of the nervous system. β1B-expressing CHL cells induce neurite outgrowth to the same degree as those expressing β1 (Patino et al., 2011). Thus, it is proposed that either β1 or the secreted splice variant β1B can interact with neuronal β1 and its associated complex to induce neurite extension. By contrast, β2-expressing CHL monolayers reduce neurite outgrowth of CGNs, and β4 subunits appear to have no effect on neurite outgrowth in this assay (Davis et al., 2004).

5.3 *Scn1b* in CNS development

From an early postnatal stage, the *Scn1b*-null mouse CNS exhibits significant morphologic abnormalities, suggesting roles for β1 and β1B in brain development (Brackenbury et al., 2008, 2013). Immunohistochemical analysis showed fewer nodes of Ranvier in the optic nerve. This defect corresponds with reduced optic nerve conduction velocity. Analysis at the ultrastructural level detected abnormalities in nodal architecture in the spinal cord, sciatic nerve, and optic nerve (Chen et al., 2004). At P14, the *Scn1b*-null cerebellar cortex exhibited defects when visualized by immunofluorescence. The external germinal layer (EGL), where CGNs first appear and proliferate, has increased thickness and increased total number of 4′,6-diamidino-2-phenylindole—positive cells compared to wild type littermates. By contrast, the number of recently divided, 5-bromo-2′-deoxyuridine—labeled cells was not different. These data suggest that *Scn1b*-null CGNs have slowed migration from the EGL inward through the cerebellar cortex to populate the granule cell layer. Axon pathfinding and parallel fiber fasciculation of CGNs are also abnormal in *Scn1b*-null cerebellum. In the normal mouse cerebellar cortex, these axons extend from the granule layer outward through the molecular layer (ML) where they bifurcate and form the tightly bundled parallel fibers. In the mutant mouse, some of these axons in the ML deviate from their normal trajectory. Most noticeably, the parallel fibers are highly defasciculated (Brackenbury et al., 2008). It is noteworthy that contactin-null mice, which are *Scn1b* hypomorphs, also have cerebellar microorganization defects, ataxia, and early death, further suggesting a link between β1 and this Ig-CAM (Berglund et al., 1999; Brackenbury et al., 2010). CNS defects are already apparent in *Scn1b*-null cerebellar cortex at P5, prior to the onset of behavioral seizures and elevated cFos expression in the cortex and hippocampus (Brackenbury et al., 2013). CGNs migrate and extend their axons along the processes of astrocytes known as Bergmann glia, which express β1, in vivo (Brackenbury et al., 2008). These cells may act as partners for β1 *trans*-homophilic binding in vivo, similar to β1-expressing CHL cells, which promote neurite outgrowth in cultured CGNs in vitro. Brain defects are not limited to the cerebellum of *Scn1b*-null mice. DiI-labeled *Scn1b*-null axons of the corticospinal tract exhibit abnormal pathfinding and defasciculation (Brackenbury et al., 2008). At P5, the *Scn1b*-null hippocampus has several abnormal features as detected by immunohistochemistry. Mutant hippocampus shows ectopic Prox1-expressing cells and increased cellular proliferation in the hilus of the dentate gyrus (DG). In addition, the *Scn1b*-null DG shows dispersion compared to wild type. The microorganization of interneurons is unusual in the DG of these mice, as well, with parvalbumin-labeled cell bodies being spread laterally rather than being tightly arranged. Axon outgrowth and pathfinding are abnormal in the *Scn1b*-null hippocampus compared to wild type (Brackenbury et al., 2013). Finally, pyramidal neurons in the *Scn1b*-null subiculum have reduced dendritic arborization (Reid et al., 2014). Zebrafish *scn1bb* morphants also have defective axon pathfinding in the olfactory nerve (Fein et al., 2008). This finding, together with the defects observed across the CNS and PNS of early postnatal *Scn1b*-null mice, and the seizures, ataxia, and early death that follow, highlight the importance of β1 and β1B for normal neurological development. Thus, it is not surprising that human mutations in *SCN1B* have disastrous, sometimes fatal consequences.

6. β SUBUNIT GENE MUTATIONS ARE LINKED TO EPILEPSY AND CARDIAC ARRHYTHMIA

6.1 Epilepsy

VGSC β subunit gene mutations are linked to epilepsy syndromes, cardiac arrhythmias, and other pathological conditions. Epilepsy is a neurological disorder characterized by recurrent, unprovoked seizures, and events of excessive hypersynchronous action potential firing in one or more brain regions. They can have many causes, but a large percentage of cases are idiopathic and likely genetic (Stafstrom & Carmant, 2015). Channelopathies, including mutations in VGSC α subunits, are responsible for many of these idiopathic epilepsy syndromes. Human mutations in *SCN1B* have been linked to generalized epilepsy with febrile seizures plus (GEFS+) and the severe pediatric epileptic encephalopathy Dravet syndrome (DS) (Audenaert et al., 2003; Escayg & Goldin, 2010; Ogiwara et al., 2012; Patino et al., 2009; Wallace et al., 2002, 1998). In DS, children in the first years of life present with frequent febrile seizures followed by afebrile seizures, cognitive decline, ataxia, and high risk of sudden unexpected death due to epilepsy (SUDEP). Apnea and cardiac arrhythmia have been proposed to be significant causes of SUDEP (Surges & Sander, 2012). *SCN1A* mutations resulting in haploinsufficiency have been determined to be responsible for most incidences of DS (Claes et al., 2001; Escayg et al., 2000; Meisler & Kearney, 2005; Shi et al., 2009; Sugawara et al., 2001). However, homozygous *SCN1B* mutations have been found in at least two patients with DS, with one of these shown to be loss-of-function (Ogiwara et al., 2012; Patino et al., 2009). In addition to GEFS+, other epilepsy syndromes, including TLE and early-onset absence epilepsy have been linked to human mutations in *SCN1B* (Audenaert et al., 2003; Scheffer et al., 2007). A β1B-specific *SCN1B* mutation, the trafficking-deficient G257R, has been discovered in two separate cases of idiopathic epilepsy (Patino et al., 2011). However, most of the epilepsy-linked mutations in *SCN1B* map to extracellular residues in the Ig domain shared by both β1 and β1B (Brackenbury & Isom, 2011). For example, the first *SCN1B* mutation to be associated with GEFS+ was the β1 (C121W) mutation that replaces one of the cysteine residues in the disulfide bridge that holds together the Ig-loop domain (Wallace et al., 1998). Further epilepsy-linked mutations in *SCN1B* include R85C, R125C, and many others (O'Malley & Isom, 2015).

β1 and β1B are multifunctional proteins with varying effects on VGSC physiology depending on developmental and cell type−specific contexts. Impaired regulation of Na^+ current likely contributes to the mechanism of *SCN1B*-linked epilepsy. In addition, all *SCN1B*-linked epilepsy mutations identified to date, except for the one specific to β1B, are located in residues of the extracellular Ig-loop (Audenaert et al., 2003; Meadows, Malhotra, et al., 2002; Patino et al., 2009; Scheffer et al., 2007; Wallace et al., 1998). Given the strength and quantity of experimental evidence demonstrating adhesive functions for β1/β1B, it is reasonable to suggest that loss of these functions might be an important contribution to the CNS neuronal network hyperexcitability underlying *SCN1B*-linked seizures and epilepsy.

6.2 Cardiac arrhythmia

Channelopathies, including mutations in VGSC α subunits, especially Nav1.5, have also been linked to cardiac arrhythmias. Such mutations are linked to long QT syndrome (LQTS), Brugada syndrome, and fatal ventricular fibrillation (VF) (Remme & Bezzina, 2010). Human mutations in *SCN1B*, *SCN2B*, and *SCN3B* have all been associated with Brugada syndrome, which carries a high risk of sudden cardiac death due to VF (Ishikawa & Takahashi, 2012; Riuró et al., 2013; Watanabe et al., 2008). Other mutations in *SCN1B*, *SCN2B*, *SCN3B*, and *SCN4B* have been linked to atrial fibrillation (Li et al., 2013; Olesen et al., 2011; Watanabe et al., 2009). Further, a mutation in *SCN4B* has been linked to LQTS, in which slowed repolarization can lead to arrhythmia and VF (Medeiros-Domingo et al., 2007). These findings suggest that all VGSC β subunits are important for healthy cardiac function in humans.

In addition to epilepsy and cardiac arrhythmias, β subunits have been implicated in a number of other pathological conditions, including sudden infant death syndrome, neuropathic pain, neurodegenerative diseases such as Huntington's disease and multiple sclerosis, and even cancer (O'Malley et al., 2009). Recent reviews provide a detailed treatment of the roles β subunits have in pathophysiology (Calhoun & Isom, 2014; O'Malley & Isom, 2015).

7. CONCLUDING REMARKS

Descendent from and sharing structural homology with a family of proteins possessing V-like Ig domains and having diverse developmental functions in the nervous system, β1, β2, β3, β4, and β1B might be thought of as Ig-superfamily CAM proteins which, in excitable cells, also regulate ion channel function. Indeed, the VGSC β subunits are expressed in nonexcitable cells like astrocytes, oligodendrocyte precursor cells, and radial glia (Cahoy et al., 2008). Like their cousins, the Ig-superfamily CAMs contactin, L1, and NCAM, these proteins, especially β1/β1B, are critical for development of the CNS. They localize in neurons at specialized intracellular structures like the AIS and nodes of Ranvier, where they link other CAMs and extracellular matrix molecules with appropriate VGSC α subunits, VGKCs, and cytoskeletal proteins. The importance of β subunits, especially those encoded by *Scn1b*, in this role is highlighted by the catastrophic pathological conditions that can occur when they are mutated. *Scn1b*-null mice, like contactin-null mice, exhibit ataxia and altered cerebellar microorganization, and both mutants have radically shortened lives. Ongoing research seeks to determine whether β1 and β1B participate in other developmental processes common to Ig-superfamily CAMs, such as synaptogenesis. Most importantly, *SCN1B* mutations are linked to serious human pathologies, including epileptic encephalopathy. Because these syndromes, especially DS, are usually caused by *SCN1A* mutations, there is a tendency to presume that *SCN1A*- and *SCN1B*-linked epileptic encephalopathies have the same underlying mechanism. However, we do not yet know whether loss of crucial adhesive functions of β1 and β1B contributes to the seizures, ataxia, and severe cognitive impairment of *SCN1B*-linked DS. VGSC β subunits are multifunctional, regulating VGSC α subunit transcription, localization, and function and contributing to key aspects of CNS development as CAMs. It might, therefore, be predicted that loss of function of *SCN1B* leads to even more severe epileptic encephalopathy than *SCN1A*-linked DS. Additional cases will need to be evaluated before such a determination can be made.

REFERENCES

Adams, M. D., Celniker, S. E., Holt, R. A., Evans, C. A., Gocayne, J. D., Amanatides, P. G., … Venter, J. C. (2000). The genome sequence of *Drosophila melanogaster*. *Science (New York, NY), 287*(5461), 2185–2195. http://doi.org/10.1126/science.287.5461.2185.

Alföldi, J., Di Palma, F., Grabherr, M., Williams, C., Kong, L., Mauceli, E., … Lindblad-Toh, K. (2011). The genome of the green anole lizard and a comparative analysis with birds and mammals. *Nature, 477*(7366), 587–591. http://doi.org/10.1038/nature10390.

Aman, T. K., Grieco-Calub, T. M., Chen, C., Rusconi, R., Slat, E. A., Isom, L. L., & Raman, I. M. (2009). Regulation of persistent Na current by interactions between beta subunits of voltage-gated Na channels. *Journal of Neuroscience : The Official Journal of the Society for Neuroscience, 29*(7), 2027–2042. http://doi.org/10.1523/JNEUROSCI.4531-08.2009.

Amemiya, C. T., Alföldi, J., Lee, A. P., Fan, S., Philippe, H., Maccallum, I., … Lindblad-Toh, K. (2013). The African coelacanth genome provides insights into tetrapod evolution. *Nature, 496*(7445), 311–316. http://doi.org/10.1038/nature12027.

Andrikopoulos, P., Fraser, S. P., Patterson, L., Ahmad, Z., Burcu, H., Ottaviani, D., … Djamgoz, M. B. A. (2011). Angiogenic functions of voltage-gated Na^+ channels in human endothelial cells: modulation of vascular endothelial growth factor (VEGF) signaling. *Journal of Biological Chemistry, 286*(19), 16846–16860. http://doi.org/10.1074/jbc.M110.187559.

Aronica, E., Troost, D., Rozemuller, A. J., Yankaya, B., Jansen, G. H., Isom, L. L., & Gorter, J. A. (2003). Expression and regulation of voltage-gated sodium channel beta1 subunit protein in human gliosis-associated pathologies. *Acta Neuropathologica, 105*(5), 515–523. http://doi.org/10.1007/s00401-003-0677-2.

Audenaert, D., Claes, L., Ceulemans, B., Löfgren, A., Van Broeckhoven, C., & De Jonghe, P. (2003). A deletion in SCN1B is associated with febrile seizures and early-onset absence epilepsy. *Neurology, 61*(6), 854–856. http://doi.org/10.1212/01.WNL.0000080362.55784.1C.

Auld, V. J., Goldin, A. L., Krafte, D. S., Marshall, J., Dunn, J. M., Catterall, W. A., ... Dunn, R. J. (1988). A rat brain Na^+ channel alpha subunit with novel gating properties. *Neuron, 1*(6), 449–461. http://dx.doi.org/10.1016/0896-6273(88)90176-6.

Azumi, K., De Santis, R., De Tomaso, A., Rigoutsos, I., Yoshizaki, F., Pinto, M. R., ... Nonaka, M. (2003). Genomic analysis of immunity in a Urochordate and the emergence of the vertebrate immune system: "waiting for Godot." *Immunogenetics, 55*(8), 570–581. http://doi.org/10.1007/s00251-003-0606-5.

Bant, J. S., & Raman, I. M. (2010). Control of transient, resurgent, and persistent current by open-channel block by Na channel beta4 in cultured cerebellar granule neurons. *Proceedings of the National Academy of Sciences of the United States of America, 107*(27), 12357–12362. http://doi.org/10.1073/pnas.1005633107.

Bean, B. P. (2005). Progress in realizing the promise of microarrays in systems neurobiology. *Neuron, 45*(2), 183–185. http://doi.org/10.1016/j.neuron.2005.01.007.

Beneski, D. A., & Catterall, W. A. (1980). Covalent labeling of protein components of the sodium channel with a photoactivable derivative of scorpion toxin. *Proceedings of the National Academy of Sciences of the United States of America,* 77(1), 639–643. http://doi.org/10.1073/pnas.77.1.639.

Bennett, P. B., Makita, N., & George, A. L. (1993). A molecular basis for gating mode transitions in human skeletal muscle Na^+ channels. *FEBS Letters, 326*(1–3), 21–24. http://dx.doi.org/10.1016/0014-5793(93)81752-L.

Berglund, E. O., Murai, K. K., Fredette, B., Sekerková, G., Marturano, B., Weber, L., ... Ranscht, B. (1999). Ataxia and abnormal cerebellar microorganization in mice with ablated contactin gene expression. *Neuron, 24*(3), 739–750. http://dx.doi.org/10.1016/S0896-6273(00)81126-5.

Brackenbury, W. J., Calhoun, J. D., Chen, C., Miyazaki, H., Nukina, N., Oyama, F., ... Isom, L. L. (2010). Functional reciprocity between Na^+ channel Nav1.6 and beta1 subunits in the coordinated regulation of excitability and neurite outgrowth. *Proceedings of the National Academy of Sciences of the United States of America, 107*(5), 2283–2288. http://doi.org/10.1073/pnas.0909434107.

Brackenbury, W. J., Davis, T. H., Chen, C., Slat, E. A., Detrow, M. J., Dickendesher, T. L., ... Isom, L. L. (2008). Voltage-gated Na^+ channel beta1 subunit-mediated neurite outgrowth requires Fyn kinase and contributes to postnatal CNS development in vivo. *Journal of Neuroscience : The Official Journal of the Society for Neuroscience, 28*(12), 3246–3256. http://doi.org/10.1523/JNEUROSCI.5446-07.2008.

Brackenbury, W. J., & Isom, L. L. (September 2011). Na channel β subunits: overachievers of the ion channel family. *Frontiers in Pharmacology, 2*, 53. http://doi.org/10.3389/fphar.2011.00053.

Brackenbury, W. J., Yuan, Y., O'Malley, H. A., Parent, J. M., & Isom, L. L. (2013). Abnormal neuronal patterning occurs during early postnatal brain development of Scn1b-null mice and precedes hyperexcitability. *Proceedings of the National Academy of Sciences of the United States of America, 110*(3), 1089–1094. http://doi.org/10.1073/pnas.1208767110.

Brette, F., & Orchard, C. (2003). T-tubule function in mammalian cardiac myocytes. *Circulation Research, 92*(11), 1182–1192. http://doi.org/10.1161/01.RES.0000074908.17214.FD.

Buffington, S. A., & Rasband, M. N. (2013). Na^+ channel-dependent recruitment of Navβ4 to axon initial segments and nodes of Ranvier. *Journal of Neuroscience : The Official Journal of the Society for Neuroscience, 33*(14), 6191–6202. http://doi.org/10.1523/JNEUROSCI.4051-12.2013.

Burgess, D. L., Kohrman, D. C., Galt, J., Plummer, N. W., Jones, J. M., Spear, B., & Meisler, M. H. (1995). Mutation of a new sodium channel gene, Scn8a, in the mouse mutant "motor endplate disease." *Nature Genetics, 10*, 196–201.

Cahoy, J. D., Emery, B., Kaushal, A., Foo, L. C., Zamanian, J. L., Christopherson, K. S., … Barres, B. A. (2008). A transcriptome database for astrocytes, neurons, and oligodendrocytes: a new resource for understanding brain development and function. *Journal of Neuroscience, 28*(1), 264–278. http://doi.org/10.1523/JNEUROSCI.4178-07.2008.

Caldwell, J. H., Schaller, K. L., Lasher, R. S., Peles, E., & Levinson, S. R. (2000). Sodium channel Na(v)1.6 is localized at nodes of ranvier, dendrites, and synapses. *Proceedings of the National Academy of Sciences of the United States of America, 97*(10), 5616–5620. http://doi.org/10.1073/pnas.090034797.

Calhoun, J. D., & Isom, L. L. (2014). The role of non-pore-forming b subunits in physiology and pathophysiology of voltage-gated sodium channels. In *Voltage-gated sodium channels* (Vol. 221, pp. 111–135). http://doi.org/10.1007/978-3-642-41588-3.

Catterall, W. A. (2000). From ionic currents to molecular mechanisms: the structure and function of voltage-gated sodium channels. *Neuron, 26*(1), 13–25. http://dx.doi.org/10.1016/s0896-6273(00)81133-2.

Chen, C., Bharucha, V., Chen, Y., Westenbroek, R. E., Brown, A., Malhotra, J. D., … Isom, L. L. (2002). Reduced sodium channel density, altered voltage dependence of inactivation, and increased susceptibility to seizures in mice lacking sodium channel beta 2-subunits. *Proceedings of the National Academy of Sciences of the United States of America, 99*(26), 17072–17077. http://doi.org/10.1073/pnas.212638099.

Chen, C., Calhoun, J. D., Zhang, Y., Lopez-Santiago, L., Zhou, N., Davis, T. H., … Isom, L. L. (2012). Identification of the cysteine residue responsible for disulfide linkage of Na^+ channel α and β2 subunits. *Journal of Biological Chemistry, 287*(46), 39061–39069. http://doi.org/10.1074/jbc.M112.397646.

Chen, C., Westenbroek, R. E., Xu, X., Edwards, C. A., Sorenson, D. R., Chen, Y., … Isom, L. L. (2004). Mice lacking sodium channel beta1 subunits display defects in neuronal excitability, sodium channel expression, and nodal architecture. *Journal of Neuroscience : The Official Journal of the Society for Neuroscience, 24*(16), 4030–4042. http://doi.org/10.1523/JNEUROSCI.4139-03.2004.

Chen, Y., Yu, F. H., Sharp, E. M., Beacham, D., Scheuer, T., & Catterall, W. A. (2008). Functional properties and differential neuromodulation of Nav1.6 channels. *Molecular and Cellular Neurosciences, 38*(4), 607–615. http://doi.org/10.1016/j.mcn.2008.05.009.

Chioni, A.-M., Brackenbury, W. J., Calhoun, J. D., Isom, L. L., & Djamgoz, M. B. A. (2009). A novel adhesion molecule in human breast cancer cells: voltage-gated Na^+ channel beta1 subunit. *International Journal of Biochemistry & Cell Biology, 41*(5), 1216–1227. http://doi.org/10.1016/j.biocel.2008.11.001.

Chopra, S. S., Watanabe, H., Zhong, T. P., & Roden, D. M. (2007). Molecular cloning and analysis of zebrafish voltage-gated sodium channel beta subunit genes: implications for the evolution of electrical signaling in vertebrates. *BMC Evolutionary Biology, 7*, 113. http://doi.org/10.1186/1471-2148-7-113.

Claes, L., Del-Favero, J., Ceulemans, B., Lagae, L., Van Broeckhoven, C., & De Jonghe, P. (2001). De novo mutations in the sodium-channel gene SCN1A cause severe myoclonic epilepsy of infancy. *American Journal of Human Genetics, 68*(6), 1327–1332. http://doi.org/10.1086/320609.

Crossin, K. L., & Krushel, L. A. (January 2000). Cellular signaling by neural cell adhesion molecules of the immunoglobulin superfamily. *Developmental Dynamics, 279*, 260–279.

Cummins, T. R., Dib-Hajj, S. D., Herzog, R. I., & Waxman, S. G. (2005). Nav1.6 channels generate resurgent sodium currents in spinal sensory neurons. *FEBS Letters, 579*(10), 2166–2170. http://doi.org/10.1016/j.febslet.2005.03.009.

Davis, T. H., Chen, C., & Isom, L. L. (2004). Sodium channel beta1 subunits promote neurite outgrowth in cerebellar granule neurons. *Journal of Biological Chemistry, 279*(49), 51424–51432. http://doi.org/10.1074/jbc.M410830200.

Deschenes, I., & Tomaselli, G. F. (2002). Modulation of Kv4. 3 current by accessory subunits. *FEBS Letters, 528*, 183–188.

Diss, J. K. J., Fraser, S. P., Walker, M. M., Patel, A., Latchman, D. S., & Djamgoz, M. B. A. (2008). Beta-subunits of voltage-gated sodium channels in human prostate cancer: quantitative in vitro and in vivo analyses of mRNA expression. *Prostate Cancer and Prostatic Diseases, 11*(4), 325–333. http://doi.org/10.1038/sj.pcan.4501012.

Du Pasquier, L., Zucchetti, I., & De Santis, R. (2004). Immunoglobulin superfamily receptors in protochordates: before RAG time. *Immunological Reviews, 198*, 233–248. http://doi.org/10.1111/j.0105-2896.2004.00122.x.

Escayg, A., & Goldin, A. L. (2010). Sodium channel SCN1A and epilepsy: mutations and mechanisms. *Epilepsia, 51*(9), 1650–1658. Retrieved from: http://www.pubmedcentral.nih.gov/articlerender.fcgi?artid=2937162&tool=pmcentrez&rendertype=abstract.

Escayg, A., MacDonald, B., Meisler, M., Baulac, S., Huberfeld, G., An-Gourfinkel, I., … Malafosse, A. (2000). Mutations of SCN1A, encoding a neuronal sodium channel, in two families with GEFS+2. *Nature Genetics, 24*(4), 343–345. http://doi.org/10.1111/j.1095-8649.2006.01157.x.

Fein, A. J., Meadows, L. S., Chen, C., Slat, E. A., & Isom, L. L. (2007). Cloning and expression of a zebrafish SCN1B ortholog and identification of a species-specific splice variant. *BMC Genomics, 8*, 226. http://doi.org/10.1186/1471-2164-8-226.

Fein, A. J., Wright, M. A., Slat, E. A., Ribera, A. B., & Isom, L. L. (2008). scn1bb, a zebrafish ortholog of SCN1B expressed in excitable and nonexcitable cells, affects motor neuron axon morphology and touch sensitivity. *Journal of Neuroscience : The Official Journal of the Society for Neuroscience, 28*(47), 12510–12522. http://doi.org/10.1523/JNEUROSCI.4329-08.2008.

Filbin, M. T., & Tennekoon, G. I. (1993). Homophilic adhesion of the myelin Po protein requires glycosylation of both molecules in the homophilic pair. *Journal of Cell Biology, 122*(2), 451–459. http://doi.org/10.1083/jcb.122.2.451.

Gallant, J. R., Traeger, L. L., Volkening, J. D., Moffett, H., Chen, P., Novina, C. D., … Albert, J. S. (2014). Genomic basis for the convergent evolution of electric organs. *Science, 344*(6191), 1522–1525.

Gibbs, R. A., Rogers, J., Katze, M. G., Bumgarner, R., Weinstock, G. M., Mardis, E. R., … Zwieg, A. S. (April 2007). Evolutionary and biomedical insights. *Science*, 222–234.

Gilchrist, J., Das, S., Van Petegem, F., & Bosmans, F. (2013). Crystallographic insights into sodium-channel modulation by the β4 subunit. *Proceedings of the National Academy of Sciences of the United States of America, 110*(51), E5016–E5024. http://doi.org/10.1073/pnas.1314557110.

Goldin, A. L. (2001). Resurgence of sodium channel research. *Annual Review of Physiology, 63*, 871–894. http://doi.org/10.1146/annurev.physiol.63.1.871/r63/1/871.

Goldin, A. L., Snutch, T., Lübbert, H., Dowsett, A., Marshall, J., Auld, V., … Dunn, R. (1986). Messenger RNA coding for only the alpha subunit of the rat brain Na channel is sufficient for expression of functional channels in *Xenopus* oocytes. *Proceedings of the National Academy of Sciences of the United States of America, 83*(19), 7503–7507. http://doi.org/10.1073/pnas.83.19.7503.

Gorter, J. A., van Vliet, E. A., da Silva, F. H. L., Isom, L. L., & Aronica, E. (2002). Sodium channel beta1-subunit expression is increased in reactive astrocytes in a rat model for mesial temporal lobe epilepsy. *European Journal of Neuroscience, 16*(2), 360–364. http://doi.org/10.1046/j.1460-9568.2002.02078.x.

Grieco, T. M., Malhotra, J. D., Chen, C., Isom, L. L., & Raman, I. M. (2005). Open-channel block by the cytoplasmic tail of sodium channel β4 as a mechanism for resurgent sodium current. *Neuron, 45*(2), 233–244. http://doi.org/10.1016/j.neuron.2004.12.035.

Hakim, P., Brice, N., Thresher, R., Lawrence, J., Zhang, Y., Jackson, A. P., … Huang, C. L. H. (2010). Scn3b knockout mice exhibit abnormal sino-atrial and cardiac conduction properties. *Acta Physiologica, 198*(1), 47–59. http://doi.org/10.1111/j.1748-1716.2009.02048.x.

Hakim, P., Gurung, I. S., Pedersen, T. H., Thresher, R., Brice, N., Lawrence, J., … Huang, C. L. H. (2008). Scn3b knockout mice exhibit abnormal ventricular electrophysiological properties. *Progress in Biophysics and Molecular Biology, 98*(2–3), 251–266. http://doi.org/10.1016/j.pbiomolbio.2009.01.005.

Hartshorne, R. P., & Catterall, W. A. (1981). Purification of the saxitoxin receptor of the sodium channel from rat brain. *Proceedings of the National Academy of Sciences of the United States of America, 78*(7), 4620–4624. http://doi.org/10.1073/pnas.78.7.4620.

Hartshorne, R. P., & Catterall, W. A. (1984). The sodium channel from rat brain: purification and subunit composition. *Journal of Biological Chemistry, 259*(3), 1667–1675.

Hartshorne, R. P., Messner, D. J., Coppersmith, J. C., & Catterall, W. A. (1982). The saxitoxin receptor of the sodium channel from rat brain. *Journal of Biological Chemistry*, 1667–1675.

Hellsten, U., Harland, R. M., Gilchrist, M. J., Hendrix, D., Jurka, J., Blitz, I. L., … Amaya, E. (April 2010). The genome of the western clawed frog *Xenopus tropicalis*. *Science, 328*, 633–636.

Hillier, L. W., Miller, W., Birney, E., Warren, W., & Hardison, R. C. (2004). Sequence and comparative analysis of the chicken genome provide unique perspectives on vertebrate evolution. *Nature, 432*(7018), 695–716. http://doi.org/10.1038/nature03394.

Hodgkin, A. L., & Huxley, A. F. (1952). A quantitative description of membrane current and its application to conduction in nerve. *Journal of Physiology (Paris)*, 500–544.

Hu, W., Tian, C., Li, T., Yang, M., Hou, H., & Shu, Y. (2009). Distinct contributions of Na(v)1.6 and Na(v)1.2 in action potential initiation and backpropagation. *Nature Neuroscience, 12*(8), 996–1002. http://doi.org/10.1038/nn.2359.

Ishikawa, T., & Takahashi, N. (2012). Novel SCN3B mutation associated with Brugada syndrome affects intracellular trafficking and function of Nav1. 5. *Circulation Journal Official Journal of the Japanese Circulation Society*, 77(4), 959–967. http://doi.org/10.1253/circj.CJ-12-0995.

Islam, R., Kristiansen, L. V., Romani, S., Garcia-Alonso, L., & Hortsch, M. (2004). Activation of EGF receptor kinase by L1-mediated homophilic cell interactions. *Molecular Biology of the Cell, 15*(4), 2003–2012. http://doi.org/10.1091/mbc.E03.

Isom, L. L. (2001). Sodium channel beta subunits: anything but auxiliary. *Neuroscientist : A Review Journal Bringing Neurobiology, Neurology and Psychiatry*, 7(1), 42–54. http://doi.org/10.1177/107385840100700108.

Isom, L. L., & Catterall, W. A. (1996). Na^+ channel subunits and Ig domains. *Nature, 383*, 307–308.

Isom, L. L., De Jongh, K. S., Patton, D. E., Reber, B. F., Offord, J., Charbonneau, H., … Catterall, W. A. (1992). Primary structure and functional expression of the beta 1 subunit of the rat brain sodium channel. *Science, 256*(5058), 839–842. Retrieved from: http://www.ncbi.nlm.nih.gov/pubmed/1375395.

Isom, L. L., Ragsdale, D. S., De Jongh, K. S., Westenbroek, R. E., Reber, B. F., Scheuer, T., & Catterall, W. A. (1995). Structure and function of the beta 2 subunit of brain sodium channels, a transmembrane glycoprotein with a CAM motif. *Cell, 83*(3), 433–442. Retrieved from: http://www.ncbi.nlm.nih.gov/pubmed/8521473.

Isom, L. L., Scheuer, T., Brownstein, A. B., Ragsdale, D. S., Murphy, B. J., & Catterall, W. A. (1995). Functional co-expresson of the b1 and type IIA a subunits of sodium channels in a mammalian cell line. *Journal of Biological Chemistry, 270*(7), 3306–3312.

Johnson, D., & Bennett, E. S. (2006). Isoform-specific effects of the b2 subunit on voltage-gated sodium channel gating. *Journal of Biological Chemistry, 281*(36), 25875–25881. http://doi.org/10.1074/jbc.M605060200.

Johnson, D., Montpetit, M. L., Stocker, P. J., & Bennett, E. S. (2004). The sialic acid component of the b1 subunit modulates voltage-gated sodium channel function. *Journal of Biological Chemistry, 279*(43), 44303–44310. http://doi.org/10.1074/jbc.M408900200.

Joho, R. H., Moorman, J. R., VanDongen, A. M., Kirsch, G. E., Silberberg, H., Schuster, G., & Brown, A. M. (1990). Toxin and kinetic profile of rat brain type III sodium channels expressed in *Xenopus* oocytes. *Brain Research. Molecular Brain Research, 7*(2), 105–113.

Kaplan, M., Cho, M., Ullian, E., & Isom, L. (2001). Differential control of clustering of the sodium channels Na(v) 1.2 and Na(v)1.6 at developing CNS nodes of Ranvier. *Neuron, 30*(1), 105–119. http://dx.doi.org/10.1016/S0896-6273(01)00266-5.

Kaufmann, S. G., Westenbroek, R. E., Maass, A. H., Lange, V., Renner, A., Wischmeyer, E., … Maier, S. K. G. (2013). Distribution and function of sodium channel subtypes in human atrial myocardium. *Journal of Molecular and Cellular Cardiology, 61*, 133–141. http://doi.org/10.1016/j.yjmcc.2013.05.006.

Kazarinova-Noyes, K., Malhotra, J. D., McEwen, D. P., Mattei, L. N., Berglund, E. O., Ranscht, B., … Xiao, Z. C. (2001). Contactin associates with Na^+ channels and increases their functional expression. *Journal of Neuroscience : The Official Journal of the Society for Neuroscience, 21*(19), 7517–7525. Retrieved from: http://www.ncbi.nlm.nih.gov/pubmed/11567041.

Kazen-Gillespie, K. A., Ragsdale, D. S., D'Andrea, M. R., Mattei, L. N., Rogers, K. E., & Isom, L. L. (2000). Cloning, localization, and functional expression of sodium channel beta1A subunits. *Journal of Biological Chemistry, 275*(2), 1079–1088. http://doi.org/10.1074/jbc.275.2.1079.

Kim, D. Y., Carey, B. W., Wang, H., Ingano, L. A. M., Binshtok, A. M., Wertz, M. H., … Kovacs, D. M. (2007). BACE1 regulates voltage-gated sodium channels and neuronal activity. *Nature Cell Biology, 9*(7), 755–764. http://doi.org/10.1038/ncb1602.

Ko, S. H., Lenkowski, P. W., Lee, H. C., Mounsey, J. P., & Patel, M. K. (2005). Modulation of Nav1.5 by β1- and β3-subunit co-expression in mammalian cells. *Pflugers Archiv European Journal of Physiology, 449*(4), 403–412. http://doi.org/10.1007/s00424-004-1348-4.

Krafte, D. S., Goldin, A. L., Auld, V. J., Dunn, R. J., Davidson, N., & Lester, H. A. (1990). Inactivation of cloned Na channels expressed in *Xenopus* oocytes. *Journal of General Physiology, 96*(1), 689–706. http://doi.org/10.1085/jgp.96.4.689.

Krämer, E., Klein, C., Boytinck, M., Kra, E., Koch, T., & Trotter, J. (1999). Compartmentation of fyn kinase with molecules in oligodendrocytes facilitates kinase activation during myelination. *Journal of Biological Chemistry, 274*(41), 29042–29049.

Li, J., Waterhouse, R. M., & Zdobnov, E. M. (2011). A remarkably stable TipE gene cluster: evolution of insect Para sodium channel auxiliary subunits. *BMC Evolutionary Biology, 11*(1), 337. http://doi.org/10.1186/1471-2148-11-337.

Li, R. G., Wang, Q., Xu, Y. J., Zhang, M., Qu, X. K., Liu, X., … Yang, Y. Q. (2013). Mutations of the SCN4B-encoded sodium channel b4 subunit in familial atrial fibrillation. *International Journal of Molecular Medicine, 32*(1), 144–150. http://doi.org/10.3892/ijmm.2013.1355.

Lin, X., O'Malley, H., Chen, C., Auerbach, D., Foster, M., Shekhar, A., … Delmar, M. (2015). Scn1b deletion leads to increased tetrodotoxin-sensitive sodium current, altered intracellular calcium homeostasis and arrhythmias in murine hearts. *Journal of Physiology, 593*(6), 1389–1407. http://doi.org/10.1113/jphysiol.2014.277699.

Liu, H., Ming-Ming, W., & Zakon, H. H. (2007). Individual variation and hormonal modulation of a sodium channel b subunit in the electric organ correlate with variation in a social signal. *Developmental Neurobiology*, 1289–1304. http://doi.org/10.1002/dneu.

Lopez-Santiago, L. F., Brackenbury, W. J., Chen, C., & Isom, L. L. (2011). Na^+ channel Scn1b gene regulates dorsal root ganglion nociceptor excitability in vivo. *Journal of Biological Chemistry, 286*(26), 22913–22923. http://doi.org/10.1074/jbc.M111.242370.

Lopez-Santiago, L. F., Meadows, L. S., Ernst, S. J., Chen, C., Malhotra, J. D., McEwen, D. P., … Isom, L. L. (2007). Sodium channel Scn1b null mice exhibit prolonged QT and RR intervals. *Journal of Molecular and Cellular Cardiology, 43*(5), 636–647. http://doi.org/10.1016/j.yjmcc.2007.07.062.

Lopez-Santiago, L. F., Pertin, M., Morisod, X., Chen, C., Hong, S., Wiley, J., … Isom, L. L. (2006). Sodium channel beta2 subunits regulate tetrodotoxin-sensitive sodium channels in small dorsal root ganglion neurons and modulate the response to pain. *Journal of Neuroscience : The Official Journal of the Society for Neuroscience, 26*(30), 7984–7994. http://doi.org/10.1523/JNEUROSCI.2211-06.2006.

Lowery, L. A., & Van Vactor, D. (2009). The trip of the tip: understanding the growth cone machinery. *Nature Reviews. Molecular Cell Biology, 10*(5), 332–343. http://doi.org/10.1038/nrm2679.

Maier, S. K. G., Westenbroek, R. E., McCormick, K. A., Curtis, R., Scheuer, T., & Catterall, W. A. (2004). Distinct subcellular localization of different sodium channel a and b subunits in single ventricular myocytes from mouse heart. *Circulation, 109*(11), 1421–1427. http://doi.org/10.1161/01.CIR.0000121421.61896.24.

Malhotra, J. D., Chen, C., Rivolta, I., Abriel, H., Malhotra, R., Mattei, L. N., … Isom, L. L. (2001). Characterization of sodium channel alpha- and beta-subunits in rat and mouse cardiac myocytes. *Circulation, 103*(9), 1303–1310. http://doi.org/10.1161/01.cir.103.9.1303.

Malhotra, J. D., Kazen-Gillespie, K., Hortsch, M., & Isom, L. L. (2000). Sodium channel β subunits mediate homophilic cell adhesion and recruit ankyrin to points of cell-cell contact. *Journal of Biological Chemistry, 275*(15), 11383–11388. http://doi.org/10.1074/jbc.275.15.11383.

Malhotra, J. D., Koopmann, M. C., Kazen-Gillespie, K. A., Fettman, N., Hortsch, M., & Isom, L. L. (2002). Structural requirements for interaction of sodium channel beta 1 subunits with ankyrin. *Journal of Biological Chemistry, 277*(29), 26681–26688. http://doi.org/10.1074/jbc.M202354200.

Malhotra, J. D., Thyagarajan, V., Chen, C., & Isom, L. L. (2004). Tyrosine-phosphorylated and nonphosphorylated sodium channel beta1 subunits are differentially localized in cardiac myocytes. *Journal of Biological Chemistry, 279*(39), 40748–40754. http://doi.org/10.1074/jbc.M407243200.

Maness, P. F., & Schachner, M. (2007). Neural recognition molecules of the immunoglobulin superfamily: signaling transducers of axon guidance and neuronal migration. *Nature Neuroscience, 10*(1), 19–26. http://doi.org/10.1038/nn1827.

Marionneau, C., Carrasquillo, Y., Norris, A. J., Townsend, R. R., Isom, L. L., Link, A. J., & Nerbonne, J. M. (2012). The sodium channel accessory subunit Nav 1 regulates neuronal excitability through modulation of repolarizing voltage-gated K^+ channels. *Journal of Neuroscience, 32*(17), 5716–5727. http://doi.org/10.1523/JNEUROSCI.6450-11.2012.

McCormick, K. A., Isom, L. L., Ragsdale, D., Smith, D., Scheuer, T., & Catterall, W. A. (1998). Molecular determinants of Na^+ channel function in the extracellular domain of the beta1 subunit. *Journal of Biological Chemistry, 273*(7), 3954–3962. http://doi.org/10.1074/jbc.273.7.3954.

McEwen, D. P., Chen, C., Meadows, L. S., Lopez-Santiago, L., & Isom, L. L. (2009). The voltage-gated Na^+ channel β3 subunit does not mediate trans homophilic cell adhesion or associate with the cell adhesion molecule contactin. *Neuroscience Letters, 462*(3), 272–275. http://doi.org/10.1016/j.neulet.2009.07.020.

McEwen, D. P., & Isom, L. L. (2004). Heterophilic interactions of sodium channel beta1 subunits with axonal and glial cell adhesion molecules. *Journal of Biological Chemistry, 279*(50), 52744–52752. http://doi.org/10.1074/jbc.M405990200.

McEwen, D. P., Meadows, L. S., Chen, C., Thyagarajan, V., & Isom, L. L. (2004). Sodium channel β1 subunit-mediated modulation of Nav1.2 currents and cell surface density is dependent on interactions with contactin and ankyrin. *Journal of Biological Chemistry, 279*(16), 16044–16049. http://doi.org/10.1074/jbc.M400856200.

Meadows, L., Malhotra, J. D., Stetzer, A., Isom, L. L., & Ragsdale, D. S. (2001). The intracellular segment of the sodium channel beta 1 subunit is required for its efficient association with the channel alpha subunit. *Journal of Neurochemistry, 76*(6), 1871–1878. http://doi.org/10.1046/j.1471-4159.2001.00192.x.

Meadows, L. S., Chen, Y. H., Powell, A. J., Clare, J. J., & Ragsdale, D. S. (2002). Functional modulation of human brain Nav1.3 sodium channels, expressed in mammalian cells, by auxiliary beta 1, beta 2 and beta 3 subunits. *Neuroscience, 114*(3), 745–753. http://dx.doi.org/10.1016/s0306-4522(02)00242-7.

Meadows, L. S., Malhotra, J., Loukas, A., Thyagarajan, V., Kazen-Gillespie, K. A., Koopman, M. C., … Ragsdale, D. S. (2002). Functional and biochemical analysis of a sodium channel beta1 subunit mutation responsible for generalized epilepsy with febrile seizures plus type 1. *Journal of Neuroscience : The Official Journal of the Society for Neuroscience, 22*(24), 10699–10709. Retrieved from: http://www.ncbi.nlm.nih.gov/pubmed/12486163.

Medeiros-Domingo, A., Kaku, T., Tester, D. J., Iturralde-Torres, P., Itty, A., Ye, B., … Ackerman, M. J. (2007). SCN4B-encoded sodium channel β4 subunit in congenital long-QT syndrome. *Circulation, 116*(2), 134–142. http://doi.org/10.1161/CIRCULATIONAHA.106.659086.

Meisler, M. H., & Kearney, J. A. (2005). Sodium channel mutations in epilepsy and other neurological disorders. *Journal of Clinical Investigation, 115*(8). http://doi.org/10.1172/JCI25466.2010.

Messner, D. J., & Catterall, W. A. (1985). The sodium channel from rat brain: separation and characterization of subunits. *Journal of Biological Chemistry, 260*(19), 1667–1675.

Mikkelsen, T. S., Wakefield, M. J., Aken, B., Amemiya, C. T., Chang, J. L., Duke, S., … Lindblad-Toh, K. (2007). Genome of the marsupial *Monodelphis domestica* reveals innovation in non-coding sequences. *Nature, 447*(7141), 167–177. http://doi.org/10.1038/nature05805.

Miller, J. A., Agnew, W. S., & Levinson, S. R. (1983). Principal glycopeptide of the tetrodotoxin/saxitoxin binding protein from *Electrophorus electricus*: isolation and partial chemical and physical characterization. *Biochemistry, 22*(2), 462–470.

Morgan, K., Stevens, E. B., Shah, B., Cox, P. J., Dixon, A. K., Lee, K., … Jackson, A. P. (2000). Beta 3: an additional auxiliary subunit of the voltage-sensitive sodium channel that modulates channel gating with distinct kinetics. *Proceedings of the National Academy of Sciences of the United States of America, 97*(5), 2308–2313. http://doi.org/10.1073/pnas.030362197.

Namadurai, S., Balasuriya, D., Rajappa, R., Wiemhöfer, M., Stott, K., Klingauf, J., … Jackson, A. P. (2014). Crystal structure and molecular imaging of the Nav channel β3 subunit indicates a trimeric assembly. *Journal of Biological Chemistry, 289*(15), 10797–10811. http://doi.org/10.1074/jbc.M113.527994.

Nelson, M., Millican-Slater, R., Forrest, L. C., & Brackenbury, W. J. (2014). The sodium channel β1 subunit mediates outgrowth of neurite-like processes on breast cancer cells and promotes tumour growth and metastasis. *International Journal of Cancer. Journal International Du Cancer, 135*, 2338–2351. http://doi.org/10.1002/ijc.28890.

Nguyen, H. M., Miyazaki, H., Hoshi, N., Smith, B. J., Nukina, N., Goldin, A. L., & Chandy, K. G. (2012). Modulation of voltage-gated K^+ channels by the sodium channel 1 subunit. *Proceedings of the National Academy of Sciences of the United States of America, 109*(45), 18577–18582. http://doi.org/10.1073/pnas.1209142109.

Niethammer, P., Delling, M., Sytnyk, V., Dityatev, A., Fukami, K., & Schachner, M. (2002). Cosignaling of NCAM via lipid rafts and the FGF receptor is required for neuritogenesis. *Journal of Cell Biology, 157*(3), 521–532. http://doi.org/10.1083/jcb.200109059.

Noda, M., Ikeda, T., Suzuki, H., Takeshima, H., Takahashi, T., Kuno, M., & Numa, S. (1986). Expression of functional sodium channels from cloned cDNA. *Nature, 322*, 826–828.

Ogiwara, I., Nakayama, T., Yamagata, T., Ohtani, H., Mazaki, E., Tsuchiya, S., … Yamakawa, K. (2012). A homozygous mutation of voltage-gated sodium channel βi gene SCN1B in a patient with Dravet syndrome. *Epilepsia, 53*(12), 200–203. http://doi.org/10.1111/epi.12040.

Oh, Y., & Waxman, S. G. (1995). Differential Na^+ channel beta 1 subunit mRNA expression in stellate and flat astrocytes cultured from rat cortex and cerebellum: a combined in situ hybridization and immunocytochemistry study. *Glia, 13*(3), 166–173. http://doi.org/10.1002/glia.440130303.

Olesen, M. S., Jespersen, T., Nielsen, J. B., Liang, B., Møller, D. V., Hedley, P., … Svendsen, J. H. (2011). Mutations in sodium channel β-subunit SCN3B are associated with early-onset lone atrial fibrillation. *Cardiovascular Research, 89*(4), 786–793. http://doi.org/10.1093/cvr/cvq348.

O'Malley, H. A., & Isom, L. L. (2015). Sodium channel β subunits: emerging targets in channelopathies. *Annual Review of Physiology,* 77(1), 481–504. http://doi.org/10.1146/annurev-physiol-021014-071846.

O'Malley, H. A., Shreiner, A. B., Chen, G.-H., Huffnagle, G. B., & Isom, L. L. (2009). Loss of Na^+ channel beta2 subunits is neuroprotective in a mouse model of multiple sclerosis. *Molecular and Cellular Neurosciences, 40*(2), 143–155. http://doi.org/10.1016/j.mcn.2008.10.001.

Patel, F., & Brackenbury, W. J. (April 2015). Dual roles of voltage-gated sodium channels in development and cancer. *International Journal of Developmental Biology*. http://doi.org/10.1387/ijdb.150171wb.

Patino, G. A., Brackenbury, W. J., Bao, Y., Lopez-Santiago, L. F., O'Malley, H. A., Chen, C., … Isom, L. L. (2011). Voltage-gated Na^+ channel β1B: a secreted cell adhesion molecule involved in human epilepsy. *Journal of Neuroscience : The Official Journal of the Society for Neuroscience, 31*(41), 14577–14591. http://doi.org/10.1523/JNEUROSCI.0361-11.2011.

Patino, G. A., Claes, L. R. F., Lopez-Santiago, L. F., Slat, E. A., Dondeti, R. S. R., Chen, C., … Isom, L. L. (2009). A functional null mutation of SCN1B in a patient with Dravet syndrome. *Journal of Neuroscience : The Official Journal of the Society for Neuroscience, 29*(34), 10764–10778. http://doi.org/10.1523/JNEUROSCI.2475-09.2009.

Peles, E., Nativ, M., Campbell, P. L., Sakurai, T., Martinez, R., Lev, S., ... Schlessinger, J. (1995). The carbonic anhydrase domain of receptor tyrosine phosphatase beta is a functional ligand for the axonal cell recognition molecule contactin. *Cell, 82*(2), 251–260. http://dx.doi.org/10.1016/0092-8674(95)90312-7.

Qin, N., D'Andrea, M. R., Lubin, M.-L., Shafaee, N., Codd, E. E., & Correa, A. M. (2003). Molecular cloning and functional expression of the human sodium channel b1B subunit, a novel splicing variant of the b1 subunit. *European Journal of Biochemistry, 270*(23), 4762–4770. http://doi.org/10.1046/j.1432-1033.2003.03878.x.

Raman, I. M., & Bean, B. P. (1997). Resurgent sodium current and action potential formation in dissociated cerebellar Purkinje neurons. *Journal of Neuroscience : The Official Journal of the Society for Neuroscience, 17*(12), 4517–4526.

Raman, I. M., Sprunger, L. K., Meisler, M. H., & Bean, B. P. (1997). Altered subthreshold sodium currents and disrupted firing patterns in Purkinje neurons of *Scn8a* mutant mice. *Neuron, 19*(4), 881–891. Retrieved from: http://www.sciencedirect.com/science/article/pii/S0896627300809691/npapers2://publication/uuid/0AF4D5FE-16DE-4CBA-B6C3-FE9218B02C85.

Ratcliffe, C. F., Qu, Y., McCormick, K. A., Tibbs, V. C., Dixon, J. E., Scheuer, T., & Catterall, W. A. (2000). A sodium channel signaling complex: modulation by associated receptor protein tyrosine phosphatase beta. *Nature Neuroscience, 3*(5), 437–444. http://doi.org/10.1038/74805.

Ratcliffe, C. F., Westenbroek, R. E., Curtis, R., & Catterall, W. A. (2001). Sodium channel beta1 and beta3 subunits associate with neurofascin through their extracellular immunoglobulin-like domain. *Journal of Cell Biology, 154*(2), 427–434. http://doi.org/10.1083/jcb.200102086.

Reid, C. A., Leaw, B., Richards, K. L., Richardson, R., Wimmer, V., Yu, C., ... Petrou, S. (2014). Reduced dendritic arborization and hyperexcitability of pyramidal neurons in a Scn1b-based model of Dravet syndrome. *Brain: A Journal of Neurology, 137*(6), 1701–1715. http://doi.org/10.1093/brain/awu077.

Remme, C. A., & Bezzina, C. R. (2010). Sodium channel (Dys)function and cardiac arrhythmias. *Cardiovascular Therapeutics, 28*(5), 287–294. http://doi.org/10.1111/j.1755-5922.2010.00210.x.

Ren, D., Navarro, B., Xu, H., Yue, L., Shi, Q., & Clapham, D. E. (December 2001). A prokaryotic voltage-gated sodium channel. *Science, 294*, 2372–2375.

Riuró, H., Beltran-Alvarez, P., Tarradas, A., Selga, E., Campuzano, O., Vergés, M., ... Brugada, R. (2013). A missense mutation in the sodium channel β2 subunit reveals SCN2B as a new candidate gene for Brugada syndrome. *Human Mutation, 34*(7), 961–966. http://doi.org/10.1002/humu.22328.

Sakurai, T., Lustig, M., Nativ, M., Hemperly, J. J., Schlessinger, J., Peles, E., & Grumet, M. (1997). Induction of neurite outgrowth through contactin and Nr-CAM by extracellular regions of glial receptor tyrosine phosphatase β. *Journal of Cell Biology, 136*(4), 907–918. http://doi.org/10.1083/jcb.136.4.907.

Scheffer, I. E., Harkin, L. A., Grinton, B. E., Dibbens, L. M., Turner, S. J., Zielinski, M. A., ... Berkovic, S. F. (2007). Temporal lobe epilepsy and GEFS+ phenotypes associated with SCN1B mutations. *Brain: A Journal of Neurology, 130*(1), 100–109. http://doi.org/10.1093/brain/awl272.

Shah, B. S., Stevens, E. B., Pinnock, R. D., Dixon, A. K., & Lee, K. (2001). Developmental expression of the novel voltage-gated sodium channel auxiliary subunit beta3, in rat CNS. *Journal of Physiology, 534*(Pt 3), 763–776. http://doi.org/10.1111/j.1469-7793.2001.t01-1-00763.x.

Shi, X., Yasumoto, S., Nakagawa, E., Fukasawa, T., Uchiya, S., & Hirose, S. (2009). Missense mutation of the sodium channel gene SCN2A causes Dravet syndrome. *Brain & Development, 31*(10), 758–762. http://doi.org/10.1016/j.braindev.2009.08.009.

Shin, S. C., Ahn, D. H., Kim, S. J., Pyo, C. W., Lee, H., Lee, H., … Park, H. (2014). The genome sequence of the Antarctic bullhead notothen reveals evolutionary adaptations to a cold environment. *Genome Biology, 15*, 1–14. http://doi.org/10.1186/s13059-014-0468-1.

Smith, J. J., Kuraku, S., Holt, C., Sauka-Spengler, T., Jiang, N., Campbell, M. S., … Li, W. (2013). Sequencing of the sea lamprey (*Petromyzon marinus*) genome provides insights into vertebrate evolution. *Nature Genetics, 45*(4), 415–421, 421e1–2 http://doi.org/10.1038/ng.2568.

Sodergren, E., Weinstock, G., Davidson, E., Cameron, R., Gibbs, R., Angerer, R., … Wright, R. (February 2006). The genome of the sea urchin. *Science, 314*, 941–952. http://doi.org/10.1126/science.1133609.

Spampanato, J., Kearney, J. A., de Haan, G., McEwen, D. P., Escayg, A., Aradi, I., … Meisler, M. H. (2004). A novel epilepsy mutation in the sodium channel SCN1A identifies a cytoplasmic domain for beta subunit interaction. *Journal of Neuroscience : The Official Journal of the Society for Neuroscience, 24*(44), 10022–10034. http://doi.org/10.1523/JNEUROSCI.2034-04.2004.

Srinivasan, J., Schachner, M., & Catterall, W. A. (1998). Interaction of voltage-gated sodium channels with the extracellular matrix molecules tenascin-C and tenascin-R. *Proceedings of the National Academy of Sciences of the United States of America, 95*(26), 15753–15757. http://doi.org/10.1073/pnas.95.26.15753.

Stafstrom, C. E., & Carmant, L. (2015). Seizures and Epilepsy: an overview for neuroscientists. *Cold Spring Harbor Perspectives in Medicine [Electronic Resource], 5*, 1–18.

Sugawara, T., Tsurubuchi, Y., Agarwala, K. L., Ito, M., Fukuma, G., Mazaki-Miyazaki, E., … Yamakawa, K. (2001). A missense mutation of the Na^+ channel alpha II subunit gene Na(v)1.2 in a patient with febrile and afebrile seizures causes channel dysssfunction. *Proceedings of the National Academy of Sciences of the United States of America, 98*(11), 6384–6389. http://doi.org/10.1073/pnas.111065098.

Surges, R., & Sander, J. (2012). Sudden unexpected death in epilepsy: mechanisms, prevalence, and prevention. *Current Opinion in Neurology, 25*(2), 201–207. http://doi.org/10.1097/WCO.0b013e3283506714.

Sutkowski, E. M., & Catterall, W. A. (1990). b1 subunits of sodium channels. *Journal of Biological Chemistry, 265*(21), 12393–12399.

Takahashi, N., Kikuchi, S., Dai, Y., Kobayashi, K., Fukuoka, T., & Noguchi, K. (2003). Expression of auxiliary β subunits of sodium channels in primary afferent neurons and the effect of nerve injury. *Neuroscience, 121*(2), 441–450. http://dx.doi.org/10.1016/S0306-4522(03)00432-9.

The C. elegans Sequencing Consortium. (1998). Genome sequence of the nematode *C. elegans*: a platform for investigating biology. *Science (New York, NY), 282*(5396), 2012–2018. http://doi.org/10.1126/science.282.5396.2012.

Theile, J. W., & Cummins, T. R. (2011). Inhibition of Nav beta4 peptide-mediated resurgent sodium currents in Nav1. 7 channels by carbamazepine, riluzole. *Molecular Pharmacology*, 724–734. http://doi.org/10.1124/mol.111.072751.

Trinidad, J. C., Barkan, D. T., Gulledge, B. F., Thalhammer, A., Sali, A., Schoepfer, R., & Burlingame, A. L. (2012). Global identification and characterization of both O-GlcNAcylation and phosphorylation at the murine synapse. *Molecular & Cellular Proteomics, 11*(8), 215–229. http://doi.org/10.1074/mcp.O112.018366.

Vaughn, D. E., & Bjorkman, P. J. (1996). The (Greek) key to structures of neural adhesion molecules. *Neuron, 16*(2), 261–273. http://dx.doi.org/10.1016/S0896-6273(00)80045-8.

Venkatesh, B., Lee, A. P., Ravi, V., Maurya, A. K., Lian, M. M., Swann, J. B., … Warren, W. C. (2014). Elephant shark genome provides unique insights into gnathostome evolution. *Nature, 505*(7482), 174–179. http://doi.org/10.1038/nature12826.

Vilella, A. J., Severin, J., Ureta-Vidal, A., Heng, L., Durbin, R., & Birney, E. (2009). EnsemblCompara GeneTrees: complete, duplication-aware phylogenetic trees in vertebrates. *Genome Research, 19*(2), 327–335. http://doi.org/10.1101/gr.073585.107.

Wallace, R. H., Scheffer, I. E., Parasivam, G., Barnett, S., Wallace, G. B., Sutherland, G. R., … Mulley, J. C. (2002). Generalized epilepsy with febrile seizures plus: mutation of the sodium channel subunit SCN1B. *Neurology, 58*(9), 1426–1429. http://doi.org/10.1212/WNL.58.9.1426.

Wallace, R. H., Wang, D. W., Singh, R., Scheffer, I. E., George, A. L., Phillips, H. A., … Mulley, J. C. (1998). Febrile seizures and generalized epilepsy associated with a mutation in the Na$^+$-channel beta1 subunit gene SCN1B. *Nature Genetics, 19*(4), 366–370. http://doi.org/10.1038/1252.

Warren, W. C., Hillier, L. W., Marshall Graves, J. A., Birney, E., Ponting, C. P., Grützner, F., … Wilson, R. K. (2008). Genome analysis of the platypus reveals unique signatures of evolution. *Nature, 453*(7192), 175–183. http://doi.org/10.1038/nature07253.

Watanabe, H., Darbar, D., Kaiser, D. W., Jiramongkolchai, K., Chopra, S., Donahue, B. S., … Roden, D. M. (2009). Mutations in sodium channel β1- and β2-subunits associated with atrial fibrillation. *Circulation: Arrhythmia and Electrophysiology, 2*(3), 268–275. http://doi.org/10.1161/CIRCEP.108.779181.

Watanabe, H., Koopmann, T. T., Le Scouarnec, S., Yang, T., Ingram, C. R., Schott, J., … Wiesfeld, A. C. P. (2008). Sodium channel β1 subunit mutations associated with Brugada syndrome and cardiac conduction disease in humans. *Structure, 118*(6), 268–275. http://doi.org/10.1172/JCI33891.2260.

Williams, A. F., & Barclay, A. N. (1988). The immunoglobulin superfamily—domains for cell surface recognition. *Annual Review of Immunology, 6*, 381–405. http://doi.org/10.1146/annurev.iy.06.040188.002121.

Wong, H. K., Sakurai, T., Oyama, F., Kaneko, K., Wada, K., Miyazaki, H., … Nukina, N. (2005). b subunits of voltage-gated sodium channels are novel substrates of b-site amyloid precursor protein-cleaving enzyme (BACE1) and g-secretase. *Journal of Biological Chemistry, 280*(24), 23009–23017. http://doi.org/10.1074/jbc.M414648200.

Xiao, Z. C., Ragsdale, D. S., Malhotra, J. D., Mattei, L. N., Braun, P. E., Schachner, M., & Isom, L. L. (1999). Tenascin R is a functional modulator of sodium channel beta subunits. *Journal of Biological Chemistry, 274*(37), 26511–26517. Retrieved from: http://www.ncbi.nlm.nih.gov/pubmed/10473612.

Yereddi, N. R., Cusdin, F. S., Namadurai, S., Packman, L. C., Monie, T. P., Slavny, P., … Jackson, A. P. (2013). The immunoglobulin domain of the sodium channel b3 subunit contains a surface-localized disulfide bond that is required for homophilic binding. *FASEB Journal, 27*(2), 568–580. http://doi.org/10.1096/fj.12-209445.

Yoshimura, T., & Rasband, M. N. (2014). Axon initial segments: diverse and dynamic neuronal compartments. *Current Opinion in Neurobiology, 27*, 96–102. http://doi.org/10.1016/j.conb.2014.03.004.

Yu, F. H., Westenbroek, R. E., Silos-Santiago, I., McCormick, K. A., Lawson, D., Ge, P., … Curtis, R. (2003). Sodium channel beta4, a new disulfide-linked auxiliary subunit with similarity to beta2. *Journal of Neuroscience : The Official Journal of the Society for Neuroscience, 23*(20), 7577–7585. http://doi.org/23/20/7577.

Zalc, B., Goujet, D., & Colman, D. (2008). The origin of the myelination program in vertebrates. *Current Biology, 18*(12), 511–512. http://doi.org/10.1016/j.cub.2008.04.010.

Zhou, J., Potts, J. F., Trimmer, J. S., Agnew, W. S., & Sigworth, F. J. (1991). Multiple gating modes and the effect of modulating factors on the μl sodium channel. *Neuron, 7*, 775–785.

CHAPTER TWELVE

Lipid Regulation of Sodium Channels

N. D'Avanzo
Université de Montréal, Montréal, QC, Canada
E-mail: nazzareno.d.avanzo@umontreal.ca

Contents

Abstract

The lipid landscapes of cellular membranes are complex and dynamic, are tissue dependent, and can change with the age and the development of a variety of diseases. Researchers are now gaining new appreciation for the regulation of ion channel proteins by the membrane lipids in which they are embedded. Thus, as membrane lipids change, for example, during the development of disease, it is likely that the ionic currents that conduct through the ion channels embedded in these membranes will also be altered. This chapter provides an overview of the complex regulation of prokaryotic and eukaryotic voltage-dependent sodium (Na_V) channels by fatty acids, sterols, glycerophospholipids, sphingolipids, and cannabinoids. The impact of lipid regulation on

Current Topics in Membranes, Volume 78
ISSN 1063-5823
http://dx.doi.org/10.1016/bs.ctm.2016.04.003

channel gating kinetics, voltage-dependence, trafficking, toxin binding, and structure are explored for Na_V channels that have been examined in heterologous expression systems, native tissue, and reconstituted into artificial membranes. Putative mechanisms for Na_V regulation by lipids are also discussed.

1. INTRODUCTION

Lipids can be broadly defined as small molecules that are hydrophobic or amphiphilic in character. Most lipids are derived from two distinct chemical scaffolds: ketoacyl and isoprene groups (Fig. 1A). Using these backbones for categorization, eight lipid classes emerge (Fahy et al., 2009). ***Fatty acids***, ***glycerolipids***, ***glycerophospholipids***, ***sphingolipids***, ***saccharolipids***, and ***polyketides*** can be derived from ketoacyl subunits, while ***sterol*** and ***prenol*** lipids can be derived from isoprene subunits. The amphiphilic nature of some lipids allows them to self-organize and form structures such as micelles, multilamellar or unilamellar liposomes, or bilayers in an aqueous environment. Which structure forms depends in part on the concentration of lipids

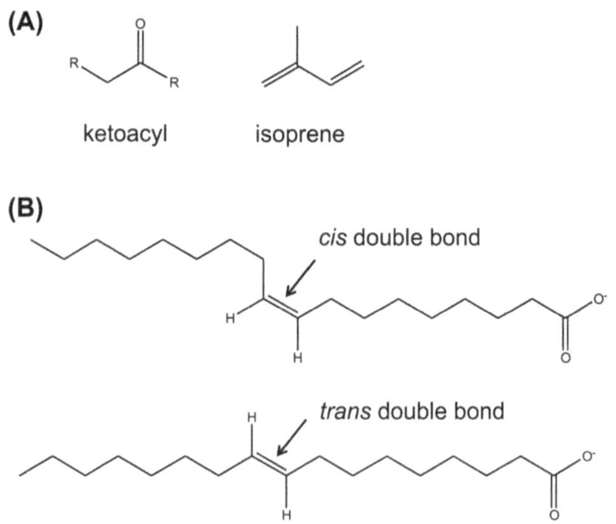

Figure 1 ***Chemistry of membrane lipids.*** (A) Most biological lipids are derived from two distinct chemical moieties: ketoacyl (*left*) and isoprene (*right*). (B) Alkyl tails of unsaturated fatty acids contain C=C double bonds that can be in the *cis* or *trans* conformations. In the *cis* conformation the R groups are on the same side, which induces a "kink" in the tail, preventing tight packing of the lipids. In the *trans* conformation, the R groups are on opposite sides of the bond, which maintains a straighter tail.

available. Of general interest in cellular systems are bilayers which are energetically favorable self-assembling structures that form when certain phospholipids are present in high enough concentrations. These lipids spontaneously arrange so that the hydrophobic acyl "tails" are isolated from the surrounding water, shielded by the hydrophilic head groups, which can interact with water and ions at the intracellular or extracellular faces. This spontaneous self-assembly is largely driven by the hydrophobic effect, which is thought to arise from the increased entropy in the system that would occur as hydrophobic regions cluster and water molecules can bond more freely to one another. The specific nature of the lipids in the bilayer system establishes a complex network of noncovalent interactions such as van der Waals, electrostatic and hydrogen bonds that enable membranes to be fluid and not rigidly fixed in space. In addition, the physiochemical properties of some lipids enable them to undergo lateral phase separations that favor the formation of stable microdomains (Boal, 2001).

Lipid bilayers form the cell membrane and act to compartmentalize different biological functions into localized substructures or organelles. This includes the cell or plasma membrane, which separates the interior and exterior environments and protects the cells from their surroundings. In most naturally occurring bilayers, the composition of the inner and outer leaflets of the membrane are different. This asymmetry is in part derived from a class of enzymes called "flippases." These enzymes transport lipids to the outer leaflet that are normally synthesized and inserted into inner leaflet (Bretscher, 1973; Rothman & Kennedy, 1977). Other lipids, such as some sphingolipids, are synthesized at the external leaflet. Once lipid asymmetry is established, it does not normally dissipate quickly because there exists a large energy barrier for spontaneous flip-flopping of lipids between leaflets.

Prokaryotes possess a lipid bilayer that makes up the plasma membrane. Gram-negative bacteria have both a plasma membrane and an outer membrane separated by periplasm. However, the outer membrane's composition is distinct from that of the inner plasma membrane. For example, the outer leaflet of the outer membrane of many gram-negative bacteria includes a complex lipopolysaccharide that acts in part as an endotoxin. Eukaryotes have a range of organelles including the nucleus, mitochondria, lysosomes, and endoplasmic reticulum, each of which surrounded by one or more lipid bilayers. Although eukaryotes use various biosynthetic pathways to both break down and synthesize lipids, some essential lipids cannot be made this way and must be obtained from the diet.

All cells maintain an electrical potential difference across the cell membrane. There also exist electrical potential differences across the membranes of many intracellular organelles. Changes in this membrane potential underlie fundamental physiological processes, such as the coordination of muscle contractions that lead to a heartbeat, the contraction of skeletal muscles, the conduction of a nerve impulse, or the release of hormones. The changing membrane potentials arise from the controlled and coordinated movements of ions across the bilayers through ion channel proteins embedded in them. The main physiological role of voltage-gated Na^+ (Na_V) channels is the initiation and propagation of action potentials in a variety of excitable cells including neurons, cardiac and skeletal muscle cells, and neuroendocrine cells (Catterall, 1986). Consequently, a variety of cardiac, skeletal, and neuronal disorders can be directly attributed to the disrupted function of Na_V channels (George, 2005). Disturbances in the function of muscle Na_V channels can affect the ability of skeletal muscle to contract or relax. Myotonia (prolonged contraction) and periodic paralysis are two symptoms characteristic of skeletal Na_V channel dysfunction (Cannon & George, 2002). Disruption of cardiac Na_V channel function causes susceptibility to impaired electrical conduction in the myocardia (Schott et al., 1999), ventricular arrhythmias (Bezzina et al., 1999; Grant et al., 2002; McNair et al., 2004; Tan et al., 2001; Wang, Viswanathan, Balser, George, & Benson, 2002), drug-induced arrhythmias (Makita et al., 2002), sudden infant death syndrome (Ackerman et al., 2001; Schwartz et al., 2000), and other forms of arrhythmia susceptibility (Splawski et al., 2002). Neuronal Na_V channels are critical for the generation and propagation of action potentials in the central and peripheral nervous systems, with most of the 13 SCN genes encoding for Na_V channels or their auxiliary subunits expressed in the brain, peripheral nerves, or both (Catterall, 1992). Mutations in neuronal Na_V channels have been linked to a variety of inherited epilepsy syndromes (Claes et al., 2001; Escayg et al., 2000; Fujiwara et al., 2003; Heron et al., 2002; Lossin et al., 2003; Rhodes, Lossin, Vanoye, Wang, & George, 2004; Sugawara et al., 2001; Wallace et al., 1998) and a rare painful neuropathy primary erythermalgia (Yang et al., 2004). Consequently, neuronal Na_V channels serve as important pharmacological targets for anticonvulsants and local anesthetic agents (Catterall, 1999; Macdonald & Greenfield, 1997).

Recently, lipidomics research has shed light on changes in membrane lipid composition during the development of many of these same diseases. Tachycardia induces numerous changes in myocardial lipid metabolism (Wojcik, Harasim, Zabielski, Chabowski, & Gorski, 2015).

Hyperthyroidism, which is associated with sinus tachycardia and a tendency to develop arrhythmias, was also found to evoke myocardial sphingolipid accumulation, with a 28% increase in ceramide, an 80% increase in sphingosine, and a 500% increase in sphingomyelin content (Miklosz et al., 2015). Cholesterol content in the sarcolemma increases with elevated serum cholesterol (Huang et al., 2004), increasing nearly 20% in the development of diabetes (Pierce & Dhalla, 1983). Changes in 38 lipid species, including phosphatidylcholine (PC), phosphatidylethanolamine (PE), and phosphatidylinositol (PI) species, were observed in isolated brains of a Drosophila epilepsy model (Kliman et al., 2010). Lipid to protein ratios are elevated in the rat hippocampus of kainic acid model of temporal epilepsy (Dal-Pizzol, Quevedo, Streck, Walz, & Moreira, 2007). Such changes in lipid profiles may affect Na_V channel function in these tissues through alterations in protein–lipid interactions, through changes in regulatory proteins or signaling pathways, or through structural changes in the lipid bilayer that influence gating.

2. REGULATION OF NA_V CHANNELS BY THE PHYSIOCHEMICAL PROPERTIES OF THE BILAYER

The functions of various membrane proteins can in some cases be regulated by the lipids through the physiochemical properties of the bilayer. These nonspecific properties include surface charge, dipole potential, lateral stress, curvature elastic energy, ordering of the lipid tails, and bilayer thickness (Andersen & Koeppe, 1992, 2007; Green & Andersen, 1991; Lundbaek, Collingwood, Ingolfsson, Kapoor, & Andersen, 2010). There is evidence that Na_V channels are regulated by some, but not all, of these nonspecific lipid properties. The dependence of conductance on ionic strength in batrachotoxin-activated Na_V channels purified from rat brain was unchanged when comparing single channel recordings performed in neutral PE versus those performed in 70% PE and 30% phosphatidylserine (PS) (Worley, French, Pailthorpe, & Krueger, 1992). These different bilayer conditions amount to a 12-fold difference in charge density. This indicates that lipid surface charges do not alter conductance in sodium channels. The lack of differences in divalent block of these channels in neutral or charged bilayers provides further evidence for this hypothesis (Worley et al., 1992). It is likely that the Na_V channel pores are too distant from the lipid bilayer head groups to be affected by their charge. However, while conductance appears to be unchanged by lipid charge, the movement of gating charges is somewhat affected (Cukierman, 1991; Cukierman, Zinkand, French, & Krueger, 1988). In the absence of Ca^{2+}, gating in these channels is unaffected by differences in the lipid bilayer charge (Cukierman et al., 1988). However, the effects of intracellular or extracellular Ca^{2+} ions on Na_V channels differ in neutral bilayers compared to charged bilayers (Cukierman et al., 1988). Furthermore, in neutral bilayers containing PE and PC, a decrease in extracellular ionic strength shifts the voltage-dependence of gating in Na_V channels to more hyperpolarized potentials by approximately −34 mV. However, in charged bilayers containing PE and PS, a decrease in ionic strength leads to an even greater hyperpolarized shift in the voltage-dependence of channel activation. Thus, it appears that the gating machinery of the sodium channel is able to sense the electric field created by surface charges on the lipid membrane. Ca^{2+} ions or high ionic strength act to shield these charges and shift the probability of opening.

The function of Na_V channels also appears to depend on membrane thickness. The addition of short-chain phospholipids such as diheptanoyl phosphatidylcholine and dioctanoyl phosphatidylcholine to giant squid axons shifts the voltage-dependence of inactivation to more depolarized potentials by +4 to +10 mV (Haydon & Urban, 1983a; Hendry, Elliott, & Haydon, 1985). Peak I_{Na} is reduced following the addition of short-chained phospholipids and shifted by +15 mV. This was largely driven by the large depolarizing shifts (+15 mV) in the voltage-dependence of activation. Notably, the long-chained phospholipid didecanoyl phosphatidylcholine has no observable effect on Na_V current (Hendry et al., 1985).

A wide variety of small molecules, volatile anesthetics, and thiazolidinedione insulin sensitizers (TZDs) have been demonstrated to regulate Na_V channels (Haydon & Kimura, 1981; Haydon & Urban, 1983a, 1983b, 1983c; Herold & Hemmings, 2012; Herold et al., 2014; Rusinova et al., 2011). Volatile anesthetics induce an inhibition of peak I_{Na} due to stabilization of the inactivated state (Ouyang, Wang, & Hemmings, 2003; Rehberg, Xiao, & Duch, 1996). There exist subtle but relevant differences in the potency between isoforms. Moreover, isoflurane has different effects on the recovery from fast-inactivation between $Na_V1.2$, $Na_V1.4$, $Na_V1.5$, and $Na_V1.6$ channels (OuYang & Hemmings, 2007; Shiraishi, Hiraiwa, & Uda, 1993), while $Na_V1.8$ channels are insensitive (Shiraishi et al., 1993). The mode of action of regulation by these compounds has been heavily examined over several decades. While some of the effects of volatile anesthetics on Na_V channel activity may be mediated through PKC phosphorylation (Cantrell & Catterall, 2001) and direct binding to the channel (Jackson, Haris, &

Chapman, 1989; Monod, Wyman, & Changeux, 1965; Raju, Barber, LeBard, Klein, & Carnevale, 2013), additional effects through modification of bilayer properties have been considered. Since many anesthetics can partition into the bilayer, their effects on Na_V channel activity have been attributed to their ability to increase membrane fluidity (Haydon & Kimura, 1981; Haydon & Urban, 1983a, 1983b, 1983c). However, at therapeutic concentrations, most anesthetics do not induce changes to lipid bilayer properties as sensed through gramicidin channels (Herold et al., 2014) despite their functional effects on Na_V channels. Moreover, any changes in membrane density and fluidity that are induced by clinical concentrations of some anesthetics can be mimicked by temperature differences as little as 1°C without the equivalent anesthetic effects (Franks & Lieb, 1982). This is in contrast to the effects of the TZD troglitazone, which effect gramicidin channels at the same concentrations as they induce hyperpolarizing shifts in Na_V channel inactivation (Rusinova et al., 2011).

Several ion channels, including Na_V channels, appear to be mechanically sensitive (Laitko, Juranka, & Morris, 2006; Laitko & Morris, 2004; Lin, Laitko, Juranka, & Morris, 2007; Morris & Juranka, 2007), that is to say they respond to changes in membrane tension, or the stress (force/unit area) that leads to deformation of the plasma membrane and membrane-to-cortex attachments (Diz-Munoz, Fletcher, & Weiner, 2013). Membrane mechanics plays a complex role in the regulation of Na_V currents. Membrane trauma, such as during the formation of a gigaohm patch seal, mechanically induces irreversible changes in the membrane nanostructure, such as de-adhesion of membrane cortex from the bilayer (known as blebbing) (Suchyna, Markin, & Sachs, 2009). This can result in a loss of lipid order, increased fluidity, bilayer thinning, and a loss in leaflet asymmetry (Tabarean, Juranka, & Morris, 1999). $Na_V1.4$ currents show different activation and inactivation kinetics depending on the degree of mechanical trauma incurred during seal formation (Tabarean et al., 1999). Membrane stretch induces an irreversible shift from slow to fast gating kinetics in the $Na_V1.4\alpha$ subunit. $Na_V1.5$ channels cells also appear sensitive to stretch in a largely reversible manner (Beyder et al., 2010). Membrane stretch shifts both the voltage-dependences of activation and inactivation to more hyperpolarized potentials with linear shifts in the kinetics of both gating transitions (Banderali, Juranka, Clark, Giles, & Morris, 2010; Beyder et al., 2010; Morris & Juranka, 2007). Interestingly, mutations of two domain IV arginine residues, R1626P and R1623Q, which are associated with LQT3 syndrome (Banderali et al., 2010; Ruan, Liu, Bloise, Napolitano, & Priori,

2007) also show a different mechanosensitivity compared to wild-type $Na_V1.5$ channels (Banderali et al., 2010). The coupled activation/inactivation of these mutant channels is less sensitive to stretch. Consequently, by remaining active longer, these mutant channels remain excitatory and their responses to stretch are decidedly proarrhythmic. Mechanistically, it appears that in general across all voltage-dependent cation channels, changes in membrane tension induced by stretch reduces the free energy barrier for movements of the voltage-sensor domains, without altering the number of gating charges required to move (Banderali et al., 2010; Laitko & Morris, 2004; Tabarean & Morris, 2002).

3. FATTY ACIDS

Fatty acids consist of a carboxyl group (−COOH) with a long aliphatic chain (tail) of carbon atoms. Under physiological conditions, this carboxyl group is typically deprotonated ($-COO^-$), and thus acidic. Since the biosynthesis of fatty acids involves the condensation of acetyl-CoA, which carries a two-carbon atom group, most naturally occurring fatty acids contain an even number of carbon atoms between 4 and 28. If all the carbon−carbon bonds in the alkyl tail are single, then the fatty acid is "saturated," while if any of the carbons are doubly or triply bonded, the fatty acid is considered "unsaturated." Fatty acids are typically found covalently bound to glycerol in the form of triglyceride, or a phospholipid. When not attached to other molecules, they are known as "free" fatty acids, which are typically insoluble in water.

Carbon atoms residing on either side of a double bond can occur in a *cis* or *trans* configuration (Fig. 1B). In the *trans* conformation, the two hydrogen atoms adjacent to the double bond are on opposite sides of the chain. The rigidity of the double bond reduces the degrees of rotational freedom and keeps the alkyl tail relatively straight. In the *cis* conformation, the two hydrogen atoms adjacent to the double bond are on the same side of the chain. The rigidity of the double bond causes a "kink" in the tail. With an increasing number of double bonds, the number of kinks increases as does the degree of bending or curvature. Unsaturated bonds in *cis* conformations prevent tight packing of lipids in bilayers, decreasing the melting temperature of the membrane and making the lipid bilayer more fluid. Additionally, *cis* unsaturation induces changes in the compressibility and the bending moduli of lipid membranes.

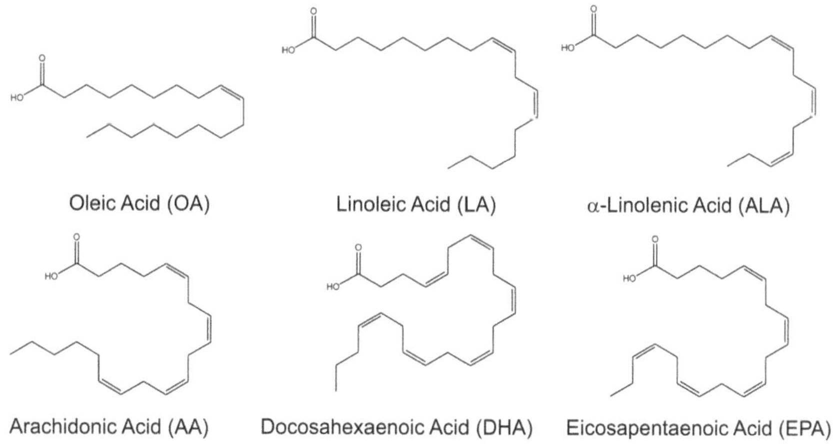

Figure 2 ***Examples of fatty acids.*** Six fatty acids of differing alkyl tail lengths and degrees of saturation are presented. The curvature of the alkyl tail increases with the number of *cis* double bonds.

Omega-3 polyunsaturated fatty acids (ω3PUFAs) contain double bonds in the alkyl tail starting with the third carbon from the end of the tail (the ω side of the lipid). ω3PUFAs are critical for the proper development and functional maintenance of several human organs, including the brain. Since mammals cannot synthesize ω3PUFAs, they are considered "essential." Three ω3PUFAs that are particularly important for human physiology are α-linolenic acid (ALA), eicosapentaenoic acid (EPA), and docosahexaenoic acid (DHA). The original source of most ω3PUFAs, both directly and indirectly, is plants, plankton, and algae. Some humans acquire most ω3PUFAs through the consumption of fish, but can also acquire them from plant oils such as soybean, flax seed, and canola (Fig. 2).

Each omega-3 fatty acid has an analogous omega-6 counterpart, which contains the same number of carbon atoms, but has the first double bond at the sixth carbon from the end of the alkyl tail. For example, the ω3 eicosapentaenoic acid (EPA) which contains 20 carbons is analogous to the ω6 arachidonic acid (AA). Some research has suggested that an ideal balance of ω6:ω3 intake is 1:1, while modern diets are often contain ratios of greater than 10:1 (Hibbeln, Nieminen, Blasbalg, Riggs, & Lands, 2006). Excessive omega-6 intake may interfere with the health benefits of omega-3 fatty acids because they compete for the same rate-limiting enzymes. For example, linoleic acid (an ω6) and α-linolenic acid (an ω3) compete for Δ6-desaturase to form AA and EPA respectively, which can effect downstream lipid production of other lipids such as prostaglandins. This can putatively shift the physiological state of tissues in prothrombotic, proinflammatory, and pro-constrictive conditions (Simopoulos, 2003).

Over the last 40 years, a large body of research has been developed examining the effect of ω3PUFAs on the cardiovascular, immune, gastrointestinal, and nervous systems. Bang, Dyerberg, and Sinclair (1980), who pioneered this domain, identified an extremely low incidence of acute myocardial infarction among the Inuit compared to their Danish neighbors, which they attributed to differences in dietary intake of ω3PUFAs. Numerous, but not all, follow-up studies over the years have suggested ω3PUFAs may prevent sudden cardiac death, reduce heart rate variability, and prevent atrial and ventricular fibrillations under certain circumstances (for a brief summary of major clinical trials on the effects of omega-3 fatty acids on the heart see (Richardson, Iaizzo, & Xiao, 2011)). Since the electrical activity of the heart is governed by the function of ion channels, the examination of the regulation of various ion channels by fatty acids has been of great interest in recent years.

3.1 Multifaceted regulation of Na_V channels by fatty acids

In both rat and human cardiac I_{Na} α-subunits, EPA, DHA, linolenic acid, and linoleic acid shift the steady-state inactivation to more hyperpolarized potentials by −20 to −30 mV (Kang, Li, & Leaf, 1997; Kang, Xiao, & Leaf, 1995; Leifert, McMurchie, & Saint, 1999; Xiao, Kang, Morgan, & Leaf, 1995). EPA also markedly accelerates the transition from the resting state to the inactivated state and slows the recovery from the inactivation (Xiao, Wright, Wang, Morgan, & Leaf, 1998). These results are similar to the effects of EPA on $Na_V1.7$ in human bronchial smooth muscle cells (Jo et al., 2005) and $Na_V1.5$ channels in human breast cancer cells MDA-MB-231 (Isbilen, Fraser, & Djamgoz, 2006). Consequently, the addition of these fatty acids causes a more negative resting membrane potential by −5 mV and increases the threshold for action potential by nearly −10 mV, with additional effects on the cycle length and rate of late depolarization (Kang et al., 1995). In adult rat ventricular myocytes, DHA, EPA, and α-linolenic acid also shift the voltage dependence of activation to more depolarized potentials by nearly +10 mV (Leifert, McMurchie, et al., 1999). Intriguingly, I_{Na} in rat ventricular myocytes is unaffected by monounsaturated and saturated fatty acids (Kang & Leaf, 1996; Kang et al., 1995; Leifert, McMurchie, et al., 1999; Xiao et al., 1995), but these lipids have similar effects on I_{Na} as PUFAs when Na_V channels are expressed in HEK293T cells

(Xiao et al., 1998). This suggests the possibility that the β-subunit of Na_V channels may play a role in regulating lipid specificity in Na_V channels. In the presence of the β-subunit, EPA and other PUFAs significantly inhibit I_{Na} by shifting the steady-state inactivation by −22 mV, and accelerating entry into the inactivated state from the closed state, and prolonging recovery from inactivation (Xiao, Wright, Wang, Morgan, & Leaf, 2000). In contrast, saturated and monounsaturated fatty acids do not affect Na_V channels in the presence of the β-subunit.

Other fatty acids have also have effects on Na_V channel function. The extracellular application of AA, oleic acid, and arachidonoyl ethanolamide (anandamide; AEA) can inhibit Na_V channels by reducing peak Na^+ currents, and shifting the voltage dependence of inactivation to more negative potentials by approximately −15 to −20 mV (Al Kury et al., 2014; Boland & Drzewiecki, 2008; Fang, Zhou, Gao, Gu, & Mei, 2011; Fraser, Hoehn, Weiss, & MacVicar, 1993; Lee, Shin, Lee, & Song, 2002; Nicholson et al., 2003; Safrany-Fark et al., 2015). Furthermore, AA accelerates entry into fast inactivation and delays recovery from inactivation in striatal neurons. These actions are independent of various AA metabolites or activation of protein kinase C (Fraser et al., 1993). Similar results are observed for $Na_V1.4$ and $Na_V1.5$ channels expressed in HEK293 cells (Bendahhou, Cummins, & Agnew, 1997). In DRG neurons, the extracellular application of AA also shifts the voltage-dependence of activation of both TTX-R and TTX-S currents to more hyperpolarized potentials by −4 mV and −10 mV, respectively (Lee et al., 2002). Similar results are observed when assessing the extracellular application of AA on cerebellar granule cells which primarily express TTX-S Na_V channels (Fang et al., 2011). Intriguingly, the intracellular application of AA to cerebellar granule cells via the patch pipette leads to a concentration-dependent increase in I_{Na} resulting from a small depolarizing shift in the steady-state activation curve (Fang et al., 2011). Intracellular AA does not shift the voltage-dependence of inactivation in these cells. Similarly, differential effects are also observed for fatty acids applied to the intracellular versus extracellular leaflet of human skeletal muscles and the murine C_2C_{12} skeletal cell line (Wieland, Fletcher, & Gong, 1992). In these cells, the intracellular application of AA increases the magnitude of peak sodium currents in a concentration-dependent manner. The percent change is similar if oleic acid or stearic acid is applied instead. On the other hand, the application of extracellular AA leads to an inhibition of peak TTX-insensitive sodium currents, which cannot be replicated by oleic or stearic acids.

In striatal neurons, AA reduces the amplitude and frequency of action potentials and depresses spontaneous inhibitory postsynaptic currents. No effect on miniature inhibitory postsynaptic currents can be observed. Together, these data implicate AA as a contributor of striatal synaptic depression by acting on Na_V channels (Fraser et al., 1993). On the other hand, AA does not inhibit evoked action potentials of the marginal mandibular motor nerve, which consists of only thick myelinated motor fibers that almost exclusively contain Na_V channels. The lack of any AA effect on the motor nerve activity may be the result of its low potency and/or fast metabolism (Safrany-Fark et al., 2015).

The application of *cis*-fatty acids such as oleate, linoleate, and linolenate to N1E-115 neuroblastoma cells, which primarily express $Na_V1.6$ and $Na_V1.3$ channels (Baroni & Moran, 2011), leads to a decrease in peak I_{Na} with no effect on steady-state activation or inactivation (Linden & Routtenberg, 1989). This attenuation is more sensitive to the intracellular application of fatty acids compared to when they are applied to the extracellular leaflet. The potency of these lipids to attenuate I_{Na} is correlated to their capacity to activate PKC and not to their capacity to alter membrane fluidization.

Olfactory mucosa contains an abundance of AA and DHA fatty acids (Russell, Evans, & Dodd, 1989). Rat olfactory receptor neurons are bipolar neurons adapted to transduce peripheral odorant signals and transmit them centrally to the olfactory bulb. These cells primarily express $Na_V1.7$ channels (Ahn et al., 2011). In these cells, AA and DHA induce a small depolarized shift in steady-state activation and a decrease in peak I_{Na} (Seebungkert & Lynch, 2002). At low concentrations, these fatty acids induce a strong transient inhibition of Na_V channels that declines within 1 min of exposure (Seebungkert & Lynch, 2002). It was suggested that this regulation may influence the encoding of odorant information by these neurons, since the release of these fatty acids would temporarily raise the action potential threshold, leading to a short-term (<1 min) decrease in excitability. However, the continued presence of AA or DHA would also mediate an inhibition of K^+ currents (Seebungkert & Lynch, 2002) that would then begin to increase excitability in these neurons. Finally, this response may be terminated by the metabolism of AA by cytochrome P-450 oxidative enzymes (Laethem, Laethem, Ding, & Koop, 1992) which is highly expressed in the olfactory epithelium (Ding, Koop, Crump, & Coon, 1986).

3.2 Putative biophysical mechanisms for fatty acid regulation of Na_V channels

Four main hypotheses can explain the effect of fatty acids on Na_V and other ion channels:

Hypothesis #1: Channels may have a specific binding site for the fatty acid, whereby the lipid acts as a ligand to modulate activity. EPA and DHA significantly reduces Na_V currents in HEK293T cells expressing either wild-type cardiac hH1α ($hNa_V1.5$) channels, or the single point mutations F1760K or Y1767K in the S6 of Domain 4 by hyperpolarizing the voltage-dependence of inactivation. However, the point mutation N406K in the S6 of Domain 1 is less sensitive to inhibition by EPA. The sensitivity to EPA is even further reduced in N406K channels coexpressed with the β-subunit (Xiao et al., 2001). The L409C/A410W mutation creates an inactivation-deficient $Na_V1.5$ channel with long-lasting persistent current. In HEK293T cells, ω3PUFAs abolishes this persistent current in this mutant channel, by shifting the steady-state inactivation to hyperpolarized potentials, effectively restoring inactivation. Peak Na^+ currents in these mutant channels are also decreased by 50% in the presence of EPA. Activation properties in these mutant channels are unaffected by ω3PUFAs (Xiao et al., 2006). These actions of EPA are similar to the effect of the antiarrhythmic drug flecainide on inactivation mutations in skeletal muscle Na^+ channels (Wang, Russell, & Wang, 2003). In addition, PUFAs that are antiarrhythmic also inhibit the binding of the radiolabeled neurotoxin [^{3}H] BTXB in a noncompetitive allosteric fashion, while monounsaturated and saturated fatty acids which are not antiarrhythmic do not inhibit toxin binding (Kang & Leaf, 1996). The effects of the β-subunit on fatty acid modulation of I_{Na} provide further support for a direct binding mechanism of fatty acids (particularly EPA and DHA) in $Na_V1.5$ channels.

Hypothesis #2: Alternatively, fatty acids may affect membrane fluidity (Jahangiri, Leifert, Patten, & McMurchie, 2000), which is correlated with the lateral pressure profiles of the membrane (Hamplova et al., 2004). ALA, EPA, and DHA all increase membrane fluidity as measured by fluorescence anisotropy using the probe TMAP-DPH, while stearic acid, oleic acid, and linoleic acid do not differ from control (Leifert, McMurchie, et al., 1999). These results correlate with the effect of these lipids on adult rat ventricular I_{Na}, which suggest these Na_V channels may be regulated by membrane fluidity. Therefore, the presence of fatty acids in the membrane may alter the lateral pressure felt by the channel, favoring the inactivated state. Similarly, insertion of fatty acids near the channel, but not bound to it, may relieve the curvature tension near the protein, altering the energy required for state transitions in the channel. The observation that nonionic detergents and benzyl alcohol mimics the inhibitory effects of PUFAs on $Na_V1.5$ channels (Leaf, Xiao, & Kang, 2002; Leifert, McMurchie, et al., 1999) provides support for this mechanism. On the other hand, the extracellular addition of the saturated fatty acid palmitic acid, which is expected to decrease membrane fluidity, has no effect on $Na_V1.4$ channels expressed in HEK293 cells (Bendahhou et al., 1997), nor does methyloleate or lysophosphatidylcholine affect Na_V currents in neuroblastoma cells (Linden & Routtenberg, 1989). This argues against changes in membrane fluidity being a common mechanism for the regulation of all Na_V channels by all PUFAs.

Hypothesis #3: Na_V channels are also located within lipid raft domains (Maguy, Hebert, & Nattel, 2006), whose composition can be altered by the presence of ω3PUFAs (Schley, Brindley, & Field, 2007). Thus, ω3PUFAs may affect Na_V channels by influencing a redistribution of these channels into nonraft domains.

Hypothesis #4: Since fatty acids, including ω3PUFAs, can act as signaling molecules, downstream intracellular changes, including metabolic changes and altered prostaglandin production, may play a role in modulation of these channels and contribute to protection against arrhythmias under ischemia. Intriguingly, DHA inhibition of I_{to} currents requires lipoperoxidation, and that DHA in its unperoxidized form has no effect (Jude et al., 2003). However, the inhibitory effects of EPA remain in the presence of inhibitors of cyclooxygenases and lipoxygenases, specifically indometacin and/or NDGA (Jo et al., 2005), suggesting this latter mechanism is likely not the primary basis for EPA regulation of Na_V channels. On the other hand, the importance of AA metabolism on Na_V channels appears to be more complicated. Inhibitors of lipoxygenase, cyclooxygenase, or cytochrome P450 epoxygenase have no effect on skeletal or cardiac Na_V channels following the extracellular application of AA (Bendahhou et al., 1997; Xiao et al., 1995), suggesting a more direct modulatory role for this fatty acid. In human and mouse skeletal muscles, the effect of intracellular AA can be mimicked by oleic and stearic acids (Wieland et al., 1992) which are not substrates for lipoxygenase or cyclooxygenase enzymes, suggesting their activation is independent of AA metabolism or the presence of an unsaturated bond. However, in DRG neurons the addition of indomethacin, a cyclooxygenase inhibitor, suppresses the hyperpolarized shift in steady-state activation induced by AA, but fails to affect steady-state inactivation (Lee et al.,

2002). On the other hand, the nonmetabolizable AA analog 5,8,11,14-Eicosatetrayonic acid (ETYA) still enables the shift in steady-state inactivation, but fails to have an effect on Na_V channel activation (Lee et al., 2002). These data suggest that in DRG neurons, the effect on Na_V channel activation is caused by metabolites of AA, while inactivation is directly regulated by the fatty acid. In cerebellar granular cells, ETYA and AA-CoA mimic the effects of extracellular AA on the voltage-dependence of activation (Fang et al., 2011), with Na_V channels directly regulated by this lipid extracellularly. On the other hand, the effects of intracellular applications of AA can be mimicked by prostaglandin E2 (an AA metabolite) but not ETYA and is affected by the inclusion of cyclooxygenase inhibitors (Fang et al., 2011). More specifically, I_{Na} modulation by intracellular AA involves PGE2-mediated activation of cAMP/PKA pathways (Fang et al., 2011). The attenuation of Na^+ currents in neuroblastoma cells by *cis*-fatty acids is more potent when applied intracellularly and is blocked in the presence of PKC inhibitors (Linden & Routtenberg, 1989), suggesting that fatty acid regulation of $Na_V1.6$ and $Na_V1.3$ channels may involve intracellular PKC signaling. The transient effect of AA and DHA exposure in olfactory receptor neurons (Seebungkert & Lynch, 2002) suggests that fatty acids may transiently elevate a second messenger which may in turn regulate $Na_V1.7$ channels.

Together, these data suggest that the mechanisms for fatty acid regulation of voltage-dependent sodium channels are complex and likely not conserved between isoforms. Alternately, if these mechanisms are conserved, there exists a relative weighting among them that differs for the various isoforms.

The fact that ω3PUFAs abolish the persistent current in inactivation-deficient $Na_V1.5$ channel, by shifting the steady-state inactivation to hyperpolarized potentials, and decreasing peak I_{Na} by 50% (Pignier et al., 2007; Xiao et al., 2006) suggests ω3PUFAs may provide a useful therapy to treat some patients with LQT3 or ischemia whose arrhythmias arise from persistent Na^+ currents. Evidence for an antiarrhythmic role for ω3PUFAs is suggested from studies of isolated neonatal and adult cardiomyocytes (Jahangiri et al., 2000; Kang & Leaf, 1995; Leifert, Jahangiri, & McMurchie, 1999, 2000). Clinical studies have even reported that PUFAs decrease the risk of sudden cardiac death from ventricular arrhythmias in humans (Marchioli et al., 2002). Dietary supplementation of adult rats with fish oils leads to an increased proportion of ω3PUFAs in ventricular sarcolemmal phospholipids and decreases the susceptibility to isoproterenol-induced asynchronous contractile activity compared to animals fed a diet enriched in saturated fats (Leifert, Jahangiri, Saint, & McMurchie, 2000). However, dietary supplemented adult rats show no change in the Na_V current—voltage relationship and only a modest +3 mV shift in the voltage-dependence of activation (Leifert, Jahangiri, Saint, et al., 2000). Thus, while dietary fish oil supplementation provides an antiarrhythmic benefit in these animals, it is not clear that this effect is driven by modulation of Na_V currents.

Dietary fat intake also contributes to the risk of developing various types of cancer (Shennan & Bishop, 1974; Wayne et al., 2006; Williams, Williams, & Weisburger, 1999). Breast cancers are known to be partially prevented by a fish-enriched diet, while prostate cancer growth is inhibited by ω3PUFAs (Bougnoux, 1999; Chavarro et al., 2007; Larsson, Kumlin, Ingelman-Sundberg, & Wolk, 2004; Leitzmann et al., 2004; Rose, 1997). It is unclear the precise mechanism by which dietary fats can regulate cancer development. However, since ion channels play essential roles in cell growth, metastasis, and secretion in many cell types, and several channels are regulated by lipids, including Na_V channels, it is conceivable they may play important roles in cancer development. 0.5 μM DHA decreases the migration of human breast cancer cells (MDA-MB-231) by 26% (Isbilen et al., 2006), which is similar to the effect of TTX-induced inhibition (Fraser et al., 2003; Isbilen et al., 2006), suggesting the effect of DHA on migration to be primarily through block of Na_V channels. DHA treatment in these cells reduces peak I_{Na} current by 43%, with no change in the voltage-dependence of activation (Isbilen et al., 2006). Intriguingly, chronic exposure of MDA-MB-231 cells to DHA induces a reduction in the expression of $Na_V1.5$ mRNA and protein levels (Isbilen et al., 2006). Similarly, prostate cancer cells (PC-3) treated with EPA for 48 h have reduced cell proliferation, invasion, and endocytosis which is linked to decreased I_{Na} and the levels of $Na_V1.6$ and $Na_V1.7$ mRNA (Nakajima et al., 2009). This suggests that I_{Na} may be necessary for endocytosis. Long-term EPA exposure also decreases $Na_V1.7$ mRNA levels in bronchial smooth muscle cells (Jo et al., 2005), but does not affect mRNA levels of $Na_V1.4$ channels in rat ventricular cells (Kang et al., 1997). The mechanism of ω3PUFAs regulation of channel expression remains to be elucidated.

3.3 Palmitoylation

Palmitoylation is the posttranslational process by which eukaryotic and viral proteins are covalently linked to the C_{16} saturated fatty acid of palmitic acid. Palmitoylation occurs on the thiol group of cysteine residues

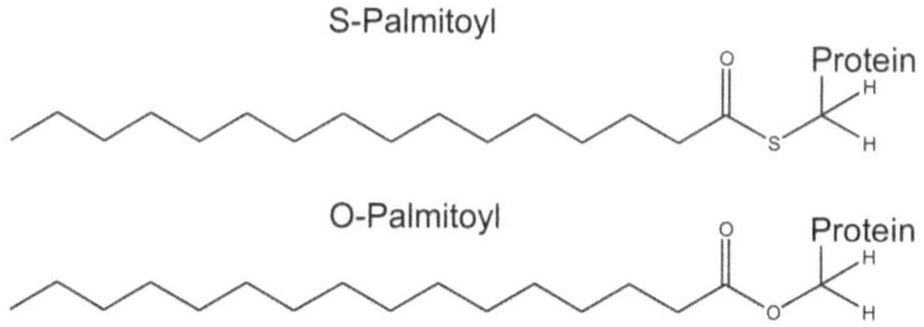

Figure 3 ***Modification of a cysteine or serine residue by a palmitoyl group.*** An internal cysteine residue in the protein can be thioester-linked to the 16-carbon saturated fatty acid, palmitic acid to give an S-palmitoyl group. Palmitic acid can also be ester-linked to serine (or threonine) residues to given an O-palmitoyl group.

(S-palmitoylation) or on the hydroxyl group of serine and threonine residues (O-palmitoylation) and is typically reversible (Fig. 3). However, the prediction of palmitoylation sites using *in silico* methods is difficult since no consensus sequence has been characterized (Smotrys & Linder, 2004). This process is largely mediated by palmitoyl acyltransferases (PATs) (Blaskovic, Blanc, & van der Goot, 2013); however, palmitate can also be added to proteins by membrane-bound O-acyltransferases (MBOATs) (Matsuda et al., 2008) and by autoacylation (Duncan & Gilman, 1996). PATs are multispanning integral membrane proteins whose active sites are present on the cytoplasmic side. While most PATs have been found in the Golgi or ER, some have also been identified to localize to the plasma membrane, endosomes, and other cellular organelles (Frenal et al., 2013; Ohno et al., 2012). Since palmitoylation is a dynamic posttranslational process, it plays a significant role in subcellular trafficking of proteins between membrane compartments and associations in lipid rafts, modulating protein–protein interactions, triggering further posttranslational modification such as ubiquitination, as well as regulating the functional properties of a variety of membrane proteins (Blaskovic et al., 2013; Charollais & Van Der Goot, 2009; Fukata & Fukata, 2010; Hayashi, Thomas, & Huganir, 2009; Iwanaga, Tsutsumi, Noritake, Fukata, & Fukata, 2009; Macdonald-Obermann & Pike, 2009; Resh, 2006; Shipston, 2011).

In rat brain neurons, palmitoylation is required for biosynthesis and processing of Na_V channels as they traffic to the plasma membrane. In these channels, N-linked glycosylation is required for palmitoylation, likely because inhibition of N-glycosylation prevents channels from being transported from the endoplasmic reticulum to the Golgi (Schmidt & Catterall, 1987). Depalmitoylation of wild-type rat $Na_V1.2$ at residue C1182 alters channel inactivation, inducing a −10 mV hyperpolarized shift in the steady-state inactivation, and a threefold slower recovery from fast inactivation (Bosmans, Milescu, & Swartz, 2011). Therefore, in addition to affecting channel expression, palmitoylation may also increase the pool of channels ready for activation at resting membrane potentials. Additionally, palmitoylation of rat $Na_V1.2$ increases the affinity of the voltage-sensor toxins PaurTx3 and ProTx-II for channels containing the G1079C mutation.

4. GLYCEROPHOSPHOLIPIDS

Glycerophospholipids are glycerol-based phospholipids derived from glycerophosphoric acid that contains at least one O-acyl, or O-alkyl, or O-alk-1′-enyl chain attached to the glycerol. These lipids are the major component of most biological membranes. Typically, each glycerophospholipid molecule consists of a small polar head group facing the aqueous phase and two long hydrophobic chains directed at one another, thereby forming a stable bilayer structure. Examples of glycerophospholipids found in biological membranes include phosphatidic acid (PA), phosphatidylcholine (also known as PC or GPCho, and lecithin), phosphatidylethanolamine (PE or GPEtn), phosphatidylserine (PS or GPSer), phosphatidylglycerol (PG), cardiolipin (CL), phosphatidylinositol (PI) and its phosphorylated derivatives (PIPs). These head groups are shown in Fig. 4 for reference. In addition to serving as a primary component of cellular membranes, intra- and intercellular proteins can bind to membranes. In eukaryotic cells, PA, PI, and PIPs can also act as precursors for second messengers, with the most notable example being the cleavage of $PI(4,5)P_2$, into the important intracellular messenger inositol triphosphate (IP3) and its membrane companion diacylglycerol (DAG).

The synthesis and degradation pathways of glycerophospholipids differ between eukaryotes and prokaryotes. Synthesis in prokaryotes involves the synthesis of phosphatidic acid and polar head groups, while in eukaryotes the Kennedy pathway is primarily responsible for the metabolism of PC and PE, while the CDP-DAG pathway generates PG, PS, and PI (including PIPs) (Fig. 5). It is important to note that the proportion of each type of phospholipid is dependent on the cell type (Table 1). Both gram-positive and gram-negative bacterial membranes primarily contain PE, PG, and CL lipids in both leaflets—although asymmetrically. PS, PA and PC lipids are also found in some bacteria (Goldfine, 1984; Jurado, Pinheiro, & Madeira, 1991; Sohlenkamp, Lopez-Lara, & Geiger, 2003; van Meer & de Kroon, 2011; van Meer, Voelker, & Feigenson, 2008). While PI lipids

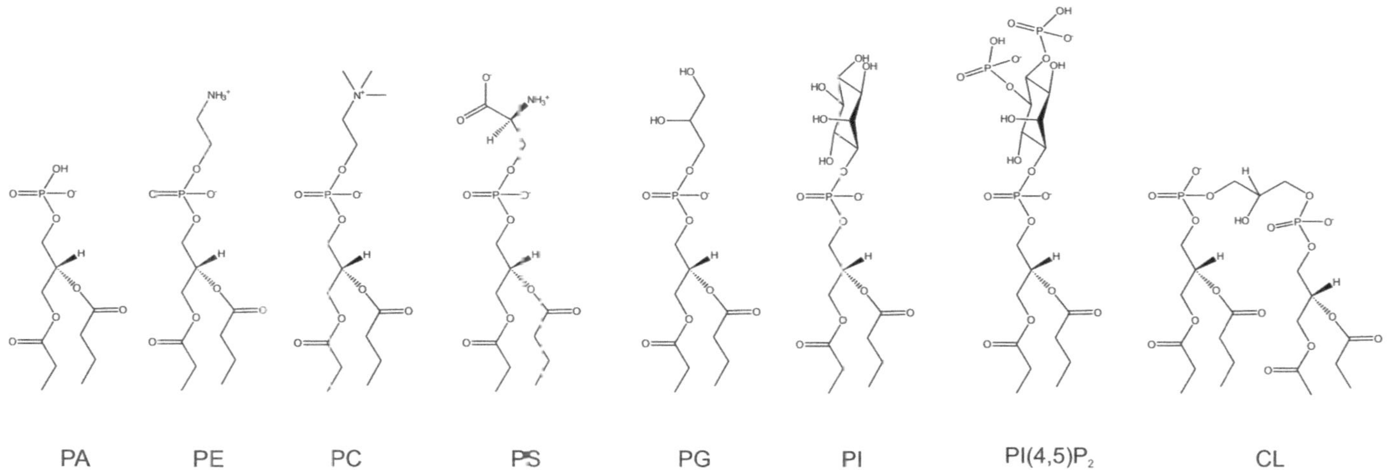

Figure 4 ***Phospholipid head groups.*** Examples of glycerophospholipids found in biological membranes include phosphatidic acid (PA), phosphatidylcholine (also known as PC or GPCho, and lecithin), phosphatidylethanolamine (PE or GPEtn), phosphatidylserine (PS or GPSer), phosphatidylglycerol (PG), cardiolipin (CL), phosphatidylinositol (PI) and its phosphorylated derivatives (PIPs).

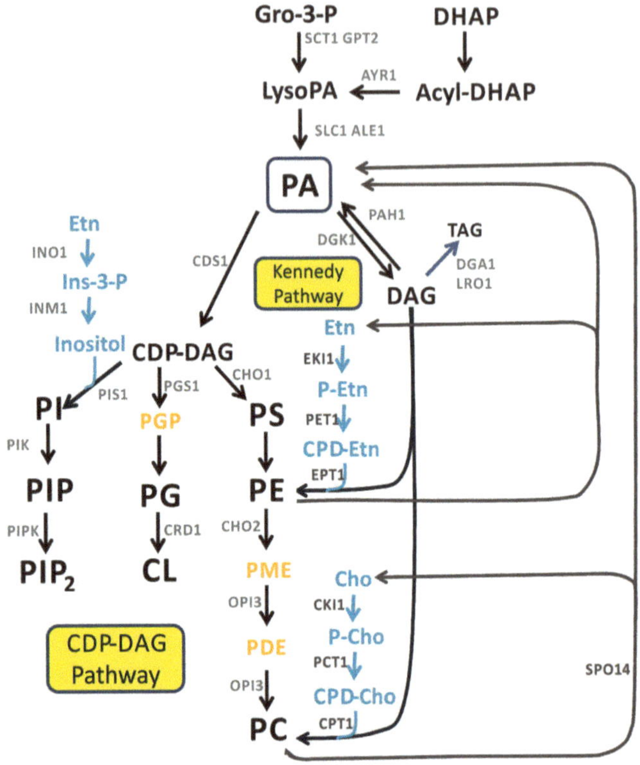

Figure 5 ***Outline of glycerophospholipid biosynthesis in eukaryotes.*** The major phospholipids (*PA*, phosphatidic acid; *PI*, phosphatidylinositol; *PIP*, phosphatidylinositol phosphate; *PS*, phosphatidylserine; *PE*, phosphatidylethanolamine; *PC*, phosphatidylcholine; *PG*, phosphatidylglycerol; *CL*, cardiolipin) are shown in bold. PGP (phosphatidylglycerophosphate), PME (phosphatidylmonomethylethanolamine), and PDE (phosphatidyldimethylethanolamine) are minor phospholipids. The Kennedy pathway is primarily responsible for the metabolism of PC and PE. The CDP-DAG pathway generates PG, PS, PI, and PIPs. Also shown are the points of entry into the pathway of the soluble phospholipid precursors such as inositol, choline, ethanolamine, and glycerol-3-phosphate (Gro-3-P). The enzymes catalyzing individual steps in the pathway are also indicated in gray.

are essential to eukaryotic cells, they are rarely found in prokaryotes and are generally confined to actinomycetes, myxobacteria, and a select few other bacteria (Sohlenkamp et al., 2003).

4.1 Glycerophospholipid regulation of bacterial and eukaryotic Na_V channels

Glycerophospholipids are an important determinant of the function of numerous ion channels (Cheng, D'Avanzo, Doyle, & Nichols, 2011;

Table 1 Percent phospholipid composition of various cell types

	Phospholipid composition					
Cell type	**PS**	**PE**	**PC**	**PI**	**SM**	**Others**
CHO-K1[a]	5.9	16.6	52.3	7.2	8.4	9.7
HEK cells						
Human placenta[b]	5.4	26.7	46.0	3.9	14.1	3.9
Human endometrium[b]	15.1	22.5	42.5	6.3	8.4	5.2
Human brain gray matter[b]	22.8	35.1	24.1	2.5	12.3	3.2
Human brain white matter[b]	19.4	35.6	24.8	2.3	15.6	2.3
Human lung[b]	7.8	18.8	53.2	4.0	11.4	4.8
Human liver[b]	3.5	28.8	45.0	9.2	5.6	7.9
Human spleen[b]	8.8	26.3	43.6	4.5	12.9	3.9
Human kidney[b]	7.2	29.6	36.2	6.4	13.9	6.7
Human cardiomyocytes[b]	3.6	28.4	42.8	6.7	5.4	13.1
Human skeletal myocytes[b]	3.5	27.0	49.0	8.5	4.5	7.5
Escherichia coli[c]	<1	75	N.D.	N.D.	N.D.	25

PI, phosphatidylinositol; *PS*, phosphatidylserine; *PE*, phosphatidylethanolamine; *PC*, phosphatidylcholine; *SM*, sphingomyelin; others include lysophosphatidylcholine, phosphatidylglycerol, phosphatidic acid, and cardiolipin.
[a]Saito, et al., JBC, 1988, *273*(27),17,199.
[b]Singh, and Swartwout, Lipids, 7(1),26.
[c]Raetz, 1990

D'Avanzo, Cheng, Doyle, & Nichols, 2010; Powl, East, & Lee, 2008a, 2008b; Schmidt, Cross, & MacKinnon, 2009; Schmidt, Jiang, & MacKinnon, 2006; Xu, Ramu, & Lu, 2008). Without question, the most widely examined phospholipids are phosphoinositides (PIPs) (Fig. 4), particularly $PI(4,5)P_2$ (Suh & Hille, 2008). PIPs comprise 0.2–1% of the phospholipids in most cell membranes and are differentially distributed in each subcellular membrane (Balla, 2013). The quantity of PIPs in cell membranes can be readily manipulated by nonspecific sequestration (eg, polylysine), specific sequestration through PH-domain containing proteins, or enzymatically by activation of phospholipases or voltage-sensitive phosphatases (VSPs). However, there is growing evidence that phospholipids possessing different head groups (Fig. 4) can also regulate the function of various ion channels.

Of 12 bacterial Na_V channel sequences cloned, ionic currents have only observed from 3 of these homologs when expressed in mammalian cells, NaChBac from *Bacillus halodurans*, Na_VSp from *Silicibacter pomeroyi*, and

Na_VPz from *Paracoccus zeaxanthinifaciens* (Chahine, Pilote, Pouliot, Takami, & Sato, 2004; Koishi et al., 2004; Kuzmenkin, Bezanilla, & Correa, 2004). These channels share only 39% sequence identity with NaChBac and approximately 60% sequence identity, suggesting a major determinant of function in these channels may reside outside the protein. It appears that this exogenous regulator of bacterial Na_V function may in part be the lipid bilayer, as is evident by examining different lipid conditions in which various homologs were found to be functional. When directly compared, the activity of purified Na_VSp, NaChBac, and Na_VMs from *M. spirillium* homologs are uniquely dependent on the lipids in which they were reconstituted (D'Avanzo et al., 2013). In a 3:1 ratio of PE and different "guest" glycerophospholipids, Na_VMs and Na_VSp show strong preferences for different negatively charged lipids (PI and PG, respectively). On the other hand, NaChBac exhibits a more modest variation with lipid type, with maximal activity observed in the presence of PG. Notably, no activity above background can be observed for Na_VMs or Na_VSp channels in proteoliposomes containing PA; however, NaChBac channels still maintain some activity under these conditions. Conversely, in proteoliposomes containing PS, activity can only be observed for Na_VMs channels, with Na_VSp and NaChBac channels showing no activity above background. NaChBac is insensitive to $PI(4,5)P_2$, while Na_VSp channels show a marked inhibition when 1% $PI(4,5)P_2$ is included in the proteoliposomes. Intriguingly, the pore-domains of bacterial Na_V channels generally remain functional in the absence of the voltage-sensing domains (D'Avanzo et al., 2013; McCusker et al., 2012; McCusker, D'Avanzo, Nichols, & Wallace, 2011; Saha et al., 2015; Shaya et al., 2011). The lipid dependence of the pore domains of Na_VMs and Na_VSp is drastically reduced compared to full-length channels (D'Avanzo et al., 2013), suggesting the lipid dependence of sodium channels may arise primarily from effects on the voltage-sensor domains, which are then transmitted to the pore. This is not to say that lipids do not affect the pore region at all. In fact, PG, PS, and PI favored Na_VSp pore opening (D'Avanzo et al., 2013). Na_VRh channels yield no measurable ionic currents when functionally assessed in insect or mammalian cells, or when purified channels and constituted into bilayers containing 3:1 POPE:POPG or DPhPC lipids (Zhang et al., 2012). However, Na_VRh channels whose selectivity filter is replaced by that of NaChBac channels are functional. Thus, it is possible that the pore-region of Na_VRh and NaChBac have differing sensitivities to these lipids. Overall, these data indicate that lipid-dependent regulation of bacterial Na_V channels varies among homologs.

The function of eukaryotic Na_V channels also appears to be regulated by the phospholipid composition. The characteristics of binding LqTx to neurotoxin-sensitive channels purified from rat brain is also lipid dependent, differing for purified channels reconstituted into PC only bilayers, PC supplemented with brain lipids, and in native membranes (Tamkun, Talvenheimo, & Catterall, 1984). For example, in native membranes, LqTX binding increases 500-fold upon depolarization, while in PC lipids supplemented with brain lipids binding affinity is voltage independent. Moreover, LqTx does not bind sodium channels in *egg* PC liposomes. Purified TTX-sensitive Na_V channels from *Electrophorus electricus* and rat brain are functional based on single-channel recordings in planar lipid bilayers of PE:PC (Hartshorne, Keller, Talvenheimo, Catterall, & Montal, 1985; Recio-Pinto, Duch, Levinson, & Urban, 1987). Batrachotoxin- and tetrodotoxin-modulated $^{22}Na^+$ influx is enhanced by the addition of up to 50% PS in the liposomes, compared to PC alone. However, negative staining EM indicates that PC/PS liposomes formed by freeze−thaw sonication method are in fact larger than PC only liposomes, and this increased influx is associated with the increased internal volume of and surface area for channel incorporation (Miguel, Balbi, Castillo, & Villegas, 1992). Purified $Na_V1.4$ channels are functional in planar black lipid membranes formed with 5% PC and 2% cholesterol in *n*-octane (Zhang, Dunlop, & Dalziel, 2007). Together, these data suggest anionic lipids may not be absolutely required for the function of eukaryotic Na_V channels.

4.2 Structural role for phospholipids in Na_V channels

The importance of phospholipids in Na_V channels is evident from recent structural studies. Several atomic resolution structures of bacterial Na_V channels in various conformations have been solved over recent years. The majority of these structures resolve electron density that can be attributed to lipid or detergent molecules. 20 or 28 molecules of 1,2-dimyristoyl-sn-glycero-3-phosphocholine (DMPC) are observed bound to the closed Na_VAb channel (PDB:3RVY and 3RVZ) (Fig. 6). The acyl chains of four of these lipids penetrate the fenestrations formed in the side of the pore domain in both Na_VAb and Na_VRh (Payandeh, Scheuer, Zheng, & Catterall, 2011; Zhang et al., 2012). Since the fenestrations were not observed in the open pore structure of Kv channels, the authors proposed that in Na_VAb channels, the lipid tails may withdraw from the fenestrations, allowing them to close in the open state. Another lipid phosphate group lies within 5 Å of a triad of polar residues (SKT) in the S1−S2 linker, and a Thr in

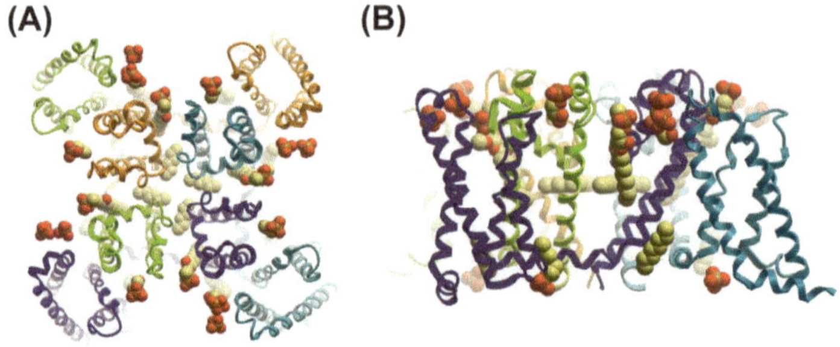

Figure 6 ***Lipids have been observed in several bacterial Na_V crystal structures.*** The crystal structure of bacterial Na_VAb from *Arcobacter butzleri* (PDB:3RVY) with 20 bound 1,2-dimyristoyl-*sn*-glycero-3-phosphocholine (DMPC) lipid molecules. (A) is a top-down view, and (B) is a lateral view. Similar arrangements of lipid and detergent molecules were observed in other bacterial Na_V channel structures.

the P-loop, and may act to help bridge the two domains. A similar arrangement of DMPC lipids can be observed in the inactivated conformations of Na_VAb (PDB:4EKW) (Payandeh, Gamal El-Din, Scheuer, Zheng, & Catterall, 2012). In the inactivated conformation of Na_VRh (PDB:4DXW), β-nonylglucoside molecules are interacting with the hydrophobic extracellular residues at the interface between two pore subunits and the S1–S2 loop of a third subunit (Zhang et al., 2012). Moreover, these channels require the presence of 3:1:1 POPC:POPE:POPG during purification and crystallization providing evidence for a structural role of membrane lipids in these channels (Zhang et al., 2012). Crystal structures of the pore-only domains of Na_VMs (McCusker et al., 2012) (PDB:4F4L) and Na_V-Ae1p (Shaya et al., 2014) (PDB:4LTO, 4LTP, 4LTR, and 4LTQ) lack bound lipids. However, in a follow-up structure the Na_VMS pore-only construct, eight HEGA-10 detergent molecules can be seen bound to each tetramer (Bagneris et al., 2014) (PDB: 4CBC), with one bound between each subunit's S5 and S6 helices and another bridging the first P-loop helix to the neighboring subunit's S6 helix. These data indicate that lipids likely play an important structural role in bacterial Na_V channels.

5. STEROLS

5.1 Cholesterol

The single channel properties of Na_V channels isolated from human cortical brain tissues and incorporated into PE/PC planar bilayers are unaffected by the addition of up to 50% w/w cholesterol (Rehberg, Urban, &

Duch, 1995). However, even at concentrations significantly lower than levels found in neuronal cells, cholesterol significantly inhibits pentobarbital-induced block of these channels in a competitive manner. At the time, it was not clear whether the effect of cholesterol on Na_V channels was due to changes in the physical properties of the bilayer or due to a direct interaction with the channel or anesthetic.

Skeletal $Na_V1.4$ channels expressed in HEK cells have reduced peak currents and inactivate at more hyperpolarized potentials by −8.9 mV when cholesterol is depleted from the cell membrane (Lundbaek et al., 2004). This effect can be partially reversed by the repletion of cholesterol. Cholesterol depletion does not appear to affect the activation properties of $Na_V1.4$ channels. On the other hand, cholesterol enrichment in HEK cells does not affect channel inactivation, but does shift the voltage-dependence of activation to more depolarized potentials by +5.5 mV. Since the effect of cholesterol depletion can also be induced by cellular exposure to 10 μM Triton X-100 or 2.5 mM β-octyl-glucoside (βOG) which alter the physical properties of the bilayer, Lundbaek and colleagues suggest that cholesterol regulation of $Na_V1.4$ channels occurs indirectly through the elastic properties of the bilayers.

Hypercholesterolemia results in a nearly twofold increase in the cholesterol content of the rabbit ventricular myocyte sarcolemma. Hypercholesterolemia also decreases peak current density of ventricular Na_V currents, with these channels inactivating at more negative potentials and recovering more slowly from inactivation (Wu et al., 1995). $Na_V1.5$ channels are implicated as the principle isoform in atrial and ventricular myocytes and contribute to the propagation of action potentials in the heart (Balijepalli & Kamp, 2008). They partition into cholesterol-rich membrane fractions and directly interact with caveolin-3 (Yarbrough, Lu, Lee, & Shibata, 2002). When this interaction is inhibited, β-adrenergic regulation of $Na_V1.5$ current through direct $G_s\alpha$-mediation is also inhibited (Palygin, Pettus, & Shibata, 2008; Yarbrough et al., 2002). Thus, caveolin-3 may directly stabilize the interaction between the channels and $G_s\alpha$ subunits (Palygin et al., 2008), and thus cholesterol regulation of $Na_V1.5$ channels results from its ability to control the integrity of caveolae.

$Na_V1.8$ channels are distributed in clusters along the neurites of DRG neurons, with 80.5% of these clusters colocalizing with the lipid raft domain marker, GM1 (Pristera, Baker, & Okuse, 2012). Moreover, $Na_V1.8$ is primarily present in fractions enriched in Flotillin-1, used to identify planar lipid rafts, rather than caveolae-type lipid rafts, which are identified by the presence of Caveolin-2. Intriguingly, in cell bodies of these same neurons, $Na_V1.8$ and GM1 do not colocalize, indicating they are distributed in non-raft domains. Treatment of DRG neurons with MβCD or 7-ketocholesterol leads to a redistribution from raft to nonraft domains, without inducing any macroscopic changes in channel distribution. Further studies in the sciatic nerve suggest that distribution into raft or nonraft domains is strongly dependent on cell type. The effects of cholesterol modulation and lipid raft disruption on $Na_V1.8$ function result in a significant decrease in the number of soma able to respond to mechanical stimulation of the axon, and a decrease in the mean speed of conduction. Since electrophysiological recordings to date have been limited to the cell soma, where $Na_V1.8$ is not found in lipid rafts, it is unclear what effects cholesterol has on $Na_V1.8$ gating.

Cholesterol depletion of *Xenopus laevis* oocytes by MβCD subtly affects the gating properties of wild-type $rNa_V1.2$ channels, but induces a 10-fold decrease in the apparent affinity of the voltage sensor toxin, PaurTx3 (Bosmans et al., 2011). The effect of cholesterol on these channels is independent of those induced by palmitoylation.

Electrophysiological recordings of the pore from Na_VSp, Na_VAe, and Na_VAb were performed in liposomes containing DPhPC and cholesterol (Shaya et al., 2011). Intriguingly, under these conditions, single-channel recording function can only be obtained from Na_VSp and Na_VAb, perhaps indicating that Na_VAe is sensitive to the presence of cholesterol or to the nature of the phospholipid in the membranes in which they are embedded.

5.2 Oxysterols

Oxysterols are oxidized derivatives of cholesterol (Fig. 7). They play important roles in cholesterol homeostasis, sphingolipid metabolism, platelet aggregation, and atherosclerosis, as well as apoptosis (Schroepfer, 2000). Little is known about how oxysterols regulate Na_V channels. Cholestane-3β,5α,6β-triol induces a hyperpolarization shift in the fast and slow steady-state inactivation curves of sodium channels and reduces Na_V current density in a concentration-dependent manner (Tang et al., 2015). Moreover, cholestane-3β,5α,6β-triol significantly increases the latency of seizure onset and attenuated the severity of seizures in kainic acid–induced seizure mouse models. Neurites of DRG neurons express $Na_V1.8$ channels which are distributed in clusters with 80.5% of these clusters in lipid raft domains (Pristera et al., 2012). Treatment of DRG neurons with 7-ketocholesterol leads to a redistribution from raft to nonraft domains, without inducing any macroscopic changes in channel distribution. This mimics the effects

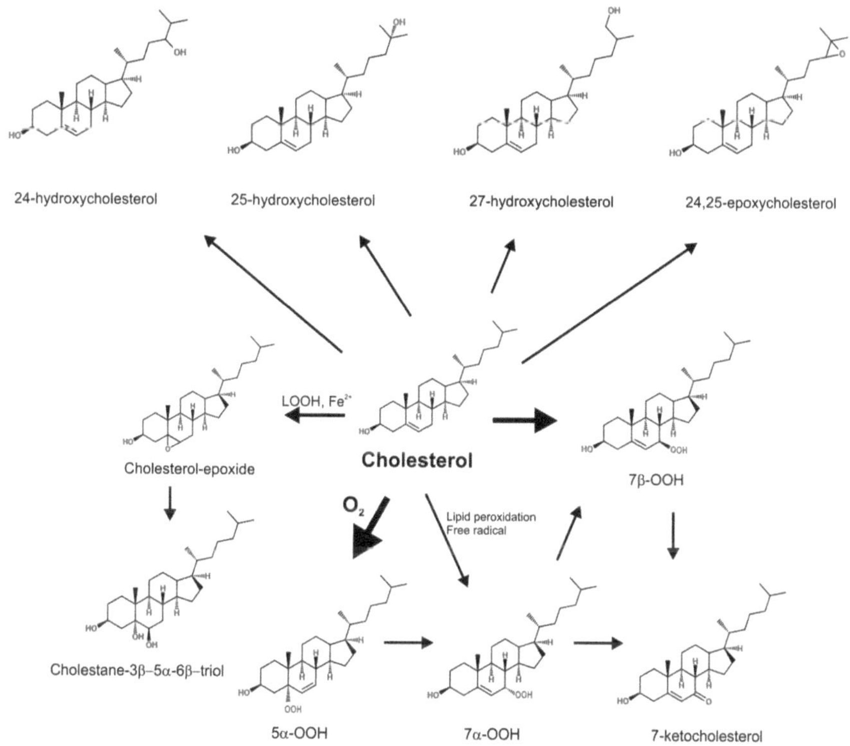

Figure 7 Cholesterol, oxysterols, and other sterol derivatives.

seen following treatment with MβCD. Additional efforts in this domain will be necessary to understand how cholesterol metabolism affects Na_V channels.

6. SPHINGOLIPIDS

Sphingolipids are a class of lipids with a backbone of aliphatic amino alcohols or sphingoid bases (or long changed bases). The sphingosine backbone is O-linked to a head group that is typically charged (for example, ethanolamine, serine, or choline) and N-linked (amide linked) to an acyl group such as a fatty acid. Sphingolipids play a central role in signal transduction and cell recognition.

Sphingosine (2-amino-4-octadecene-1,3-diol) is an amino alcohol with an 18-carbon unsaturated alkyl chain, forming a primary component of other sphingolipids. This lipid is synthesized from palmitoyl CoA and serine

(Fig. 8) via the formation of dehydrosphingosine and dihydrosphingosine intermediates. Sphingosine can then be phosphorylated to shingosine-1-phosphate by sphingosine kinase type 1 or type 2.

Sphingosine incubation of rat ventricular myocytes induces a ninefold reduction of Na^+ currents, with peak currents shifting toward more depolarized potentials (Yasui & Palade, 1996). Time to peak is nearly double in cells treated with sphingosine, while inactivation is 2.5-fold slower. The authors of this study suggest that the inhibition of Na^+ current by sphingosine may contribute to the reduced excitability in muscles during ischemic insult or skeletal muscle fatigue, as well as sphingosine blocked depolarization-induced Ca^{2+} influx in synaptosomes of the nervous system.

6.1 Sphingomyelin

Sphingomyelin, a type of sphingolipid found in animal membranes, particularly the myelin sheath, is comprised of phosphatidylcholine (PC) or phosphatidylethanolamine (PE) head group linked to a ceramide (sphingosine backbone and fatty acid). It is synthesized in the ER, and at the *trans* Golgi and is enriched in the plasma membrane with a greater concentration in the outer leaflet (Testi, 1996). In humans, sphingomyelins comprise nearly 85% of all sphingolipids and 10–20 mol% of the total plasma membrane lipids. Sphingomyelins comprise 4–18% of all sarcolemma membrane lipids (Hamplova et al., 2004; Post, Verkleij, & Langer, 1995), with 93% present in the outer leaflet (Post et al., 1995; Testi, 1996). In Chinese hamster ovary (CHO) cells, sphingomyelins comprise only 12–14% of the plasma membrane phospholipids (Hermansson, Uphoff, Kakela, & Somerharju, 2005; Pike, Han, & Gross, 2005), but in the plasma membranes of *X. laevis* oocytes they can comprise as much as 25% of the total phospholipids (Hermansson et al., 2005; Hill et al., 2005; Pike et al., 2005; Stith et al., 2000). Thus, it is important to consider the cellular context in which sphingomyelin regulation of membrane proteins is studied.

Sphingomyelins are nearly cylindrical, though their alkyl tails often differ in the number of carbons with a high degree of saturation (Barenholz & Thompson, 1999). Their high transition temperature and greater tendency to form intermolecular hydrogen bonds compared to similar phospholipids can lead to lateral heterogeneity and the spontaneous formation of microdomains (Barenholz & Thompson, 1999; Massey, 2001). Moreover, intermolecular hydrogen bonding between sphingomyelin and cholesterol may enable tight packing in the liquid-ordered phase of "rafts" and favoring their formation (Barenholz, 2004).

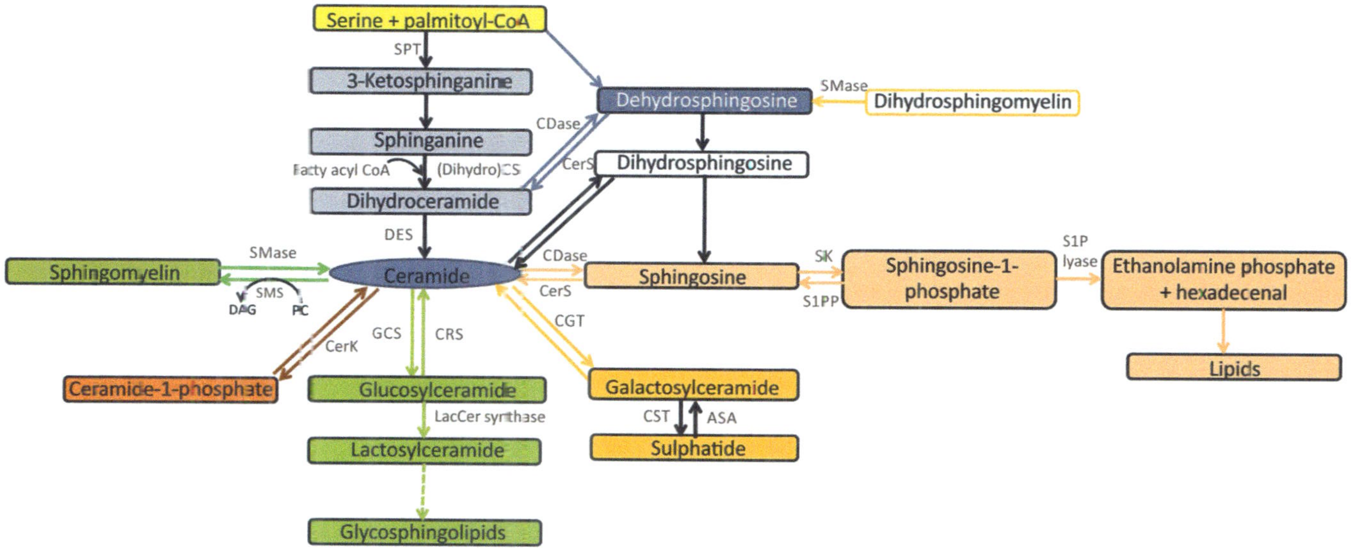

Figure 8 ***Pathways for sphingolipid and ceramide metabolism.*** Sphingolipids and ceramides are generated primarily from serine and palmitoyl-CoA. Ceramides lie at the center of sphingolipid biosynthesis. The enzymes that catalyze individual steps in the pathway are indicated in gray.

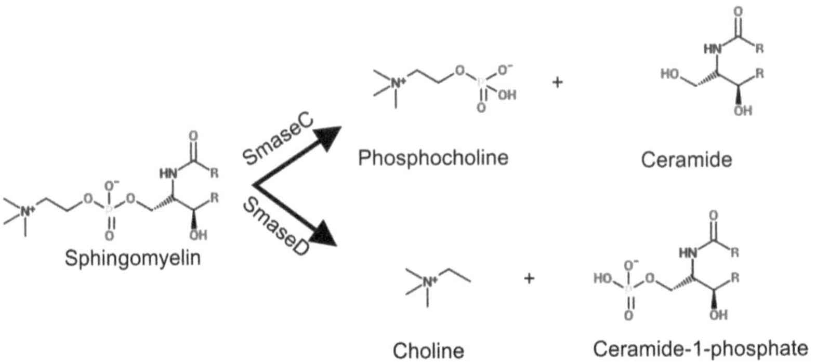

Figure 9 ***Lipid hydrolysis by sphingomyelinases C and D.*** Cleavage of sphingomyelin lipids by SMaseC generates a phospholipid (PC) and a ceramide molecule. Sphingomyelin cleavage by SMaseD generates a choline molecule and a nonphospholipid (ceramide-1-phosphate). Treatment of cells with SMases has been a useful in uncovering the regulation of several ion channels by sphingomyelin.

Sphingomyelinase (SMase) D is the active toxin in brown spider venom that cleaves the positively charged choline moiety from the phosphoryl group of sphingomyelin leaving ceramide-1-phosphate to diffuse through the membrane (Fig. 9). It is also produced by some bacteria including *Corynebacterium pseudotuberculosis* (Tambourgi, Goncalves-de-Andrade, & van den Berg, 2010). Since this enzyme can be readily purified as a recombinant protein in *E. coli*, it has quickly developed into a rather useful tool to assess the role of sphingomyelin to regulate membrane proteins (Combs, Shin, Xu, Ramu, & Lu, 2013; Ramu, Xu, & Lu, 2006, 2007; Xu et al., 2008). The use of SMaseD on Na_V channels has uncovered a complex relationship between these lipids and the properties of activation and inactivation (Combs et al., 2013). The treatment of *Xenopus* oocytes expressing skeletal muscle $rNa_V1.4$-IR, which lacks inactivation, with SMaseD induces a hyperpolarizing shift in the voltage-dependence of activation. Intriguingly, this corresponds to similar stimulation observed for Kv and Cav channels. SMaseD treatment slows the inactivation of $rNa_V1.2$, $rNa_V1.4$ and $hNa_V1.4$ channels expressed in *Xenopus* oocytes, but only when they were coexpressed with the $Na_V\beta1$ subunit. On the other hand, the inactivation of human $Na_V1.5$ channels expressed in oocytes is slowed by SMaseD treatment in the presence or absence of the $Na_V\beta1$ subunit. This suggests that $Na_V1.4$ interactions with sphingomyelin that enable fast inactivation are enhanced in the presence of the $Na_V\beta1$ subunit, but these interactions are already optimized for $Na_V1.5$ channels. Intriguingly,

SMaseD treatment does not slow the inactivation kinetics of native Na_V channels in Chinese hamster ovary (CHO) cells nor Neuro-2A (N2A) cells even in the presence of overexpressed $Na_V\beta1$. Moreover, the sensitivity of Na_V channel activation to SMaseD depends on the paddle motif, while the effect of SMaseD on inactivation is conferred on the channel by the S5–S6 linkers. The authors of this study suggest that sphingomyelin occupies two different sites; one that affects inactivation, and the other channel activation. In oocytes, both sites may be bound by sphingomyelin on account of the high concentrations of this lipid in the membrane, while in mammalian cells, the site related to channel inactivation is likely bound by other lipids.

Sphingosylphosphorycholine (SPC) is synthesized from sphingomyelin via the sphingomyelin deacylase. SPC is implicated as a promoter of mitosis in a variety of cell types (Desai et al., 1993; Desai & Spiegel, 1991). The extracellular application of SPC induces a rapid increase in intracellular calcium in a variety of cell types (Chen, Chin, & Chueh, 1998; Chin & Chueh, 1998; Desai et al., 1993; Meyer zu Heringdrof, van Koppen, Windorfer, Himmel, & Jakobs, 1996; Okajima & Kondo, 1995; Orlati, Porcelli, Hrelia, Lorenzini, & Rugolo, 1998; Tornquist & Ekokoski, 1994), cytoskeletal rearrangements (Seufferlein & Rozengurt, 1995), alters cell migration (Boguslawski, Lyons, Harvey, Kovala, & English, 2000; Rumenapp et al., 2000), and activates acetyl-choline sensitive potassium currents ($_{IK,ACh}$) in atrial myocytes (Bunemann et al., 1996) and Ca^{2+} currents of ventricular myocytes (Yasui & Palade, 1996). SPC has little to no effect on Na^+ currents measured from isolated rat ventricular myocytes (Yasui & Palade, 1996).

6.2 Ceramides

Ceramides are a family of waxy lipids composed of a sphingosine backbone and a fatty acid. They are the products of sphingomyelin cleavage. Ceramides enhance membrane rigidity and can stabilize the formation of lipid raft domains (Catapano et al., 2011). In addition to their structural and organizational role in the cell membrane, ceramides also act as novel lipid second messengers to mediate a diversity of signaling pathways, including apoptosis, cell growth arrest, cell differentiation, cell senescence, cell migration, and adhesion (Hannun & Obeid, 2008; Testi, 1996).

Neuronal Growth Factor (NGF) which is elevated during inflammation and known to activate the sphingomyelin signaling pathway (Dobrowsky, Werner, Castellino, Chao, & Hannun, 1994) augments the number of action potentials evoked over a 20 min period in embryonic and adult sensory neurons (Zhang, Vasko, & Nicol, 2002). Intracellular perfusion of a neutral SMase as well as the direct perfusion of 1 μM C_2-ceramide have similar effects. While inhibition of neutral SMase by glutathione completely blocks the NGF-induced augmentation of action potential firing, dithiothreitol (DTT) which inhibits acidic SMase, has no effect. Part of the increase in action potential firing in these neurons can be attributed to enhanced peak currents for tetrodotoxin (TTX)-resistant Na_v channels induced by the presence of NGF or C_2-ceramide. While it is unclear what is the effector system by which ceramide modulates neuronal Na_v channels, it may be through direct stimulation of the serine/threonine type ceramide-activated protein kinase (CAPK) (Joseph, Byun, Bittman, & Kolesnick, 1993; Liu, Mathias, Yang, & Kolesnick, 1994; Mathias, 1993).

Ceramides have also been reported to induce apoptosis of cerebellar granule neurons (Taniwaki, Yamada, Asahara, Ohyagi, & Kira, 1999; Toman et al., 2002), possibly through activation of executioner caspase 3 (Vaudry et al., 2003). The application of 10 μM C_2- or C_6-ceramide to rat cerebellar granule neurons rapidly inhibit I_{Na} by over 15%, which can be recovered after its removal from the bath solution (Liu et al., 2008). C_6-ceramide has no effect on the steady-state activation or inactivation properties of the Na_v channels; however, peak I_{Na} are reduced. Ceramide inhibition of I_{Na} in these cells appears to be mediated through G-protein/phospholipase C—dependent and ryanodine-sensitive receptor-mediated Ca^{2+} release, which is independent of apoptosis.

In myoblasts, a Na^+-activated K^+ current ($I_{K,Na}$) may play an important role in fusion and are altered in development (Zhou, Liu, Hu, Zhang, & Mei, 2004). This current is directly correlated with I_{Na} from TTX-sensitive Na_v channels, which is the main current in prefusion myoblasts (Lu, Liu, Liao, Zhang, & Mei, 2005). Extracellular application of C_6-ceramide induces a dose-dependent inhibition of peak I_{Na} amplitude with no change in the voltage-dependence of steady-state activation (Liu, Xu, Zhang, Fang, & Mei, 2007). Additionally, C_6-ceramide induces a −16 mV hyperpolarizing shift in the voltage-dependence of steady-state inactivation. This effect can be reproduced by C_2-ceramide, but not the inactive dihydro-C_6-ceramide. Inhibition of I_{Na} in rat myoblasts by ceramides is mediated by IP_3 receptor—dependent Ca^{2+} release, downstream of the G-protein/PLC pathway. In summary, ceramides affect voltage-gated sodium channels primarily through the activation of intracellular signaling pathways.

7. CANNABINOIDS

Cannabinoids are a diverse class of molecules that act to repress neurotransmitter release in the brain and periphery. These molecules primarily act as ligands for the two cannabinoid receptors identified thus far, CB1 and CB2, though there is mounting evidence for the existence of more such receptors (Brown, 2007). Ligands can be divided into three categories: (1) *endocannabinoids* produced naturally by the body of humans and other animals, (2) *phytocannabinoids* found in some plants, and (3) *synthetic cannabinoids* artificially manufactured for a variety of purposes. All endocannabinoids are considered *eicosanoids* (or *icosanoids*) (Pertwee, 2006), a class of signaling molecules synthesized via the oxidation of 20-carbon omega-3 or omega-6 fatty acids. The chemical structures of several cannabinoids used in studies of Na_V channels are shown in Fig. 10.

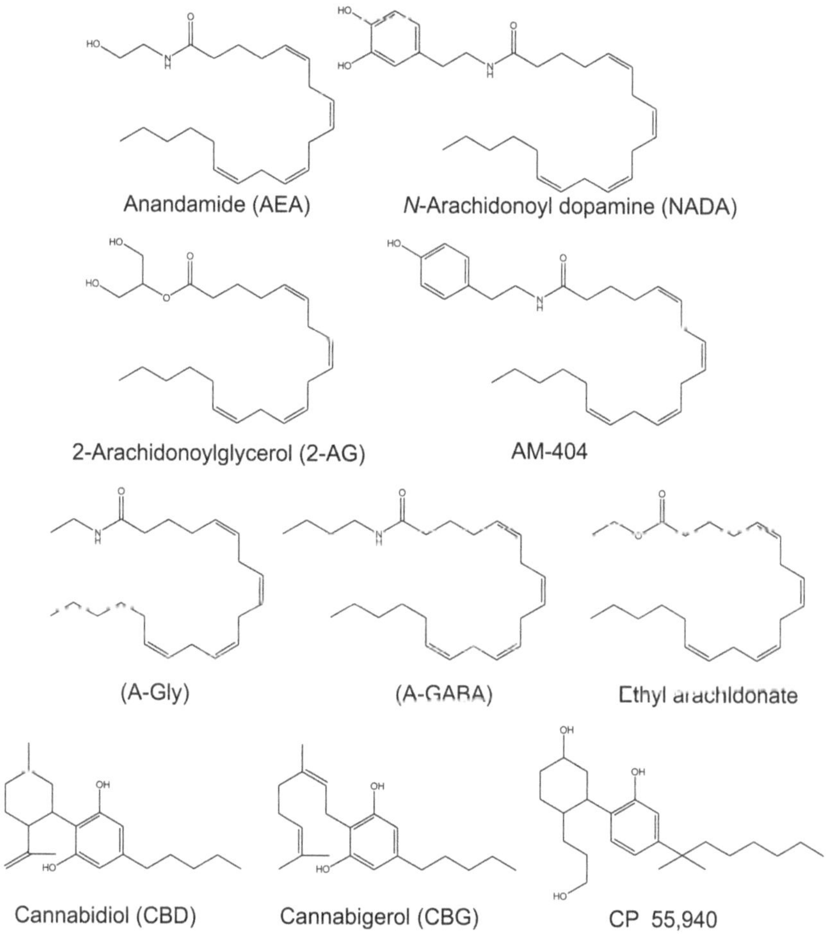

Figure 10 Endocannabinoids, phytocannabinoids, and cannabimimetics.

Arachidonoyl ethanolamide (anandamide; AEA), the first endocannabinoid isolated from the brain, is the ethanolamide of AA. As a partial agonist of the cannabinoid receptor 1 (CB1), several groups have shown anandamide can act as an analgesic in acute or persistent inflammatory and neuropathic pain models. However, since anandamide was also found to interact with other ion channels present in the periaqueductal gray, amygdala, and spinal trigeminal tract of the central nervous system (CNS), and dorsal root ganglion (DRG) neurons of the peripheral nervous system (PNS), the complete mechanism of the analgesic effects remains unclear. Na^+ currents can be categorized as those sensitive to tetrodotoxin (TTX-S) and those that are resistant (TTX-R). Anandamide and cannabimimetics AM404 and WIN55,212-2 inhibit veratridine-induced depolarization of synaptoneurosomes from cortical neurons (Nicholson et al., 2003). Anandamide also blocks TTX-S sustained repetitive firing of cultured cortical neurons in a dose-dependent manner (Nicholson et al., 2003). In DRG neurons, anandamide also suppresses TTX-S and TTX-R Na^+ currents in a use-dependent manner, with K_d values of 5.4 and 38.4 μM, respectively (Kim et al., 2005). Here, anandamide does not affect the voltage-dependence of activation I_{Na}, but increases the rate of activation in TTX-R channels. Anandamide induces shifts in the voltage-dependence of inactivation toward more negative potentials (by −6.8 mV in TTX-S and by −8.3 mV for TTX-R) and increases the rate of inactivation in TTX-R channels alone. At high concentrations, anandamide also causes a slight inhibition of action potentials of the marginal mandibular motor nerve, which consists of only thick myelinated motor fibers that almost exclusively contain Na_v channels (Safrany-Fark et al., 2015).

$Na_v1.7$, $Na_v1.8$, and $Na_v1.9$ channels are expressed in DRG and are believed to play critical roles in inflammatory and neuropathic pain (Cummins, Sheets, & Waxman, 2007; Decosterd, Ji, Abdi, Tate, & Woolf, 2002; Wang, Gu, Li, & Tao, 2011; Wood, Boorman, Okuse, & Baker, 2004). Anandamide suppresses the activity of $Na_v1.2$, $Na_v1.6$, $Na_v1.7$, and $Na_v1.8$ channels (Table 2) in a concentration-dependent manner and is more potent from the inactivated state (at $V_{1/2}$ holding potential) than from the resting state (V_{max} holding potential) (Okura et al., 2014). This endocannabinoid shifts the midpoint of activation in these channels to slightly more (+4 mV) positive potentials while shifting their midpoints of inactivation to somewhat more (−5 mV) negative potentials. While $Na_v1.2$, $Na_v1.6$, and $Na_v1.7$ channels show use-dependent block by anandamide, $Na_v1.8$ channels do not. However, $Na_v1.8$ channels are

Table 2 Effects of anandamide on Na_v currents

	$Na_v1.2$	$Na_v1.6$	$Na_v1.7$	$Na_v1.8$
% inhibition at $V_{1/2}$	46 ± 4[a]	49 ± 3[a]	37 ± 2[a]	27 ± 2[a]
% inhibition at V_{max}	7 ± 2	6 ± 1	9 ± 1	21 ± 5[a]
IC_{50} (μmol/L)	17 ± 3	12 ± 1	27 ± 3	40 ± 14
Hill slope (μmol/L)	0.74 ± 0.04	0.79 ± 0.08	0.52 ± 0.06	0.71 ± 0.10
$\Delta V_{1/2}$ of activation from $V_{H,max}$ (mV)	+1.9[a]	+2.1[a]	+2.4[a]	+2.6[a]
$\Delta V_{1/2}$ of activation from $V_{H,1/2}$ (mV)	+3.2[a]	+3.8[a]	+2.3[a]	+3.3[a]
$\Delta V_{1/2}$ of inactivation from $V_{H,max}$ (mV)	−5.2[a]	−5.0[a]	−4.1[a]	−6.3[a]
% reduction of plateau amplitude	11	25	22	0

[a]Significant compared to control recordings in the absence of anandamide.
Data summarized from Okura, D., Horishita, T., Ueno, S., Yanagihara, N., Sudo, Y., Uezono, Y., & Sata, T. (2014). The endocannabinoid anandamide inhibits voltage-gated sodium channels $Na_v1.2$, $Na_v1.6$, $Na_v1.7$, and $Na_v1.8$ in *Xenopus* oocytes. *Anesthesia and Analgesia, 118*(3), 554–562.

significantly inhibited at V_{max} holding potential, while the other channels are not. Together these data indicate that resting-state block is an important mechanism for $Na_v1.8$ channels, while $Na_v1.2$, $Na_v1.6$, and $Na_v1.7$ may be more favorably inhibited in the open state. Since these experiments were performed in oocytes, anandamide inhibition of sodium channels appears independent of cannabinoid receptor–dependent signaling. Anandamide suppression of sodium currents in DRG neurons even in the presence of CB1, CB2, and TRPV1 antagonists provides further support for this hypothesis (Kim et al., 2005).

In rat ventricular myocytes, the application of 10 μM AEA inhibits peak I_{Na} with no change in steady-state activation (Al Kury et al., 2014). AEA speeds up the rate of inactivation and shifts the voltage-dependence of inactivation by −11.4 mV. Anandamide belongs to a group of lipids called long-chain N-acylethanolamines (NAEs), which are produced in large quantities during necrosis or cellular stress (Berger et al., 2004; Hansen, Moesgaard, Hansen, & Petersen, 2000) and accumulate during myocardial infarction in canines with ligated coronary arteries (Epps, Mandel, & Schwartz, 1982; Epps, Schmid, Natarajan, & Schmid, 1979; Schmid & Berdyshev, 2002). Although anandamide only constitutes approximately 1–3% of the NAEs (Schmid & Berdyshev, 2002), at resting membrane potentials, fewer Na^+ channels would be available to open causing a reduction in the amplitude and upstroke rate of the action potential which may contribute to changes in ventricular contractility in response to ischemia or cellular stress (Berger et al., 2004; Hansen et al., 2000; Schmid & Berdyshev, 2002).

Other endocannabinoids have also been shown to regulate the function of Na_V channels. NADA, AGE, and 2-AG inhibit veratridine-dependent depolarization of synaptoneurosomes driven by TTX-S channels with increasing low micromolar affinity (Duan, Zheng, & Nicholson, 2008). Ethyl arachidonate, A-GABA, and A-Gly are less effective at inhibiting these currents, with incomplete block even at high concentrations (Catterall, Morrow, Daly, & Brown, 1981; Charollais & Van Der Goot, 2009; Chen et al., 1998; Chin & Chueh, 1998; Combs et al., 2013; Desai et al., 1993; Duan, Liao, Jain, & Nicholson, 2008; Duan, Zheng, et al., 2008). *Cis*-9,10-octadecenoamide (cOA or oleamide) accumulates in the cerebrospinal fluid during sleep deprivation and induces sleep in mammals. Oleamide stereoselectively suppresses sustained repetitive firing in cultured pyramidal neurons by inhibiting TTX-S Na_V channels (Nicholson, Zheng, Ganellin, Verdon, & Lees, 2001). Similar to other cannabinoids, oleamide reduces peak I_{Na} by shifting the inactivation properties to more negative potentials by −15.4 mV without altering activation properties. Furthermore, oleamide slows the recovery from inactivation nearly twofold, but do not show any frequency-dependent block. These results suggest a possible novel approach to regulating sleep and arousal through the manipulation of fatty acid amides in the brain.

Phytocannabinoids such as cannabidiol (CBD) have diverse pharmacological effects and are currently utilized in the treatment of epilepsies (Cunha et al., 1980). Intriguingly, CBD does not interact with CB1 receptors (Cunha et al., 1980; Hill, Williams, Whalley, & Stephens, 2012), but does interact with several ion channels including T-type Ca^{2+} channels; TRPV1, 2, and 3; TRPA1; and TRPM8 channels (De Petrocellis et al., 2011, 2012; Qin et al., 2008; Ross, Napier, & Connor, 2008). CBD reduces the spike frequency of CA1 pyramidal neurons stimulated by current injection, increases the first and second spike duration, and increases the minimum stimulus required to evoke spiking, which are all consistent with actions on Na_V channels (Hill et al., 2014). In fact, 10 μM CBD decreases peak I_{Na} by ~75% in human neuroblastoma SH-SY5Y cells and isolated mouse cortical neurons. Moreover, CBD inhibits veratridine activation of human $Na_V1.1$, $Na_V1.2$, and $Na_V1.5$ with EC_{50}'s in the range of 37–59 μM. A structurally similar phytocannabinoid, cannabigerol (CBG), has nearly equivalent impacts on Na_V channel function, however, lacks the anticonvulsant properties of other cannabinoids including CBD, when screened against the pentylenetetrazol (ptz)-induced kindling model of generalized seizures. This suggests that block of Na_V channels by CBD is not the primary mechanism of the anticonvulsant action of CBD (Hill et al., 2014).

7.1 Possible mechanisms for cannabinoid regulation of Na_V channels

As described previously, endocannabinoid inhibition of sodium channels appears independent of cannabinoid receptor—dependent signaling or downstream effects of TRPV1 channels (Al Kury et al., 2014; Duan, Zheng, et al., 2008; Kim et al., 2005; Nicholson et al., 2003). Cannabinoids and other cannabimimetics displace the binding of radiolabeled batrachotoxin ([^{3}H] BTX-B) (Al Kury et al., 2014; Duan, Zheng, et al., 2008; Nicholson et al., 2001, 2003) which selectively binds to site 2 of the sodium channel (Catterall et al., 1981). Anandamide, 2-AG, AGE, CP 55,940, ethyl arachidonate, and NADA do so in a noncompetitive fashion, while cOA, AM404, and WIN55,212-2 displacement involved competitive action. The sites of interaction for Anandamide, 2-AG, AGE, CP 55,940, ethyl arachidonate, and NADA remain to be determined.

Overall, electrophysiological and biochemical analysis suggest a preferential interaction of cannabinoids with the inactivated or toxin-modified conformations. These actions parallel those of Class I antiarrhythmics and neuroactive drugs that inhibit sodium channel function, including anticonvulsants, local and general anesthetics, and neuroprotectants (Hill, Duff, & Sheldon, 1989; Ratnakumari & Hemmings, 1996; Sheldon, Duff, Thakore, & Hill, 1994; Sheldon, Hill, Cannon, & Duff, 1989). Anandamide is the ethanolamide of AA, which also affects Na^+ currents. However, since AA shifts the voltage-dependences of both activation and inactivation, while AEA affects only inactivation, it is unlikely that free AA generated via the hydrolysis of AEA is the mediator of anandamide regulation of Na^+ currents (Kim et al., 2005), although this hypothesis may be challenged by the subtle changes in activation observed when examining specific Na_V isoforms (Okura et al., 2014).

8. CONCLUSIONS

It is clear from lipidomics studies that lipid profiles are altered during development and pathogenesis. Moreover, lipid profiles can also be altered by environmental influences such as dietary intake. Importantly, altered electrical activity through Na_V channels is also reported with development, changes in diet, and the pathogenesis of cancer as well as a variety of neuronal and cardiac disorders. However, it is only in recent years that researchers are gaining a greater appreciation for the interplay between integral membrane proteins, and the lipids which form the membrane in which they are embedded. It is evident from the data presented in this chapter that voltage-dependent sodium channels (Na_V) channels are regulated by a wide variety of lipids, including sphingolipids, fatty acids, glycerophospholipids, sterols, and cannabinoids. In some cases, these interactions involve intimate protein–lipid interactions, which are both structurally and functionally important. In other cases, lipid regulation takes place indirectly through changes in the bilayer properties, or through intracellular signaling. Interestingly, the nature of how lipids regulate Na_V channels depends both on the specific type of lipid and the isoform in question, as well as the cellular system being examined. This complex relationship between Na_V channels and the lipids in which they are embedded contributes to the cell-specific distribution and variations in the currents observed for Na_V channels. Further research into this clearly important mode of regulation in these channels will undoubtedly lead to a greater understanding of the development of a variety of diseases. Even more intriguing is the possibility of targeted lipid modulation as a novel approach for therapeutic developments.

REFERENCES

Ackerman, M. J., Siu, B. L., Sturner, W. Q., Tester, D. J., Valdivia, C. R., Makielski, J. C., & Towbin, J. A. (2001). Postmortem molecular analysis of SCN5A defects in sudden infant death syndrome. *JAMA, 286*(18), 2264–2269. pii:jci10024.

Ahn, H. S., Black, J. A., Zhao, P., Tyrrell, L., Waxman, S. G., & Dib-Hajj, S. D. (2011). Nav1.7 is the predominant sodium channel in rodent olfactory sensory neurons. *Molecular Pain, 7*, 32. http://dx.doi.org/10.1186/1744-8069-7-32. pii:1744-8069-7-32.

Al Kury, L. T., Voitychuk, O. I., Yang, K. H., Thayyullathil, F. T., Doroshenko, P., Ramez, A. M., … Oz, M. (2014). Effects of the endogenous cannabinoid anandamide on voltage-dependent sodium and calcium channels in rat ventricular myocytes. *British Journal of Pharmacology, 171*(14), 3485–3498. http://dx.doi.org/10.1111/bph.12734.

Andersen, O. S., & Koeppe, R. E., 2nd (1992). Molecular determinants of channel function. *Physiological Reviews, 72*(4 Suppl.), S89–S158. Retrieved from: http://www.ncbi.nlm.nih.gov/pubmed/1279736.

Andersen, O. S., & Koeppe, R. E., 2nd (2007). Bilayer thickness and membrane protein function: an energetic perspective. *Annual Review of Biophysics and Biomolecular Structure, 36*, 107–130. http://dx.doi.org/10.1146/annurev.biophys.36.040306.132643.

Bagneris, C., DeCaen, P. G., Naylor, C. E., Pryde, D. C., Nobeli, I., Clapham, D. E., & Wallace, B. A. (2014). Prokaryotic NavMs channel as a structural and functional model for eukaryotic sodium channel antagonism. *Proceedings of the National Academy of Sciences of the United States of America, 111*(23), 8428–8433. http://dx.doi.org/10.1073/pnas.1406855111.

Balijepalli, R. C., & Kamp, T. J. (2008). Caveolae, ion channels and cardiac arrhythmias. *Progress in Biophysics and Molecular Biology, 98*(2–3), 149–160. http://dx.doi.org/10.1016/j.pbiomolbio.2009.01.012.

Balla, T. (2013). Phosphoinositides: tiny lipids with giant impact on cell regulation. *Physiological Reviews, 93*(3), 1019–1137. http://dx.doi.org/10.1152/physrev.00028.2012.

Banderali, U., Juranka, P. F., Clark, R. B., Giles, W. R., & Morris, C. E. (2010). Impaired stretch modulation in potentially lethal cardiac sodium channel mutants. *Channels (Austin), 4*(1), 12–21. Retrieved from: http://www.ncbi.nlm.nih.gov/pubmed/20090423.

Bang, H. O., Dyerberg, J., & Sinclair, H. M. (1980). The composition of the Eskimo food in north western Greenland. *American Journal of Clinical Nutrition, 33*(12), 2657–2661. Retrieved from: http://www.ncbi.nlm.nih.gov/pubmed/7435433.

Barenholz, Y. (2004). Sphingomyelin and cholesterol: from membrane biophysics and rafts to potential medical applications. *Subcellular Biochemistry, 37*, 167–215. Retrieved from: http://www.ncbi.nlm.nih.gov/pubmed/15376621.

Barenholz, Y., & Thompson, T. E. (1999). Sphingomyelin: biophysical aspects. *Chemistry and Physics of Lipids, 102*(1–2), 29–34. Retrieved from: http://www.ncbi.nlm.nih.gov/pubmed/11001558.

Baroni, D., & Moran, O. (2011). Molecular differential expression of voltage-gated sodium channel alpha and beta subunit mRNAs in five different mammalian cell lines. *Journal of Bioenergetics and Biomembranes, 43*(6), 729–738. http://dx.doi.org/10.1007/s10863-011-9399-7.

Bendahhou, S., Cummins, T. R., & Agnew, W. S. (1997). Mechanism of modulation of the voltage-gated skeletal and cardiac muscle sodium channels by fatty acids. *American Journal of Physiology, 272*(2 Pt 1), C592–C600. Retrieved from: http://www.ncbi.nlm.nih.gov/entrez/query.fcgi?cmd=Retrieve&db=PubMed&dopt=Citation&list_uids=9124303.

Berger, C., Schmid, P. C., Schabitz, W. R., Wolf, M., Schwab, S., & Schmid, H. H. (2004). Massive accumulation of N-acylethanolamines after stroke. Cell signalling in acute cerebral ischemia? *Journal of Neurochemistry, 88*(5), 1159–1167. Retrieved from: http://www.ncbi.nlm.nih.gov/pubmed/15009671.

Beyder, A., Rae, J. L., Bernard, C., Strege, P. R., Sachs, F., & Farrugia, G. (2010). Mechanosensitivity of Nav1.5, a voltage-sensitive sodium channel. *Journal of Physiology, 588*(Pt 24), 4969–4985. http://dx.doi.org/10.1113/jphysiol.2010.199034.

Bezzina, C., Veldkamp, M. W., van Den Berg, M. P., Postma, A. V., Rook, M. B., Viersma, J. W., … Wilde, A. A. (1999). A single Na(+) channel mutation causing both long-QT and Brugada syndromes. *Circulation Research, 85*(12), 1206–1213. Retrieved from: http://www.ncbi.nlm.nih.gov/entrez/query.fcgi?cmd=Retrieve&db=PubMed&dopt=Citation&list_uids=10590249.

Blaskovic, S., Blanc, M., & van der Goot, F. G. (2013). What does S-palmitoylation do to membrane proteins? *FEBS Journal, 280*(12), 2766–2774. http://dx.doi.org/10.1111/febs.12263.

Boal, D. (2001). *Mechanics of the cell*. Cambridge: Cambridge University Press.

Boguslawski, G., Lyons, D., Harvey, K. A., Kovala, A. T., & English, D. (2000). Sphingosylphosphorylcholine induces endothelial cell migration and morphogenesis. *Biochemical and Biophysical Research Communications, 272*(2), 603–609. http://dx.doi.org/10.1006/bbrc.2000.2822.

Boland, L. M., & Drzewiecki, M. M. (2008). Polyunsaturated fatty acid modulation of voltage-gated ion channels. *Cell Biochemistry and Biophysics, 52*(2), 59–84 http://dx.doi.org/10.1007/s12013-008-9027-2.

Bosmans, F., Milescu, M., & Swartz, K. J. (2011). Palmitoylation influences the function and pharmacology of sodium channels. *Proceedings of the National Academy of Sciences of the United States of America, 108*(50), 20213–20218. http://dx.doi.org/10.1073/pnas.1108497108.

Bougnoux, P. (1999). n-3 polyunsaturated fatty acids and cancer. *Current Opinion in Clinical Nutrition and Metabolic Care, 2*(2), 121–126. Retrieved from: http://www.ncbi.nlm.nih.gov/pubmed/10453342.

Bretscher, M. S. (1973). Membrane structure: some general principles. *Science, 181*(4100), 622–629. Retrieved from: http://www.ncbi.nlm.nih.gov/pubmed/4724478.

Brown, A. J. (2007). Novel cannabinoid receptors. *British Journal of Pharmacology, 152*(5), 567–575. http://dx.doi.org/10.1038/sj.bjp.0707481.
Bunemann, M., Liliom, K., Brandts, B. K., Pott, L., Tseng, J. L., Desiderio, D. M., … Tigyi, G. (1996). A novel membrane receptor with high affinity for lysosphingomyelin and sphingosine 1-phosphate in atrial myocytes. *EMBO Journal, 15*(20), 5527–5534. Retrieved from: http://www.ncbi.nlm.nih.gov/pubmed/8896446.
Cannon, S. C., & George, A. L. (2002). Pathophysiology of myotonia and periodic paralysis. In A. K. Asbury, G. M. Mckhann, W. I. McDonald, P. J. Goadsby, & J. C. McArthur (Eds.), *Diseases of the nervous system: Clinical neuroscience and therapeutic principles* (pp. 1183–1206). Cambridge: Cambridge University Press.
Cantrell, A. R., & Catterall, W. A. (2001). Neuromodulation of Na^+ channels: an unexpected form of cellular plasticity. *Nature Reviews. Neuroscience, 2*(6), 397–407. http://dx.doi.org/10.1038/35077553.
Catapano, E. R., Arriaga, L. R., Espinosa, G., Monroy, F., Langevin, D., & Lopez-Montero, I. (2011). Solid character of membrane ceramides: a surface rheology study of their mixtures with sphingomyelin. *Biophysical Journal, 101*(11), 2721–2730. http://dx.doi.org/10.1016/j.bpj.2011.10.049.
Catterall, W. A. (1986). Molecular properties of voltage-sensitive sodium channels. *Annual Review of Biochemistry, 55*, 953–985. http://dx.doi.org/10.1146/annurev.bi.55.070186.004513.
Catterall, W. A. (1992). Cellular and molecular biology of voltage-gated sodium channels. *Physiological Reviews, 72*(4 Suppl.), S15–S48. Retrieved from: http://www.ncbi.nlm.nih.gov/pubmed/1332090.
Catterall, W. A. (1999). Molecular properties of brain sodium channels: an important target for anticonvulsant drugs. *Advances in Neurology, 79*, 441–456. Retrieved from: http://www.ncbi.nlm.nih.gov/pubmed/10514834.
Catterall, W. A., Morrow, C. S., Daly, J. W., & Brown, G. B. (1981). Binding of batrachotoxinin A 20-alpha-benzoate to a receptor site associated with sodium channels in synaptic nerve ending particles. *Journal of Biological Chemistry, 256*(17), 8922–8927. Retrieved from: http://www.ncbi.nlm.nih.gov/pubmed/6114956.
Chahine, M., Pilote, S., Pouliot, V., Takami, H., & Sato, C. (2004). Role of arginine residues on the S4 segment of the *Bacillus halodurans* Na^+ channel in voltage-sensing. *Journal of Membrane Biology, 201*(1), 9–24. Retrieved from: http://www.ncbi.nlm.nih.gov/entrez/query.fcgi?cmd=Retrieve&db=PubMed&dopt=Citation&list_uids=15635808.
Charollais, J., & Van Der Goot, F. G. (2009). Palmitoylation of membrane proteins (Review). *Molecular Membrane Biology, 26*(1), 55–66. http://dx.doi.org/10.1080/09687680802620369.
Chavarro, J. E., Stampfer, M. J., Li, H., Campos, H., Kurth, T., & Ma, J. (2007). A prospective study of polyunsaturated fatty acid levels in blood and prostate cancer risk. *Cancer Epidemiology, Biomarkers & Prevention: A Publication of the American Association for Cancer Research, Cosponsored By the American Society of Preventive Oncology, 16*(7), 1364–1370. http://dx.doi.org/10.1158/1055-9965.EPI-06-1033.
Chen, P. F., Chin, T. Y., & Chueh, S. H. (1998). Ca^{2+} signaling induced by sphingosylphosphorylcholine and sphingosine 1-phosphate via distinct mechanisms in rat glomerular mesangial cells. *Kidney International, 54*(5), 1470–1483. http://dx.doi.org/10.1046/j.1523-1755.1998.00162.x.
Cheng, W. W., D'Avanzo, N., Doyle, D. A., & Nichols, C. G. (2011). Dual-mode phospholipid regulation of human inward rectifying potassium channels. *Biophysical Journal, 100*(3), 620–628. http://dx.doi.org/10.1016/j.bpj.2010.12.3724.
Chin, T. Y., & Chueh, S. H. (1998). Sphingosylphosphorylcholine stimulates mitogen-activated protein kinase via a Ca^{2+}-dependent pathway. *American Journal of Physiology,*

275(5 Pt 1), C1255–C1263. Retrieved from: http://www.ncbi.nlm.nih.gov/pubmed/9814974.
Claes, L., Del-Favero, J., Ceulemans, B., Lagae, L., Van Broeckhoven, C., & De Jonghe, P. (2001). De novo mutations in the sodium-channel gene SCN1A cause severe myoclonic epilepsy of infancy. *American Journal of Human Genetics, 68*(6), 1327–1332. http://dx.doi.org/10.1086/320609. pii:S0002-9297(07)61043-X.
Combs, D. J., Shin, H. G., Xu, Y., Ramu, Y., & Lu, Z. (2013). Tuning voltage-gated channel activity and cellular excitability with a sphingomyelinase. *Journal of General Physiology, 142*(4), 367–380. http://dx.doi.org/10.1085/jgp.201310986.
Cukierman, S. (1991). Asymmetric electrostatic effects on the gating of rat brain sodium channels in planar lipid membranes. *Biophysical Journal, 60*(4), 845–855. http://dx.doi.org/10.1016/S0006-3495(91)82118-X.
Cukierman, S., Zinkand, W. C., French, R. J., & Krueger, B. K. (1988). Effects of membrane surface charge and calcium on the gating of rat brain sodium channels in planar bilayers. *Journal of General Physiology, 92*(4), 431–447. Retrieved from: http://www.ncbi.nlm.nih.gov/pubmed/2849628.
Cummins, T. R., Sheets, P. L., & Waxman, S. G. (2007). The roles of sodium channels in nociception: implications for mechanisms of pain. *Pain, 131*(3), 243–257. http://dx.doi.org/10.1016/j.pain.2007.07.026.
Cunha, J. M., Carlini, E. A., Pereira, A. E., Ramos, O. L., Pimentel, C., Gagliardi, R., … Mechoulam, R. (1980). Chronic administration of cannabidiol to healthy volunteers and epileptic patients. *Pharmacology, 21*(3), 175–185. Retrieved from: http://www.ncbi.nlm.nih.gov/pubmed/7413719.
D'Avanzo, N., Cheng, W. W., Doyle, D. A., & Nichols, C. G. (2010). Direct and specific activation of human inward rectifier K^+ channels by membrane phosphatidylinositol 4,5-bisphosphate. *Journal of Biological Chemistry, 285*(48), 37129–37132. http://dx.doi.org/10.1074/jbc.C110.186692.
D'Avanzo, N., McCusker, E. C., Powl, A. M., Miles, A. J., Nichols, C. G., & Wallace, B. A. (2013). Differential lipid dependence of the function of bacterial sodium channels. *PLoS One, 8*(4), e61216. http://dx.doi.org/10.1371/journal.pone.0061216. pii:PONE-D-12-40087.
Dal-Pizzol, F., Quevedo, J., Streck, E., Walz, R., & Moreira, J. C. (2007). Changes in lipid composition in hippocampus early and late after status epilepticus induced by kainic acid in Wistar rats. *Metabolic Brain Disease, 22*(1), 25–29. http://dx.doi.org/10.1007/s11011-006-9033-2.
De Petrocellis, L., Ligresti, A., Moriello, A. S., Allara, M., Bisogno, T., Petrosino, S., … Di Marzo, V. (2011). Effects of cannabinoids and cannabinoid-enriched Cannabis extracts on TRP channels and endocannabinoid metabolic enzymes. *British Journal of Pharmacology, 163*(7), 1479–1494. http://dx.doi.org/10.1111/j.1476-5381.2010.01166.x.
De Petrocellis, L., Orlando, P., Moriello, A. S., Aviello, G., Stott, C., Izzo, A. A., & Di Marzo, V. (2012). Cannabinoid actions at TRPV channels: effects on TRPV3 and TRPV4 and their potential relevance to gastrointestinal inflammation. *Acta Physiologica (Oxford), 204*(2), 255–266. http://dx.doi.org/10.1111/j.1748-1716.2011.02338.x.
Decosterd, I., Ji, R. R., Abdi, S., Tate, S., & Woolf, C. J. (2002). The pattern of expression of the voltage-gated sodium channels Na(v)1.8 and Na(v)1.9 does not change in uninjured primary sensory neurons in experimental neuropathic pain models. *Pain, 96*(3), 269–277. Retrieved from: http://www.ncbi.nlm.nih.gov/pubmed/11972999.
Desai, N. N., Carlson, R. O., Mattie, M. E., Olivera, A., Buckley, N. E., Seki, T., … Spiegel, S. (1993). Signaling pathways for sphingosylphosphorylcholine-mediated mitogenesis in Swiss 3T3 fibroblasts. *Journal of Cell Biology, 121*(6), 1385–1395. Retrieved from: http://www.ncbi.nlm.nih.gov/pubmed/8389770.

Desai, N. N., & Spiegel, S. (1991). Sphingosylphosphorylcholine is a remarkably potent mitogen for a variety of cell lines. *Biochemical and Biophysical Research Communications, 181*(1), 361–366. Retrieved from: http://www.ncbi.nlm.nih.gov/pubmed/1958205.

Ding, X. X., Koop, D. R., Crump, B. L., & Coon, M. J. (1986). Immunochemical identification of cytochrome P-450 isozyme 3a (P-450ALC) in rabbit nasal and kidney microsomes and evidence for differential induction by alcohol. *Molecular Pharmacology, 30*(4), 370–378. Retrieved from: http://www.ncbi.nlm.nih.gov/entrez/query.fcgi?cmd=Retrieve&db=PubMed&dopt=Citation&list_uids=3762523.

Diz-Munoz, A., Fletcher, D. A., & Weiner, O. D. (2013). Use the force: membrane tension as an organizer of cell shape and motility. *Trends in Cell Biology, 23*(2), 47–53. http://dx.doi.org/10.1016/j.tcb.2012.09.006.

Dobrowsky, R. T., Werner, M. H., Castellino, A. M., Chao, M. V., & Hannun, Y. A. (1994). Activation of the sphingomyelin cycle through the low-affinity neurotrophin receptor. *Science, 265*(5178), 1596–1599. Retrieved from: http://www.ncbi.nlm.nih.gov/pubmed/8079174.

Duan, Y., Liao, C., Jain, S., & Nicholson, R. A. (2008). The cannabinoid receptor agonist CP-55,940 and ethyl arachidonate interfere with [(3)H]batrachotoxinin A 20 alpha-benzoate binding to sodium channels and inhibit sodium channel function. *Comparative Biochemistry and Physiology. Toxicology & Pharmacology: CBP, 148*(3), 244–249. http://dx.doi.org/10.1016/j.cbpc.2008.06.002.

Duan, Y., Zheng, J., & Nicholson, R. A. (2008). Inhibition of [3H]batrachotoxinin A-20alpha-benzoate binding to sodium channels and sodium channel function by endocannabinoids. *Neurochemistry International, 52*(3), 438–446. http://dx.doi.org/10.1016/j.neuint.2007.08.004.

Duncan, J. A., & Gilman, A. G. (1996). Autoacylation of G protein alpha subunits. *Journal of Biological Chemistry, 271*(38), 23594–23600. Retrieved from: http://www.ncbi.nlm.nih.gov/entrez/query.fcgi?cmd=Retrieve&db=PubMed&dopt=Citation&list_uids=8798571.

Epps, D. E., Mandel, F., & Schwartz, A. (1982). The alteration of rabbit skeletal sarcoplasmic reticulum function by N-acylethanolamine, a lipid associated with myocardial infarction. *Cell Calcium, 3*(6), 531–543. Retrieved from: http://www.ncbi.nlm.nih.gov/pubmed/6219743.

Epps, D. E., Schmid, P. C., Natarajan, V., & Schmid, H. H. (1979). N-Acylethanolamine accumulation in infarcted myocardium. *Biochemical and Biophysical Research Communications, 90*(2), 628–633. Retrieved from: http://www.ncbi.nlm.nih.gov/pubmed/508325.

Escayg, A., MacDonald, B. T., Meisler, M. H., Baulac, S., Huberfeld, G., An-Gourfinkel, I., … Malafosse, A. (2000). Mutations of SCN1A, encoding a neuronal sodium channel, in two families with GEFS+2. *Nature Genetics, 24*(4), 343–345. http://dx.doi.org/10.1038/74159.

Fahy, E., Subramaniam, S., Murphy, R. C., Nishijima, M., Raetz, C. R., Shimizu, T., … Dennis, E. A. (2009). Update of the LIPID MAPS comprehensive classification system for lipids. *Journal of Lipid Research, 50*(Suppl.), S9–S14. http://dx.doi.org/10.1194/jlr.R800095-JLR200.

Fang, Y. J., Zhou, M. H., Gao, X. F., Gu, H., & Mei, Y. A. (2011). Arachidonic acid modulates Na^+ currents by non-metabolic and metabolic pathways in rat cerebellar granule cells. *Biochemical Journal, 438*(1), 203–215. http://dx.doi.org/10.1042/BJ20110569. pii: BJ20110569.

Franks, N. P., & Lieb, W. R. (1982). Molecular mechanisms of general anaesthesia. *Nature, 300*(5892), 487–493. Retrieved from: http://www.ncbi.nlm.nih.gov/pubmed/6755267.

Fraser, D. D., Hoehn, K., Weiss, S., & MacVicar, B. A. (1993). Arachidonic acid inhibits sodium currents and synaptic transmission in cultured striatal neurons. *Neuron, 11*(4), 633–644. pii:0896-6273(93)90075-3.

Fraser, S. P., Salvador, V., Manning, E. A., Mizal, J., Altun, S., Raza, M., … Djamgoz, M. B. (2003). Contribution of functional voltage-gated Na^+ channel expression to cell behaviors involved in the metastatic cascade in rat prostate cancer: I. Lateral motility. *Journal of Cellular Physiology, 195*(3), 479–487. http://dx.doi.org/10.1002/jcp.10312.

Frenal, K., Tay, C. L., Mueller, C., Bushell, E. S., Jia, Y., Graindorge, A., … Soldati-Favre, D. (2013). Global analysis of apicomplexan protein S-acyl transferases reveals an enzyme essential for invasion. *Traffic, 14*(8), 895–911. http://dx.doi.org/10.1111/tra.12081.

Fujiwara, T., Sugawara, T., Mazaki-Miyazaki, E., Takahashi, Y., Fukushima, K., Watanabe, M., … Inoue, Y. (2003). Mutations of sodium channel alpha subunit type 1 (SCN1A) in intractable childhood epilepsies with frequent generalized tonic-clonic seizures. *Brain, 126*(Pt 3), 531–546. Retrieved from: http://www.ncbi.nlm.nih.gov/entrez/query.fcgi?cmd=Retrieve&db=PubMed&dopt=Citation&list_uids=12566275.

Fukata, Y., & Fukata, M. (2010). Protein palmitoylation in neuronal development and synaptic plasticity. *Nature Reviews. Neuroscience, 11*(3), 161–175. http://dx.doi.org/10.1038/nrn2788.

George, A. L., Jr. (2005). Inherited disorders of voltage-gated sodium channels. *Journal of Clinical Investigation, 115*(8), 1990–1999. http://dx.doi.org/10.1172/JCI25505.

Goldfine, H. (1984). Bacterial membranes and lipid packing theory. *Journal of Lipid Research, 25*(13), 1501–1507. Retrieved from: http://www.ncbi.nlm.nih.gov/pubmed/6530599.

Grant, A. O., Carboni, M. P., Neplioueva, V., Starmer, C. F., Memmi, M., Napolitano, C., & Priori, S. (2002). Long QT syndrome, Brugada syndrome, and conduction system disease are linked to a single sodium channel mutation. *Journal of Clinical Investigation, 110*(8), 1201–1209. http://dx.doi.org/10.1172/JCI15570.

Green, W. N., & Andersen, O. S. (1991). Surface charges and ion channel function. *Annual Review of Physiology, 53*, 341–359. http://dx.doi.org/10.1146/annurev.ph.53.030191.002013.

Hamplova, B., Pelouch, V., Novakova, O., Skovranek, J., Hucin, B., & Novak, F. (2004). Phospholipid composition of myocardium in children with normoxemic and hypoxemic congenital heart diseases. *Physiological Research, 53*(5), 557–560. Retrieved from: http://www.ncbi.nlm.nih.gov/entrez/query.fcgi?cmd=Retrieve&db=PubMed&dopt=Citation&list_uids=15479135.

Hannun, Y. A., & Obeid, L. M. (2008). Principles of bioactive lipid signalling: lessons from sphingolipids. *Nature Reviews. Molecular Cell Biology, 9*(2), 139–150. http://dx.doi.org/10.1038/nrm2329.

Hansen, H. S., Moesgaard, B., Hansen, H. H., & Petersen, G. (2000). N-Acylethanolamines and precursor phospholipids – relation to cell injury. *Chemistry and Physics of Lipids, 108*(1–2), 135–150. Retrieved from: http://www.ncbi.nlm.nih.gov/pubmed/11106787.

Hartshorne, R. P., Keller, B. U., Talvenheimo, J. A., Catterall, W. A., & Montal, M. (1985) Functional reconstitution of the purified brain sodium channel in planar lipid bilayers. *Proceedings of the National Academy of Sciences of the United States of America, 82*(1), 240–244. Retrieved from: http://www.ncbi.nlm.nih.gov/pubmed/2578662.

Hayashi, T., Thomas, G. M., & Huganir, R. L. (2009). Dual palmitoylation of NR2 subunits regulates NMDA receptor trafficking. *Neuron, 64*(2), 213–226. http://dx.doi.org/10.1016/j.neuron.2009.08.017.

Haydon, D. A., & Kimura, J. E. (1981). Some effects of n-pentane on the sodium and potassium currents of the squid giant axon. *Journal of Physiology, 312*, 57–70. Retrieved from: http://www.ncbi.nlm.nih.gov/pubmed/6267268.

Haydon, D. A., & Urban, B. W. (1983a). The action of alcohols and other non-ionic surface active substances on the sodium current of the squid giant axon. *Journal of Physiology, 341*, 411–427. Retrieved from: http://www.ncbi.nlm.nih.gov/pubmed/6312030.

Haydon, D. A., & Urban, B. W. (1983b). The action of hydrocarbons and carbon tetrachloride on the sodium current of the squid giant axon. *Journal of Physiology, 338*, 435–450. Retrieved from: http://www.ncbi.nlm.nih.gov/pubmed/6875965.

Haydon, D. A., & Urban, B. W. (1983c). The effects of some inhalation anaesthetics on the sodium current of the squid giant axon. *Journal of Physiology, 341*, 429–439. Retrieved from: http://www.ncbi.nlm.nih.gov/pubmed/6312031.

Hendry, B. M., Elliott, J. R., & Haydon, D. A. (1985). Further evidence that membrane thickness influences voltage-gated sodium channels. *Biophysical Journal, 47*(6), 841–845. http://dx.doi.org/10.1016/S0006-3495(85)83988-6.

Hermansson, M., Uphoff, A., Kakela, R., & Somerharju, P. (2005). Automated quantitative analysis of complex lipidomes by liquid chromatography/mass spectrometry. *Analytical Chemistry, 77*, 2166–2175.

Herold, K. F., & Hemmings, H. C., Jr. (2012). Sodium channels as targets for volatile anesthetics. *Frontiers in Pharmacology, 3*, 50. http://dx.doi.org/10.3389/fphar.2012.00050.

Herold, K. F., Sanford, R. L., Lee, W., Schultz, M. F., Ingolfsson, H. I., Andersen, O. S., & Hemmings, H. C., Jr. (2014). Volatile anesthetics inhibit sodium channels without altering bulk lipid bilayer properties. *Journal of General Physiology, 144*(6), 545–560. http://dx.doi.org/10.1085/jgp.201411172.

Heron, S. E., Crossland, K. M., Andermann, E., Phillips, H. A., Hall, A. J., Bleasel, A., … Scheffer, I. E. (2002). Sodium-channel defects in benign familial neonatal-infantile seizures. *Lancet, 360*(9336), 851–852. http://dx.doi.org/10.1016/S0140-6736(02)09968-3. pii:S0140-6736(02)09968-3.

Hibbeln, J. R., Nieminen, L. R., Blasbalg, T. L., Riggs, J. A., & Lands, W. E. (2006). Healthy intakes of n-3 and n-6 fatty acids: estimations considering worldwide diversity. *American Journal of Clinical Nutrition, 83*(6 Suppl.), 1483S–1493S. Retrieved from: http://www.ncbi.nlm.nih.gov/pubmed/16841858.

Hill, A. J., Jones, N. A., Smith, I., Hill, C. L., Williams, C. M., Stephens, G. J., & Whalley, B. J. (2014). Voltage-gated sodium (NaV) channel blockade by plant cannabinoids does not confer anticonvulsant effects per se. *Neuroscience Letters, 566*, 269–274. http://dx.doi.org/10.1016/j.neulet.2014.03.013.

Hill, A. J., Williams, C. M., Whalley, B. J., & Stephens, G. J. (2012). Phytocannabinoids as novel therapeutic agents in CNS disorders. *Pharmacology & Therapeutics, 133*(1), 79–97. http://dx.doi.org/10.1016/j.pharmthera.2011.09.002.

Hill, R. J., Duff, H. J., & Sheldon, R. S. (1989). Class I antiarrhythmic drug receptor: biochemical evidence for state-dependent interaction with quinidine and lidocaine. *Molecular Pharmacology, 36*(1), 150–159. Retrieved from: http://www.ncbi.nlm.nih.gov/pubmed/2546048.

Hill, W. G., Southern, N. M., MacIver, B., Potter, E., Apodaca, G., Smith, C. P., & Zeidel, M. L. (2005). Isolation and characterization of the *Xenopus* oocyte plasma membrane: a new method for studying activity of water and solute transporters. *American Journal of Physiology. Renal Physiology, 289*(1), F217–F224. http://dx.doi.org/10.1152/ajprenal.00022.2005.

Huang, Y., Walker, K. E., Hanley, F., Narula, J., Houser, S. R., & Tulenko, T. N. (2004). Cardiac systolic and diastolic dysfunction after a cholesterol-rich diet. *Circulation, 109*(1), 97–102. http://dx.doi.org/10.1161/01.CIR.0000109213.10461.F6.

Isbilen, B., Fraser, S. P., & Djamgoz, M. B. (2006). Docosahexaenoic acid (omega-3) blocks voltage-gated sodium channel activity and migration of MDA-MB-231 human breast cancer cells. *International Journal of Biochemistry & Cell Biology, 38*(12), 2173–2182. http://dx.doi.org/10.1016/j.biocel.2006.06.014.

Iwanaga, T., Tsutsumi, R., Noritake, J., Fukata, Y., & Fukata, M. (2009). Dynamic protein palmitoylation in cellular signaling. *Progress in Lipid Research, 48*(3–4), 117–127. http://dx.doi.org/10.1016/j.plipres.2009.02.001.

Jackson, M., Haris, P. I., & Chapman, D. (1989). Fourier transform infrared spectroscopic studies of lipids, polypeptides and proteins. *Journal of Molecular Structure, 214*, 329–355.

Jahangiri, A., Leifert, W. R., Patten, G. S., & McMurchie, E. J. (2000). Termination of asynchronous contractile activity in rat atrial myocytes by n-3 polyunsaturated fatty acids. *Molecular and Cellular Biochemistry, 206*(1–2), 33–41. Retrieved from: http://www.ncbi.nlm.nih.gov/entrez/query.fcgi?cmd=Retrieve&db=PubMed&dopt=Citation&list_uids=10839192.

Jo, T., Iida, H., Kishida, S., Imuta, H., Oonuma, H., Nagata, T., … Nakajima, T. (2005). Acute and chronic effects of eicosapentaenoic acid on voltage-gated sodium channel expressed in cultured human bronchial smooth muscle cells. *Biochemical and Biophysical Research Communications, 331*(4), 1452–1459. http://dx.doi.org/10.1016/j.bbrc.2005.04.062.

Joseph, C. K., Byun, H. S., Bittman, R., & Kolesnick, R. N. (1993). Substrate recognition by ceramide-activated protein kinase. Evidence that kinase activity is proline-directed. *Journal of Biological Chemistry, 268*(27), 20002–20006. Retrieved from: http://www.ncbi.nlm.nih.gov/pubmed/8376361.

Jude, S., Bedut, S., Roger, S., Pinault, M., Champeroux, P., White, E., & Le Guennec, J. Y. (2003). Peroxidation of docosahexaenoic acid is responsible for its effects on I TO and I SS in rat ventricular myocytes. *British Journal of Pharmacology, 139*(4), 816–822. http://dx.doi.org/10.1038/sj.bjp.0705308.

Jurado, A. S., Pinheiro, T. J., & Madeira, V. M. (1991). Physical studies on membrane lipids of *Bacillus stearothermophilus* temperature and calcium effects. *Archives of Biochemistry and Biophysics, 289*(1), 167–179. Retrieved from: http://www.ncbi.nlm.nih.gov/pubmed/1898060.

Kang, J. X., & Leaf, A. (1995). Prevention and termination of beta-adrenergic agonist-induced arrhythmias by free polyunsaturated fatty acids in neonatal rat cardiac myocytes. *Biochemical and Biophysical Research Communications, 208*(2), 629–636. http://dx.doi.org/10.1006/bbrc.1995.1385. pii:S0006-291X(85)71385-X.

Kang, J. X., & Leaf, A. (1996). Evidence that free polyunsaturated fatty acids modify Na^+ channels by directly binding to the channel proteins. *Proceedings of the National Academy of Sciences of the United States of America, 93*(8), 3542–3546. Retrieved from: http://www.ncbi.nlm.nih.gov/entrez/query.fcgi?cmd=Retrieve&db=PubMed&dopt=Citation&list_uids=8622972.

Kang, J. X., Li, Y., & Leaf, A. (1997). Regulation of sodium channel gene expression by class I antiarrhythmic drugs and n-3 polyunsaturated fatty acids in cultured neonatal rat cardiac myocytes. *Proceedings of the National Academy of Sciences of the United States of America, 94*(6), 2724–2728. Retrieved from: http://www.ncbi.nlm.nih.gov/pubmed/9122264.

Kang, J. X., Xiao, Y. F., & Leaf, A. (1995). Free, long-chain, polyunsaturated fatty acids reduce membrane electrical excitability in neonatal rat cardiac myocytes. *Proceedings of the National Academy of Sciences of the United States of America, 92*(9), 3997–4001. Retrieved from: http://www.ncbi.nlm.nih.gov/entrez/query.fcgi?cmd=Retrieve&db=PubMed&dopt=Citation&list_uids=7732020.

Kim, H. I., Kim, T. H., Shin, Y. K., Lee, C. S., Park, M., & Song, J. H. (2005). Anandamide suppression of Na^+ currents in rat dorsal root ganglion neurons. *Brain Research, 1062*(1–2), 39–47. http://dx.doi.org/10.1016/j.brainres.2005.09.004.

Kliman, M., Vijayakrishnan, N., Wang, L., Tapp, J. T., Broadie, K., & McLean, J. A. (2010). Structural mass spectrometry analysis of lipid changes in a *Drosophila* epilepsy model brain. *Molecular Biosystems, 6*(6), 958–966. http://dx.doi.org/10.1039/b927494d.

Koishi, R., Xu, H., Ren, D., Navarro, B., Spiller, B. W., Shi, Q., & Clapham, D. E. (2004). A superfamily of voltage-gated sodium channels in bacteria. *Journal of Biological*

Chemistry, 279(10), 9532–9538. http://dx.doi.org/10.1074/jbc.M313100200. pii: M313100200.

Kuzmenkin, A., Bezanilla, F., & Correa, A. M. (2004). Gating of the bacterial sodium channel, NaChBac: voltage-dependent charge movement and gating currents. *Journal of General Physiology, 124*(4), 349–356. http://dx.doi.org/10.1085/jgp.200409139. pii: jgp.200409139.

Laethem, R. M., Laethem, C. L., Ding, X., & Koop, D. R. (1992). P-450-dependent arachidonic acid metabolism in rabbit olfactory microsomes. *Journal of Pharmacology and Experimental Therapeutics, 262*(1), 433–438. Retrieved from: http://www.ncbi.nlm.nih.gov/entrez/query.fcgi?cmd=Retrieve&db=PubMed&dopt=Citation&list_uids=1625211.

Laitko, U., Juranka, P. F., & Morris, C. E. (2006). Membrane stretch slows the concerted step prior to opening in a Kv channel. *Journal of General Physiology, 127*(6), 687–701. http://dx.doi.org/10.1085/jgp.200509394.

Laitko, U., & Morris, C. E. (2004). Membrane tension accelerates rate-limiting voltage-dependent activation and slow inactivation steps in a Shaker channel. *Journal of General Physiology, 123*(2), 135–154. http://dx.doi.org/10.1085/jgp.200308965.

Larsson, S. C., Kumlin, M., Ingelman-Sundberg, M., & Wolk, A. (2004). Dietary long-chain n-3 fatty acids for the prevention of cancer: a review of potential mechanisms. *American Journal of Clinical Nutrition, 79*(6), 935–945. Retrieved from: http://www.ncbi.nlm.nih.gov/pubmed/15159222.

Leaf, A., Xiao, Y. F., & Kang, J. X. (2002). Interactions of n-3 fatty acids with ion channels in excitable tissues. *Prostaglandins, Leukotrienes, and Essential Fatty Acids, 67*(2–3), 113–120. pii:S0952327802904072.

Lee, G. Y., Shin, Y. K., Lee, C. S., & Song, J. H. (2002). Effects of arachidonic acid on sodium currents in rat dorsal root ganglion neurons. *Brain Research, 950*(1–2), 95–102. pii:S0006899302030081.

Leifert, W. R., Jahangiri, A., & McMurchie, E. J. (1999). Antiarrhythmic fatty acids and antioxidants in animal and cell studies. *Journal of Nutritional Biochemistry, 10*(5), 252–267. pii:S0955-2863(99)00011-X.

Leifert, W. R., Jahangiri, A., & McMurchie, E. J. (2000). Membrane fluidity changes are associated with the antiarrhythmic effects of docosahexaenoic acid in adult rat cardiomyocytes. *Journal of Nutritional Biochemistry, 11*(1), 38–44. pii:S0955-2863(99)00069-8.

Leifert, W. R., Jahangiri, A., Saint, D. A., & McMurchie, E. J. (2000). Effects of dietary n-3 fatty acids on contractility, Na^+ and K^+ currents in a rat cardiomyocyte model of arrhythmia. *Journal of Nutritional Biochemistry, 11*(7–8), 382–392. pii:S0955-2863(00)00094-2.

Leifert, W. R., McMurchie, E. J., & Saint, D. A. (1999). Inhibition of cardiac sodium currents in adult rat myocytes by n-3 polyunsaturated fatty acids. *Journal of Physiology, 520*(Pt 3), 671–679. Retrieved from: http://www.ncbi.nlm.nih.gov/pubmed/10545135.

Leitzmann, M. F., Stampfer, M. J., Michaud, D. S., Augustsson, K., Colditz, G. C., Willett, W. C., & Giovannucci, E. L. (2004). Dietary intake of n-3 and n-6 fatty acids and the risk of prostate cancer. *American Journal of Clinical Nutrition, 80*(1), 204–216. Retrieved from: http://www.ncbi.nlm.nih.gov/pubmed/15213050.

Lin, W., Laitko, U., Juranka, P. F., & Morris, C. E. (2007). Dual stretch responses of mHCN2 pacemaker channels: accelerated activation, accelerated deactivation. *Biophysical Journal, 92*(5), 1559–1572. http://dx.doi.org/10.1529/biophysj.106.092478.

Linden, D. J., & Routtenberg, A. (1989). cis-Fatty acids, which activate protein kinase C, attenuate Na^+ and Ca^{2+} currents in mouse neuroblastoma cells. *Journal of Physiology, 419*, 95–119. Retrieved from: http://www.ncbi.nlm.nih.gov/pubmed/2559978.

Liu, J., Mathias, S., Yang, Z., & Kolesnick, R. N. (1994). Renaturation and tumor necrosis factor-alpha stimulation of a 97-kDa ceramide-activated protein kinase. *Journal of*

Biological Chemistry, 269(4), 3047–3052. Retrieved from: http://www.ncbi.nlm.nih.gov/pubmed/8300638.

Liu, Z., Fei, X. W., Fang, Y. J., Shi, W. J., Zhang, Y. Q., & Mei, Y. A. (2008). PLC-dependent intracellular Ca^{2+} release was associated with C6-ceramide-induced inhibition of Na^{+} current in rat granule cells. *Journal of Neurochemistry, 106*(6), 2463–2475. http://dx.doi.org/10.1111/j.1471-4159.2008.05562.x.

Liu, Z., Xu, J. G., Zhang, H., Fang, Y. J., & Mei, Y. A. (2007). C(6)-ceramide inhibited Na(+) currents by intracellular Ca(2+) release in rat myoblasts. *Journal of Cellular Physiology, 213*(1), 151–160. http://dx.doi.org/10.1002/jcp.21106.

Lossin, C., Rhodes, T. H., Desai, R. R., Vanoye, C. G., Wang, D., Carniciu, S., … George, A. L., Jr. (2003). Epilepsy-associated dysfunction in the voltage-gated neuronal sodium channel SCN1A. *Journal of Neuroscience: The Official Journal of the Society for Neuroscience, 23*(36), 11289–11295. pii:23/36/11289.

Lu, B. X., Liu, L. Y., Liao, L., Zhang, Z. H., & Mei, Y. A. (2005). Inhibition of Na^{+} channel currents in rat myoblasts by 4-aminopyridine. *Toxicology and Applied Pharmacology, 207*(3), 275–282. http://dx.doi.org/10.1016/j.taap.2005.01.009.

Lundbaek, J. A., Birn, P., Hansen, A. J., Sogaard, R., Nielsen, C., Girshman, J., … Andersen, O. S. (2004). Regulation of sodium channel function by bilayer elasticity: the importance of hydrophobic coupling. Effects of Micelle-forming amphiphiles and cholesterol. *Journal of General Physiology, 123*(5), 599–621. http://dx.doi.org/10.1085/jgp.200308996.

Lundbaek, J. A., Collingwood, S. A., Ingolfsson, H. I., Kapoor, R., & Andersen, O. S. (2010). Lipid bilayer regulation of membrane protein function: gramicidin channels as molecular force probes. *Journal of the Royal Society, Interface, 7*(44), 373–395. http://dx.doi.org/10.1098/rsif.2009.0443.

Macdonald, R. L., & Greenfield, L. J., Jr. (1997). Mechanisms of action of new antiepileptic drugs. *Current Opinion in Neurology, 10*(2), 121–128. Retrieved from: http://www.ncbi.nlm.nih.gov/pubmed/9147003.

Macdonald-Obermann, J. L., & Pike, L. J. (2009). Palmitoylation of the EGF receptor impairs signal transduction and abolishes high-affinity ligand binding. *Biochemistry, 48*(11), 2505–2513. http://dx.doi.org/10.1021/bi802249x.

Maguy, A., Hebert, T. E., & Nattel, S. (2006). Involvement of lipid rafts and caveolae in cardiac ion channel function. *Cardiovascular Research, 69*(1), 798–807. http://dx.doi.org/10.1016/j.cardiores.2005.11.013.

Makita, N., Horie, M., Nakamura, T., Ai, T., Sasaki, K., Yokoi, H., … Kitabatake, A. (2002). Drug-induced long-QT syndrome associated with a subclinical SCN5A mutation. *Circulation, 106*(10), 1269–1274. Retrieved from: http://www.ncbi.nlm.nih.gov/entrez/query.fcgi?cmd=Retrieve&db=PubMed&dopt=Citation&list_uids=12208804.

Marchioli, R., Barzi, F., Bomba, E., Chieffo, C., Di Gregorio, D., Di Mascio, R., … Investigators, G. I.-P. (2002). Early protection against sudden death by n-3 polyunsaturated fatty acids after myocardial infarction: time-course analysis of the results of the Gruppo Italiano per lo Studio della Sopravvivenza nell'Infarto Miocardico (GISSI)-Prevenzione. *Circulation, 105*(16), 1897–1903. Retrieved from: http://www.ncbi.nlm.nih.gov/pubmed/11997274.

Massey, J. B. (2001). Interaction of ceramides with phosphatidylcholine, sphingomyelin and sphingomyelin/cholesterol bilayers. *Biochimica et Biophysica Acta, 1510*(1–2), 167–184. Retrieved from: http://www.ncbi.nlm.nih.gov/pubmed/11342156.

Mathias, C. J. (1993). Sympathetic nervous system disorders in man. *Baillière's Clinical Endocrinology and Metabolism, 7*(2), 465–490. Retrieved from: http://www.ncbi.nlm.nih.gov/pubmed/8489487.

Matsuda, S., Inoue, T., Lee, H. C., Kono, N., Tanaka, F., Gengyo-Ando, K., … Arai, H. (2008). Member of the membrane-bound O-acyltransferase (MBOAT) family encodes a lysophospholipid acyltransferase with broad substrate specificity. *Genes Cells, 13*(8), 879–888. http://dx.doi.org/10.1111/j.1365-2443.2008.01212.x. pii:GTC1212.

McCusker, E. C., Bagneris, C., Naylor, C. E., Cole, A. R., D'Avanzo, N., Nichols, C. G., & Wallace, B. A. (2012). Structure of a bacterial voltage-gated sodium channel pore reveals mechanisms of opening and closing. *Nature Communications, 3*, 1102. http://dx.doi.org/10.1038/ncomms2077.

McCusker, E. C., D'Avanzo, N., Nichols, C. G., & Wallace, B. A. (2011). Simplified bacterial "pore" channel provides insight into the assembly, stability, and structure of sodium channels. *Journal of Biological Chemistry, 286*(18), 16386–16391. http://dx.doi.org/10.1074/jbc.C111.228122. pii:C111.228122.

McNair, W. P., Ku, L., Taylor, M. R., Fain, P. R., Dao, D., Wolfel, E., & Mestroni, L. (2004). SCN5A mutation associated with dilated cardiomyopathy, conduction disorder, and arrhythmia. *Circulation, 110*(15), 2163–2167. http://dx.doi.org/10.1161/01.CIR.0000144458.58660.BB. pii:01.CIR.0000144458.58660.BB.

van Meer, G., & de Kroon, A. I. (2011). Lipid map of the mammalian cell. *Journal of Cell Science, 124*(Pt 1), 5–8. http://dx.doi.org/10.1242/jcs.071233.

van Meer, G., Voelker, D. R., & Feigenson, G. W. (2008). Membrane lipids: where they are and how they behave. *Nature Reviews. Molecular Cell Biology, 9*(2), 112–124. http://dx.doi.org/10.1038/nrm2330.

Meyer zu Heringdrof, D., van Koppen, C. J., Windorfer, B., Himmel, H. M., & Jakobs, K. H. (1996). Calcium signalling by G protein-coupled sphingolipid receptors in bovine aortic endothelial cells. *Naunyn-schmiedeberg's Archives of Pharmacology, 354*(4), 397–403. Retrieved from: http://www.ncbi.nlm.nih.gov/pubmed/8897440.

Miguel, V., Balbi, D., Castillo, C., & Villegas, R. (1992). Reconstitution of sodium channels in large liposomes formed by the addition of acidic phospholipids and freeze-thaw sonication. *Journal of Membrane Biology, 129*(1), 37–47. Retrieved from: http://www.ncbi.nlm.nih.gov/pubmed/1328646.

Miklosz, A., Lukaszuk, B., Chabowski, A., Rogowski, F., Kurek, K., & Zendzian-Piotrowska, M. (2015). Hyperthyroidism evokes myocardial ceramide accumulation. *Cellular Physiology and Biochemistry: International Journal of Experimental Cellular Physiology, Biochemistry, and Pharmacology, 35*(2), 755–766. http://dx.doi.org/10.1159/000369735.

Monod, J., Wyman, J., & Changeux, J. P. (1965). On the nature of allosteric transitions: a plausible model. *Journal of Molecular Biology, 12*, 88–118. Retrieved from: http://www.ncbi.nlm.nih.gov/pubmed/14343300.

Morris, C. E., & Juranka, P. F. (2007). Nav channel mechanosensitivity: activation and inactivation accelerate reversibly with stretch. *Biophysical Journal, 93*(3), 822–833. http://dx.doi.org/10.1529/biophysj.106.101246.

Nakajima, T., Kubota, N., Tsutsumi, T., Oguri, A., Imuta, H., Jo, T., … Nagata, T. (2009). Eicosapentaenoic acid inhibits voltage-gated sodium channels and invasiveness in prostate cancer cells. *British Journal of Pharmacology, 156*(3), 420–431. http://dx.doi.org/10.1111/j.1476-5381.2008.00059.x.

Nicholson, R. A., Liao, C., Zheng, J., David, L. S., Coyne, L., Errington, A. C., … Lees, G. (2003). Sodium channel inhibition by anandamide and synthetic cannabimimetics in brain. *Brain Research, 978*(1–2), 194–204. Retrieved from: http://www.ncbi.nlm.nih.gov/pubmed/12834914.

Nicholson, R. A., Zheng, J., Ganellin, C. R., Verdon, B., & Lees, G. (2001). Anesthetic-like interaction of the sleep-inducing lipid oleamide with voltage-gated sodium channels in mammalian brain. *Anesthesiology, 94*(1), 120–128. Retrieved from: http://www.ncbi.nlm.nih.gov/pubmed/11135731.

Ohno, Y., Kashio, A., Ogata, R., Ishitomi, A., Yamazaki, Y., & Kihara, A. (2012). Analysis of substrate specificity of human DHHC protein acyltransferases using a yeast expression system. *Molecular Biology of the Cell, 23*(23), 4543–4551. http://dx.doi.org/10.1091/mbc.E12-05-0336. pii:mbc.E12-05-0336.

Okajima, F., & Kondo, Y. (1995). Pertussis toxin inhibits phospholipase C activation and Ca^{2+} mobilization by sphingosylphosphorylcholine and galactosylsphingosine in HL60 leukemia cells. Implications of GTP-binding protein-coupled receptors for lysosphingolipids. *Journal of Biological Chemistry, 270*(44), 26332–26340. Retrieved from: http://www.ncbi.nlm.nih.gov/pubmed/7592844.

Okura, D., Horishita, T., Ueno, S., Yanagihara, N., Sudo, Y., Uezono, Y., & Sata, T. (2014). The endocannabinoid anandamide inhibits voltage-gated sodium channels Nav1.2, Nav1.6, Nav1.7, and Nav1.8 in Xenopus oocytes. *Anesthesia and Analgesia, 118*(3), 554–562. http://dx.doi.org/10.1213/ANE.0000000000000070.

Orlati, S., Porcelli, A. M., Hrelia, S., Lorenzini, A., & Rugolo, M. (1998). Intracellular calcium mobilization and phospholipid degradation in sphingosylphosphorylcholine-stimulated human airway epithelial cells. *Biochemical Journal, 334*(Pt 3), 641–649. Retrieved from: http://www.ncbi.nlm.nih.gov/pubmed/9729473.

OuYang, W., & Hemmings, H. C., Jr. (2007). Isoform-selective effects of isoflurane on voltage-gated Na^+ channels. *Anesthesiology, 107*(1), 91–98. http://dx.doi.org/10.1097/01.anes.0000268390.28362.4a.

Ouyang, W., Wang, G., & Hemmings, H. C., Jr. (2003). Isoflurane and propofol inhibit voltage-gated sodium channels in isolated rat neurohypophysial nerve terminals. *Molecular Pharmacology, 64*(2), 373–381. http://dx.doi.org/10.1124/mol.64.2.373.

Palygin, O. A., Pettus, J. M., & Shibata, E. F. (2008). Regulation of caveolar cardiac sodium current by a single Gsalpha histidine residue. *American Journal of Physiology. Heart and Circulatory Physiology, 294*(4), H1693–H1699. http://dx.doi.org/10.1152/ajpheart.01337.2007.

Payandeh, J., Gamal El-Din, T. M., Scheuer, T., Zheng, N., & Catterall, W. A. (2012). Crystal structure of a voltage-gated sodium channel in two potentially inactivated states. *Nature, 486*(7401), 135–139. http://dx.doi.org/10.1038/nature11077.

Payandeh, J., Scheuer, T., Zheng, N., & Catterall, W. A. (2011). The crystal structure of a voltage-gated sodium channel. *Nature, 475*(7356), 353–358. http://dx.doi.org/10.1038/nature10238.

Pertwee, R. G. (2006). The pharmacology of cannabinoid receptors and their ligands: an overview. *International Journal of Obesity (London), 30*(Suppl. 1), S13–S18. http://dx.doi.org/10.1038/sj.ijo.0803272.

Pierce, G. N., & Dhalla, N. S. (1983). Sarcolemmal Na^+-K^+-ATPase activity in diabetic rat heart. *American Journal of Physiology, 245*(3), C241–C247. Retrieved from: http://www.ncbi.nlm.nih.gov/pubmed/6137147.

Pignier, C., Revenaz, C., Rauly-Lestienne, I., Cussac, D., Delhon, A., Gardette, J., & Le Grand, B. (2007). Direct protective effects of poly-unsaturated fatty acids, DHA and EPA, against activation of cardiac late sodium current: a mechanism for ischemia selectivity. *Basic Research in Cardiology, 102*(6), 553–564. http://dx.doi.org/10.1007/s00395-007-0676-x

Pike, L. J., Han, X., & Gross, R. W. (2005). Epidermal growth factor receptors are localized to lipid rafts that contain a balance of inner and outer leaflet lipids: a shotgun lipidomics study. *Journal of Biological Chemistry, 280*(29), 26796–26804. http://dx.doi.org/10.1074/jbc.M503805200

Post, J. A., Verkleij, A. J., & Langer, G. A. (1995). Organization and function of sarcolemmal phospholipids in control and ischemic/reperfused cardiomyocytes. *Journal of Molecular and Cellular Cardiology, 27*(2), 749–760. Retrieved from: http://www.ncbi.nlm.nih.gov/pubmed/7776380.

Powl, A. M., East, J. M., & Lee, A. G. (2008a). Anionic phospholipids affect the rate and extent of flux through the mechanosensitive channel of large conductance MscL. *Biochemistry, 47*(14), 4317–4328. http://dx.doi.org/10.1021/bi702409t.

Powl, A. M., East, J. M., & Lee, A. G. (2008b). Importance of direct interactions with lipids for the function of the mechanosensitive channel MscL. *Biochemistry, 47*(46), 12175–12184. http://dx.doi.org/10.1021/bi801352a.

Pristera, A., Baker, M. D., & Okuse, K. (2012). Association between tetrodotoxin resistant channels and lipid rafts regulates sensory neuron excitability. *PLoS One,* 7(8), e40079. http://dx.doi.org/10.1371/journal.pone.0040079.

Qin, N., Neeper, M. P., Liu, Y., Hutchinson, T. L., Lubin, M. L., & Flores, C. M. (2008). TRPV2 is activated by cannabidiol and mediates CGRP release in cultured rat dorsal root ganglion neurons. *Journal of Neuroscience: The Official Journal of the Society for Neuroscience, 28*(24), 6231–6238. http://dx.doi.org/10.1523/JNEUROSCI.0504-08.2008.

Raju, S. G., Barber, A. F., LeBard, D. N., Klein, M. L., & Carnevale, V. (2013). Exploring volatile general anesthetic binding to a closed membrane-bound bacterial voltage-gated sodium channel via computation. *PLoS Computational Biology, 9*(6), e1003090. http://dx.doi.org/10.1371/journal.pcbi.1003090.

Ramu, Y., Xu, Y., & Lu, Z. (2006). Enzymatic activation of voltage-gated potassium channels. *Nature, 442*(7103), 696–699. http://dx.doi.org/10.1038/nature04880.

Ramu, Y., Xu, Y., & Lu, Z. (2007). Inhibition of CFTR Cl- channel function caused by enzymatic hydrolysis of sphingomyelin. *Proceedings of the National Academy of Sciences of the United States of America, 104*(15), 6448–6453. http://dx.doi.org/10.1073/pnas.0701354104.

Ratnakumari, L., & Hemmings, H. C., Jr. (1996). Inhibition by propofol of [3H]-batrachotoxinin-A 20-alpha-benzoate binding to voltage-dependent sodium channels in rat cortical synaptosomes. *British Journal of Pharmacology, 119*(7), 1498–1504. Retrieved from: http://www.ncbi.nlm.nih.gov/pubmed/8968561.

Recio-Pinto, E., Duch, D. S., Levinson, S. R., & Urban, B. W. (1987). Purified and unpurified sodium channels from eel electroplax in planar lipid bilayers. *Journal of General Physiology, 90*(3), 375–395. Retrieved from: http://www.ncbi.nlm.nih.gov/pubmed/2443607.

Rehberg, B., Urban, B. W., & Duch, D. S. (1995). The membrane lipid cholesterol modulates anesthetic actions on a human brain ion channel. *Anesthesiology, 82*(3), 749–758. Retrieved from: http://www.ncbi.nlm.nih.gov/pubmed/7879943.

Rehberg, B., Xiao, Y. H., & Duch, D. S. (1996). Central nervous system sodium channels are significantly suppressed at clinical concentrations of volatile anesthetics. *Anesthesiology, 84*(5), 1223–1233. Discussion 1227A. Retrieved from: http://www.ncbi.nlm.nih.gov/pubmed/8624017.

Resh, M. D. (2006). Palmitoylation of ligands, receptors, and intracellular signaling molecules. *Science STKE, 2006*(359), re14. http://dx.doi.org/10.1126/stke.3592006re14.

Rhodes, T. H., Lossin, C., Vanoye, C. G., Wang, D. W., & George, A. L., Jr. (2004). Noninactivating voltage-gated sodium channels in severe myoclonic epilepsy of infancy. *Proceedings of the National Academy of Sciences of the United States of America, 101*(30), 11147–11152. http://dx.doi.org/10.1073/pnas.0402482101. pii:0402482101.

Richardson, E. S., Iaizzo, P. A., & Xiao, Y. F. (2011). Electrophysiological mechanisms of the anti-arrhythmic effects of omega-3 fatty acids. *Journal of Cardiovascular Translational Research, 4*(1), 42–52. http://dx.doi.org/10.1007/s12265-010-9243-1.

Rose, D. P. (1997). Effects of dietary fatty acids on breast and prostate cancers: evidence from in vitro experiments and animal studies. *American Journal of Clinical Nutrition, 66*(6 Suppl.), 1513S–1522S. Retrieved from: http://www.ncbi.nlm.nih.gov/pubmed/9394709.

Ross, H. R., Napier, I., & Connor, M. (2008). Inhibition of recombinant human T-type calcium channels by Delta9-tetrahydrocannabinol and cannabidiol. *Journal of Biological Chemistry, 283*(23), 16124–16134. http://dx.doi.org/10.1074/jbc.M707104200.

Rothman, J. E., & Kennedy, E. P. (1977). Rapid transmembrane movement of newly synthesized phospholipids during membrane assembly. *Proceedings of the National Academy of Sciences of the United States of America, 74*(5), 1821–1825. Retrieved from: http://www.ncbi.nlm.nih.gov/pubmed/405668.

Ruan, Y., Liu, N., Bloise, R., Napolitano, C., & Priori, S. G. (2007). Gating properties of SCN5A mutations and the response to mexiletine in long-QT syndrome type 3 patients. *Circulation, 116*(10), 1137–1144. http://dx.doi.org/10.1161/CIRCULATIONAHA.107.707877.

Rumenapp, U., Lummen, G., Virchow, S., Hanske, J., Meyer zu Heringdorf, D., & Jakobs, K. H. (2000). Sphingolipid receptor signaling and function in human bladder carcinoma cells: inhibition of LPA- but enhancement of thrombin-stimulated cell motility. *Naunyn-schmiedeberg's Archives of Pharmacology, 361*(1), 1–11. Retrieved from: http://www.ncbi.nlm.nih.gov/pubmed/10651140.

Rusinova, R., Herold, K. F., Sanford, R. L., Greathouse, D. V., Hemmings, H. C., Jr., & Andersen, O. S. (2011). Thiazolidinedione insulin sensitizers alter lipid bilayer properties and voltage-dependent sodium channel function: implications for drug discovery. *Journal of General Physiology, 138*(2), 249–270. http://dx.doi.org/10.1085/jgp.201010529.

Russell, Y., Evans, P., & Dodd, G. H. (1989). Characterization of the total lipid and fatty acid composition of rat olfactory mucosa. *Journal of Lipid Research, 30*(6), 877–884. Retrieved from: http://www.ncbi.nlm.nih.gov/entrez/query.fcgi?cmd=Retrieve&db=PubMed&dopt=Citation&list_uids=2529336.

Safrany-Fark, A., Petrovszki, Z., Kekesi, G., Liszli, P., Benedek, G., Keresztes, C., & Horvath, G. (2015). In vivo potency of different ligands on voltage-gated sodium channels. *European Journal of Pharmacology, 762*, 158–164. http://dx.doi.org/10.1016/j.ejphar.2015.05.053.

Saha, S. C., Henderson, A. J., Powl, A. M., Wallace, B. A., de Planque, M. R., & Morgan, H. (2015). Characterization of the prokaryotic sodium channel NavSp pore with a microfluidic bilayer platform. *PLoS One, 10*(7), e0131286. http://dx.doi.org/10.1371/journal.pone.0131286.

Schley, P. D., Brindley, D. N., & Field, C. J. (2007). (n 3) PUFA alter raft lipid composition and decrease epidermal growth factor receptor levels in lipid rafts of human breast cancer cells. *Journal of Nutrition, 137*(3), 548–553. Retrieved from: http://www.ncbi.nlm.nih.gov/pubmed/17311938.

Schmid, H. H., & Berdyshev, E. V. (2002). Cannabinoid receptor-inactive N-acylethanolamines and other fatty acid amides: metabolism and function. *Prostaglandins, Leukotrienes, and Essential Fatty Acids, 66*(2–3), 363–376. http://dx.doi.org/10.1054/plef.2001.0348.

Schmidt, D., Cross, S. R., & MacKinnon, R. (2009). A gating model for the archeal voltage-dependent K(+) channel KvAP in DPhPC and POPE: POPG decane lipid bilayers. *Journal of Molecular Biology, 390*(5), 902–912. http://dx.doi.org/10.1016/j.jmb.2009.05.062.

Schmidt, D., Jiang, Q. X., & MacKinnon, R. (2006). Phospholipids and the origin of cationic gating charges in voltage sensors. *Nature, 444*(7120), 775–779. http://dx.doi.org/10.1038/nature05416.

Schmidt, J. W., & Catterall, W. A. (1987). Palmitoylation, sulfation, and glycosylation of the alpha subunit of the sodium channel. Role of post-translational modifications in channel assembly. *Journal of Biological Chemistry, 262*(28), 13713–13723. Retrieved from: http://www.ncbi.nlm.nih.gov/pubmed/2443496.

Schott, J. J., Alshinawi, C., Kyndt, F., Probst, V., Hoorntje, T. M., Hulsbeek, M., … Le Marec, H. (1999). Cardiac conduction defects associate with mutations in SCN5A. *Nature Genetics, 23*(1), 20–21. http://dx.doi.org/10.1038/12618.

Schroepfer, G. J., Jr. (2000). Oxysterols: modulators of cholesterol metabolism and other processes. *Physiological Reviews, 80*(1), 361–554. Retrieved from: http://www.ncbi.nlm.nih.gov/pubmed/10617772.

Schwartz, P. J., Priori, S. G., Dumaine, R., Napolitano, C., Antzelevitch, C., Stramba-Badiale, M., … Bloise, R. (2000). A molecular link between the sudden infant death syndrome and the long-QT syndrome. *New England Journal of Medicine, 343*(4), 262–267. http://dx.doi.org/10.1056/NEJM200007273430405. pii:MJBA-430405.

Seebungkert, B., & Lynch, J. W. (2002). Effects of polyunsaturated fatty acids on voltage-gated K^+ and Na^+ channels in rat olfactory receptor neurons. *European Journal of Neuroscience, 16*(11), 2085–2094. pii:2288.

Seufferlein, T., & Rozengurt, E. (1995). Sphingosylphosphorylcholine rapidly induces tyrosine phosphorylation of p125FAK and paxillin, rearrangement of the actin cytoskeleton and focal contact assembly. Requirement of p21rho in the signaling pathway. *Journal of Biological Chemistry, 270*(41), 24343–24351. Retrieved from: http://www.ncbi.nlm.nih.gov/pubmed/7592646.

Shaya, D., Findeisen, F., Abderemane-Ali, F., Arrigoni, C., Wong, S., Nurva, S. R., … Minor, D. L., Jr. (2014). Structure of a prokaryotic sodium channel pore reveals essential gating elements and an outer ion binding site common to eukaryotic channels. *Journal of Molecular Biology, 426*(2), 467–483. http://dx.doi.org/10.1016/j.jmb.2013.10.010.

Shaya, D., Kreir, M., Robbins, R. A., Wong, S., Hammon, J., Bruggemann, A., & Minor, D. L., Jr. (2011). Voltage-gated sodium channel (NaV) protein dissection creates a set of functional pore-only proteins. *Proceedings of the National Academy of Sciences of the United States of America, 108*(30), 12313–12318. http://dx.doi.org/10.1073/pnas.1106811108. pii:1106811108.

Sheldon, R. S., Duff, H. J., Thakore, E., & Hill, R. J. (1994). Class I antiarrhythmic drugs: allosteric inhibitors of [3H] batrachotoxinin binding to rat cardiac sodium channels. *Journal of Pharmacology and Experimental Therapeutics, 268*(1), 187–194. Retrieved from: http://www.ncbi.nlm.nih.gov/pubmed/8301556.

Sheldon, R. S., Hill, R. J., Cannon, N. J., & Duff, H. J. (1989). Amiodarone: biochemical evidence for binding to a receptor for class I drugs associated with the rat cardiac sodium channel. *Circulation Research, 65*(2), 477–482. Retrieved from: http://www.ncbi.nlm.nih.gov/pubmed/2546697.

Shennan, D. H., & Bishop, O. S. (1974). Diet and mortality from malignant disease in 32 countries. *West Indian Medical Journal, 23*(1), 44–53. Retrieved from: http://www.ncbi.nlm.nih.gov/pubmed/4456820.

Shipston, M. J. (2011). Ion channel regulation by protein palmitoylation. *Journal of Biological Chemistry, 286*(11), 8709–8716. http://dx.doi.org/10.1074/jbc.R110.210005.

Shiraishi, T., Hiraiwa, M., & Uda, Y. (1993). Effects of cyclodextrins on the hydrolysis of ganglioside GM1 by acid beta-galactosidases. *Glycoconjugate Journal, 10*(2), 170–174. Retrieved from: http://www.ncbi.nlm.nih.gov/pubmed/8400826.

Simopoulos, A. P. (2003). Importance of the ratio of omega-6/omega-3 essential fatty acids: evolutionary aspects. *World Review of Nutrition and Dietetics, 92*, 1–22. Retrieved from: http://www.ncbi.nlm.nih.gov/pubmed/14579680.

Smotrys, J. E., & Linder, M. E. (2004). Palmitoylation of intracellular signaling proteins: regulation and function. *Annual Review of Biochemistry, 73*, 559–587. http://dx.doi.org/10.1146/annurev.biochem.73.011303.073954.

Sohlenkamp, C., Lopez-Lara, I. M., & Geiger, O. (2003). Biosynthesis of phosphatidylcholine in bacteria. *Progress in Lipid Research, 42*(2), 115–162. Retrieved from: http://www.ncbi.nlm.nih.gov/pubmed/12547654.

Splawski, I., Timothy, K. W., Tateyama, M., Clancy, C. E., Malhotra, A., Beggs, A. H., … Keating, M. T. (2002). Variant of SCN5A sodium channel implicated in risk of cardiac arrhythmia. *Science, 297*(5585), 1333–1336. http://dx.doi.org/10.1126/science.1073569. pii:297/5585/1333.

Stith, B. J., Hall, J., Ayres, P., Waggoner, L., Moore, J. D., & Shaw, W. A. (2000). Quantification of major classes of *Xenopus* phospholipids by high performance liquid chromatography with evaporative light scattering detection. *Journal of Lipid Research, 41*(9), 1448–1454. Retrieved from: http://www.ncbi.nlm.nih.gov/pubmed/10974052.

Suchyna, T. M., Markin, V. S., & Sachs, F. (2009). Biophysics and structure of the patch and the gigaseal. *Biophysical Journal, 97*(3), 738–747. http://dx.doi.org/10.1016/j.bpj.2009.05.018.

Sugawara, T., Tsurubuchi, Y., Agarwala, K. L., Ito, M., Fukuma, G., Mazaki-Miyazaki, E., … Yamakawa, K. (2001). A missense mutation of the Na^+ channel alpha II subunit gene Na(v)1.2 in a patient with febrile and afebrile seizures causes channel dysfunction. *Proceedings of the National Academy of Sciences of the United States of America, 98*(11), 6384–6389. http://dx.doi.org/10.1073/pnas.111065098. pii:98/11/6384.

Suh, B. C., & Hille, B. (2008). PIP2 is a necessary cofactor for ion channel function: how and why? *Annual Review of Biophysics, 37*, 175–195. http://dx.doi.org/10.1146/annurev.biophys.37.032807.125859.

Tabarean, I. V., Juranka, P., & Morris, C. E. (1999). Membrane stretch affects gating modes of a skeletal muscle sodium channel. *Biophysical Journal, 77*(2), 758–774. http://dx.doi.org/10.1016/S0006-3495(99)76930-4.

Tabarean, I. V., & Morris, C. E. (2002). Membrane stretch accelerates activation and slow inactivation in Shaker channels with S3-S4 linker deletions. *Biophysical Journal, 82*(6), 2982–2994. http://dx.doi.org/10.1016/S0006-3495(02)75639-7.

Tambourgi, D. V., Goncalves-de-Andrade, R. M., & van den Berg, C. W. (2010). Loxoscelism: from basic research to the proposal of new therapies. *Toxicon, 56*(7), 1113–1119. http://dx.doi.org/10.1016/j.toxicon.2010.01.021.

Tamkun, M. M., Talvenheimo, J. A., & Catterall, W. A. (1984). The sodium channel from rat brain. Reconstitution of neurotoxin-activated ion flux and scorpion toxin binding from purified components. *Journal of Biological Chemistry, 259*(3), 1676–1688. Retrieved from: http://www.ncbi.nlm.nih.gov/pubmed/6319406.

Tan, H. L., Bink-Boelkens, M. T., Bezzina, C. R., Viswanathan, P. C., Beaufort-Krol, G. C., van Tintelen, P. J., … Balser, J. R. (2001). A sodium-channel mutation causes isolated cardiac conduction disease. *Nature, 409*(6823), 1043–1047. http://dx.doi.org/10.1038/35059090.

Tang, L., Wang, Y., Leng, T., Sun, H., Zhou, Y., Zhu, W., … Yan, G. (2015). Cholesterol metabolite cholestane-3beta,5alpha,6beta-triol suppresses epileptic seizures by negative modulation of voltage-gated sodium channels. *Steroids, 98*, 166–172. http://dx.doi.org/10.1016/j.steroids.2014.12.025.

Taniwaki, T., Yamada, T., Asahara, H., Ohyagi, Y., & Kira, J. (1999). Ceramide induces apoptosis to immature cerebellar granule cells in culture. *Neurochemical Research, 24*(5), 685–690. Retrieved from: http://www.ncbi.nlm.nih.gov/pubmed/10344598.

Testi, R. (1996). Sphingomyelin breakdown and cell fate. *Trends in Biochemical Sciences, 21*(12), 468–471. Retrieved from: http://www.ncbi.nlm.nih.gov/pubmed/9009829.

Toman, R. E., Movsesyan, V., Murthy, S. K., Milstien, S., Spiegel, S., & Faden, A. I. (2002). Ceramide-induced cell death in primary neuronal cultures: upregulation of ceramide levels during neuronal apoptosis. *Journal of Neuroscience Research, 68*(3), 323–330. http://dx.doi.org/10.1002/jnr.10190.

Tornquist, K., & Ekokoski, E. (1994). Effect of sphingosine derivatives on calcium fluxes in thyroid FRTL-5 cells. *Biochemical Journal, 299*(Pt 1), 213–218. Retrieved from: http://www.ncbi.nlm.nih.gov/pubmed/8166643.

Vaudry, D., Falluel-Morel, A., Basille, M., Pamantung, T. F., Fontaine, M., Fournier, A., … Gonzalez, B. J. (2003). Pituitary adenylate cyclase-activating polypeptide prevents C2-ceramide-induced apoptosis of cerebellar granule cells. *Journal of Neuroscience Research, 72*(3), 303–316. http://dx.doi.org/10.1002/jnr.10530.

Wallace, R. H., Wang, D. W., Singh, R., Scheffer, I. E., George, A. L., Jr., Phillips, H. A., … Mulley, J. C. (1998). Febrile seizures and generalized epilepsy associated with a mutation in the Na^+-channel beta1 subunit gene SCN1B. *Nature Genetics, 19*(4), 366–370. http://dx.doi.org/10.1038/1252.

Wang, D. W., Viswanathan, P. C., Balser, J. R., George, A. L., Jr., & Benson, D. W. (2002). Clinical, genetic, and biophysical characterization of SCN5A mutations associated with atrioventricular conduction block. *Circulation, 105*(3), 341–346. Retrieved from: http://www.ncbi.nlm.nih.gov/entrez/query.fcgi?cmd=Retrieve&db=PubMed&dopt=Citation&list_uids=11804990.

Wang, G. K., Russell, C., & Wang, S. Y. (2003). State-dependent block of wild-type and inactivation-deficient Na^+ channels by flecainide. *Journal of General Physiology, 122*(3), 365–374. http://dx.doi.org/10.1085/jgp.200308857. pii:jgp.200308857.

Wang, W., Gu, J., Li, Y. Q., & Tao, Y. X. (2011). Are voltage-gated sodium channels on the dorsal root ganglion involved in the development of neuropathic pain? *Molecular Pain, 7*, 16. http://dx.doi.org/10.1186/1744-8069-7-16.

Wayne, S. J., Baumgartner, K., Baumgartner, R. N., Bernstein, L., Bowen, D. J., & Ballard-Barbash, R. (2006). Diet quality is directly associated with quality of life in breast cancer survivors. *Breast Cancer Research and Treatment, 96*(3), 227–232. http://dx.doi.org/10.1007/s10549-005-9018-6.

Wieland, S. J., Fletcher, J. E., & Gong, Q. H. (1992). Differential modulation of a sodium conductance in skeletal muscle by intracellular and extracellular fatty acids. *American Journal of Physiology, 263*(2 Pt 1), C308–C312. Retrieved from: http://www.ncbi.nlm.nih.gov/entrez/query.fcgi?cmd=Retrieve&db=PubMed&dopt=Citation&list_uids=1514578.

Williams, G. M., Williams, C. L., & Weisburger, J. H. (1999). Diet and cancer prevention: the fiber first diet. *Toxicological Sciences: An Official Journal of the Society of Toxicology, 52*(2 Suppl.), 72–86. Retrieved from: http://www.ncbi.nlm.nih.gov/pubmed/10630594.

Wojcik, B., Harasim, E., Zabielski, P., Chabowski, A., & Gorski, J. (2015). Effect of tachycardia on lipid metabolism and expression of fatty acid transporters in heart ventricles of the rat. *Journal of Physiology and Pharmacology: An Official Journal of the Polish Physiological Society, 66*(5), 691–699. Retrieved from: http://www.ncbi.nlm.nih.gov/pubmed/26579575.

Wood, J. N., Boorman, J. P., Okuse, K., & Baker, M. D. (2004). Voltage-gated sodium channels and pain pathways. *Journal of Neurobiology, 61*(1), 55–71. http://dx.doi.org/10.1002/neu.20094.

Worley, J. F., 3rd, French, R. J., Pailthorpe, B. A., & Krueger, B. K. (1992). Lipid surface charge does not influence conductance or calcium block of single sodium channels in planar bilayers. *Biophysical Journal, 61*(5), 1353–1363. http://dx.doi.org/10.1016/S0006-3495(92)81942-2.

Wu, C. C., Su, M. J., Chi, J. F., Chen, W. J., Hsu, H. C., & Lee, Y. T. (1995). The effect of hypercholesterolemia on the sodium inward currents in cardiac myocyte. *Journal of Molecular and Cellular Cardiology, 27*(6), 1263–1269. Retrieved from: http://www.ncbi.nlm.nih.gov/pubmed/8531208.

Xiao, Y. F., Kang, J. X., Morgan, J. P., & Leaf, A. (1995). Blocking effects of polyunsaturated fatty acids on Na^+ channels of neonatal rat ventricular myocytes. *Proceedings of the National Academy of Sciences of the United States of America, 92*(24), 11000–11004. Retrieved from: http://www.ncbi.nlm.nih.gov/entrez/query.fcgi?cmd=Retrieve&db=PubMed&dopt=Citation&list_uids=7479925.

Xiao, Y. F., Ke, Q., Wang, S. Y., Auktor, K., Yang, Y., Wang, G. K., … Leaf, A. (2001). Single point mutations affect fatty acid block of human myocardial sodium channel alpha subunit Na^+ channels. *Proceedings of the National Academy of Sciences of the United States of*

America, 98(6), 3606–3611. http://dx.doi.org/10.1073/pnas.061003798. pii: 061003798.

Xiao, Y. F., Ma, L., Wang, S. Y., Josephson, M. E., Wang, G. K., Morgan, J. P., & Leaf, A. (2006). Potent block of inactivation-deficient Na^+ channels by n-3 polyunsaturated fatty acids. *American Journal of Physiology. Cell Physiology, 290*(2), C362–C370. http://dx.doi.org/10.1152/ajpcell.00296.2005. pii:00296.2005.

Xiao, Y. F., Wright, S. N., Wang, G. K., Morgan, J. P., & Leaf, A. (1998). Fatty acids suppress voltage-gated Na^+ currents in HEK293t cells transfected with the alpha-subunit of the human cardiac Na^+ channel. *Proceedings of the National Academy of Sciences of the United States of America, 95*(5), 2680–2685. Retrieved from: http://www.ncbi.nlm.nih.gov/entrez/query.fcgi?cmd=Retrieve&db=PubMed&dopt=Citation&list_uids=9482947.

Xiao, Y. F., Wright, S. N., Wang, G. K., Morgan, J. P., & Leaf, A. (2000). Coexpression with beta(1)-subunit modifies the kinetics and fatty acid block of hH1(alpha) Na(+) channels. *American Journal of Physiology. Heart and Circulatory Physiology, 279*(1), H35–H46. Retrieved from: http://www.ncbi.nlm.nih.gov/entrez/query.fcgi?cmd=Retrieve&db=PubMed&dopt=Citation&list_uids=10899039.

Xu, Y., Ramu, Y., & Lu, Z. (2008). Removal of phospho-head groups of membrane lipids immobilizes voltage sensors of K^+ channels. *Nature, 451*(7180), 826–829. http://dx.doi.org/10.1038/nature06618.

Yang, Y., Wang, Y., Li, S., Xu, Z., Li, H., Ma, L., … Shen, Y. (2004). Mutations in SCN9A, encoding a sodium channel alpha subunit, in patients with primary erythermalgia. *Journal of Medical Genetics, 41*(3), 171–174. Retrieved from: http://www.ncbi.nlm.nih.gov/pubmed/14985375.

Yarbrough, T. L., Lu, T., Lee, H. C., & Shibata, E. F. (2002). Localization of cardiac sodium channels in caveolin-rich membrane domains: regulation of sodium current amplitude. *Circulation Research, 90*(4), 443–449. Retrieved from: http://www.ncbi.nlm.nih.gov/pubmed/11884374.

Yasui, K., & Palade, P. (1996). Sphingolipid actions on sodium and calcium currents of rat ventricular myocytes. *American Journal of Physiology, 270*(2 Pt 1), C645–C649. Retrieved from: http://www.ncbi.nlm.nih.gov/pubmed/8779930.

Zhang, X., Ren, W., DeCaen, P., Yan, C., Tao, X., Tang, L., … Yan, N. (2012). Crystal structure of an orthologue of the NaChBac voltage-gated sodium channel. *Nature, 486*(7401), 130–134. http://dx.doi.org/10.1038/nature11054.

Zhang, Y. H., Vasko, M. R., & Nicol, G. D. (2002). Ceramide, a putative second messenger for nerve growth factor, modulates the TTX-resistant Na(+) current and delayed rectifier K(+) current in rat sensory neurons. *Journal of Physiology, 544*(Pt 2), 385–402. Retrieved from: http://www.ncbi.nlm.nih.gov/pubmed/12381813.

Zhang, Y. L., Dunlop, J., & Dalziel, J. E. (2007). Recombinant human voltage-gated skeletal muscle sodium channels are pharmacologically functional in planar lipid bilayers. *Biosensors & Bioelectronics, 22*(6), 1006–1012. http://dx.doi.org/10.1016/j.bios.2006.04.009.

Zhou, M., Liu, Z., Hu, C., Zhang, Z., & Mei, Y. (2004). Developmental regulation of a Na(+)-activated fast outward K^+ current in rat myoblasts. *Cellular Physiology and Biochemistry: International Journal of Experimental Cellular Physiology, Biochemistry, and Pharmacology, 14*(4–6), 225–230. http://dx.doi.org/10.1159/000080331.

CHAPTER THIRTEEN

Mechanism of Inactivation in Voltage-Gated Na^+ Channels

V.S. Gawali and H. Todt[1]
Medical University of Vienna, Vienna, Austria
[1]Corresponding author: E-mail: hannes.todt@meduniwien.ac.at

Contents

Current Topics in Membranes, Volume 78
ISSN 1063-5823
http://dx.doi.org/10.1016/bs.ctm.2016.07.004

Abstract

Voltage-gated Na^+ channels (VGSCs) initiate action potentials thereby giving rise to rapid transmission of electrical signals along cell membranes and between cells. Depolarization of the cell membrane causes VGSCs to open but also gives rise to a nonconducting state termed inactivation. Inactivation of VGSCs serves a critical physiologic function as it determines the extent of excitability of neurons and of muscle cells. Depending on the time course of development and removal of inactivation both "fast-" and "slow"-inactivated states have been described. Evidence from mutagenesis studies suggests that fast inactivation is produced by a block of the internal vestibule by a tethered inactivation particle that has been mapped to the internal linker between domains III and IV. The motion of this linker may be regulated by parts of the internal C-terminus. The molecular mechanism of slow inactivation is less clear. However, aside from a high number of mutagenesis studies, the recent availability of 3D structures of crystallized prokaryotic VGSCs offers insights into the molecular motions associated with slow inactivation. One possible scenario is that slow movements of the voltage sensors are transmitted to the external vestibule giving rise to a conformational change of this region. This molecular rearrangement is transmitted to the S6 segments giving rise to collapse of the internal vestibule.

List of Abbreviations

(μ-CTX) μ-Conotoxin GIIIA
DI–IV Domain I–IV
FGF Fibroblast growth factor
FHF Fibroblast growth factor homologous factors
IFM motif Isoleucine, phenylalanine, and methionine
NaAb Bacterial voltage-gated sodium channel from *Arcobacter butzleri*
Nav Voltage-gated sodium channels
NavRh Bacterial voltage-gated sodium channel from the marine alpha proteobacterium HIMB114
VGSC Voltage-gated sodium channels

1. MOLECULAR ARCHITECTURE OF VOLTAGE-GATED Na^+ CHANNELS

Voltage-gated Na^+ channels (VGSCs) are necessary for rapid transmission of excitation along cell membranes, between cells and throughout cell networks. This spread of excitation forms the basis of function of skeletal muscle, the heart, and the nervous system. VGSCs consist of one pore forming α subunit, and several accessory β subunits (Isom et al., 1992; Sutkowski & Catterall, 1990). Heterologous expression of α subunits alone is sufficient to form functional VGSCs (Goldin et al., 1986; Noda et al., 1984). The α-subunit is a large polypeptide of about 2000 amino acids containing four homologous domains of 300–400 amino acids. Each domain (DI–DIV) consists of six membrane spanning α-helical (S1–S6) segments. The three intracellular linker loops as well as the N- and C-termini of the channel are cytoplasmic. The four S6 segments line the internal part of the conduction pathway ("internal vestibule"). The external part of the permeation pathway is considered to be composed of the external loops connecting the upper ends of the S5 and S6 segments (P-loops). These loops fold into the channel pore and form the selectivity filter at their innermost turn. The S1–S4 segments of each domain are arranged laterally to the S5–S6 pore forming unit and are considered to constitute the voltage sensor which, in response to depolarization of the membrane, moves in the extracellular direction thereby opening the channel (Catterall, Goldin, & Waxman, 2005). Nine VGSC isoforms are expressed in mammalian tissues, Nav1.1, Nav1.2, Nav1.3, Nav1.6, Nav1.7, Nav1.9 are found in neurons; Nav1.4 in skeletal muscle; and Nav1.5 in heart muscle (Catterall et al., 2005).

2. THE KINETIC STATES OF VOLTAGE-GATED Na^+ CHANNELS

The fact that the sodium permeability of the axon membrane rises rapidly and then decays during a step depolarization led Hodgkin and Huxley to suggest that sodium permeability first activates and then inactivates (Hodgkin & Huxley, 1952a, 1952b). Activation is a rapid process that opens VGSCs during a depolarization, whereas inactivation causes channels to enter a nonconducting state. Once inactivated, channels must be repolarized, or hyperpolarized for a specific time period, to remove inactivation ("recovery from inactivation"). Inactivation is critical for control of excitability in nerve and muscle cells. Depending on the time spent at depolarized potentials, channels may enter into a number of different inactivated states (Nuss et al., 1996). Thus, if adult rat skeletal muscle VGSC (rNav1.4), heterologously expressed in *Xenopus* oocytes, are inactivated for up to 5 s duration and then repolarized, the channels recover from inactivation with three distinct exponential phases. Each phase can be characterized by a distinct time constant, which is believed to reflect the time course of recovery from a distinct inactivated state. These time constants are in the order of several milliseconds, several hundred milliseconds, and several thousand milliseconds (Nuss et al., 1996). Accordingly the respective inactivated states from which recovery occurs are referred to as "fast inactivation," "intermediate inactivation," and "slow inactivation" (Kambouris et al., 1998). When inactivated for even longer periods, rNav1.4 channels recover with a time constants of >30 s (Cummins & Sigworth, 1996; Featherstone, Richmond, & Ruben, 1996; Todt, Dudley, Kyle, French, & Fozzard, 1999; Wang & Wang, 1997). The inactivated state from which recovery occurs with such very large time constants has been termed "ultraslow inactivation" (Fox, 1976). However, this terminology is not generally accepted and in many publications all non—fast-inactivated states are referred to as "slow inactivation."

3. FAST INACTIVATION

3.1 Mechanism of fast inactivation

3.1.1 The fast-inactivation gate

As first described by Hodgkin and Huxley (1952b) fast inactivation in VGSCs occurs within a few milliseconds of opening. Based on the observation that intracellular application of proteases removes fast inactivation, Armstrong and Bezanilla originally proposed a ball and chain model to describe VGSC fast inactivation (Armstrong, 1981; Armstrong & Bezanilla, 1977). This notion is supported by the finding that fast inactivation can be abolished by antibodies directed against an intracellular loop connecting DIII and DIV of VGSCs (Vassilev et al., 1989; Vassilev, Scheuer, & Catterall, 1988) and that fast inactivation modulates the accessibility of this site for antibody binding (Vassilev et al., 1989, 1988). Stühmer et al. (1989) performed cuts between the individual domains of the rNav1.2 channel. Cuts between DI and DII did not result in sufficient current expression. Whereas cleaving the link between DII and DIII did not result in substantial functional changes, cleaving the link between DIII and DIV strongly reduced the rate of inactivation. West et al. (1992) then found that a hydrophobic triad of isoleucine, phenylalanine, and methionine ("IFM motif") in the DIII—DIV linker is required for fast inactivation (West et al., 1992). Furthermore, peptides containing the IFM motif are able to restore inactivation of VGSCs with disabled fast inactivation (Eaholtz, Scheuer, & Catterall, 1994).

Kellenberger, Scheuer, and Catterall (1996) used the substituted cysteine accessibility method to test the availability of a cysteine residue substituted for F1489 (rNav1.2 numbering, unless stated otherwise, amino acid numbers in the following sections will refer to this isoform) of the IFM motif in rNav1.2 to modification by methane thiosulfonate reagents applied from the cytoplasmic side. This modification slowed inactivation from both closed and open states and destabilized the inactivated state. Furthermore, the voltage dependence of the reaction rate of modification correlated closely with steady-state inactivation confirming the voltage-dependent movement of the inactivation gate (Kellenberger et al., 1996).

Glycine residues confer flexibility to polypeptides, whereas proline side chains induce formation of bends (Creighton, 1993). Mutations of glycine and proline residues in the DIII–DIV linker, flanking the IFM motif, slow the time course of macroscopic inactivation suggesting that these amino acids may serve as molecular hinges to allow closure of the inactivation gate like a hinged lid (Kellenberger, West, Catterall, & Scheuer, 1997).

What is the nature of the interaction of the inactivation lid with the internal mouth of the channel? To explore this question, F1489 was substituted by a series of amino acids with different physicochemical properties. These constructs disrupted inactivation to different extents and the degree of disruption was closely correlated with the hydrophilicity of the amino acid at position 1489. Furthermore, the precise location of F1489 is critical for its function because substituting a phenylalanine residue at position 1488 or 1490 in mutants lacking F1489 does not restore normal inactivation. Hence, during inactivation the inactivation particle most likely forms a hydrophobic interaction with an inactivation gate receptor (Kellenberger, West, Scheuer, & Catterall, 1997).

Using standard homo- and heteronuclear methods, the NMR, the solution structure of a protein fragment corresponding to the III–IV linker has been determined (Rohl et al., 1999). According to this analysis, the linker is composed of a rigid α-helix that is capped on its N-terminal side by two turns, the second of which contains the IFM motif. The hydrophobic F1489, which is considered to mainly interact with the gate receptor is "forced" to project into the solvent by an interaction of the adjacent M1490 with two tyrosine side chains in the α-helix. T1491 also points towards the solvent to interact with the gate receptor as suggested by experimental evidence (Kellenberger, West, Scheuer, et al., 1997) ("IFMT motif").

3.1.2 Modulation of fast inactivation by the C-terminus of the alpha subunit

In addition to the IFM motif the DIII–DIV linker contains a number of negatively and positively charged amino acids. Mutagenesis studies suggest that these charged residues may play a role in channel deactivation and coupling of fast inactivation to activation (Groome, Dice, Fujimoto, & Ruben, 2007; Groome, Fujimoto, & Ruben, 2003; Miller, Patel, John, Mounsey, & Moorman, 2000). However, the DIII–DIV linker is not the only determinant of the fast inactivation process. A number of studies suggest that the carboxy terminus of the VGSC modulates the action of the DIII–DIV linker. This interaction may account for some of the isoform-dependent modulation of the fast inactivation (Desch, Chen, Kallen, & Chahine, 1998; Deschenes, Trottier, & Chahine, 2001; Glaaser, Bankston, Liu, Tateyama, & Kass, 2006; Hartmann, Tiedeman, Chen, Brown, & Kirsch, 1994; Kass, 2006; Lee & Goldin, 2008; Mantegazza, Yu, Catterall, & Scheuer, 2001; Motoike et al., 2004; Nguyen & Goldin, 2010). Recently, crystal structures of the carboxy terminus have become available alone in combination with (Sarhan, Tung, Van, & Ahern, 2012) and without (Wang et al., 2014; Gabelli, Yoder, Tomaselli, & Amzel, 2015) the DIII–IV linker.

A model has been proposed whereby a single Ca^{2+}/CaM bridges the C-terminal IQ motif of the carboxy terminus to the DIII–IV linker. However, a recent study does not confirm a direct interaction. Instead, the carboxy terminus itself may be a determinant of fast inactivation, perhaps via the DIV S6 segment (Wang et al., 2014). Ca^{2+} regulation of fast inactivation of VGSC has been a matter of considerable controversy (Ben-Johny et al., 2014; Van, Lobo, & Ahern, 2012). A recent study elegantly confirms conspicuous Ca^{2+} dependent inactivation in Nav1.4 but not in Nav1.5 channels (Ben-Johny et al., 2014). Prior studies found that Ca^{2+} elevation shifts steady-state inactivation curves by up to 10 mV, to depolarized potentials in the case of Nav1.5 (Biswas, DiSilvestre, Tian, Halperin, & Tomaselli, 2009; Potet et al., 2009; Sarhan et al., 2012; Shah et al., 2006; Wingo et al., 2004), and to hyperpolarized potentials for Nav1.4 (Biswas et al., 2008; Deschenes et al., 2002).

In these studies the intracellular concentration of Ca^{2+} was changed via the dialyzing pipette solution when studying cells in the whole-cell mode. By contrast, Ben-Johny et al. employed the technique of rapid elevation of intracellular calcium levels by photouncaging and by calcium entry via neighboring Ca^{2+} channels. In this study, intracellular rise in Ca^{2+} did not shift the steady-state inactivation curve for fast inactivation but rather scaled down the steady-state inactivation curve without shift along the voltage axis. Thus, Ca^{2+}-dependent inactivation reflects decreased channel open probability, separate from fast inactivation. As mentioned above, this effect was observed in Nav1.4 but not in Nav1.5 channels (Ben-Johny et al., 2014). Further studies using this technique are needed to investigate possible Ca^{2+}-dependent inactivation in other VGSC isoforms (Catterall et al., 2005).

3.1.3 The receptor for the inactivation gate

The intracellular receptor for the inactivation lid has been mapped to DIIIS4-S5 loop. Mutating the conserved A1329 in this loop disrupts inactivation. Combination of a charged mutation at 1329 with replacement of F1489 in the inactivation gate with an amino acid carrying an opposite charge partially restores inactivation of the channel (Smith & Goldin, 1997). Interestingly the DIIIS4-S5 linker may also be implicated in slow inactivation (Casini et al., 2007). Other studies suggest that the DIVS4-S5 loop may be part of the inactivation gate receptor (Filatov, Nguyen, Kraner, & Barchi, 1998; Lerche et al., 1997; McPhee, Ragsdale, Scheuer, & Catterall, 1998; Tang, Chehab, Wieland, & Kallen, 1998). Furthermore amino acids of the S6 segments which line the internal channel vestibule may bind the inactivation lid thereby blocking the permeation pathway (Bai, Glaaser, Sawanobori, & Sunami, 2003; McPhee, Ragsdale, Scheuer, & Catterall, 1995; Wang, Bonner, Russell, & Wang, 2003).

3.1.4 Role of the voltage-gated Na^+ channel β subunits in the regulation of fast inactivation

The VGSC β subunits share a broad spectrum of functions (Chapter: Developmental and regulatory functions of Na^+ channel non-pore-forming β subunits by Winters & Isom, 2016): an immunoglobulin loop enables them to function as cell adhesion molecules (Brackenbury & Isom, 2011; Calhoun & Isom, 2014) which is important for the development of the central nervous system and possibly the peripheral nervous system and the heart (Brackenbury & Isom, 2011; Calhoun & Isom, 2014). Furthermore, these molecules contribute to the regulation of VGSC α subunit gene expression by acting as substrates for sequential cleavage by β-secretases and γ-secretases (Wong et al., 2005). Here, we focus on their function as modulators of ion channel gating of VGSCs. However, it has to be mentioned that modulation of voltage-gated K^+ channels by β subunits has been reported (Deschenes, Armoundas, Jones, & Tomaselli, 2008; Marionneau et al., 2012; Nguyen et al., 2012).

To date five VGSC β subunit proteins have been found which are encoded by a family of four genes (O'Malley & Isom, 2015). With the exception of β1B, β subunits consist of a single polypeptide chain with an extracellular N terminus, a single transmembrane-spanning segment, and an intracellular C terminus.

β1 and β3, form noncovalent interactions with α subunits via their N and C termini (McCormick et al., 1998; Meadows, Malhotra, Stetzer, Isom, & Ragsdale, 2001; Spampanato et al., 2004). β1B is a secreted protein that selectively associates with Nav1.5, but not with Nav1.1 or Nav1.3, in heterologous systems (Patino et al., 2011). β2 and β4 bind covalently to α subunits via a single N-terminal cysteine in the extracellular Ig loop (Buffington & Rasband, 2013; Messner & Catterall, 1985; Yu et al., 2003).

All four subunits shift voltage-dependent gating and alter Na^+ current kinetics expressed by various VGSC α subunits in heterologous expression systems, although effects are inconsistent between different studies (for review see Adelman, 1995; Brackenbury & Isom, 2008; Catterall, 1992, 2000; Isom, De Jongh, & Catterall, 1994). Thus, when co-expressed in *Xenopus* oocytes with Nav1.2, β1 and β2 both increase the peak Na^+ current, accelerate inactivation, and shift the voltage dependence of activation and inactivation to more negative potentials (Isom et al., 1992, 1995). When examined in mammalian expression cell lines these effects are similar, but more subtle (Isom et al., 1994).

The physical mechanism of interaction between α and β subunits is not well defined. Early studies implicated the NH_2-terminus in the regulation of Nav1.4 and Nav1.2 channels (Chen & Cannon, 1995; McCormick et al., 1998; Qu et al., 1999). On the other hand, if 34 amino acids are deleted from the C-terminus of the β1 subunit the ability of β1 to modulate Nav1.2 α-subunits expressed in Chinese hamster lung cells is lost. Furthermore, co-immunoprecipitation data suggest that this loss of functional modulation is caused by inefficient association of the mutant β1 with the α-subunit (Meadows et al., 2001). In addition, an epilepsy-causing mutation, D1866Y, in the C-terminal cytoplasmic domain of Nav1.1 disrupts modulation of Na^+ currents by β1. Spampanato et al. (2004) suggested that the molecular basis for the biophysical alterations by this mutation impairs the interaction between the cytoplasmic C-termini of the α and β1 subunits. Also, only β1/β2 chimeras retaining the COOH-terminal domain of the β1-subunit are able to give rise to hyperpolarizing shifts in gating Nav1.8 (Zhao, O'Leary, & Chahine, 2011). By contrast in Nav1.5 a part of membrane-spanning domain coupled with additional interactions with either the intracellular or extracellular domain of the β1-subunit are required for regulation (Zimmer & Benndorf, 2002). Hence, the mechanism of functional interaction between α and β subunits may be isoform specific (Zhao et al., 2011).

The use of genetically engineered mouse models and viral-mediated gene transfer shows that β subunit modulation of sodium current may occur in vivo and may be critical for normal electrical excitability in the nervous system and heart. However, the modulation is largely cell type specific. Recordings from brain slices of mice lacking the β1 subunit (Scn1b null mice) demonstrate hyperexcitability in the CA3 region of the hippocampus as well as epileptiform activity in the hippocampus and cortex (Brackenbury, Yuan, O'Malley, Parent, & Isom, 2013; Patino et al., 2009). This neuronal hyperexcitability may contribute to the severe seizures observed in these mice (Chen et al., 2004; Patino et al., 2009). On the other hand, excitability of cerebellar granule neurons in Scn1b null mice is reduced (Brackenbury et al., 2010) which may contribute to the ataxic phenotype of these animals (Chen et al., 2004).

Also, the β2 subunit has modulates channel function in vivo. Hence, the voltage dependency of inactivation of sodium currents in hippocampal neurons is shifted to negative potentials in mice lacking the β2 subunit (Scn2b null) compared to wild-type animals (Chen et al., 2002). Modulation of expression of β4 in primary neurons suggests that this subunit promotes persistent sodium current (Aman et al., 2009; Bant & Raman, 2010).

Resurgent sodium current occurs if channels reopen during repolarization (Bean, 2005). This phenomenon is considered to allow for high frequency firing of some neurons (Burgess et al., 1995; Raman, Sprunger, Meisler, & Bean, 1997). The intracellular domain of β4 is considered to block the open channel, thereby allowing for resurgent sodium current (Bant & Raman, 2010; Grieco, Malhotra, Chen, Isom, & Raman, 2005; Theile & Cummins, 2011). Thus, β-subunits are important modulatory factors of fast inactivation.

3.1.5 Other modulatory factors of fast inactivation

Apart from β-subunits a number of other modulator factors such as post-translational modification, pH, interaction with intracellular factors, phosphorylation modulate fast inactivation. A discussion of these modulations is beyond the scope of this chapter and the reader is referred to recently published reviews (Abriel, 2010; Chen-Izu et al., 2015; Ednie & Bennett, 2012;

Herren, Bers, & Grandi, 2013; Jones & Ruben, 2014; Lampert, Eberhardt, & Waxman, 2014; Marionneau & Abriel, 2015).

4. LONG-TERM INACTIVATION

Fast and slow inactivated states are characterized by rapid and prolonged development and recovery, respectively. Recently a new kinetic state has been reported that bears resemblance to both fast and slow inactivation (Dover, Solinas, D'Angelo, & Goldfarb, 2010). "Long-term inactivation" develops within milliseconds following membrane depolarization, similar to fast inactivation. However, recovery upon repolarization may be 100-fold slower than recovery from fast inactivation, similar to slow inactivation. In contrast to fast inactivation, long-term inactivation has a higher voltage threshold closer to the $V_{1/2}$ of channel activation (Dover et al., 2010). This process has been suggested to occur by closure of the internal vestibule of VGSCs by the N-terminal part of A-type fibroblast growth factor homologous factors (FHFs), which compete with the channel's intrinsic fast inactivation particle (IFM-motif, see above; Dover et al., 2010). Although FHFs and fibroblast growth factors (FGFs) are similar in terms of sequence and structure, the biochemical and functional properties of FHFs are largely, if not totally, unrelated to those of FGFs. Whereas FGFs bind to the extracellular domains of FGF receptors, FHFs associate with the intracellular domains of VGSCs and with a neuronal MAP kinase scaffold protein, islet-brain-2 (IB2) (Goldfarb, 2005). Such interaction between FHFs and VGSCs has been shown for a number of neuronal isoforms and the cardiac isoform (Liu, Dib-Hajj, & Waxman, 2001; Liu, Dib-Hajj, Renganathan, Cummins, & Waxman, 2003; Lou et al., 2005; Wang, Wang, Hoch, & Pitt, 2011). In hippocampal CA1 pyramidal neurons use-dependent accumulating internal channel blockade by FHFs progressively elevates spike voltage threshold and lengthens interspike intervals and may contribute to spike accommodation which may play a role in perceptual processing and learning (Venkatesan, Liu, & Goldfarb, 2014).

5. SLOW INACTIVATION

The process of slow inactivation has been subject of a number of reviews (Bagneris, Naylor, McCusker, & Wallace, 2015; Bahring & Covarrubias, 2011; Catterall, 2014; Goldfarb, 2012; Goldin, 2003;

Payandeh & Minor, 2015; Silva, 2015; Ulbricht, 2005; Vilin & Ruben, 2001). Here we will provide a short account of the available literature and summarize our own work on a special form of slow inactivation which we refer to as "ultraslow inactivation." Furthermore, we will provide an overview over recent evidence regarding the physical mechanism of slow-inactivation states from recently published crystal structures of bacterial VGSCs.

5.1 Slow inactivation in native tissues

Slow recovery from inactivation was first observed in squid giant axons (Adelman & Palti, 1969; Chandler & Meves, 1970a, 1970b, 1970c; Narahashi, 1964). Rudy demonstrated that this process is resistant to internal perfusion with pronase (Rudy, 1978). Slow inactivation was observed if the membrane was depolarized for long time periods, ie, several seconds to minutes (Chandler & Meves, 1970c; Rudy, 1978). Subsequently, this phenomenon was also detected in many other excitable tissues such as Myxicola axons (Schauf, Pencek, & Davis, 1976), neuroblastoma cells (Quandt, 1987), myelinated nerve fibers (Brismar, 1977), hippocampal granule cells (Ellerkmann, Riazanski, Elger, Urban, & Beck, 2001), skeletal (Collins, Rojas, & Suarez, 1982; Hayward, Brown, & Cannon, 1997; Simoncini & Stühmer, 1987) and heart muscles (Clarkson, Matsubara, & Hondeghem, 1984; Saikawa & Carmeliet, 1982; Schneider, Proebstle, Hombach, Hannekum, & Rudel, 1994).

5.2 Mechanism of slow inactivation—low inactivation in mutated voltage-gated Na^+ channels

5.2.1 Role of the outer vestibule

The first data acquired in recombinant channels suggested that amino acids located in the outer mouth of the Na^+ channel may not only be involved in the control of ion permeation, but may also contribute to slower processes of inactivation, similar to C-type inactivation in K^+ channels (Baukrowitz & Yellen, 1995; Choi, Aldrich, & Yellen, 1991; Grissmer & Cahalan, 1989; Lopez-Barneo, Hoshi, Heinemann, & Aldrich, 1993; Ogielska & Aldrich, 1999; Yellen, Sodickson, Chen, & Jurman, 1994).

In 1996, Balser et al. (1996) reported that the mutation W402C, located in the external vestibule of the rNav1.4 channel altered the inactivation properties of rNav1.4 channels (Balser et al., 1996; Tomaselli et al., 1995). Subsequently, a high number of mutations in the external vestibule were found to modulate slow inactivation (Bendahhou, Cummins, Kula, Fu, &

Ptacek, 2002; Brancati et al., 2003; Cummins & Sigworth, 1996; Hayward, Sandoval, & Cannon, 1999; Kambouris et al., 1998; Ong, Tomaselli, & Balser, 2000; O'Reilly, Wang, & Wang, 2001; Vedantham & Cannon, 2000; Webb & Cannon, 2008). Vilin et al. demonstrated that the differences in the probability of slow inactivation between human skeletal muscle and human heart isoforms are determined by a single amino acid in the DII-P-loop (Vilin, Fujimoto, & Ruben, 2001; Vilin, Makita, George, & Ruben, 1999). Recently, it has been demonstrated that amino acids of the outer vestibule not only control slow inactivation but also mediate proton modulation of slow inactivation and isoform specific proton modulation of inactivation by high-frequency pulses (Jones, Peters, Allard, Claydon, & Ruben, 2013; Jones, Peters, Tolhurst, Claydon, & Ruben, 2011; Vilin, Peters, & Ruben, 2012).

As to the molecular mechanism of slow inactivation a closure of the extracellular mouth of the channel has been proposed to account for this kinetic process (Balser et al., 1996; Hilber et al., 2001, 2002; Kambouris et al., 1998; Ong et al., 2000; Todt et al., 1999; Tomaselli et al., 1995; Townsend & Horn, 1997; Vilin et al., 2001; Xiong et al., 2006; Xiong, Li, Tian, & Tomaselli, 2003; Wu, Ye, Sengupta, & Zakon, 2013).

On the other hand, Struyk and Cannon (2002) found the rates of sulfhydryl modification of cysteines engineered to several sites of the external vestibule to be not different between slow and non—slow-inactivated states, a finding that argues against a pore closure as mechanism of slow inactivation.

5.2.2 The selectivity filter is involved in ultraslow inactivation

The selectivity filter of the channel of eukaryotic VGSCs is considered to be formed by the amino acids D400, E755, K1237, and A1529 ("DEKA motif"; amino acid numbering corresponds to rNav1.4; unless stated otherwise, amino acid numbering in the following sections will correspond to this isoform) which are located at the innermost turn of the P-loops connecting S5 and S6 segments of each domain (Chiamvimonvat, Perez Garcia, Tomaselli, & Marban, 1996; Favre, Moczydlowski, & Schild, 1996; Heinemann, Terlau, Stühmer, Imoto, & Numa, 1992; Lipkind & Fozzard, 1994; Schlief, Schoenherr, Imoto, & Heinemann, 1996; Terlau et al., 1991; Tsushima, Li, & Backx, 1997a).

The amino acid DIII-K1237 of the DEKA motif is critical for selectivity of ion permeation through the VGSC (Chiamvimonvat et al., 1996; Favre et al., 1996; Heinemann et al., 1992; Perez Garcia, Chiamvimonvat, Marban, & Tomaselli, 1996; Tsushima et al., 1997a). If DIII-K1237E channels are expressed in *Xenopus laevis* oocytes and depolarized for several minutes, these mutant channels enter a state of ultraslow inactivation (Todt et al., 1999). Specifically, as shown in Fig. 1, the channels are inactivated by a 300-s prepulse to −30 mV. After reestablishment of a negative holding potential of −120 mV by two-electrode voltage clamp, the peak Na^+ current assessed by consecutive short test pulses increases to a stable level over a several minute period.

Fig. 1B compares the normalized time course of recovery from a 300 s inactivating prepulse to −30 mV between wild-type rNav1.4 and the constructs DIII-K1237R, DIII-K1237S, and DIII-K1237E. Peak Na^+ currents of wild-type rNav1.4 and charge conserving mutant DIII-K1237R channels recover almost completely within ~60 s. In contrast to rNav1.4 and DIII-K1237R, channels carrying the charge altering mutations DIII-K1237S and DIII-K1237E do not completely recover until more than 300 s had elapsed. The time course of recovery of normalized peak inward currents is best fit with a double exponential function. The calculated time constants of recovery were 9−17 s (τ_1), and 75−130 s for (τ_2). τ_1 and τ_2 most likely represent the time constants of recovery from two distinct inactivated states. τ_2, the longer time constant of recovery, corresponded to a component representing ~20% of the overall time course of recovery of rNav1.4 and DIII-K1237R channels. This component is dramatically increased to ~80% in channels carrying the mutations DIII-K1237S and DIII-K1237E (Todt et al., 1999).

Because of its long duration τ_2 may reflect recovery from "ultraslow" inactivation (Fox, 1976) which was first observed in the squid giant axon (Adelman & Palti, 1969; Rudy, 1978). The faster component of recovery accounting for ~80% of recovery of rNav1.4 and DIII-K1237R channels, and for only 15−20% of recovery of DIII-K1237S and DIII-K1237E mutants has a time constant of ~10 s. This time constant most likely reflects the predominant state of slow inactivation which can be induced in wild-type channels (Featherstone et al., 1996; Kambouris et al., 1998). In summary, the mutations DIII-K1237S and DIII-K1237E appear to favor entry into a state of ultraslow inactivation (Todt et al., 1999). However, mutating residue K1237 not only alters channel gating, but also causes a substantial loss of ionic selectivity (Favre et al., 1996; Heinemann et al., 1992). This dual effect of amino acid replacements at site 1237 raises the question whether ultraslow inactivation was produced by variations in species or in number of permeating ions rather than by an alteration in channel structure per se. Thus, lowering the external Na^+ concentration enhances entry into slow

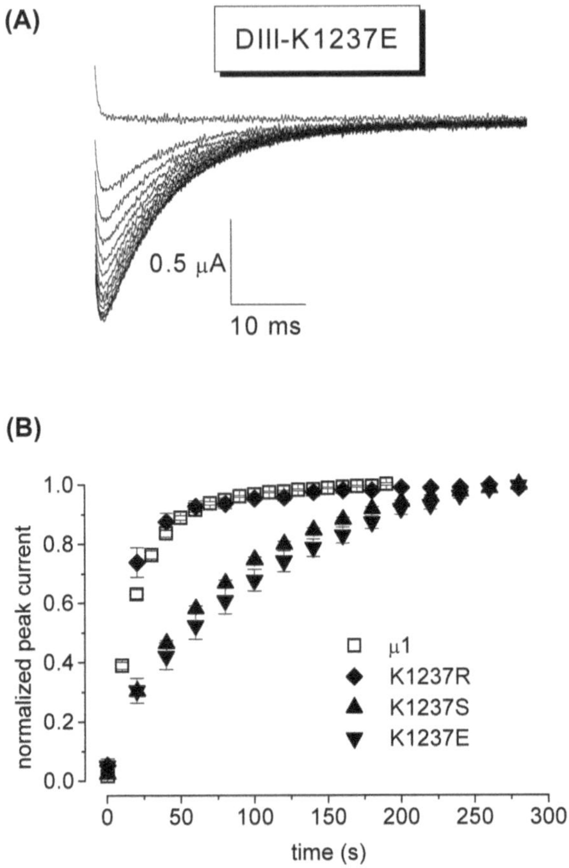

Figure 1 ***Time course of recovery from ultraslow inactivation in wild-type rNav1.4 ("μ1") and by rNav1.4 channels harboring amino acid replacements of DIII-K1237E in the selectivity filter.*** The constructs were expressed in *Xenopus laevis* oocytes as described previously (Todt et al., 1999) and examined by means of two-electrode voltage clamp. (A) Typical original currents traces from one oocyte. From a holding potential of −120 mV, the channels were inactivated by a 300 s depolarizing step to −30 mV. After returning to −120 mV, recovery from inactivation was monitored by repetitive test pulses to −10 mV at 20 s intervals. Shown are slowly increasing inward currents through DIII-K1237E channels elicited by test pulses after the holding potential was returned to −120 mV. Test pulse duration was 100 ms. (B) Comparison of the time courses of recovery from ultraslow inactivation in native rNav1.4 and in constructs with mutations at the DIII-K1237 site (n = 6 for each data point). Maximum inward currents from experiments such as in (A) were normalized to the respective maximum and plotted as a function of time after the 300 s inactivating prepulse. The mutations in domains III, in which the charge of the residue was altered, exhibited a substantial amount of slow recovery from inactivation, whereas wild-type rNav1.4 and the charge conserving mutation in DIII, K1237R, recover much faster. *From Todt, H., Dudley, S. C., Jr, Kyle, J. W., French, R. J., & Fozzard, H. A. (1999). Ultra-slow inactivation in μ1 Na^+ channels is produced by a structural rearrangement of the outer vestibule.* Biophysical Journal, *76, 1335–1345 with permission.*

inactivated states that have faster kinetic properties than ultraslow inactivation (Chen et al., 2000; Townsend & Horn, 1997). However, the finding that the charge conserving mutation K1237R is not associated with substantial amounts of ultraslow inactivation (Todt et al., 1999), despite having dramatic effects on ionic selectivity (Favre et al., 1996), argues against changes in permeation properties as underlying mechanism. Further evidence for the independence of the generation of ultraslow inactivation from changes in ionic selectivity is provided by the fact that mutations of two other positions of the DEKA ring, ie, DIV-A1529D and DI-D400A also produce significant amounts of ultraslow inactivation despite having no or only minor effects on ionic selectivity (Hilber et al., 2001, 2005). Furthermore, mutation DII-E755A does not give rise to ultraslow inactivation despite significant loss of selectivity (Hilber et al., 2005). These results indicate that ultraslow inactivation occurs by a conformational change at the selectivity filter.

5.2.3 Ultraslow inactivation is modulated by blocking the external vestibule

Further studies showed that this conformational change could be modulated by binding of a large molecule to the outer vestibule. μ-Conotoxin GIIIA (μ-CTX) is a 22 amino acid peptide toxin that completely blocks the skeletal muscle VGSC outer vestibule with high affinity (Becker et al., 1992; Cruz et al., 1985; Dudley, Todt, Lipkind, & Fozzard, 1995; French et al., 1996; Moczydlowski, Olivera, Gray, & Strichartz, 1986). The toxin mutant μ-CTX R13Q only partially occludes the channel, reducing but not abolishing single channel current (Becker et al., 1992; Chang, French, Lipkind, Fozzard, & Dudley, 1998; French et al., 1996). This partial block allows for assessment of channel gating kinetics in the toxin-bound state. The K_d of μ-CTX R13Q binding to the mutant DIII-K1237E is about 10 μM (Todt et al., 1999). Fig. 2 shows the effect of a near-saturating ($\sim 3 \times K_d$) concentration of 27 μM μ-CTX R13Q on the time course of recovery from inactivation by the mutant channel DIII-K1237E. Binding of μ-CTX R13Q markedly reduces the amplitude of ultraslow recovery from inactivation by >30%, from 0.81 ± 0.05 under control conditions to 0.49 ± 0.09 during perfusion with 27 μM μ-CTX R13Q ($p < 0.01$, $n = 6$, Fig. 2 inset). The time constant of ultraslow inactivation is reduced from 112.6 ± 16.8 s at control to 66.3 ± 13.6 s with μ-CTX R13Q ($p < 0.05$; Fig. 2 inset). Hence, this molecule dramatically reduces entry of the VGSC with the K1237E mutation into its ultraslow inactivated state.

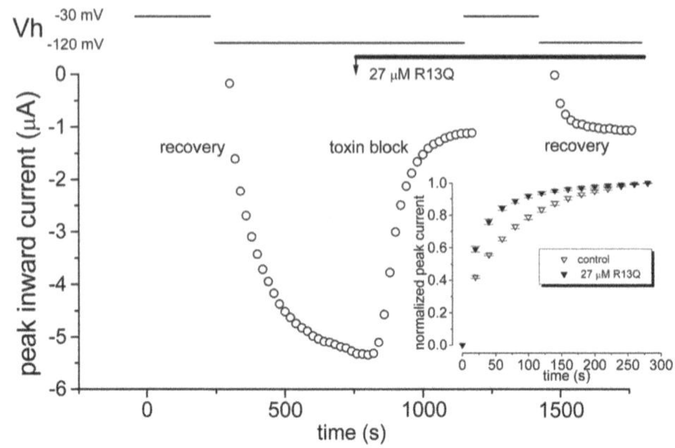

Figure 2 ***Recovery from ultraslow inactivation in DIII-K1237E (rNav1.4) is modulated by a mutant μ-conotoxin GIIIA (R13Q) known to bind at the outer vestibule.*** The constructs were expressed in *Xenopus laevis* oocytes as described previously (Todt et al., 1999) and examined by means of two-electrode voltage clamp. Data points represent peak inward currents from a typical experiment which were elicited by 20 ms step depolarizations to −10 mV from the indicated holding potential (Vh). During the control period, DIII-K1237E channels were depolarized to −30 mV for 300 s and recovery from ultraslow inactivation was monitored for repetitive test pulses ("recovery"). The following superfusion of 27 μM μ-CTX R13Q reduced peak inward currents by 75–80% (*arrow*, "27 μM R13Q"; "toxin block"). Clearly, during superfusion with μ-CTX R13Q, recovery from ultraslow inactivation had a much faster time course ("recovery") than during control. The inset summarizes six experiments. The data points were normalized to the final level of ultraslow recovery, and the time courses of ultraslow recovery with and without superfusion with μ-CTX R13Q were fitted with two exponentials. Clearly, application of μ-CTX R13Q substantially accelerated recovery. *From Todt, H., Dudley, S. C., Jr, Kyle, J. W., French, R. J., & Fozzard, H. A. (1999). Ultra-slow inactivation in μ1 Na^+ channels is produced by a structural rearrangement of the outer vestibule.* Biophysical Journal, 76, *1335–1345 with permission.*

This effect was independent of the voltage of the prepulse and recovery potential, ruling out a simple electrostatic influence of the charged toxin on the nearby voltage-sensing systems (French et al., 1996; Todt et al., 1999). Similar results were observed in channels carrying mutations in the selectivity filter in domains I (D400A) (Hilber et al., 2001) and IV (A1529D) (Hilber et al., 2005). These data suggest that the toxin may interfere with those structural changes in the Na^+ channel outer vestibule that generate ultraslow inactivation, thereby underscoring the importance of conformational changes at this site in the generation of this kinetic state. μ-CTX R13Q may act as a "splint" in the outer vestibule, stabilizing the structure of this region of the channel. Recent evidence for the notion that block of the outer vestibule alters slow ion channel gating is provided by the observation that the pore blocker tetrodotoxin impairs the motions of voltage sensors in domains I and II associated with slow inactivation (Silva & Goldstein, 2013). Also, use-dependent pore block by tetrodotoxin and saxitoxin may arise from coupling of voltage-sensor movement to the outer vestibule (Huang, Schild, & Moczydlowski, 2012).

5.2.4 Connection between ultraslow inactivation and fast inactivation

To enter into the ultraslow inactivated state channels have to be depolarized for several minutes. Therefore, channels have to transit through faster inactivated states on the way to ultraslow inactivation. This raises the question in which way fast inactivation is related to ultraslow inactivation. As shown in Fig. 3, the voltage dependence of ultraslow inactivation closely matches the voltage dependence of fast inactivation. This suggests that the two kinetic states are closely related, ie, fast inactivation and ultraslow inactivation emerges from similar voltage-dependent conformational states.

On the other hand, the voltage dependencies of ultraslow inactivation strongly depend on the type of underlying mutation. Thus, the mutations DI-D400A and DIV-A1529D (Fig. 4) produce an ultraslow inactivated state whose voltage-dependency is U-shaped, ie, the fraction of channels recovering with an ultraslow time course following a 300-s depolarization is greatest at approximately −60 mV (Fig. 4). Depolarization to voltages more negative and more positive to −60 mV elicits substantially less amounts of ultraslow inactivation (Hilber et al., 2002, 2005). Thus, the greater the fraction of channels entering into fast inactivation during short depolarizations to a given voltage, the lower the fraction of channels entering into ultraslow inactivation upon long lasting depolarizations to the same voltage.

This suggests that fast inactivation protects channels from entry into the ultraslow inactivated state. This notion is supported by two additional observations: (1) Coexpression of α subunits with the $\beta1$ (which stabilizes the fast inactivated state) reduces the propensity of DIV-A1529 channels to enter into ultraslow inactivation; (2) Destabilization of the fast inactivated state by additional mutations in the inactivation particle enhances entry into ultraslow inactivation and abolishes the U-shaped voltage dependency of ultraslow inactivation (Hilber et al., 2002). These experimental results raise the question of which molecular mechanisms could account for the protective effect of fast inactivation regarding entry into ultraslow inactivation. As mentioned above, fast inactivation is generated by an occlusion of the

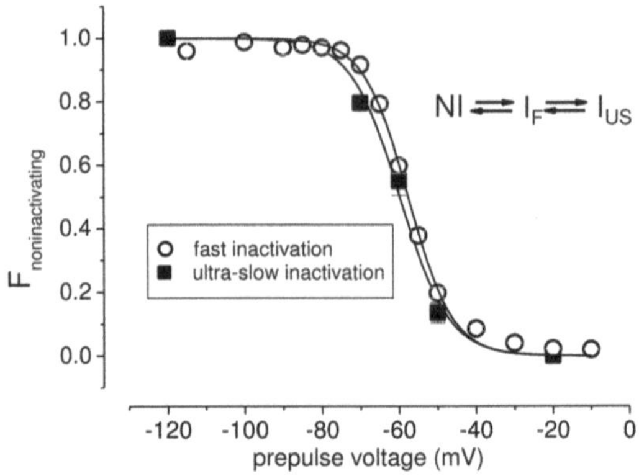

Figure 3 ***Voltage dependence of fast inactivation and ultraslow inactivation in K1237E (rNav1.4).*** The construct was expressed in *Xenopus laevis* oocytes as described previously (Todt et al., 1999) and examined by means of two-electrode voltage clamp. To test for the relationship between fast inactivation and ultraslow inactivation we examined the voltage dependence of both kinetic states during steady-state conditions. The voltage dependence of fast inactivation was tested by depolarizing the membranes to a variety of voltages for 50 ms after which a short test pulse to −20 mV was applied to open noninactivated channels. The voltage dependence of ultraslow inactivation was examined by depolarizing the membranes for 600 s after the potential was returned to −120 mV. The time course of recovery from each depolarizing pulse was then assessed by sequential short test pulses which were applied at 20 s intervals. The time course of recovery of inward currents after each prepulse was then fitted with a double exponential function, which yielded the time constants and amplitudes of slow inactivation and ultraslow inactivation. The voltage dependence of ultraslow inactivation was derived by plotting the amplitudes of ultraslow inactivation as a function of the respective prepulse voltage. The data points were fit with a Boltzmann equation. The fitted $V_{1/2}$ values and k (slope factor, mV/e-fold change) values were -57.6 ± 0.3 mV and 5.7 ± 0.3 for steady-state fast inactivation and -59 ± 0.8 mV and 6.2 ± 0.8 for steady-state ultraslow inactivation ($n = 6$; $P =$ n.s). $F_{noninactivating}$ = -fraction of channels that is not inactivated either by fast or by ultraslow inactivation. Inset: A schematic model illustrating the kinetic transitions (*arrows*) that may underlie the similar voltage-dependency of fast inactivation and ultraslow inactivation. Upon depolarization channels are assumed to first move from the noninactivated state (NI) to the fast inactivated state (I_F). If the depolarization is maintained for several seconds, fast inactivated channels enter into an ultraslow inactivated state (I_{US}). Thus the voltage dependence of ultraslow inactivation is a reflection of the voltage dependence of fast inactivation. Unpublished data.

internal vestibule by the inactivation lid located to the intracellular linker between DIII and DIV (Featherstone et al., 1996; Hartmann et al., 1994; Kellenberger et al., 1996; Patton, West, Catterall, & Goldin, 1992; Stühmer et al., 1989; West et al., 1992). It has been shown that slow inactivation

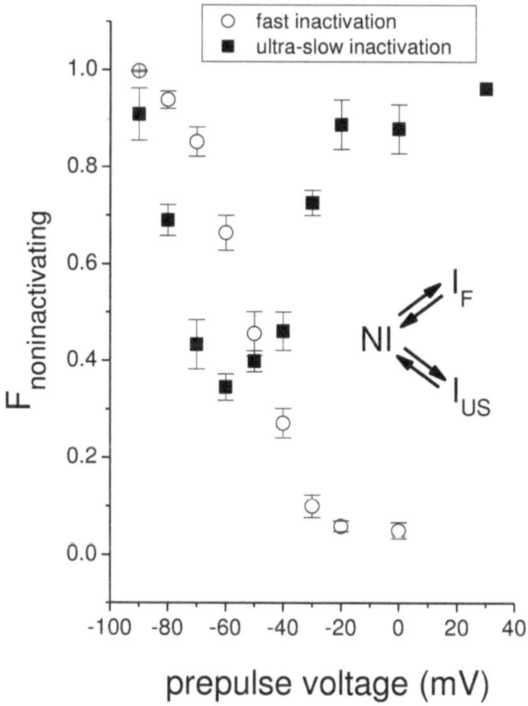

Figure 4 ***Voltage dependence of ultraslow inactivation and steady-state fast inactivation in the mutation DIV-A1529D (rNav1.4).*** The construct was expressed in *Xenopus laevis* oocytes as described previously (Todt et al., 1999) and examined by means of two-electrode voltage clamp. A1529 is the amino acid in DIV of the DEKA-motif in the eukaryotic VGSC selectivity filter. To assess the voltage dependence of ultraslow inactivation, the membrane potential was depolarized from −120 mV to the indicated prepulse voltages for 300 s, and the time course of recovery at −120 mV was evaluated as in Fig. 3. Steady-state fast inactivation was assessed by delivery of 20-ms test pulses to −20 mV after 50-ms prepulses to the indicated potentials. The respective test pulse current was normalized to the current under fully available conditions. Whereas steady-state fast inactivation became more complete at prepulse voltages positive to −60 mV, ultraslow inactivation had a U-shaped voltage-dependence, reaching a local minimum at −60 mV. Inset: A schematic model illustrating the kinetic transitions (*arrows*) that may account for the differences between the voltage dependence of fast inactivation and ultraslow inactivation. In this scheme, channels can move from the noninactivated state (NI) either to the fast inactivated state (I_F) or to an ultraslow inactivated state (I_{US}). Thus, at voltages positive to the $V_{1/2}$ of fast inactivation the fraction of fast inactivated channel increases resulting in a decrease in the fraction of ultraslow inactivated channels. *This research was originally published in Hilber, K., Sandtner, W., Kudlacek, O., Schreiner, B., Glaaser, I., Schutz, W.…Todt, H. (2002). Interaction between fast and ultraslow inactivation in the voltage-gated sodium channel. Does the inactivation gate stabilize the channel structure?* Journal of Biological Chemistry, 277, *37105–37115. © The American Society for Biochemistry and Molecular Biology.*

remains intact if the fast inactivation gate is disabled by pronase (Rudy, 1978), trypsin (Starkus & Shrager, 1978), α-chymotrypsin (Valenzuela & Bennett, 1994), or N-bromoacetamide (Salgado, Yeh, & Narahashi, 1985). As mentioned earlier, replacement of the IFM motif in the DIII-IV linker by glutamine residues (QQQ) completely removes fast inactivation (West et al., 1992), but leaves slow inactivation intact (Featherstone et al., 1996). As a matter of fact, in both hNav1.4 and hNav1.5 the propensity to enter into slow inactivation increases when fast inactivation is removed (Featherstone et al., 1996; Richmond, Featherstone, Hartmann, & Ruben, 1998). On the other hand, slow inactivation does not affect recovery from fast inactivation in rNav1.4 channels expressed in Xenopus oocytes (Vedantham & Cannon, 1998). It has been suggested that the fast inactivation gate, in its position bound to the internal vestibule, prevents the pore of the channel from collapsing during slow inactivation (Vilin & Ruben, 2001). Because the S6 segments of all four domains line the internal vestibule, it is likely that fast inactivation is associated with conformational changes of some or all S6 segments. Indeed, this notion is supported by substantial experimental evidence.

5.2.5 The S6 segments as determinants of both fast and slow inactivated states

Studying the binding sites of the local anesthetic lidocaine in the S6 segment, Ragsdale, McPhee, Scheuer, and Catterall (1994) noted that mutations I1761A, V1767A, and N1769A (rNav1.2) shift the half voltage of fast inactivation negatively by 4–13 mV, ie, stabilizes the fast inactivated state. Conversely, F1764A (rNav1.2) shifts the half voltage of fast inactivation by 8 mV in the positive direction consistent with a destabilization of the fast inactivated state. I1760A (rNav1.2) increases the recovery rate from local anesthetic block perhaps by creating an escape route for the local anesthetic. Hence, this mutation must produce a substantial change in the structure of the inner vestibule (Ragsdale et al., 1994).

McPhee et al. reported a number of mutations in DIV-S6 (rNav1.2), which cause incomplete inactivation resulting in sustained currents at the end of a 30-ms-long depolarization. The mutants F1764A and V1774A exhibit the largest sustained currents, whereas S1763A, Y1771A, I1775A, and L1776A have smaller sustained currents. The double mutant F1764A/V1774A almost completely abolishes fast inactivation. Single channel analysis revealed continued reopening late in 40-ms pulses, indicating that the stability of the inactivated state is substantially impaired compared with wild type. A peptide containing the IFM motif restored inactivation to the F1764A/V1774A mutant indicating that the binding site for the IFM motif remains intact in these mutants (McPhee et al., 1995). Thus, the mechanism by which these mutants generate sustained current remains obscure. The triple mutant VIL1774-6AAA (rNav1.2), located at the cytoplasmic end of DIV-S6 has >85% non inactivating current. Again, single channel analysis revealed continued reopening during the depolarizing pulse indicating that inactivation is substantially impaired (McPhee, Ragsdale, Scheuer, & Catterall, 1994). Sustained currents during a depolarizing pulse have been reported in the mutations I1598 W/C and I1590 W/C in rNav1.4, homologous to V1774 and I1775 in rNav1.2. These mutants also exhibit an enhanced slow inactivation phenotype (Wang et al., 2003; slow inactivation being defined as inactivation produced by a 10-s-conditioning pulse). The midpoint of fast inactivation is shifted to more positive potentials in I1589 W/C, I1590 W/C, L1591 W/C, and E1592 W/C (Nav1.4; Wang et al., 2003). Similar results have been obtained in the homologous residues in rNav1.2 (McPhee et al., 1995). In rNav1.4 substitution of Y1586 by lysine or arginine, but not by alanine or aspartic acid, produces an atypical intermediate inactivated state, ie, an inactivated state from which the channels recover with a time constant 10-fold slower than in wild-type (O'Reilly, Wang, & Wang, 2000). The authors concluded that a conformational change during depolarization alters the relative position of 1586 K, and that the atypical inactivated state occurs by an electrostatic interaction near the inner pore region. The mutation F1579C (rNav1.4) has been reported to inhibit both fast and slow inactivation, whereas Y1586C and I1575C enhances both fast and slow inactivation (slow inactivation defined as inactivation produced by a 30-s-conditioning pulse; Bai et al., 2003). Furthermore, a possible role of the inner vestibule in the generation of slow inactivated states is suggested by the fact that a number of other mutations in S6 segments affect slow inactivation (O'Reilly et al., 2001, Takahashi & Cannon, 1999). If the valine at position 1583, ie, in the middle of the DIV-S6 helix is replaced by a cysteine, then this site is accessible for 2-aminoethyl methane thiosulfonate applied from the cytoplasmic side. The rate of modification by 2-aminoethyl methane thiosulfonate increases when the exposure occurs during brief depolarizing pulses, but decreases when the exposures occurs at the end of prolonged depolarizations, revealing two conformational changes near site 1583, one coupled to fast inactivation and one tightly associated with slow inactivation (Vedantham & Cannon, 2000; slow inactivation being defined as inactivation produced by a 2.5-s conditioning pulse). In Nav1.5 and Nav1.7, residues in DII-S6 are involved in the modulation of slow inactivation and some mutations at these sites are associated with human channelopathies (Chancey, Shockett, & O'Reilly, 2007; Cheng et al., 2011; O'Reilly & Shockett, 2006). In addition, G protein coupled receptor- and protein kinase-dependent reductions in Na^+ channel availability have been shown to be mediated by modulation of slow inactivation (Carr et al., 2003), consistent with a significant role of the internal parts of the Na^+ channel in slow inactivation. Taken together, the available data indicate that DIV-S6 has a critical role in the generation of both fast and slow inactivated states. However the existing studies have not resulted in a clear structure—function correlation of amino acids in DIV-S6.

5.2.6 Role of the internal vestibule in ultraslow inactivation

Local anesthetics are small molecules that modulate gating of Na^+ channels by binding to specific sites in the internal vestibule (Ragsdale et al., 1994). In the mutation DIII-K1237E lidocaine inhibits entry into ultraslow inactivation in a concentration-dependent manner. A model that assumes that the lidocaine-bound inactivated state and the ultraslow inactivated state are mutually exclusive and can numerically reproduce the mode of drug-induced inhibition of entry into ultraslow inactivation. We suggested a molecular model in which, similar to slow inactivation (Vilin & Ruben, 2001), the ultraslow inactivated state occurs by a collapse of the internal vestibule. Lidocaine, bound to the internal vestibule would then inhibit entry into the ultraslow inactivated state by acting as a "foot-in-the door" (Sandtner et al., 2004). A similar model could be invoked to explain the inhibition of some forms of ultraslow inactivation by fast inactivation (see above; Hilber et al., 2002, 2005) which is reflected by a U-shaped voltage-dependence (Fig. 4). Here, binding of the inactivation particle to the internal channel pore during fast inactivation may stabilize the internal vestibule thereby inhibiting the collapse of this structure that gives rise to ultraslow inactivation. This idea is supported by the fact that removal of fast inactivation abolishes the U-shaped voltage dependency of ultraslow inactivation and allows for entry into ultraslow inactivation at strongly depolarized potentials, whereas co-expression of the β1 subunit stabilizes fast inactivation and has the opposite effect on ultraslow inactivation (Hilber et al., 2002) and on slow inactivation (Webb, Wu, & Cannon, 2009). But if this mechanism holds true, why does it not apply to ultraslow inactivation in the mutation DIII-K1237E that has a monotonic voltage-dependence (Fig. 3)? It is remarkable that the DIII-K1237 mutations drive a much large channel fraction into ultraslow inactivation, than mutations in DIV-A1529D and DI-D400A are considerably reduced. Perhaps the DIII-K1237 mutations more severely destabilize the pore structure than mutations DIV-A1529D and DI-D400A, and this may finally result in the inability of the bound inactivation gate to prevent the collapse of the internal vestibule in DIII-K1237E.

As mentioned above, ultraslow inactivation is generated by mutations in the outer vestibule of the channel. However, if ultraslow inactivation is produced by a conformational change involving the S6 segments, then mutations in the S6 segments could be expected to give rise to ultraslow inactivation. To explore this question we performed cysteine-scanning mutations in the S6 segment domain IV of rNav1.4 channels between positions 1575 and 1591. This region is predicted by homology modeling to line the internal vestibule (Lipkind & Fozzard, 2000). With the notable exception of site 1575, all cysteine replacements result in enhancement of recovery from ultraslow inactivation compared to wild-type channels supporting a significant role of the S6 segment of domain IV in the generation of the ultraslow inactivated state (Zarrabi et al., 2010). The fact that mutations of amino acids located both in the external and in the internal vestibule are able to enhance entry into ultraslow inactivation raises the question whether there is a functional connection between these two regions of the channel.

5.2.7 A functional link between the outer and the inner vestibule of the channel

A systematic analysis of possible interactions between outer and inner vestibule of the VGSC has been performed by Zarrabi et al. (2010). The study was inspired by a structural homology model of the Nav1.4 channel, based on the crystal structure of the bacterial K channel KcsA (Lipkind & Fozzard, 2000). In this model, I1575 in DIV-S6 is in close proximity to DIII-K1237 of the selectivity filter. The mutation DIII-K1237E has been shown to increase the permeability for large cations (Favre et al., 1996; Sunami, Glaaser, & Fozzard, 2001). Hence, this mutation most likely increases the diameter of the selectivity filter, perhaps by virtue of a lateral movement of E1237. Such "swinging out" of the side chain of E1237 has been suggested by Moczydlowski to explain the permeation of rNav1.4 channels by the large organic molecule spermidine at high voltages (Huang & Moczydlowski, 2001). The assumption of such a molecular movement is reasonable because the P-loops have been shown to be flexible structures in studies involving paired cysteine substitutions that allow disulfide bond formation (Tsushima, Li, & Backx, 1997b; Xiong et al., 2003). The lateral motion of E1237 may then alter the nature of the interaction with the neighboring S6 residue I1575. To test this hypothesis, Zarrabi et al. investigated a series of double mutations in which the mutation DIII-K1237E was combined with serial cysteine replacements of amino acids in the DIV-S6 segment. The authors found that the double mutation DIII-K1237E/DIV-I1575C is unique among the tested constructs in that it has a significantly smaller propensity to enter into the ultraslow inactivated state than the single mutation DIII-K1237E. Moreover, adding the mutation DIV-I1575C to DIII-K1237C partially restores the loss of ionic selectivity of permeation for Na ion produced by the single mutation DIII-K1237E. These data support the notion that the ultraslow inactivated state is associated with molecular rearrangements both in the external and internal vestibules, and that the regions in which the internal turn of the DIII-P-loop contacts the DIV-S6 segment may serve as site for transmission of molecular movements between the external and the internal vestibules. In this model, ultraslow inactivation is associated with a widening of the external vestibule as opposed to a collapse (see above). This is consistent with the finding that the aqueous accessibility of the external vestibule is not altered by long depolarizing pulses (Struyk & Cannon, 2002).

5.3 Insights into the mechanism of slow inactivation from crystal structures of bacterial voltage-gated Na^+ channels

5.3.1 The architecture of bacterial voltage-gated Na^+ channels

Recently, a number of crystal structures of bacterial VGSCs have become available. As opposed to eukaryotic VGSCs, bacterial VGSCs are homotetramers of identical subunits (Ren et al., 2001). They lack the linker between domains III and IV, which is required for fast inactivation of eukaryotic VGSCs, but they retain a slow inactivation process (Pavlov et al., 2005). These crystal structures enhanced our understanding of the molecular underpinnings of some gating transitions. Owing to these structures we now have 3D "snapshots" of channels in presumably closed (Payandeh, Scheuer, Zheng, & Catterall, 2011; Shaya et al., 2014; Tsai et al., 2013), inactivated (Payandeh, Gamal El-Din, Scheuer, Zheng, & Catterall, 2012; Zhang et al., 2012), and open (Bagneris et al., 2013; McCusker et al., 2012; Tsai et al., 2013) conformations. If viewed from the extracellular side of the membrane, the available structures show a central pore surrounded by four pore-forming modules composed of S5 and S6 segments and the intervening pore loops. The four voltage-sensing modules composed of S1—S4 segments are symmetrically associated with the outer rim of the pore module. Each voltage-sensing module is closely associated with the pore-forming module of its neighboring subunit. The S6 segments line the permeation pathway of a large central cavity that forms the internal vestibule. An internal crossing of S6 segments forms the intracellular activation gate. At the extracellular side the two transmembrane helices, S5 and S6 of each subunit are bridged by the pore helices P1 and P2, which are linked by the selectivity filter. As mentioned earlier, the selectivity filter of eukaryotic VGSCs is considered to be formed by a ring of P-loop amino acids, the DEKA motif, whereby each domain contributes one amino acid to the ring (Chiamvimonvat et al., 1996; Favre et al., 1996; Heinemann et al., 1992; Lipkind & Fozzard, 1994; Schlief et al., 1996; Terlau et al., 1991; Tsushima et al., 1997a). In the first crystallized bacterial VGSC, NavAb, from the gram-negative bacterium *Arcobacter butzleri* the four E177 side chains form an orifice of 4.6 × 4.6 Å (Payandeh et al., 2011). Thus, unlike eukaryotic VGSCs the bacterial counterparts form a nearly symmetric selectivity filter ("EEEE motif" as compared to "DEKA" in eukaryotic VGSCs). E177 forms the extracellular entrance of the selectivity filter followed by the backbone carbonyls of L176 and T175. The P1-helix ends with the conserved T175, which accepts a hydrogen bond from the conserved W179 in the outer vestibule of a neighboring subunit. This interaction stabilizes adjacent subunits at the selectivity filter. Another stabilizing interaction is the hydrogen bond between Q172 from the P-helix and the carbonyl of E177 of the selectivity filter (Payandeh et al., 2011).

5.3.2 Molecular mechanism of slow inactivation in bacterial voltage-gated Na^+ channels

Three crystal structures of NavAb are available. The first reported structure of NaVAb-I217C represents a potentially preopen state with four activated voltage-sensor modules and a closed pore module (Payandeh et al., 2011). The NaVAb-I217C structures were obtained by introducing a cysteine at two positions near the intracellular end of S6 (I217C or M221C). These mutations may have stabilized the activated state and prevented inactivation of the protein during the crystallization process at 0 mV. The other two available structures, NavAb-AB and NavAb-CD, are wild-type dimers and are considered to represent slow inactivated states of the channel (Payandeh et al., 2012). Thus, the whole structure represents a dimer of these dimers. When compared with the preopen NavAb, the inactivated structures display changes both in the selectivity filter region and in the central cavity. With regard to the selectivity filter region the slow inactivated structures have a distorted and narrowed T175—L176 carbonyl funnel. Furthermore, the key intersubunit interactions T175—W179 are lost as is the Q172 side chain-E177 backbone interaction (Payandeh et al., 2012). These changes may substantially decrease the stability of the outer vestibule structure.

With regard to the central cavity, compared to the preopen NavAb, the S6 segments from NavAb-AB have moved closer to the central pore axis, while the adjacent S6 segments of NavAb-CD are displaced laterally, giving rise to an asymmetric pore collapse of the internal vestibule. Another available crystal structure considered to represent an inactivated state is NavRh. When compared to the preopen NavAb, NavRh has a collapsed selectivity filter and a closed internal gate (Zhang et al., 2012). Interestingly, the closure at the internal gate occurs in NavRh one helical turn above the equivalent site in the preopen NavAb. This may indicate a rearrangement of the S6 segments during inactivation in NavRh, similar to the inactivated structures of NavAb.

As mentioned above, the interaction T175—W179 acts to stabilize the selectivity filter region and loss of this interaction is associated with entry into a slow inactivated state. W179 is part of a ring of tryptophans that is highly conserved among both eukaryotic and prokaryotic voltage-gated Na and Ca^{2+} channels (Tikhonov & Zhorov, 2011). This ring of tryptophans is located two positions C-terminal (or external) to the DEKA or EEEE motif comprising the selectivity filter in eukaryotic voltage-gated Na and Ca^{2+} channels, respectively (the only exception is the P-loop in DII of VGSCs in which the respective tryptophan is located next to the selectivity filter glutamic acid).

Regarding the position of the tryptophan in DIV of VGSC, W1531 (rNav1.4 numbering), we recently compared earlier structural homology models of eukaryotic VGSCs, which were based on the crystal structures of the bacterial K^+ channels KcsA (Lipkind & Fozzard, 2000) and MthK (Tikhonov & Zhorov, 2005) with the recently published crystal structures of bacterial VGSCs (McCusker et al., 2012; Payandeh et al., 2011; Zhang et al., 2012) and a revised homology model (Tikhonov & Zhorov, 2012), based on the structure of the NavAb (Payandeh et al., 2011). We found that, only in the NavAb-based model, W1531 intersects between the DIII-P-loop and the adjacent DIV-S6 segment. This arrangement was confirmed by the fact that removal of the side chain of W1531 by the mutation W1531G opens a pathway for external access of the permanently charged local anesthetic QX-222 (Lukacs et al., 2014). As mentioned earlier, it has been suggested that ultraslow inactivation in VGSCs is produced by a conformational change of both the external and the internal vestibules that is controlled by the interaction of the DIII-P-loop and the DIV-S6-segement (Zarrabi et al., 2010). The comparison of the preopen and the inactivated structures of NavAb supports the idea of an involvement of both the external and the internal vestibules in the generation of slow inactivated states. The position of W1531 at the contact surface between the external and the internal vestibules suggests that this tryptophan may be involved in the modulation of inactivated states. Indeed mutations within the ring of tryptophans alter both fast and slow inactivation gating (Benitah, Chen, Balser, Tomaselli, & Marban, 1999; Kambouris et al., 1998; Lukacs et al., 2014; Tomaselli et al., 1995; Tsang, Tsushima, Tomaselli, Li, & Backx, 2005; Xiong et al., 2003; Yang, Hsieh, & Kuo, 2009). Furthermore, if both DIV-W1531 and DIII-K1237 are replaced by cysteines then the rates of catalyzed internal disulfide formation between these cysteines are dependent on the duration of a depolarizing pulse, with a maximum at 200 ms, suggesting that the occupancy of an intermediate-inactivation state facilitates pore motions (Benitah et al., 1999). Hence the ring of tryptophans may serve to coordinate molecular motions of the external and the internal vestibules.

When NavAb (wild-type) is expressed in insect cells, it activates at very negative potentials, with a half-activation potential of −97.8 mV. Channel inactivation during 50-ms depolarizations also has unusually negative voltage dependence, with a half-inactivation potential of −119 mV. During repolarization, after a single 100-ms depolarization the time course of recovery has three kinetic phases: early ($\tau = 103$ ms), intermediate ($\tau = 3.8$ s), and late ($\tau > 1$ min) (Payandeh et al., 2012). Perhaps the two discrete conformations of NaVAb-WT provide a structural basis for some of these functional properties. Furthermore, repetitive 7-ms pulses at 0.2 Hz cause a late phase of slow inactivation that is dependent on the frequency of depolarization and nears completion in 600 s. Interestingly, if the DIV-alanine of the DEKA motif in the selectivity filter of rNav1.4 is mutated to a negatively charged aspartate (in part mimicking the negatively charged selectivity filter of NavAb) a similar ultraslow cumulative inactivation can be observed (Hilber et al., 2001). These data again underscore the importance of the selectivity filter in the modulation of slow inactivation gating.

In NavAb mutation N49K in the S2 segment in the extracellular negative cluster of the voltage sensor abolishes the late phase of slow inactivation (Gamal El-Din, Martinez, Payandeh, Scheuer, & Catterall, 2013). One of the gating charges of the S4 voltage sensor interacts with N49 in the crystal structure of NavAb, and mutation of this residue to cysteine also removes the late phase of slow inactivation. For eukaryotic VGSCs the voltage sensors of DII, DIII, and DIV have been implicated in both fast and slow inactivation (Capes, rcisio-Miranda, Jarecki, French, & Chanda, 2012; Cha, Ruben, George, Fujimoto, & Bezanilla, 1999; Mitrovic, George, & Horn, 2000; Nguyen, Wang, Rhodes, & George, 2008) and in Nav1.4 channels the voltage sensor in DIV couples to the selectivity filter region thereby modulating slow inactivation (Capes et al., 2012). In K channels conformational changes in the pore domain are triggered and stabilized by interaction with the voltage sensor S4 segment (Loots & Isacoff, 1998). There may be an electrostatic interaction between the S4 segment and the pore domain (Laine et al., 2003; Loots & Isacoff, 2000; Soler-Llavina, Chang, & Swartz, 2006) perhaps involving the uppermost S4 arginine residue (Laine et al., 2003).

With regard to eukaryotic VGSCs, Silva and Goldstein recently investigated the role of the voltage sensors in slow inactivation in rNav1.4 channels (Silva & Goldstein, 2013). Using voltage clamp fluorimetry (Cha et al., 1999; Chanda & Bezanilla, 2002), these authors found that onset and recovery from inactivation are correlated with voltage sensor movements 10,000-fold slower than the voltage sensor movements associated with activation. Developments of slow inactivation are best correlated with immobilization of voltage sensors in DI and DII, whereas recovery from slow inactivation was associated with slow restoration of mobility of the sensor in DIII.

From these studies a hypothetical mechanism for slow inactivation may be suggested: long depolarizations give rise to slow movements of the voltage sensor domains which is transmitted to the pore domains resulting in a collapse of the selectivity filter and the internal vestibule (Gamal El-Din et al., 2013). Clearly, much work is still needed to verify this model, to resolve the exact sequence of conformational changes and to examine its applicability to the various forms of slow inactivation in eukaryotic VGSCs.

6. SUMMARY AND OUTLOOK

Since the first description of "inactivation of sodium permeability" by Hodgkin and Huxley (Hodgkin & Huxley, 1952a, 1952b), our understanding of the mechanisms of inactivation has increased enormously. We have learned that there are several types of inactivated states with distinct biophysical properties. Furthermore, these states are subject to differential modulation by high a number of factors, eg, accessory subunits, the C-terminus, cytoskeletal proteins, pH, or drugs. The availability of crystal structures of bacterial VGSCs allowed for first insights into the molecular rearrangements that may underlie slow inactivation. Nevertheless, the exact structural underpinnings of the various forms of inactivation in eukaryotic Na channels are yet to be defined. Given the substantial importance of the process of inactivation with regard to pathophysiology and pharmacology, a deeper knowledge of the mechanisms of inactivation will be of great medical benefit. Recently, the structure of a mammalian voltage-gated Ca^{2+} channel has been solved by means of single-particle cryo-electron microscopy (Wu et al., 2015). This raises the hope that high-resolution structures of eukaryotic VGSCs will be available in the near future, which may then allow for further insights into the mechanism of inactivation.

REFERENCES

Abriel, H. (2010). Cardiac sodium channel Na(v)1.5 and interacting proteins: physiology and pathophysiology. *Journal of Molecular and Cellular Cardiology, 48*, 2–11.

Adelman, J. P. (1995). Proteins that interact with the pore-forming subunits of voltage- gated ion channels. *Current Opinion in Neurobiology, 5*, 286–295.

Adelman, W. J., & Palti, Y. (1969). The effects of external potassium and long duration voltage conditioning on the amplitude of sodium currents in the giant axon of the squid, *Loglio pealei. Journal of General Physiology, 54*, 589–606.

Aman, T. K., Grieco-Calub, T. M., Chen, C., Rusconi, R., Slat, E. A., Isom, L. L., & Raman, I. M. (2009). Regulation of persistent Na current by interactions between beta subunits of voltage gated Na channels. *Journal of Neuroscience: The Official Journal of the Society for Neuroscience, 29*, 2027–2042.

Armstrong, C. M. (1981). Sodium channels and gating currents. *Physiological Reviews, 61*, 644–683.

Armstrong, C. M., & Bezanilla, F. (1977). Inactivation of the sodium channel. II. Gating current experiments. *Journal of General Physiology, 70*, 567–590.

Bagneris, C., Decaen, P. G., Hall, B. A., Naylor, C. E., Clapham, D. E., Kay, C. W., & Wallace, B. A. (2013). Role of the C terminal domain in the structure and function of tetrameric sodium channels. *Nature Communications, 4*, 2465.

Bagneris, C., Naylor, C. E., McCusker, E. C., & Wallace, B. A. (2015). Structural model of the open closed-inactivated cycle of prokaryotic voltage-gated sodium channels. *Journal of General Physiology, 145*, 5–16.

Bahring, R., & Covarrubias, M. (2011). Mechanisms of closed-state inactivation in voltage-gated ion channels. *Journal of Physiology, 589*, 461–479.

Bai, C. X., Glaaser, I. W., Sawanobori, T., & Sunami, A. (2003). Involvement of local anesthetic binding sites on IVS6 of sodium channels in fast and slow inactivation. *Neuroscience Letters, 337*, 41–45.

Balser, J. R., Nuss, H. B., Chiamvimonvat, N., Perez-Garcia, M. T., Marban, E., & Tomaselli, G. F. (1996). External pore residue mediates slow inactivation in µ1 rat skeletal muscle sodium channels. *Journal of Physiology, 494*, 431–442.

Bant, J. S., & Raman, I. M. (2010). Control of transient, resurgent, and persistent current by open-channel block by Na^+ channel beta4 in cultured cerebellar granule neurons. *Proceedings of the National Academy of Sciences of the United States of America, 107*, 12357–12362.
Baukrowitz, T., & Yellen, G. (1995). Modulation of K^+ current by frequency and external [K^+]: a tale of two inactivation mechanisms. *Neuron, 15*, 951–960.
Bean, B. P. (2005). The molecular machinery of resurgent sodium current revealed. *Neuron, 45*, 185–187.
Becker, S., Prusak-Sochaczewski, E., Zamponi, G., Beck-Sickinger, A. G., Gordon, R. D., & French, R. J. (1992). Action of derivatives of μ–conotoxin GIIIA on sodium channels. Single amino acid substitutions in the toxin separately affect association and dissociation rates. *Biochemistry, 31*, 8229–8238.
Bendahhou, S., Cummins, T. R., Kula, R. W., Fu, Y. H., & Ptacek, L. J. (2002). Impairment of slow inactivation as a common mechanism for periodic paralysis in DIIS4-S5. *Neurology, 58*, 1266–1272.
Benitah, J. P., Chen, Z., Balser, J. R., Tomaselli, G. F., & Marban, E. (1999). Molecular dynamics of the sodium channel pore vary with gating: interactions between P-segment motions and inactivation. *Journal of Neuroscience, 19*, 1577–1585.
Ben-Johny, M., Yang, P. S., Niu, J., Yang, W., Joshi-Mukherjee, R., & Yue, D. T. (2014). Conservation of Ca^{2+}/calmodulin regulation across Na and Ca^{2+} channels. *Cell, 157*, 1657–1670.
Biswas, S., Deschenes, I., DiSilvestre, D., Tian, Y., Halperin, V. L., & Tomaselli, G. F. (2008). Calmodulin regulation of Nav1.4 current: role of binding to the carboxyl terminus. *Journal of General Physiology, 131*, 197–209.
Biswas, S., DiSilvestre, D., Tian, Y., Halperin, V. L., & Tomaselli, G. F. (2009). Calcium-mediated dual-mode regulation of cardiac sodium channel gating. *Circulation Research, 104*, 870–878.
Brackenbury, W. J., Calhoun, J. D., Chen, C., Miyazaki, H., Nukina, N., Oyama, F., … Isom, L. L. (2010). Functional reciprocity between Na^+ channel Nav1.6 and beta1 subunits in the coordinated regulation of excitability and neurite outgrowth. *Proceedings of the National Academy of Sciences of the United States of America, 107*, 2283–2288.
Brackenbury, W. J., & Isom, L. L. (2008). Voltage-gated Na^+ channels: potential for beta subunits as therapeutic targets. *Expert Opinion on Therapeutic Targets, 12*, 1191–1203.
Brackenbury, W. J., & Isom, L. L. (2011). Na^+ channel beta subunits: overachievers of the ion channel family. *Frontiers in Pharmacology, 2*, 53.
Brackenbury, W. J., Yuan, Y., O'Malley, H. A., Parent, J. M., & Isom, L. L. (2013). Abnormal neuronal patterning occurs during early postnatal brain development of Scn1b-null mice and precedes hyperexcitability. *Proceedings of the National Academy of Sciences of the United States of America, 110*, 1089–1094.
Brancati, F., Valente, E. M., Davies, N. P., Sarkozy, A., Sweeney, M. G., Lomonaco, M., … Dallapiccola, B. (2003). Severe infantile hyperkalaemic periodic paralysis and paramyotonia congenita: broadening the clinical spectrum associated with the T704M mutation in SCN4A. *Journal of Neurology, Neurosurgery, and Psychiatry, 74*, 1339–1341.
Brismar, T. (1977). Slow mechanism for sodium permeability inactivation in myelinated nerve fibre of *Xenopus laevis*. *Journal of Physiology, 270*, 283–297.
Buffington, S. A., & Rasband, M. N. (2013). Na^+ channel-dependent recruitment of Navbeta4 to axon initial segments and nodes of Ranvier. *Journal of Neuroscience: The Official Journal of the Society for Neuroscience, 33*, 6191–6202.
Burgess, D. L., Kohrman, D. C., Galt, J., Plummer, N. W., Jones, J. M., Spear, B., & Meisler, M. H. (1995). Mutation of a new sodium channel gene, Scn8a, in the mouse mutant 'motor endplate disease'. *Nature Genetics, 10*, 461–465.

Calhoun, J. D., & Isom, L. L. (2014). The role of non-pore-forming beta subunits in physiology and pathophysiology of voltage-gated sodium channels. *Handbook of Experimental Pharmacology, 221*, 51–89.

Capes, D. L., rcisio-Miranda, M., Jarecki, B. W., French, R. J., & Chanda, B. (2012). Gating transitions in the selectivity filter region of a sodium channel are coupled to the domain IV voltage sensor. *Proceedings of the National Academy of Sciences of the United States of America, 109*, 2648–2653.

Carr, D. B., Day, M., Cantrell, A. R., Held, J., Scheuer, T., Catterall, W. A., & Surmeir, D. J. (2003). Transmitter modulation of slow, activity-dependent alterations in sodium channel availability endows neurons with a novel form of cellular plasticity. *Neuron, 39*, 793–806.

Casini, S., Tan, H. L., Bhuiyan, Z. A., Bezzina, C. R., Barnett, P., Cerbai, E., … Veldkamp, M. W. (2007). Characterization of a novel SCN5A mutation associated with Brugada syndrome reveals involvement of DIIIS4-S5 linker in slow inactivation. *Cardiovascular Research, 76*, 418–429.

Catterall, W. A. (1992). Cellular and molecular biology of voltage-gated sodium channels. *Physiological Reviews, 72*, S15–S48.

Catterall, W. A. (2000). From ionic currents to molecular mechanisms: the structure and function of voltage-gated sodium channels. *Neuron, 26*, 13–25.

Catterall, W. A. (2014). Structure and function of voltage-gated sodium channels at atomic resolution. *Experimental Physiology, 99*, 35–51.

Catterall, W. A., Goldin, A. L., & Waxman, S. G. (2005). International Union of Pharmacology. XLVII. Nomenclature and structure-function relationships of voltage-gated sodium channels. *Pharmacological Reviews, 57*, 397–409.

Cha, A., Ruben, P. C., George, A. L. J., Fujimoto, E., & Bezanilla, F. (1999). Voltage sensors in domains III and IV, but not I and II, are immobilized by Na^+ channel fast inactivation. *Neuron, 22*, 73–87.

Chancey, J. H., Shockett, P. E., & O'Reilly, J. P. (2007). Relative resistance to slow inactivation of human cardiac Na^+ channel hNav1.5 is reversed by lysine or glutamine substitution at V930 in D2-S6. *American Journal of Physiology. Cell Physiology, 293*, C1895–C1905.

Chanda, B., & Bezanilla, F. (2002). Tracking voltage-dependent conformational changes in skeletal muscle sodium channel during activation. *Journal of General Physiology, 120*, 629–645.

Chandler, W. K., & Meves, H. (1970a). Evidence for two types of sodium conductance in axons perfused with sodium fluoride solution. *Journal of Physiology, 211*, 653–678.

Chandler, W. K., & Meves, H. (1970b). Rate constants associated with changes in sodium conductance in axons perfused with sodium fluoride. *Journal of Physiology, 211*, 679–705.

Chandler, W. K., & Meves, H. (1970c). Slow changes in membrane permeability and long-lasting action potentials in axons perfused with fluoride solutions. *Journal of Physiology, 211*, 707–728.

Chang, N. S., French, R. J., Lipkind, G. M., Fozzard, H. A., & Dudley, S. (1998). Predominant interactions between μ–conotoxin Arg-13 and the skeletal muscle Na^+ channel localized by mutant cycle analysis. *Biochemistry, 37*, 4407–4419.

Chen, C., Bharucha, V., Chen, Y., Westenbroek, R. E., Brown, A., Malhotra, J. D., … Isom, L. L. (2002). Reduced sodium channel density, altered voltage dependence of inactivation, and increased susceptibility to seizures in mice lacking sodium channel beta 2-subunits. *Proceedings of the National Academy of Sciences of the United States of America, 99*, 17072–17077.

Chen, C., & Cannon, S. C. (1995). Modulation of Na^+ channel inactivation by the beta 1 subunit: a deletion analysis. *Pflügers Archiv, 431*, 186–195.

Chen, C., Westenbroek, R. E., Xu, X., Edwards, C. A., Sorenson, D. R., Chen, Y., ... Isom, L. L. (2004). Mice lacking sodium channel beta1 subunits display defects in neuronal excitability, sodium channel expression, and nodal architecture. *Journal of Neuroscience, 24*, 4030–4042.

Chen, Z., Ong, B. H., Kambouris, N. G., Marban, E., Tomaselli, G. F., & Balser, J. R. (2000). Lidocaine induces a slow inactivated state in rat skeletal muscle sodium channels. *Journal of Physiology, 524*(Pt 1), 37–49.

Cheng, X., Dib-Hajj, S. D., Tyrrell, L., te Morsche, R. H., Drenth, J. P., & Waxman, S. G. (2011). Deletion mutation of sodium channel $Na_V1.7$ in inherited erythromelalgia: enhanced slow inactivation modulates dorsal root ganglion neuron hyperexcitability. *Brain: A Journal of Neurology, 134*, 1972–1986.

Chen-Izu, Y., Shaw, R. M., Pitt, G. S., Yarov-Yarovoy, V., Sack, J. T., Abriel, H., ... Bers, D. M. (2015). Na^+ channel function, regulation, structure, trafficking and sequestration. *Journal of Physiology, 593*, 1347–1360.

Chiamvimonvat, N., Perez Garcia, M. T., Tomaselli, G. F., & Marban, E. (1996). Control of ion flux and selectivity by negatively charged residues in the outer mouth of rat sodium channels. *Journal of Physiology, 491*, 51–59.

Choi, K. L., Aldrich, R. W., & Yellen, G. (1991). Tetraethylammonium blockade distinguishes two inactivation mechanisms in voltage-activated K^+ channels. *Proceedings of the National Academy of Sciences of the United States of America, 88*, 5092–5095.

Clarkson, C. W., Matsubara, T., & Hondeghem, L. M. (1984). Slow inactivation of Vmax in guinea pig ventricular myocardium. *American Journal of Physiology, 247*, H645–H654.

Collins, C. A., Rojas, E., & Suarez, I. B. (1982). Fast charge movements in skeletal muscle fibres from *Rana temporaria*. *Journal of Physiology, 324*, 319–345.

Creighton, T. E. (1993). *Proteins. Structures and molecular properties* (2 ed.). New York: W.H.Freeman and Company.

Cruz, L. J., Gray, W. R., Olivera, B. M., Zeikus, R. D., Kerr, L., Yoshikami, D., & Moczydlowski, E. (1985). *Conus geographus* toxins that discriminate between neuronal and muscle sodium channels. *Journal of Biological Chemistry, 260*, 9280–9288.

Cummins, T. R., & Sigworth, F. J. (1996). Impaired slow inactivation in mutant sodium channels. *Biophysical Journal, 71*, 227–236.

Desch, I., Chen, Q., Kallen, R. G., & Chahine, M. (1998). Electrophysiological study of chimeric sodium channels from heart and skeletal muscle. *Journal of Membrane Biology, 164*, 25–34.

Deschenes, I., Armoundas, A. A., Jones, S. P., & Tomaselli, G. F. (2008). Post-transcriptional gene silencing of KChIP2 and Navbeta1 in neonatal rat cardiac myocytes reveals a functional association between Na and Ito currents. *Journal of Molecular and Cellular Cardiology, 45*, 336–346.

Deschenes, I., Neyroud, N., DiSilvestre, D., Marban, E., Yue, D. T., & Tomaselli, G. F. (2002). Isoform-specific modulation of voltage-gated Na^+ channels by calmodulin. *Circulation Research, 90*, E49–E57.

Deschenes, I., Trottier, E., & Chahine, M. (2001). Implication of the C-terminal region of the alpha-subunit of voltage-gated sodium channels in fast inactivation. *Journal of Membrane Biology, 183*, 103–114.

Dover, K., Solinas, S., D'Angelo, E., & Goldfarb, M. (2010). Long-term inactivation particle for voltage-gated sodium channels. *Journal of Physiology, 588*, 3695–3711.

Dudley, S. C., Jr., Todt, H., Lipkind, G., & Fozzard, H. A. (1995). A μ–conotoxin-insensitive Na^+ channel mutant: possible localization of a binding site at the outer vestibule. *Biophysical Journal, 69*, 1657–1665.

Eaholtz, G., Scheuer, T., & Catterall, W. A. (1994). Restoration of inactivation and block of open sodium channels by an inactivation gate peptide. *Neuron, 12*, 1041–1048.

Ednie, A. R., & Bennett, E. S. (2012). Modulation of voltage-gated ion channels by sialylation. *Comprehensive Physiology, 2*, 1269–1301.
Ellerkmann, R. K., Riazanski, V., Elger, C. E., Urban, B. W., & Beck, H. (2001). Slow recovery from inactivation regulates the availability of voltage-dependent Na^+ channels in hippocampal granule cells, hilar neurons and basket cells. *Journal of Physiology, 532*, 385–397.
Favre, I., Moczydlowski, E., & Schild, L. (1996). On the structural basis for ionic selectivity among Na^+, K^+, and Ca^{2+} in the voltage-gated sodium channel. *Biophysical Journal, 71*, 3110–3125.
Featherstone, D. E., Richmond, J. E., & Ruben, P. C. (1996). Interaction between fast and slow inactivation in Skm1 sodium channels. *Biophysical Journal, 71*, 3098–3109.
Filatov, G. N., Nguyen, T. P., Kraner, S. D., & Barchi, R. L. (1998). Inactivation and secondary structure in the D4/S4-5 region of the SkM1 sodium channel. *Journal of General Physiology, 111*, 703–715.
Fox, J. M. (1976). Ultra-slow inactivation of the ionic currents through the membrane of myelinated nerve. *Biochimica et Biophysica Acta, 426*, 232–244.
French, R. J., Prusak-Sochaczewski, E., Zamponi, G. W., Becker, S., Kularatna, A. S., & Horn, R. (1996). Interactions between a pore-blocking peptide and the voltage sensor of the sodium channel: an electrostatic approach to channel geometry. *Neuron, 16*, 407–413.
Gabelli, S. B., Yoder, J. B., Tomaselli, G. F., & Amzel, L. M. (2015). Calmodulin and Ca^{2+} control of voltage gated Na^+ channels. *Channels, 10*, 45–54.
Gamal El-Din, T. M., Martinez, G. Q., Payandeh, J., Scheuer, T., & Catterall, W. A. (2013). A gating charge interaction required for late slow inactivation of the bacterial sodium channel NavAb. *Journal of General Physiology, 142*, 181–190.
Glaaser, I. W., Bankston, J. R., Liu, H., Tateyama, M., & Kass, R. S. (2006). A carboxyl-terminal hydrophobic interface is critical to sodium channel function. Relevance to inherited disorders. *Journal of Biological Chemistry, 281*, 24015–24023.
Goldfarb, M. (2005). Fibroblast growth factor homologous factors: evolution, structure, and function. *Cytokine & Growth Factor Reviews, 16*, 215–220.
Goldfarb, M. (2012). Voltage-gated sodium channel-associated proteins and alternative mechanisms of inactivation and block. *Cellular and Molecular Life Sciences: CMLS, 69*, 1067–1076.
Goldin, A. L. (2003). Mechanisms of sodium channel inactivation. *Current Opinion in Neurobiology, 13*, 284–290.
Goldin, A. L., Snutch, T., Lubbert, H., Dowsett, A., Marshall, J., Auld, V., … Dunn, R. (1986). Messenger RNA coding for only the alpha subunit of the rat brain Na channel is sufficient for expression of functional channels in *Xenopus* oocytes. *Proceedings of the National Academy of Sciences of the United States of America, 83*, 7503–7507.
Grieco, T. M., Malhotra, J. D., Chen, C., Isom, L. L., & Raman, I. M. (2005). Open-channel block by the cytoplasmic tail of sodium channel beta4 as a mechanism for resurgent sodium current. *Neuron, 45*, 233–244.
Grissmer, S., & Cahalan, M. (1989). TEA prevents inactivation while blocking open K^+ channels in human T lymphocytes. *Biophysical Journal, 55*, 203–206.
Groome, J. R., Dice, M. C., Fujimoto, E., & Ruben, P. C. (2007). Charge immobilization of skeletal muscle Na^+ channels: role of residues in the inactivation linker. *Biophysical Journal, 93*, 1519–1533.
Groome, J. R., Fujimoto, E., & Ruben, P. C. (2003). Negative charges in the DIII-DIV linker of human skeletal muscle Na^+ channels regulate deactivation gating. *Journal of Physiology, 548*, 85–96.
Hartmann, H. A., Tiedeman, A. A., Chen, S. F., Brown, A. M., & Kirsch, G. E. (1994). Effects of III-IV linker mutations on human heart Na^+ channel inactivation gating. *Circulation Research, 75*, 114–122.

Hayward, L. J., Brown, R. H., Jr., & Cannon, S. C. (1997). Slow inactivation differs among mutant Na channels associated with myotonia and periodic paralysis. *Biophysical Journal, 72*, 1204–1219.
Hayward, L. J., Sandoval, G. M., & Cannon, S. C. (1999). Defective slow inactivation of sodium channels contributes to familial periodic paralysis. *Neurology, 52*, 1447–1453.
Heinemann, S. H., Terlau, H., Stühmer, W., Imoto, K., & Numa, S. (1992). Calcium channel characteristics conferred on the sodium channel by single mutations. *Nature, 356*, 441–443.
Herren, A. W., Bers, D. M., & Grandi, E. (2013). Post-translational modifications of the cardiac Na channel: contribution of CaMKII-dependent phosphorylation to acquired arrhythmias. *American Journal of Physiology - Heart and Circulatory Physiology, 305*, H431–H445.
Hilber, K., Sandtner, W., Kudlacek, O., Glaaser, I. W., Weisz, E., Kyle, J. W., … Todt, H. (2001). The selectivity filter of the voltage-gated sodium channel is involved in channel activation. *Journal of Biological Chemistry, 276*, 27831–27839.
Hilber, K., Sandtner, W., Kudlacek, O., Schreiner, B., Glaaser, I., Schutz, W., … Todt, H. (2002). Interaction between fast and ultra-slow inactivation in the voltage-gated sodium channel. Does the inactivation gate stabilize the channel structure? *Journal of Biological Chemistry, 277*, 37105–37115.
Hilber, K., Sandtner, W., Zarrabi, T., Zebedin, E., Kudlacek, O., Fozzard, H. A., & Todt, H. (2005). Selectivity filter residues contribute unequally to pore stabilization in voltage-gated sodium channels. *Biochemistry, 44*, 13874–13882.
Hodgkin, A. L., & Huxley, A. F. (1952a). The components of membrane conductance in the giant axon of *Loglio*. *Journal of Physiology, 116*, 473–496.
Hodgkin, A. L., & Huxley, A. F. (1952b). The dual effect of membrane potential on sodium conductance in the giant axon of *Loglio*. *Journal of Physiology, 116*, 497–506.
Huang, C. J., & Moczydlowski, E. (2001). Cytoplasmic polyamines as permeant blockers and modulators of the voltage-gated sodium channel. *Biophysical Journal, 80*, 1262–1279.
Huang, C. J., Schild, L., & Moczydlowski, E. G. (2012). Use-dependent block of the voltage-gated Na(+) channel by tetrodotoxin and saxitoxin: effect of pore mutations that change ionic selectivity. *Journal of General Physiology, 140*, 435–454.
Isom, L. L., De Jongh, K. S., & Catterall, W. A. (1994). Auxiliary subunits of voltage-gated ion channels. *Neuron, 12*, 1183–1194.
Isom, L. L., De Jongh, K. S., Patton, D. E., Reber, B. F., Offord, J., Charbonneau, H., … Catterall, W. A. (1992). Primary structure and functional expression of the beta 1 subunit of the rat brain sodium channel. *Science, 256*, 839–842.
Isom, L. L., Ragsdale, D. S., De Jongh, K. S., Westenbroek, R. E., Reber, B. F., Scheuer, T., & Catterall, W. A. (1995). Structure and function of the beta 2 subunit of brain sodium channels, a transmembrane glycoprotein with a CAM motif. *Cell, 83*, 433–442.
Jones, D. K., Peters, C. H., Allard, C. R., Claydon, T. W., & Ruben, P. C. (2013). Proton sensors in the pore domain of the cardiac voltage-gated sodium channel. *Journal of Biological Chemistry, 288*, 4782–4791.
Jones, D. K., Peters, C. H., Tolhurst, S. A., Claydon, T. W., & Ruben, P. C. (2011). Extracellular proton modulation of the cardiac voltage-gated sodium channel, Nav1.5. *Biophysical Journal, 101*, 2147–2156.
Jones, D. K., & Ruben, P. C. (2014). Proton modulation of cardiac I Na: a potential arrhythmogenic trigger. *Handbook of Experimental Pharmacology, 221*, 169–181.
Kambouris, N. G., Hastings, L. A., Stepanovic, S., Marban, E., Tomaselli, G. F., & Balser, J. R. (1998). Mechanistic link between lidocaine block and inactivation probed by outer pore mutations in the rat μ1 skeletal muscle sodium channel. *Journal of Physiology, 512*, 693–705.

Kass, R. S. (2006). Sodium channel inactivation in heart: a novel role of the carboxy-terminal domain. *Journal of Cardiovascular Electrophysiology, 17*(Suppl. 1), S21–S25.
Kellenberger, S., Scheuer, T., & Catterall, W. A. (1996). Movement of the Na^+ channel inactivation gate during inactivation. *Journal of Biological Chemistry, 271*, 30971–30979.
Kellenberger, S., West, J. W., Catterall, W. A., & Scheuer, T. (1997). Molecular analysis of potential hinge residues in the inactivation gate of brain type IIA Na^+ channels. *Journal of General Physiology, 109*, 607–617.
Kellenberger, S., West, J. W., Scheuer, T., & Catterall, W. A. (1997). Molecular analysis of the putative inactivation particle in the inactivation gate of brain type IIA Na^+ channels. *Journal of General Physiology, 109*, 589–605.
Laine, M., Lin, M. C., Bannister, J. P., Silverman, W. R., Mock, A. F., Roux, B., & Papazian, D. M. (2003). Atomic proximity between S4 segment and pore domain in Shaker potassium channels. *Neuron, 39*, 467–481.
Lampert, A., Eberhardt, M., & Waxman, S. G. (2014). Altered sodium channel gating as molecular basis for pain: contribution of activation, inactivation, and resurgent currents. *Handbook of Experimental Pharmacology, 221*, 91–110.
Lee, A., & Goldin, A. L. (2008). Role of the amino and carboxy termini in isoform-specific sodium channel variation. *Journal of Physiology, 586*, 3917–3926.
Lerche, H., Peter, W., Fleischhauer, R., Pika-Hartlaub, U., Malina, T., Mitrovic, N., & Lehmann-Horn, F. (1997). Role in fast inactivation of the IV/S4-S5 loop of the human muscle Na^+ channel probed by cysteine mutagenesis. *Journal of Physiology, 505*, 345–352.
Lipkind, G. M., & Fozzard, H. A. (1994). A structural model of the tetrodotoxin and saxitoxin binding site of the Na^+ channel. *Biophysical Journal, 66*, 1–13.
Lipkind, G. M., & Fozzard, H. A. (2000). KcsA crystal structure as framework for a molecular model of the Na^+ channel pore. *Biochemistry, 39*, 8161–8170.
Liu, C., Dib-Hajj, S. D., & Waxman, S. G. (2001). Fibroblast growth factor homologous factor 1B binds to the C terminus of the tetrodotoxin-resistant sodium channel rNav1.9a (NaN). *Journal of Biological Chemistry, 276*, 18925–18933.
Liu, C. J., Dib-Hajj, S. D., Renganathan, M., Cummins, T. R., & Waxman, S. G. (2003). Modulation of the cardiac sodium channel Nav1.5 by fibroblast growth factor homologous factor 1B. *Journal of Biological Chemistry, 278*, 1029–1036.
Loots, E., & Isacoff, E. Y. (1998). Protein rearrangements underlying slow inactivation of the Shaker K^+ channel. *Journal of General Physiology, 112*, 377–389.
Loots, E., & Isacoff, E. Y. (2000). Molecular coupling of S4 to a K(+) channel's slow inactivation gate. *Journal of General Physiology, 116*, 623–636.
Lopez-Barneo, J., Hoshi, T., Heinemann, S. H., & Aldrich, R. W. (1993). Effects of external cations and mutations in the pore region on C-type inactivation of Shaker potassium channels. *Receptor Channel, 1*, 61–71.
Lou, J. Y., Laezza, F., Gerber, B. R., Xiao, M., Yamada, K. A., Hartmann, H., ... Ornitz, D. M. (2005). Fibroblast growth factor 14 is an intracellular modulator of voltage-gated sodium channels. *Journal of Physiology, 569*, 179–193.
Lukacs, P., Gawali, V. S., Cervenka, R., Ke, S., Koenig, X., Rubi, L., ... Todt, H. (2014). Exploring the structure of the voltage-gated Na^+ channel by an engineered drug access pathway to the receptor site for local anesthetics. *Journal of Biological Chemistry, 289*, 21770–21781.
Mantegazza, M., Yu, F. H., Catterall, W. A., & Scheuer, T. (2001). Role of the C-terminal domain in inactivation of brain and cardiac sodium channels. *Proceedings of the National Academy of Sciences of the United States of America, 98*, 15348–15353.
Marionneau, C., & Abriel, H. (2015). Regulation of the cardiac Na^+ channel NaV1.5 by post-translational modifications. *Journal of Molecular and Cellular Cardiology, 82*, 36–47.

Marionneau, C., Carrasquillo, Y., Norris, A. J., Townsend, R. R., Isom, L. L., Link, A. J., & Nerbonne, J. M. (2012). The sodium channel accessory subunit Navbeta1 regulates neuronal excitability through modulation of repolarizing voltage-gated K^+ channels. *Journal of Neuroscience: The Official Journal of the Society for Neuroscience, 32*, 5716–5727.
McCormick, K. A., Isom, L. L., Ragsdale, D., Smith, D., Scheuer, T., & Catterall, W. A. (1998). Molecular determinants of Na^+ channel function in the extracellular domain of the beta1 subunit. *Journal of Biological Chemistry, 273*, 3954–3962.
McCusker, E. C., Bagneris, C., Naylor, C. E., Cole, A. R., D'Avanzo, N., Nichols, C. G., & Wallace, B. A. (2012). Structure of a bacterial voltage-gated sodium channel pore reveals mechanisms of opening and closing. *Nature Communications, 3*, 1102.
McPhee, J. C., Ragsdale, D. S., Scheuer, T., & Catterall, W. A. (1994). A mutation in segment IVS6 disrupts fast inactivation of sodium channels. *Proceedings of the National Academy of Sciences of the United States of America, 91*, 12346–12350.
McPhee, J. C., Ragsdale, D. S., Scheuer, T., & Catterall, W. A. (1995). A critical role for transmembrane segment IVS6 of the sodium channel alpha subunit in fast inactivation. *Journal of Biological Chemistry, 270*, 12025–12034.
McPhee, J. C., Ragsdale, D. S., Scheuer, T., & Catterall, W. A. (1998). A critical role for the S4-S5 intracellular loop in domain IV of the sodium channel alpha-subunit in fast inactivation. *Journal of Biological Chemistry, 273*, 1121–1129.
Meadows, L., Malhotra, J. D., Stetzer, A., Isom, L. L., & Ragsdale, D. S. (2001). The intracellular segment of the sodium channel beta 1 subunit is required for its efficient association with the channel alpha subunit. *Journal of Neurochemsitry, 76*, 1871–1878.
Messner, D. J., & Catterall, W. A. (1985). The sodium channel from rat brain. Separation and characterization of subunits. *Journal of Biological Chemistry, 260*, 10597–10604.
Miller, J. R., Patel, M. K., John, J. E., Mounsey, J. P., & Moorman, J. R. (2000). Contributions of charged residues in a cytoplasmic linking region to Na channel gating. *Biochimica et Biophysica Acta, 1509*, 275–291.
Mitrovic, N., George, A. L., & Horn, R. (2000). Role of domain 4 in sodium channel slow inactivation. *Journal of General Physiology, 115*, 707–718.
Moczydlowski, E., Olivera, B. M., Gray, W. R., & Strichartz, G. R. (1986). Discrimination of muscle and neuronal Na^+-channel subtypes by binding competition between [3H] saxitoxin and μ–conotoxins. *Proceedings of the National Academy of Sciences of the United States of America, 83*, 5321–5325.
Motoike, H. K., Liu, H., Glaaser, I. W., Yang, A. S., Tateyama, M., & Kass, R. S. (2004). The Na^+ channel inactivation gate is a molecular complex: a novel role of the COOH-terminal domain. *Journal of General Physiology, 123*, 155–165.
Narahashi, T. (1964). Restoration of action potential by anodal polarization in lobster giant axons. *Journal of Cellular and Comparative Physiology, 64*, 73–96.
Nguyen, H. M., & Goldin, A. L. (2010). Sodium channel carboxyl-terminal residue regulates fast inactivation. *Journal of Biological Chemistry, 285*, 9077–9089.
Nguyen, H. M., Miyazaki, H., Hoshi, N., Smith, B. J., Nukina, N., Goldin, A. L., & Chandy, K. G. (2012). Modulation of voltage-gated K^+ channels by the sodium channel beta1 subunit. *Proceedings of the National Academy of Sciences of the United States of America, 109*, 18577–18582.
Nguyen, T. P., Wang, D. W., Rhodes, T. H., & George, A. L., Jr. (2008). Divergent biophysical defects caused by mutant sodium channels in dilated cardiomyopathy with arrhythmia. *Circulation Research, 102*, 364–371.
Noda, M., Shimizu, S., Tanabe, T., Takai, T., Kayano, T., Ikeda, T., … Minamino, N. (1984). Primary structure of *Electrophorus electricus* sodium channel deduced from cDNA sequence. *Nature, 312*, 121–127.

Nuss, H. B., Balser, J. R., Orias, D. W., Lawrence, J. H., Tomaselli, G. F., & Marban, E. (1996). Coupling between fast and slow inactivation revealed by analysis of a point mutation (F1304Q) in mu 1 rat skeletal muscle sodium channels. *Journal of Physiology, 494*, 411–429.

Ogielska, E. M., & Aldrich, R. W. (1999). Functional consequences of a decreased potassium affinity in a potassium channel pore. Ion interactions and C-type inactivation. *Journal of General Physiology, 113*, 347–358.

O'Malley, H. A., & Isom, L. L. (2015). Sodium channel beta subunits: emerging targets in channelopathies. *Annual Review of Physiology, 77*, 481–504.

Ong, B. H., Tomaselli, G. F., & Balser, J. R. (2000). A structural rearrangement in the sodium channel pore linked to slow inactivation and use dependence. *Journal of General Physiology, 116*, 653–662.

O'Reilly, J. P., & Shockett, P. E. (2006). Slow-inactivation induced conformational change in domain 2-segment 6 of cardiac Na^+ channel. *Biochemical and Biophysical Research Communication, 345*, 59–66.

O'Reilly, J. P., Wang, S. Y., & Wang, G. K. (2000). A point mutation in domain 4-segment 6 of the skeletal muscle sodium channel produces an atypical inactivation state. *Biophysical Journal, 78*, 773–784.

O'Reilly, J. P., Wang, S. Y., & Wang, G. K. (2001). Residue-specific effects on slow inactivation at V787 in D2-S6 of $Na_V1.4$ sodium channels. *Biophysical Journal, 81*, 2100–2111.

Patino, G. A., Brackenbury, W. J., Bao, Y., Lopez-Santiago, L. F., O'Malley, H. A., Chen, C., … Isom, L. L. (2011). Voltage-gated Na^+ channel beta1B: a secreted cell adhesion molecule involved in human epilepsy. *Journal of Neuroscience: The Official Journal of the Society for Neuroscience, 31*, 14577–14591.

Patino, G. A., Claes, L. R., Lopez-Santiago, L. F., Slat, E. A., Dondeti, R. S., Chen, C., … Isom, L. L. (2009). A functional null mutation of SCN1B in a patient with Dravet syndrome. *Journal of Neuroscience: The Official Journal of the Society for Neuroscience, 29*, 10764–10778.

Patton, D. E., West, J. W., Catterall, W. A., & Goldin, A. L. (1992). Amino acid residues required for fast Na^+-channel inactivation: charge neutralizations and deletions in the III-IV linker. *Proceedings of the National Academy of Sciences of the United States of America, 89*, 10905–10909.

Pavlov, E., Bladen, C., Winkfein, R., Diao, C., Dhaliwal, P., & French, R. J. (2005). The pore, not cytoplasmic domains, underlies inactivation in a prokaryotic sodium channel. *Biophysical Journal, 89*, 232–242.

Payandeh, J., Gamal El-Din, T. M., Scheuer, T., Zheng, N., & Catterall, W. A. (2012). Crystal structure of a voltage-gated sodium channel in two potentially inactivated states. *Nature, 486*, 135–139.

Payandeh, J., & Minor, D. L., Jr. (2015). Bacterial voltage-gated sodium channels ($BacNa_Vs$) from the soil, sea, and salt lakes enlighten molecular mechanisms of electrical signaling and pharmacology in the brain and heart. *Journal of Molecular Biology, 427*, 3–30.

Payandeh, J., Scheuer, T., Zheng, N., & Catterall, W. A. (2011). The crystal structure of a voltage-gated sodium channel. *Nature, 475*, 353–358.

Perez Garcia, M. T., Chiamvimonvat, N., Marban, E., & Tomaselli, G. F. (1996). Structure of the sodium channel pore revealed by serial cysteine mutagenesis. *Proceedings of the National Academy of Sciences of the United States of America, 93*, 300–304.

Potet, F., Chagot, B., Anghelescu, M., Viswanathan, P. C., Stepanovic, S. Z., Kupershmidt, S., … Balser, J. R. (2009). Functional interactions between distinct sodium channel cytoplasmic domains through the action of calmodulin. *Journal of Biological Chemistry, 284*, 8846–8854.

Qu, Y., Rogers, J. C., Chen, S. F., McCormick, K. A., Scheuer, T., & Catterall, W. A. (1999). Functional roles of the extracellular segments of the sodium channel alpha subunit in voltage-dependent gating and modulation by beta1 subunits. *Journal of Biological Chemistry, 274*, 32647–32654.

Quandt, F. N. (1987). Burst kinetics of sodium channels which lack fast inactivation in mouse neuroblastoma cells. *Journal of Physiology (London), 392*, 563–585.

Ragsdale, D. S., McPhee, J. C., Scheuer, T., & Catterall, W. A. (1994). Molecular determinants of state-dependent block of Na^+ channels by local anesthetics. *Science, 265*, 1724–1728.

Raman, I. M., Sprunger, L. K., Meisler, M. H., & Bean, B. P. (1997). Altered subthreshold sodium currents and disrupted firing patterns in Purkinje neurons of Scn8a mutant mice. *Neuron, 19*, 881–891.

Ren, D., Navarro, B., Xu, H., Yue, L., Shi, Q., & Clapham, D. E. (2001). A prokaryotic voltage-gated sodium channel. *Science, 294*, 2372–2375.

Richmond, J. E., Featherstone, D. E., Hartmann, H. A., & Ruben, P. C. (1998). Slow inactivation in human cardiac sodium channels. *Biophysical Journal, 74*, 2945–2952.

Rohl, C. A., Boeckman, F. A., Baker, C., Scheuer, T., Catterall, W. A., & Klevit, R. E. (1999). Solution structure of the sodium channel inactivation gate. *Biochemistry, 38*, 855–861.

Rudy, B. (1978). Slow inactivation of the sodium conductance in squid giant axons. Pronase resistance. *Journal of Physiology (London), 283*, 1–21.

Saikawa, T., & Carmeliet, E. (1982). Slow recovery of the maximal rate of rise (Vmax) of the action potential in sheep cardiac Purkinje fibers. *Pflügers Archiv, 394*, 90–93.

Salgado, V. L., Yeh, J. Z., & Narahashi, T. (1985). Voltage-dependent removal of sodium inactivation by N-bromoacetamide and pronase. *Biophysical Journal, 47*, 567–571.

Sandtner, W., Szendroedi, J., Zarrabi, T., Zebedin, E., Hilber, K., Glaaser, I., … Todt, H. (2004). Lidocaine: a foot in the door of the inner vestibule prevents ultra-slow inactivation of a voltage-gated sodium channel. *Molecular Pharmacology, 66*, 648–657.

Sarhan, M. F., Tung, C. C., Van, P. F., & Ahern, C. A. (2012). Crystallographic basis for calcium regulation of sodium channels. *Proceedings of the National Academy of Sciences of the United States of America, 109*, 3558–3563.

Schauf, C. L., Pencek, T. L., & Davis, F. A. (1976). Slow sodium inactivation in Myxicola axons. Evidence for a second inactive state. *Biophysical Journal, 16*, 771–778.

Schlief, T., Schoenherr, R., Imoto, K., & Heinemann, S. H. (1996). Pore properties of rat brain II sodium channels mutated in the selectivity filter domain. *European Biophysics Journal, 25*, 75–91.

Schneider, M., Proebstle, T., Hombach, V., Hannekum, A., & Rudel, R. (1994). Characterization of the sodium currents in isolated human cardiocytes. *Pflügers Archiv, 428*, 84–90.

Shah, V. N., Wingo, T. L., Weiss, K. L., Williams, C. K., Balser, J. R., & Chazin, W. J. (2006). Calcium-dependent regulation of the voltage-gated sodium channel hH1: intrinsic and extrinsic sensors use a common molecular switch. *Proceedings of the National Academy of Sciences of the United States of America, 103*, 3592–3597.

Shaya, D., Findeisen, F., Abderemane-Ali, F., Arrigoni, C., Wong, S., Nurva, S. R., … Minor, D. L., Jr. (2014). Structure of a prokaryotic sodium channel pore reveals essential gating elements and an outer ion binding site common to eukaryotic channels. *Journal of Molecular Biology, 426*, 467–483.

Silva, J. R. (2015)P. C. Ruben (Ed.), Slow Inactivation of Na^+ Channels. *Handbook of Experimental Pharmacology, 221*, 33–49.

Silva, J. R., & Goldstein, S. A. (2013). Voltage-sensor movements describe slow inactivation of voltage-gated sodium channels I: wild-type skeletal muscle $Na_V1.4$. *Journal of General Physiology, 141*, 309–321.

Simoncini, L., & Stühmer, W. (1987). Slow sodium channel inactivation in rat fast-twitch muscle. *Journal of Physiology, 383*, 327–337.

Smith, M. R., & Goldin, A. L. (1997). Interaction between the sodium channel inactivation linker and domain III S4-S5. *Biophysical Journal, 73*, 1885–1895.

Soler-Llavina, G. J., Chang, T. H., & Swartz, K. J. (2006). Functional interactions at the interface between voltage-sensing and pore domains in the Shaker K(v) channel. *Neuron, 52*, 623–634.

Spampanato, J., Kearney, J. A., De Haan, G., McEwen, D. P., Escayg, A., Aradi, I., … Meisler, M. H. (2004). A novel epilepsy mutation in the sodium channel SCN1A identifies a cytoplasmic domain for beta subunit interaction. *Journal of Neuroscience: The Official Journal of the Society for Neuroscience, 24*, 10022–10034.

Starkus, J. G., & Shrager, P. (1978). Modification of slow sodium inactivation in nerve after internal perfusion with trypsin. *American Journal of Physiology, 235*, C238–C244.

Struyk, A. F., & Cannon, S. C. (2002). Slow inactivation does not block the aqueous accessibility to the outer pore of voltage-gated Na^+ channels. *Journal of General Physiology, 120*, 509–516.

Stühmer, W., Conti, F., Suzuki, H., Wang, X. D., Noda, M., Yahagi, N., … Numa, S. (1989). Structural parts involved in activation and inactivation of the sodium channel. *Nature, 339*, 597–603.

Sunami, A., Glaaser, I. W., & Fozzard, H. A. (2001). Structural and gating changes of the sodium channel induced by mutation of a residue in the upper third of IVS6, creating an external access path for local anesthetics. *Molecular Pharmacology, 59*, 684–691.

Sutkowski, E. M., & Catterall, W. A. (1990). Beta 1 subunits of sodium channels. Studies with subunit- specific antibodies. *Journal of Biological Chemistry, 265*, 12393–12399.

Takahashi, M. P., & Cannon, S. C. (1999). Enhanced slow inactivation by V445M: a sodium channel mutation associated with myotonia. *Biophysical Journal, 76*, 861–868.

Tang, L., Chehab, N., Wieland, S. J., & Kallen, R. G. (1998). Glutamine substitution at alanine1649 in the S4-S5 cytoplasmic loop of domain 4 removes the voltage sensitivity of fast inactivation in the human heart sodium channel. *Journal of General Physiology, 111*, 639–652.

Terlau, H., Heinemann, S. H., Stühmer, W., Pusch, M., Conti, F., Imoto, K., & Numa, S. (1991). Mapping the site of block by tetrodotoxin and saxitoxin of sodium channel II. *FEBS Letters, 293*, 93–96.

Theile, J. W., & Cummins, T. R. (2011). Inhibition of Navbeta4 peptide-mediated resurgent sodium currents in Nav1.7 channels by carbamazepine, riluzole, and anandamide. *Molecular Pharmacology, 80*, 724–734.

Tikhonov, D. B., & Zhorov, B. S. (2005). Modeling P-loops domain of sodium channel: homology with potassium channels and interaction with ligands. *Biophysical Journal, 88*, 184–197.

Tikhonov, D. B., & Zhorov, B. S. (2011). Possible roles of exceptionally conserved residues around the selectivity filters of sodium and calcium channels. *Journal of Biological Chemistry, 286*, 2998–3006.

Tikhonov, D. B., & Zhorov, B. S. (2012). Architecture and pore block of eukaryotic voltage-gated sodium channels in view of NavAb bacterial sodium channel structure. *Molecular Pharmacology, 82*, 97–104.

Todt, H., Dudley, S. C., Jr., Kyle, J. W., French, R. J., & Fozzard, H. A. (1999). Ultra-slow inactivation in μ1 Na^+ channels is produced by a structural rearrangement of the outer vestibule. *Biophysical Journal, 76*, 1335–1345.

Tomaselli, G. F., Chiamvimonvat, N., Nuss, H. B., Balser, J. R., Perez-Garcia, M. T., Xu, R. H., … Marban, E. (1995). A mutation in the pore of the sodium channel alters gating. *Biophysical Journal, 68*, 1814–1827.

Townsend, C., & Horn, R. (1997). Effect of alkali metal cations on slow inactivation of cardiac Na^+ channels. *Journal of General Physiology, 110*, 23–33.

Tsai, C. J., Tani, K., Irie, K., Hiroaki, Y., Shimomura, T., McMillan, D. G., ... Li, X. D. (2013). Two alternative conformations of a voltage-gated sodium channel. *Journal of Molecular Biology, 425*, 4074–4088.

Tsang, S. Y., Tsushima, R. G., Tomaselli, G. F., Li, R. A., & Backx, P. H. (2005). A multifunctional aromatic residue in the external pore vestibule of Na^+ channels contributes to the local anesthetic receptor. *Molecular Pharmacology, 67*, 424–434.

Tsushima, R. G., Li, R. A., & Backx, P. H. (1997a). Altered ionic selectivity of the sodium channel revealed by cysteine mutations within the pore. *Journal of General Physiology, 109*, 463–475.

Tsushima, R. G., Li, R. A., & Backx, P. H. (1997b). P-loop flexibility in Na^+ channel pores revealed by single- and double-cysteine replacements. *Journal of General Physiology, 110*, 59–72.

Ulbricht, W. (2005). Sodium channel inactivation: molecular determinants and modulation. *Physiological Reviews, 85*, 1271–1301.

Valenzuela, C., & Bennett, P. B., Jr. (1994). Gating of cardiac Na^+ channels in excised membrane patches after modification by alpha-chymotrypsin. *Biophysical Journal, 67*, 161–171.

Van, P. F., Lobo, P. A., & Ahern, C. A. (2012). Seeing the forest through the trees: towards a unified view on physiological calcium regulation of voltage-gated sodium channels. *Biophysical Journal, 103*, 2243–2251.

Vassilev, P., Scheuer, T., & Catterall, W. A. (1989). Inhibition of inactivation of single sodium channels by a site- directed antibody. *Proceedings of the National Academy of Sciences of the United States of America, 86*, 8147–8151.

Vassilev, P. M., Scheuer, T., & Catterall, W. A. (1988). Identification of an intracellular peptide segment involved in sodium channel inactivation. *Science, 241*, 1658–1661.

Vedantham, V., & Cannon, S. C. (1998). Slow inactivation does not affect movement of the fast inactivation gate in voltage-gated Na^+ channels. *Journal of General Physiology, 111*, 83–93.

Vedantham, V., & Cannon, S. C. (2000). Rapid and slow voltage-dependent conformational changes in segment IVS6 of voltage-gated Na^+ channels. *Biophysical Journal, 78*, 2943–2958.

Venkatesan, K., Liu, Y., & Goldfarb, M. (2014). Fast-onset long-term open-state block of sodium channels by A-type FHFs mediates classical spike accommodation in hippocampal pyramidal neurons. *Journal of Neuroscience: The Official Journal of the Society for Neuroscience, 34*, 16126–16139.

Vilin, Y. Y., Fujimoto, E., & Ruben, P. C. (2001). A single residue differentiates between human cardiac and skeletal muscle Na^+ channel slow inactivation. *Biophysical Journal, 80*, 2221–2230.

Vilin, Y. Y., Makita, N., George, A. L. J., & Ruben, P. C. (1999). Structural determinants of slow inactivation in human cardiac and skeletal muscle sodium channels. *Biophysical Journal, 77*, 1384–1393.

Vilin, Y. Y., Peters, C. H., & Ruben, P. C. (2012). Acidosis differentially modulates inactivation in $Na_V1.2$, $Na_V1.4$, and $Na_V1.5$ channels. *Frontiers in Pharmacology, 3*, 109.

Vilin, Y. Y., & Ruben, P. C. (2001). Slow inactivation in voltage-gated sodium channels — molecular substrates and contributions to channelopathies. *Cell Biochemistry and Biophysics, 35*, 171–190.

Wang, C., Chung, B. C., Yan, H., Wang, H. G., Lee, S. Y., & Pitt, G. S. (2014). Structural analyses of Ca^{2+}/CaM interaction with Na_V channel C-termini reveal mechanisms of calcium-dependent regulation. *Nature Communications, 5*, 4896.

Wang, C., Wang, C., Hoch, E. G., & Pitt, G. S. (2011). Identification of novel interaction sites that determine specificity between fibroblast growth factor homologous factors and voltage-gated sodium channels. *Journal of Biological Chemistry, 286*, 24253–24263.

Wang, S., & Wang, G. K. (1997). A mutation in segment I-S6 alters slow inactivation of sodium channels. *Biochemical Journal, 72*, 1633–1640.

Wang, S. Y., Bonner, K., Russell, C., & Wang, G. K. (2003). Tryptophan scanning of D1S6 and D4S6 C-termini in voltage-gated sodium channels. *Biophysical Journal, 85*, 911–920.

Webb, J., & Cannon, S. C. (2008). Cold-induced defects of sodium channel gating in atypical periodic paralysis plus myotonia. *Neurology, 70*, 755–761.

Webb, J., Wu, F. F., & Cannon, S. C. (2009). Slow inactivation of the NaV1.4 sodium channel in mammalian cells is impeded by co-expression of the beta1 subunit. *Pflugers Archiv, 457*, 1253–1263.

West, J. W., Patton, D. E., Scheuer, T., Wang, Y., Goldin, A. L., & Catterall, W. A. (1992). A cluster of hydrophobic amino acid residues required for fast Na^+-channel inactivation. *Proceedings of the National Academy of Sciences of the United States of America, 89*, 10910–10914.

Wingo, T. L., Shah, V. N., Anderson, M. E., Lybrand, T. P., Chazin, W. J., & Balser, J. R. (2004). An EF-hand in the sodium channel couples intracellular calcium to cardiac excitability. *Nature Structural & Molecular Biology, 11*, 219–225.

Wingo, J. J., & Isom, L. L. (2016). Developmental and regulatory functions of Na^+ channel non-pore-forming β subunits. In R. J. French, & S. Noskov (Eds.), *Na Channel* (Vol. 78).

Wong, H. K., Sakurai, T., Oyama, F., Kaneko, K., Wada, K., Miyazaki, H., … Nukina, N. (2005). Beta Subunits of voltage-gated sodium channels are novel substrates of beta-site amyloid precursor protein-cleaving enzyme (BACE1) and gamma-secretase. *Journal of Biological Chemistry, 280*, 23009–23017.

Wu, J., Yan, Z., Li, Z., Yan, C., Lu, S., Dong, M., & Yan, N. (2015). Structure of the voltage-gated calcium channel Cav1.1 complex. *Science, 350*, aad2395.

Wu, M., Ye, N., Sengupta, B., & Zakon, H. H. (2013). A naturally occurring amino acid substitution in the voltage-dependent sodium channel selectivity filter affects channel gating. *Journal of Comparative Physiology. A, Neuroethology, Sensory, Neural, and Behavioral Physiology, 199*, 829–842.

Xiong, W., Farukhi, Y. Z., Tian, Y., DiSilvestre, D., Li, R. A., & Tomaselli, G. F. (2006). A conserved ring of charge in mammalian Na^+ channels: a molecular regulator of the outer pore conformation during slow inactivation. *Journal of Physiology, 576*, 739–754.

Xiong, W., Li, R. A., Tian, Y., & Tomaselli, G. F. (2003). Molecular motions of the outer ring of charge of the sodium channel: do they couple to slow inactivation? *Journal of General Physiology, 122*, 323–332.

Yang, Y. C., Hsieh, J. Y., & Kuo, C. C. (2009). The external pore loop interacts with S6 and S3-S4 linker in domain 4 to assume an essential role in gating control and anticonvulsant action in the Na^+ channel. *Journal of General Physiology, 134*, 95–113.

Yellen, G., Sodickson, D., Chen, T. Y., & Jurman, M. E. (1994). An engineered cysteine in the external mouth of a K^+ channel allows inactivation to be modulated by metal binding. *Biophysical Journal, 66*, 1068–1075.

Yu, F. H., Westenbroek, R. E., Silos-Santiago, I., McCormick, K. A., Lawson, D., Ge, P., … Curtis, R. (2003). Sodium channel beta4, a new disulfide-linked auxiliary subunit with similarity to beta2. *Journal of Neuroscience, 23*, 7577–7585.

Zarrabi, T., Cervenka, R., Sandtner, W., Lukacs, P., Koenig, X., Hilber, K., … Todt, H. (2010). A molecular switch between the outer and the inner vestibules of the voltage-gated Na^+ channel. *Journal of Biological Chemistry, 285*, 39458–39470.

Zhang, X., Ren, W., DeCaen, P., Yan, C., Tao, X., Tang, L., … Yan, N. (2012). Crystal structure of an orthologue of the NaChBac voltage-gated sodium channel. *Nature, 486*, 130–134.

Zhao, J., O'Leary, M. E., & Chahine, M. (2011). Regulation of Nav1.6 and Nav1.8 peripheral nerve Na^+ channels by auxiliary beta-subunits. *Journal of Neurophysiology, 106*, 608–619.

Zimmer, T., & Benndorf, K. (2002). The human heart and rat brain IIA Na^+ channels interact with different molecular regions of the beta1 subunit. *Journal of General Physiology, 120*, 887–895.

CHAPTER FOURTEEN

Current—Voltage Relationship for Late Na^+ Current in Adult Rat Ventricular Myocytes

R.B. Clark and W.R. Giles[1]
University of Calgary, Calgary, AB, Canada
[1]Corresponding author: E-mail: wgiles@ucalgary.ca

Contents

Abstract

It is now well established that the slowly inactivating component of the Na^+ current ($I_{Na\text{-}L}$) in the mammalian heart is a significant regulator of the action potential waveform. This insight has led to detailed studies of the role of $I_{Na\text{-}L}$ in a number of important and challenging pathophysiological settings. These include genetically based ventricular arrhythmias (LQT 1, 2, and 3), ventricular arrhythmias arising from progressive cardiomyopathies (including diabetic), and proarrhythmic abnormalities that develop during local or global ventricular ischemia. Inhibition of $I_{Na\text{-}L}$ may also be a useful strategy for management of atrial flutter and fibrillation.

Many important biophysical parameters that characterize $I_{Na\text{-}L}$ have been identified; and $I_{Na\text{-}L}$ as an antiarrhythmia drug target has been studied extensively. However,

Current Topics in Membranes, Volume 78
ISSN 1063-5823
http://dx.doi.org/10.1016/bs.ctm.2016.04.002

relatively little information is available regarding (1) the ion transfer or current–voltage relationship for $I_{Na\text{-}L}$ or (2) the time course of its reactivation at membrane potentials similar to the resting or diastolic membrane potential in mammalian ventricle.

This chapter is based on our preliminary findings concerning these two very important physiological/biophysical descriptors for $I_{Na\text{-}L}$. Our results were obtained using whole-cell voltage clamp methods applied to enzymatically isolated rat ventricular myocytes. A chemical agent, BDF 9148, which was once considered to be a drug candidate in the Na^+-dependent inotropic agent category has been used to markedly enhance $I_{Na\text{-}L}$ current. BDF acts in a potent, selective, and reversible fashion. These BDF 9148 effects are compared and contrasted with the prototypical activator of $I_{Na\text{-}L}$, a sea anemone toxin, ATX II.

List of Abbreviations

APD	Action potential duration
ATX II	Sea anemone toxin
BDF 9148	Na^+-dependent inotropic agent
DMSO	Dimethyl sulfoxide
$I_{Na\text{-}L}$	Late sodium current
$[Na^+]_i$	Intracellular sodium
TTX	Tetrodotoxin

1. INTRODUCTION

In mammalian heart tissues there are several different indirect cellular mechanisms by which contractility may be enhanced (Aronsen, Swift, & Sejersted, 2013; Murphy & Eisner, 2009). All of these ultimately involve an increase in intracellular calcium $[Ca^{2+}]_i$ or altered sensitivity to Ca^{2+} of intracellular target proteins (Baczko, Giles, & Light, 2003; Shattock & Bers, 1989). For example, alkalinization of intracellular pH, which enhances the Ca^{2+}-sensitivity of the myofilaments, can significantly increase the strength of contraction. Increases in intracellular Na^+, ($[Na^+]_i$), can also result in augmented contractility of mammalian heart tissue (Aronsen et al., 2013; Baczko et al., 2003) and within the last 10–15 years this mechanism has received significant attention (Bers, Barry, & Despa, 2003; Murphy & Eisner, 2009). The related positive inotropic effects in ventricular tissue in vitro develop with little or no change in heart rate, which suggested that therapeutic agents designed to enhance $[Na^+]_i$ could be effective in the treatment of heart failure (Doggrell, Bishop, & Brosch, 1995; Flesch & Erdmann, 2001; Hoey, Amos, & Ravens, 1994b; Makielski & Farley, 2006; Müller-Ehmsen, Brixius, & Schwinger, 1998; Müller-Ehmsen, Näbauer, & Schwinger, 1999; Schiffmann et al., 2003; Schwinger, Böhm, Mittmann, La

Rosée, & Erdmann, 1991; Schwinger, Müller-Ehmsen, Böhm, & Erdmann, 1996).

Several classes of compounds are known to increase $[Na^+]_i$. One such agent, veratridine, augments Na^+ influx by slowing the rate of fast inactivation of the tetrodotoxin (TTX)-sensitive Na^+ current that in adult myocytes is generated mainly by $Na_V1.5$ (Doggrell et al., 1995; Vassalle, Bocchi, & Du, 2007; Verdonck, Bielen, & Ver Donck, 1991). In addition, in mammalian hearts a number of different toxins, such as Anthopleura toxin A, markedly enhance $[Na^+]_i$ (Belardinelli, Giles, Rajamani, Karagueuzian, & Shryock, 2015; Belardinelli, Shryock, & Fraser, 2006; Benndorf, 1994; Yang et al., 2015; Zaza, Belardinelli, & Shryock, 2008). A pharmacological agent, DPI 201-106, that also dramatically slows the rate of inactivation of cardiac Na^+ channels and thus lengthens action potential duration (APD), was shown to result in significant increases of $[Na^+]_i$ in rat, guinea pig, and human cardiac tissue (Cheng, & Incardona, 2009; Herzig, Lilienthal, & Mohr, 1991; Hoey et al., 1994a; Wang, Lin, Lin, Peng, & Wu, 2008). A structurally related compound, BDF 9148, was reported to have similar effects, although a full analysis of the biophysical mechanism(s) of responsible for its actions on mammalian cardiac Na^+ channels has not been published previously (Hoey et al., 1994a; Müller-Ehmsen et al., 1998, 1999; Nand & Doggrell, 1999; Schwinger et al., 1996; Yuill, Convery, Dooley, Doggrell, & Hancox, 2000).

Thus, even a decade ago it was apparent that agents which slow the rate of inactivation of the Na^+ current can cause a positive inotropic effect (Attwell, Cohen, Eisner, Ohba, & Ojeda, 1979, Carmeliet, 1987; Gintant, Datyner, & Cohen, 1984; Kiyouse & Arita, 1989; Makielski & Farley, 2006; Saint, Ju, & Gage, 1992). However, the cellular mechanism(s) remained unclear. Both classical data and much more recent findings have focused on a putative increase in $[Na^+]_i$, and related, secondary reduction in the activity of the Na^+/Ca^{2+} exchanger as being key mechanistic steps in ultimately increasing $[Ca^{2+}]_i$ (Baczko et al., 2003; Bouchard, Clark, & Giles, 1993b; Clark, Bouchard, & Giles, 1996; Ramirez, Sah, Liu, Rose, & Backx, 2011). The presently accepted rationale for the positive inotropic actions of an increase in $[Na^+]_i$ can be summarized as follows: by reducing the Na^+ electrochemical gradient an increase of $[Na^+]_i$ can decrease the forward mode, or possibly even reverse Na^+/Ca^{2+} exchange, transport, thus resulting in an increase of intracellular $[Ca^{2+}]_i$. Indeed, several comprehensive studies have implicated the electrogenic Na^+/Ca^{2+} exchanger as the main mechanism by which agents that slow the inactivation of the Na^+ current can prolong APD and cause a significant positive inotropic effect (Benndorf, 1994; Bouchard, Clark, & Giles, 1993a, 1995; Fernandes et al., 2014).

Within the past 15 years electrophysiological studies in mammalian hearts (including human) have focused on a small component of the Na^+ current that exhibits very slow inactivation (Attwell et al., 1979; Carmeliet, 1987; Ju, Saint, & Gage, 1994; Liu, DeFelice, & Mazzanti, 1992; Patlak & Ortiz, 1985; Saint et al., 1992; Valdivia et al., 2005). This late Na^+ current $I_{Na\text{-}L}$ is known to increase significantly in a broad range of genetically induced or acquired pathologies (Belardinelli et al., 2015, 2006; Nagatomo et al., 1998; Sossalla et al., 2008). The corresponding enhanced Na^+ influx lengthens the action potential and often results in a proarrhythmic substrate, perhaps due to the increase in $[Na^+]_i$ and then altered intracellular $[Ca^{2+}]$ homeostasis (Belardinelli et al., 2006; Benndorf, 1994; Carmeliet, 2006a; Clark et al., 1996). The appearance of alternans in APD (Carmeliet, 2006b)—and/or abnormal early or delayed afterdepolarizations (EADs or DADs) in both atria and ventricles is common (Fedida, Peter, Hesketh & Ezrin, 2006; Makielski & Farley, 2006; Stump, Wallace, Gilberto, Gehret, & Lynch, 2000; Yang et al., 2015).

To better understand the cellular mechanism(s) by which $I_{Na\text{-}L}$ is proarrhythmic in mammalian ventricular myocytes, it is necessary to have detailed information concerning its ion transfer or current—voltage (*I*- *V*) relationship over the entire physiological range of membrane potentials. In addition, the duty cycle and kinetic properties of this current need to be studied in detail. That is, it is necessary to characterize not only the steady-state inactivation of $I_{Na\text{-}L}$, but also to have information concerning its kinetic properties of recovery at membrane potentials similar to the resting or diastolic membrane potential in mammalian ventricle (Baczko et al., 2003; Carmeliet, 1987; Kleber, 1983; Makielski & Farley, 2006).

In the experiments in isolated myocytes from adult rat ventricles that form the basis for this chapter, the effects of BDF 9148 on the action potential waveform and unloaded cell shortening have been studied in detail. Rat ventricle preparations were chosen since the electrophysiological and mechanical consequences of changes in action potential waveform and Na^+/Ca^{2+} exchange in these myocytes are well characterized (Baczko et al., 2003; Bouchard et al., 1993b; Kiyouse & Arita, 1989; Pandit, Clark, Giles, & Demir, 2001). Additionally, comprehensive sets of data are available on the cellular mechanism(s) that modulate intracellular Ca^{2+} transients. Based on this information, our experiments were designed to (1) confirm that BDF 9148 acted on TTX-sensitive Na^+ current, (2) characterize the main changes in the macroscopic Na^+ current waveforms, (3) record the ion transfer or I-V relation for the I_{Na-L} over the entire physiological range of membrane potentials, and (4) analyze the reactivation kinetics of I_{Na-L} at the resting or diastolic membrane potential of mammalian ventricular myocytes. These findings provide important new insights into (1) whether the positive inotropic effect of BDF 9148 results mainly from an augmentation of $[Na^+]_i$, (2) from the BDF-induced action potential prolongation, or (3) both of these factors.

2. METHODS

2.1 Solutions and chemicals

Standard HEPES-buffered Tyrode solution contained (mM): NaCl, 140; KCl, 5; $CaCl_2$, 1; $MgCl_2$ 1, HEPES, 10; glucose, 5.5. "Low calcium" solution was prepared by omitting $CaCl_2$ from this standard Tyrode solution. The "enzymatic" solution used for cell dissociation contained (mM): NaCl, 130; KCl, 5, $MgCl_2$, 1; HEPES, 10; glucose, 5.5; collagenase (Type I; Sigma Chemical Co., St. Louis, MO, USA), 0.8 mg/mL; protease (type XIV; Sigma Chem. Co.) 60 μg/mL and taurine, 10. The "washout" solution was a modification of standard Tyrode solution in which $CaCl_2$ was reduced to 100 μM and 1 mg/mL bovine serum albumin (type V; Sigma Chem. Co.) was added. All solutions were gassed with 100% O_2, and pH was adjusted to 7.4 by addition of appropriate amounts of 1.0 M NaOH.

The external solution used in most action potential and voltage-clamp recordings was the same as standard HEPES-buffered Tyrode, except where noted in figure legends. In some experiments, tetrodotoxin (TTX; Sigma Chem. Co.) was added at a concentration of 15 μM. The patch pipette solution contained (mM): K-aspartate, 120; KCl, 30; $MgSO_4$, 1; HEPES, 10; Na-ATP, 5. The pH of the solution was adjusted to 7.2 with appropriate amounts of 1 M KOH. In some experiments, K-aspartate and KCl in the pipette solution were replaced by Cs-aspartate (120 mM) and CsCl (30 mM); pH was adjusted to 7.2 with CsOH. In these experiments, KCl in the external solution was also removed and replaced with 3 mM CsCl.

The compound BDF 9148 was obtained in 2002 from Dr. M. Steinberg then, of Eli Lilly Research Laboratories, Indianapolis, USA. A stock solution (10^{-3} M) was prepared using 100% dimethyl sulfoxide (DMSO) as a solvent. At a final concentration of 1 μM BDF 9148, the DMSO concentration was

0.1%. This concentration of DMSO alone had no effect on the electrophysiology properties of those ventricular myocytes.

2.2 Myocyte dissociation procedure

The dissociation method was similar to that described by Clark et al (Bouchard et al., 1993b, 1995; Bouchard, Clark, Juhasz, & Giles, 2004; Clark et al., 1996). Each rat weighing 250–350 g was killed by cervical dislocation. The heart was immediately removed and mounted on a modified Langendorff perfusion system for retrograde perfusion of the coronary circulation. Temperature was maintained at $36 \pm 1°C$ throughout all perfusion steps. An initial 5-min period of perfusion with standard HEPES-buffered Tyrode solution was used to remove as much blood as possible from the heart. Three further steps of perfusion were as follows: 7 min with "low Ca" solution, 15 min with "enzymatic" solution, and 5 min with "washout" solution. The heart was then removed from the perfusion system, and small pieces of ventricle were cut from the heart and placed in a flask containing "washout" solution. These tissue fragments were gently stirred until they were dissociated into single myocytes.

2.3 Electrophysiological methods

A drop of myocyte suspension was placed in a 200-μL superfusion chamber mounted on the stage of an inverted microscope (Diaphot; Nikon, Tokyo, Japan), and then were left for 5–10 min to allow them to settle onto the bottom of the chamber. Thereafter, the myocytes were superfused (at about 1 mL/min) at room temperature (19–21°C) with either control or BDF-containing solutions.

Electrophysiological recordings were made from single ventricular myocytes using standard whole-cell patch-clamp methods. Patch electrodes with resistances of 1–2 MΩ were made from borosilicate glass (World Precision Inst., Sarasota, FL, USA). Voltage and current-clamp experiments were carried out using a conventional patch-clamp amplifier (EPC-7; List Electronic, Medical Systems Corp., Greenvale, NY, USA). Membrane potentials were corrected by −10 mV to compensate for liquid junction potentials between external and pipette solutions. Series resistance in the whole-cell mode was 5–8 MΩ; 60–80% of this resistance was compensated electronically. Cell capacitance was usually not compensated since no transient transmembrane currents were studied. Data acquisition was carried out with a 12-bit A/D convertor board (DT2801A; Data Translation, Marlborough, MA, USA), which was controlled by custom software ("CELLSOFT," Mr. D. Bergman,

University of Calgary), and acquired data were stored on hard disk in a microcomputer for later analysis. A commercial graphing program ("Sigmaplot," Jandel Scientific, Corta Madera, CA, USA) was used for data plotting and nonlinear curve fitting.

In some experiments the contractions of one end of the myocyte ("unloaded cell shortening") were measured using a video edge detector device (Steadman, Moore, Spitzer, & Bridge, 1988), and these cell shortening results were digitized together with membrane potential and current records.

3. RESULTS

3.1 Effects of BDF 9148 on rat ventricular myocytes

Fig. 1A shows a representative example of the effect of BDF 9148 (1 μM) on the action potential, and accompanying unloaded cell shortening, of an isolated rat ventricular myocyte stimulated at 0.1 Hz. As expected for this myocyte preparation, the action potential in control conditions is short, with durations at 50% and 90% repolarization of about 18 and 55 ms, respectively. After application of BDF, the action potential waveform developed a

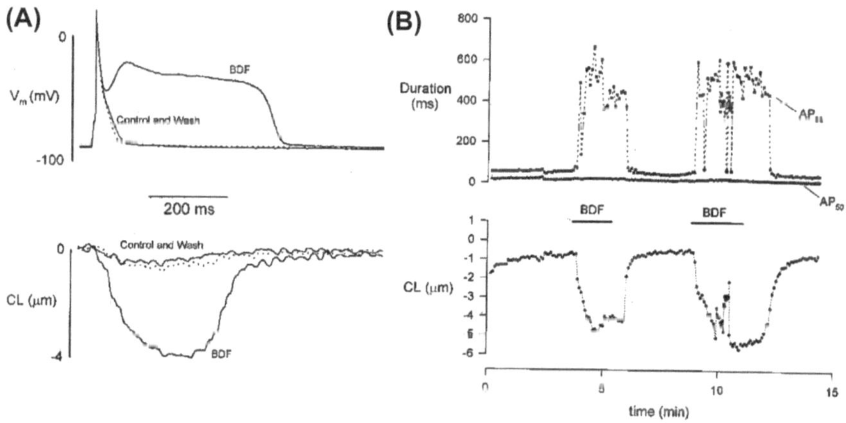

Figure 1 Effect of BDF 9148 (1 μM) on rat ventricular myocyte action potentials and contractions measured as unloaded cell shortening. Stimuli (5 ms, 0.9 nA) were applied at 0.2 Hz. Note that the action potentials were prolonged markedly after 75 s exposure to BDF (Panel A, top), and a large positive inotropic response "accompanied" this action potential prolongation (Panel A, bottom). These very significant drug-induced effects were reversible on washout of BDF (*dashed lines*). Time course of changes in action potential duration (50% and 90% repolarization: top) and unloaded cell shortening (bottom) on exposure to and washout of 1 μM BDF 9148 are shown in Panel B.

pronounced "spike and dome" configuration. Note that the initial phase of repolarization of the action potential was altered only slightly by BDF. In contrast, the final phase of repolarization was prolonged markedly. The resulting action potential had a very long plateau at membrane potentials in the range −20 to −40 mV. For the example shown in Fig. 1, the APD at 50% repolarization increased by only 15−20% after application of BDF, while the 90% duration increased by about 10-fold. Resting membrane potential was not significantly (<1 mV) changed by BDF application. Importantly, a large positive inotropic response, as judged by unloaded cell shortening, accompanied this increase in APD (Fig. 1). These effects of BDF on both action potential shape and unloaded cell shortening were reversible on washout of this compound. Fig. 1B illustrates the time course of the changes in APD and associated unloaded cell shortening produced by BDF. As shown, onset of the effects of BDF on APD occurred rapidly (<30 s) after introduction of the drug. In contrast, the increases in the amplitude of unloaded cell shortening developed more slowly than the change in the lengthening of the APD. This positive inotropic effect also declined (washed out) much more slowly than APD, after removal of BDF (see Fig. 6).

The exceptionally long plateau of the action potential in the presence of BDF suggested that BDF might produce a "persistent" inward Na^+ current, as had been reported previously in this preparation (Hoey et al., 1994b; Trenor et al., 2012). This was confirmed in the voltage-clamp experiments, shown in Fig. 2. This ventricular myocyte was voltage-clamped at holding potential of −80 mV. Based on studies of BDF 9148 done in guinea pig myocytes (Wang et al., 2008), 350 ms steps over a range of potentials from −110 to +50 mV were applied at 0.1 Hz. Fig. 2A shows examples of membrane currents produced by three different voltage-clamp steps (to −50, −40, and −20 mV) in control conditions, and then about 2 min after application of 1 μM BDF 9148. As expected, in control conditions, each of the voltage-clamp steps produced a large, rapidly activating inward Na^+ current that completely inactivated in less than 30 ms. In contrast, in the presence of BDF, the Na^+ current in response to the same series of voltage-clamp steps exhibited of two very distinct phases: (1) an initial rapidly activating and inactivating component, and the following (2) slowly inactivating inward component which decayed in amplitude by about 50% during the 350 ms voltage-clamp steps. Fig. 2B shows the current−voltage (*I-V*) relation for net membrane current measured at the end of the 350 ms voltage-clamp steps, over the potential range −110 to +50 mV. Note that

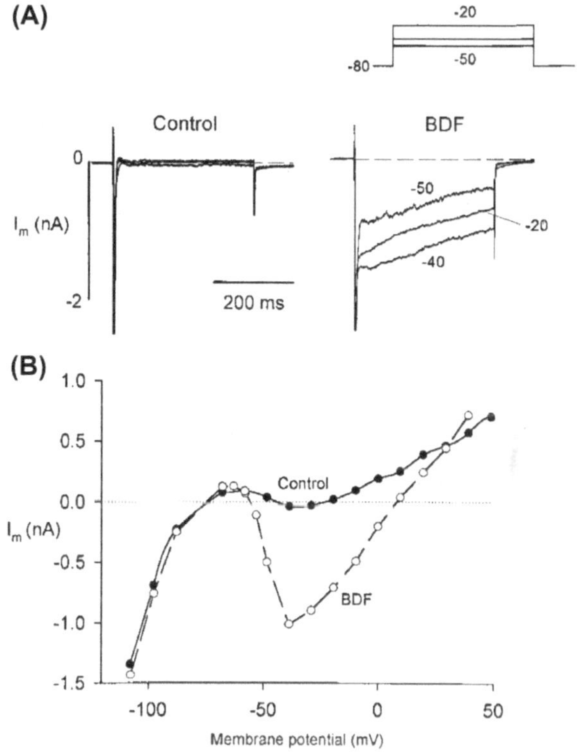

Figure 2 Effects of BDF 9148 (1 μM) on membrane currents recorded from a voltage-clamped rat ventricular myocyte. In Panel A the voltage-clamp protocol is shown in the inset; voltage steps were applied at 0.1 Hz. BDF induced a large "late" inward current component. Note that the large initial transient inward component of I_{Na} is truncated in this and all subsequent plots. Panel B shows the current–voltage relations for the control and BDF-induced currents. The myocyte was held at −80 mV, and 350 ms voltage-clamp steps from −110 mV to +40 mV were applied at 0.1 Hz. Membrane current was measured at the end of each voltage-clamp step.

BDF did not change membrane currents at potentials negative to about −60 mV; this current is primarily the inward rectifier K^+ current, I_{K1} (Bouchard et al., 2004). Lack of change in this current is consistent with the absence of BDF effects on resting membrane potential (Baczko et al., 2003; Bouchard et al., 1995; Yang et al., 2015). At potentials positive to −60 mV, BDF induced a pronounced and very slowly inactivating inward current that reached its peak magnitude at about −40 mV. In summary, the *I-V* in Fig. 2B shows that the BDF-induced current is relatively large in the same range of membrane potentials at which the very long plateau of the action potentials was observed.

Fig. 3A illustrates an example of the time course of development of this "late" inward Na^+ current (denoted $I_{Na\text{-}L}$) produced by BDF in a rat ventricular myocyte. $I_{Na\text{-}L}$ was measured at the end of 100 ms voltage-clamp steps to −30 mV. This increase developed completely within about 2 min after application of BDF (1 μM), while reversal of these BDF effects following removal of the drug was somewhat slower. The selected current records in Fig. 3B, which were taken from the data of Fig. 3A, show that addition of 15 μM TTX (in the presence of BDF) blocked both the initial fast and "late" components of membrane current. This consistent result confirms that the slowly inactivating component is also a Na^+ current.

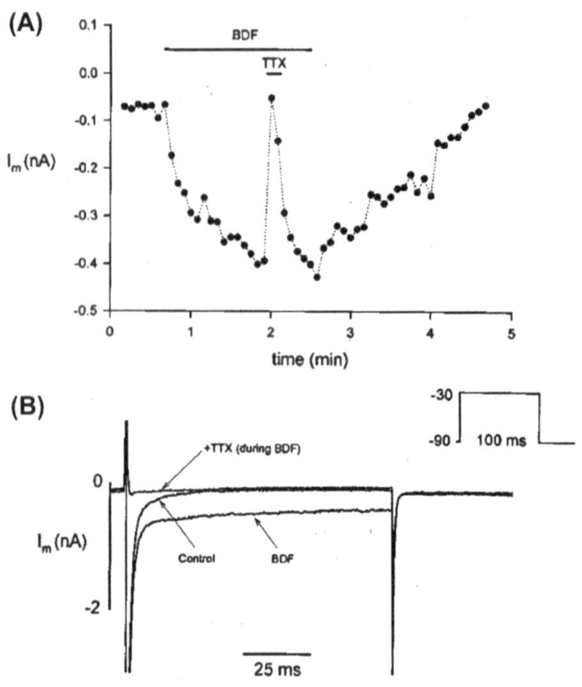

Figure 3 Time course of development of BDF-induced membrane current and demonstration that this Na^+ current is blocked by tetrodotoxin (TTX). In Panel A, the current at the end of each 100 ms voltage-clamp step to −30 mV (holding potential −90 mV) is plotted versus time. Application of 15 μM TTX in the presence of BDF completely abolished the BDF-induced current. Note also that this BDF-induced current returned toward baseline values after removal of this compound. Panel B shows representative $I_{Na\text{-}L}$ current records taken before application of BDF (*control*), and immediately before (*BDF*) and during application of TTX (+*TTX*). As in previous figures the initial rapid transient inward current component is truncated in this plot. Pipette solution contained Cs^+ in place of K^+, and external KCl was replaced with 3 mM CsCl (see Section 2).

3.2 Inactivation and reactivation (recovery) of BDF 9148-induced $I_{Na\text{-}L}$

Despite the very slow inactivation of the BDF-induced component of Na^+ current (compared with its rapid inactivation in the absence of this drug candidate), "steady-state" inactivation of the BDF-induced current occurred over a very similar range of membrane potentials as the inactivation of the fast component of I_{Na}. This is illustrated by the superimposed current records as illustrated in Fig. 4. In this experiment the ventricular myocyte was exposed to BDF for about 2 min to allow its effects to fully develop. The myocyte was then voltage-clamped or "held" for 30 s at each of a series of different holding potentials, after which the voltage was stepped to −50 mV. At holding potential of −80 mV, the step to −50 mV produced a two component inward current. However, at holding potentials of both −70 and −60 mV, both components of the Na^+ current were strongly reduced. In fact, both were almost completely abolished, as expected from the well-established steady-state relationship for voltage-dependent inactivation of I_{Na} in rat ventricle.

Recovery of this BDF-induced current from inactivation was very slow, as illustrated in Fig. 5. The voltage-clamp protocol used in this experiment was similar to that shown in Fig. 4. This BDF-treated myocyte was

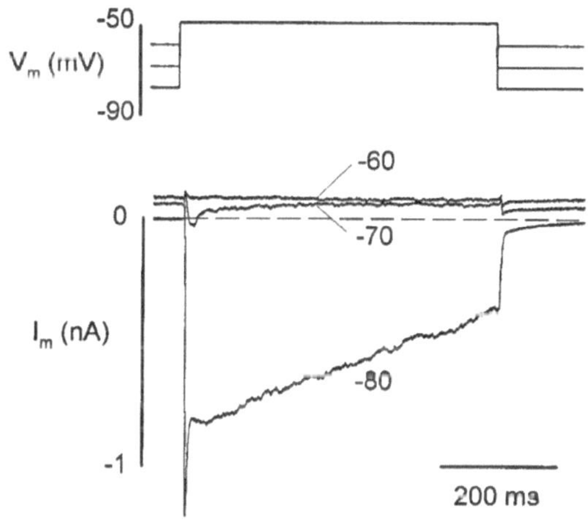

Figure 4 Inactivation of BDF 9148 induced $I_{Na\text{-}L}$ current. Voltage clamp records of $I_{Na\text{-}L}$ demonstrating that selected changes of the holding potential in the range that approximates the known diastolic membrane potentials in mammalian ventricular myocytes can markedly reduce (inactivate) $I_{Na\text{-}L}$. Again, the peak Na^+ current records are truncated. See text for details.

voltage-clamped at holding potential of −80 mV, and 150 ms test steps to −50 mV were applied at a rate of 0.1 Hz. Next, the holding potential was abruptly switched to −60 mV for about 80 s. Note that the BDF-induced current measured at the end of the 150 ms voltage-clamp steps ("test current" in Fig. 5A) was completely abolished at this holding potential (cf. Fig. 4). At the end of the 80 s depolarization to −60 mV period, the holding potential was returned to −80 mV. As demonstrated by the response to subsequent voltage-clamp steps, the magnitude of the BDF-induced current fully recovered to its initial value very slowly, taking approximately 3 min. This time course of recovery was fitted to a single exponential function with a time constant of 72 s (at −80 mV, ie, during diastole). The sample records shown in Fig. 5B, taken before the holding

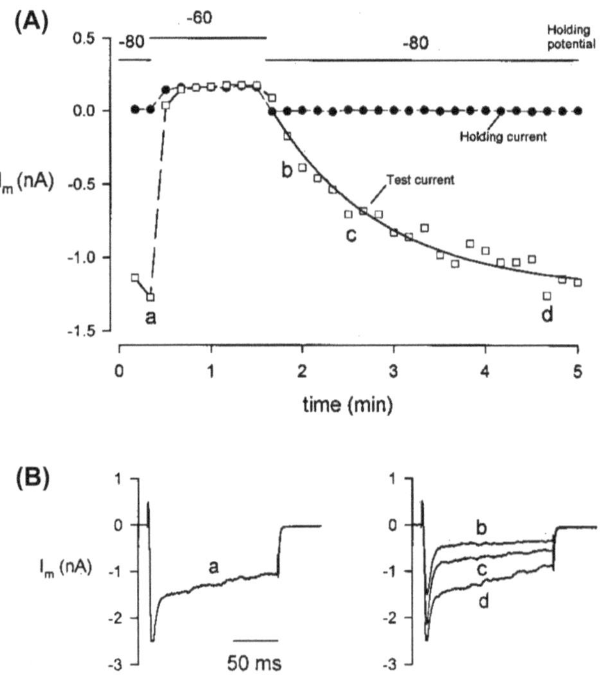

Figure 5 Time course of recovery of (1 μM) BDF-induced currents from inactivation. As shown in Panel A, the holding potential in the BDF-treated myocyte was changed from −80 mV to −60 mV. Holding currents, and current at the end of 150 ms voltage-clamp steps to −50 mV, are plotted versus time. The solid line denotes the time course of recovery from inactivation. It is a best-fit single exponential with a time constant of 72 s at −80 mV. Panel B shows representative voltage-clamp current records from Panel A.

potential was changed to −60 mV (a), and during recovery (b−d) show that the fast component of this Na^+ current recovered from inactivation considerably more quickly than the BDF-induced component.

3.3 Is BDF 9148-induced contractility entirely due to an increase in $[Na^+]_i$?

As shown in Fig. 1 the large positive inotropic effect that accompanied the BDF-induced changes in action potential develops slowly, so that most of this effect was observed well after prolongation of the action potential (Fig. 1B). Since this prolongation in APD resulted from the slowly inactivating or late component of Na^+ current that was induced by BDF, one possible explanation for this positive inotropic effect is that it resulted indirectly from an increase in internal Na^+ concentration $[Na^+]_i$ which may have occurred during this series of prolonged action potentials. Such an increase in $[Na^+]_i$ would decrease the electrochemical reversal potential for Na^+, and hence diminish the effectiveness of the Na^+/Ca^{2+} exchanger to extrude Ca^{2+} from these myocytes. In turn, this would be expected to result in increased loading of the sarcoplasmic reticulum with Ca^{2+} (Bouchard et al., 1993b, 1995, 2004; Despa & Bers, 2003; Murphy & Eisner, 2009; Shattock & Bers, 1989).

The experimental results shown in Fig. 6 suggest that this conventional explanation for the positive inotropic effect of BDF 9148 response is unlikely to be correct, at least for exposures to BDF, that last 1−5 min. In this experiment, the myocyte was voltage-clamped at holding potential of −90 mV, and 100 ms steps to −30 mV were applied at 0.2 Hz. As illustrated in previous figures, exposure to BDF caused activation of $I_{Na\text{-}L}$ during the voltage-clamp step. The two superimposed transmembrane Na^+ currents in Fig. 6 were recorded before, and 75 s after exposure of the myocyte to BDF. Note that despite the marked increase in inward current, both the magnitude and time course of unloaded cell shortening were unchanged. This pattern of results reveals that the positive inotropic effect (Fig. 1) was due to the increase in duration of the action potential rather than exclusively due to an increase in $[Na^+]_i$ (Bouchard et al., 1993b, 1995, 2004; Makielski & Farley, 2006; Murphy & Eisner, 2009).

3.4 Comparison of BDF 9148 and ATX II-induced $I_{Na\text{-}L}$

Results that we have obtained using BDF 9148 can be related to ongoing, multidisciplinary work on the late Na^+ current as an antiarrhythmic drug target by comparing electrophysiological findings in BDF 9148 with those

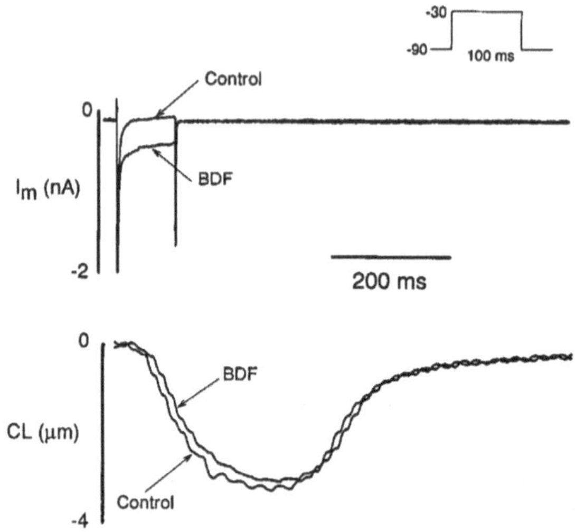

Figure 6 Effect of BDF 9148 (1 μM) on membrane current and unloaded cell shortening recorded from a voltage-clamped rat ventricular myocyte. The inset in the top panel shows the voltage-clamp protocol. Comparison of the superimposed voltage-clamp current and unloaded cell shortening records reveals that although BDF (75 s exposure) induced a substantial I_{Na-L}, unloaded cell shortening was unchanged during the time course of this experiment. Pipette solution contained Cs^+ in place of K^+; external solution contained 3 mM CsCl (see text for more detail).

obtained in the presence and after sea anemone toxin (ATX II) applications. As shown in Fig. 7 [and expected from previous published work (cf. (Belardinelli et al., 2015, 2006)], ATX II superfusion (even at 1–5 nM concentrations) markedly prolongs the rat ventricular action potential. In addition, ATX also induced both EADs and DADs (Belardinelli et al., 2006; Verdonck et al., 1991). These ATX II effects were reversible with washout of this toxin.

The superimposed action potential waveforms in Fig. 7 show/confirm even at 1.0 nM level the marked effects on the rat ventricular action potential of ATX. Note also that ATX results in pronounced EAD activity at the plateau of the action potential, and also some DADs that occur infrequently at the resting or diastolic membrane potential (Baczko et al., 2003; Belardinelli et al., 2015; Kleber, 1983; Trenor et al., 2012; Vassalle et al., 2007; Yang et al., 2015). These effects can be reversed on washout.

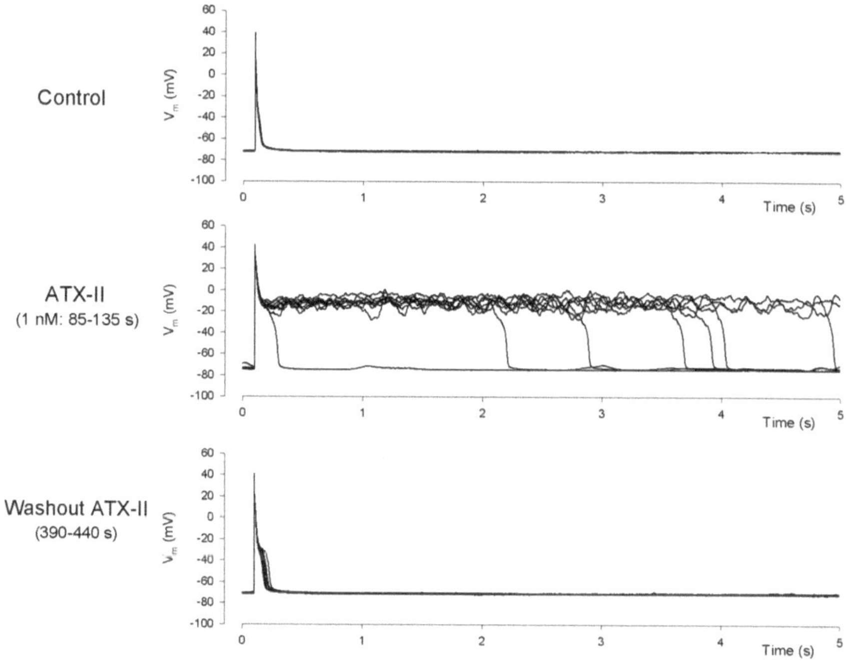

Figure 7 Effects of addition of sea anemone toxin (ATX II, 1 nM) on the action potential waveform recorded from an isolated rat ventricular myocyte stimulated at 0.01 Hz. Control (top) action potential and action potentials after washout (bottom) are shown; along with the superimposed markedly prolonged action potential waveforms recorded between 85 and 135 s after ATX application. Note that ATX induced prominent early afterdepolarization activity and also that some delayed afterdepolarizations were observed (see text for more details).

The data in Fig. 8 are from a similar ATX experiment. In this case, ranolazine (20 μM) a well-known, but perhaps a nonselective, inhibitor of $I_{Na\text{-}L}$ was used to block the effects of ATX (1 nM).

The results in Fig. 9 show $I_{Na\text{-}L}$ records in ATX (0.5 nM, Panel A) and in ATX (0.5 nM) plus ranolazine (10 μM). Note the marked activation of $I_{Na\text{-}L}$ by ATX and its partial inhibition by ranolazine (Panel C). The two superimposed *I-V* curves in Panel B illustrate the ion transfer relationship for the ATX-induced $I_{Na\text{-}L}$. The ATX II data in Fig. 9 can be compared directly with the BDF 9148 results in Fig. 2. Note that ATX induces a very similar $I_{Na\text{-}L}$ to that activated by BDF 9148. Panel B of Fig. 9 shows the ATX induced late current and its inhibition by ranolazine in the form of superimposed raw current records. Once again, note that the very large and rapid transient inward component of I_{Na} has been truncated.

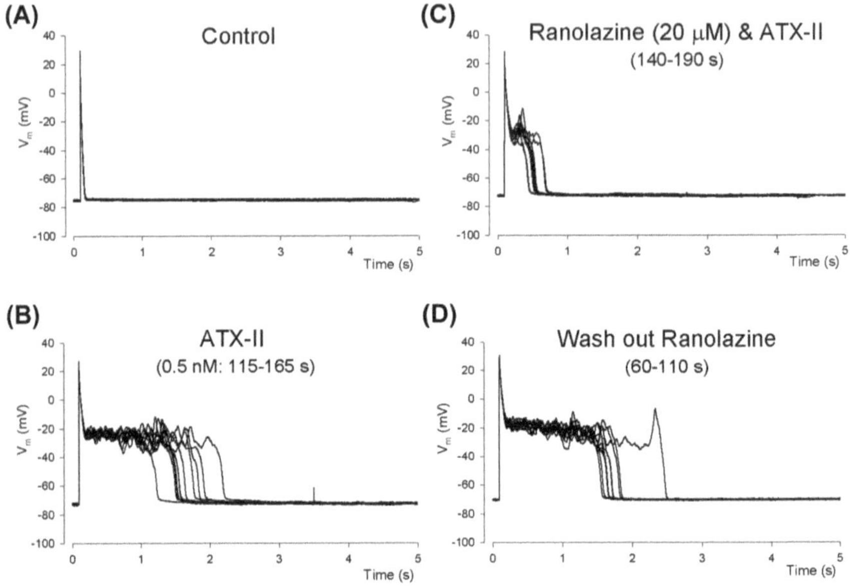

Figure 8 Confirmation that ranolazine, an antiarrhythmic drug that blocks $I_{Na\text{-}L}$ in mammalian heart can markedly inhibit the action potential prolonging effects of ATX. Numbers in parenthesis in each panel indicate time (in seconds) after toxin or drug application or washout. In this experiment, after baseline or control data were collected (A) ATX II (0.5 nM) was applied (B) and this was followed by ranolazine (20 μM) application in the presence of (ATX); (C) and then by washout of ranolazine (D).

4. DISCUSSION

In this study, we have been successful in using an agent that activates a substantial $I_{Na\text{-}L}$; and as a result we have been able to record and define the ion transfer or current–voltage (*I-V*) relationship for this Na^+ flux. As shown in Fig. 2 when $I_{Na\text{-}L}$ is separated from the much larger and faster transient inward Na^+ current (by making measurements after the transient Na^+ current has completely inactivated), a full data set defining the relative size and the voltage dependence of $I_{Na\text{-}L}$ as a function of transmembrane voltages can be obtained when either BDF 9148 or ATX II is used as the agonist. Interestingly, the resulting *I-V* curve for the $I_{Na\text{-}L}$ closely resembles that for the peak Na^+ current; and (importantly), the $I_{Na\text{-}L}$ can be inhibited completely by TTX whether it is induced by BDF (Fig. 3) or by ATX (data not shown).

Somewhat similar results have been published previously, based on classical studies of Na^+ "window current" in cardiac Purkinje fibers (Attwell

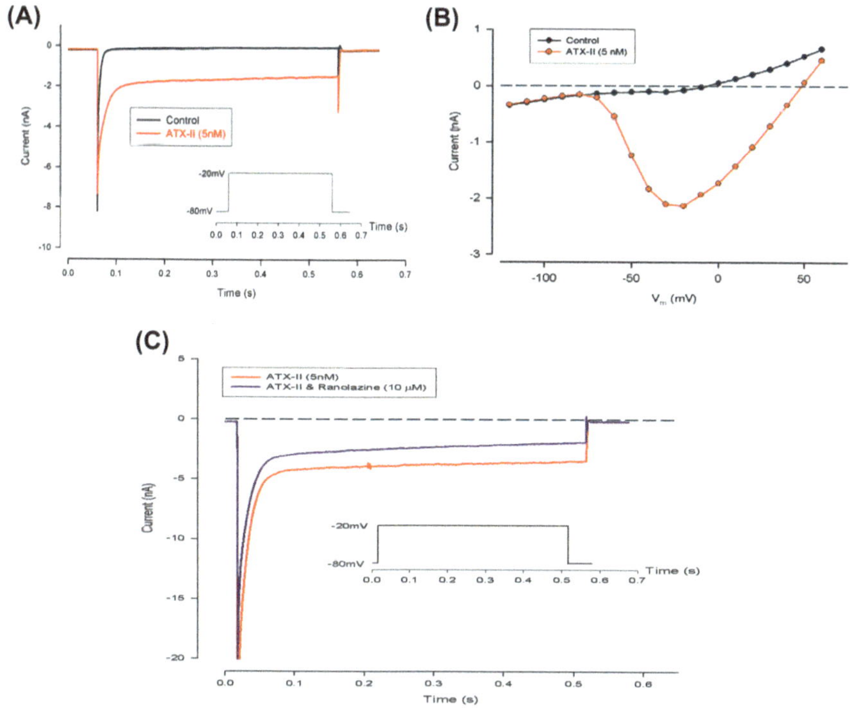

Figure 9 Illustration of block of ATX-induced $I_{Na\text{-}L}$ by ranolazine. Panel A confirms that ATX induces a significant $I_{Na\text{-}L}$. Panel B consists of two *I-V* curves constructed from measurements of Na^+ current at the end of a series of 500 ms voltage-clamp depolarizations from a holding potential of −80 mV. In these two isochronal *I-V* curves, the filled circles represent the baseline (no toxin) data; and the data denoted by open circles was obtained in the presence of ATX (5 nM). In Panel C, the two superimposed I_{Na} records show the raw data at −20 mV and illustrate the measurement of $I_{Na\text{-}L}$ and partial block by ranolazine (10 μM).

et al., 1979; Carmeliet, 1987, 2006b; Fedida et al., 2006; Gintant et al., 1984; Verdonck et al., 1991). The Carmeliet work also includes the suggestion that the reactivation or recovery processes for the peak versus the $I_{Na\text{-}L}$ may be separate and distinct from the kinetic process involved in the regulation of peak I_{Na} (Carmeliet, 1987). Since this topic is complex and remains controversial, one of our main goals was to devise a protocol, that at a minimum, would reveal the predominant recovery kinetics of the $I_{Na\text{-}L}$ at membrane potentials that are very close to the physiological diastolic potential of the tissue of our choice, rat ventricle. As shown in Fig. 5 we have succeeded in doing this. Indeed, the recovery of the $I_{Na\text{-}L}$ is approximately two orders of magnitude slower than the peak Na^+ current: the predominant time constant of recovery at −80 mV is approximately 70 s. As a consequence, this current would be expected to change very significantly if an abrupt change in heart rate occurred in any mammalian species; and this effect would be prominent in those species where heart rate was approximately 1 Hz.

More than 30 years ago, studies of the Na^+ window current in cardiac Purkinje fiber preparations also drew attention to the importance of reactivation of the Na^+ current that is now denoted $I_{Na\text{-}L}$. Specifically, these papers drew attention to the consequence of this slow reactivation of Na^+ current on APD at both short times and during prolonged stimulus trains where quasi-steady-state conditions would be achieved (Carmeliet, 2006a, 2006b). Later work, directed toward understanding the mechanism of the genetically induced cardiac rhythm disorder denoted LQT3, represents an interesting and potentially important application of this reasoning. Thus, this mammalian ventricle Na^+ current(s) (particularly when augmented by hyperactivity of the sympathetic nervous system) shows very marked frequency dependence. This change in current contributes to the induction of a significant ventricular arrhythmia when the subject or animal transitions from baseline conditions to a more active, or flight or fight situation (Nagatomo et al., 1998; Qi et al., 2015). A very relevant example in Human Physiology is the response to a sleep—awake cycle where it is well known that there is a significant increase in the incidence of cardiac arrhythmias during the time period shortly after adult humans wake up.

Another important insight which our data concerning the *I*–*V* relationship and its dynamics (recovery) has yielded concerns the effectiveness of $I_{Na\text{-}L}$ as an intrinsic current source. Since the $I_{Na\text{-}L}$ recorded in the presence of either BDF 9148 or ATX II (as a first approximation) shows little or no time dependence during the time course of the murine, rat, rabbit, or the human action potential, one can consider it to be a steady inward background Na^+ current. This information, in conjunction with the voltage profile of the action potential in the tissue of interest, becomes an interesting and important parameter. It provides information concerning when (and how) this Na^+ current would be expected to be an effective proarrhythmic factor (Qu, Karaguenzian, Garfinkel, & Weiss, 2003; Trenor et al., 2012, 2013; Zaza et al., 2008). Inspection of the *I*–*V* relationship shows that whether the plateau level of the action potential is quite hyperpolarized, about −20 mV as it is in Purkinje fibers or relatively depolarized, approximately +20 mV as it is in the guinea pig, rabbit, or human ventricular action potential $I_{Na\text{-}L}$ still (at all times) generates a significant inward current. This would bias the background I-V relationship toward being a net inward current. In other preparations, this consideration applied in conjunction with knowledge of the connexon (gap junction) mediated cell to cell network that is characteristic of atrium or ventricle gives rise to the possibility of this steady current being a significant factor in augmenting intrinsic tissue oscillations: a phenomenon, sometimes denoted stochastic resonance (Clancy, Tateyama, Liu, Wehrens, & Kass, 2003).

4.1 Can $I_{Na\text{-}L}$ increase $[Na^+]_i$?

It is well known that pronounced and or prolonged increases in heart rate can result in small but significant increases in $[Na^+]_i$ (Carmeliet, 2006a; Clancy et al., 2015; Hirn, Shapovalov, Petermann, Roulet, & Ruegg, 2008; Kleber, 1983; Liu, Liu, & Dudley, 2010; Murphy & Eisner, 2009; Pandit et al., 2001). These changes are on the order of 3 mM from baseline values of 8–10 mM. In addition, both experimental evidence and theoretical analyses have provided evidence that $[Na^+]_i$ may be increased under conditions where the $I_{Na\text{-}L}$ is enhanced. This possibility may appear unlikely; however, Carmeliet, Makielski (Makielski & Farley, 2006), and others have argued that although the endogenous or agonist-induced $I_{Na\text{-}L}$ is quite small, this Na^+ influx is active for prolonged periods of time, eg, during the entire cardiac action potential plateau. The data we have obtained using either BDF 9148 or ATX II provided a meaningful test based of this working hypothesis.

Interestingly the data in Fig. 6 show that under conditions where BDF 9148 markedly augments the $I_{Na\text{-}L}$ (increases it by a factor of 2–2.5) there is *no significant* change in cellular contractility as measured by unloaded cell shortening. Thus, the immediate effect of increasing $I_{Na\text{-}L}$ (although it is known to be strongly proarrhythmic) must derive its effects from action potential shape changes, secondary changes in other currents, for example; the L-type Ca^{2+}, current, or a significant change in the electrophysiological duty cycle of the ventricular action potential due to its lengthening. We acknowledge, however, that under conditions where the $I_{Na\text{-}L}$ is markedly enhanced, and this effect is maintained, $[Na^+]_i$ can increase. This will have important secondary consequences that have been well documented (Despa & Bers, 2003; Makielski & Farley, 2006; Murphy & Eisner, 2009). For example, $[Na^+]_i$ can increase to a level that causes the Na^+/Ca^{2+} exchanger to be compromised or reversed.

Since augmentation of the $I_{Na\text{-}L}$ under most circumstances would be expected to lengthen APD, it is important to remember the classical setting in which the increases in $[Na^+]_i$ have been well documented. Predominantly, these are the responses of ventricular myocardium to the long-lasting high-frequency stimulus trains would *not apply* to any physiological or pathophysiological augmentation of $I_{Na\text{-}L}$. It is also worth noting that if $[Na^+]_i$ did increase, there would be a number of significant consequences in addition to the change in the Na^+ gradient. These would include compromise of the ability of the myocyte to buffer pH due to reduced Na^+/H^+ ion activity and Na^+/HCO_3^- exchange activity (Murphy & Eisner, 2009). In addition, and perhaps more importantly, any significant or maintained increase in $[Na^+]_i$ would strongly activate the electrogenic Na^+/K^+ pump (Despa & Bers, 2003; Kleber, 1983). This would result in increased ATP utilization by the myocyte. The expected reduction in intracellular ATP is a known, significant proarrhythmic factor. Careful inspection of our action potential records reveals that there is no such activation in the Na^+/K^+ pump under any of our experimental circumstances since this would result in a small but detectable hyperpolarization of the resting potential.

One way that the data in Figs. 1–6 of this paper can be put in the context of commonly occurring heart pathophysiology is to recall from previous publications the effects the redox mediator hydrogen peroxide, H_2O_2, on the rat ventricle action potential waveform (Song, Shryock, Wagner, Maier, & Belardinelli, 2006; Ward & Giles, 1997). As shown in Fig. 10 and published previously (Ward & Giles, 1997), H_2O_2 markedly prolongs the APD while also giving rise to prominent EAD oscillations and relatively frequent diastolic depolarizations (DADs). However, in contrast to the effects of both BDF 9148 and ATX II, these H_2O_2 effects on Na^+ current reversed or washed out only very slowly. In fact these changes did not return completely to the baseline or control levels. This may be a result of the PKC-mediated intracellular signaling that is involved in these H_2O_2 effects (Ward & Giles, 1997). Alternatively, the lack of reversibility may be a consequence of the nonselective actions of H_2O_2: $I_{Na\text{-}L}$ is augmented markedly, but H_2O_2 at 200 μM also alters (1) the L-type Ca^{2+} current, (2) one or more of the repolarizing K^+ current, I_{Kr}.

In summary, although it is acknowledged that an increase in $I_{Na\text{-}L}$ and a consequent rise in $[Ca^{2+}]_i$ are commonly invoked as the main overall electrophysiological mechanism (or scenario) secondary to an increase in $I_{Na\text{-}L}$, our experimental results do not strongly support this hypothesis (Belardinelli et al., 2015, 2006; Makielski & Farley, 2006; Sossalla et al.,

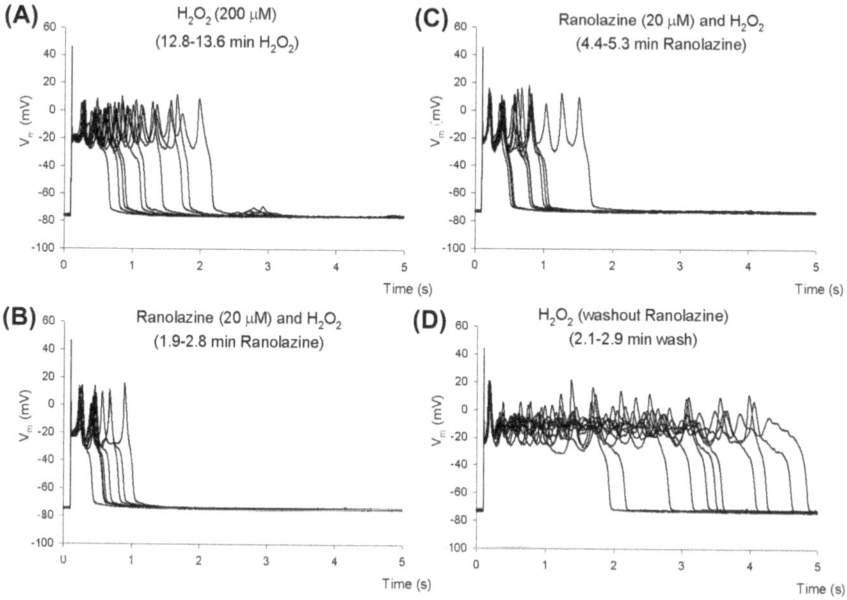

Figure 10 Illustration of the effects of a redox challenge on the action potential waveform in rat ventricle. H_2O_2 (200 μM) effects are shown in Panel A, while Panels B and C show time-dependent changes after ranolazine was added to the H_2O_2—containing superfusate. Panel D illustrates data obtained when ranolazine was removed from the H_2O_2—containing superfusate (see text for more detail).

2008; Yang et al., 2015) as an explanation for short-term or transient proarrhythmic effects. In fact, detailed inspection of the literature suggests that the increase in $[Na^+]_i$ does not occur in a volume of distribution corresponding to the myoplasm per se; rather, it may develop in a small "restricted diffusion" space just beneath the sarcolemma of the myocyte. This has been denoted the "fuzzy space". We accept the important secondary consequences of this rationale and phenomenon and have not made any significant measurements of or analysis about this possibility. Nonetheless, the data in Fig. 4 speak for themselves: under circumstances where the I_{Na-L} has been increased substantially, there has been no measurable increase in myocyte contractility.

4.2 I_{Na-L} is enhanced by redox challenges

Much of our previous work on I_{Na-L} is based on our studies of the effects of H_2O_2 as a prototypical "free radical" in the setting of redox challenge within the compromised myocardium. We have also studied the mechanisms of action of sterile inflammation in some detail (Poon, Ward, Giles, & Kubes, 1999a, 1999b; Ward, Bazzazi, Clark, Nygren, & Giles, 2006), see also Dib-Haji, Black, Cummins, & Waxman, 2002; Maingret, et al., 2008. Our results have shown that one of the primary localized chemical reactions involved in the interaction between emigrated neutrophils and ventricular myocytes also involves the production of H_2O_2. A typical response of an isolated rat ventricular myocyte to bath superfusion with a pathophysiological level of H_2O_2 (approximately 200 μM) is shown in Fig. 10. In many ways, the changes in action potential waveform and proarrhythmic activity resemble the actions of either BDF 9148 or ATX II. Detailed studies of this effect have led our group and others to conclude that this H_2O_2 effect has an obligatory dependence on intracellular protein kinase C, and others have established the importance of calmodulin as an intracellular "messenger" that can modulate the cardiac Na^+ channel (Abriel, 2010; Xie, Chen, Karagueuzian, & Weiss, 2009).

The similarities in BDF 9148, ATX, and H_2O_2 effects on $I_{Na\text{-}L}$ have yielded useful insights. In fact, however, it is also informative to look in some detail at the consistent differences in these responses. For example, the BDF effect is readily reversible, the ATX II washout is slow and incomplete; and the H_2O_2 effect is extremely difficult to reverse. Although the reason for these differences is not known, it is likely that their respective binding affinities and/or differences in their intracellular second messenger pathways can combine to result in this phenotype. Perhaps more importantly, there are also marked differences in the actions of these compounds on the proarrhythmic activity that is observed even in settings where it can be documented that the size of the induced $I_{Na\text{-}L}$ is comparable. Thus, BDF 9148 elicits only small EADs and generates very few if any DADs. In marked contrast and as shown in Fig. 9 and other publications, effective concentrations of ATX consistently exhibited a substantial spiking at the plateau or EAD activity together with less frequent, but still significant DADs, seen as small depolarizations from the resting potential. Somewhat similar to ATX, H_2O_2 induced prominent EAD activity as well as DADs and sustained runs of arrhythmic activity. Interestingly, in all cases these effects could be blocked by ranolazine which until recently was the prototypical blocker of $I_{Na\text{-}L}$ (Bonatti et al., 2014).

$I_{Na\text{-}L}$ can be elicited consistently by a number of different agonist compounds. Marked enhancement of the $I_{Na\text{-}L}$ broadens the action potential and gives rise to abnormal or proarrhythmic activity in the form of EADs or DADs (Bouchard et al., 1995; Fedida et al., 2006; Makielski & Farley,

2006; Ramirez et al., 2011; Sakmann, Spindler, Bryan, Linz, & Noble, 2000). With respect to the main goal or motivation for the development of these compounds as "sodium-dependent inotropes," one can readily see that although the principle of increasing $[Na^+]_i$ and thus augmenting diastolic Ca^{2+} is valid as a principle, in fact, doing so has at least two very significant negative consequences. First, the $I_{Na\text{-}L}$ essentially removes repolarization reserve resulting in abnormally prolonged action potentials and significant, perhaps dangerous rhythm disturbances. Second, any tendency to increase global $[Na^+]_i$ would significantly diminish the essential ability of the cell to buffer H^+; and perhaps equally importantly would augment Na^+/K^+ pump activity and therefore increase energy utilization which in the hypodynamic myocardium is strongly contraindicated (Aronsen et al., 2013; Makielski & Farley, 2006; Sossalla et al., 2008; Trenor et al., 2012; Valdivia et al., 2005; Yang et al., 2015).

4.3 Limitations

1. All of our experiments were done using single enzymatically isolated rat ventricular myocytes. It is important to acknowledge that the relatively high heart rate in an adult rat, and the consequent quite short APD, gives rise to a situation where it would appear to be somewhat unlikely that findings generated using this preparation are directly applicable to the human ventricular myocardium. However, the early repolarization is a phase of the ventricular action potential that is of growing importance. The currents which underlie the repolarization process in rat ventricle between membrane potentials, approximately +20 mV and −60 mV are very similar to those in human heart. Accordingly, the findings in this chapter may be very significant for understanding the "repolarization reserve" of early, as supposed to late repolarization in the mammalian myocardium.
2. Our findings clearly demonstrate that BDF 9148 can dramatically enhance the $I_{Na\text{-}L}$ and reveal that this late phase of the Na^+ current has a very slow time course of recovery (reactivation) at diastolic membrane potentials. We recognize, however, that BDF 9148 has been reported to have small but significant effects on other currents that contribute to the cardiac action potential in a variety of species; and also that there are enantiomer-specific effects of this and similar chemical compounds on human ventricle (Schwinger et al., 1996; Wang et al., 2008).
3. As mentioned previously, approximately 15 years ago there was a very significant interest in attempting to increase the Ca^{2+} sensitivity of the

contractile proteins in the hypodynamic or failing mammalian heart. At that time, it was correctly reasoned that, in principle, one way that this may be able to be brought about was to increase $[Na^+]_i$ and thus decrease the electrochemical gradient for Na^+ and secondarily of increased intracellular $[Ca^{2+}]_i$. Although in principle this is true, it is clear from the work in this chapter as well as other more extensive work with ATX and other $I_{Na\text{-}L}$ ligands that a global increase in $[Na^+]_i$ has definite proarrhythmic effects. Moreover, and as mentioned previously, this type of change in the Na^+ electrochemical gradient would be expected to compromise critically important H^+ homeostasis mechanisms. It would require increased ATP as a result of Na^+ dependent increased Na^+/K^+ pump activity. Both are strongly contradicted.

4. This chapter presents quite limited data with respect to inhibition of $I_{Na\text{-}L}$ by ranolazine. This was not a major focus of our experimental design. In part, this is because the compound that is the topic of this chapter, BDF 9148, has actions those are readily reversible; therefore, its study does not require inhibition by compounds such as ranolazine. We are, however, very interested in further defining the molecular mechanisms of action of ranolazine and newer compounds that are ranolazine derivatives but exhibit considerable selectivity for the $I_{Na\text{-}L}$ (Yang et al., 2015).

ACKNOWLEDGMENTS

We are grateful for operating grants from the Canadian Institutes for Health Research, the Heart and Stroke Foundations of Alberta and Canada; and salary support from the Alberta Heritage Foundation for Medical Research (now Alberta Innovates Health Solutions).

REFERENCES

Abriel, H. (2010). Cardiac sodium channel $Na_v1.5$ and interacting proteins: physiology and pathophysiology. *Journal of Molecular and Cellular Cardiology, 48*, 2–11.

Aronsen, J. M., Swift, F., & Sejersted, O. M. (2013). Cardiac sodium transport and excitation – contraction coupling. *Journal of Molecular and Cellular Cardiology, 61*, 11–19.

Attwell, D., Cohen, I., Eisner, D., Ohba, M., & Ojeda, C. (1979). The steady state TTX-sensitive ("window") sodium current in cardiac Purkinje fibres. *Pflügers Archiv: European Journal of Physiology, 379*, 137–142.

Baczko, I., Giles, W. R., & Light, P. E. (2003). Resting membrane potential regulates Na^+-Ca^{2+} exchange-mediated Ca^{2+} overload during hypoxia-reoxygenation in rat ventricular myocytes. *Journal of Physiology, 550*, 889–898.

Belardinelli, L., Giles, W. R., Rajamani, S., Karagueuzian, H. S., & Shryock, J. C. (2015). Cardiac late Na^+ current: proarrhythmic effects, roles in long QT syndromes, and pathological relationship to CAMKII and oxidative stress. *Heart Rhythm, 12*, 440–448.

Belardinelli, L., Shryock, J. C., & Fraser, H. (2006). Inhibition of the late sodium current a potential cardioprotective principle: effects of the late sodium current inhibitor ranolazine. *Heart, 92*(Suppl. IV), iv6–iv14.

Benndorf, K. (1994). Properties of single cardiac Na channels at 35°C. *Journal of General Phsyiology, 104*(5), 801–820.

Bers, D. M., Barry, W. H., & Despa, S. (2003). Intracellular Na^+ regulation in cardiac myocytes. *Cardiovascular Research, 57*, 897–912.

Bonatti, R., Silva, A. F. G., Batatinha, J. A. P., Sobrado, L. F., Machado, A. D., Varone, B. B., … Verrier, R. L. (2014). Selective late sodium current blockade with GS-458967 markedly reduces ischemia-induced atrial and ventricular repolarization alternans and ECG heterogeneity. *Heart Rhythm, 11*(10), 1827–1835.

Bouchard, R. A., Clark, R. B., & Giles, W. R. (1993a). Regulation of unloaded cell shortening by sarcolemmal sodium-calcium exchange in isolated rat ventricular myocytes. *Journal of Physiology, 469*, 583–599.

Bouchard, R. A., Clark, R. B., & Giles, W. R. (1993b). Role of sodium-calcium exchange in activation of contraction in rat ventricle. *Journal of Physiology, 472*, 391–413.

Bouchard, R. A., Clark, R. B., & Giles, W. R. (1995). Effects of action potential duration on excitation-contraction coupling in rat ventricular myocytes. Action potential voltage-clamp measurements. *Circulation Research, 76*, 790–801.

Bouchard, R., Clark, R. B., Juhasz, A. E., & Giles, W. R. (2004). Changes in extracellular K^+ concentration modulate contractility of rat and rabbit cardiac myocytes via the inward rectifier K^+ current I_{K1}. *Journal of Physiology, 556*, 773–790.

Carmeliet, E. (1987). Slow inactivation of the sodium current in rabbit cardiac Purkinje fibres. *Pflügers Archiv: European Journal of Physiology, 408*, 18–26.

Carmeliet, E. (2006a). Action potential duration, rate of stimulation, and intracellular sodium. *Journal of Cardiovascular Electrophysiology, 17*, S2–S7.

Carmeliet, E. (2006b). Electrical alternans: membrane-limited and subcellular components. *Journal of Cardiovascular Electrophysiology, 17*, 94–96.

Cheng, H. C., & Incardona, J. (2009). Models of Torsades de pointes: effects of FPL64176, DPI 201106, dofetilide and chromanol 293B in isolated rabbit and guinea pig hearts. *Journal of Pharmacological Toxicology Methods, 60*, 174–184.

Clancy, C. E., Chen-Izu, Y., Bers, D. M., Belardinelli, L., Boyden, P. A., Csernoch, L., … Zaza, A. (2015). Deranged sodium to sudden death. *Journal of Physiology, 593*(6), 1331–1345.

Clancy, C. E., Tateyama, M., Liu, H., Wehrens, X. H., & Kass, R. S. (2003). Non-equilibrium gating in cardiac Na^+ channels: an original mechanism of arrhythmia. *Circulation, 107*(17), 2233–2237.

Clark, R. B., Bouchard, R. A., & Giles, W. R. (1996). Action potential duration modulates calcium influx, Na^+-Ca^{2+} exchange, and intracellular calcium release in rat ventricular myocytes. *Annals of New York Academy of Science, 779*, 417–429.

Despa, S., & Bers, D. M. (2003). Na/K pump current and $[Na]_i$ in rabbit ventricular myocytes: local $[Na]_i$ depletion and Na buffering. *Biophysical Journal, 84*, 4157–4166.

Dib-Haji, S., Black, J. A., Cummins, T. R., & Waxman, S. G. (2002). NaN/Na_v 1.9: a sodium channel with unique properties. *Trends in Neurosciences, 25*(5), 253–259.

Doggrell, S. A., Bishop, B. E., & Brosch, S. (1995). The effects of veratridine and BDF 9148 on the action potentials and contractility of the rat right ventricle. *General Pharmacology, 26*(3), 593–601.

Fedida, D., Peter, M. R., Hesketh, C., & Ezrin, A. M. (2006). The role of late I_{Na} and antiarrhythmic drugs in EAD formation and termination in Purkinje fibers. *Journal of Cardiovascular Electrophysiology, 17*(Suppl. 1), S71–S78.

Fernandes, S., Hoyer, K., Liu, G., Wang, W. Q., Dhalla, A. K., Belardinelli, L., & Rajamani, S. (2014). Selective inhibition of the late sodium current has no adverse effect on electrophysiological or contractile function of the normal heart. *Cardiovascular Pharmacology, 63*, 512–519.

Flesch, M., & Erdmann, E. (2001). Na^+ channel activators as positive intropic agents for the treatment of chronic heart failure. *Cardiovascular Drugs and Therapy, 15*, 379–386.

Gintant, G. A., Datyner, N. B., & Cohen, I. S. (1984). Slow inactivation of a tetrodotoxin-sensitive current in canine Purkinje fibers. *Biophysical Journal, 45*, 509–512.
Herzig, S., Lilienthal, E., & Mohr, K. (1991). The positive inotropic drugs DPI 201 – 106, BDF 9148, and veratridine increase ouabain toxicity and (3H)-ouabain binding in guinea pig heart. *Journal of Cardiovascular Pharmacology, 18*(2), 182–189.
Hirn, C., Shapovalov, G., Petermann, O., Roulet, E., & Ruegg, U. T. (2008). Na_v 1.4 deregulation in dystrophic skeletal muscle leads to Na^+ overload and enhanced cell death. *Journal of General Physiology, 132*(2), 199–208.
Hoey, A., Amos, G. J., & Ravens, U. (1994a). Differential effects of BDF 9148 and DPI 201-106 on action potential and contractility in rat and guinea pig myocardium. *Journal of Cardiovascular Pharmacology, 23*(6), 907–915.
Hoey, A., Amos, G. J., & Ravens, U. (1994b). Comparison of the action potential prolonging and positive inotropic activity of DPI 201-106 and BDF 9148 in human ventricular myocardium. *Journal of Molecular and Cellular Cardiology, 26*(8), 985–994.
Ju, Y.-K., Saint, D. A., & Gage, P. W. (1994). Inactivation-resistant channels underlying the persistent sodium current in rat ventricular myocytes. *Proceedings of Biological Science, 256*, 163–168.
Kiyouse, T., & Arita, M. (1989). Late sodium current and its contribution to action potential configuration in guinea pig ventricular myocytes. *Circulation Research, 64*(2), 389–397.
Kleber, A. G. (1983). Resting membrane potential, extracellular potassium activity, and intracellular sodium activity during acute global ischemia in isolated perfused guinea pig hearts. *Circulation Research, 52*, 442–450.
Liu, M., Liu, H., & Dudley, S. C., Jr. (2010). Reactive oxygen species originating from mitochondria regulate the cardiac sodium channel. *Circulation Research, 107*, 967–974.
Liu, Y. M., DeFelice, L. J., & Mazzanti, M. (1992). Na channels that remain open throughout the cardiac action potential plateau. *Biophysical Journal, 63*, 654–662.
Maingret, F., Coste, B., Padilla, F., Clerc, N., Crest, M., Korogod, S. M., & Delmas, P. (2008). Inflammatory mediators increase Na_v1.9 current and excitability in nociceptors through a coincident detection mechanism. *Journal of General Physiology, 131*(3), 211–225.
Makielski, J. C., & Farley, A. L. (2006). Na^+ current in human ventricle: implications for sodium loading and homeostasis. *Cardiovascular Electrophysiology, 17*(Suppl. 1), S15–S20.
Müller-Ehmsen, J., Brixius, K., & Schwinger, R. H. (1998). Positive inotropic effects of the novel Na^+-channel modulator BDF 9198 in human non failing myocardium. *Journal of Cardiovascular Pharmacology, 32*(5), 684–689.
Müller-Ehmsen, J., Näbauer, M., & Schwinger, R. H. (1999). Na^+-channel modulating effect of the inotropic compound S(-)BDF 9196 in human myocardium. *Naunyn Schmiedebergs Archives of Pharmacology, 359*(1), 60–64.
Murphy, E., & Eisner, D. A. (2009). Regulation of intracellular and mitochondrial sodium in health and disease. *Circulation Research, 104*, 292–303.
Nagatomo, T., Ye, B., Valdivia, C. R., Fan, Z., January, C. T., & Makielski, J. C. (1998). Frequency-dependent reduction of late Na current by ultra-slow recovery in the LQT3 DKPQ mutant Na channel. *Circulation, 98*, 469.
Nand, V., & Doggrell, S. A. (1999). Effects of BDF 9148 on the action potentials and contractions of left ventricles from normo- and hypertensive rats. *Clinical and Experimental Pharmacology and Physiology, 26*(3), 212–219.
Pandit, S. V., Clark, R. B., Giles, W. R., & Demir, S. S. (2001). A mathematical model of action potential heterogeneity in adult rat left ventricular myocytes. *Biophysical Journal, 81*, 3029–3051.
Patlak, J. B., & Ortiz, M. (1985). Slow currents through single sodium channels of the adult rat heart. *Journal of General Physiology, 86*(1), 89–104.

Poon, B. Y., Ward, C. A., Giles, W. R., & Kubes, P. (1999a). Emigrated neutrophils regulate ventricular contractility via α4 integrin. *Circulation Research, 84*, 1245–1251.

Poon, B. Y., Ward, C. A., Giles, W. R., & Kubes, P. (1999b). Emigrated neutrophils reduce ventricular contractility via α4 integrin. *FASEB Journal, 13*, A178.

Qi, D., Yang, Z., Robinson, V. M., Li, J., Gao, C., Guo, D., ... Yan, G. X. (2015). Heterogeneous distribution of I_{Na-L} determines interregional differences in rate adaptation of repolarization. *Heart Rhythm, 12*(6), 1295–1303.

Qu, Z., Karaguenzian, H. S., Garfinkel, A., & Weiss, J. N. (2003). Effects of Na^+ channel and cell coupling abnormalities on vulnerability to reentry: a simulation study. *American Journal of Physiology Heart and Circulatory Physiology, 286*, H1310–H1321.

Ramirez, R. J., Sah, R., Liu, J., Rose, R. A., & Backx, P. H. (2011). Intracellular [Na^+] modulates synergy between Na^+/Ca^{2+} exchanger and L-type Ca^{2+} current in cardiac excitation – contraction coupling during action potentials. *Basic Research in Cardiology, 106*, 967–977.

Saint, D. A., Ju, Y.-K., & Gage, P. W. (1992). A persistent sodium current in rat ventricular myocytes. *Journal of Physiology, 453*, 219–231.

Sakmann, B. F., Spindler, A. J., Bryan, S. M., Linz, K. W., & Noble. (2000). Distribution of a persistent sodium current across the ventricular wall in guinea pigs. *Circulation Research, 87*, 910–914.

Schiffmann, H., Rizouli, V., Luers, F., Hackmann, F., Hoebel, D., Pfahlberg, A., & Hellige, G. (2003). Na^+-channel modulation, a new principle of inotropic intervention: effects on hemodynamic and myocardial energetics in the immature rabbit heart. *Pediatric Research, 54*(6), 875–884.

Schwinger, R. H., Böhm, M., Mittmann, C., La Rosée, K., & Erdmann, E. (1991). Evidence for a sustained effectiveness of sodium-channel activators in failing human myocardium. *Journal of Molecular and Cellular Cardiology, 23*(4), 461–471.

Schwinger, R. H., Müller-Ehmsen, J., Böhm, M., & Erdmann, E. (1996). Enantioselective inotropic actions of the Na+-channel activators BDF 9148, BDF 9196 and BDF 9167 in human failing and nonfailing myocardium. *Journal of Pharmacology and Experimental Therapeutics, 276*(3), 1180–1188.

Shattock, M. J., & Bers, D. M. (1989). Rat vs. rabbit ventricle: Ca flux and intracellular Na assessed by ion-selective microelectrodes. *American Journal of Physiology, 256*, C813–C822.

Song, Y., Shryock, J. C., Wagner, S., Maier, L. S., & Belardinelli, L. (2006). Blocking late sodium current reduces hydrogen peroxide-induced arrhythmogenic activity and contractile dysfunction. *Journal of Pharmacology and Experimental Therapies, 318*, 214–222.

Sossalla, S., Wagner, S., Rasenack, E. C. L., Ruff, H., Weber, S. L., Schöndube, F. A., ... Maier, L. S. (2008). Ranolazine improves diastolic dysfunction in isolated myocardium from failing human hearts – role of late sodium current and intracellular ion accumulation. *Journal of Molecular and Cellular Cardiology, 45*, 32–43.

Steadman, B. W., Moore, K. B., Spitzer, K. W., & Bridge, J. H. (1988). A video system for measuring motion in contracting heart cells. *IEEE Transactions on Biomedical Engineering, 35*, 264–272.

Stump, G. L., Wallace, A. A., Gilberto, D. B., Gehret, J. R., & Lynch, J. J., Jr. (2000). Arrhythmogenic potential of positive inotropic agents. *Basic Research in Cardiology, 95*(3), 186–198.

Trenor, B., Cardona, K., Gomez, J. F., Rajamani, S., Ferrero, J. M., Jr., Belardinelli, & Saiz, J. (2012). Simulation and mechanistic investigation of the arrhythmogenic role of the late sodium current in human heart failure. *PLoS One, 7*(3), e32659.

Trenor, B., Gomis-Tena, J., Cardona, K., Romero, L., Rajamani, S., Belardineli, L., … Saiz, J. (2013). In silico assessment of drug safety in human heart applied to late sodium current blockers. *Channels, 7*(4), 1–14.

Valdivia, C. R., Chu, W. W., Pu, J., Foell, J. D., Haworth, R. A., Wolff, M. R., … Makielski, J. C. (2005). Increased late sodium current in myocytes from a canine heart failure model and from failing human heart. *Journal of Molecular and Cellular Cardiology, 38*, 475–483.

Vassalle, M., Bocchi, L., & Du, F. (2007). A slowly inactivating sodium current (I_{Na2}) in the plateau range in canine cardiac Purkinje single cells. *Experimental Physiology, 92*(1), 161–173.

Verdonck, F., Bielen, F. V., & Ver Donck, L. (1991). Preferential block of the veratridine-induced, non-inactivating Na^+ current by R56865 in single cardiac Purkinje cells. *European Journal of Pharmacology, 203*, 371–378.

Wang, Y. J., Lin, M. W., Lin, A. A., Peng, H., & Wu, S. N. (2008). Evidence for state-dependent block of DPI 201-106 a synthetic inhibitor of Na^+ channel inactivation on delayed rectifier K^+ current in pituitary tumor (GH3) cells. *Journal of Physiology Pharmacology, 59*, 409–423.

Ward, C. A., Bazzazi, H., Clark, R. B., Nygren, A., & Giles, W. R. (2006). Actions of emigrated neutrophils on Na^+ and K^+ currents in rat ventricular myocytes. *Progress in Biophysics and Molecular Biology, 90*, 249–269.

Ward, C. A., & Giles, W. R. (1997). Ionic mechanism of the effects of hydrogen peroxide in rat ventricular myocytes. *Journal of Physiology, 500*, 631–642.

Xie, L.-H., Chen, F., Karagueuzian, H. S., & Weiss, J. N. (2009). Oxidative stress-induced afterdepolarizations and calmodulin kinase II signaling. *Circulation Research, 104*, 79–86.

Yang, P.-C., Song, Y., Giles, W. R., Horvath, B., Chen-Izu, Y., Belardinelli, L., … Clancy, C. E. (2015). A computational modeling approach combined with cellular electrophysiology data provides insights into the therapeutic benefit of targeting the late Na^+ current. *Journal of Physiology, 593*, 1429–1442.

Yuill, K. H., Convery, M. K., Dooley, P. C., Doggrell, S. A., & Hancox, J. C. (2000). Effects of BDF 9198 on action potentials and ionic currents from guinea-pig isolated ventricular myocytes. *British Journal of Pharmacology, 130*(8), 1753–1766.

Zaza, A., Belardinelli, L., & Shryock, J. C. (2008). Pathophysiology and pharmacology of the cardiac "late sodium current". *Pharmacology and Therapeutics, 119*, 326–339.

CHAPTER FIFTEEN

Physiology and Pathophysiology of Sodium Channel Inactivation

M-R. Ghovanloo, K. Aimar[a], R. Ghadiry-Tavi[a], A. Yu[a] and P.C. Ruben[1]
Department of Biomedical Physiology and Kinesiology, Simon Fraser University, Burnaby, BC, Canada
[1]Corresponding author: E-mail: pruben@sfu.ca

Contents

Abstract

Voltage-gated sodium channels are present in different tissues within the human body, predominantly nerve, muscle, and heart. The sodium channel is composed of four similar domains, each containing six transmembrane segments. Each domain can be functionally organized into a voltage-sensing region and a pore region. The sodium channel may exist in resting, activated, fast inactivated, or slow inactivated states. Upon depolarization, when the channel opens, the fast inactivation gate is in its open state. Within the time frame of milliseconds, this gate closes and blocks the channel pore from conducting any more sodium ions. Repetitive or continuous stimulations of sodium channels result in a rate-dependent decrease of sodium current. This process may continue until the channel fully shuts down. This collapse is known as slow inactivation. This chapter reviews

[a] Indicates these authors contributed equally.

Current Topics in Membranes, Volume 78
ISSN 1063-5823
http://dx.doi.org/10.1016/bs.ctm.2016.04.001

what is known to date regarding, sodium channel inactivation with a focus on various mutations within each Na_V subtype and with clinical implications.

1. PHYSIOLOGY AND STRUCTURAL BASIS OF INACTIVATION

Voltage-gated ion channels (VGICs) are transmembrane proteins that are crucial in cellular electrical signaling. The membrane potential regulates the activity of VGICs. According to the type of ions conducted, these VGICs can be classified as calcium, chloride, potassium, or sodium channels. Hodgkin and Huxley first discovered the voltage-gated sodium currents in 1952 by studying the squid giant axon (Hodgkin & Huxley, 1952a, 1952b). The sodium channels responsible for these currents are heteromultimeric proteins that are composed of a large pore-forming α-subunit and smaller β subunits. The α-subunit is organized into four domains (DI–DIV). Each domain consists of six transmembrane helices (S1–S6) that are linked together by intra- and extracellular loops (Catterall, 2000). Segments S1–S4 of each domain of the α-subunit form the voltage-sensing domain (VSD). Voltage sensing is mainly mediated through the lysine and arginine residues in the S4 helices (Bezanilla, 2008; Pless et al., 2014). The S5–S6 and a membrane reentrant p-loop form the channel pore. The VSDs are located around the central pore domain (PD) (Shaya et al., 2011). The PD and VSD are connected via the linker between S4 and S5 (Payandeh, Gamal El-Din, Scheuer, Zheng, & Catterall, 2012). In contrast, β subunits are composed of single transmembrane domain whose prominent feature is a large extracellular domain (Catterall, 2000) (Fig. 1). The α-subunit is capable of conducting currents on its own, and it is expressed in a tissue specific manner. $Na_V1.1$, $Na_V1.2$, and $Na_V1.3$ paralogs are expressed in the central nervous system (CNS). $Na_V1.4$ and $Na_V1.5$ are predominantly expressed in skeletal and cardiac muscles, respectively. $Na_V1.6$ is expressed in both CNS and the peripheral nervous system (PNS); $Na_V1.7$, $Na_V1.8$, and $Na_V1.9$ are primarily found in the PNS (Catterall, Goldin, & Waxman, 2005).

Sodium channels are the engines behind the generation and propagation of action potentials. In a simplified model, sodium channels can exist in three functional states: resting (closed), activated (open), and inactivated (closed) (Catterall et al., 2005). During activation, depolarization causes a shift of the positively charged residues on the S4 segments toward the extracellular side of the membrane. This conformational change in the VSD pulls on the connecting linker, which in turn pulls open the pore leading to the conduction of sodium ions. The process of inactivation can exist in two separate states: fast inactivation and slow inactivation. Within milliseconds of

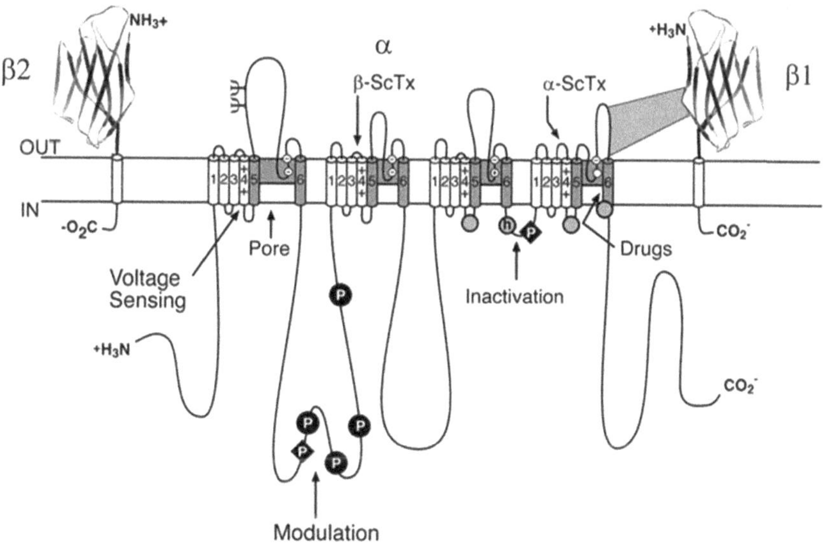

Figure 1 The primary sequence of the voltage-gated sodium channel. The alpha-subunit is a single protein with 24 transmembrane segments arranged in four domains. The S5 and S6 segments of each domain as well as the S5–S6 linkers form the pore of the channel. The alpha-subunit is also associated with beta-subunits.

activation the fast inactivation gate closes. The fast inactivation gate is composed of a very rigid α-helix, two turns, and four key amino acid residues: I, F, M, and T in the DIII-DIV cytoplasmic linker (Payandeh et al., 2012; Rohl et al., 1999). These amino acids are essentially the latch that holds the inactivation gate closed (Goldin, 2003). Repetitive or prolonged stimulation can result in slow inactivation. During slow inactivation, voltage sensors undergo an additional movement, secondary to that of activation (Silva & Goldstein, 2013a, 2013b), and two of the opposite S6 segments move toward the pore axis with the other two pointing outward. This asymmetry causes a structural collapse leading to a slowly reversible inactivation (Gamal El-Din, Martinez, Payandeh, Scheuer, & Catterall, 2013). The pore and voltage-sensing α-subunit coassociates with one or two β subunits that are known to modify the gating properties of the α-subunit. As in the α-subunit, mutations in the β subunits are related to pathophysiology arising from inactivation defects. For the purposes of this review, however, we have limited our discussion to α-subunit mutations.

The work of Hodgkin and Huxley (1952a, 1952b) indicated that activation and inactivation are independent processes, solely linked together on the basis of membrane potential. However, later findings such as the

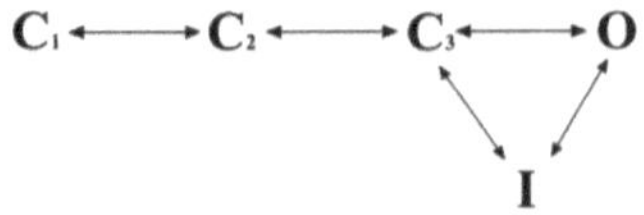

Figure 2 The five-state model.

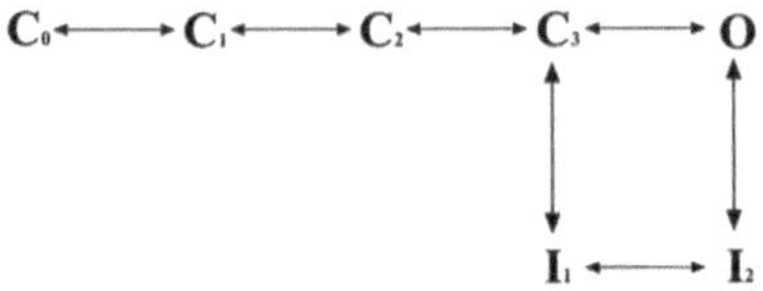

Figure 3 The seven-state model.

presence of multiple inactivation components and single-channel recordings led to the emergence of more extended models. The vast majority of such models suggest that the voltage dependence of inactivation is derived from activation (Correa & Bezanilla, 1994; Patlak, 1991). One of these models represents a five-state scheme that allows the channel to undergo inactivation from both open and closed states, but there can only be a single inactivated state (Horn & Vandenberg, 1984) (Fig. 2). Another more complicated seven-state model that accounts for gating currents, suggests that although the channel can enter inactivation from both open and closed states, there are two inactivated states (Patlak, 1991) (Fig. 3). One problem with these models is that they do not distinguish the difference between the fast and slow inactivated states. We propose a kinetic model resembling the shape of a Toblerone chocolate bar, which suggests that the sodium channel may exist in one open state, and multiple distinctive closed, fast inactivated, and slow inactivated states (Fig. 4). In this model, the sodium channel may enter fast inactivation or slow inactivation from any state along the activation pathway (Groome, Fujimoto, George, & Ruben, 1999; Groome, Fujimoto, & Ruben, 2000; Groome, Fujimoto, & Ruben, 2003; Groome, Fujimoto, & Ruben, 2005). Furthermore, fast and slow inactivation are recognized in this model as separate states, as has been demonstrated experimentally (eg, Featherstone, Richmond, & Ruben, 1996).

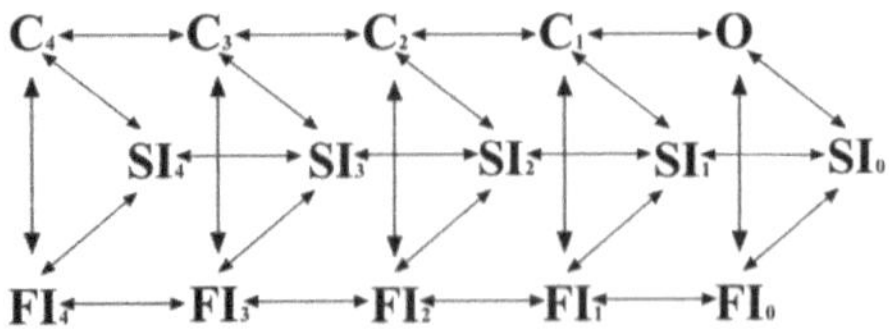

Figure 4 The Toblerone scheme. During the course of activation pathway, the sodium channel may enter fast inactivation or slow inactivation from any state.

2. $Na_V1.1$

The *SCN1A* gene, located at position 2q24.3 on chromosome 2, encodes the pore-forming $Na_V1.1$ α-subunit (Malo et al., 1994). This channel is predominantly expressed in the somata of neurons and axon initial segments within the CNS (Lorincz & Nusser, 2008; Whitaker et al., 2001). Given their somatodendritic abundance, $Na_V1.1$ channels may control the generation and propagation of action potentials to the axon through the summation of synaptic inputs. $Na_V1.1$ is essential for normal electrical activity of neuronal networks since defects in this channel result in serious neurological conditions. Missense mutations in *SCN1A* have been shown to alter the kinetics of $Na_V1.1$ fast inactivation and are associated with epilepsy and migraine disorders (Dichgans et al., 2005; Lossin, 2009; Mulley et al., 2005).

Mutations in *SCN1A* were the first to be causally linked to inherited epilepsy in humans (Escayg et al., 2000). GEFS+ is a distinct form of familial epilepsy in which approximately 10% of cases are attributed to *SCN1A* mutations (Marini et al., 2007). GEFS+ patients present with convulsions associated with febrile or afebrile seizures beyond the age of 6 years, when febrile seizures typically abate (Scheffer & Berkovic, 1997). Mutations that impair fast inactivation gating, characterized by persistent sodium current during the depolarizing phase of the action potential, may lead to neuronal hyperexcitability (Lossin et al., 2003). Escayg et al. (2000) were the first to describe GEFS+-associated missense mutations in the *SCN1A* gene. The mutations, T875M and R1648H, are located in the domain I and domain IV voltage-sensing S4 segments of the $Na_V1.1$ α-subunit, respectively (Escayg et al., 2000; Ulbricht, 2005). Lossin, Wang, Rhodes, Vanoye, and George (2002) characterized whole-cell current properties of the GEFS+ mutants and determined that T875M and R1648H (Fig. 5) produced persistent sodium current not observed in wild type-SCN1A. A fractional increase of inward sodium current depolarizes the membrane potential toward the action potential initiation threshold, thereby enhancing the susceptibility of neurons to repetitive, pathophysiological firing (Stafstrom, 2007). The persistent current generated by the two mutations is expected to be a consequence of pronounced alterations in fast inactivation (Lossin et al., 2002). T875M mutants had a greater probability of entering the activated state at any given membrane potential compared to wild-type SCN1A, as indicated by a depolarizing shift in the voltage dependence of channel availability (Lossin et al., 2002). Thus, enhanced likelihood of T875M channel activation reflects a reduction in steady-state fast inactivation (SSFI). In mutant R1648C channels, impaired fast inactivation increases the probability of late openings (Stafstrom, 2007). Single-channel recordings revealed that

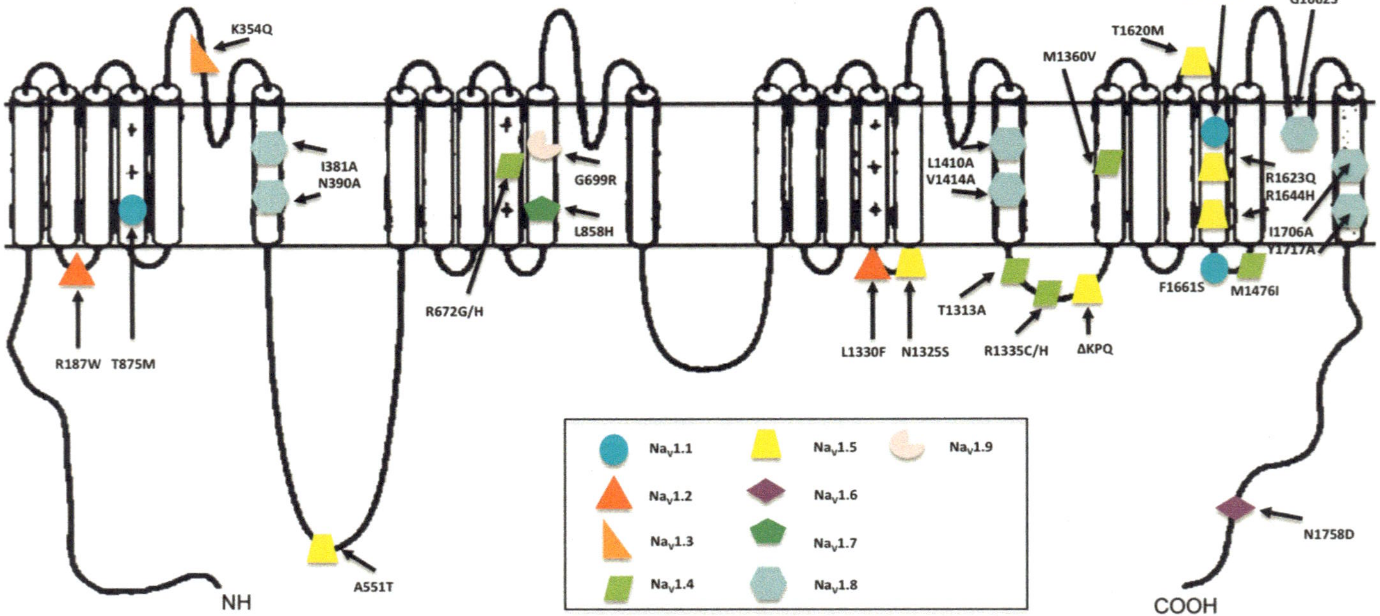

Figure 5 Location of point mutations in a generic sodium channel α-subunit identified in various Na_V subtypes. The shapes corresponding to each subtype mutants are indicated in the box.

R1648H mutants exhibited multiple channel openings occurring throughout a 200 ms test depolarization (Lossin et al., 2002). In contrast, wild-type SCN1A channel openings occurred within the first millisecond of the test depolarization, after which openings were rare (Lossin et al., 2002). Mutant channels with persistent sodium current are thought to enhance neuronal susceptibility to epileptic firing (Stafstrom, 2007). In contrast, SCN1A loss-of-function mutations in inhibitory GABAergic neurons also have been implicated in GEFS+ (Martin et al., 2010).

A second form of epilepsy known to be associated with mutations in *SCN1A* is SMEI, commonly referred to as Dravet syndrome. The SMEI phenotype is characterized by febrile or afebrile seizures within the first year of life, followed by marked developmental delay and progressive cognitive deterioration and motor impairment (Dravet, 2011). Of the SMEI-associated *SCN1A* mutations, approximately half are nonsense or frameshift mutations resulting in truncated, nonfunctional $Na_V1.1$ α-subunits and one-third are missense mutations, some of which result in channels with gain-of-function properties (Volkers et al., 2011). Biophysically characterizing mutants F1661S and R1648C, Rhodes, Lossin, Vanoye, Wang, and George (2004) were the first to demonstrate that SMEI can arise from enhanced sodium channel function (Wallace, 2005). F1661S is located between the S4 and S5 segments of DIV (Wallace, 2005) whereas R1648C, affecting the same amino acid as the R1648H GEFS+ allele, is located within the voltage-sensing S4 segment of DIV (Escayg et al., 2000). Both F1661S and R1648C channels generated noninactivating current when coexpressed with human $\beta1$ and $\beta2$ accessory subunits in mammalian cells (Rhodes et al., 2004). The persistent current produced by the SMEI mutants is caused by slowed fast inactivation and accelerated recovery from inactivation compared with wild-type SCN1A (Rhodes et al., 2004). Prolonged channel opening and more rapid reactivation are direct consequences of impaired fast inactivation which results in gain-of-function effects (Wallace, 2005). In GEFS+ and SMEI alike, increased sodium influx through mutant $Na_V1.1$ predisposes CNS neurons to hyperexcitable states leading to seizure activity (Stafstrom, 2007).

3. $Na_V1.2$

$Na_V1.2$ channels harboring missense mutations have been implicated in GEFS+ and benign familial neonatal infantile seizures (BFNIS) (Escayg et al., 2000; Liao, Deprez, Maljevic, Pitsch, & Claes, 2010; Scalmani et al., 2006). Normally, $Na_V1.2$ ensures the propagation of excitatory stimuli through brain and spinal cord neurons by the selective expression of α-subunits in fiber tracts and nerve terminals and aggregations of these channels at developing nodes of Ranvier (Whitaker et al., 2001). The distribution of this channel was found to be widespread, occurring in various regions of the brain (Gong, Rhodes, Bekele-Arcuri, & Trimmer, 1999; Westenbroek, Merrick, & Catterall, 1989; Whitaker et al., 2001), which may allow for an expanse of cortical hyperexcitability that occurs in episodes of generalized seizure.

The *SCN2A* gene occupies the chromosomal position adjacent to *SCN1A* and, together, these genes comprise a GEFS+ susceptibility locus. The R187W missense mutation in *SCN2A* affects the cytoplasmic linker between segments S2 and S3 in domain I (Sugawara et al., 2001). Impaired $Na_V1.2$ channel inactivation is the molecular basis of the disease phenotype, which is characterized by frequent episodes of febrile and afebrile seizures (Sugawara et al., 2001). Electrophysiological recordings of wild-type and mutant rat—derived $Na_V1.2$ channels revealed a prolongation in the time to reach peak current and a slowing of the time constant of current decay in the R187W channels compared to wild type (Sugawara et al., 2001). These changes reflect a decreased rate of inactivation and presumably induce convulsive activity through the excessive influx of sodium ions during neuronal depolarization (Sugawara et al., 2001). As the first GEFS+-associated missense mutation to be discovered in *SCN2A*, analysis of R187W biophysical properties provided early evidence that inactivation deficits in $Na_V1.2$ contribute to disorders of familial epilepsy.

The phenotype more commonly associated with the $Na_V1.2$ sodium channel subtype is the BFNIS, caused exclusively by *SCN2A* mutations (Scalmani et al., 2006). First described by Kaplan and Lacey (1983), BFNIS describes the sudden onset of afebrile seizures during the neonatal period or very early infancy and remission of such seizures by 12 months. Similar to GEFS+, neuronal hyperexcitability caused by noninactivating sodium current is thought to contribute to the pathogenic mechanism of BFNIS (Scalmani et al., 2006). The L1330F $Na_V1.2$ mutant in particular exhibits a depolarizing shift in the voltage dependence of inactivation (Scalmani et al., 2006). The increased sodium current is greater for mutant channels compared with wild-type channels for a given voltage, reflecting impairments in fast inactivation. The L1330F mutation affects the cytoplasmic linker between transmembrane segments S4 and S5 in DIII (Scalmani et al., 2006). The DIII S4–S5 loop likely serves as a receptor for the inactivation particle within the DIII–DIV cytoplasmic loop (Smith & Goldin, 1997). When a charged residue replaces alanine-1329 in the putative receptor site, the inactivation kinetics of the mutant channel are significantly slower than those of the wild-type channel (Smith & Goldin, 1997). Disrupting the interaction between the inactivation gate and docking site may reduce the rate of inactivation, resulting in increased inward sodium current through $Na_V1.2$ and network hyperexcitability (Heron et al., 2002). Although the mechanisms of BFNIS pathogenicity and remission have yet to be elucidated, there is a clear basis for impaired fast inactivation in $Na_V1.2$ to cause excessive cortical excitability and generalized seizures.

4. $Na_V1.3$

Positioned within the cluster of neuronal voltage-gated sodium channel genes on chromosome 2, *SCN3A* encodes the $Na_V1.3$ α-subunit and alterations in this gene have been linked to epilepsy (Estacion, Gasser, Dib-Hajj, & Waxman, 2010; Estacion, Waxman, & Dib-Hajj, 2010). $Na_V1.3$ is found in the developing brain with particular localization in the cerebellum (Schaller & Caldwell, 2003). In the adult brain, $Na_V1.3$ is found in the neocortex, primarily in association with somata and dendrites (Vacher, Mohapatra, & Trimmer, 2008). A missense mutation in *SCN3A* has been implicated in cryptogenic pediatric partial epilepsy (Holland et al., 2008). Patients with the K354Q mutation experienced infantile spasms with focal cerebral electrical abnormalities (Holland et al., 2008) (Fig. 5). The replacement of an evolutionarily conserved lysine by glutamine in the pore-forming S5–S6 loop of domain I conferred a gain-of-function effect on the $Na_V1.3$ mutant channels (Estacion, Gasser, Dib-Hajj, & Waxman, 2010; Estacion, Waxman, & Dib-Hajj, 2010). Compared to wild-type $Na_V1.3$, K354Q mutant channels generated a twofold increase in noninactivating persistent current (Estacion, Gasser, Dib-Hajj, & Waxman, 2010; Estacion, Waxman, & Dib-Hajj, 2010). Since $Na_V1.3$ modulates neuronal excitability through its contribution to the subthreshold membrane potential (Yu & Catterall, 2003), increased sodium ion inflow is expected to depolarize the membrane potential toward the action potential threshold and predispose the affected neurons to epileptic firing (Stafstrom, 2007). Persistent current is likely the consequence of mutation-induced fast inactivation deficits. In support of this notion, the onset of fast inactivation is delayed in the mutant channel (Holland et al., 2008). Also, K354Q-expressing hippocampal neurons have a propensity for firing spontaneous bursts of action potentials in a background of resting membrane potential fluctuations (Estacion, Gasser, Dib-Hajj, & Waxman, 2010; Estacion, Waxman, & Dib-Hajj, 2010). Impaired inactivation and enhanced persistent current are characteristics of neuronal voltage-gated sodium channel mutations resulting in epilepsy disorders.

5. $Na_V1.4$

The *SCN4A* gene encodes the α-subunit of the voltage-gated sodium channel $Na_V1.4$, located in position 17q23.3 on the 17th chromosome. $Na_V1.4$ is primarily found in the sarcolemma and T-tubule membranes of skeletal muscle fibers (Goldin, 2001). A depolarizing current through the channel pore initiates the skeletal muscle action potential and leads to contraction.

Mutations in *SCN4A* are linked to muscular diseases including periodic paralysis and congenital or induced myotonia (Ulbricht, 2005). These mutations delay the onset of fast inactivation and cause a sustained depolarized state that is reflective of the diseases' phenotypical symptoms.

Paramyotonia congenita (PMC) is a muscular disorder characterized by prolonged muscle tension following voluntary contraction (Gutmann & Phillips, 1991). To date, over 30 *SCN4A* mutations have been causally linked to PMC (Cannon, 2006; Cannon & Bean, 2010). PMC can be triggered by both decreases in core body temperature and increases in extracellular potassium. One cold-induced myotonia mutant, M1476I, leads to a two- to threefold slower time constant of fast inactivation (τ_h) and a depolarizing shift in the midpoint of SSFI compared to the wild type (Zhao, Duprè, Puymirat, & Chahine, 2012). These changes result in a greater fraction of mutant channels being active at any given voltage. The mutant also exhibited incomplete fast inactivation, which is reflected by an increase in persistent Na^+ current. This lowers the threshold for further firing of action potentials, leading to the onset of muscular tension characteristic of PMC (Zhao et al., 2012). Destabilized fast inactivation is shared with other *SCN4A* mutants. A mutant in the IFMT motif, T1313A, behaves similarly to M1476I, with a two- to fourfold slower τ_h and a depolarizing shift in SSFI (Bouhours et al., 2004). The mutant displayed a varied rate of fast inactivation during inactivation-triggering potentials in comparison to the wild type. At positive potentials, τ_h decreased in the wild-type channel, but exhibited a slight increase in the mutant channel, signaling a destabilization of inactivation at positive potentials (Bouhours et al., 2004). This is consistent with both the electrophysiological measurements indicating a depolarizing shift in SSFI, and the clinical phenotypes of cold-induced myotonia. Interestingly, the T1313A mutant exhibited no marked increase in persistent Na^+ current. This is unlike a different point mutation of the same threonine residue, T1313M (Dice et al., 2004). Hayward, Brown, and Cannon (1996) reported similar effects on the fast inactivation time constant and SSFI in addition to a large increase in persistent current. The change in polarity and hydrophilicity between threonine and alanine or methionine proximal to the IFM motif may explain the disruption of fast inactivation (Bouhours et al., 2004).

Potassium-aggravated myotonia (PAM) is induced by potassium ingestion and has a similar phenotype to PMC (Heine, Pika, & Lehmann-Horn, 1993). A PAM mutation, Q1633E, impairs fast inactivation by depolarizing its onset and slowing the time constant (Kubota et al., 2009). The depolarizing shift in the onset of fast inactivation results in an increased sodium ion flow through the mutant channels relative to wild type in both potassium- and cold-induced myotonia. An increase in extracellular potassium concentration augments the electrochemical gradient, causing an increased sodium influx (Cannon, 2006), while cold temperatures further slow the kinetics of an already delayed fast inactivation (Bouhours et al., 2004).

The periodic paralysis phenotype is characterized by extreme muscle weakness, fatigue, and prevention of normal contraction (Burge & Hanna, 2012). Two well-documented variations of periodic paralysis are hypokalemic periodic paralysis (HypoPP) and hyperkalemic periodic paralysis (HyperPP).

HypoPP symptoms are most severe during phases of low extracellular potassium, which causes paradoxical muscle fiber depolarizations, leading to the inactivation and inexcitability of Na^+ channels that underlie muscle weakness (Jurkat-Rott et al., 2000). Jurkat-Rott et al. (2000) described two mutations, R672H and R672G, in five families with HypoPP. In low-potassium extracellular solution, individual muscle fibers expressing mutant channels displayed spontaneous depolarization and weakness. Whole-cell patch clamp experiments revealed a hyperpolarizing shift in SSFI for both mutants, indicating an onset of fast inactivation at more negative potentials. Both displayed significantly slower recoveries from fast inactivation, decreasing the probability of channel availability during transitional voltages between inactivated and activated states (Jurkat-Rott et al., 2009). A hyperpolarizing shift in SSFI and a slower recovery rate were also exhibited by R1135H and R1135C mutants in cases of HypoPP (Groome et al., 2014). These mutations are localized to arginine residues on S4 voltage sensors in domains II and III respectively, indicating that the pathophysiological disturbances in fast inactivation are at least partially due to the movement of the voltage sensor. Indeed, gating current experiments performed on *Xenopus* oocytes have shown impeded inward translocation of the S4 segments, leading to the observed slower recovery from the fast inactivated state (Groome et al., 2014).

HyperPP symptoms are very similar to those of HypoPP, but differ substantially in its molecular mechanisms. HyperPP is commonly triggered by periods of rest after exercise or the consumption of potassium-rich foods, with the increased extracellular potassium thought to initially trigger, and then disrupt Na^+ channel activation (Venance et al., 2006). Missense mutations leading to HyperPP are present in highly conserved regions of the $Na_V1.4$ channel (Rojas et al., 1991), the S4—S5 linker (Bendahhou, Cummins, Tawil, Waxman, & Ptácek, 1999) and the S4 voltage sensor (Ptacek et al., 1991). Electrophysiological data collected for these mutations tend to be inconsistent. A mutation in the S5 segment, I1495F, resulted in a slightly lower τ_h and a hyperpolarizing shift of both steady-state activation and fast inactivation (Bendahhou et al., 1999). In contrast, the HyperPP mutation T704M similarly impacted activation but did not exhibit the same hyperpolarization of fast inactivation (Bendahhou et al., 1999). The shifts in activation result in sustained sodium current over an increased range of physiological voltages, hastening the onset of fast inactivation and triggering periodic paralysis (Bendahhou et al., 1999). Mutations in families presenting features of HyperPP can also display symptoms of PMC and cold-induced weakness (Wagner et al., 1997). A single mutation in the S1 segment of domain IV, M1360V, exhibits these particular overlapping phenotypes. The mutant channels displayed a hyperpolarizing shift in SSFI resembling that of HyperPP, but τ_h was slower than the wild type, resembling that of PMC (Wagner et al., 1997).

Although many mutations in *SCN4A* have been causally linked to inherited skeletal myopathies, the molecular mechanisms that underlie these diseases remain poorly understood. Environmental and extracellular conditions produce a shift in the rate of fast inactivation, rendering the muscle fibers hyper- or hypoexcitable and resulting in clinical phenotypes of muscle tension or weakness. Current pharmaceutical agents such as Mexiletine are used to effectively diminish symptoms, but serious cases still require continuous long-term therapy (Jurkat-Rott, Holzherr, Fauler, Lehmann-Horn, 2010), and further research is required to more rationally design drugs to treat these diseases.

6. $Na_V1.5$

$Na_V1.5$ is the predominant cardiac voltage-dependent sodium channel subtype, which consists of a transmembrane pore-forming α-subunit protein associated with a modulatory β subunit and several regulatory proteins (Brugada, Brugada, & Brugada, 1998; Brugada, Towbin, et al., 1998). The α-subunit is encoded by the *SCN5A* gene located on chromosome 3p21. Its overall structure is similar to those channel paralogs already described. The influx of sodium ions through voltage-gated cardiac sodium channels is responsible for the initial fast upstroke of the cardiac action potential and the conduction of the electrical impulse throughout the heart (Catterall, 2012). $Na_V1.5$ plays a crucial role in ensuring that the heart contracts regularly as a functional syncytium. However, the proper conduction of the electrical impulse throughout the heart is dependent on the availability of the cardiac sodium channels, which is a product of the dynamic interplay of channel activation and inactivation.

A wide array of cardiac arrhythmias have been linked to mutations in *SCN5A*. $Na_V1.5$ mutants in channel regions directly controlling inactivation are those most commonly implicated in cardiac channelopathies. When these arrhythmias occur, cardiac output is diminished, peripheral pulse and blood pressure are absent, and death is imminent if the arrhythmia is not immediately controlled (Trappe et al., 1988). In a subset of these mutations, the clinical manifestations are commonly due to the impaired inactivation of $Na_V1.5$ channels. These include but are not limited to mutations that cause long QT syndrome type 3 (LQT3) (Meisler & Kearney, 2005) and Brugada syndrome (Wang, Yuan, Khaiche, & Zhang, 2009).

SCN5A mutations have been described to cause LQT3 in both the inherited and the sporadic forms of the disease (Zimmer & Surber, 2008). Patients with these mutations exhibit conduction abnormalities as evidenced by prolonged QT intervals on ECG resulting from sustained inward Na^+ currents (Wang, Yazawa, George, & Bennett, 1996). In the case of LQT3, the symptoms are due to $Na_V1.5$ dysfunction and dysregulation, leading to life-threatening polymorphic ventricular tachycardias, in particular torsades de pointes. Compared to other LQT3 subtypes, LQT3 patients are at an increased risk for sudden cardiac death, and cardiac arrest is often the first clinical event (Zareba, Sattari, Rosero, Couderc, & Moss, 2001). The *SCN5A* mutations ΔKPQ1505 to 1507, N1325S, and R1644H have been shown to cause the inherited form of LQT3 and coincidentally exhibit striking phenotypic similarities (Dumaine et al., 1996) (Fig. 5). The N1325S and R1644H mutations are near docking sites for the inactivation gate, whereas ΔKPQ1505 resides on the linker itself. Biophysical characterization has revealed that the rapid decay of sodium current is preserved in these mutant channels but that a persistent plateau of inward sodium current follows the decay (Kambouris, Hastings, et al., 1998; Kambouris, Nuss, et al., 1998). Single-channel recordings have revealed inactivation is compromised, allowing reopenings to occur throughout depolarization. These channel reopenings result in the delayed afterdepolarization events that contribute to torsades de pointes (Bennett, Yazawa, Makita, & George, 1995).

The R1623Q mutation is an example of an *SCN5A* mutation that causes the sporadic and a more severe form of LQT3 (Oginosawa et al., 2005). Structurally, this mutation has replaced a charged arginine residue with a neutral glutamine at an external position in the S4 segment of domain IV (Yamagishi et al., 1998). In contrast with the inherited LQT3 mutations, which do not slow the current decay (Bennet et al., 1995), Kambouris, Hastings, et al. (1998) and Kambouris, Nuss, et al. (1998) showed that R1623Q channels exhibit a slower rate of macroscopic current decay. Single-channel recording of R1623Q channels have shown that the inactivated state is impaired, allowing multiple, prolonged openings to occur early in the depolarization period. These early channel openings functionally result in the prolongation in the time-course of the depolarizing sodium current leading to a delay in repolarization, which ultimately results in the prolongation of the cardiac action potential.

Several *SCN5A* mutations have been causally linked to Brugada syndrome since the first description of the disease in 1992 (Rivolta et al., 2001). Brugada syndrome is a life-threatening familial disorder that is characterized by ventricular arrhythmias and sudden cardiac death occurring in otherwise healthy individuals at a relatively young age (<40 years) (Brugada, Campuzano, Sarquella-Brugada, Brugada, & Brugada, 2014). This disease is clinically manifested by distinctive ST-segment elevations in the right precordial ECG leads V1—V3, indicating abnormal electrical conduction in the right ventricle (Brugada, Brugada, & Brugada, 1998; Brugada, Towbin, et al.,

1998). Approximately 20% of Brugada syndrome cases have been attributed to *SCN5A* mutations (Sheikh & Ranjan, 2014). The $Na_V1.5$ mutation, A551T, has been found in patients with Brugada syndrome and is located in the cytoplasmic loop linking domains I and II. A551T mutant channels exhibit a smaller current density relative to that of the wild type, when coexpressed with human β_1 accessory subunits in HEK293T cells (Chiang et al., 2009). In addition, at the resting membrane potential, the A551T mutation was shown to increase the fraction of channels entering fast inactivation from the closed state and to decrease the rate and the fraction of channel recovery from inactivation at −90 mV. These electrophysiological changes by A551T could potentially result in significant sodium channel current reduction during phase 1 of the cardiac action potential. Reduced sodium current in turn decreases action potential upstroke velocity, leading to slower atrial and ventricular conduction, clinically indicated by prolonged PR and QRS intervals in the ECG.

Dumaine et al. (1999) delineated the functional consequences of the first identified missense mutation, producing T1620M, in patients with Brugada syndrome. This mutant is located on the extracellular loop between S3 and S4 segments of DIV, and therefore it is not closely associated with structures that contribute to permeation. Studies on cultured mammalian cells expressing T1620M in the presence of $h\beta_1$ revealed that a small but significant shift in the voltage dependence of SSFI (Wang, Makita, Kitabatake, Balser, & George, 2000). A novel molecular mechanism for Brugada syndrome was first described by Veldkamp et al. (2000), in which *SCN5A* mutations enhance the slow inactivation process with intermediate kinetics. The intermediate inactivation process in $Na_V1.5$ channels was operationally defined as a form of inactivation occurring with a 10- to 100-ms time constant (Kambouris, Hastings, et al., 1998; Kambouris, Nuss, et al., 1998). Some studies described that certain *SCN5A* mutations (T1620M, 1795insD, and others) associated with Brugada syndrome functionally reduced cardiac sodium current in the myocardium by enhancing the entry process of the channel into an intermediate inactivated state (Wang et al., 2000). In addition, some investigators revealed that S4 segment of the domain IV was responsible for fast and slow inactivation of sodium channels (Mitrovic, George, & Horn, 2000). DIV-S4 plays a critical role in the inactivation of sodium channels, and the R1629Q mutation is located in the DIV-S4. Zeng et al. (2013) demonstrated that the R1629Q mutant alters the availability and intermediate inactivation of $Na_V1.5$ channels. Thus, it has been argued that enhanced intermediated inactivation might be an additional biophysical mechanism by which Na^+ channels have reduced function in the setting of Brugada syndrome.

7. $Na_V1.6$

The *SCN8A* gene codes for the α-subunit of the voltage-gated sodium channel $Na_V1.6$, located in the position 12q13.13 on the 12th chromosome. This gene is widely expressed in the CNS and PNS, and is the most abundantly expressed sodium channel isoform in the brain (Goldin, 2001).

The channel is associated and predominantly expressed with two β subunits, β_1 and β_2, both of which modulate persistent current and inactivation in $Na_V1.6$ (Smith, Smith, Plummer, Meisler, & Goldin, 1998). Although individual effects of the subunits seem to be inconsistent (He & Soderlund, 2014; Tan, 2011), expression of both β subunits along with the α-subunit hyperpolarize activation onset and increase the rate of fast inactivation (Tan, 2011).

Due to its important role and wide distribution in the CNS, $Na_V1.6$ is predicted to be involved in diseases related to the development and control of cognitive and motor processes. Missense mutations have been causally linked to cerebellar ataxia (Catterall et al., 2005; Kohrman et al., 1996) and early onset epileptic encephalopathy (EOEE) (Vaher et al., 2013), while *SCN8A* truncations underlie cerebellar atrophy and mental retardation (Trudeau, Dalton, Day, Ranum, & Meisler, 2006).

Of these disorders, only mutations linked to EOEE have been shown to result in changes to the fast inactivation process of $Na_V1.6$. EOEEs are a class of severe neonatal disorders characterized by impaired cognitive, motor, and sensory development due to recurrent seizures (Beal, Cherian, & Moshe, 2012). Although a hyperpolarizing shift in the voltage dependence of activation is the most prevalent cause of neuronal excitability resulting in seizures (Estacion et al., 2014), mutations disrupting fast inactivation have also been documented (Veeramah et al., 2012). N1768D was the first EOEE-associated mutant to demonstrate a significant change in fast inactivation. Electrophysiological data revealed a depolarizing shift in SSFI, along with a small depolarizing shift in activation (Veeramah et al., 2012). The mutant channel also displayed incomplete inactivation, resulting in a persistent Na^+ current five times greater than in wild type (Veeramah et al., 2012). Both the shifts in fast inactivation and exacerbation of persistent Na^+ elevate intracellular sodium concentrations, leaving individuals vulnerable to epileptic seizures. These findings stand out as a rare variation of EOEE; other EOEE mutations in *SCN8A* do not result in mutants that shift the voltage dependence of fast inactivation (Estacion et al., 2014; de Kovel et al., 2014).

At least 13 mutations in *SCN8A* have been linked to EOEE (Estacion et al., 2014; de Kovel et al., 2014; Ohba et al., 2014; Vaher et al., 2013; Veeramah et al., 2012); however, few have been biophysically characterized. Consequently, our understanding of the molecular basis for the disease is limited. Further work is required to delineate its phenotypes from those connected to more prominent epilepsy disorders and to better understand the impact $Na_V1.6$ mutants have on fast inactivation.

8. $Na_V1.7$

The *SCN9A* gene codes for the α-subunit of the voltage-gated sodium channel $Na_V1.7$, located in the position 2q24 on chromosome 2. The channel is primarily found in nociceptive neurons in dorsal root ganglia and sympathetic ganglia (Raymond et al., 2004; Huang et al., 2014) and has been detected in breast tumor cells, prostate tumor cells, and immune cells (Dib-Hajj, Yang, Black, & Waxman, 2012). $Na_V1.7$ functions as a "threshold channel" that rapidly activates and inactivates to amplify the generator potentials created at nociceptor endings, thereby modulating the transmission of pain signals to the brain (Dib-Hajj et al., 2012).

Because $Na_V1.7$ is located on nociceptors, missense mutations in the channel tend to result in pain-related disorders, including inherited erythromelalgia (IEM) (Dib-Hajj et al., 2005; Krafte & Bannon, 2008) and paroxysmal extreme pain disorder (PEPD) (Waxman, 2011). Less common mutations can cause indifference to pain (Cox et al., 2010) or febrile seizures (Eberhardt et al., 2014). The majority of these syndromes can be linked to a depolarizing shift in the voltage sensitivity of fast inactivation, resulting in the hyperexcitability of neurons responsible for pain or motor function (Rush et al., 2006).

IEM is a relatively common *SCN9A*-related disorder characterized by burning pain in the extremities in response to moderate exercise or warmth (Dib-Hajj et al., 2005). Missense mutations resulting in a gain of function mutation in $Na_V1.7$ affect a variety of regions in the protein and alter different aspects of the channel's biophysical properties (Han et al., 2009). A series of point mutations affecting the alanine residue at position 1632 on the S5 segment of domain IV uncovered a uniform substantial depolarizing shift in fast inactivation (Eberhardt et al., 2014). The onset of activation is hyperpolarized for some of the mutants, but not all, suggesting that residue charge and hydrophilicity plays a role in determining pathophysiological attributes. Further studies investigating the kinetics of IEM mutant channels found a dramatic tenfold increase in the time constant of deactivation and a comparatively similar time constant of fast inactivation, in the mutant L858H (Cummins, Dib-Hajj, & Waxman, 2004). Each of these changes is likely to create the increased excitability of nociceptors thought to underlie the symptoms of IEM.

PEPD is another major disease related to mutations in *SCN9A* and is characterized by diffuse pain and flushed skin. Attacks of pain begin around the rectum in infancy and progress to face and jaw area during adulthood, often triggered by sudden changes in temperature, emotional distress, and bowel movements (Estacion et al., 2011). Depolarizing shifts in fast inactivation and enhanced persistent Na^+ current are trademark traits of PEPD mutants, both of which incite the hypersensitivity of nociceptors known to underlie the phenotypes of extreme pain (Fertleman et al., 2006). Functional analyses performed on two PEPD mutants proximal to the IMF motif and a third in the S4–S5 linker of domain IV revealed a depolarizing shift in SSFI (Fertleman et al., 2006). The two inactivation gate mutants also demonstrated increased persistent currents and incomplete inactivation (Fertleman et al., 2006). Further studies have confirmed these findings and in addition, have found the time constants of fast inactivation to be much larger in mutant channels (Dib-Hajj et al., 2008; Jarecki, Sheets, Jackson, & Cummins, 2008). When expressed with β_1 and β_2 subunits, known to modify SCN9A's biophysical properties, results revealed a decreased persistent current but no change in the shifts to fast inactivation (Jarecki et al., 2008; Zhang et al., 2013).

Findings indicate that *SCN9A*-related pain disorders might be on a continuum, as a novel mutant was associated with a mixed phenotype with characteristics of both IEM and PEPD (Eberhardt et al., 2014). The A1632E mutant induces channel pathophysiology that resembles both diseases: hyperpolarization of activation onset and slowing of deactivation as observed in IEM mutants, and depolarization of fast inactivation, slowing of fast inactivation, and presence of persistent Na^+ current as observed in PEPD mutants (Eberhardt et al., 2014). These shared characteristics may be what allow nonselective channel blocking drugs such as ranolazine to attenuate the pain and associated phenotypes of both IEM and PEPD

(Estacion et al., 2011). Further research and clinical testing is required to elucidate these findings and sharpen our understanding of *SCN9A* channelopathies.

9. $Na_V1.8$

$Na_V1.8$ is a slowly inactivating channel that is preferentially expressed in dorsal root ganglion (DRG) and trigeminal ganglion (TG) neurons (Sangameswaran et al., 1996). The *SCN10A* gene, located at position 3p22.2 on chromosome 3, encodes the sodium channel isoform $Na_V1.8$ (Bezzina et al., 2013). Human studies have shown that chronic neurogenic pain is correlated with increased $Na_V1.8$ expression levels in close proximities to the sites of injury. $Na_V1.8$ has a depolarized voltage dependence of inactivation, with $V_{1/2}$ values ranging from −30 to −50 mV. Local anesthetics and other chemically related agents bind selectively to the inactivated state of this Na_V isoform, which may lead to a use-dependent block during periods of sustained depolarization (Hille, 1977). One such compound is A-803467 which potently blocks tetrodotoxin-resistant ($Na_V1.8$ and $Na_V1.9$ are sensitive to 1000-fold higher concentrations of TTX than other subtypes) currents (Jarvis et al., 2007). Mutation-based studies of A-803467 have indicated that this compound primarily interacts with the DIV-S6 segment of $Na_V1.8$, thereby modulating its inactivation (Browne, Blaney, Yusaf, Clare, & Wray, 2009).

The $Na_V1.8$ channel is linked to pain signaling following noxious and cold stimulation. The role of $Na_V1.8$ has been well established in studies in *SCN10A*-knockout mice (Lu, Ikeda, & Puhl, 2015). The absence of $Na_V1.8$ channels has been reported to impair the murine pain response. Mutational studies in the channel paralogs $Na_V1.2$ to $Na_V1.5$ have indicated that there are a number of key residues that play vital roles in drug-mediated blockage in the S6 segments of domains I, III, and IV (Catterall, 2002; Nau & Wang, 2004). Browne et al. (2009) performed mutational analysis on the corresponding S6 residues in $Na_V1.8$, some of which shifted the voltage dependence of inactivation. The mutation of I381 and N390 in DI, V1414 in DIII, and L1410, I1706, and Y1717 in DIV to alanine hyperpolarizes $Na_V1.8$ inactivation. The wild-type $Na_V1.8$ channel shows incomplete inactivation even at positive potentials, and the DIV-S6 mutants L1410A, I1706A, and Y1717A display further decreases in steady-state inactivation. Additionally, during depolarizing pulses, these mutations lead to a decrease of the inactivation time constants; thus, the mutated channels enter inactivation from open state faster than wild type (Browne et al., 2009).

Painful small fiber neuropathy (SFN) is a significant health problem (Han et al., 2014). Recent findings indicate that a small percentage of SFNs are due to gain-of-function mutations in $Na_V1.8$. Han et al. (2014) characterized the novel G1662S mutation isolated from two female patients diagnosed with idiopathic SNF, located in the highly conserved P-loop region within the extracellular linker between S5 and S6 of DIV. The two patients were described to suffer from continuous tingling and cramp-like pain in both feet; moreover, these conditions were aggravated during exercise and in cold temperatures. The G1662S channel displays a depolarized fast inactivation and an accelerated recovery from inactivation. These changes contribute to an increased excitability in DRG neurons, which may lead to SFN phenotypes (Han et al., 2014).

10. $Na_V1.9$

Similar to $Na_V1.8$, $Na_V1.9$ is also mostly expressed in DRG and TG neurons, as such abnormalities in its properties may result in SFN. The human *SCN11A*, encoding $Na_V1.9$, has been mapped to the conserved linkage group on chromosome 3p21-p24, close to the locations of the human *SCN5A* and *SCN10A* genes (Dib-Hajj et al., 1999). Despite containing all of the structural motifs as the rest of the Na_V superfamily, $Na_V1.9$ is the least conserved sodium channel (Dib-Hajj, Tyrrell, Black, & Waxman, 1998; Dib-Hajj et al., 1999). Functionally, $Na_V1.9$ is distinguished by its unique ultraslow inactivation kinetics, depolarized inactivation, and hyperpolarized activation. The large overlap of the activation and fast inactivation in this channel lead to substantial window and persistent currents (Dib-Hajj et al., 1999).

Earlier studies of $Na_V1.4$ and $Na_V1.7$ have determined that the DII S4–S5 linker is crucial to these channels' activation kinetics. The characterization of G699R shows that this mutation results in a gain-of-function alteration to $Na_V1.9$, leading to a greater depolarization of SSFI and greater hyperpolarization of steady-state activation. Furthermore, these changes in the biophysical properties of $Na_V1.9$ depolarize the resting membrane potential and result in increased firing frequency in response to small levels of stimulation (Han et al., 2015). These events mediate the appearance of SFN phenotypes in diagnosed patients.

11. DRUG INTERACTIONS

Fast inactivation is partly responsible for the depolarization phase of action potentials. The majority of the toxins targeting sodium channels interfere with the inactivation process. Neurotoxins are biomolecules that have lethal effects in the nanogram range on neurons. Through evolution many organisms have developed neurotoxins that bind with various affinities to sodium channels, resulting in loss- or gain-of-function phenotypes. These phenotypes range from disruption in sensory messages such as pain perception to motor function failures such as paralysis (Ulbricht, 2005).

Sea anemones produce powerful venoms in their nematocyte cells. These venoms act as powerful toxic cocktails used for hunting. The sea anemone toxins consist of a large family of polypeptides, one of which are proteins that are 3–5 kilodaltons in size, and modify the sodium channel gating mechanisms (Catterall & Beress, 1978). The anemone toxin (ATX) has been suggested to bind the DIVS3–S4 loop and DIVS3 segment; therefore, ATX tremendously slows down the sodium channel fast inactivation by impairing the outward movement of S4 segments (Campos, Chanda, Beirao, & Bezanilla, 2008; Gordon & Zlotkin, 1993; Moran et al., 2007). Mutational studies in $Na_V1.2$ suggest that the ATX subtype-II (ATX-II) tends to interact close to Glu-1613 on the DIVS3-S4 extracellular loop (Rogers, Qu, Tanada, Scheuer, & Catterall, 1996). ATX-II has also been implicated to disrupt the completeness of SSFI in $Na_V1.5$. This toxin is sometimes used as a laboratory reagent to induce persistent currents in $Na_V1.5$ (Wu et al., 2008).

Apart from various toxins, numerous drugs have been developed to target sodium channels. Carbamazepine is a potent antiepileptic drug that has been described to reduce inward sodium currents. This effect has mainly attributed to a shift in the steady-state inactivation curve to hyperpolarizing potentials. Thus, the net result of carbamazepine is reducing the availability of channels at resting membrane potentials (Kuo, Chen, Lu, & Chen, 1997; Reckziegel, Beck, Schramm, Urban, & Elger, 1999; Vreugdenhil & Wadman, 1999). Diazepam is another potent drug against epilepsy. Despite its recognition as a γ-amino-butyric acid (GABA) inhibitor, studies in mouse neuroblastoma cells indicate that diazepam reduces sodium currents by 40%. This reduction in sodium currents seems to be voltage dependent and results in left-shifting the steady-state inactivation curve (Backus, Pflimlin, & Trube, 1991; Wakamori, Kaneda, Oyama, & Akaike, 1989). Mutations in the rat $Na_V1.2$ indicate that both of these drugs bind the inner lumen of the channel pore, which is the same as most other local anesthetics (Ragsdale, McPhee, Scheuer, & Catterall, 1996).

12. CONCLUDING REMARKS

Inactivation is a highly important feature of sodium channel function. Fast inactivation allows for the repolarization of excitable membranes during action potentials. Fast inactivation further permits the duration of the inward transient sodium current to be fine-tuned resulting in a tunable action potential duration. Slow inactivation allows the number of sodium channels available for activation to be adjusted as a function of membrane potential and underlies adjustments in cell excitability.

Extensive recent research in the sodium channel field has produced an enormous body of knowledge on the structure and physiological roles of fast inactivation and slow inactivation leading to a variety of disorders. For instance, in CNS, understanding the role of gain-of-function mutations in $Na_V1.1$ has led us to better differentiate between $GEFS^+$ and SMEI epilepsies. Interest in seizures has led to identifying more than 10 missense mutations in $Na_V1.2$, which all result in BFNIS. Furthermore, Holland et al. (2008) were able to characterize K354Q, which was the first ever epilepsy-linked mutation in $Na_V1.3$.

In skeletal muscles, $Na_V1.4$ is responsible for voluntary muscular contraction. Loss-of-function mutations in $Na_V1.4$ have been shown to delay the onset of fast inactivation, causing diseases including PMC and periodic paralysis. These diseases typically feature involuntary muscular tightness or muscular weakness following exercise or a change in various environmental factors. In cardiac muscles, mutations in $Na_V1.5$ have been involved in an array of different conditions namely LQT3 and Brugada syndrome.

Although $Na_V1.6$ is the most abundant sodium channel in the brain, it has been found to function in both the CNS and PNS. Missense mutations in $Na_V1.6$ can lead to depolarizing shifts in fast inactivation and have been causally linked to EOEE. A gain of function in activation combined with a loss of function in fast inactivation results in neuronal hyperexcitability that leads to frequent seizures.

Among the PNS channels, mutations in $Na_V1.7$ result in the slowing of fast inactivation and often also render inactivation incomplete. This can lead to the pain-related diseases such as IEM and PEPD, which are characterized by attacks of pain during changes in environmental conditions or physiological processes. Similarly, various mutations in DIV-S6 and P-loop region of $Na_V1.8$ can disrupt fast inactivation, resulting in SFN. $Na_V1.9$ represents the least conserved member of the sodium channel family. Although this channel is not as thoroughly studied as the rest of the sodium channels, mutations in its DII S4–S5 have been shown to induce SFN phenotypes.

Through evolution, many organisms have developed toxins that specifically target sodium channels. These chemicals represent powerful arsenals against predators and to also feed on prey. Many toxins disrupt the inactivation processes of sodium channel, as it seems to be the most potent way of rendering the channel, and therefore the animal, to become incapacitated. Following this pattern, many drugs have been developed that function on similar principles as many natural toxins. A better understanding of fast and slow inactivation, including their voltage-dependent and kinetic properties, and their structural basis, will lead to more rational drug design and solutions to important and pressing medical problems.

REFERENCES

Backus, K. H., Pflimlin, P., & Trube, G. (1991). Action of diazepam on the voltage-dependent Na^+ current. Comparison with the effects of phenytoin, carbamazepine, lidocaine and flumazenil. *Brain Research, 548*, 41–49.

Beal, J. C., Cherian, K., & Moshe, S. L. (2012). Early-onset epileptic encephalopathies: Ohtahara syndrome and early myoclonic encephalopathy. *Pediatric Neurology, 47*, 317–323.

Bendahhou, S., Cummins, T. R., Tawil, R., Waxman, S. G., & Ptácek, L. J. (1999). Activation and inactivation of the voltage-gated sodium channel: role of segment S5 revealed by a novel hyperkalaemic periodic paralysis mutation. *Journal of Neuroscience, 19*, 4762–4771.

Bennett, P. B., Yazawa, K., Makita, N., & George, A. L., Jr. (1995). Molecular mechanism for an inherited cardiac arrhythmia. *Nature, 376*, 683–685.

Bezanilla, F. (2008). How membrane proteins sense voltage. *Nature Reviews Molecular Cell Biology, 9*, 323–332.

Bezzina, C. R., Barc, J., Mizusawa, Y., Remme, C. A., Gourraud, J. B., Simonet, F., & Redon, R. (2013). Common variants at SCN5A-SCN10A and HEY2 are associated with Brugada syndrome, a rare disease with high risk of sudden cardiac death. *Nature Genetics, 45*, 1044–1049.

Bouhours, M., Sternberg, D., Davoine, C. S., Ferrer, X., Willer, J. C., Fontaine, B., & Tabti, N. (2004). Functional characterization and cold sensitivity of T1313A, a new mutation of the skeletal muscle sodium channel causing paramyotonia congenita in humans. *Journal of Physiology, 554*, 635–647.

Browne, L. E., Blaney, F. E., Yusaf, S. P., Clare, J. J., & Wray, D. (2009). Structural determinants of drugs acting on the Nav1.8 channel. *Journal of Biological Chemistry, 284*, 10523–10536.

Brugada, J., Brugada, R., & Brugada, P. (1998). Right bundle branch block, ST segment elevation in leads V1–V3: a marker for sudden death in patients without demonstrable structural heart disease. *Circulation, 97*, 457–460.

Brugada, J., Towbin, J. A., Kirsch, G. E., Wang, Z., Antzelevitch, C., Moya, A., & Potenza, D. (1998). Genetic basis and molecular mechanism for idiopathic ventricular fibrillation. *Nature, 392*, 293–296.
Brugada, R., Campuzano, O., Sarquella-Brugada, G., Brugada, J., & Brugada, P. (2014). Brugada syndrome. *Methodist DeBakey Cardiovascular Journal, 10*, 25.
Burge, J. A., & Hanna, M. G. (2012). Novel insights into the pathomechanisms of skeletal muscle channelopathies. *Current Neurology and Neuroscience Reports, 12*, 62–69.
Campos, F. V., Chanda, B., Beirao, P. S., & Bezanilla, F. (2008). Alpha-scorpion toxin impairs a conformational change that leads to fast inactivation of muscle sodium channels. *Journal of General Physiology, 132*, 251–263.
Cannon, S. C. (2006). Pathomechanisms in channelopathies of skeletal muscle and brain. *Annual Review of Neuroscience, 29*, 387–415.
Cannon, S. C., & Bean, B. P. (2010). Sodium channels gone wild: resurgent current from neuronal and muscle channelopathies. *Journal of Clinical Investigation, 120*, 80–83.
Catterall, W. A. (2000). From ionic currents to molecular Mechanisms: the structure and function of voltage-gated sodium channels. *Neuron, 26*, 13–25.
Catterall, W. A. (2002). Molecular mechanisms of gating and drug block of sodium channels. *Novartis Foundation Symposium, 241*, 206–218.
Catterall, W. A. (2012). Voltage-gated sodium channels at 60: structure, function and pathophysiology. *Journal of Physiology, 590*, 2577–2589.
Catterall, W. A., & Beress, L. (1978). Sea anemone toxin and scorpion toxin share a common receptor site associated with the action potential sodium ionophore. *Journal of Biological Chemistry, 253*, 7393–7396.
Catterall, W. A., Goldin, A. L., & Waxman, S. G. (2005). International union of pharmacology. XLVII. nomenclature and structure-function relationships of voltage-gated sodium channels. *Pharmacological Reviews, 57*, 397–409.
Chiang, C., Shen, W. K., Wang, T. T., Lin, Y. H., Hsu, E. L., & Dai, S. M. (2009). A novel amino acid substitution in a voltage-gated sodium channel is associated with knockdown resistance to permethrin in *Aedes aegypti*. *Insect Biochemistry and Molecular Biology, 39*, 272–278.
Correa, A. M., & Bezanilla, F. (1994). Gating of the squid sodium channel at positive potentials: II. single channels reveal two open states. *Biophysical Journal, 66*, 1864–1878.
Cox, J. J., Sheynin, J., Shorer, Z., Reimann, F., Nicholas, A. K., Zubovic, L., & Parvari, R. (2010). Congenital insensitivity to pain: novel SCN9A missense and in-frame deletion mutations. *Human Mutation, 31*, 1670–1686.
Cummins, T. R., Dib-Hajj, S. D., & Waxman, S. G. (2004). Electrophysiological properties of mutant Nav1.7 sodium channels in a painful inherited neuropathy. *Journal of Neuroscience, 24*, 8232–8236.
Dib-Hajj, S. D., Estacion, M., Jarecki, B. W., Tyrrell, L., Fischer, T. Z., Lawden, M., & Waxman, S. G. (2008). Paroxysmal extreme pain disorder M1627K mutation in human Nav1.7 renders DRG neurons hyperexcitable. *Molecular Pain, 4*, 37.
Dib-Hajj, S. D., Rush, A. M., Cummins, T. R., Hisama, F. M., Novella, S., Tyrrell, L., & Waxman, S. G. (2005). Gain-of-function mutation in Nav1.7 in familial erythromelalgia induces bursting of sensory neurons. *Brain: A Journal of Neurology, 128*(Pt 8), 1847–1854. pii:awh514.
Dib-Hajj, S. D., Tyrrell, L., Black, J. A., & Waxman, S. G. (1998). NaN, a novel voltage-gated Na channel, is expressed preferentially in peripheral sensory neurons and down-regulated after axotomy. *Proceedings of the National Academy of Sciences of the United States of America, 95*, 8963–8968.
Dib-Hajj, S. D., Tyrrell, L., Escayg, A., Wood, P. M., Meisler, M. H., & Waxman, S. G. (1999). Coding sequence, genomic organization, and conserved chromosomal localization of the mouse gene Scn11a encoding the sodium channel NaN. *Genomics, 59*, 309–318.

Dib-Hajj, S. D., Yang, Y., Black, J. A., & Waxman, S. G. (2012). The NaV1.7 sodium channel: from molecule to man. *Nature, 14*, 1–14.
Dice, M. S., Abbruzzese, J. L., Wheeler, J. T., Groome, J. R., Fujimoto, E., & Ruben, P. C. (2004). Temperature-sensitive defects in paramyotonia congenita mutants R1448C and T1313M. *Muscle & Nerve, 30*, 277–288.
Dichgans, M., Freilinger, T., Eckstein, G., Babini, E., Lorenz-Depiereux, B., Biskup, S., … Storm, T. M. (2005). Mutation in the neuronal voltage-gated sodium channel SCN1A in familial hemiplegic migraine. *Lancet, 366*, 371–377.
Dravet, C. (2011). The core Dravet syndrome phenotype. *Epilepsia, 52*, 3–9.
Dumaine, R., Towbin, J. A., Brugada, P., Vatta, M., Nesterenko, D. V., Nesterenko, V. V., & Antzelevitch, C. (1999). Ionic mechanisms responsible for the electrocardiographic phenotype of the Brugada syndrome are temperature dependent. *Circulation Research, 85*, 803–809.
Dumaine, R., Wang, Q., Keating, M. T., Hartmann, H. A., Schwartz, P. J., Brown, A. M., & Kirsch, G. E. (1996). Multiple mechanisms of Na1 channel–linked long-QT syndrome. *Circulation Research, 78*, 916–924.
Eberhardt, M., Nakajima, J., Klinger, A. B., Neacsu, C., Hühne, K., O'Reilly, A. O., & Lampert, A. (2014). Inherited pain sodium channel Nav1.7 A1632T mutation causes erythromelalgia due to a shift of fast inactivation. *Journal of Biological Chemistry, 289*, 1971–1980.
Escayg, A., MacDonald, B. T., Meisler, M. H., Baulac, S., Huberfeld, G., An-Gourfinkel, I., … Malafosse, A. (2000). Mutations of SCN1A, encoding a neuronal sodium channel, in two families with GEFS+2. *Nature Genetics, 24*, 343–345.
Estacion, M., Gasser, A., Dib-Hajj, S. D., & Waxman, S. G. (2010). A sodium channel mutation linked to epilepsy increases ramp and persistent current of Nav1.3 and induces hyperexcitability in hippocampal neurons. *Experimental Neurology, 224*, 362–368.
Estacion, M., Han, C., Choi, J.-S., Hoeijmakers, J. G., Lauria, G., Drenth, J. P., & Waxman, S. G. (2011). Intra- and interfamily phenotypic diversity in pain syndromes associated with a gain-of-function variant of NaV1.7. *Molecular Pain, 7*, 92.
Estacion, M., O'Brien, J. E., Conravey, A., Hammer, M. F., Waxman, S. G., Dib-Hajj, S. D., & Meisler, M. H. (2014). A novel de novo mutation of SCN8A (Nav1.6) with enhanced channel activation in a child with epileptic encephalopathy. *Neurobiology of Disease, 69*, 117–123.
Estacion, M., Waxman, S. G., & Dib-Hajj, S. D. (2010). Effects of ranolazine on wild-type and mutant hNav1.7 channels and on DRG neuron excitability. *Molecular Pain, 6*, 35.
Featherstone, D. E., Richmond, J. E., & Ruben, P. D. (1996). Interaction between fast and slow inactivation in skm1 sodium channels. *Biophysical Journal, 71*, 3098–3901.
Fertleman, C. R., Baker, M. D., Parker, K. A., Moffatt, S., Elmslie, F. V., Abrahamsen, B., & Rees, M. (2006). SCN9A mutations in paroxysmal extreme pain disorder: allelic variants underlie distinct channel defects and phenotypes. *Neuron, 52*, 767–774.
Gamal El-Din, T. M., Martinez, G. Q., Payandeh, J., Scheuer, T., & Catterall, W. A. (2013). A gating charge interaction required for late slow inactivation of the bacterial sodium channel NavAb. *Journal of General Physiology, 142*, 181–190.
Goldin, A. L. (2001). Resurgence of sodium channel research. *Annual Review of Physiology, 63*, 871–894.
Goldin, A. L. (2003). Mechanisms of sodium channel inactivation. *Current Opinion in Neurobiology, 13*, 284–290.
Gong, B., Rhodes, K. J., Bekele-Arcuri, Z., & Trimmer, J. S. (1999). Type I and type II Na^+ channel alpha-subunit polypeptides exhibit distinct spatial and temporal patterning, and association with auxiliary subunits in rat brain. *Journal of Comparative Neurology, 412*, 342–352.
Gordon, D., & Zlotkin, E. (1993). Binding of alpha scorpion toxin to insect sodium channels is not dependent on membrane potential. *FEBS Letters, 315*, 125–128.

Groome, J. R., Fujimoto, E., George, A. L., & Ruben, P. C. (1999). Differential effects of homologous S4 mutations in human skeletal muscle sodium channels on deactivation gating from open and inactivated states. *Journal of Physiology, 516*, 687–698.

Groome, J. R., Fujimoto, E., & Ruben, P. C. (2000). The delay in recovery from fast inactivation in skeletal muscle sodium channels is deactivation. *Cellular and Molecular Neurobiology, 20*, 521–527.

Groome, J. R., Fujimoto, E., & Ruben, P. C. (2003). Negative charges in the DIII-DIV linker of human skeletal muscle Na^+ channels regulate deactivation gating. *Journal of Physiology, 548*, 85–96.

Groome, J. R., Fujimoto, E., & Ruben, P. C. (2005). K-aggravated myotonia mutations at residue G1306 differentially alter deactivation gating of human skeletal muscle sodium channels. *Cellular and Molecular Neurobiology, 25*, 1075–1092.

Groome, J. R., Lehmann-Horn, F., Fan, C., Wolf, M., Winston, V., Merlini, L., & Jurkat-Rott, K. (2014). NaV1.4 mutations cause hypokalaemic periodic paralysis by disrupting IIIS4 movement during recovery. *Brain, 137*, 998–1008.

Gutmann, L., & Phillips, L. H. (1991). Myotonia congenita. *Seminars in Neurology, 11*, 244–248.

Han, C., Dib-Hajj, S. D., Lin, Z., Li, Y., Eastman, E. M., Tyrrell, L., & Waxman, S. G. (2009). Early- and late-onset inherited erythromelalgia: genotypephenotype correlation. *Brain, 132*, 1711–1722.

Han, C., Vasylyev, D., Macala, L. J., Gerrits, M. M., Hoeijmakers, J. G., Bekelaar, K. J., & Waxman, S. G. (2014). The G1662S NaV1.8 mutation in small fibre neuropathy: impaired inactivation underlying DRG neuron hyperexcitability. *Journal of Neurology, Neurosurgery, and Psychiatry, 85*, 499–505.

Han, C., Yang, Y., de Greef, B. T., Hoeijmakers, J. G., Gerrits, M. M., Verhamme, C., & Waxman, S. G. (2015). The domain II S4–S5 linker in Nav1.9: a missense mutation enhances activation, impairs fast inactivation, and produces human painful neuropathy. *Neuromolecular Medicine, 17*, 158–169.

Hayward, L. J., Brown, R. H., & Cannon, S. C. (1996). Inactivation defects caused by myotonia-associated mutations in the sodium channel III-IV linker. *Journal of General Physiology, 107*, 559–576.

He, B., & Soderlund, D. M. (2014). Functional expression of rat Nav1.6 voltage-gated sodium channels in HEK293 cells: modulation by the auxiliary β1 subunit. *PLoS One, 9*.

Heine, R., Pika, U., & Lehmann-Horn, F. (1993). A novel SCN4A mutation causing myotonia aggravated by cold and potassium. *Human Molecular Genetics, 2*, 1349–1353.

Heron, S. E., Crossland, K. M., Andermann, E., Phillips, H. A., Hall, A. J., Bleasel, A., … Scheffer, I. E. (2002). Sodium-channel defects in benign familial neonatal-infantile seizures. *Lancet, 360*, 851–852.

Hille, B. (1977). Local anesthetics: hydrophilic and hydrophobic pathways for the drug-receptor reaction. *Journal of General Physiology, 69*, 497–515.

Hodgkin, A. L., & Huxley, A. F. (1952a). A quantitative description of membrane current and its application to conduction and excitation in nerve. *Journal of Physiology, 117*, 500–544.

Hodgkin, A. L., & Huxley, A. F. (1952b). The dual effect of membrane potential on sodium conductance in the giant axon of loligo. *Journal of Physiology, 116*, 497–506.

Holland, K. D., Kearney, J. A., Glauser, T. A., Buck, G., Keddache, M., Blankston, J. R., … Meisler, M. H. (2008). Mutation of sodium channel SCN3A in a patient with cryptogenic pediatric partial epilepsy. *Neuroscience Letters, 433*, 65–70.

Horn, R., & Vandenberg, C. A. (1984). Statistical properties of single sodium channels. *Journal of General Physiology, 84*, 505–534.

Huang, J., Yang, Y., Dib-Hajj, S. D., van Es, M., Zhao, P., Salomon, J., & Waxman, S. G. (2014). Depolarized inactivation overcomes impaired activation to produce DRG neuron hyperexcitability in a Nav1.7 mutation in a patient with distal limb pain. *Journal of Neuroscience, 34*, 12328–12340.

Jarecki, B. W., Sheets, P. L., Jackson, J. O., & Cummins, T. R. (2008). Paroxysmal extreme pain disorder mutations within the D3/S4–S5 linker of Nav1.7 cause moderate destabilization of fast inactivation. *Journal of Physiology, 586*, 4137–4153.

Jarvis, M. F., Honore, P., Shieh, C. C., Chapman, M., Joshi, S., Zhang, X. F., & Krafte, D. S. (2007). A-803467, a potent and selective Nav1.8 sodium channel blocker, attenuates neuropathic and inflammatory pain in the rat. *Proceedings of the National Academy of Sciences of the United States of America, 104*, 8520–8525.

Jurkat-Rott, K., Mitrovic, N., Hang, C., Kouzmekine, A., Iaizzo, P., Herzog, J., & Lehmann-Horn, F. (2000). Voltage-sensor sodium channel mutations cause hypokalemic periodic paralysis type 2 by enhanced inactivation and reduced current. *Proceedings of the National Academy of Sciences of the United States of America, 97*, 9549–9554.

Jurkat-Rott, K., Weber, M. A., Fauler, M., Guo, X. H., Holzherr, B. D., Paczulla, A., & Lehmann-Horn, F. (2009). K+-dependent paradoxical membrane depolarization and Na^+ overload, major and reversible contributors to weakness by ion channel leaks. *Proceedings of the National Academy of Sciences of the United States of America, 106*, 4036–4041.

Jurkat-Rott, K., Holzherr, B., Fauler, M., & Lehmann-Horn, F. (2010). Sodium channelopathies of skeletal muscle result from gain or loss of function. *Pflugers Archives, 460*(2), 239–248.

Kambouris, N. G., Hastings, L. A., Stepanovic, S., Marban, E., Tomaselli, G. F., & Balser, J. R. (1998). Mechanistic link between lidocaine block and inactivation probed by outer pore mutations in the rat m1 skeletal muscle sodium channel. *Journal of Physiology, 512*, 693–705.

Kambouris, N. G., Nuss, H. B., Johns, D. C., Tomaselli, G. F., Marban, E., & Balser, J. R. (1998). Phenotypic characterization of a novel long-QT syndrome mutation (R1623Q) in the cardiac sodium channel. *Circulation, 97*, 640–644.

Kaplan, R. E., & Lacey, D. J. (1983). Benign familial neonatal-infantile seizures. *American Journal of Medical Genetics, 16*, 598–599.

Kohrman, D. C., Smith, M. R., Goldin, A. L., Harris, J., & Meisler, M. H. (1996). A missense mutation in the sodium channel Scn8a is responsible for cerebellar ataxia in the mouse mutant jolting. *Journal of Neuroscience, 16*, 5993–5999.

de Kovel, C. G., Meisler, M. H., Brilstra, E. H., van Berkestijn, F. M., van 't Slot, R., van Lieshout, S., … Koeleman, B. P. (2014). Characterization of a de novo SCN8A mutation in a patient with epileptic encephalopathy. *Epilepsy Research, 109*, 1511–1518.

Krafte, D. S., & Bannon, A. W. (2008). Sodium channels and nociception: recent concepts and therapeutic opportunities. *Current Opinions in Pharmacology, 8*, 50–56.

Kubota, T., Kinoshita, M., Sasaki, R., Aoike, F., Takahashi, M. P., Sakoda, S., & Hirose, K. (2009). New mutation of the Na channel in the severe form of potassium-aggravated myotonia. *Muscle & Nerve, 39*, 666–673.

Kuo, C. C., Chen, R. S., Lu, L., & Chen, R. C. (1997). Carbamazepine inhibition of neuronal Na^+ currents: quantitative distinction from phenytoin and possible therapeutic implications. *Molecular Pharmacology, 51*, 1077–1083.

Liao, Y., Deprez, L., Maljevic, S., Pitsch, J., & Claes, L. (2010). Molecular correlates of age-dependent seizures in an inherited neonatal-infantile epilepsy. *Brain, 133*, 1403–1414.

Lorincz, A., & Nusser, Z. (2008). Cell-type-dependent molecular composition of the axon initial segment. *Journal of Neuroscience, 28*, 14329–14340.

Lossin, C. (2009). A catalog of SCN1A variants. *Brain & Development, 31*, 114–130.
Lossin, C., Rhodes, T. H., Desai, R. R., Vanoye, C. G., Wang, D., Carniciu, C., ... George, A. L., Jr. (2003). Epilepsy-associated dysfunction in the voltage-gated neuronal sodium channel SCN1A. *Journal of Neuroscience, 23*, 11289–11295.
Lossin, C., Wang, D. W., Rhodes, T. H., Vanoye, C. G., & George, A. L. (2002). Molecular basis of an inherited epilepsy. *Neuron, 34*, 877–884.
Lu, V. B., Ikeda, S. R., & Puhl, H. L., 3rd (2015). A 3.7 kb fragment of the mouse Scn10a gene promoter directs neural crest but not placodal lineage EGFP expression in a transgenic animal. *Journal of Neuroscience, 35*, 8021–8034.
Malo, M. S., Blanchard, B. J., Andresen, J. M., Srivastava, K., Chen, X. N., Li, X., ... Ingram, V. M. (1994). Localization of a putative human brain sodium channel gene (SCN1A) to chromosome band 2q24. *Cytogenetics and Cell Genetics, 67*, 178–186.
Marini, C., Mei, D., Temudo, T., Ferrari, A. R., Buti, D., Dravet, C., ... Guerrini, R. (2007). Idiopathic epilepsies with seizures precipitated by fever and SCN1A abnormalities. *Epilepsia, 48*, 1678–1685.
Martin, M. S., Dutt, K., Papale, L. A., Dubé, C. M., Dutton, S. B., de Haan, G., ... Escayg, A. (2010). Altered function of the SCN1A voltage-gated sodium channel leads to gamma-aminobutyric acid-ergic (GA Aergic) interneuron abnormalities. *Journal of Biological Chemistry, 285*, 9823–9834.
Meisler, M. H., & Kearney, J. A. (2005). Sodium channel mutations in epilepsy and other neurological disorders. *Journal of Clinical Investigation, 115*, 2010–2017.
Mitrovic, N., George, A. L., Jr., & Horn, R. (2000). Role of domain 4 in sodium channel slow inactivation. *Journal of General Physiology, 115*, 707–718.
Moran, Y., Kahn, R., Cohen, L., Gur, M., Karbat, I., Gordon, D., & Gurevitz, M. (2007). Molecular analysis of the sea anemone toxin Av3 reveals selectivity to insects and demonstrates the heterogeneity of receptor site-3 on voltage-gated Na^+ channels. *Biochemical Journal, 406*, 41–48.
Mulley, J. C., Scheffer, I. E., Petrou, S., Dibbens, L. M., Berkovic, S. F., & Harkin, L. A. (2005). SCN1A mutations and epilepsy. *Human Mutation, 25*, 535–542.
Nau, C., & Wang, G. K. (2004). Interactions of local anesthetics with voltage-gated Na^+ channels. *Journal of Membrane Biology, 201*, 1–8.
Oginosawa, Y., Nagatomo, T., Abe, H., Makita, N., Makielski, J. C., & Nakashima, Y. (2005). Intrinsic mechanism of the enhanced rate-dependent QT shortening in the R1623Q mutant of the LQT3 syndrome. *Cardiovascular Research, 65*, 138–147.
Ohba, C., Kato, M., Takahashi, S., Lerman-Sagie, T., Lev, D., Terashima, H., & Saitsu, H. (2014). Early onset epileptic encephalopathy caused by de novo SCN8A mutations. *Epilepsia, 55*, 994–1000.
Patlak, J. (1991). Molecular kinetics of voltage-dependent Na^+ channels. *Physiological Reviews, 71*, 1047–1080.
Payandeh, J., Gamal El-Din, T. M., Scheuer, T., Zheng, N., & Catterall, W. A. (2012). Crystal structure of a voltage-gated sodium channel in two potentially inactivated states. *Nature, 486*, 135–139.
Pless, S. A., Elstone, F. D., Niciforovic, A. P., Galpin, J. D., Yang, R., Kurata, H. T., & Ahern, C. A. (2014). Asymmetric functional contributions of acidic and aromatic side chains in sodium channel voltage-sensor domains. *Journal of General Physiology, 143*, 645–656.
Ptacek, L. J., George, A. L., Barchi, R. L., Griggs, R. C., Riggs, J. E., Robertson, M., & Leppert, M. F. (1991). Mutations in an S4 segment of the adult skeletal muscle sodium channel cause paramyotonia congenital. *Neuron, 8*, 891–897.
Ragsdale, D. S., McPhee, J. C., Scheuer, T., & Catterall, W. A. (1996). Common molecular determinants of local anesthetic, antiarrhythmic, and anticonvulsant block of voltage-gated Na^+ channels. *Proceedings of the National Academy of Sciences of the United States of America, 93*, 9270–9275.

Raymond, C. K., Castle, J., Garrett-Engele, P., Armour, C. D., Kan, Z., Tsinoremas, N., & Johnson, J. M. (2004). Expression of alternatively spliced sodium channel subunit genes. Unique splicing patterns are observed in dorsal root ganglia. *Journal of Biological Chemistry, 279*, 46234–46241.

Reckziegel, G., Beck, H., Schramm, J., Urban, B. W., & Elger, C. E. (1999). Carbamazepine effects on Na^+ currents in human dentate granule cells from epileptogenic tissue. *Epilepsia, 40*, 401–407.

Rhodes, T. H., Lossin, C., Vanoye, C. G., Wang, D. W., & George, A. L. (2004). Noninactivating voltage-gated sodium channels in severe myoclonic epilepsy of infancy. *Proceedings of the National Academy of Sciences of the United States of America, 101*, 11147–11152.

Rivolta, I., Abriel, H., Tateyama, M., Liu, H., Memmi, M., Vardas, P., … Kass, R. S. (2001). Inherited Brugada and LQT-3 syndrome mutations of a single residue of the cardiac sodium channel confer distinct channel and clinical phenotypes. *Journal of Biological Chemistry, 276*, 30623–30630.

Rogers, J. C., Qu, Y., Tanada, T. N., Scheuer, T., & Catterall, W. A. (1996). Molecular determinants of high affinity binding of alpha-scorpion toxin and sea anemone toxin in the S3–S4 extracellular loop in domain IV of the Na^+ channel alpha subunit. *Journal of Biological Chemistry, 271*, 15950–15962.

Rohl, C. A., Boeckman, F. A., Backer, C., Scheuer, T., Catterall, W. A., & Klevit, R. L. (1999). Solution structure of the sodium channel inactivation gate. *Biochemistry, 38*, 855–861.

Rojas, C. V., Wang, J. Z., Schwartz, L. S., Hoffman, E. P., Powell, B. R., & Brown, R. H., Jr. (1991). A Met-to-Val mutation in the skeletal muscle Na^+ channel alpha-subunit in hyperkalaemic periodic paralysis. *Nature, 354*, 387–389.

Rush, A. M., Dib-Hajj, S. D., Liu, S., Cummins, T. R., Black, J. A., & Waxman, S. G. (2006). A single sodium channel mutation produces hyper- or hypoexcitability in different types of neurons. *Proceedings of the National Academy of Sciences of the United States of America, 103*, 8245–8250.

Sangameswaran, L., Delgado, S. G., Fish, L. M., Koch, B. D., Jakeman, L. B., Stewart, G. R., & Herman, R. C. (1996). Structure and function of a novel voltage-gated, tetrodotoxin-resistant sodium channel specific to sensory neurons. *Journal of Biological Chemistry, 271*, 5953–5956.

Scalmani, P., Rusconi, R., Armatura, E., Zara, F., Avanzini, G., Franceschetti, S., & Mantegazza, M. (2006). Effects in neocortical neurons of mutations of the Nav1.2 Na^+ channel causing benign familial neonatal-infantile seizures. *Journal of Neuroscience, 26*, 10100–10109.

Schaller, K. L., & Caldwell, J. H. (2003). Expression and distribution of voltage-gated sodium channels in the cerebellum. *Cerebellum, 2*, 2–9.

Scheffer, I. E., & Berkovic, S. F. (1997). Generalized epilepsy with febrile seizures plus. A genetic disorder with heterogeneous clinical phenotypes. *Brain, 120*, 479–490.

Shaya, D., Kreir, M., Robbins, R. A., Wong, S., Hammon, J., Bruggemann, A., & Minor, D. L., Jr. (2011). Voltage-gated sodium channel (NaV) protein dissection creates a set of functional pore-only proteins. *Proceedings of the National Academy of Sciences of the United States of America, 108*, 12313–12318.

Sheikh, A. S., & Ranjan, K. (2014). Brugada syndrome: a review of the literature. *Clinical Medicine, 14*, 482–509.

Silva, J. R., & Goldstein, S. A. (2013a). Voltage sensor movements describe slow inactivation of voltage-gated sodium channels I. wild-type skeletal muscle Na(V)1.4. *Journal of General Physiology, 141*, 309–321.

Silva, J. R., & Goldstein, S. A. (2013b). Voltage sensor movements describe slow inactivation of voltage-gated sodium channels II. a periodic paralysis mutation in Na(V)1.4. *Journal of General Physiology, 141*, 321–334.

Smith, M. R., & Goldin, A. L. (1997). Interaction between the sodium channel inactivation linker and domain III S4–S5. *Biophysical Journal, 73*, 1885–1895.

Smith, M. R., Smith, R. D., Plummer, N. W., Meisler, M. H., & Goldin, A. L. (1998). Functional analysis of the mouse Scn8a sodium channel. *Journal of Neuroscience, 18*, 6093–6102.

Stafstrom, C. E. (2007). Persistent sodium current and its role in epilepsy. *Epilepsy Currents, 7*, 15–22.

Sugawara, T., Tsurubuchi, Y., Agarwala, K. L., Ito, M., Fukuma, G., Mazaki-Miyazaki, E., … Yamakawa, K. (2001). A missense mutation of the Na^+ channel αII subunit gene Nav1.2 in a patient with febrile and afebrile seizures causes channel dysfunction. *Proceedings of the National Academy of Sciences of the United States of America, 98*, 6384–6389.

Tan, J. (2011). Independent and joint modulation of rat Nav1,6 voltage-gated sodium channels by coexpression with the auxiliary b1 and b2 subunits. *Changes, 29*, 997–1003.

Trappe, H. J., Brugada, P., Talajic, M., Della Bella, P., Lezaun, R., Mulleneers, R., & Wellens, H. J. (1988). Prognosis of patients with ventricular tachycardia and ventricular fibrillation: role of the underlying etiology. *Journal of the American College of Cardiology, 12*, 166–174.

Trudeau, M. M., Dalton, J. C., Day, J. W., Ranum, L. P. W., & Meisler, M. H. (2006). Heterozygosity for a protein truncation mutation of sodium channel SCN8A in a patient with cerebellar atrophy, ataxia, and mental retardation. *Journal of Medical Genetics, 43*, 527–530.

Ulbricht, W. (2005). Sodium channel inactivation: molecular determinants and modulation. *Physiological Reviews, 85*, 1271–1301.

Vacher, H., Mohapatra, D. P., & Trimmer, J. S. (2008). Localization and targeting of voltage-dependent ion channels in mammalian central neurons. *Physiological Reviews, 88*, 1407–1447.

Vaher, U., Nõukas, M., Nikopensius, T., Kals, M., Annilo, T., Nelis, M., & Talvik, T. (2013). De novo SCN8A mutation identified by whole-exome sequencing in a boy with neonatal epileptic encephalopathy, multiple congenital anomalies, and movement disorders. *Journal of Child Neurology, 29*, 202–206.

Veeramah, K. R., O'Brien, J. E., Meisler, M. H., Cheng, X., Dib-Hajj, S. D., Waxman, S. G., & Hammer, M. F. (2012). De novo pathogenic SCN8A mutation identified by whole-genome sequencing of a family quartet affected by infantile epileptic encephalopathy and SUDEP. *American Journal of Human Genetics, 90*, 502–510.

Veldkamp, M. W., Viswanathan, P. C., Bezzina, C., Baartscheer, A., Wilde, A. A., & Balser, J. R. (2000). Two distinct congenital arrhythmias evoked by a multidysfunctional Na^+ channel. *Circulation Research, 86*, 91–97.

Venance, S. L., Cannon, S. C., Fialho, D., Fontaine, B., Hanna, M. G., Ptacek, L. J., & Griggs, R. C. (2006). The primary periodic paralyses: diagnosis, pathogenesis and treatment. *Brain, 129*, 8–17.

Volkers, L., Kahlig, K. M., Verbeek, N. E., Das, J. H., van Kempen, M. J. A., Stroink, H., … Rook, M. B. (2011). Nav1.1 dysfunction in genetic epilepsy with febrile seizures-plus or Dravet syndrome. *European Journal of Neuroscience, 34*, 1268–1275.

Vreugdenhil, M., & Wadman, W. J. (1999). Modulation of sodium currents in rat CA1 neurons by carbamazepine and valproate after kindling epileptogenesis. *Epilepsia, 40*, 1512–1522.

Wagner, S., Lerche, H., Mitrovic, N., Heine, R., George, A. L., & Lehmann-Horn, F. (1997). A novel sodium channel mutation causing a hyperkalemic paralytic and paramyotonic syndrome with variable clinical expressivity. *Neurology, 49*, 1018–1025.

Wakamori, M., Kaneda, M., Oyama, Y., & Akaike, N. (1989). Effects of chlordiazepoxide, chlorpromazine, diazepam, diphenylhydantoin, flunitrazepam and haloperidol on the voltage-dependent sodium current of isolated mammalian brain neurons. *Brain Research, 494*, 374–378.

Wallace, R. (2005). A plethora of SCN1A mutations: what can they tell us? *Epilepsy Currents, 5*, 17–20.

Wang, D. W., Makita, N., Kitabatake, A., Balser, J. R., & George, A. L., Jr. (2000). Enhanced Na(+) channel intermediate inactivation in Brugada syndrome. *Circulation Research, 87*, 37–43.

Wang, D. W., Yazawa, K., George, A. L., Jr., & Bennett, P. B. (1996). Characterization of human cardiac Na^+ channel mutations in the congenital long QT syndrome. *Proceedings of the National Academy of Sciences of the United States of America, 93*, 13200–13205.

Wang, K., Yuan, Y., Kharche, S., & Zhang, H. (2009). The E1784K mutation in SCN5A and phenotypic overlap of type 3 long QT syndrome and Brugada syndrome: a simulation study. *Computational Cardiology, 36*, 301.

Waxman, S. G. (2011). Neuroscience: channelopathies have many faces. *Nature, 472*, 173–174.

Westenbroek, R. E., Merrick, D. K., & Catterall, W. A. (1989). Differential subcellular localization of the RI and RII Na^+ channel subtypes in central neurons. *Neuron, 3*, 6384–6389.

Whitaker, W. R., Faull, R. L., Waldvogel, H. J., Plumpton, C. J., Emson, P. C., & Clare, J. J. (2001). Comparative distribution of voltage-gated sodium channel proteins in human brain. *Molecular Brain Research, 88*, 37–53.

Wu, L., Rajamani, S., Shryock, J. C., Li, H., Ruskin, J., Antzelevitch, C., & Belardinelli, L. (2008). Augmentation of late sodium current unmasks the proarrhythmic effects of amiodarone. *Cardiovascular Research*, 77, 481–488.

Yamagishi, H., Furutani, M., Kamisago, M., Morikawa, Y., Kojima, Y., Hino, Y., & Matsuoka, R. (1998). A de novo missense mutation (R1623Q) of the SCN5A gene in a Japanese girl with sporadic long QT syndrome. *Human Mutation, 11*, 481.

Yu, F. H., & Catterall, W. A. (2003). Overview of the voltage-gated sodium channel family. *Genome Biology, 4*, 207.

Zareba, W., Sattari, M. N., Rosero, S., Couderc, J. P., & Moss, A. J. (2001). Altered atrial, atrioventricular, and ventricular conduction in patients with the long QT syndrome caused by the KPQ SCN5A sodium channel gene mutation. *American Journal of Cardiology, 88*, 1311–1314.

Zeng, Z., Zhou, J., Hou, Y., Liang, X., Zhang, Z., Xu, X., ... Huang, Z. (2013). Electrophysiological characteristics of a SCN5A voltage sensors mutation R1629Q associated with Brugada syndrome. *PLoS One, 8*, e78382.

Zhang, M. M., Wilson, M. J., Azam, L., Gajewiak, J., Rivier, J. E., Bulaj, G., & Yoshikami, D. (2013). Co-expression of NaVβ subunits alters the kinetics of inhibition of voltage-gated sodium channels by pore-blocking μ-conotoxins. *British Journal of Pharmacology, 168*, 1597–1610.

Zhao, J., Duprè, N., Puymirat, J., & Chahine, M. (2012). Biophysical characterization of M1476I, a sodium channel founder mutation associated with cold-induced myotonia in French Canadians. *Journal of Physiology, 590*, 2629–2644.

Zimmer, T., & Surber, R. (2008). SCN5A channelopathies – an update on mutations and mechanisms. *Progress in Biophysics and Molecular Biology, 98*, 120–136.

SECTION FIVE

Sodium Channels Under Stress and in Disease

CHAPTER SIXTEEN

Cardiac Sodium Channel Mutations: Why so Many Phenotypes?

M. Liu, K.-C. Yang and S.C. Dudley, Jr.[1]
The Warren Alpert Medical School of Brown University, Providence, RI, United States
[1]Corresponding author: E-mail: samuel_dudley@brown.edu

Contents

ISSN 1063-5823
http://dx.doi.org/10.1016/bs.ctm.2015.12.004

Abstract

The cardiac Na^+ channel ($Na_V1.5$) conducts a depolarizing inward Na^+ current that is responsible for the generation of the upstroke Phase 0 of the action potential. In heart tissue, changes in Na^+ currents can affect conduction velocity and impulse propagation. The cardiac $Na_V1.5$ is also involved in determination of the action potential duration, since some channels may reopen during the plateau phase, generating a persistent or late inward current. Mutations of cardiac $Na_V1.5$ can induce gain or loss of channel function because of an increased late current or a decrease of peak current, respectively. Gain-of-function mutations cause Long QT syndrome type 3 and possibly atrial fibrillation, while loss-of-function channel mutations are associated with a wider variety of phenotypes, such as Brugada syndrome, cardiac conduction disease, dilated cardiomyopathy, and sick sinus node syndrome. The penetrance and phenotypes resulting from $Na_V1.5$ mutations also vary with age, gender, body temperature, circadian rhythm, and between regions of the heart. This phenotypic variability makes it difficult to correlate genotype–phenotype. We propose that mutations are only one contributor to the phenotype and additional modifications on $Na_V1.5$ lead to the phenotypic variability. Possible modifiers include other genetic variations and alterations in the life cycle of $Na_V1.5$ such as gene transcription, RNA processing, translation, posttranslational modifications, trafficking, complex assembly, and degradation. In this chapter, we summarize potential modifiers of cardiac $Na_V1.5$ that could help explain the clinically observed phenotypic variability. Consideration of these modifiers could help improve genotype–phenotype correlations and lead to new therapeutic strategies.

1. INTRODUCTION

The cardiac Na^+ channel ($Na_V1.5$) is a macromolecular complex consisting of an α subunit encoded by *SCN5A* and accessory proteins, including β subunits and Na^+ channel interacting proteins (Fig. 1A). Cardiac $Na_V1.5$ is responsible in cardiomyocytes for the initiation of the depolarizing phase 0 of the cardiac action potential by conducting rapid depolarizing inward Na^+ currents (I_{Na}) (Fig. 1B). In addition to intercellular communication via gap junctions, the upstroke velocity of the cardiac action potential is a major determinant of impulse conduction velocity in heart tissue. The importance of $Na_V1.5$ for stable cardiac electrical conduction is highlighted by lethal arrhythmias that results from inherited genetic defects in the $Na_V1.5$ channel associated with sudden cardiac death syndromes such as Brugada syndrome (BrS) and Long QT syndrome type 3 (LQT3). Alternations in cardiac $Na_V1.5$ have also been implicated in arrhythmic risk associated with acquired heart disease (Gao et al., 2011; Liu et al., 2009).

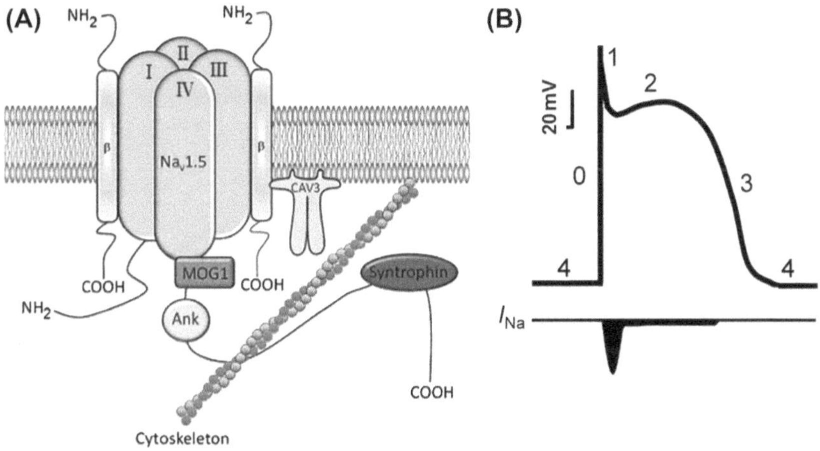

Figure 1 (A) The macromolecular complex of cardiac Na^+ channel: the channel α subunit interacts with multiple cellular proteins, including accessory proteins such as β subunits, ankyrin, CAV3, MOG1, syntrophin, and cytoskeleton at the cell membrane, forming a macromolecular complex. $Na_v1.5$ channel activity can be modified by the altered expression or function of the components of the macromolecular protein complex. Abbreviations: CAV3, caveolin-3; MOG1, Ran guanine nucleotide release factor. (B) A schematic representation of the cardiac action potential with cardiac $Na_v1.5$ channel currents. *Modified from Liu, M., Yang, K.-C., & Dudley, S.C., Jr. (2014). Cardiac sodium channel mutations: why so many phenotypes? Nature Review Cardiology, 11, 607–615.*

Despite a relatively straightforward role in cardiac conduction, mutations of *SCN5A* have been shown to produce a variety of phenotypes that are not easily anticipated. In gain-of-function mutations, an increase in the persistent late I_{Na} (I_{NaL}) during the action potential plateau, rather than an increase in peak I_{Na}, is generally thought to be responsible for this phenotype, which results in a prolonged action potential and QT intervals (Fig. 2A). Conversely, loss-of-function mutations in *SCN5A* lead to a decrease in peak I_{Na}, which decreases the action potential upstroke speed and confers arrhythmic risk by virtue of slowing conduction velocity or shortening the action potential duration (APD), conditions favoring reentrant-type tachycardias. Occasionally, mutations can cause both an elevation in I_{NaL} and a simultaneous reduction in peak I_{Na}, thereby having both gain- and loss-of-function properties.

Different *SCN5A* mutations can generate different clinical syndromes. These syndromes can manifest in a particular chamber or region of the heart. Additionally, the phenotype of a particular *SCN5A* mutation can show a variable penetrance between individuals and can vary within an individual

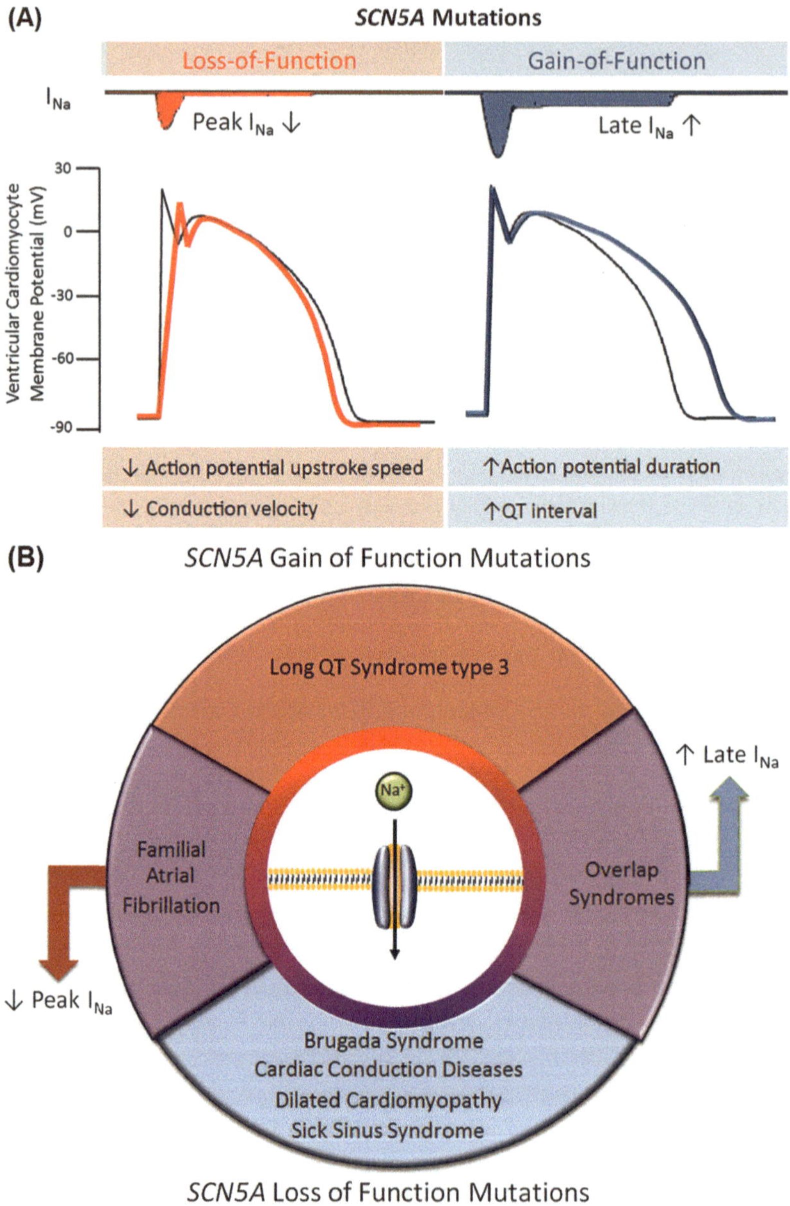

Figure 2 (A) Altered I_{Na} by *SCN5A* mutations, either peak I_{Na} or late I_{Na} can change the cardiac action potential waveforms of ventricular cardiomyocytes. Gain-of-function *SCN5A* mutations can result in increased late I_{Na}, leading to delayed repolarization of ventricular action potentials and prolonged QT intervals on the surface ECG. In contrast, loss-of-function *SCN5A* mutations can reduce peak I_{Na}, leading to decreased action potential upstroke velocity and slowed cardiac electrical conduction velocity. (B)

as a result of factors such as age, time of day, heart rate, and body temperature. The source of this phenotypic variation is unclear, and this degree of variability makes genotype—phenotype correlations difficult. Furthermore, the lack of correlation complicates medical decision-making in patients with known mutations.

In this chapter, we will discuss some potential modifiers that regulate the life cycle of the cardiac $Na_V1.5$ channel and that could help explain the clinically observed phenotypic variability. We propose that phenotypic variability, not ascribed to mutation-dependent changes in channel biophysical properties, might be the result of additional modulations of channel behavior by modifiers (Liu, Yang, & Dudley, 2014) such as additional genetic variants in the *SCN5A* gene. Other potential modifiers include intrinsic modifiers that alter the expression or function of a mutated channel, and extrinsic modifiers that change the expression or function of other channels or genes that can modify the overall biophysical properties and electrical effects of a mutated channel. Consideration of these modifiers could help improve genotype—phenotype correlations and lead to new therapeutic strategies.

2. THE CARDIAC $Na_V1.5$

One contributor to the complexity of various phenotypes is the complexity of the channel itself. In response to the changes in membrane potential, the α subunit of $Na_V1.5$ changes its structural conformation on a timescale of milliseconds, a process termed voltage-dependent gating. The α subunit is composed of intracellular N and C termini, and four homologous domains (I—IV) that form a principal pore to conduct Na^+ ions across the cell membrane (Fig. 1A) (Gellens et al., 1992). The four domains are attached to one another by cytoplasmic linker sequences with important residues for channel gating and posttranslational modifications

Phenotypic variability of $Na_V1.5$ mutations: gain-of-function mutations of *SCN5A* can cause long QT syndrome type 3 and possibly atrial fibrillation (AFib); loss-of-function mutations of *SCN5A* can cause Brugada syndrome, sick sinus syndrome, cardiac conduction diseases, familial lone AFib, and dilated cardiomyopathy. Some *SCN5A* mutations have also been shown to cause overlap syndromes. *(Modified from Liu, M., Yang, K.-C., & Dudley, S.C., Jr. (2014). Cardiac sodium channel mutations: why so many phenotypes? Nature Review Cardiology, 11, 607—615.)*

such as phosphorylation. Each domain contains six transmembrane segments (S1−S6). The four positively charged S4 segments constitute the voltage sensor and are responsible for increased channel permeability during membrane depolarization (Fozzard & Hanck, 1996). The S5 and S6 segments of all four domains form the ion-conducting pore, and the interconnecting P-loops make up the channel's selectivity filter (Balser, 2001; Kass, 2006). As shown in Fig. 1A, the α subunit of $Na_V1.5$ interacts with accessory proteins, including β subunits and Na^+ channel interacting proteins that modulate the channel expression and function, and forms a macromolecular complex on the cell membrane (for a recent review of accessory proteins, see Shy, Gillet, & Abriel, 2013). The $Na_V1.5$ channel has three states: activated (open), inactivated, and deactivated (rested). Transitions among these states depend on the membrane potential and modifier functions. Under physiological conditions, the three-state properties of wild-type $Na_V1.5$ are tightly regulated, while under pathological condition the gating properties and current kinetics of $Na_V1.5$ are altered possibly because of changes in composition of the macromolecular complex. The complexity of gating and the nature of the channel as a macromolecular complex introduce the opportunity for differential modulation of $Na_V1.5$ mutation phenotypes among patients and within areas of the heart under different disease states.

3. PHENOTYPIC VARIABILITY OF CARDIAC $Na_V1.5$ MUTATIONS

The first mutants of *SCN5A* were reported by Chen et al. (1998) in association with BrS. In a number of families with BrS or idiopathic ventricular fibrillation, the R1232W and T1620M mutants of human cardiac $Na_V1.5$ show loss of channel function. Since then, many more *SCN5A* mutations have been found to be responsible for different syndromes with both gain and loss of channel function (Adsit, Vaidyanathan, Galler, Kyle, & Makielski, 2013; Amin, Asghari-Roodsari, & Tan, 2010; Remme, 2013; Ruan, Liu, & Priori, 2009; Zimmer & Surber, 2008). As shown in Fig. 2B, gain-of-function mutations of *SCN5A* can cause LQT3 and possibly atrial fibrillation (AFib). Loss-of-function mutations of *SCN5A* can cause BrS, sick sinus syndrome (SSS), cardiac conduction diseases (CCDs), AFib, and possibly dilated cardiomyopathy (DCM). Some *SCN5A* mutations have also been shown to cause overlap syndromes or been linked with familial lone AFib.

3.1 Phenotypes with $Na_V1.5$ gain-of-function mutations

3.1.1 Long QT syndrome type 3

Gain-of-function mutations of cardiac $Na_V1.5$ can cause LQT3, which is characterized by abnormally prolonged cardiac repolarization phase (Schwartz, Priori, & Napolitano, 2003). LQT3 is the substrate for fatal polymorphic ventricular arrhythmia known as torsades de pointes. The majority of LQT3 patients with a *SCN5A* mutation experience cardiac events or sudden cardiac death during sleep (Fig. 3) (Beaufort-Krol et al., 2005). Most of the *SCN5A* mutations linked to LQT3 show defective inactivation (such as M1766L (Valdivia et al., 2002), E1784K (Wei et al., 1999), D1790G (Abriel, Wehrens, Benhorin, Kerem, & Kass, 2000; An et al., 1998)) or repeated channel reopening (such as ΔKPQ, N1325S, and R1644H (Bennett, Yazawa, Makita, & George, 1995; Dumaine et al., 1996; Wang, Yazawa, George, & Bennett, 1996)) that evoke increased I_{NaL}. This current further depolarizes the cell membrane during phase 1 and 2 of the cardiac action potential plateau (Fig. 2A), resulting in prolongation of the APD and the QT interval. Elevated I_{NaL} leads to increased dispersion of repolarization and early afterdepolarizations that predispose to ventricular arrhythmias (Cranefield & Aronson, 1991; Ming, Aronson, & Nordin, 1994; Surawicz, 1989).

3.1.2 Atrial fibrillation

As one of the most common cardiac arrhythmias, particularly for elder people, AFib is characterized by a disorganized, rapid electrical activation of the

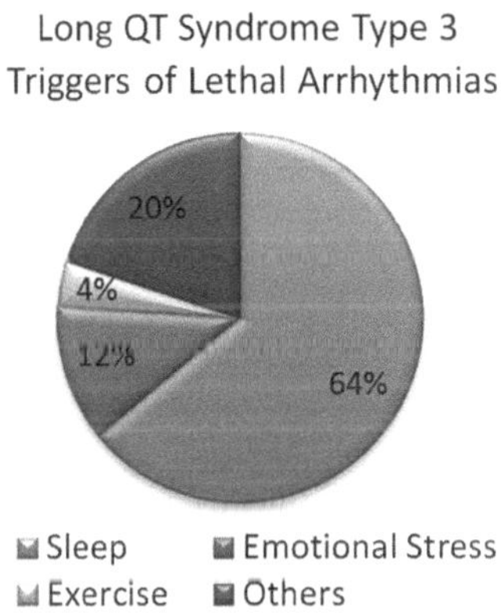

Figure 3 In Long QT syndrome type 3 patients, the events triggering lethal ventricular arrhythmias are variable, ranging from sleep (64%), emotional stress (12%), and exercise (4%) (Beaufort-Krol et al., 2005).

atria without P waves on the electrocardiogram (ECG). Gain-of-function *SCN5A* mutations have been linked to familial AFib with structurally normal hearts (lone AFib) (Li et al., 2009; Makiyama et al., 2008). Reported in lone AFib patients, mutations M1875T and K1493R both show positively shifted steady-state inactivation without elevated I_{NaL}. Consequently, M1875T produces a larger peak I_{Na} (Makiyama et al., 2008), while K1493R has a large ramp I_{Na} at the resting membrane potential (Li et al., 2009). Increased I_{Na} can cause early afterdepolarizations or repolarization impairment to induce AFib. It can also increase the conduction velocity, which favors the maintenance of the fibrillation wave or induces vulnerability to AFib. Nevertheless, family members with the same mutation M1875T exhibit different phenotypes as shown in Fig. 4. While most of

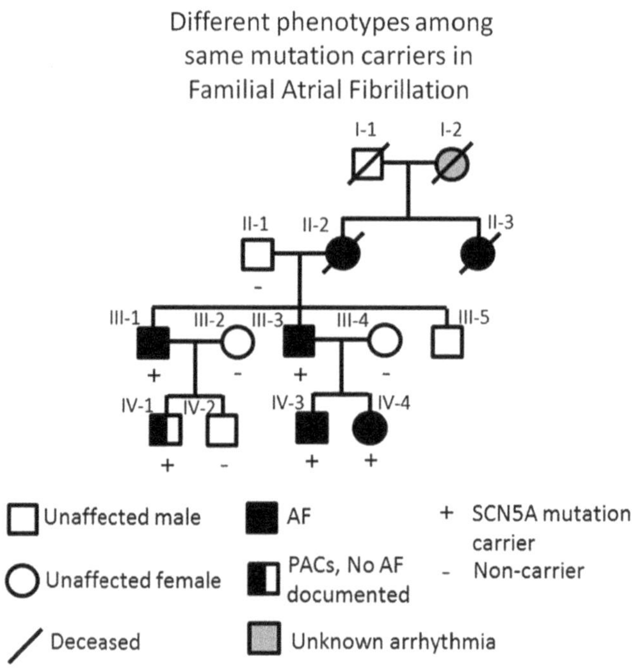

Figure 4 Phenotypic variability and incomplete penetrance have been reported among family members with *SCN5A* mutations causing familial AFib (Makiyama et al., 2008). When most of the family members carrying M1875T showed AFib, member IV–1 had premature atrial contractions instead of AFib. Abbreviation: *AFib*, atrial fibrillation; *PACs*, premature atrial contractions. *Reprinted from Makiyama, T., Akao, M., Shizuta, S., Doi, T., Nishiyama, K., Oka, Y, … Horie, M. (2008). A novel SCN5A gain-of-function mutation M1875T associated with familial atrial fibrillation. Journal of the American College of Cardiology, 52, 1326–1334.*

the family members carrying M1875T showed AFib, member IV-1 has only premature atrial contractions instead of AFib (Makiyama et al., 2008).

3.2 Phenotypes with $Na_V1.5$ loss-of-function mutations

Clinical manifestations vary more with loss-of-function $Na_V1.5$ mutations, which present as decreased or undetectable I_{Na}. Decreased I_{Na} is thought to lead to an abnormally short action potential (Fig. 2A) or even no action potential and confer arrhythmic risk by virtue of slowing conduction and favoring reentry. In the case of loss-of-function mutations, clinically relevant arrhythmias can involve any part of the cardiac conduction system and present as sinus node dysfunction, atrial arrhythmias, CCDs, ventricular arrhythmias, and combinations thereof.

3.2.1 Brugada syndrome

Mutations in *SCN5A* account for ~20% of subjects with BrS. There are over 100 loss-of-function *SCN5A* mutations linked to BrS to date (Amin, Verkerk, Bhuiyan, Wilde, & Tan, 2005; Baroudi, Acharfi, Larouche, & Chahine, 2002; Baroudi, Napolitano, Priori, Del Bufalo, & Chahine, 2004; Baroudi et al., 2001; Vatta et al., 2002). Decreased peak I_{Na} prompts premature repolarization, especially in the right ventricle, causing an action potential shortening and ST-segment elevation in the right precordial leads (Fig. 5). I_{Na} reduction results from less or no functional channel membrane expression (such as nonfunctional G1408R with no trafficking defect (Kyndt et al., 2001) and R1432G being trapped in the endoplasmic reticulum (ER) (Baroudi et al., 2001)) and channel gating defects (such as R1232W and T1620M (Chen et al., 1998) with more rapid recovery from inactivation, and G351V (Vatta et al., 2002) showing a slow decay of currents). With a constant genetic background, there is phenotypic variability in the ECG manifestations of the disease and the timing of fatal events. A spontaneous BrS ECG can be triggered by fever and pharmaceutical medications such as antiarrhythmic $Na_V1.5$ channel blockers (Dumaine et al., 1999; Junttila et al., 2008).

3.2.2 Sick sinus syndrome

SSS is characterized by various forms of arrhythmia including sinus bradycardia, sinoatrial arrest/block, and alternating bradycardia—tachycardia. Although $Na_V1.5$ is not essential for the pacemaking activity of sinoatrial node, it does contribute to cardiac automaticity. At least 20 *SCN5A* mutations induced inherited SSS and showed decreased or nondetectable I_{Na}

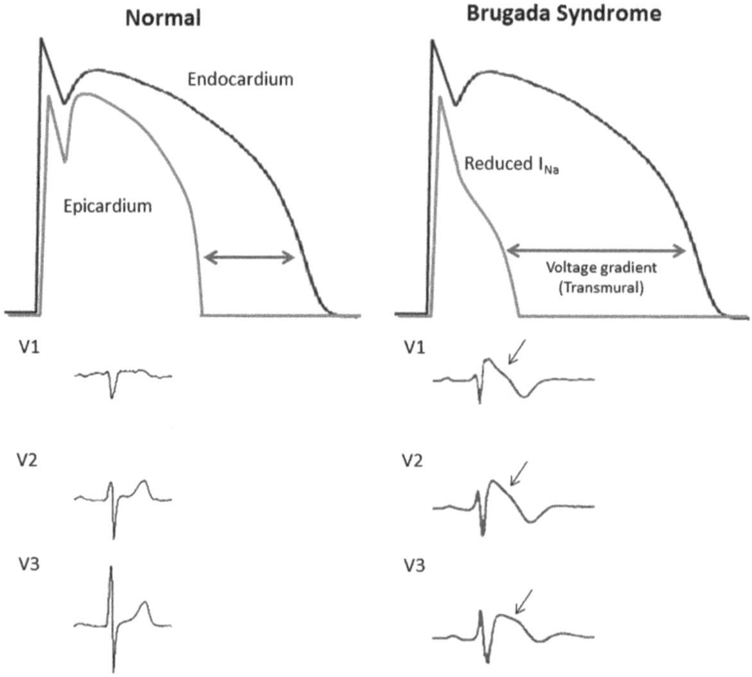

Figure 5 Schematic epi- and endocardial action potential waveforms and ECG recordings from normal and Brugada syndrome patients. In Brugada syndrome, reduced I_{Na} during phase 1 of the action potential leads to markedly abbreviated action potential waveforms in the epicardium, resulting in a widened transmural voltage gradient and the characteristic ST-segment elevation in V1 through V3 on the ECG recording. *Adapted from Fig. 1 of Nademanee, K., Veerakul, G., Nimmannit, S., Chaowakul, V., Bhuripanyo, K., Likittanasombat, K., …Tatsanavivat, P. (1997). Arrhythmogenic marker for the sudden unexplained death syndrome in Thai men. Circulation, 96, 2595–2600.*

(such as E161K (Smits et al., 2005), T187I, and R1623X (Makiyama et al., 2005)). In these cases, reduction in I_{Na} results from decreased membrane expression of the mutant channel proteins or altered gating properties such as delayed activation and earlier inactivation. Studies suggest that SSS induced by I_{Na} reduction involves reduced automaticity and conduction slowing or conduction exit block from the sinoatrial node to the surrounding atrial muscle (Lei et al., 2005; Lei, Huang, & Zhang, 2008).

3.2.3 *Cardiac conduction diseases*

CCD can be caused by loss-of-function $Na_V1.5$ mutations with reduced action potential upstroke velocity and therefore reduced conduction velocity. Such defects can lead to atrial standstill (such as L212P (Makita et al.,

2005) and D1275N (Groenewegen et al., 2003)), atrioventricular block or bundle branch block. Progressive CCDs (PCCDs) present as progressive prolongation of the conduction parameters and bundle branch blocks without ST-segment elevation or QT interval prolongation (Amin et al., 2010). A complete atrioventricular block may occur with PCCD and cause syncope or sudden death. PCCD is degenerative and worsens with age. Several mutations of *SCN5A* showing reduced or undetectable I_{Na} have been found in inherited PCCD. The reduction in I_{Na} can be the result of nonfunctional channels (in-frame skipping mutation IVS.22 + 2 T→C (Schott et al., 1999) and G1406R (Kyndt et al., 2001)) or altered gating properties (G514C (Tan et al., 2001)). Nonprogressive conduction diseases induced by *SCN5A* mutations show defects from birth. The nucleotide deletion at position 5280 of *SCN5A* has been reported in a Dutch family with congenital cardiac conduction disorder (Herfst et al., 2003; Schott et al., 1999). This mutation results in a coding frame shift and a premature stop codon in the channel C terminus. Transient expression of this mutated channel shows that the protein does not express on the plasma membrane but accumulates in the perinuclear ER region.

3.2.4 *Atrial fibrillation*

Loss-of-function genetic variants in *SCN5A* have also been linked to lone AFib. One common nonsynonymous polymorphism of *SCN5A*, H558R, present in over one-third of the population, has been shown to increase the susceptibility to lone AFib (Ackerman et al., 2004; Chen, Ballew, Herron, Rodeheffer, & Olson, 2007). H558R has been shown to reduce I_{Na} significantly with no trafficking defect and no changes of steady-state activation and inactivation (Chen et al., 2007; Makielski et al., 2003; Viswanathan, Benson, & Balser, 2003), which indicates the primary defect is a loss of single channel conductance. Another loss-of-function mutation in *SCN5A*, N1986K, causes familial AFib by negatively shifting steady-state inactivation without affecting the peak I_{Na}, resulting in reduced functional availability of $Na_V1.5$ (Ellinor et al., 2008). Reduced I_{Na} decreases the atrial conduction velocity, shortens the atrial reentry wavelength, promotes reentry, and favors AFib.

3.2.5 *Dilated cardiomyopathy*

DCM is characterized by ventricular dilation and systolic dysfunction, which may proceed into symptoms of congestive heart failure. DCM has been reported in patients carrying *SCN5A* missense and truncation mutations. For

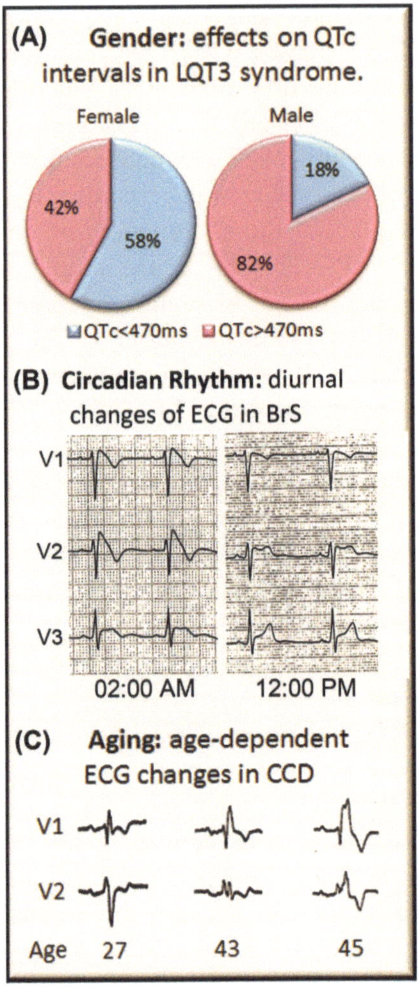

Figure 6 Phenotypic variability is often observed among patients with mutations in $Na_V1.5$ channel gene SCN5A. (A) Sex can affect the phenotype of long QT syndrome type 3. In these patients, men tend to have a longer QTc interval than women (Locati et al., 1998). *(Modified from Liu, M., Yang, K.-C., & Dudley, S.C., Jr. (2014). Cardiac sodium channel mutations: why so many phenotypes? Nature Review Cardiology, 11, 607–615.)* (B) Circadian changes in electrocardiogram patterns and the susceptibility to lethal ventricular fibrillation has been shown in Brugada syndrome. An electrocardiogram that is typical of Brugada syndrome (with ST-segment elevation in the precordial leads) is more frequently observed in the midnight-to-early morning period during sleep than during the day (Mizumaki et al., 2004). *(Reprinted from Mizumaki, K., Fujiki, A., Tsuneda, T., Sakabe, M., Nishida, K., Sugao, M., & Inoue, H. (2004). Vagal activity modulates spontaneous augmentation of ST elevation in the daily life of patients with Brugada syndrome. Journal of Cardiovascular Electrophysiology, 15, 667–673.)* (C) *SCN5A* mutations have been shown to cause progressive cardiac conduction defects, including deterioration

example, mutations R814W and D1595H are associated with DCM patients with atrial and ventricular arrhythmia (Nguyen, Wang, Rhodes, & George, 2008). Biophysical properties of these mutations show loss of channel function associated with diverse changes in gating properties without trafficking defects (Remme & Bezzina, 2010; Nguyen et al., 2008). The mechanism whereby *SCN5A* mutations cause DCM is unclear, but since $Na_V1.5$ forms a macromolecular complex with other accessory proteins, it is speculated that *SCN5A* mutations may disrupt the normal organization of cytoskeletal proteins that are essential for the contractile function and architecture of cardiomyocytes (Amin et al., 2010; Gavillet et al., 2006). Alternatively, the mechanisms underlying *SCN5A* mutation-induced DCM may result from longstanding arrhythmias.

3.2.6 Combinations of phenotypes

Combined phenotypes with a single *SCN5A* mutation are common. Patients with the T1620K mutation (Surber et al., 2008) show concomitant phenotypes of LQT3 and CCD. Patients with E161K exhibit SSS, bradycardia, generalized conduction disease, BrS, or combinations of these phenotypes (Smits et al., 2005). Patients with E1784K exhibit LQT3, BrS, and sinus node dysfunction (Makita et al., 2008). D1275N is associated with age-dependent, variably expressed phenotypes of DCM, AFib, impaired automaticity, and conduction delay (Olson et al., 2005). Patients with the truncated mutant R1860GfsX12 (R at 1860 replaced by G with a reading frame shift and a premature stop code at 1871) have induced SSS, AFib, atrial flutter, and atrioventricular block (Clatot et al., 2013). The family members with 1795insD exhibit not only LQT3 but also BrS (Bezzina et al., 1999). The mutant 1795insD (an insertion of aspartic acid at 1795) gives rise to earlier inactivation and delayed activation, and results in both I_{Na} gain- and loss-of-function: significantly increased I_{NaL} and decreased peak I_{Na} (Bezzina et al., 1999).

The clinical phenotypic penetrance and manifestation of a *SCN5A* mutation also vary with gender, circadian rhythm, and aging (Fig. 6). Gender has been found to influence the clinical phenotype of an individual mutation

in ventricular conduction and widening of the QRS complex, with aging (Probst et al., 2003). *(Modified from Probst, V., Kyndt, F., Potet, F., Trochu, J.N., Mialet, G., Demolombe, S., … Le Marec, H. (2003). Haploinsufficiency in combination with aging causes SCN5A-linked hereditary Lenègre disease. Journal of the American College of Cardiology, 41, 643–652.)* Abbreviation: *QTc*, corrected QT interval.

for many cases. In patients with LQT3, men tend to have longer corrected QT intervals than women (Fig. 6A) (Locati et al., 1998). Furthermore, BrS affects far more men than women (Alings & Wilde, 1999; Kyndt et al., 2001; Priori et al., 2000). Within a single family, four male members carrying a loss-of-function mutation in *SCN5A* (G1406R) had BrS, whereas seven other family members (six of whom were women) carrying the same mutation had CCD (Kyndt et al., 2001). The mechanism of this gender-dependent effect is still uncertain. Sex hormones such as 5-dehydrotestosterone, estrogen, and progesterone might cause differences in QT intervals and gene expression of cardiac ion channels (Kääb, Pfeufer, Hinterseer, äbauer, & hulze-Bahr, 2004). For female patients, hormonal changes owing to menses and pregnancy might induce QT prolongation and increase susceptibility to arrhythmias (Rashba et al., 1996). Premenopausal females have shown a reduced incidence of cardiovascular disease compared with males of same ages. Estrogen plays cardioprotective roles against ischemia reperfusion injury through activation of estrogen receptor (reviewed in (Deschamps, Murphy, & Sun, 2010; Meyer, Haas, & Barton, 2006)). On a channel level, estrogen has been found to induce increased *S*-nitrosylation of the L-type Ca^{2+} channel ($Ca_V1.2$) and result in a cardioprotective phenotype in mouse female hearts (Sun et al., 2006). Serum- and glucocorticoid-induced kinase 1 (SGK1) that is regulated by gonadotropins (Alliston, Gonzalez-Robayna, Buse, Firestone, & Richards, 2000; Richards et al., 1995) has been found to alter $Na_V1.5$ membrane expression and current levels (Boehmer et al., 2003; Das et al., 2012), and therefore could modulate $Na_V1.5$ function and alter cardiac function during pregnancy. An antiestrogen agent, clomiphene, has been found to inhibit I_{Na}, L-type Ca^{2+} current, and some K^+ currents (Borg, Yuill, Hancox, Spencer, & Kozlowski, 2002).

Within an individual patient with an *SCN5A* mutation associated with BrS, phenotypic variability exists in the ECG manifestations of the disease as a function of time. As shown in Fig. 6B, a patient with BrS can show circadian changes (Mizumaki et al., 2004) in ECG patterns and susceptibility to lethal ventricular fibrillation, and ECG manifestations can worsen with fever (Junttila et al., 2008). The phenotype of *SCN5A* mutations can be age-dependent, as well. For example, CCDs can be progressive and worsen with age, as shown in Fig. 6C (Probst et al., 2003). Furthermore, despite carrying the mutation since birth, patients with BrS do not typically develop an arrhythmic phenotype until aged approximately 34–53 years (Gehi, Duong, Metz, Gomes, & Mehta, 2006). Interestingly, the male predominance of an arrhythmic phenotype is not observed in children with BrS (Probst et al., 2007). Age-dependent hormonal changes have been proposed to explain why the risk of spontaneous arrhythmias in men increases after puberty (Matsuo, Akahoshi, Seto, & Yano, 2003; Probst et al., 2007).

In summary, mutations in the cardiac $Na_V1.5$ channel can give rise to multiple clinical phenotypes that can vary between patients, over time, or with different body temperature. This suggests that there are modulating factors that influence the effect of mutations to increase phenotypic variability. In this chapter, we will discuss our hypothesis that numerous modifiers of $Na_V1.5$ are responsible for the various phenotypes. Variability in these modifiers may help explain how a single mutation can cause multiple phenotypes, how a specific region of the heart can be affected, why there can be variable penetrance of a phenotype within a family, or why a phenotype varies with time. In this hypothesis, these modifiers act as second-hits to further modify the $Na_V1.5$ current and function and eventually determine the final phenotype(s) exhibited.

4. SOURCES OF PHENOTYPIC VARIABILITY

For the two groups of potential $Na_V1.5$ modifiers we cataloged, intrinsic modifiers alter the expression or function of a mutated $Na_V1.5$ channel. Extrinsic modifiers change the expression or function of other channels or genes that can modify the overall biophysical properties and electrical effect of a $Na_V1.5$ mutation.

4.1 Intrinsic modifiers of $Na_V1.5$ function

The life cycle of cardiac $Na_V1.5$ includes gene transcription, posttranscription RNA processing, translation (protein synthesis and assembly), posttranslational trafficking and modifications, complex assembly, and degradation, as demonstrated in Fig. 7 (Cusdin, Clare, & Jackson, 2008; Herfst, Rook, & Jongsma, 2004). The half-life of $Na_V1.5$ protein turnover has been reported to be 17–35 h (Maltsev, Kyle, Mishra, & Undrovinas, 2008; Sherman, Chrivia, & Catterall, 2000), considerably longer than most other channels that have shorter half-lives of 1–8 h (Chen-Izu et al., 2015). The life cycle of $Na_V1.5$ is a dynamic process, which can lead to diurnal changes in $Na_V1.5$ expression temporally (channel quantity) and spatially (channel localization) that contribute to the diverse phenotypes and penetration of disease-related $Na_V1.5$ mutations. The life cycle of $Na_V1.5$ is regulated by many intrinsic

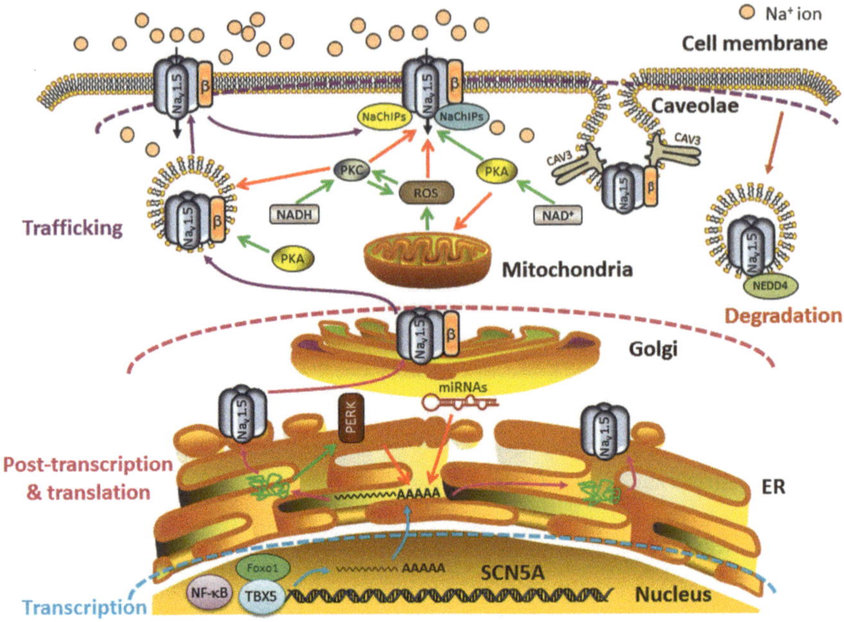

Figure 7 The life cycle of cardiac $Na_V1.5$ and potential phenotypic modifiers. The life cycle starts with *SCN5A* gene transcription (below the *blue dash line*) in the nucleus, which can be regulated by transcription factors, such as FOXO1, NF-κB, and TBX5. The next steps are protein translation and assembly in the endoplasmic reticulum and Golgi apparatus (between the *pink* and *blue dash lines*). Appropriately folded $Na_V1.5$ protein is trafficked to the cell membrane, whereas misfolded protein (as a result of mutations and splicing variants) can activate the PERK pathway of the unfolded protein response to downregulate the *SCN5A* mRNA level. microRNAs also regulate the *SCN5A* mRNA level. Channel trafficking (left and middle parts between the *purple* and *pink dash lines*) can be modulated by PKA, PKC, and caveolae. After being expressed on the cell membrane, NaChIPs join the α and β subunits to form a macromolecular complex, which can be regulated by PKA, PKC, oxidative stress (ROS), and metabolic states (NADH and NAD^+). Eventually, the channel will go through ubiquitin-mediated degradation (the right part between the *purple* and *pink dash lines*), which can be regulated by NEDD4. *Green arrows* indicate activation, and *red arrows* indicate inhibition. *Blue arrows* indicate activities in transcription, *pink arrows* in translation, *purple arrows* in trafficking, and *brown arrows* in degradation. Abbreviations: *ER*, endoplasmic reticulum; *FOXO1*, forkhead box protein O1; *NaChIPs*, Na^+ channel interacting proteins; *NEDD4*, E3 ubiquitin-protein ligase NEDD4; *NF-κB*, nuclear factor NF-κB; *PERK*, eukaryotic translation initiation factor 2α kinase 3; *PKA*, cAMP-dependent protein kinase (protein kinase A); *PKC*, protein kinase C; *ROS*, reactive oxygen species; *TBX5*, T-box transcription factor TBX5. *Modified from Liu, M., Yang, K.-C., & Dudley, S.C., Jr. (2014). Cardiac sodium channel mutations: why so many phenotypes? Nature Review Cardiology, 11, 607–615.*

modifiers, including genetic background (extra *SCN5A* mutations) and any potential modifiers that regulate every stage of the channel's life cycle. These modifiers can amplify, localize, or minimize a mutation-mediated phenotype. Modifiers such as transcription factors, microRNAs (miRNAs), and kinases regulate these processes and result in functional changes of cardiac $Na_V1.5$. These modifiers do not have established roles in modifying inherited $Na_V1.5$ channelopathies, but they can vary with age, gender, clinical state, and between regions of the heart. Conceivably, therefore, variability in these modifiers might contribute to the overall phenotypic variability observed. Moreover, these modifiers are candidates to explain regional localization of $Na_V1.5$ mutation effects.

4.1.1 Multiple mutations of SCN5A

The variability of phenotypes associated with an *SCN5A* mutation can result from another genetic variant that coexists within the gene. For example, two BrS-associated *SCN5A* variants T1620M and R1232W are known to cause minor changes in $Na_V1.5$ channel properties when they exist independently (Chen et al., 1998). When they coexist, however, the T1620M/R1232W double-mutated channel is trapped in the ER and therefore, produces no sodium current (Baroudi et al., 2002). Individuals carrying heterozygous loss-of-function mutations W156X or R225W do not have a clinical phenotype (Bezzina et al., 2003). Compound heterozygosity for W156X and R225W mutations, however, leads to severe conduction disease, DCM, and death at young age (Bezzina et al., 2003). These examples demonstrate that the coexistence of a genetic variant within *SCN5A* can affect the severity of the phenotype associated with another *SCN5A* mutation. Additionally, the *SCN5A* H558R polymorphism, which we discussed before, is another important disease-modifying genetic variation. It is considered as a risk factor for lone AFib since it is more common in patients with lone AFib (Chen et al., 2007). It has also been found to be able to rescue not only a gain-of-function *SCN5A* mutation P2006A by decreasing I_{NaL} but also a loss-of-function *SCN5A* mutation R282H by increasing peak I_{Na}, respectively, as shown in Fig. 8 (Shinlapawittayatorn, Du, et al., 2011; Shinlapawittayatorn, Dudash, et al., 2011). These observations provide a potential mechanism to account for the variable penetrance that is often observed with a *SCN5A* mutation. Taken together, the presence of genetic modifiers within *SCN5A* may contribute to the phenotypic variability, even in familial cases sharing a single mutation.

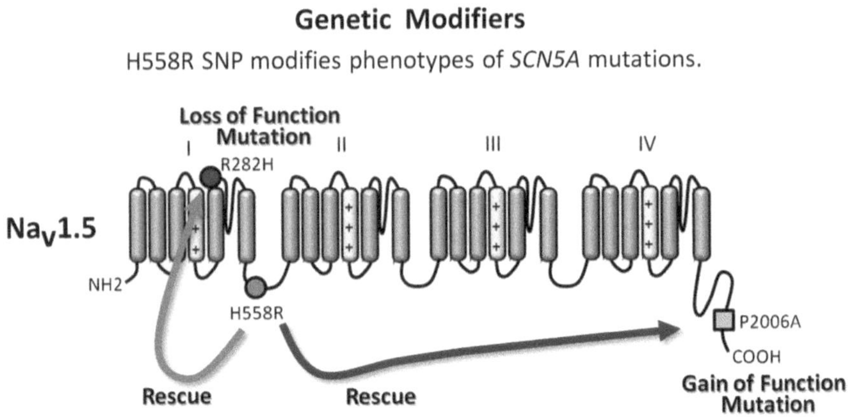

Figure 8 Genetic variants within or outside the gene can alter the phenotype of a *SCN5A* mutation. For example, the *SCN5A* H558R polymorphism has been shown to rescue the phenotype of P2006A and R282H mutations in *SCN5A*. P2006A induced a decreased peak I_{Na} that was restored with H558R (the *green arrow* (light gray in print versions)), while R282H resulted in an increased I_{NaL} that was decreased by H558R (the *red arrow* (dark gray in print versions)) (Shinlapawittayatorn, Du, et al., 2011; Shinlapawittayatorn, Dudash, et al., 2011). *Modified from Liu, M., Yang, K.-C., & Dudley, S.C., Jr. (2014). Cardiac sodium channel mutations: why so many phenotypes? Nature Review Cardiology, 11, 607–615.*

4.1.2 Transcriptional modifiers

Transcriptional regulation of *SCN5A* (lower part of Fig. 7) can lead to either up- or downregulation of Na$_V$1.5 membrane expression and macroscopic I_{Na}, which will modify the upstroke phase 0 of the cardiac action potential. This change in current can enhance or mask the phenotype of a mutation. Moreover, regional differences in transcriptional regulation could explain how various areas of the heart and conduction system are most affected.

4.1.2.1 Transcriptional factors of SCN5A

Some transcriptional factors play essential roles in signaling normal *SCN5A* expression, dysregulation of which can disrupt *SCN5A* expression, affect channel protein membrane expression, and therefore, cause phenotypic variability. T-box transcription factor 5 (TBX5) is essential for Na$_V$1.5 expression in the ventricular conduction system, removal of which causes a dramatic reduction in membrane expression of Na$_V$1.5 and gap junction α5 protein (also known as connexin 40), leading to conduction velocity slowing, cardiac arrhythmias, and increased animal mortality (Arnolds et al., 2012). Alterations in TBX5 expression could modulate I_{Na} in the conduction system and contribute to conduction-specific phenotypes. *ARNTL* (also called *BMAL1*), encoding a core molecular clock transcription factor, also plays an important role in *SCN5A* expression and cardiac arrhythmia susceptibility. Cardiomyocyte-specific deletion of *Arntl* in mice causes slowed heart rate, prolonged electrocardiographic RR and QRS intervals, and increased episodes of arrhythmia. These effects result from loss of *Scn5a* expression (Schroder et al., 2013). Alterations in *ARNTL* may help explain diurnal variations in arrhythmic risk in the presence of a given mutation.

The *SCN5A* promoter region contains consensus binding sites for several transcription factors, such as nuclear factor-κB (NF-κB) and Forkhead box O 1 (FOXO1). When activated by angiotensin II or oxidative stress, NF-κB and FOXO1 are upregulated and induce a significant decrease in $Na_v1.5$ expression and I_{Na} (Mao et al., 2012; Shang et al., 2008). Interestingly, transforming growth factor β1 can counteract FOXO1's effect and increase the *SCN5A* mRNA level and I_{Na} (Kaur et al., 2013). This suggests that the phenotypic penetrance of a mutation could vary as a result of changes in these transcription factors. Additionally, single nucleotide polymorphisms in the promoter region of *SCN5A* have been shown to modulate the transcriptional activity of the gene (Yang, Kupershmidt, & Roden, 2004) and to affect the phenotypic severity in heterozygous carriers of an *SCN5A* loss-of-function mutation (1936delC) (Park, Martin, Zhang, Jegga, & Benson, 2012). These data suggest that genetic variants in the *SCN5A* promoter region can contribute to the phenotypic variability and penetrance of *SCN5A* mutations through regulating *SCN5A* transcription.

4.1.2.2 mRNA processing and regulation

The phenotypes of $Na_v1.5$ mutations can be modified at the stage of mRNA processing. Recently, in human heart failure, our group has shown that angiotensin II and hypoxia can regulate *SCN5A* mRNA splicing and generate truncated, nonfunctional Na^+ channels (Gao et al., 2011). These abnormally spliced, truncated channels accumulate in the ER and activate the unfolded protein response (UPR) sensor, eukaryotic translation initiation factor 2α kinase 3 (also known as PERK; Fig. 7, lower middle on the ER membrane) (Gao et al., 2013). As shown in Fig. 9A, the protein level of PERK is elevated significantly in AngII treatment and in human failing heart tissue (Gao et al., 2013). As shown in Fig. 9B–D, activated PERK shows a dominant negative effect on the full-length *SCN5A* mRNA transcript and leads to a significant reduction of *SCN5A* mRNA, $Na_v1.5$ protein

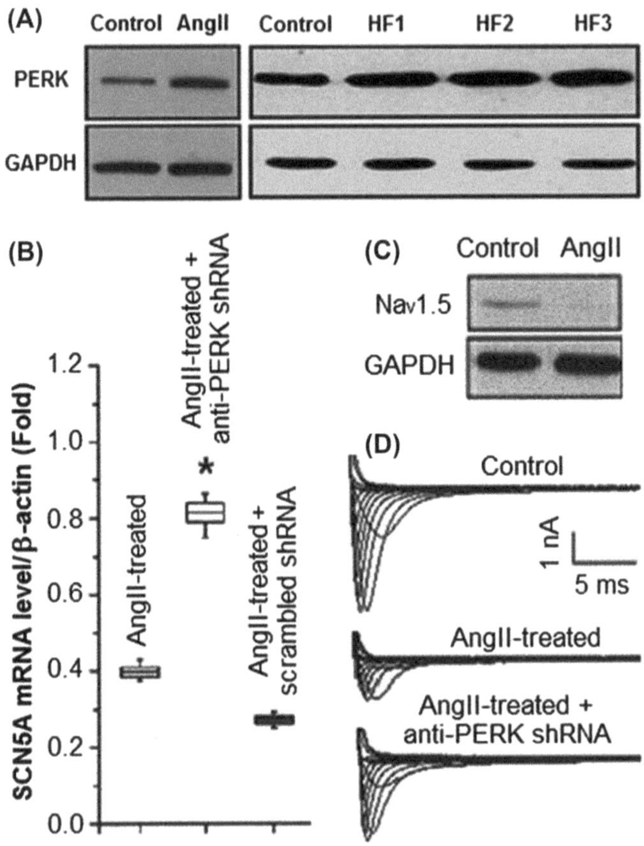

Figure 9 Activated PERK in human heart failure and its downregulation on $Na_v1.5$. (A) PERK is upregulated in hiPSC-CMs with AngII treatment and in human HF tissue samples (from three patients), confirmed by Western blot quantification. AngII (200 nM 24 h treatment) decreases (B) mRNA, (C) protein, and (D) I_{Na} levels of $Na_v1.5$ in hiPSC-CMs, which can be restored by preinfection of anti-PERK short hair RNA. *Panel A, B, and D are modified from Gao, G., Xie, A., Zhang, J., Herman, A.M., Jeong, E.M., Gu, L., ... Dudley, S.C. (2013). Unfolded protein response regulates cardiac sodium current in systolic human heart failure. Circulation: Arrhythmia and Electrophysiology, 6, 1018–1024. Panel C is modified from Gao, G., Xie, A., Huang, S.C., Zhou, A., Zhang, J., Herman, A.M. ..., Dudley, S.C., Jr. (2011). Role of RBM25/LUC7L3 in abnormal cardiac sodium channel splicing regulation in human heart failure/clinical perspective. Circulation, 124, 1124–1131.*

membrane expression, and I_{Na} (Gao et al., 2011, 2013). This could mask a gain-of-function *SCN5A* mutation or amplify a loss-of-function *SCN5A* mutation, leading to phenotypic variability. Meanwhile, angiotensin II can be differentially regulated in local regions of the heart. For example, angiotensin II is elevated in ventricular cells but not in atrial cells of type 1 diabetic rats (Shimoni, Hunt, Chen, Emmett, & Kargacin, 2006). Combining local regulation with induction of abnormal mRNA processing could contribute to a phenotype that varies with clinical states.

miRNAs are endogenous noncoding small ribonucleic acids emerging as important regulators for gene expression by either restricting translation or inducing degradation of the target mRNA. Altered expression of miRNAs, mainly upregulation, has been widely reported to reduce the expression of cardiac ion channels in the pathological studies of cardiovascular diseases (Luo, Zhang, Xiao, & Wang, 2010; Yang, Lu, & Wang, 2008; Zicha, Maltsev, Nattel, Sabbah, & Undrovinas, 2004). In heart failure, upregulated miR-125a/b is associated with suppression of $Na_v1.5$ protein expression and reduction of I_{Na} (Luo et al., 2010). A stress-inducible miRNA, miR-195, which is upregulated in cardiac hypertrophy and heart failure (van Rooij et al., 2006), can downregulate *SCN5A*, leading to I_{Na} reduction (Yang et al., 2008). In myocardial infarction, upregulation of let-7f miRNA precursor and miR-378 also cause I_{Na} reduction (Luo et al., 2010). Therefore, changes in miRNAs at different cardiac disease stages may contribute to phenotypic variability. A recent study reveals several miRNAs modulating *Scn5a* transcription in mouse HL-1 cells: miR-98, miR-106, miR-200, miR-125, and miR-153 decrease *Scn5a* expression level; and miR-219 increases *Scn5a* level (Daimi et al., 2015). Any changes or unbalanced changes of these miRNAs (most of which downregulate *Scn5a* transcription while miR-219 upregulates it) may change the phenotype and penetrance of the mutation carrier patients.

Other miRNAs may have indirect effects on the $Na_v1.5$. For example, miR-199a is downregulated after induction of ischemia, leading to an upregulation of its target hypoxia-inducible factor 1 α. This factor can activate RBM25 (Zhou, Ou, Cho, Benz, & Huang, 2008) and enhance abnormal *SCN5A* mRNA splicing to generate truncated, nonfunctional Na^+ channels (Gao et al., 2011, 2013). In human AFib and ventricular arrhythmia, miRNA-1 is decreased, which can induce upregulation of the Kir2.1 potassium channel (Girmatsion et al., 2009). This channel conducts an inward potassium current (I_{K1}) that controls the resting membrane potential. An increase in I_{K1} can lead to membrane hyperpolarization, which can decrease the availability of activated $Na_v1.5$. This can either amplify the reduction in I_{Na} induced by a *SCN5A* loss-of-function mutation or minimize the increase in I_{NaL} induced by a *SCN5A* gain-of-function mutation. In a canine tachy-paced model of atrial fibrillation, the level of miRNA-328 was increased, and led to a reduction in L-type Ca^{2+} current and a shortened atrial APD (Lu et al., 2010). These changes might combine with those of $Na_V1.5$ loss-of-function mutations to worsen arrhythmic risk.

4.1.3 Translational modifiers

At the translational level, many modifiers regulate cardiac $Na_V1.5$ protein synthesis, folding, trafficking, and assembling with accessory subunits and interacting proteins. For example, activated mitogen-activated protein kinases (MAPKs) such as p38-MAPK and MAPK8 in human HF (Cook, Sugden, & Clerk, 1999; Muslin, 2008) can affect translation of $Na_V1.5$ channel and connexin 43 and contribute to arrhythmia (van Bemmelen et al., 2004; Gasser et al., 2010; Yanagita et al., 2003).

4.1.3.1 Modifiers of $Na_V1.5$ protein translation, folding, and assembling in the endoplasmic reticulum

$Na_V1.5$ protein folding in the ER can be regulated by the UPR. When the ER homeostasis is disturbed, unfolded, or misfolded proteins accumulate in the ER lumen, triggering the UPR (Fig. 7, lower middle part). The UPR is involved in ischemia/reperfusion (Doroudgar, Thuerauf, Marcinko, Belmont, & Glembotski, 2009; Thuerauf et al., 2006) and human heart failure (Gao et al., 2013) and can either protect or impair heart function (reviewed in (Liu & Dudley, 2014)). As discussed in Section 4.1.2.2, our group has recently found that activated PERK downregulates *SCN5A* mRNA transcription and leads to a significant reduction of $Na_V1.5$ membrane expression and I_{Na} in human failing heart (Fig. 9) (Gao et al., 2013). This could lead to phenotypic variability by ameliorating the phenotype of a gain-of-function *SCN5A* mutation or enhancing the phenotype of a loss-of-function *SCN5A* mutation. A recent study also indicates that protein misfolding and oligomerization are associated with infiltrative structural damages in AFib (Sidorova et al., 2015).

4.1.3.2 Modifiers of $Na_V1.5$ trafficking to the plasma membrane

After *SCN5A* is translated to protein and assembled in the ER, the channel is transited to the Golgi complex, where $Na_V1.5$ channel trafficking begins. The Golgi vesicle containing the $Na_V1.5$ channel is then transported to specific subdomains of the plasma membrane such as the intercalated discs, T-tubules, and lateral membrane (Lin et al., 2011). Gap junction α-1 protein (also known as connexin 43) and ankyrin (Ank) has been recently reported to help $Na_V1.5$ localize at the intercalated discs (Jansen et al., 2012; Mohler et al., 2004). Interactions between the $Na_V1.5$ α subunit and proteins such as syntrophins and dystrophin has recently been found to be important for $Na_V1.5$ to localize at the lateral membrane area (Rougier, Gavillet, & Abriel, 2013; Shy et al., 2014). Any changes of these modifiers will affect the $Na_V1.5$ distribution and may affect the phenotype and penetrance of a $Na_V1.5$ mutation.

The channel trafficking efficiency to the plasma membrane can be regulated by modifiers such as microtubule plus-end trafficking proteins (Smyth et al., 2010), actin cytoskeletal proteins (Lu et al., 2009; Rafizadeh et al., 2014), cAMP-dependent protein kinase (also known as protein kinase A; PKA) and protein kinase C (PKC) (Fig. 7, upper left part). For instance, PKC activation that is reported in HF (Konishi et al., 1997; Liu et al., 2009) can decrease $Na_V1.5$ channel trafficking to the plasma membrane, while PKA can play an opposite role (Hallaq et al., 2012; Zhou et al., 2002). This suggests that a patient's clinical state can potentially affect the penetrance of a $Na_V1.5$ mutation.

There are small invaginations of the cell membrane called caveolae (Fig. 7, upper right part on the cell membrane). Caveolae are enriched in signaling molecules, such as caveolin-3, PKC, PKA, Src family tyrosine kinases, and G proteins (Razani, Woodman, & Lisanti, 2002; Rybin, Xu, Lisanti, & Steinberg, 2000; Smart et al., 1999). Colocalization of the $Na_V1.5$ protein and caveolin-3 in caveolae is observed in ventricular tissues (Yarbrough, Lu, Lee, & Shibata, 2002), indicating caveolae are potential $Na_V1.5$ modifiers. Caveolae constitute a reservoir of I_{Na} that can be modulated by opening or closing of the caveolar neck (Levin & Page, 1980). Meanwhile, the other signaling molecules in caveolae regulate $Na_V1.5$ expression and function. A complex formed with PKA, caveolin-3 and $Gs\alpha$ in caveolae modulates $Na_V1.5$ function (Palygin, Pettus, & Shibata, 2008; Rybin et al., 2000). PKA and PKC regulate $Na_V1.5$ trafficking (Hallaq et al., 2012; Zhou et al., 2002) and gating properties (Liu, Hanck, & Dudley, 2013; Liu, Liu, & Dudley, 2010; Liu et al., 2009) by phosphorylation. A caveolin-3 mutation results in an approximately three-fold increase of late I_{Na}, which could enhance the penetrance of LQT3 and AFib induced by $Na_V1.5$ gain-of-function mutations (Vatta et al., 2006). Any changes of these molecules could affect $Na_V1.5$ membrane availability and alter the phenotype of $Na_V1.5$ mutations.

$Na_V1.5$ is also found to autoregulate its own trafficking. Coexpression of wild-type and mutant $Na_V1.5$ has shown reduced $Na_V1.5$ membrane expression and I_{Na} (Chen-Izu et al., 2015). Possible explanation for this phenomenon is that trafficking of a cluster of the wild-type α subunits of $Na_v1.5$ channel is more efficient than that of a cluster of mutant/wild-type α subunits. This may explain the different penetrance among the same $Na_v1.5$ mutation carrier patients but with homo- or heterozygous mutations.

4.1.3.3 Modifiers of the channel complex

The α subunit of $Na_v1.5$ interacts with accessory proteins (β subunits and interacting proteins) and forms a macromolecular complex on the cell membrane (Fig. 1A) (reviewed in Shy et al., 2013). Mutations and misregulation of these accessory proteins can disturb $Na_v1.5$ membrane expression and function and have been reported to cause cardiac death syndromes. For example, patients with BrS and mutations in the channel β1 and β3 subunits (encoded by *SCN1B* and *SCN3B*) show impaired $Na_v1.5$ channel function (Amin et al., 2010; Hu et al., 2009; Watanabe et al., 2008). A recent study with *Scn1b* null mice showed that loss of *Scn1b* induces an increase of *Scn3a* mRNA, disturbed Ca^{2+} homeostasis, prolonged APD, delayed afterdepolarizations, and increased susceptibility to polymorphic ventricular arrhythmias at the heart level, all of which will modify $Na_v1.5$ mutation-induced phenotypes (Lin et al., 2015). The expression level of the β4 subunit affects the severity of CCD in mice by modulating $Na_v1.5$ channel activation (Remme et al., 2009).

$Na_v1.5$ interacting proteins such as Ank, Ran guanine nucleotide release factor (MOG1), and syntrophin interact with the α and β subunits of Na^+ channel. Ank-G and Ank-B regulate $Na_v1.5$ localization. Ank-B also regulates the trafficking and/or membrane targeting of Na^+-Ca^{2+} exchanger (NCX), $Ca_v1.3$, the Na^+/K^+-ATPase, and inositol-1,4,5-trisphosphate receptors (reviewed in Cunha & Mohler, 2011). A mutation in Ank-B causes LQT syndrome type 4, resulting from disruption of Ank-B interactions with these channels and receptors (Mohler et al., 2003). Mutations in the MOG1 gene *RANGRF* are found in patients with BrS (E83D) (Kattygnarath et al., 2011) and AFib (E61X) (Olesen et al., 2011). The A257G syntrophin mutation induces LQT3 because of a gain-of-function modulation of $Na_v1.5$ gating properties (Wu et al., 2008). Fibroblast growth factor homologous factor 13 has been found recently to bind directly to $Na_v1.5$ in the sarcolemma membrane of adult mouse ventricular myocytes, and knockdown of this factor causes significant reduction of $Na_v1.5$ membrane expression and I_{Na} (Wang et al., 2011). Therefore, differential expression and regulation of accessory proteins may modulate the $Na_v1.5$ mutation-induced phenotypes.

4.1.4 Posttranslational modifiers

Posttranslational modifications of $Na_V1.5$ have been investigated broadly due to their critical modulations on $Na_V1.5$ functions under both the physiological and pathological conditions, such as phosphorylation, redox modifications, methylation, and glycosylation (for a recent review, see Marionneau & Abriel, 2015).

4.1.4.1 Protein kinase A and protein kinase C

The most extensively studied posttranslational modification of cardiac $Na_V1.5$ is channel phosphorylation by protein kinases such as PKA, PKC, Ca^{2+}/calmodulin-dependent protein kinase II (CaMKII), serum- and glucocorticoid-induced kinases (SGK), and phosphatidylinositol 3-kinase (PI3K). CaMKII and PI3K have been reported to regulate I_{NaL} although with some controversial results (reviewed in Ballou, Lin, & Cohen, 2015; Marionneau & Abriel, 2015). SGK has been found to increase $Na_V1.5$ membrane expression and peak I_{Na} by suppressing $Na_V1.5$ degradation (Boehmer et al., 2003). SGK also plays a role on regulating I_{NaL}, mutations of which (S433D and K127M) leads to altered I_{NaL} (Das et al., 2012). Here we will focus on PKA and PKC, which regulate the $Na_V1.5$ channel membrane expression and gating properties (Fig. 7, upper middle part) (Hallaq et al., 2012; Konishi et al., 1997; Liu, Hanck, et al., 2013; Liu et al., 2009; Zhou et al., 2002). Summated effects of these modifiers with the baseline genetic defect may explain phenotypes of $Na_V1.5$ mutations that vary with time, patient states, and regions of the heart.

PKCs are a family of serine/threonine protein kinases with at least 15 isoforms in humans, with the PKCα, δ, and ε isoforms being most intensively investigated in cardiomyocytes (Poole, Pula, Hers, Crosby, & Jones, 2004). PKC has been found to decrease the cardiac I_{Na} by channel redistribution away from plasma membrane (Hallaq et al., 2012) and by reducing channel conductance (Liu, Hanck, et al., 2013). A PKC activator phorbol 12 myristate 13-acetate (PMA) induced significant I_{Na} reduction of human cardiac $Na_V1.5$ (Liu et al., 2009). S1503 has been identified as a PKC phosphorylation site (West, Numann, Murphy, Scheuer, & Catterall, 1991), mutation of which blocks PKC effect on both I_{NaL} (Tateyama, Kurokawa, Terrenoire, Rivolta, & Kass, 2003) and peak I_{Na} (our unpublished data). This site has also been found to be very important in PKC stimulation of LQT3-associated mutations of *SCN5A* (Y1795C, Y1795H, and ΔKPQ). PKC activation significantly decreased I_{NaL} in the mutants but not in the wild-type channels (Tateyama, Kurokawa, et al., 2003). Recently, our group found that the isoform PKCδ (our unpublished data) reduces I_{Na} by inducing overproduction of mitochondrial reactive oxygen species (ROS) (Liu, Gu, et al., 2013; Liu, Hanck, et al., 2013; Liu et al., 2010, 2009,), which in turn decreases the single channel conductance (Liu, Hanck, et al., 2013). This effect occurs in minutes with no change in *SCN5A* mRNA abundance or channel membrane expression, suggesting these effects are in addition to any effect on channel trafficking. In summary, PKC activation generally reduces I_{Na} and could exacerbate the phenotype of loss-of-function *SCN5A* mutations or mitigate the phenotype of gain-of-function *SCN5A* mutations.

PKA is a tetrameric serine/threonine kinase with broad specificity that is known to upregulate $Na_v1.5$ through changing the channel trafficking (Zhou et al., 2002) and conductance (Liu, Hanck, et al., 2013). A PKA activator forskolin can restore peak I_{Na} decreased by NADH or mitochondrial ROS overproduction (Liu et al., 2010, 2009). PKA phosphorylation at S36 and S525 enhances I_{NaL} of mutant channel D1790G associated with LQT3 (Tateyama, Rivolta, Clancy, & Kass, 2003). Important sites of PKA regulation have also been identified at S528, R526, and RRR533-535 on human cardiac $Na_v1.5$ (Aiba et al., 2014; Hallaq et al., 2006; Zhou et al., 2002; Zhou, Yi, Hu, George, Jr., & Murray, 2000), mutations at which block PKA effects on $Na_v1.5$ upregulation. Our group has found that RRR533-535 is important for the PKA-dependent increase and NADH-dependent decrease in peak I_{Na} (our unpublished data and (Liu, Hanck, et al., 2013). PKA may upregulate $Na_v1.5$ through minimizing mitochondrial ROS overproduction, which downregulates $Na_v1.5$ (Liu et al., 2010, 2009). Furthermore, PKA activation of connexin 43 stimulates gap junctional intercellular communication and alters the electrical distance over which $Na_v1.5$ mutations have their effects (Atkinson et al., 1995; Chanson, White, & Garber, 1996; Darrow, Fast, Kleber, Beyer, & Saffitz, 1996).

Cardiac ion channel dephosphorylation is regulated by serine/threonine protein phosphatases (PPs). It has not been identified which subtype of PPs regulates $Na_v1.5$ dephosphorylation and the regulating mechanisms are still unknown. Nevertheless, a phosphatase inhibitor okadaic acid can increase peak I_{Na} in canine ventricular myocytes (Baba, Dun, & Boyden, 2004), indicating PPs modulation of $Na_v1.5$. PP type-2A (PP2A) has been found to associate with rat brain Na^+ channels (Chen, Law, Kondratyuk, & Rossie, 1995). PP type-1 (PP1) and PP2A are major subtypes expressed ubiquitously in cardiac tissue and have been shown to regulate $Ca_v1.2$, K_vLQT1, NCX, the Na^+/K^+-ATPase (NKA), connexin 43, and some Ca^{2+} handling proteins such as sarco/endoplasmic reticulum Ca^{2+}-ATPase, ryanodine receptor 2, and phospholemman (Christ et al., 2004; El-Armouche et al., 2006; Heijman, Dewenter, El-Armouche, & Dobrev, 2013). PP1 expression is increased in HF (Nicolaou & Kranias, 2009) and in chronic AFib (Christ et al., 2004; El-Armouche et al., 2006), which could alter phosphorylation of a $Na_V1.5$ mutation and therefore the phenotype. Both up- and downregulation of PP2A have been reported in cardiomyopathy (Heijman et al., 2013), which may also affect a $Na_V1.5$ mutation function and modify any mutation's phenotype.

4.1.4.2 Reactive oxygen species downregulation on $Na_V1.5$

ROS elevation is a common feature shared by various cardiac diseases (Dhalla & Singal, 1994; Hill & Singal, 1997). Increasing evidence suggests that elevated ROS alter the cardiac ion channels (such as $Na_V1.5$, $K_V1.4$, $K_V4.2$, $K_V4.3$, and $Ca_V1.2$), Ca^{2+} handling proteins, and gap junction proteins (Choudhary & Dudley, 2002; Jeong et al., 2011; Liu et al., 2010, 2009). For $Na_V1.5$, ROS can downregulate the channel at the transcriptional level by reducing the *SCN5A* gene expression (Gao et al., 2013; Shang et al., 2008) and at the posttranslational level probably by affecting channel phosphorylation and conductance (Liu, Gu, et al., 2013; Liu, Hanck, et al., 2013). Therefore, fluctuation in ROS is another possible contributor to the phenotypic variability of ion channel mutations. Major sources of cardiac ROS include mitochondrial electron transport chain, uncoupled nitric oxide synthase, NAD(P)H oxidase, cyclooxygenase, and xanthine oxidase. Mitochondrial ROS overproduction is observed in a heart failure mouse model and leads to $Na_V1.5$ downregulation and I_{Na} reduction (Fig. 7, upper middle part) (Liu, Gu, et al., 2013). Mitochondrial ROS can also activate PKC and CaMKII, which can decrease peak I_{Na} by phosphorylation (Liu, Gu, et al., 2013) or impairing channel trafficking, and increase I_{NaL}, respectively (Valdivia, Ueda, Ackerman, & Makielski, 2009; Wagner et al., 2011). Antimycin A, a complex III inhibitor of the mitochondrial electron transport chain that induces mitochondrial ROS overproduction, decreases peak I_{Na} significantly and that a mitochondrial targeted scavenger mitoTEMPO can restore peak I_{Na} decreased by elevated NADH or PKC activation (by PMA). Hydrogen peroxide treatment has been found to decrease peak I_{Na} and increase I_{NaL} (Shang et al., 2008; Wagner et al., 2011). The changes of currents could lead to APD prolongation. ROS-mediated lipid peroxidation products are also potent modifiers of $Na_V1.5$ (reviewed in (Boyden et al., 2007)). For instance, isoketals, which are formed as products of arachidonic acid peroxidation and accumulated in the epicardial border zone of canine infarcted hearts, induce similar downregulation on $Na_V1.5$ and I_{Na} as hydrogen peroxide by changing channel gating function (Fukuda et al., 2005). Therefore, ROS can influence $Na_V1.5$ mutation behavior. Enzymes such as superoxide dismutase, glutathione peroxidase, and catalase form the defensive system against oxidative stress. Upregulation of these enzymes could counteract ROS downregulation on $Na_V1.5$. Our group has found that superoxide dismutase and mitochondrial-targeted scavenger mitoTEMPO can block the downregulation of mitochondrial ROS on $Na_V1.5$ and I_{Na} (Liu, Gu, et al., 2013; Liu et al., 2010, 2009). A summary scheme of effects of PKC, PKA, and mitochondrial ROS on cardiac $Na_V1.5$ is shown in Fig. 10.

4.1.4.3 Alterations in metabolism

HF and cardiac injury from many causes are associated with altered metabolism and downregulated $Na_V1.5$ (Pillai, Isbatan, Imai, & Gupta, 2005; Pu & Boyden, 1997; Ufret-Vincenty et al., 2001; Valdivia et al., 2005). NADH is known to oscillate with myocardial ischemia, and mitochondrial injury is associated with increased NADH and ROS levels (Aon, Cortassa, Marban, & O'Rourke, 2003; Di, Menabo, Canton, Barile, & Bernardi, 2001). Elevated NADH can result in PKC activation, mitochondrial ROS overproduction, and I_{Na} reduction (Fig. 10). Many such alterations could

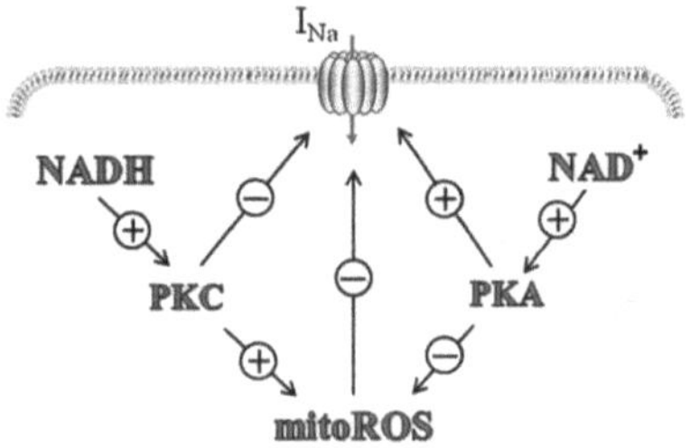

Figure 10 A scheme of the signaling pathways of NADH/NAD^+ regulation on cardiac $Na_V1.5$ by modulating PKC/PKA activation and mitochondrial ROS generation (Liu et al., 2010, 2009). NADH downregulates $Na_V1.5$ via PKC activation that induces mitochondrial ROS overproduction. On the other hand, NAD^+ upregulates $Na_V1.5$ via PKA activation that inhibits mitochondrial ROS overproduction. In addition, both kinases act directly on the channel to mediate their effects. *PKA*, protein kinase A; *PKC*, protein kinase C; *ROS*, reactive oxygen species

be local, following ischemic segments, for example. In an *ex vivo* study, NADH elevation by pyruvate/lactate buffer perfusion (1:10 mmol/L) induces ventricular tachycardia in mouse hearts, which can be reversible after the buffer is washed out (Liu et al., 2009). NAD^+ antagonizes the I_{Na} reduction seen with a rise of internal NADH by activating PKA and reducing mitochondrial ROS level (Fig. 10). NAD^+ perfusion can correct the ventricular tachycardia induced in heterozygous knockout *SCN5A*$^{+/-}$ mouse hearts (Liu et al., 2009) and improve the decreased conduction velocity in human failing heart tissue (Liu, Gu, et al., 2013). Elevated free fatty acids as can occur in mitochondrial dysfunction in HF can constitute ionophores on the cell membrane and allow extracellular Na^+ ions to enter the cell (Nickel, Löffler, & Maack, 2013), which may generate current flow similar to I_{NaL} and increase intracellular Na^+ concentration ($[Na^+]_i$). This current could either exacerbate the phenotype of gain-of-function *SCN5A* mutations or mitigate the phenotype of loss-of-function *SCN5A* mutations. Therefore, any alteration in cardiac metabolism could alter the phenotype of $Na_V1.5$ mutations.

4.1.4.4 Arginine methylation of $Na_V1.5$

Recently, arginine methylation of $Na_V1.5$ has been reported by Dr. Brugada's group (Beltran-Alvarez, Pagans, & Brugada, 2011) at R513, R526, and R680. This dynamic modulation may vary I_{Na} and the phenotype of SCN5A mutations. R526 methylation is observed in end-stage heart failure human heart (Beltran-Alvarez et al., 2014), a condition associated with increased sudden death. Coexpression of protein arginine methylatransferase 3 and 5, which catalyze arginine methylation, induces increased $Na_V1.5$ channel membrane expression (Beltran-Alvarez et al., 2013).

4.1.5 Degradation pathways of $Na_V1.5$

Degradation is the last step of the $Na_V1.5$ channel's life cycle. A balance between synthesis and degradation of the channel determines the available channel amount to generate current. Ubiquitination of membrane proteins by ubiquitin-protein ligases is common for protein internalization and degradation. E3 ubiquitin-protein ligase NEDD4 is an ubiquitin-protein ligase that regulates the degradation of cardiac ion channels such as $Na_V1.5$, K_VLQT1, and $Ca_V1.2$ (Fig. 7, upper right part). In *Xenopus* oocytes, Nedd4 significantly reduces these channels' membrane expression levels and attenuates ion currents to 40–75% (Abriel, Kamynina, Horisberger, &

Staub, 2000; van Bemmelen et al., 2004; Jespersen et al., 2007; Rougier, Albesa, Abriel, & Viard, 2011; Rougier et al., 2005). An inactive Nedd4 mutant can increase the $Na_v1.5$ density on the membrane and elevate I_{Na} (Abriel, Kamynina, et al., 2000). NEDD4 also regulates the degradation of neuronal Na^+ channels ($Na_v1.2$, $Na_v1.3$, $Na_v1.7$, and $Na_v1.8$), which are expressed in cardiomyocytes and affect the cardiac function (as discussed later) (Fotia et al., 2004; Rougier et al., 2005, 2011). NEDD4 can be regulated by PKA, MAPKs, serine/threonine-protein kinase 1 (SGK1), and 14-3-3 (van Bemmelen et al., 2004; Boehmer et al., 2003; Gasser et al., 2010; Snyder, 2009; Yanagita et al., 2003; Yang & Kumar, 2010). Any changes of NEDD4 and its modifiers can affect the membrane expression of cardiac ion channels, which will give rise to various phenotypes when combined with $Na_v1.5$ mutations. For instance, SGK1 activation has been found to increase $Na_v1.5$ membrane expression, peak I_{Na} and I_{NaL} via decreasing NEDD4-2 binding to $Na_v1.5$ and therefore decreasing $Na_v1.5$ degradation, which is similar to the effect of pressure overload—induced HF in mice on $Na_v1.5$ (Das et al., 2012). SGK1 itself binds to $Na_v1.5$ physically and likely alters $Na_v1.5$ conformation to decrease NEDD's binding to the channel (Das et al., 2012).

4.2 Extrinsic modifiers of $Na_v1.5$ function

Extrinsic modifiers affect the expression or function of other channels or genes that can modify the overall biophysical properties and electrical effect of a $Na_v1.5$ mutation. Here, we will discuss other voltage-gated ion channels including other isoforms of Na^+ channels, K^+ and Ca^{2+} channels. Changes of these channels may affect the plasma membrane potential and conduction velocity. There are also proteins that regulate intracellular Na^+ homeostasis that affect the driving force for Na^+ entering the myocytes during cardiac action potential. We will present that modifiers extrinsic to the Na^+ channel could contribute to the phenotypic variability of a $Na_v1.5$ mutation.

4.2.1 Other sodium channel isoforms

In addition to cardiac $Na_v1.5$, the neuronal and skeletal isoforms of voltage-gated sodium channels ($Na_v1.1$ to $Na_v1.8$) are also expressed in the heart. These tetrodotoxin-sensitive channel isoforms are found to mainly populate the midsection transverse tubules of myocytes but remain inactivated under normal rest conditions (Lin et al., 2011; Maier et al., 2004). In adult mouse hearts, $Na_v1.5$ is responsible for 92% of I_{Na}, whereas the other isoforms account for 8% (Haufe et al., 2005). $Na_V1.1$ and $Na_V1.6$ participate in pacemaking in the sinoatrial node (Lei et al., 2004) and in excitation–contraction coupling (Noujaim et al., 2012). In dog heart ventricular myocytes $Na_V1.1$ contributes to 60% of I_{NaL} at −60 mV (Mishra et al., 2015). In ventricular cardiomyocytes isolated from human ischemic and dog failing hearts, $Na_V1.1$ upregulation and $Na_V1.6$ downregulation were observed, changes which could affect the phenotypic variability of a $Na_V1.5$ mutation (Mishra et al., 2015).

Mutations in $Na_V1.8$, encoded by *SCN10A*, have been associated with prolonged cardiac conduction and an increased risk of heart block (Chambers et al., 2010). $Na_V1.8$ is mainly expressed in intracardiac neurons of the heart and has an important role in regulating the frequency of action potential firing in mice (Verkerk et al., 2012). This channel also contributes to I_{NaL}, and its inhibition might be antiarrhythmic (Yang et al., 2012). An association between common variants in *SCN5A*, *SCN10A*, and a transcriptional regulator gene *HEY2* was reported in a genome-wide association study of patients with BrS.[104] The cumulative effect of the three loci significantly decreased cardiac conduction velocity and increased susceptibility to BrS. Differential regulation of these neuronal and skeletal isoforms of sodium channels might modulate the phenotype of any $Na_V1.5$ mutation.

4.2.2 Other cardiac ion channels

Mutations in genes encoding cardiac ion channels other than $Na_V1.5$ are also associated with BrS and LQT3, which suggests that many different channels can be involved in generating one clinical phenotype. This idea implies that changes in these other ion channels can affect the phenotype of any $Na_V1.5$ mutation. For example, the *SCN5A* mutation D1275N has been shown to cause atrial standstill if this mutation is combined with polymorphisms in the atrial-specific connexin 40 (Groenewegen et al., 2003). The potassium current I_{to} (conducted by $K_V4.2$ and $K_V4.3$) and I_{K1} (conducted by Kir2.1-Kir2.3) have been shown to modify I_{Na} and affect the APD and conduction velocity. I_{to} counteracts the depolarizing currents during the early phase of the action potential plateau, resulting in an action potential notch (phase 1) and a "spike-and-dome" morphology (Fig. 1B). In a canine heart model, increased I_{to} at the early phase of action potential was found to cause conduction slowing (Hoogendijk et al., 2010). When combined with a I_{Na} reduction, I_{to} that is larger in subepicardial than in subendocardial myocytes can lead to loss of the action potential dome and early completion of repolarization in the right ventricular subepicardium but not the subendocardium. This situation can result in ST-segment elevation being evident on the ECG in patients with BrS. Therefore, alterations or mutations of $K_v4.3$ and its accessory proteins could amplify early afterdepolarizations or mitigate early repolarization in BrS.

I_{K1} controls the resting membrane potential, which determines voltage-gated $Na_v1.5$ availability. $I_{K1}-I_{Na}$ interactions have been shown to involve reciprocal modulation of their respective channel expressions within a macromolecular complex (Milstein et al., 2012). An increase in I_{K1} expression can lead to membrane hyperpolarization, which can either amplify a reduction in I_{Na} induced by *SCN5A* loss-of-function mutations, or minimize an increase in I_{NaL} induced by *SCN5A* gain-of-function mutations.

4.2.3 Modulators of intracellular Na^+ homeostasis

Intracellular Na^+ homeostasis is not only a determinant of driving force for Na^+ entering the cells during cardiac action potential but also coupled with intracellular Ca^{2+} homeostasis and cytosol pH values. Several proteins transport Na^+ ions through the plasma membrane, including the NCX, the NKA, the Na^+-H^+ exchanger (NHE), the Na^+-bicarbonate cotransporter (NBC), and connexins (reviewed in Bay, Kohlhaas, & Maack, 2013; Clancy et al., 2015; Despa & Bers, 2013; Wagner, Maier, & Bers, 2015; Yang, Kyle, Makielski, & Dudley, 2015). The NCX and NKA also contribute to the membrane potential by transporting one net positive charge inward and outward, respectively. Any changes of these transporters in a disease state may disturb the intracellular Na^+ and Ca^{2+} homeostasis, which in turn may affect the manifestation of a $Na_v1.5$ mutation. For instance, elevated $[Na^+]_i$ often observed in heart failure and ischemia can trigger the reverse mode transport of the NCX with Ca^{2+} influx, leading to Ca^{2+} overload and mitochondrial dysfunction with decreased ATP production (Liu & O'Rourke, 2008) and increased oxidative stress (Kohlhaas et al., 2010), which downregulates $Na_v1.5$ (Liu et al., 2010).

4.2.4 Genetic and genomic background

Genetic background can contribute to the variable phenotypes observed with *SCN5A* mutations. A study in families with BrS carrying *SCN5A* mutations suggested that the presence of the Brugada phenotype is determined not only by *SCN5A* mutation but also by the genetic background of an individual (Probst et al., 2009). The *Scn5a*$^{1798insD/+}$ mutation has been found to generate a more severe phenotype of conduction slowing in mice with a 129P2 genetic background than in those with an FVB/N background (Remme et al., 2009). A genome-wide association study in patients with BrS showed that polymorphisms in *SCN10A* and *HEY2* can have a large effect on the phenotype of BrS (Bezzina et al., 2013). Mutations of glycerol-3-phosphate dehydrogenase 1-like have been also reported in patients of BrS and sudden infant death syndrome by changing cellular NADH levels that decrease I_{Na} (Liu et al., 2009; London et al., 2007; Van Norstrand et al., 2007). Therefore, multiple genetic mutations in genes other than *SCN5A* can act to alter I_{Na} and affect the phenotypes, which highlights the importance of considering genetic and genomic modifiers when undertaking genotype—phenotype correlations (reviewed in Hedley et al., 2009).

5. SUMMARY

Mutations in the cardiac $Na_v1.5$ result in variable clinical syndromes. Some of this phenotypic variability can be explained by the direct effect of the mutation on channel availability and biophysical properties. Conversely, it is difficult to explain various manifestations of a mutation in a particular region or chamber of the heart, circadian variation, age-dependent arrhythmic episodes, incomplete penetrance in families, and variations according to clinical state within an affected individual. We propose that this variability might be the result of secondary modifiers of the $Na_v1.5$ channel. We have divided these modifiers into those within the *SCN5A* gene, those intrinsic to the life cycle of $Na_v1.5$, and those extrinsic to $Na_v1.5$. Most of the modifiers discussed above are summarized in Fig. 11. Much of the evidence for modifiers of $Na_v1.5$ expression and function has been derived from the literature on heart failure. Further experimental evidence will be required to support and generalize this "second-hit" hypothesis of phenotypic variability to the field of arrhythmia syndromes associated with *SCN5A* mutations.

$Na_v1.5$ channel modifiers that reduce I_{Na} may alleviate gain-of-function mutation effects and exacerbate loss-of-function mutation effects, while the modifiers that increase I_{Na} may have an opposite effect. Changes of localized modifier may explain regional expression of mutation effects. Stable versus continuous changes in modifiers may explain why some conduction disease phenotypes progress and others do not. Variations in modifiers may explain why ECG phenotypes and arrhythmic risk varies with time of day and with age. These considerations may also explain why there is less phenotypic

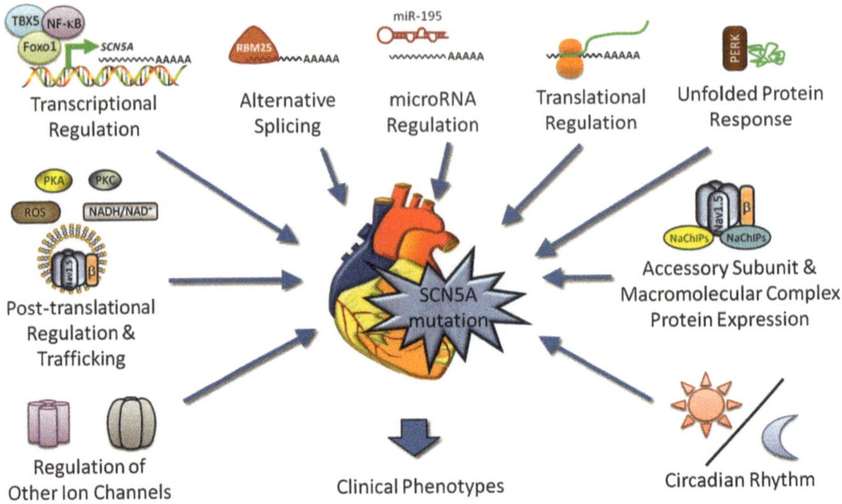

Figure 11 Modifiers, which modulate $Na_V1.5$ expression and function and may lead to clinical phenotype variability of *SCN5A* mutations, are summarized, including intrinsic modifiers such as transcriptional regulators (TBX5, Foxo1, NF-κB, etc.), alternative splicing regulators of SCN5A (RBM25 and LUC7L3), microRNA regulation, translational regulators, the PERK branch of unfolded protein response, post-translational modifies (PKA, PKC, ROS, $NADH/NAD^+$, etc.), and accessory proteins that form the macromolecular complex with $Na_V1.5$, and extrinsic modifiers such as other ion channels and circadian rhythm. *PKA*, protein kinase A; *PKC*, protein kinase C; *ROS*, reactive oxygen species.

variability in gain-of-function mutations than in loss-of-function mutations, because fewer modifiers exist that affect only inactivation than affect overall channel current. Nevertheless, the net effect of a gain-of-function mutation can be modulated by the total number of available channels, which might explain the variable penetrance of a phenotype with time or between regions of the heart.

In summary, considering these "second-hit" modifiers may improve genotype-phenotype correlation and risk prediction algorithms. Moreover, targeting these modifiers may prove to be an effective strategy to prevent diseases associated with $Na_V1.5$ mutations.

6. DISCLOSURES

SCD is the inventor on patents or patent applications: (1) *SCN5A* Splice Variants for Use in Methods Relating to Sudden Cardiac Death and Need for Implanted Cardiac Defibrillators, PCT/US.2012/20564, (2) 13/291,826, *SCN5A* Splicing Factors and Splice Variants For Use in Diagnostic and Prognostic Methods, (3) 8,003,324, Modulation of sodium channels by nicotinamide adenine dinucleotide, (4) 11/707,882, Human sodium channel isoforms, (5) 12/929,786, Method for modulating or controlling sodium channel current by reactive oxygen species, (6) 13/032,629 0, Activation of the Renin-Angiotensin System (RAS) and Sudden Cardiac Death, (7) 13/067,953, Modulation of sodium channels by nicotinamide adenine dinucleotide, and (8) 61/786,882, Method for Detecting Brugada Syndrome.

ACKNOWLEDGMENTS

This work was funded by an AHA Midwest Affiliation Postdoctoral Fellowship AHA13-POST14380029 (KCY); National Institutes of Health (NIH) grants RO1 HL104025 (SCD), RO1 HL106592, and R41 HL112355 (SCD); and a Veterans Affairs MERIT grant BX000859 (SCD).

REFERENCES

Abriel, H., Kamynina, E., Horisberger, J. D., & Staub, O. (2000). Regulation of the cardiac voltage-gated Na^+ channel (H1) by the ubiquitin-protein ligase Nedd4. *FEBS Letters, 466*, 377–380.

Abriel, H., Wehrens, X. H. T., Benhorin, J., Kerem, B., & Kass, R. S. (2000). Molecular pharmacology of the sodium channel mutation D1790G linked to the long-QT syndrome. *Circulation, 102*, 921–925.

Ackerman, M., Splawski, I., Makielski, J. C., Tester, D. J., Will, M. L., Timothy, K. W.... Curran, M. E. (2004). Spectrum and prevalence of cardiac sodium channel variants among black, white, Asian, and Hispanic individuals: implications for arrhythmogenic susceptibility and Brugada/long QT syndrome genetic testing. *Heart Rhythm, 1*, 600–607.

Adsit, G. S., Vaidyanathan, R., Galler, C. M., Kyle, J. W., & Makielski, J. C. (2013). Channelopathies from mutations in the cardiac sodium channel protein complex. *Journal of Molecular and Cellular Cardiology, 61*, 34–43.

Aiba, T., Farinelli, F., Kostecki, G., Hesketh, G., Edwards, D., Biswas, S.... Tomaselli, G. F. (2014). A mutation causing Brugada syndrome identifies a mechanism for altered autonomic and oxidant regulation of cardiac sodium currents. *Circulation: Cardiovascular Genetics, 7*, 249–256.

Alings, M., & Wilde, A. (1999). "Brugada" syndrome: clinical data and suggested pathophysiological mechanism. *Circulation, 99*, 666–673.

Alliston, T. N., Gonzalez-Robayna, I. J., Buse, P., Firestone, G. L., & Richards, J. S. (2000). Expression and localization of serum/glucocorticoid-induced kinase in the rat ovary: relation to follicular growth and differentiation. *Endocrinology, 141*, 385–395.

Amin, A., Asghari-Roodsari, A., & Tan, H. (2010). Cardiac sodium channelopathies. *Pflugers Archiv European Journal of Physiology, 460*, 223–237.

Amin, A. S., Verkerk, A. O., Bhuiyan, Z. A., Wilde, A. A. M., & Tan, H. L. (2005). Novel Brugada syndrome-causing mutation in ion-conducting pore of cardiac Na^+ channel does not affect ion selectivity properties. *Acta Physiologica Scandinavica, 185*, 291–301.

An, R. H., Wang, X. L., Kerem, B., Benhorin, J., Medina, A., Goldmit, M., & Kass, R. S. (1998). Novel LQT-3 mutation affects Na^+ channel activity through interactions between α- and β1-subunits. *Circulation Research, 83*, 141–146.

Aon, M. A., Cortassa, S., Marban, E., & O'Rourke, B. (2003). Synchronized whole cell oscillations in mitochondrial metabolism triggered by a local release of reactive oxygen species in cardiac myocytes. *Journal of Biological Chemistry, 278*, 44735–44744.

Arnolds, D. E., Liu, F., Fahrenbach, J. P., Kim, G. H., Schillinger, K. J., Smemo, S. … Moskowitz, I. P. (2012). TBX5 drives Scn5a expression to regulate cardiac conduction system function. *Journal of Clinical Investigation, 122*, 2509–2518.

Atkinson, M. M., Lampe, P. D., Lin, H. H., Kollander, R., Li, X. R., & Kiang, D. T. (1995). Cyclic AMP modifies the cellular distribution of connexin43 and induces a persistent increase in the junctional permeability of mouse mammary tumor cells. *Journal of Cell Sciences, 108*, 3079–3090.

Baba, S., Dun, W., & Boyden, P. A. (2004). Can PKA activators rescue Na^+ channel function in epicardial border zone cells that survive in the infarcted canine heart? *Cardiovascular Research, 64*, 260–267.

Ballou, L. M., Lin, R. Z., & Cohen, I. S. (2015). Control of cardiac repolarization by phosphoinositide 3-kinase signaling to ion channels. *Circulation Research, 116*, 127–137.

Balser, J. R. (2001). The cardiac sodium channel: gating function and molecular pharmacology. *Journal of Molecular and Cellular Cardiology, 33*, 599–613.

Baroudi, G., Acharfi, S., Larouche, C., & Chahine, M. (2002). Expression and intracellular localization of an SCN5A double mutant R1232W/T1620M implicated in Brugada syndrome. *Circulation Research, 90*, e11–e16.

Baroudi, G., Napolitano, C., Priori, S. G., Del Bufalo, A., & Chahine, M. (2004). Loss of function associated with novel mutations of the SCN5A gene in patients with Brugada syndrome. *Canadian Journal of Cardiology, 20*, 425–430.

Baroudi, G., Pouliot, V., Denjoy, I., Guicheney, P., Shrier, A., & Chahine, M. (2001). Novel mechanism for Brugada syndrome: defective surface localization of an SCN5A mutant (R1432G). *Circulation Research, 88*, e78–e83.

Bay, J., Kohlhaas, M., & Maack, C. (2013). Intracellular Na^+ and cardiac metabolism. *Journal of Molecular and Cellular Cardiology, 61*, 20–27.

Beaufort-Krol, G. C. M., van den Berg, M. P., Wilde, A. A. M., van Tintelen, J. P., Viersma, J. W., Bezzina, C. R., & Bink-Boelkens, M. T. (2005). Developmental aspects of long QT syndrome type 3 and Brugada syndrome on the basis of a single SCN5A mutation in childhood. *Journal of the American College of Cardiology, 46*, 331–337.

Beltran-Alvarez, P., Espejo, A., Schmauder, R., Beltran, C., Mrowka, R., Linke, T. … Brugada, R. (2013). Protein arginine methyl transferases-3 and -5 increase cell surface expression of cardiac sodium channel. *FEBS Letters, 587*, 3159–3165.

Beltran-Alvarez, P., Pagans, S., & Brugada, R. (2011). The cardiac sodium channel is post-translationally modified by arginine methylation. *Journal of Proteome Research, 10*, 3712–3719.

Beltran-Alvarez, P., Tarradas, A., Chiva, C., Pérez-Serra, A., Batlle, M., Pérez-Villa, F. … Pagans, S. (2014). Identification of N-terminal protein acetylation and arginine methylation of the voltage-gated sodium channel in end-stage heart failure human heart. *Journal of Molecular and Cellular Cardiology, 76*, 126–129.

van Bemmelen, M. X., Rougier, J. S., Gavillet, B., Apotheloz, F., Daidie, D., Tateyama, M. … Abriel, H. (2004). Cardiac voltage-gated sodium channel $Na_V1.5$ is regulated by Nedd4-2 mediated ubiquitination. *Circulation Research, 95*, 284–291.

Bennett, P. B., Yazawa, K., Makita, N., & George, A. L. (1995). Molecular mechanism for an inherited cardiac arrhythmia. *Nature, 376*, 683–685.

Bezzina, C. R., Barc, J., Mizusawa, Y., Remme, C. A., Gourraud, J., Simonet, F. … Redon, R. (2013). Common variants at SCN5A-SCN10A and HEY2 are associated with Brugada syndrome, a rare disease with high risk of sudden cardiac death. *Nature Genetics, 45*, 1044–1049.

Bezzina, C. R., Rook, M. B., Groenewegen, W. A., Herfst, L. J., van der Wal, A. C., Lam, J. … Mannens, M. M. (2003). Compound heterozygosity for mutations (W156X and R225W) in SCN5A associated with severe cardiac conduction disturbances and degenerative changes in the conduction system. *Circulation Research, 92*, 159–168.

Bezzina, C., Veldkamp, M. W., van den Berg, M. P., Postma, A. V., Rook, M. B., Viersma, J. W. … Wilde, A. A. (1999). A single Na^+ channel mutation causing both long-QT and Brugada syndromes. *Circulation Research, 85*, 1206–1213.

Boehmer, C., Wilhelm, V., Palmada, M., Wallisch, S., Henke, G., Brinkmeier, H., & Lang, F. (2003). Serum and glucocorticoid inducible kinases in the regulation of the cardiac sodium channel SCN5A. *Cardiovascular Research, 57*, 1079–1084.

Borg, J. J., Yuill, K. H., Hancox, J. C., Spencer, I. C., & Kozlowski, R. Z. (2002). Inhibitory effects of the antiestrogen agent clomiphene on cardiac sarcolemmal anionic and cationic currents. *Journal of Pharmacology and Experimental Therapeutics, 303*, 282–292.

Boyden, P. A., Divies, S., Viswanathan, P. C., Amarnath, V., Balser, J. R., & Roberts, L. J., III (2007). Potential role of isoketals formed via the isoprostane pathway of lipid peroxidation in ischemic arrhythmias. *Journal of Cardiovascular Pharmacology, 50*, 480–486.

Chambers, J., Zhao, J., Terracciano, C., Bezzina, C. R., Zhang, W., Kaba, R. … Kooner, J. S. (2010). Genetic variations in SCN10A influences cardiac conduction. *Nature Genetics, 42*, 149–152.

Chanson, M., White, M. M., & Garber, S. S. (1996). cAMP promotes gap junctional coupling in T84 cells. *American Journal of Physiology — Cell Physiology, 271*, C533–C539.

Chen-Izu, Y., Shaw, R. M., Pitt, G. S., Yarov-Yarovoy, V., Sack, J. T., Abriel, H. … Bers, D. M. (2015). Na^+ channel function, regulation, structure, trafficking and sequestration. *The Journal of Physiology, 593*, 1347–1360.

Chen, L., Ballew, J. D., Herron, K. J., Rodeheffer, R. J., & Olson, T. M. (2007). A common polymorphism in SCN5A is associated with lone atrial fibrillation. *Clinical Pharmacology & Therapeutics, 81*, 35–41.

Chen, Q., Kirsch, G. E., Zhang, D., Brugada, R., Brugada, J., Brugada, P. … Wang, Q. (1998). Genetic basis and molecular mechanism for idiopathic ventricular fibrillation. *Nature, 392*, 293–296.

Chen, T. C., Law, B., Kondratyuk, T., & Rossie, S. (1995). Identification of soluble protein phosphatases that dephosphorylate voltage-sensitive sodium channels in rat brain. *Journal of Biological Chemistry, 270*, 7750–7756.

Choudhary, G., & Dudley, S. C., Jr. (2002). Heart failure, oxidative stress, and ion channel modulation. *Congestive Heart Failure, 8*, 148–155.

Christ, T., Boknik, P., Wöhrl, S., Wettwer, E., Graf, E. M., Bosch, R. F. … Dobrev, D. (2004). L-type Ca^{2+} current downregulation in chronic human atrial fibrillation is associated with increased activity of protein phosphatases. *Circulation, 110*, 2651–2657.

Clancy, C. E., Chen-Izu, Y., Bers, D. M., Belardinelli, L., Boyden, P. A., Csernoch, L. … Zaza, A. (2015). Deranged sodium to sudden death. *The Journal of Physiology, 593*, 1331–1345.

Clatot, J., Ziyadeh-Isleem, A., Duchatelet, S., Gandjbakhch, E., Denjoy, I., Hidden-Lucet, F. … Guichiney, P. (2013). Different atrial and ventricular resting membrane potentials may explain the phenotypical variability of a truncating SCN5A mutation. *Heart Rhythm, 10*, 1746.

Cook, S. A., Sugden, P. H., & Clerk, A. (1999). Activation of c-Jun N-terminal kinases and p38-mitogen-activated protein kinases in human heart failure secondary to ischaemic heart disease. *Journal of Molecular and Cellular Cardiology, 31*, 1429–1434.

Cranefield, P. F., & Aronson, R. S. (1991). Torsades de pointes and early afterdepolarizations. *Cardiovascular Drugs and Therapy, 5*, 531–537.

Cunha, S. R., & Mohler, P. J. (2011). Ankyrin-based cellular pathways for cardiac ion channel and transporter targeting and regulation. *Seminars in Cell & Developmental Biology, 22*, 166–170.

Cusdin, F. S., Clare, J. J., & Jackson, A. P. (2008). Trafficking and cellular distribution of voltage-gated sodium channels. *Traffic, 9*, 17–26.

Daimi, H., Lozano-Velasco, E., Haj Khelil, A., Chibani, J. B. E., Barana, A., Amorós, I. … Franco, D. (2015). Regulation of SCN5A by microRNAs: miR-219 modulates SCN5A transcript expression and the effects of flecainide intoxication in mice. *Heart Rhythm, 12*(6), 1333–1342.

Darrow, B. J., Fast, V. G., Kleber, A. G., Beyer, E. C., & Saffitz, J. E. (1996). Functional and structural assessment of intercellular communication: increased conduction velocity and enhanced connexin expression in dibutyryl cAMP-treated cultured cardiac myocytes. *Circulation Research, 79*, 174–183.

Das, S., Aiba, T., Rosenberg, M., Hessler, K., Xiao, C., Quintero, P. A. … Rosenzweig, A. (2012). Pathological role of serum- and glucocorticoid-regulated kinase 1 in adverse ventricular remodeling. *Circulation, 126*, 2208–2219.

Deschamps, A. M., Murphy, E., & Sun, J. (2010). Estrogen receptor activation and cardioprotection in ischemia reperfusion injury. *Trends in Cardiovascular Medicine, 20*, 73–78.

Despa, S., & Bers, D. M. (2013). Na^+ transport in the normal and failing heart — Remember the balance. *Journal of Molecular and Cellular Cardiology, 61*, 2–10.

Dhalla, A. K., & Singal, P. K. (1994). Antioxidant changes in hypertrophied and failing guinea pig hearts. *American Journal of Physiology — Heart and Circulatory Physiology, 266*, H1280–H1285.

Di, L. F., Menabo, R., Canton, M., Barile, M., & Bernardi, P. (2001). Opening of the mitochondrial permeability transition pore causes depletion of mitochondrial and cytosolic NAD^+ and is a causative event in the death of myocytes in postischemic reperfusion of the heart. *Journal of Biological Chemistry, 276*, 2571–2575.

Doroudgar, S., Thuerauf, D. J., Marcinko, M. C., Belmont, P. J., & Glembotski, C. C. (2009). Ischemia activates the ATF6 branch of the endoplasmic reticulum stress response. *Journal of Biological Chemistry, 284*, 29735–29745.

Dumaine, R., Towbin, J. A., Brugada, P., Vatta, M., Nesterenko, D. V., Nesterenko, V. V. … Antzelevitch, C. (1999). Ionic mechanisms responsible for the electrocardiographic phenotype of the Brugada syndrome are temperature dependent. *Circulation Research, 85*, 803–809.

Dumaine, R., Wang, Q., Keating, M. T., Hartmann, H. A., Schwartz, P. J., Brown, A. M., & Kirsch, G. E. (1996). Multiple mechanisms of Na^+ channel-linked long-QT syndrome. *Circulation Research, 78*, 916–924.

El-Armouche, A., Boknik, P., Eschenhagen, T., Carrier, L., Knaut, M., Ravens, U., & Dobrev, D. (2006). Molecular determinants of altered Ca^{2+} handling in human chronic atrial fibrillation. *Circulation, 114*, 670–680.

Ellinor, P. T., Nam, E. G., Shea, M. A., Milan, D. J., Ruskin, J. N., & MacRae, C. A. (2008). Cardiac sodium channel mutation in atrial fibrillation. *Heart Rhythm, 5*, 99–105.

Fotia, A. B., Ekberg, J., Adams, D. J., Cook, D. I., Poronnik, P., & Kumar, S. (2004). Regulation of neuronal voltage-gated sodium channels by the ubiquitin-protein ligases Nedd4 and Nedd4-2. *Journal of Biological Chemistry, 279*, 28930–28935.

Fozzard, H. A., & Hanck, D. A. (1996). Structure and function of voltage-dependent sodium channels: comparison of brain II and cardiac isoforms. *Physiological Reviews, 76*, 887–926.

Fukuda, K., Davies, S. S., Nakajima, T., Ong, B. H., Kupershmidt, S., Fessel, J. … Balser, J. R. (2005). Oxidative mediated lipid peroxidation recapitulates proarrhythmic effects on cardiac sodium channels. *Circulation Research, 97*, 1262–1269.

Gao, G., Xie, A., Huang, S. C., Zhou, A., Zhang, J., Herman, A. M. ... Dudley, S. C., Jr. (2011). Role of RBM25/LUC7L3 in abnormal cardiac sodium channel splicing regulation in human heart failure/clinical perspective. *Circulation, 124*, 1124–1131.

Gao, G., Xie, A., Zhang, J., Herman, A. M., Jeong, E. M., Gu, L. ... Dudley, S. C. (2013). Unfolded protein response regulates cardiac sodium current in systolic human heart failure. *Circulation: Arrhythmia and Electrophysiology, 6*, 1018–1024.

Gasser, A., Cheng, X., Gilmore, E. S., Tyrrell, L., Waxman, S. G., & Dib-Hajj, S. D. (2010). Two Nedd4-binding motifs underlie modulation of sodium channel $Na_V1.6$ by p38 MAPK. *Journal of Biological Chemistry, 285*, 26149–26161.

Gavillet, B., Rougier, J. S., Domenighetti, A. A., Behar, R., Boixel, C., Ruchat, P. ... Abriel, H. (2006). Cardiac sodium channel $Na_V1.5$ is regulated by a multiprotein complex composed of syntrophins and dystrophin. *Circulation Research, 99*, 407–414.

Gehi, A., Duong, T., Metz, L., Gomes, J., & Mehta, D. (2006). Risk stratification of individuals with the Brugada electrocardiogram: a meta-analysis. *Journal of Cardiovascular Electrophysiology, 17*, 577–583.

Gellens, M. E., George, A. L., Chen, L. Q., Chahine, M., Horn, R., Barchi, R. L., & Kallen, R. G. (1992). Primary structure and functional expression of the human cardiac tetrodotoxin-insensitive voltage-dependent sodium channel. *Proceedings of the National Academy of Sciences of the United States of America, 89*, 554–558.

Girmatsion, Z., Biliczki, P., Bonauer, A., Wimmer-Greinecker, G., Scherer, M., Moritz, A. ... Ehrlich, J. R. (2009). Changes in microRNA-1 expression and I_{K1} up-regulation in human atrial fibrillation. *Heart Rhythm, 6*, 1802–1809.

Groenewegen, W. A., Firouzi, M., Bezzina, C. R., Vliex, S., van Langen, I. M., Sandkuijl, L. ... Wilde, A. A. (2003). A cardiac sodium channel mutation cosegregates with a rare connexin 40 genotype in familial atrial standstill. *Circulation Research, 92*, 14–22.

Hallaq, H., Wang, D. W., Kunic, J. D., George, A. L., Wells, K. S., & Murray, K. T. (2012). Activation of protein kinase C alters the intracellular distribution and mobility of cardiac Na^+ channels. *American Journal of Physiology – Heart and Circulatory Physiology, 302*, H782–H789.

Hallaq, H., Yang, Z., Viswanathan, P. C., Fukuda, K., Shen, W., Wang, D. W. ... Murray, K. T. (2006). Quantitation of protein kinase A-mediated trafficking of cardiac sodium channels in living cells. *Cardiovascular Research, 72*, 250–261.

Haufe, V., Camacho, J. A., Dumaine, R., Günther, B., Bollensdorff, C., von Banchet, G. S. ... Zimmer, T. (2005). Expression pattern of neuronal and skeletal muscle voltage-gated Na^+ channels in the developing mouse heart. *The Journal of Physiology, 564*, 683–696.

Hedley, P. L., Jorgensen, P., Schlamowitz, S., Moolman-Smook, J., Kanters, J. K., Corfield, V. A., & Christiansen, M. (2009). The genetic basis of Brugada syndrome: a mutation update. *Human Mutation, 30*, 1256–1266.

Heijman, J., Dewenter, M., El-Armouche, A., & Dobrev, D. (2013). Function and regulation of serine/threonine phosphatases in the healthy and diseased heart. *Journal of Molecular and Cellular Cardiology, 64*, 90–98.

Herfst, L. J., Potet, F., Bezzina, C. R., Groenewegen, W. A., Le Marec, H., Hoorntje, T. M. ... Rook, M. B. (2003). Na^+ channel mutation leading to loss of function and non-progressive cardiac conduction defects. *Journal of Molecular and Cellular Cardiology, 35*, 549–557.

Herfst, L., Rook, M. B., & Jongsma, H. J. (2004). Trafficking and functional expression of cardiac Na^+ channels. *Journal of Molecular and Cellular Cardiology, 36*, 185–193.

Hill, M. F., & Singal, P. K. (1997). Right and left myocardial antioxidant responses during heart failure subsequent to myocardial infarction. *Circulation, 96*, 2414–2420.

Hoogendijk, M. G., Opthof, T., Postema, P. G., Wilde, A. A. M., de Bakker, J. M. T., & Coronel, R. (2010). The Brugada ECG pattern: a marker of channelopathy, structural heart disease, or neither? Toward a unifying mechanism of the Brugada syndrome. *Circulation: Arrhythmia and Electrophysiology, 3*, 283–290.

Hu, D., Barajas-Martinez, H., Burashnikov, E., Springer, M., Wu, Y., Varro, A.... Antzelevitch, C. (2009). A mutation in the β3 subunit of the cardiac sodium channel associated with brugada ECG phenotype. *Circulation: Cardiovascular Genetics, 2*, 270–278.

Jansen, J. A., Noorman, M., Musa, H., Stein, M., de Jong, S., van der Nagel, R.... van Rijen, H. V. (2012). Reduced heterogeneous expression of Cx43 results in decreased $Na_V1.5$ expression and reduced sodium current that accounts for arrhythmia vulnerability in conditional Cx43 knockout mice. *Heart Rhythm, 9*, 600–607.

Jeong, E. M., Liu, M., Sturdy, M., Gao, G., Sovari, A. A., & Dudley, J. (2011). Metabolic stress, reactive oxygen species, and arrhythmia. *Journal of Molecular and Cellular Cardiology, 52*, 454–463.

Jespersen, T., Membrez, M., Nicolas, C. S., Pitard, B., Staub, O., Olesen, S. P.... Abriel, H. (2007). The KCNQ1 potassium channel is down-regulated by ubiquitylating enzymes of the Nedd4/Nedd4-like family. *Cardiovascular Research, 74*, 64–74.

Junttila, M. J., Gonzalez, M., Lizotte, E., Benito, B., Vernooy, K., Sarkozy, A.... Brugada, R. (2008). Induced Brugada-type electrocardiogram, a sign for imminent malignant arrhythmias. *Circulation, 117*, 1890–1893.

Kääb, S., Pfeufer, A., Hinterseer, M., äbauer, M., & hulze-Bahr, E. (2004). Long QT syndrome. *Zeitschrift fur Kardiologie, 93*, 641–645.

Kass, R. S. (2006). Sodium channel inactivation in heart: a novel role of the carboxy-terminal domain. *Journal of Cardiovascular Electrophysiology, 17*, S21–S25.

Kattygnarath, D., Maugenre, S., Neyroud, N., Balse, E., Ichai, C., Denjoy, I.... Guicheney, P. (2011). MOG1: a new susceptibility gene for Brugada syndrome. *Circulation: Cardiovascular Genetics, 4*, 261–268.

Kaur, K., Zarzoso, M., Ponce-Balbuena, D., Guerrero-Serna, G., Hou, L., Musa, H., & Jalife, J. (2013). TGF-β1, released by myofibroblasts, differentially regulates transcription and function of sodium and potassium channels in adult rat ventricular myocytes. *PLoS One, 8*, e55391.

Kohlhaas, M., Liu, T., Knopp, A., Zeller, T., Ong, M. F., Böhm, M.... Maack, C. (2010). Elevated cytosolic Na^+ increases mitochondrial formation of reactive oxygen species in failing cardiac myocytes. *Circulation, 121*, 1606–1613.

Konishi, H., Tanaka, M., Takemura, Y., Matsuzaki, H., Ono, Y., Kikkawa, U., & Nishizuka, Y. (1997). Activation of protein kinase C by tyrosine phosphorylation in response to H_2O_2. *Proceedings of the National Academy of Sciences of the United States of America, 94*, 11233–11237.

Kyndt, F., Probst, V., Potet, F., Demolombe, S., Chevallier, J. C., Baro, I.... Le Marec, H. (2001). Novel SCN5A mutation leading either to isolated cardiac conduction defect or Brugada Syndrome in a large French family. *Circulation, 104*, 3081–3086.

Lei, M., Goddard, C., Liu, J., Léoni, A. L., Royer, A., Fung, S. S. M.... Huang, C. L. (2005). Sinus node dysfunction following targeted disruption of the murine cardiac sodium channel gene Scn5a. *The Journal of Physiology, 567*, 387–400.

Lei, M., Huang, C. L. H., & Zhang, Y. (2008). Genetic Na^+ channelopathies and sinus node dysfunction. *Progress in Biophysics and Molecular Biology, 98*, 171–178.

Lei, M., Jones, S. A., Liu, J., Lancaster, M. K., Fung, S. S. M., Dobrzynski, H.... Boyett, M. R. (2004). Requirement of neuronal- and cardiac-type sodium channels for murine sinoatrial node pacemaking. *The Journal of Physiology, 559*, 835–848.

Levin, K. R., & Page, E. (1980). Quantitative studies on plasmalemmal folds and caveolae of rabbit ventricular myocardial cells. *Circulation Research, 46*, 244–255.

Li, Q., Huang, H., Liu, G., Lam, K., Rutberg, J., Green, M. S. ... Gollob, M. H. (2009). Gain-of-function mutation of $Na_V1.5$ in atrial fibrillation enhances cellular excitability and lowers the threshold for action potential firing. *Biochemical and Biophysical Research Communications, 380*, 132–137.

Lin, X., Liu, N., Lu, J., Zhang, J., Anumonwo, J. M. B., Isom, L. L. ... Delmar, M. (2011). Subcellular heterogeneity of sodium current properties in adult cardiac ventricular myocytes. *Heart Rhythm, 8*, 1923–1930.

Lin, X., O'Malley, H., Chen, C., Auerbach, D., Foster, M., Shekhar, A. ... Delmar, M. (2015). Scn1b deletion leads to increased tetrodotoxin-sensitive sodium current, altered intracellular calcium homeostasis and arrhythmias in murine hearts. *The Journal of Physiology, 593*, 1389–1407.

Liu, M., & Dudley, S. C. (2014). Targeting the unfolded protein response in heart diseases. *Expert Opinion on Therapeutic Targets, 18*, 719–723.

Liu, M., Gu, L., Sulkin, M. S., Liu, H., Jeong, E. M., Greener, I. ... Dudley, S. C., Jr. (2013). Mitochondrial dysfunction causing cardiac sodium channel downregulation in cardiomyopathy. *Journal of Molecular and Cellular Cardiology, 54*, 25–34.

Liu, M., Hanck, D. A., & Dudley, S. C., Jr. (2013). The molecular mechanism of cardiac sodium channel current reduction with metabolic stress. *Heart Rhythm, 10*, 1742–1743.

Liu, M., Liu, H., & Dudley, S. C., Jr. (2010). Reactive oxygen species originating from mitochondria regulate the cardiac sodium channel. *Circulation Research, 107*, 967–974.

Liu, T., & O'Rourke, B. (2008). Enhancing mitochondrial Ca^{2+} uptake in myocytes from failing hearts restores energy supply and demand matching. *Circulation Research, 103*, 279–288.

Liu, M., Sanyal, S., Gao, G., Gurung, I. S., Zhu, X., Gaconnet, G. ... Dudley, S. C., Jr. (2009). Cardiac Na^+ current regulation by pyridine nucleotides. *Circulation Research, 105*, 737–745.

Liu, M., Yang, K.-C., & Dudley, S. C., Jr. (2014). Cardiac sodium channel mutations: why so many phenotypes? *Nature Review Cardiology, 11*, 607–615.

Locati, E. H., Zareba, W., Moss, A. J., Schwartz, P. J., Vincent, G. M., Lehmann, M. H. ... Hall, W. J. (1998). Age- and sex-related differences in clinical manifestations in patients with congenital long-QT syndrome: findings from the international LQTS registry. *Circulation, 97*, 2237–2244.

London, B., Michalec, M., Mehdi, H., Zhu, X., Kerchner, L., Sanyal, S. ... Dudley, S. C., Jr. (2007). Mutation in glycerol-3-phosphate dehydrogenase 1-like gene (GPD1-L) decreases cardiac Na^+ current and causes inherited arrhythmias. *Circulation, 116*, 2260–2268.

Lu, L., Timofeyev, V., Li, N., Rafizadeh, S., Singapuri, A., Harris, T. R., & Chiamvimonvat, N. (2009). α-Actinin2 cytoskeletal protein is required for the functional membrane localization of a Ca^{2+}-activated K^+ channel (SK2 channel). *Proceedings of the National Academy of Sciences of the United States of America, 106*, 18402–18407.

Lu, Y., Zhang, Y., Wang, N., Pan, Z., Gao, X., Zhang, F. ... Yang, B. (2010). MicroRNA-328 contributes to adverse electrical remodeling in atrial fibrillation. *Circulation, 122*, 2378–2387.

Luo, X., Zhang, H., Xiao, J., & Wang, Z. (2010). Regulation of human cardiac ion channel genes by microRNAs: theoretical perspective and pathophysiological implications. *Cellular Physiology and Biochemistry, 25*, 571–586.

Maier, S. K. G., Westenbroek, R. E., McCormick, K. A., Curtis, R., Scheuer, T., & Catterall, W. A. (2004). Distinct subcellular localization of different sodium channel α and β subunits in single ventricular myocytes from mouse heart. *Circulation, 109*, 1421–1427.

Makielski, J. C., Ye, B., Valdivia, C. R., Pagel, M. D., Pu, J., Tester, D. J., & Ackerman, M. J. (2003). A ubiquitous splice variant and a common polymorphism affect heterologous expression of recombinant human SCN5A heart sodium channels. *Circulation Research, 93*, 821–828.

Makita, N., Behr, E., Shimizu, W., Horie, M., Sunami, A., Grotti, L. ... Roden, D. M. (2008). The E1784K mutation in SCN5A is associated with mixed clinical phenotype of type 3 long QT syndrome. *Journal of Clinical Investigation, 118*, 2219–2229.

Makita, N., Sasaki, K., Groenewegen, W. A., Yokota, T., Yokoshiki, H., Murakami, T., & Tsutsui, H. (2005). Congenital atrial standstill associated with coinheritance of a novel SCN5A mutation and connexin 40 polymorphisms. *Heart Rhythm, 2*, 1128–1134.

Makiyama, T., Akao, M., Shizuta, S., Doi, T., Nishiyama, K., Oka, Y. ... Horie, M. (2008). A novel SCN5A gain-of-function mutation M1875T associated with familial atrial fibrillation. *Journal of the American College of Cardiology, 52*, 1326–1334.

Makiyama, T., Akao, M., Tsuji, K., Doi, T., Ohno, S., Takenaka, K. ... Horie, M. (2005). High risk for bradyarrhythmic complications in patients with Brugada syndrome caused by SCN5A gene mutations. *Journal of the American College of Cardiology, 46*, 2100–2106.

Maltsev, V. A., Kyle, J. W., Mishra, S., & Undrovinas, A. (2008). Molecular identity of the late sodium current in adult dog cardiomyocytes identified by $Na_v1.5$ antisense inhibition. *American Journal of Physiology – Heart and Circulatory Physiology, 295*, H667–H676.

Mao, W., You, T., Ye, B., Li, X., Dong, H., Hill, J. ... Xu, H. (2012). Reactive oxygen species suppress cardiac $Na_v1.5$ expression through Foxo1. *PLoS One, 7*, e32738.

Marionneau, C., & Abriel, H. (2015). Regulation of the cardiac Na^+ channel $Na_v1.5$ by post-translational modifications. *Journal of Molecular and Cellular Cardiology, 82*, 36–47.

Matsuo, K., Akahoshi, M., Seto, S., & Yano, K. (2003). Disappearance of the Brugada-type electrocardiogram after surgical castration: a role for testosterone and an explanation for the male preponderance? *Pacing and Clinical Electrophysiology, 26*, 1551–1553.

Meyer, M. R., Haas, E., & Barton, M. (2006). Gender differences of cardiovascular disease: new perspectives for estrogen receptor signaling. *Hypertension, 47*, 1019–1026.

Milstein, M. L., Musa, H., Balbuena, D. P., Anumonwo, J. M. B., Auerbach, D. S., Furspan, P. B. ... Jalife, J. (2012). Dynamic reciprocity of sodium and potassium channel expression in a macromolecular complex controls cardiac excitability and arrhythmia. *Proceedings of the National Academy of Sciences of the United States of America, 109*, E2134–E2143.

Ming, Z., Aronson, R., & Nordin, C. (1994). Mechanism of current-induced early afterdepolarizations in guinea pig ventricular myocytes. *American Journal of Physiology, 267*, H1419–H1428.

Mishra, S., Reznikov, V., Maltsev, V. A., Undrovinas, N. A., Sabbah, H. N., & Undrovinas, A. (2015). Contribution of sodium channel neuronal isoform $Na_v1.1$ to late sodium current in ventricular myocytes from failing hearts. *The Journal of Physiology, 593*, 1409–1427.

Mizumaki, K., Fujiki, A., Tsuneda, T., Sakabe, M., Nishida, K., Sugao, M., & Inoue, H. (2004). Vagal activity modulates spontaneous augmentation of ST elevation in the daily life of patients with Brugada syndrome. *Journal of Cardiovascular Electrophysiology, 15*, 667–673.

Mohler, P. J., Rivolta, I., Napolitano, C., LeMaillet, G., Lambert, S., Priori, S. G., & Bennett, V. (2004). $Na_v1.5$ E1053K mutation causing Brugada syndrome blocks binding to ankyrin-G and expression of $Na_v1.5$ on the surface of cardiomyocytes. *Proceedings of the National Academy of Sciences United States of America, 101*, 17533–17538.

Mohler, P. J., Schott, J. J., Gramolini, A. O., Dilly, K. W., Guatimosim, S., duBell, W. H. ... Bennett, V. (2003). Ankyrin-B mutation causes type 4 long-QT cardiac arrhythmia and sudden cardiac death. *Nature, 421*, 634–639.

Muslin, A. (2008). MAPK signaling in cardiovascular health and disease: molecular mechanisms and therapeutic targets. *Clinical Science, 115*, 203–218.

Nademanee, K., Veerakul, G., Nimmannit, S., Chaowakul, V., Bhuripanyo, K., Likittanasombat, K. ... Tatsanavivat, P. (1997). Arrhythmogenic marker for the sudden unexplained death syndrome in Thai men. *Circulation, 96*, 2595–2600.

Nguyen, T. P., Wang, D. W., Rhodes, T. H., & George, A. L. (2008). Divergent biophysical defects caused by mutant sodium channels in dilated cardiomyopathy with arrhythmia. *Circulation Research, 102*, 364–371.

Nickel, A., Löffler, J., & Maack, C. (2013). Myocardial energetics in heart failure. *Basic Research in Cardiology, 108*, 1–20.

Nicolaou, P., & Kranias, E. G. (2009). Role of PP1 in the regulation of Ca cycling in cardiac physiology and pathophysiology. *Frontiers in Bioscience, 14*, 3571–3585.

Noujaim, S. F., Kaur, K., Milstein, M., Jones, J. M., Furspan, P., Jiang, D. ... Jalife, J. (2012). A null mutation of the neuronal sodium channel $Na_V1.6$ disrupts action potential propagation and excitation-contraction coupling in the mouse heart. *The FASEB Journal, 26*, 63–72.

Olesen, M. S., Jensen, N. F., Holst, A. G., Nielsen, J. B., Tfelt-Hansen, J., Jespersen, T. ... Svendsen, J. H. (2011). A novel nonsense variant in $Na_V1.5$ cofactor MOG1 eliminates its sodium current increasing effect and may increase the risk of arrhythmias. *Canadian Journal of Cardiology, 27*, 523.e17–523.e23.

Olson, T. M., Michels, V. V., Ballew, J. D., Reyna, S. P., Karst, M. L., Herron, K. J. ... Anderson, J. L. (2005). Sodium channel mutations and susceptibility to heart failure and atrial fibrillation. *Journal of American Medical Association, 293*, 447–454.

Palygin, O. A., Pettus, J. M., & Shibata, E. F. (2008). Regulation of caveolar cardiac sodium current by a single Gsα histidine residue. *American Journal of Physiology - Heart and Circulatory Physiology, 294*, H1693–H1699.

Park, J. K., Martin, L. J., Zhang, X., Jegga, A. G., & Benson, D. W. (2012). Genetic variants in SCN5A promoter are associated with arrhythmia phenotype severity in patients with heterozygous loss-of-function mutation. *Heart Rhythm, 9*, 1090–1096.

Pillai, J. B., Isbatan, A., Imai, S.i., & Gupta, M. P. (2005). Poly(ADP-ribose) polymerase-1-dependent cardiac myocyte cell death during heart failure is mediated by NAD^+ depletion and reduced Sir2α deacetylase activity. *Journal of Biological Chemistry, 280*, 43121–43130.

Poole, A. W., Pula, G., Hers, I., Crosby, D., & Jones, M. L. (2004). PKC-interacting proteins: from function to pharmacology. *Trends in Pharmacological Sciences, 25*, 528–535.

Priori, S. G., Napolitano, C., Gasparini, M., Pappone, C., Della Bella, P., Brignole, M. ... Schwartz, P. J. (2000). Clinical and genetic heterogeneity of right bundle branch block and ST-segment elevation syndrome: a prospective evaluation of 52 families. *Circulation, 102*, 2509–2515.

Probst, V., Denjoy, I., Meregalli, P. G., Amirault, J. C., Sacher, F., Mansourati, J. ... Wilde, A. A. (2007). Clinical aspects and prognosis of Brugada syndrome in children. *Circulation, 115*, 2042–2048.

Probst, V., Kyndt, F., Potet, F., Trochu, J. N., Mialet, G., Demolombe, S. ... Le Marec, H. (2003). Haploinsufficiency in combination with aging causes SCN5A-linked hereditary Lenègre disease. *Journal of the American College of Cardiology, 41*, 643–652.

Probst, V., Wilde, A. A. M., Barc, J., Sacher, F., Babuty, D., Mabo, P. ... Schott, J. J. (2009). SCN5A mutations and the role of genetic background in the pathophysiology of Brugada syndrome. *Circulation: Cardiovascular Genetics, 2*, 552–557.

Pu, J., & Boyden, P. A. (1997). Alterations of Na^+ currents in myocytes from epicardial border zone of the infarcted heart: a possible ionic mechanism for reduced excitability and postrepolarization refractoriness. *Circulation Research, 81*, 110–119.

Rafizadeh, S., Zhang, Z., Woltz, R. L., Kim, H. J., Myers, R. E., Lu, L. … Chiamvimonvat, N. (2014). Functional interaction with filamin A and intracellular Ca^{2+} enhance the surface membrane expression of a small-conductance Ca^{2+}-activated K^+ (SK2) channel. *Proceedings of the National Academy of Sciences of the United States of America, 111*, 9989–9994.

Rashba, E., Zareba, W., Moss, A., Hall, W. J., Robinson, J., Locati, E. H. … Andrews, M. (1996). Does pregnancy cause an increased risk of cardiac events in long QT syndrome patients? *Circulation, 94*. I–203.

Razani, B., Woodman, S. E., & Lisanti, M. P. (2002). Caveolae: from cell biology to animal physiology. *Pharmacological Reviews, 54*, 431–467.

Remme, C. A. (2013). Cardiac sodium channelopathy associated with SCN5A mutations: electrophysiological, molecular and genetic aspects. *The Journal of Physiology, 591*, 4099–4116.

Remme, C. A., & Bezzina, C. R. (2010). Sodium channel (dys)function and cardiac arrhythmias. *Cardiovascular Therapeutics, 28*, 287–294.

Remme, C. A., Scicluna, B. P., Verkerk, A. O., Amin, A. S., van Brunschot, S., Beekman, L. … Bezzinna, C. R. (2009). Genetically determined differences in sodium current characteristics modulate conduction disease severity in mice with cardiac sodium channelopathy. *Circulation Research, 104*, 1283–1292.

Richards, J., Fitzpatrick, S., Clemens, J., Morris, J., Alliston, T., & Sirois, J. (1995). Ovarian cell differentiation: a cascade of multiple hormones, cellular signals, and regulated genes. *Recent Progress in Hormone Research, 50*, 223–254.

van Rooij, E., Sutherland, L. B., Liu, N., Williams, A. H., McAnally, J., Gerard, R. D. … Olson, E. N. (2006). A signature pattern of stress-responsive microRNAs that can evoke cardiac hypertrophy and heart failure. *Proceedings of the National Academy of Sciences of the United States of America, 103*, 18255–18260.

Rougier, J. S., Albesa, M., Abriel, H., & Viard, P. (2011). Neuronal precursor cell-expressed developmentally down-regulated 4-1 (NEDD4-1) controls the sorting of newly-synthesized $Ca_v1.2$ calcium channels. *Journal of Biological Chemistry, 286*, 8829–8838.

Rougier, J. S., van Bemmelen, M. X., Bruce, M. C., Jespersen, T., Gavillet, B., Apotheloz, F. … Abriel, H. (2005). Molecular determinants of voltage-gated sodium channel regulation by the Nedd4/Nedd4-like proteins. *American Journal of Physiology – Cell Physiology, 288*, C692–C701.

Rougier, J. S., Gavillet, B., & Abriel, H. (2013). Proteasome inhibitor (MG132) rescues $Na_v1.5$ protein content and the cardiac sodium current in dystrophin-deficient mdx5cv mice. *Frontiers in Physiology, 4*, 51.

Ruan, Y., Liu, N., & Priori, S. G. (2009). Sodium channel mutations and arrhythmias. *Nature Review Cardiology, 6*, 337–348.

Rybin, V. O., Xu, X., Lisanti, M. P., & Steinberg, S. F. (2000). Differential targeting of β-adrenergic receptor subtypes and adenylyl cyclase to cardiomyocyte caveolae. *Journal of Biological Chemistry, 275*, 41447–41457.

Schott, J. J., Alshinawi, C., Kyndt, F., Probst, V., Hoorntje, T., Hulsbeek, M. … Le Marec, H. (1999). Cardiac conduction defects associate with mutations in SCN5A. *Nature Genetics, 23*, 20–21.

Schroder, E. A., Lefta, M., Zhang, X., Bartos, D. C., Feng, H. Z., Zhao, Y. … Delisle, B. P. (2013). The cardiomyocyte molecular clock, regulation of Scn5a, and arrhythmia susceptibility. *American Journal of Physiology – Cell Physiology, 304*, C954–C965.

Schwartz, P. J., Priori, S. G., & Napolitano, C. (2003). How really rare are rare diseases? *Journal of Cardiovascular Electrophysiology, 14*, 1120–1121.

Shang, L. L., Sanyal, S., Pfahnl, A. E., Jiao, Z., Allen, J., Liu, H., & Dudley, S. C., Jr. (2008). NF-κB-dependent transcriptional regulation of the cardiac scn5a sodium channel by angiotensin II. *American Journal of Physiology – Cell Physiology, 294*, C372–C379.

Sherman, S., Chrivia, J., & Catterall, W. A. (2000). Cyclic adenosine 3':5'-monophosphate and cytosolic calcium exert opposing effects on biosynthesis of tetrodotoxin-sensitive sodium channels in rat muscle cells. *The Journal of Neurosciences, 5*(6), 1570–1576.

Shimoni, Y., Hunt, D., Chen, K., Emmett, T., & Kargacin, G. (2006). Differential autocrine modulation of atrial and ventricular potassium currents and of oxidative stress in diabetic rats. *American Journal of Physiology – Heart and Circulatory Physiology, 290*, H1879–H1888.

Shinlapawittayatorn, K., Du, X. X., Liu, H., Ficker, E., Kaufman, E. S., & Deschênes, I. (2011). A common SCN5A polymorphism modulates the biophysical defects of SCN5A mutations. *Heart Rhythm, 8*, 455–462.

Shinlapawittayatorn, K., Dudash, L. A., Du, X. X., Heller, L., Poelzing, S., Ficker, E., & Deschenes, I. (2011). A novel strategy using cardiac sodium channel polymorphic fragments to rescue trafficking-deficient SCN5A mutations. *Circulation: Cardiovascular Genetics, 4*, 500–509.

Shy, D., Gillet, L., & Abriel, H. (2013). Cardiac sodium channel $Na_V1.5$ distribution in myocytes via interacting proteins: the multiple pool model. *Biochimica et Biophysica Acta, 1833*, 886–894.

Shy, D., Gillet, L., Ogrodnik, J., Albesa, M., Verkerk, A. O., Wolswinkel, R. … Abriel, H. (2014). PDZ domain-binding motif regulates cardiomyocyte compartment-specific $Na_V1.5$ channel expression and function. *Circulation, 130*, 147–160.

Sidorova, T. N., Yermalitskaya, L. V., Mace, L. C., Wells, K. S., Boutaud, O., Prinsen, J. K. … Murray, K. T. (2015). Reactive γ-ketoaldehydes promote protein misfolding and preamyloid oligomer formation in rapidly-activated atrial cells. *Journal of Molecular and Cellular Cardiology, 79*, 295–302.

Smart, E. J., Graf, G. A., McNiven, M. A., Sessa, W. C., Engelman, J. A., Scherer, P. E. … Lisanti, M. P. (1999). Caveolins, liquid-ordered domains, and signal transduction. *Molecular and Cellular Biology, 19*, 7289–7304.

Smits, J. P. P., Koopmann, T. T., Wilders, R., Veldkamp, M. W., Opthof, T., Bhuiyan, Z. A. … Wilde, A. A. (2005). A mutation in the human cardiac sodium channel (E161K) contributes to sick sinus syndrome, conduction disease and Brugada syndrome in two families. *Journal of Molecular and Cellular Cardiology, 38*, 969–981.

Smyth, J. W., Hong, T. T., Gao, D., Vogan, J. M., Jensen, B. C., Fong, T. S. … Shaw, R. M. (2010). Limited forward trafficking of connexin 43 reduces cell cell coupling in stressed human and mouse myocardium. *The Journal of Clinical Investigation, 120*, 266–279.

Snyder, P. M. (2009). Down-regulating destruction: phosphorylation regulates the E3 ubiquitin ligase Nedd4-2. *Science Signaling, 2*, e41.

Sun, J., Picht, E., Ginsburg, K. S., Bers, D. M., Steenbergen, C., & Murphy, E. (2006). Hypercontractile female hearts exhibit Increased S-nitrosylation of the L-type Ca^{2+} Channel α1 subunit and reduced ischemia/reperfusion injury. *Circulation Research, 98*, 403–411.

Surawicz, B. (1989). Electrophysiologic substrate of torsade de pointes: dispersion of repolarization or early afterdepolarizations? *Journal of the American College of Cardiology, 14*, 172–184.

Surber, R., Hensellek, S., Prochnau, D., Werner, G. S., Benndorf, K., Figulla, H. R., & Zimmer, T. (2008). Combination of cardiac conduction disease and long QT syndrome caused by mutation T1620K in the cardiac sodium channel. *Cardiovascular Research, 77*, 740–748.

Tan, H. L., Bink-Boelkens, M. T. E., Bezzina, C. R., Viswanathan, P. C., Beaufort-Krol, G. C. M., van Tintelen, P. J. … Balser, J. R. (2001). A sodium-channel mutation causes isolated cardiac conduction disease. *Nature, 409*, 1043–1047.

Tateyama, M., Kurokawa, J., Terrenoire, C., Rivolta, I., & Kass, R. S. (2003). Stimulation of protein kinase C inhibits bursting in disease-linked mutant human cardiac sodium channels. *Circulation, 107*, 3216–3222.

Tateyama, M., Rivolta, I., Clancy, C. E., & Kass, R. S. (2003). Modulation of cardiac sodium channel gating by protein kinase A can be altered by disease-linked mutation. *Journal of Biological Chemistry, 278*, 46718–46726.

Thuerauf, D. J., Marcinko, M., Gude, N., Rubio, M., Sussman, M. A., & Glembotski, C. C. (2006). Activation of the unfolded protein response in infarcted mouse heart and hypoxic cultured cardiac myocytes. *Circulation Research, 99*, 275–282.

Ufret-Vincenty, C. A., Baro, D. J., Lederer, W. J., Rockman, H. A., Quinones, L. E., & Santana, L. F. (2001). Role of sodium channel deglycosylation in the genesis of cardiac arrhythmias in heart failure. *Journal of Biological Chemistry, 276*, 28197–28203.

Valdivia, C. R., Ackerman, M., Tester, D. J., Wada, T., McCormack, J., Ye, B., & Makielski, J. C. (2002). A novel SCN5A arrhythmia mutation, M1766L, with expression defect rescued by mexiletine. *Cardiovascular Research, 55*, 279–289.

Valdivia, C. R., Chu, W. W., Pu, J., Foell, J. D., Haworth, R. A., Wolff, M. R., & Makielski, J. C. (2005). Increased late sodium current in myocytes from a canine heart failure model and from failing human heart. *Journal of Molecular and Cellular Cardiology, 38*, 475–483.

Valdivia, C. R., Ueda, K., Ackerman, M., & Makielski, J. C. (2009). GPD1L links redox state to cardiac excitability by PKC-dependent phosphorylation of the sodium channel SCN5A. *American Journal of Physiology – Heart and Circulatory Physiology, 297*, H1446–H1452.

Van Norstrand, D. W., Valdivia, C. R., Tester, D. J., Ueda, K., London, B., Makielski, J. C., & Ackerman, M. J. (2007). Molecular and functional characterization of novel glycerol-3-phosphate dehydrogenase 1 like gene (GPD1-L) mutations in sudden infant death syndrome. *Circulation, 116*, 2253–2259.

Vatta, M., Ackerman, M., Ye, B., Makielski, J. C., Ughanze, E. E., Taylor, E. W. … Towbin, J. A. (2006). Mutant caveolin-3 induces persistent late sodium current and is associated with long-QT syndrome. *Circulation, 114*, 2104–2112.

Vatta, M., Dumaine, R., Antzelevitch, C., Brugada, R., Li, H., Bowles, N. E. … Towbin, J. A. (2002). Novel mutations in domain I of SCN5A cause Brugada syndrome. *Molecular Genetics and Metabolism, 75*, 317–324.

Verkerk, A. O., Remme, C. A., Schumacher, C. A., Scicluna, B. P., Wolswinkel, R., de Jonge, B. … Veldkamp, M. W. (2012). Functional $Na_v1.8$ channels in intracardiac neurons: the link between SCN10A and cardiac electrophysiology. *Circulation Research, 111*, 333–343.

Viswanathan, P. C., Benson, D. W., & Balser, J. R. (2003). A common SCN5A polymorphism modulates the biophysical effects of an SCN5A mutation. *Journal of Clinical Investigation, 111*, 341–346.

Wagner, S., Maier, L. S., & Bers, D. M. (2015). Role of sodium and calcium dysregulation in tachyarrhythmias in sudden cardiac death. *Circulation Research, 116*, 1956–1970.

Wagner, S., Ruff, H. M., Weber, S. L., Bellmann, S., Sowa, T., Schulte, T. … Maier, L. S. (2011). Reactive oxygen species – activated Ca/Calmodulin kinase IIδ is required for late I_{Na} augmentation leading to cellular Na and Ca overload. *Circulation Research, 108*, 555–565.

Wang, C., Hennessey, J. A., Kirkton, R. D., Wang, C., Graham, V., Puranam, R. S. … Pitt, G. S. (2011). Fibroblast growth factor homologous factor 13 regulates Na^+ channels and conduction velocity in murine hearts. *Circulation Research, 109*, 775–782.

Wang, D., Yazawa, K., George, A., & Bennett, P. (1996). Characterization of human cardiac Na^+ channel mutations in the congenital long QT syndrome. *Proceedings of the National Academy of Sciences United States of America, 93*, 13200–13205.

Watanabe, H., Koopmann, T. T., Le Scouarnec, S., Yang, T., Ingram, C. R., Schott, J. J. … Bezzina, C. R. (2008). Sodium channel β1 subunit mutations associated

with Brugada syndrome and cardiac conduction disease in humans. *Journal of Clinical Investigation, 118*, 2260.
Wei, J., Wang, D. W., Alings, M., Fish, F., Wathen, M., Roden, D. M., & George, A. L. (1999). Congenital long-QT syndrome caused by a novel mutation in a conserved acidic domain of the cardiac Na^+ channel. *Circulation, 99*, 3165–3171.
West, J. W., Numann, R., Murphy, B. J., Scheuer, T., & Catterall, W. A. (1991). A phosphorylation site in the Na^+ channel required for modulation by protein kinase C. *Science, 254*, 866–868.
Wu, G., Ai, T., Kim, J., Mohapatra, B., Xi, Y., Li, Z.… Vatta, M. (2008). α-1-syndrophin mutation and the long-QT syndrome: a disease of sodium channel disruption. *Circulation: Arrhythmia and Electrophysiology, 1*, 193–201.
Yanagita, T., Kobayashi, H., Uezono, Y., Yokoo, H., Sugano, T., Saitoh, T.… Wada, A. (2003). Destabilization of $Na_V1.7$ sodium channel α-subunit mRNA by constitutive phosphorylation of extracellular signal-regulated kinase: negative regulation of steady-state level of cell surface functional sodium channels in adrenal chromaffin cells. *Molecular Pharmacology, 63*, 1125–1136.
Yang, B., Lu, Y., & Wang, Z. (2008). Control of cardiac excitability by microRNAs. *Cardiovascular Research, 79*, 571–580.
Yang, B., & Kumar, S. (2010). Nedd4 and Nedd4-2: closely related ubiquitin-protein ligases with distinct physiological functions. *Cell Death and Differentiation, 17*, 68.
Yang, K. C., Kyle, J. W., Makielski, J. C., & Dudley, S. C. (2015). Mechanisms of sudden cardiac death: oxidants and metabolism. *Circulation Research, 116*, 1937–1955.
Yang, P., Kupershmidt, S., & Roden, D. M. (2004). Cloning and initial characterization of the human cardiac sodium channel (SCN5A) promoter. *Cardiovascular Research, 61*, 56–65.
Yang, T., Atack, T. C., Stroud, D. M., Zhang, W., Hall, L., & Roden, D. M. (2012). Blocking Scn10a channels in heart reduces late sodium current and is antiarrhythmic. *Circulation Research, 111*, 322–332.
Yarbrough, T. L., Lu, T., Lee, H. C., & Shibata, E. F. (2002). Localization of cardiac sodium channels in caveolin-rich membrane domains: regulation of sodium current amplitude. *Circulation Research, 90*, 443–449.
Zhou, A., Ou, A. C., Cho, A., Benz, E. J., & Huang, S. C. (2008). Novel splicing factor RBM25 modulates Bcl-x pre-mRNA 5' splice site selection. *Molecular and Cellular Biology, 28*, 5924–5936.
Zhou, J., Shin, H. G., Yi, J., Shen, W., Williams, C. P., & Murray, K. T. (2002). Phosphorylation and putative ER retention signals are required for protein kinase A-mediated potentiation of cardiac sodium current. *Circulation Research, 91*, 540–546.
Zhou, J., Yi, J., Hu, N., George, A. L., Jr., & Murray, K. T. (2000). Activation of protein kinase A modulates trafficking of the human cardiac sodium channel in *Xenopus* oocytes. *Circulation Research, 87*, 33–38.
Zicha, S., Maltsev, V. A., Nattel, S., Sabbah, H. N., & Undrovinas, A. I. (2004). Post-transcriptional alterations in the expression of cardiac Na^+ channel subunits in chronic heart failure. *Journal of Molecular and Cellular Cardiology, 37*, 91–100.
Zimmer, T., & Surber, R. (2008). SCN5A channelopathies – an update on mutations and mechanisms. *Progress in Biophysics and Molecular Biology, 98*, 120–136.

CHAPTER SEVENTEEN

Nav Channels in Damaged Membranes

C.E. Morris*[,1] and B. Joos[§]
*Ottawa Hospital Research Institute, Ottawa, ON, Canada
[§]University of Ottawa, Ottawa, ON, Canada
[1]Corresponding author: E-mail: cmorris@uottawa.ca

Contents

Current Topics in Membranes, Volume 78
ISSN 1063-5823
http://dx.doi.org/10.1016/bs.ctm.2016.06.001

Abstract

Sick excitable cells (ie, Nav channel—expressing cells injured by trauma, ischemia, inflammatory, and other conditions) typically exhibit "acquired sodium channelopathies" which, we argue, reflect bleb-damaged membranes rendering their Nav channels "leaky." The situation is excitotoxic because untreated Nav leak exacerbates bleb damage. Fast Nav inactivation (a voltage-independent process) is so tightly coupled, kinetically speaking, to the inherently voltage-dependent process of fast activation that when bleb damage accelerates and thus left-shifts macroscopic fast activation, fast inactivation accelerates to the same extent. The coupled *g(V)* and availability*(V)* processes and their window conductance regions consequently left-shift by the same number of millivolts. These damage-induced hyperpolarizing shifts, whose magnitude increases with damage intensity, are called coupled left shift (CLS). Based on past work and modeling, we discuss how to test for Nav-CLS, emphasizing the virtue of sawtooth ramp clamp. We explain that it is the inherent mechanosensitivity of Nav activation that underlies Nav-CLS. Using modeling of excitability, we show the known process of Nav-CLS is sufficient to predict a wide variety of "sick excitable cell" phenomena, from hyperexcitability through to depolarizing block. When living cells are mimicked by inclusion of pumps, mild Nav-CLS produces a wide array of burst phenomena and subthreshold oscillations. Dynamical analysis of mild damage scenarios shows how these phenomena reflect changes in spike thresholds as the pumps try to counteract the leaky Nav channels. Smart Nav inhibitors designed for sick excitable cells would target bleb-damaged membrane, buying time for cell-mediated removal or repair of Nav-bearing membrane that has become bleb-damaged (ie, detached from the cytoskeleton).

1. THE PROBLEM: ACQUIRED SODIUM CHANNELOPATHIES—A FEATURE OF SICK EXCITABLE CELLS

Ischemic, traumatic, and inflammatory conditions produce "sick excitable cells," as do assorted chemical reagents and inherited impairments of membrane skeleton (Morris, Boucher, & Joós, 2012). In other words, their genetically normal Na^+-selective voltage-gated channels (VGCs) or Nav channels, notwithstanding excitable cells (those designed to fire action potential using Nav channels), have a high propensity to "acquire" sodium channelopathies when the tissues they reside in experience ischemia, trauma, inflammation, and so on (eg, Douglas, Callaway, Sword, Kirov, & Andrew, 2011; Latronico & Bolton, 2011; Merkies, Faber, & Lauria, 2015; Wagner, Maier, & Bers, 2015). Unlike inherited sodium channelopathies, whose dysfunctions often provide mechanistic insight for normal function, but for which the clinical load is small, the acquired ones are numerous but have been mechanistically baffling.

Issues of cause-and-effect between acquired sodium channelopathies and membrane damage in "sick" Nav-rich cells have been confusing. If effective therapeutic strategies are to be developed, they need to be seen clearly. Our view is as follows. *Cause*: some primary insult subjects an excitable cell to bleb damage. *Effect*: Nav gating is modified where there is bleb damage (Wang et al., 2009) because Nav channel protein embedded in a bleb-damaged plasma membrane bilayer activates at abnormally hyperpolarized membrane potentials (V_m). *Secondary cause*: the resultant "Nav leak" (evident as tetrodotoxin-sensitive stretch-induced Na^+-loading (Wang et al., 2009)), stresses ATP-dependent ion homeostatic mechanisms (White, Brisson, & Andrew, 2012), and (Secondary effect) ATP depletion triggers excitotoxic feedbacks that worsen bleb damage (Zou et al., 2013), increase the Nav leak, and so on round the loop (Fig. 1) (Morris, 2011a). With the excitotoxic cascade operating, the malfunctioning Nav channels are indeed a *cause* of worsening conditions in the sick excitable cell.

Bleb damage as initial cause of acquired channelopathies was overlooked, perhaps, because nascent bleb damage, though measureable via mechanical compliance (Sheetz, Sable, & Döbereiner, 2006), is invisible by light microscopy. Once blebbing is conspicuous at the light microscope level, multiple excitotoxic-type dysfunctions, secondary injuries, chief among them the bleb-acquired Nav leak, are rendering the excitable cell necrotic. Secondary excitable cell damage can thus be mitigated/prevented by early or prior blockade of Nav channels that initially "acquire" their "channelopathy" in a bleb-inducing primary injury.

Assuming this is correct, efforts to find cell- and tissue-specific mechanistic explanations for the diverse manifestations of acquired sodium channelopathies would be difficult if the causal (primary) source of bleb damage is bypassed. A search for molecular explanations of Nav dysfunctions in critical illness myopathies and polyneuropathies (Friedrich, 2008; Novak et al., 2009) illustrate this point. Both states develop under the powerfully "blebbogenic" condition (Celes et al., 2013), namely sepsis with inflammation. Bacterial endotoxins, a key player in sepsis, elicit neuronal blebbing (Picot et al., 2003) and reduce Nav availability (Haeseler et al., 2008). Carefully performed loose-patch V-clamp using a septic muscle model reveals left-shifted steady-state inactivation (Filatov & Rich, 2004), precisely the expected outcome from severe bleb damage. This "acquired sodium

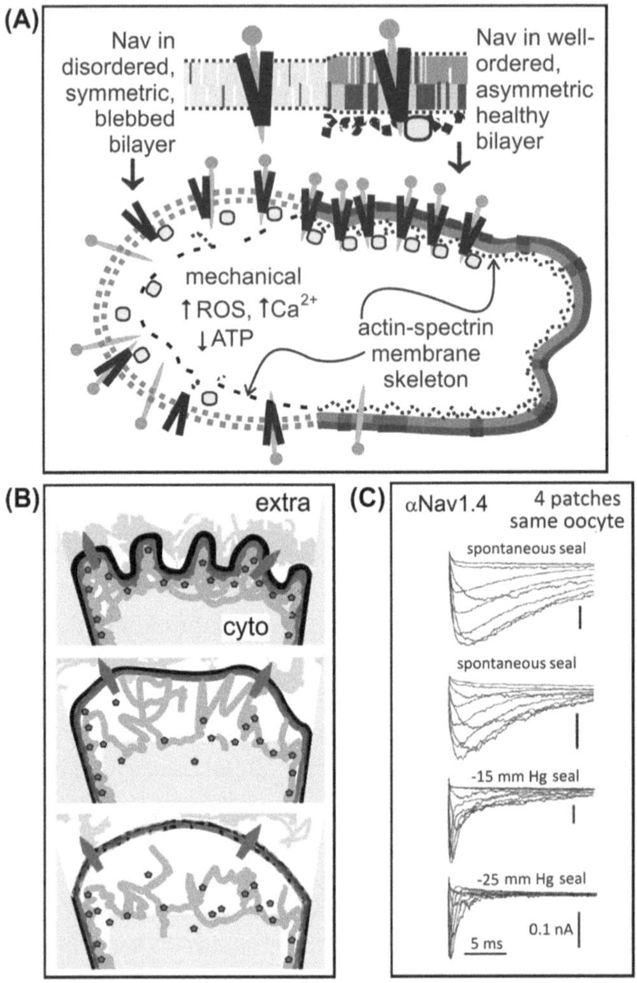

Figure 1 ***Bleb-damaged membranes.*** (A) Cartoon highlighting Nav channels in healthy versus bleb-damaged membrane (several blebbogenic conditions are listed) (see Morris, 2011a; Morris, Boucher, et al., 2012; Morris, Juranka, et al., 2012; Morris et al., 2015; Zou et al., 2013). (B) cartoon depicting a cell-attached patch (blue (gray in print versions) = pipette walls) with extracellular, cytoplasmic, and membrane disruptions expected after a gentle gigaohm-seal forms (top) then after mild (middle) and severe (bottom) bleb damage due to pipette aspiration. Though the bilayer's structure denatures (see text) relative to its intact plasma membrane state, membrane proteins (*green arrow-like* (light gray in print versions)) do not denature (C) in intact oocyte plasma membrane, the kinetics of heterologously expressed αNav1.4 channels exhibit anomalously slow right-shifted kinetics but, unexpectedly, as illustrated by these traces, damaging cell-attached oocyte membrane patches causes αNav1.4 channels to exhibit normal-seeming fast-mode gating indistinguishable from that seen when αNav1.4 is coexpressed with β subunits. *Traces are modified from Figure 5 of Tabarean, I. V., Juranka, P., & Morris, C. E. (1999). Membrane stretch affects gating modes of a skeletal muscle sodium channel.* Biophysical Journal, 77, *758–774. (As per Morris, C. E., & Juranka, P. F. (2007b). Lipid stress at play: mechanosensitivity of voltage-gated channels.* Current Topics in Membranes, 59, *297–338.).*

channelopathy" readily explains the muscle weakness attendant on critical illness neuropathy. Similar shifts observed for peripheral neurons (Novak et al., 2009) would explain sepsis-related loss of excitability there.

Likewise, the bleb damage associated with mechanical trauma and ischemia and ischemia/reperfusion and inherited membrane skeleton "dystrophy" (Hirn, Shapovalov, Petermann, Roulet, & Ruegg, 2008) situations would, in those situations, be primary cause for the acquired sodium channelopathies. In experimental model systems for diverse sick excitable cell conditions, Nav channel antagonists improve outcomes (see Morris, Boucher, et al., 2012; Morris, Juranka, & Joós, 2012).

Thus conditions as seemingly unrelated as sepsis, traumatic brain injury, brain and cardiac and diabetic ischemias, membrane skeleton impairments, various degenerative diseases such as ALS, muscular dystrophy and many others, have a common factor. These are excitable (i.e., V-dependent) gradient-dissipating ion channels embedded in bleb-damaged ("denatured") bilayer, albeit the embedded Nav channels are by no means denatured. "Denatured bilayer" refers to the irreversible loss, in blebbing plasma membrane bilayer, of higher order bilayer-structural features (eg, leaflet asymmetries, lateral inhomogeneities in packing order). For sick excitable cells, severe pathological consequences appear to result from the propensity of damaged plasma membrane to retain low their level bilayer structure plus functional Nav channels even as higher level bilayer structure decays.

2. MODELING: UNDERLYING SIMPLICITY FOR DIVERSE EXCITABILITY PATHOLOGIES

A biophysical model of Nav dysfunction has been developed (Boucher, Joós, & Morris, 2012) based on recombinant Nav1.6 channel behavior measured in membrane mechanically traumatized by pipette aspiration (Fig. 2A) (Wang et al., 2009). Fig. 3 summarizes the principal results from simulating excitability dysfunctions predicted by this model, "Nav-CLS" [CLS: coupled left-shift coupled hyperpolarizing or left-shift of the Nav channel availability and activation curves, see, for example, Fig. 4B]; these results will be revisited as we proceed. While our work has been generic, Nav-CLS has also been used in a computational model of white matter axons where it successfully mimics the acute electrophysiological consequences of concussive injury (Volman & Ng, 2013), a mild traumatic brain injury that can yield devastating diffuse axonal injury.

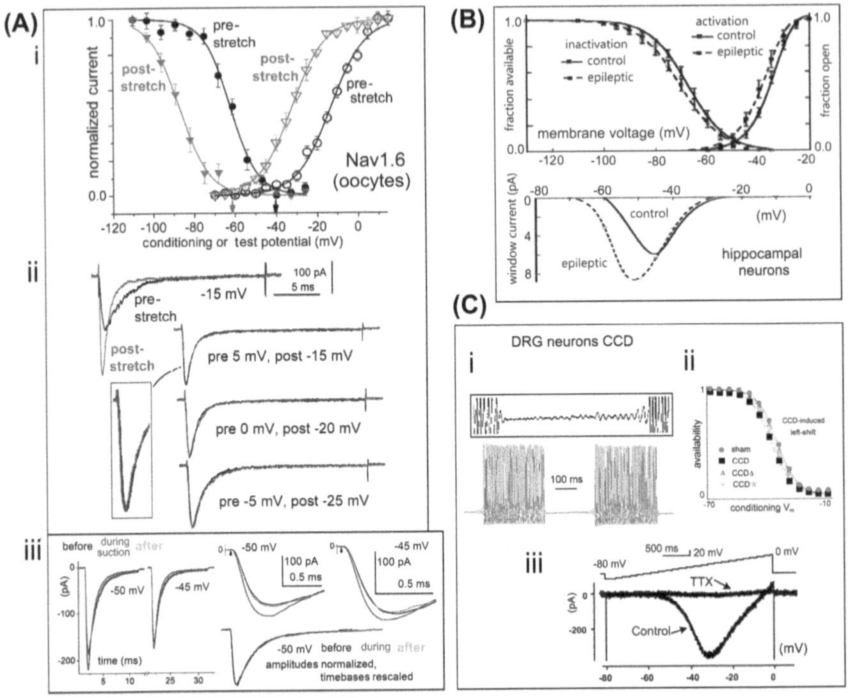

Figure 2 ***Damaged membranes and I_{Na}.*** (Ai–iii) Recombinant Nav1.6, heterologously expressed in oocytes, cell-attached patch data, explanations as previously (for i and ii, see also legend to Figure 3 in Morris, Boucher, et al., 2012). (B) Left-shifted I_{window} in epilepsy model. Chronic crush damage (CCD); (Ci) trace fragments illustrate crush-related spontaneous bursts and subthreshold oscillations, phenomena they showed to be associated with a "mild" Nav-CLS-like left-shift of availability(V). This point is depicted here (Cii) via a cartoon summary of their findings for CCD (plus treatment variants). (Ciii) $\mathbf{I_{Na}(V)}$ from slow depolarizing ramp clamp unavoidably leaves ambiguous modal (slow? fast?) contributions to $I_{Na\text{-}persistent}$, especially at near-threshold V_m values (see Fig. 6A). *DRG*, dorsal root ganglion. *(Ai-iii) Modified from Wang, J. A., Lin, W., Morris, T., Banderali, U., Juranka, P. F., & Morris, C. E. (2009). Membrane trauma and Na^+ leak from Nav1.6 channels.* American Journal of Physiology and Cell Physiology, 297, *C823–C834, (B) Modified from Sun, G. C., Werkman, T. R., & Wadman, W. J. (2006). Kinetic changes and modulation by carbamazepine on voltage-gated sodium channels in rat CA1 neurons after epilepsy.* Acta Pharmacological Sinica, 12, *1537–1546, (Ci) Modified from Song, X. S., Huang, Z. J., & Song, X. J. (2009). Thiamine suppresses thermal hyperalgesia, inhibits hyperexcitability, and lessens alterations of sodium currents in injured, dorsal root ganglion neurons in rats.* Anesthesiology, 110, *387–400, and (Ciii) Modified from Song, Y., Li, H. M., Xie, R. G., Yue, Z. F., Song, X. J., Hu, S. J., & Xing, J.L. (2012). Evoked bursting in injured Aβ dorsal root ganglion neurons: a mechanism underlying tactile allodynia.* Pain, 153, *657–665.*

3. THERAPEUTIC AGENTS: A NEED TO RECOGNIZE BLEB-DAMAGED MEMBRANE

Nav inhibitors suitable for prolonged systemic administration are needed for sick excitable cell conditions. Inhibitor molecules (eg, tetrodotoxin) that do not distinguish in their bind preferences between fast- and slow-mode channel Nav conformations, though remarkably protective in tissue/cell models (Hirn et al., 2008; Iwata et al., 2004; White et al., 2012), are lethal when systemically applied, while safer slow-mode inhibitors have generally met with only "small-but-significant" clinical success (Nagoshi, Nakashima, & Fehlings, 2015; Rognoni et al., 2013). Where there is bleb damage—induced Nav dysfunction, all-modes blockade may be needed, with ideal inhibitors accumulating exclusively into the leaflets of Nav-bearing bleb-damaged bilayer. Lipophilicity, an attribute of clinically effective Nav-inhibitors (Lenkey, Karoly, Epresi, Vizi, & Mike, 2011; Lenkey et al., 2010), fosters bilayer accumulation of inhibitors. Natural amphiphilic V-sensor toxins accumulate at the bilayer's outer leaflet interface, burying themselves a few hydrophobic residues deep into the membrane interior (Jung et al., 2005; Mihailescu et al., 2014). Insofar as voltage-sensor/toxin interactions are responsive to the identity of the lipid components of the bilayer (Milescu et al., 2009) or to the palmitoylation state of the channel (Nav1.2 (Bosmans, Milescu, & Swartz, 2011)), and insomuch as the toxin's inhibitory binding interaction produces local bilayer thinning (Mihailescu et al., 2014), it is perhaps not surprising that the inhibitory efficacy of a V-sensor toxins can show a dramatic increase in bleb-damaged versus intact oocyte bilayers (Schmidt & MacKinnon, 2008) (see further discussion of this striking finding by Schmidt and MacKinnon (2008) in Morris (2011b)). An immense diversity of spider venom peptides target Nav channels; there may be untapped tools from these sources (Klint et al., 2012). Targeting inhibitors exclusively to bleb-damaged Nav-rich bilayers will be a challenge. A counterapproach (Shankarappa et al., 2012) illustrates the state of affairs: systemically toxic nonlipophilic saxitoxin is microencapsulated in liposomes for delivery application to the discrete small volume of imposed nerve damage. While this peripheral treatment reduces longer term central impacts of peripheral nerve damage, it is applicable only for precisely localized injuries. In a fascinatingly different approach that reduced cardiac reperfusion injury, leaky Nav channels residing in cardiac monocyte/macrophages were targeted by having the monocytes engulf a liposome encapsulated nonselective Nav inhibitor (phenytoin) with the aim of exploiting the cells' phagocytosis for delivery-to-target (putatively,

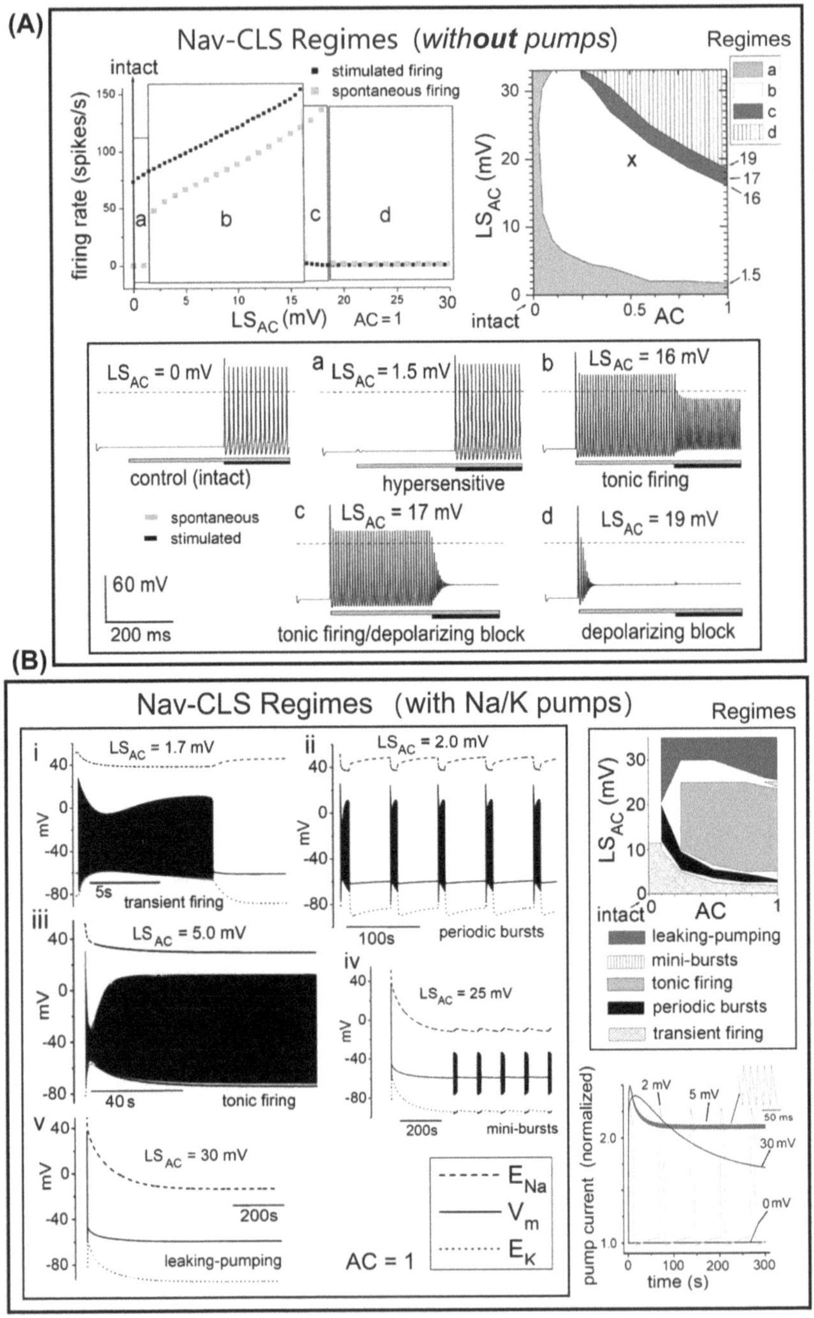

Figure 3 Computed excitability with Nav-CLS injury, as labeled. Regime plots are alluded to throughout the text. In (A), there are no pumps (in voltage plots, *dashed lines* indicate 0 mV). In (B), Na/K pumps are included with a surface *s* to inner volume vol_i ratio, 20 $cm^2/\mu L$, small enough to observe neuropathic behaviors over conveniently

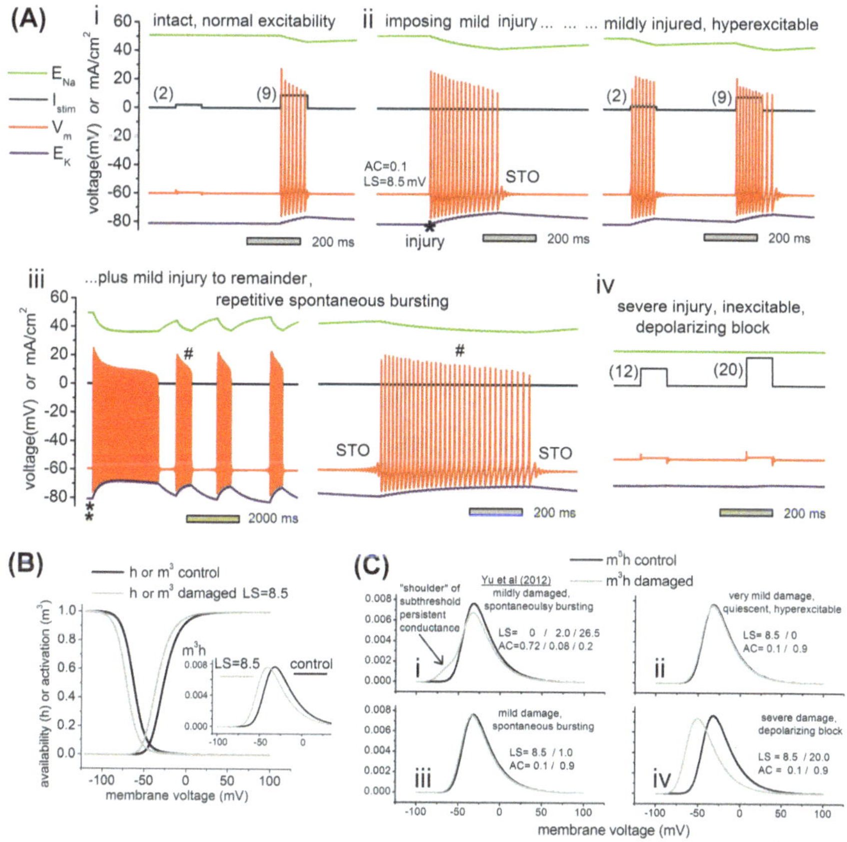

Figure 4 ***Modeling Nav-CLS injury in an excitable cell.*** (A) For a computational axon, time courses for V_m prior to (i) then after (ii–iv) worsening Nav-CLS injury depicted at three discrete stages (see text). (B) The biophysical underpinnings, in Hodgkin–Huxley terms, of an 8.5 mV Nav-CLS injury; availability(V) (ie, $h_\infty(V)$) and $g(V)$ (ie, $m^3_\infty(V)$) curves are left-shifted 8.5 mV with respect to the intact condition. Inset, $m^3_\infty h_\infty(V)$ for both conditions. (Ci–iv) Steady-state Nav conductances (ie, $(m^3_\infty h_\infty(V))$ for intact and injured Nav-bearing membranes as described in the text. *CLS*, coupled left-shift; *STO*, subthreshold oscillation.

short simulation times. The specific burst patterns vary with the out/in volume ratio here 4:1. Other smaller ratios down to 0.1:1 are shown in Figure 11 of Boucher et al. (2012). With respect to the ratio s/vol_i, 20 $cm^2/\mu L$ was used in Boucher et al. (2012), Yu et al. (2012) and Lachance et al. (2014) (Parameter Tables in Boucher et al., 2012 and Yu et al., 2012 have a typo: "m^3" instead of liters, L, for volumes). For all simulations, $vol_i = 3 \times 10^{-15}$ L $= 3 \times 10^{-9}$ μL. *CLS*, coupled left-shift. *Modified from Boucher, P. A., Joós, B., & Morris, C. E. (2012). Coupled left-shift of Nav channels: modeling the Na⁺-loading and dysfunctional excitability of damaged axons.* Journal of Computational Neuroscience, *33, 301–319.*

endomembrane Nav channels) (Zhou et al., 2013). An intriguing possibility not considered by Zhou et al. (2013) is that these bleb-sensing white blood cells also deliver their onboard Nav drugs to the damaged excitable cell membranes associated with the infarcted tissue these cells seek out. If they could be developed, "smart" Nav reagents for sick excitable cells would generally need to be delivered systemically (possibly microencapsulated) using delivery systems that recognize blebbed membrane and, in those locales only, inhibit both fast- and slow-mode Nav channels.

Upon injury to neurons and muscle fibers, which are large, long-lived, terminally differentiated cells, buying time for the membrane repair machinery is crucial. Thus, recent reports that sick excitable cells show improved survival when treated with surface-active block copolymers are very encouraging; block polymer molecules in question adhere to bilayer surfaces in a manner that reduces the likelihood of bilayer rupture (see Martindale & Metzger, 2014). Examples of improved outcomes include cardiomyocytes subjected to ischemia/reperfusion injury (Martindale & Metzger, 2014), neural cells subjected to mechanical injury (Rad, Mobasheri, Najafi, & Rezaei, 2014), hippocampal neurons subjected to oxygen-glucose deprivation (Shelat, Plant, Wang, Lee, & Marks, 2013). Perhaps coadministering protective, membrane-stabilizing block copolymers (Jin, 2014; Kilinc, Gallo, & Barbee, 2008) together with highly lipophilic Nav inhibitors would yield synergisms. Stanching excitotoxicity (Zou et al., 2013) during prolonged application of both agents at low levels would be expected to foster excitable cell repair.

4. IRREVERSIBLE BLEB-TYPE DAMAGE IN MEMBRANE PATCHES

A bilayer that forms part of a healthy plasma membrane is not self-organized. A plasma membrane bilayer has asymmetries (Manno, Takakuwa, & Mohandas, 2002), curvatures, and subdomains that are actively generated and actively maintained. As detachment of the bilayer from adherent filamentous membrane skeleton proteins initiated blebbing (Boulbitch et al., 2000), a plasma membrane bilayer would begin degrading closer to its minimal free energy self-organized (low enthalpy, high entropy—"dead") state, consistent with the situation described by Schwille and colleagues (Sezgin et al., 2015) who "speculate…that in live cells" (in contrast to cytoskeleton-free giant unilamellar vesicles prepared from living cells, and whose degraded features they measure) "multiple domains of different compositions could coexist within a single membrane due to active, local processes like membrane traffic, lipid turnover, enzymatic activity." Electron microcopy of stretch-damaged white matter shows axolemmal bilayer detached from its Nav-organizing spectrin (Maxwell, 1996). Issues of repair/maintenance of excitable cell membranes are receiving increasing attention (eg, Wang et al., 2010; Hinman, Rasband, & Carmichael, 2013; Zhang et al., 2012), and recently key ion transporters' postischemia capabilities were rigorously examined in cardiomyocytes (Lin et al., 2013). Imposing membrane injury of a known intensity is problematic, but for ion channels, pipette aspiration of gently sealed cell-attached patches provides a progressively worsening (albeit not quantified) damage scenario (Fig. 1B). Irreversibly changing gating behaviors after aspiration (suction, "stretch") are attributed to the imposed bleb damage (Morris, Juranka, Lin, Morris, & Laitko, 2006). Changes can be abrupt or develop over minutes (eg, Morris, Boucher, et al., 2012; Small & Morris, 1994; Wan, Juranka, & Morris, 1999; Wang et al., 2009).

From the earliest gigaohm-seal recordings on oocytes (Methfessel et al., 1986), it was acknowledged that patches formed using gentle aspiration will have bleb damage. To steer the focus toward putative mechanophysiology, stretch-activated channels' irreversibly altered responses to stretch after such damage (Morris & Horn, 1991; Morris, 2011a) are often downplayed, and terms such as "hysteresis," "patch history," "exercising the patch" (Suchyna, Markin, & Sachs, 2009) are used to acknowledge what we, instead, allude to as membrane trauma (ie, Wan et al., 1999; Wang et al., 2009). Fortunately, in oocytes (and molluscan neurons) demonstrably "gentle" seals can be formed. In whole oocytes, gating of αNav1.4 is slow-mode dominated, and as Fig. 1C illustrates, "gentle patches" can largely retain those intact-membrane characteristics. Oocyte's endogenous MS-cation channels also gate differently pre/postsuction, a finding by Hamill & McBride (1992) eventually linked to bleb damage. To compare different native MS-channels, Small and Morris (1994) reproduced the Hamill-lab "patch history" effect on oocyte MS channels then turned to molluscan neuron stretch-activated K channels, finding there that successive fixed-intensity "hits" (on-off steps of pipette suction) progressively increase the channels' susceptibility to stretch. Other blebbogenic pretreatments act similarly (Wan et al., 1999). Irreversible changes in bilayer structure might be a factor, but loss of the "mechanoprotection" against mechanical load (ie, bilayer tension) normally provided by adhesion of the bilayer to intact membrane skeleton (Morris, 2001) sufficed to explain these data.

Thus, the body of findings on MS channels (ie, the non-VGC ones) are reminder that, while reversible ("elastic") MS gating changes imply reversible changes in membrane structure (ie, in the mechanical characteristics of the bilayer), irreversible ones could have irreversible bilayer mechanical and/or chemical explanations (see Morris, Juranka, et al., 2012). Complicating matters, patch tension can increase irreversibly via increased adhesion to the glass pipette, due to so-called "patch creep." As discussed recently (Morris, Prikryl, & Joos, 2015), if procedures known to prevent creep of oocyte patch preparations (Gil, Silberberg, & Magleby, 1999; Morris et al., 2006) are omitted or prove ineffective, experimentally irreversible gating changes could in fact reflect an inherently "elastic" (reversible) MS response. For example, does the excised outside-out patch phenomenon reported by Schmidt, del Mármol, and MacKinnon (2013), a phenomenon they call "conversion" of Kv, reflect a reversible process that appeared irreversible because patch creep had irreversibly increased membrane tension?

5. Nav1.4 IN DAMAGED OOCYTE MEMBRANES—TRAUMA-INDUCED SWITCH FROM SLOW TO FAST MODE

As reported independently from two labs (Shcherbatko, Ono, Mandel, & Brehm, 1999; Tabarean et al., 1999), skeletal muscle αNav1.4 (expressed in oocytes without β subunits) was the first identified channel whose irreversible membrane damage–induced gating changes were studied. Damage progressively hyperpolarizes αNav1.4 activation and steady-state inactivation (availability) by ~20 mV. Strikingly, the channels in progressively damaged membrane also progressively and irreversibly switch from slow-mode into fast-mode gating. Eventually the kinetics of αNav1.4-alone become indistinguishable from αNav1.4-plus-β (co-expressed) in intact membrane. For α-with-β subunits, fast-mode dominates, and activation and availability midpoints are >20 mV more hyperpolarized than for αNav1.4-alone. The slow-mode preference of αNav1.4 in intact oocyte membrane and the damage-induced switch constitute an unexplained anomaly of that isoform, not observed, for example, with Nav1.5 channels expressed in oocytes (Banderali, Juranka, Clark, Giles, & Morris, 2010; Morris & Juranka, 2007a). With the determinants of the speed of Nav activation sensor motions' speeds having been identified as particular residues in the DI–DIII voltage sensors (Lacroix, Campos, Frezza, & Bezanilla, 2013), it would be interesting to test whether WT (αNav1.4) sensitivity to membrane damage (ie, damage-induced stabilization of fast-mode gating) is retained in extra-slow and extra-fast Nav mutants designed in that study.

Simultaneous megapatch recordings/videos of αNav1.4-alone before, during, and after visibly damaging suction (Shcherbatko et al., 1999), show that only during applied suction is patch tension high enough to activate endogenous MS channels. While the oocyte patch is curved at all stages (indicating some Laplace tension), conditions were not such that patch-glass adhesion tension produced an inward creeping "drumhead" (ie, planar tensed) (Suchyna et al., 2009) that would indicate inexorably increasing patch tension (Ursell, Agrawal, & Phillips, 2011). Thus, while it is appropriate in the case of Kv "conversion" (an excised outside-out patch phenomenon as discussed above) to invoke the (unintended) irreversible imposition of high tension to explain a gating change, there are not grounds for doing so to explain the irreversible mode switch of αNav1.4 in cell-attached oocyte patches. Also ruled out (by the findings of Moran, Tammaro, Nizzari, and Conti (2000) who studied Nav1.4 kinetics using cytoskeleton reagents including colchicine) is the notion (Shcherbatko et al., 1999) of mechanically destroyed microtubule/Nav1.4 channel links as the cause for altered gating characteristics (see also Morris & Juranka, 2007b). A likelier conjecture is that the changed bilayer—protein interface is the "site of action" that renders αNav1.4 anomalously more stable in its slow-mode configurations in the "oocyte-organized" (albeit heterologous) bilayer than in the bleb-damaged ("self-organized") bilayer.

6. Nav1.6 IN DAMAGED OOCYTE MEMBRANES

For Nav1.6 (in oocytes, with or without coexpressed β subunits), fast-mode gating dominates and the response to progressive bleb damage is straightforward: pipette suction yields irreversible hyperpolarizing shifts of *g(V)* and availability*(V)* (Wang et al., 2009) (Fig. 2Ai). Because of patch-rupture, "left-shift" maxima were not quantitated. In some experiments, however, "saturation" of the irreversible change was noted and at that point, fully reversible MS-gating changes were measured by comparing "before, during, and after" currents (Fig. 2Aiii). Under tension, activation and inactivation accelerate to the same extent, as evidenced by complete $I_{Na}(t)$ time course overlap following amplitude normalization and extension of the "during stretch" time base (multiple examples in Banderali et al., 2010; Morris &

Juranka, 2007a). The increased amplitudes of the "during" $I_{Na}(t)$ traces are consistent with left-shifted g(V)s. While signal averaging is needed to resolve small reversible changes, single sweeps suffice to reveal progressive irreversible change (until saturation approaches; see Figure 1Ai in Wang et al., 2009). Reversible MS changes were also observed (oocyte patches) for Nav1.4 and Nav1.2 with or without β (Morris & Juranka, unpublished observations) and again, a small fraction of the response is fully reversible. Thus, all oocyte-expressed Nav channels tested exhibit reversible MS modulation; Nav1.5 is the only isoform for which the reversible MS response dominates and, in this regard, is the best characterized (Banderali et al., 2010; Morris & Juranka, 2007a). MS modulation of Nav1.5 currents is evident in rat ventricular myocytes (Banderali et al., 2010) and also in mammalian cell-line patches (with reversibility/irreversibility less clear) (Beyder et al., 2010).

For Nav1.6, with regard to membrane damage, Fig. 2Aii shows an example of irreversible left-shift that happens to be exactly 20 mV. Note that, with amplitudes normalized, $I_{Na}(t)$ overlaps perfectly for pre/posttraces 20 mV apart. This shows that activation/inactivation time courses accelerate to the same extent for irreversible MS changes, just as we saw for reversible MS changes. Evidently, at this coupled transition (about which more will be said later), elevated tension and bilayer damage are both essentially "surrogates" for depolarization. For a gNa(V) with an e-fold per 6 mV slope, an MS acceleration factor of say 1.4-fold (or 1.9-fold) would yield a 2 mV (or 4 mV) hyperpolarizing shift of the activation curve (see Morris & Juranka, 2007a), more than enough to render a membrane hyperexcitable (eg, see Boucher et al., 2012).

Functionally, it follows that bilayer perturbations for reversible and irreversible g(V) shifts are not distinguishable. From the point of view of Nav energetics, damage changes would include reduced packing order, reduced leaflet asymmetry, thinning. Outright loss of lipids (fatty acids, inositol lipids (Rusinova, Hobart, Koeppe, & Andersen, 2013)) might be a factor. Inevitably, the magnitude of change possible going healthy-to-blebbed would exceed that for already-blebbed membrane going low-to-high tension. While unintended tension elevations might cause some overestimation of irreversible components, the fundamental message is: irreversible MS Nav left-shifts are expected where bleb damage irreversibly thins the bilayer and diminishes its packing order. Nano-level specifics of bleb damage in excitable cells were broached two decades ago by Maxwell (1996) but have not to our knowledge received serious attention since then. This issue warrants experimental attention.

7. MS MODULATION OF SQUID AXON VOLTAGE-GATED CHANNELS

Hydrostatic pressures and surface active agents that increase bilayer lipid packing order (decreased area/lipid) make bilayers thicker, and membrane stretch and surface active agents that decrease packing order make the bilayer thinner (see Rawicz, Smith, McIntosh, Simon, & Evans, 2008 and references in Finol-Urdaneta, McArthur, Juranka, French, & Morris, 2010). Well before VGC mechanosensitivity was examined in patches, reversible exposure to intense hydrostatic pressures on the classic channels squid axon Kv and Nav channels (see Conti, Inoue, Kukita, & Stuhmer, 1984) was shown to reversibly shift activation in the depolarizing direction, as expected of these MS channels (see discussion in Morris & Juranka, 2007a). The original interpretation was in terms of intraprotein volume changes, but subsequent high-resolution Kv structures (eg, references in Schmidt & MacKinnon, 2008) showing that VGC activation involves repacking at the lateral protein—lipid interface implicates the embedding bilayer structure as a factor in VGC gating energetics. Of course when changes in area-per-lipid are imposed (through stretch or hydrostatic pressure), boundary lipid impacts on the behavior of a multistate protein might well evince complexities related to, eg, hydrophobic mismatch (Kinnun, Mallikarjunaiah, Petrache & Brown, 2015).

8. COUPLED LEFT-SHIFT: KINETIC LINKAGE OF ACTIVATION TO INACTIVATION

How damage-modified $g_{Na}(t,V)$ (Wang et al., 2009) would be expected to affect excitability will be addressed below (See Section 10) after a look at the inherent MS kinetics underlying Nav dysfunction.

At least one Nav transition must be MS (some VGCs have more) (Laitko & Morris, 2004; Lin, Laitko, Juranka, & Morris, 2007). MS-VGCs do not necessarily exhibit MS left-shift (Cav and Kv3 channels do not) (Morris, 2011b, 2011c), but MS-left-shift will result if the rate-limiting V-dependent transition along the activation path accelerates with stretch (as, evidently, in Navs). For Shaker-type Kvs, the dominant MS transition (as seen via $I_K(t)$) involves a concerted V-dependent "lateral pull" preceding the channel's V-independent opening (Morris et al., 2015). Essentially, as envisaged by Yifrach and MacKinnon (2002) and by Pathak, Kurtz, Tombola, and Isacoff (2005), from structure/function studies geared at understanding channel opening, a VGC is so designed so that "vertical" (transmembrane) electric field energy can be transformed into "horizontal" (in the plane of the bilayer) mechanical work. Once this transformation has "spring-loaded" the appropriate region of the VGC, thermal energy is sufficient for pore-opening to occur.

Though Nav structure is unknown (unlike Kv), functionally, Nav opening is V-dependent (Capes, Goldschen-Ohm, Arcisio-Miranda, Bezanilla, & Chanda, 2013). As shown recently (Figure 8 in Morris et al., 2015), an opening transition that was MS but not V-dependent (Schmidt et al., 2013) could have nontrivial consequences for excitability. That work shows how an MS-VGC's response to the bilayer-mechanical state will depend on (1) the VGC's kinetic scheme and (2) which of that scheme's transitions is MS.

Nav channels behave as if their rate-limiting (slowest) V-dependent activation transition is also an MS transition (Banderali et al., 2010). Reciprocal Nav responses to hydrostatic pressure and increased membrane tension could suggest a forward/backward V-dependent transition involves a lateral expansion/compaction in the plane of the bilayer (see Banderali et al., 2010; Morris & Juranka, 2007a). However, since preliminary tests on Nav1.5 deactivation showed no effect of stretch on this transition (Morris, unpublished), the transition state for MS-V-dependent opening could be very close (in terms of conformational area and stiffness) to the open state (see Figure 2 of Morris et al., 2015).

The molecular transition for Nav channel fast-inactivation gating is not V-dependent (Capes et al., 2013) (inactivation's V-dependence as characterized from macroscopic I_{Na} is a separate issue, discussed in the next section); at the membrane's cytoplasmic face, beyond the electric field, beyond the bilayer, binding of a small hinged "lid" motif occludes permeation. However, the open state facilitates binding of this lid motif strongly enough that, kinetically, macroscopic inactivation couples to opening. This tight linkage simultaneously renders inactivation gating both pseudo-V-sensitive and pseudo-MS (Banderali et al., 2010): if a Nav channel's activation time course decelerates or accelerates X-fold (due bilayer compaction or expansion respectively), the inactivation time course also decelerates or accelerates X-fold.

Such kinetics predict coupled *de*polarizing shifts of activation (g(V)) and availability(V) curves, which we could call "coupled right-shift" (CRS) and the coupled hyperpolarizing shifts of g(V) and availability(V) curves seen with membrane damage that we have termed "coupled left-shift" (CLS) (Boucher et al., 2012).

Slow-mode activation in Nav1.6 has been observed making MS left-shifts (Figure 4 in Morris, Boucher, et al., 2012), but for slow-mode inactivation (which, mechanistically, is unrelated to fast inactivation), there are no data. In any case, CLS (and CRS) are terms exclusively applicable to fast-mode Nav gating. Unless explicitly indicated, "activation" and "inactivation" refer to the fast-mode processes.

Thus, for more compressed/thickened/orderly and for more stretched/thinned/disordered membranes, outcomes are respectively, Nav-CRS and Nav-CLS.

9. DOES THE DOMAIN-4 V-SENSOR, THE SITE OF "COUPLING MUTANTS," PLAY A ROLE IN Nav-CLS?

The Nav channel biophysics term "activation/inactivation coupling" has been associated with several inherited sodium channelopathies (Chen, Santarelli, Horn, & Kallen, 1996), and we expected MS-CLS behavior of "coupling mutants" to be abnormal. The idea is that stretch will not accelerate their activation and inactivation time courses in a coupled fashion. However, as explained below, this idea was wrong (Banderali et al., 2010).

Nav-CLS happens because Nav opening (monitored as g(V,t)), which occurs only after three of the channel's four V-sensors respond to depolarization (Lacroix et al., 2013), strongly facilitates inactivation. Nothing is implied about absolute positions of $g_{infinity}(V)$ and $h_{infinity}(V)$ nor the position/steepness of $\tau_{inactivation}(V)$ relative to g(V). V-dependent motions of Nav's domain-4 V-sensor regulate a $\tau_{inactivation}(V)/g(V)$ "coupling" relationship, a relationship that weakens or fails in "coupling mutants" (Chen et al., 1996). Banderali et al. (2010) examined the MS responses of two coupling mutants, each lacking one S4-arginine in the domain-4 V-sensor, thus prolonging inactivation and causing "long QT" syndrome. $\tau_{inactivation}(V)$ is entirely flat in one and overly shallow in the other.

In the flat $\tau_{inactivation}(V)$ case, ie, Nav1.5-R1626P, $\tau_{inactivation}(V)$ is fully uncoupled from g(V). Nevertheless, just as for wild-type Nav1.5, membrane stretch reversibly accelerates the onset and decay $I_{Na}(t)$ to the same extent. Normal, reversible Nav-CLS remains in effect. Likewise for Nav1.5-R1623Q, the $\tau_{inactivation}(V)$-too-shallow mutant. This argues that in both mutants, binding of the inactivation motif proceeds normally after channel opening. We conclude, therefore, that MS Nav-CLS is solely dependent on the biomechanics of the rate-limiting MS- and V-dependent transition in the activation pathway; the V-sensor in domain-4 seems not to come into play for this process.

10. USING THE HODGKIN–HUXLEY FORMULATION TO SIMULATE THE COUPLING OF g(V) AND AVAILABILITY(V) IN MS Nav-CLS

While not "molecularly" accurate regarding inactivation (which it depicts as an inherently V-dependent process occurring independent of activation) the Hodgkin–Huxley (HH) (Hodgkin & Huxley, 1952) formulation can be used in a very straightforward way to mimic the key feature of Nav-CLS (Fig. 4B and Figure 2 in Boucher et al., 2012). An HH depiction of Nav-CLS simply shifts $m^3{}_\infty(V)$ (the nondimensional g(V) curve) by a given number of millivolts then repeats that shift for $h_\infty(V)$, the HH version of steady-state availability(V). Because HH depicts inactivation as a fully independent process, HH-based Nav-CLS would depart, albeit in a minor way, from experimental data for time course overlaps (eg, that Fig. 2Aii and iii). HH is mathematically tractable and provides a direct, easily grasped link from model to experimentally observed MS shifts of g(V) and availability(V). Detailed Markov schemes featuring putatively accurate transitions are difficult to parameterize even for simpler kinetic cases (Morris et al., 2015); HH is the sensible "better the devil you know" choice for modeling sick cell excitability.

11. COMPUTATIONS TO ILLUSTRATE THE PATHOEXCITABILITY SPECTRUM OF Nav-CLS

Fig. 4 shows Nav-CLS computations, with intact (healthy, predamage) parameter values chosen as previously (Boucher et al., 2012) but with the computations sequenced to mimic current clamp of a sick excitable cell with progressive bleb injury. Unlike the situations depicted in Fig. 3A where pump activity was not included, the Fig. 4 action potential computations have pumps (and thus relate to injury regimes depicted in Fig. 3B), and thus include simple cyto-geometry specifics such as (inside and outside volumes plus a cellular surface/volume ratio). Left-shift (LS) values are in mV, and the fraction of affected channels (ACs) (referring only to Nav channels) is nondimensional. Different fractions of membrane have damage as given by AC, LS pairs. Concurrent with I_{stim} and V_m "traces," Fig. 4A plots $E_{Na}(t)$ and $E_K(t)$. Given the membrane surface area, the 4:1 out/in volume ratio yields $E_{Na}(t)$ changes quickly enough to generate neuropathic behaviors over conveniently short times (Figure 11 in Boucher et al., 2012 shows a range of other out/in ratios). Fig. 4B and C depicts, in HH terms, the Nav-CLS injury mechanism at various intensities.

12. AN ASIDE ON EXCITABILITY THRESHOLDS AND ION HOMEOSTASIS

Changing thresholds are critical in excitability dysfunction (eg, Park et al., 2009), so to before dissecting the Fig. 4 scenario, we first briefly address the issue of "excitability threshold." Dysfunctional excitability due to Nav-CLS injury needs to be seen together with ion homeostatic mechanisms that defend normal excitation thresholds (Fig. 3, Regimes). In the context of Nav-CLS injury, Yu, Morris, Joos, and Longtin (2012) modeled dynamical interactions between channel-based ion fluxes and restorative Na/K pump activity showing how activity-dependent changes in E_{Na} and E_K affect thresholds for Nav-based excitability. Accordingly, excitability dysfunctions, eg, as depicted in Fig. 4, could be associated with quite different Nav-CLS values for say different intra/extracellular volumes or if the pump was impaired (for Fig. 4, pump intensity is at a "healthy" level). Addressing changing threshold values for the simplest possible AP (a biphasic plateau AP) due to a noninactivating $g_{Ca}(V)$ and a g_{leak} (Morris & Lecar, 1981), Fig. 5A–C is provided as a primer on dynamical approaches to threshold where prolonged channel activity runs down ion gradient (E_{Ca} in this case) because pump activity is inadequate to the task. The system's simplicity allows for a fully analytical approach in identifying how this affects system threshold (unstable) points. For Nav-based excitability examined by Yu et al. (2012), threshold points are much more complex, but (with extensive numerical integration) can be identified from graphical analyses, as seen in Fig. 5D.

In Fig. 5D we will see "STOs" (subthreshold oscillations). These are an indicator of Na/K pump homeostatic activity "spontaneously" bringing V_m nearer an unstable threshold region, as in Fig. 3; recordings from sick excitable cells are typically often rife with STOs (see, for example, Fig. 2C (Costigan, Scholz, & Woolf, 2009; Kovalsky, Amir, & Devor, 2009; Novak et al., 2009; Song et al., 2012; Xie, Strong, Kim, Shahrestani, & Zhang, 2012)).

Yu et al. (2012) demonstrate how STOs lead in and out of spontaneous AP bursting as a particular Nav-CLS injury stresses ion homeostasis. Since Nav-CLS damage is simply an unwanted shift in the Nav channels'

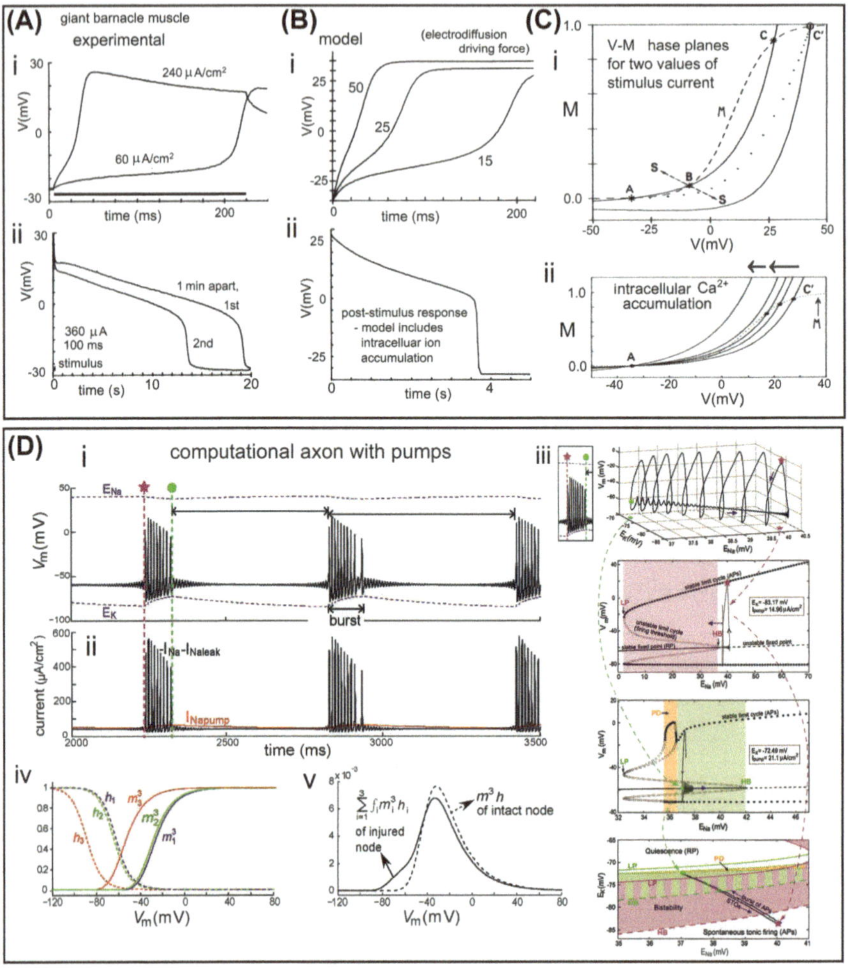

Figure 5 ***Excitation thresholds vary as driving forces change.*** (Ai, ii) Threshold for bistable Ca^{2+}-plateau APs changes with ΔE_{Ca} (barnacle muscle, noninactivating Cav channels, g_{leak}, Kv channels inhibited, stimuli as labeled). This behavior is modeled (Bi, ii) and explored via dynamical analysis (Ci, ii). In (A) prolonged I_{Ca} yields a plateau V_m whose drop is putatively due to intracellular Ca^{2+} accumulation (hence, pump inadequacy). For the "up/down" AP jumps, thresholds differ (compare Ai and ii). Consistent with diminished E_{Ca}, the initial V-plateau on a second AP (ii) is diminished. (Bi) These phenomena are modeled for various stimulus intensities ("M" = HH-type activation variable for g_{Ca}). To give an I_{Ca}-dependent ΔE_{Ca}, cell volume is added ("intracellular Ca^{2+} accumulation," driving force nonohmic (electrodiffusive)) (Bii). A phase plane (dynamical) analysis (Ci, ii) reveals unstable points B ("bifurcation") that signify thresholds, and stable points A, C, C′ for V_{rest} and $V_{plateau}$. (D) This computational work illustrates a central theme for Nav-based ectopic (spontaneous) excitability (i, ii) in damaged membrane, namely that firing thresholds change. For Nav-based

operational range, pathological spontaneous bursting with STOs understandably resembles the healthy-cell Nav physiology of some sensory cells and interneurons (see Yu et al., 2012 for refs). Without stimuli, V_m should be quiescent in axons and skeletal muscle, hyperexcitability is troublesome, and ectopic spontaneous excitation is a pathological disaster, as is the next stage, depolarizing block (see the plots labeled "Regimes" in Fig. 3). Sick excitable cells stumble through parameter ranges they would normally not encounter. Studying the responses to electrical stimuli of isolated crayfish axons and isolated barnacle muscle fibers, Hodgkin (1948) and Morris and Lecar (1981), respectively, were both categorizing epiphenomena of moribund (ie, isolated and thus dying) excitable cells. While their interest was healthy excitable cell action potential behaviors, the preparations were undergoing run-down, so unintentionally they were describing behaviors of (mildly) sick excitable cells. The important overlap in what constitutes fully physiological versus mildly pathological excitability patterns for various classes of excitable membranes (eg, axons, axon initial segments, pacemaker neurons, skeletal muscle, cardiac muscle subtypes) is evident in the fact that Prescott, De Koninck, and Sejnowski (2008) successfully modeled

excitability, analytical solutions are not possible but, as illustrated, "brute-force" graphical analysis (iii) enables dynamical analysis of thresholds, (iv and v) give the HH underpinnings of this particular Nav-CLS injury. This spontaneous bursting activity (no stimulation) is pathological and ectopic. Driving forces (see $E_{Na}(t)$ and $E_K(t)$ in Di) oscillate slowly, yielding instabilities evident first as subthreshold oscillations then as AP bursts. Reference points (*purple star, green dot*) for the first V_m burst (replotted as a *boxed inset*) are also found in the three-dimensional trajectory (*blue time-direction arrows*) depiction of this burst at right (top box, iii), below which are "snapshot" phase-plane analyses (the "instantaneous" values for E_K and I_{pump} are listed; note different X,Y scales). The bottom plot emphasizes that an excitable system will exhibit various stable and unstable behaviors, including some chaoslike unstable zones (see Yu et al., 2012 for explanation), as the pump attempts to control E_K:E_{Na} in the face of Nav-CLS injury. If the contributing channels or the pump efficacy or the relevant volumes (all held constant here) were instead allowed to vary, the system would exhibit similar complexities. These pathological behavior patterns are not fundamentally different from healthy cell patterns; it is the context that is wrong. *HH*, Hodgkin–Huxley; *CLS*, coupled left-shift. *(Ci, ii) Modified from Morris, C. E., & Lecar, H. (1981). Voltage oscillations in the barnacle giant muscle fiber.* Biophysical Journal, 35, *193–213 and (D) Modified from Yu, N., Morris, C. E., Joós, B., & Longtin, A. (2012). Spontaneous excitation patterns computed for axons with injury-like impairments of sodium channels and Na/K pumps.* PLoS Computational Biology, 8, *e1002664.*

the major activity patterns of healthy mammalian interneurons based on the activity patterns of crayfish axons and barnacle fibers in variously moribund states.

13. AN EXCITABLE CELL'S INJURY WORSENS—MODELED

We now return to the simulated "I-clamp experiment" which we set up two sections ago. It begins with calibration of intact-cell excitability (Fig. 4Ai); 2 mA/cm^2 has no effect, 9 mA/cm^2 elicits a train of APs, and (not shown) hyperpolarizing steps (eg, $-9\,mA/cm^2$) elicit single overshooting "anode-break" APs. Next (Fig. 4Aii and Cii), 10% of the membrane (AC = 0.1) is "injured" (asterisk) by imposing an 8.5 mV Nav-CLS (LS = 8.5); the relative severity of this injury can be gauged via coordinates (AC = 0.1, LS = 8.5) in a "Regimes" plot (see Fig. 3Bii). Applying the CLS injury (*asterisk*, Fig. 4Aii) destabilizes V_m; an AP train ensues, with the AP amplitudes falling as ion gradients decay. The train ends spontaneously via an STO, another class of V_m instability. After the single prolonged burst, the system approaches a new V_{rest} or quiescent steady state (computation of many seconds—not shown—confirmed the lack of further bursts) but, unlike its healthy cell counterpart, this V_{rest} is compromised. Though a physiologist would be challenged to directly measure the very small V_{rest} or E_{Na} or E_K changes here, pathology would be evident via the test current injections; a previously subthreshold stimulus of 2 mA/cm^2 now elicits APs (Fig. 4Aii, right) and 9 mA/cm^2 now triggers an AP train of increased frequency and duration. The mild injury has produced a hyperexcitable cell.

After the "primary localized" injury (AC = 0.1, LS = 8.5), there follows a secondary injury that mildly damages the rest of the excitable membrane (ie, AC = 0.9; LS = 1 mV; *double asterisk* Fig. 4Aiii, corresponds to Fig. 4Ciii). With the augmented Nav-leak, pump strength is inadequate to return the system to a stable steady state. Instead, the prolonged injury-induced burst is followed by a series of shorter AP bursts, each initiated and terminated by STOs. In this simple excitable system (only two VGCs, a nonspecific leak, pump), bursting is a mathematically robust form of dynamic instability, evident over wide-ranging system parameters. In sick excitable cells, unsurprisingly, bursting is typically irregular, since inter- and intraburst characteristics would vary with ATP depletion, tissue edema, use-dependent changes in channel characteristics, and so on. In fact, simply introducing current noise along with Nav-CLS injury is sufficient to elicit chaotic, paroxysmal burst firing (Yu et al., 2012). This reflects observations in injured dorsal root ganglion (DRG) neurons (Wang et al., 2011; Xie et al., 2012).

As conditions that worsen bleb damage came into play in a sick excitable cell (Fig. 1A), increasing its Nav-CLS injury, two further distinctive phenomena, ie, tonic firing then depolarizing block, could be expected. The former would not likely be short-lived given the stress it would place on pumps. In the simulation, however, after ~10 s with at AC = 0.1, LS = 8.5 plus AC = 0.9, LS = 15 mV, our system settles stably into tonic firing (not shown, but see Fig. 3, Regimes) since we did not allow for ATP depletion. Finally (Fig. 4Aiv and Civ) with bleb-induced Nav-CLS at AC = 0.1, LS = 8.5 plus AC = 0.9, LS = 20 mV, depolarizing block prevails. Again there is quiescence, but now at an easily measurable diminished V_{rest} (~-50 mV) and current injection reveals an inexcitable membrane (even stimuli >9 mA/cm^2 are ineffective). We noted that large hyperpolarizing currents injected at this stage can elicit complex STOs (not shown). In current clamp, anode break excitation is used because, without having to inject depolarizing current, it tests for AP-producing VGCs. Similarly, the ability of sustained hyperpolarizations to elicit STOs could be used to probe sick cells for the presence of activatable Nav channels (eg, after Nav proteolysis is suspected) (von Reyn, Mott, Siman, Smith, & Meaney, 2012).

Fig. 4B shows, in HH terms, a Nav-CLS exemplar. Steady-state activation (m^3) and availability (h) are shown for intact membrane and for LS = 8.5 mV. As all Nav channels (AC = 1) undergo 8.5 mV of coupled LS (m^3(V) and h(V) are displaced by −8.5 mV), the steady-state or "window conductance" $m^3_\infty h_\infty(V)$ (inset) follows. Note that the y-axis for $m^3_\infty h_\infty(V)$ is <1% that of the $m^3_\infty(V)$ and $h_\infty(V)$. The shifted $m^3_\infty h_\infty$ "conductance" is, effectively, a "subthreshold persistent conductance"; multiplied by $g_{Na,max}$ and V_m-E_{Na}, it passes the "Nav-leak" current.

The left-most $m^3_\infty h_\infty(V)$ plot in Fig. 4C (plot Ci) applies not to the "I-clamp experiment" calculations, but to the ectopic bursting Nav-CLS condition mentioned above (Yu et al., 2012), where Nav-CLS injury to an axon (72% intact, 8% mildly injured, 20% severely injured) was designed to give a $m^3_\infty h_\infty(V)$ with a left-shifted "shoulder" similar to what is seen in healthy pacemaking interneurons driven by a fast-mode "subthreshold persistent current" (Taddese & Bean, 2002).

As the next two graphs emphasize, however, mild injury could be very challenging to detect by V-clamp, distinctive V(t) dysfunctions notwithstanding. Consider the hyperexcitabilty exemplar [Fig. 4A; (AC = 0.1, LS = 8.5 mV), (AC = 0.9, LS = 0 mV)]; plotted $m^3_\infty h_\infty(V)$ changes (Fig. 4Cii) are almost invisible, and even with enough CLS to engender repetitive spontaneous bursting (Ciii), the window conductance change is barely discernible. Though $I_{Na}(V)$, which multiplies $g_{Na}(V)$ by the Na^+-driving force, would be an improvement over the Δg_{Na} shown, for mild yet clinically important injury, it might be a lost cause given the noise inherent in experimental data sets. However, since mild manifestations (eg, hyperexcitability, ectopicity) of acquired sodium channelopathies presage the severer ones (inexcitability), experimentalists could design V-clamp investigations that compare healthy cells against sick cells at the severe end of their injury spectrum. Where depolarizing block prevails (Civ), V-shifts would likely be quantifiable experimentally (Novak et al., 2009). Indeed, left-shifted window conductance is reported for hippocampal neurons metabolically injured by prolonged epileptic barrage stimulation (see Fig. 2C) (Sun et al., 2006). We reemphasize (Morris, Boucher, et al., 2012; Morris, Juranka, et al., 2012) the value, for molecular pharmacologists, of using V-steps plus "sawtooth" clamp (as opposed to standard ramps) to dissect window current (fast-mode) from slow-mode changes. Additionally, experimentalists should watch for shallow-seeming g(V) curves (which are precisely the expectation from heterogeneous ("smeared out")) left-shift injury (see Figure 5 in Yu et al., 2012) that get steeper as injury intensifies (eg, for a left-shifting gK, see Schmidt & MacKinnon, 2008).

14. TACTILE ALLODYNIA (HYPERSENSITIVITY TO STIMULI) AND LEFT-SHIFTED Nav CURRENTS

To model tactile allodynia, a phenomenon linked to hyperexcitable DRG neurons, chronic compression of DRG neurons [chronic crush damage (CCD)] has been used (Song et al., 2009, 2012). CCD yields bursting discharges associated with STOs in response to peripheral stimulation (see Fig. 2C). Burst durations increase with depolarized V_m, and more depolarized V_m elicits spontaneous bursting; this is associated with a larger "persistent sodium current."

The DRG ramp current, we suggest, may not be due solely to slow mode; one could compare it, for example, with recombinant Nav1.6 slow-mode ramp currents (irreversibly increased and left-shifted after patch suction, Fig. 6A), as discussed previously (Morris, Boucher, et al., 2012). Further investigation could determine if the pre-postinjury availability(V) curves indicate any Nav-CLS.

Congruent with the view (Song et al., 2012) that ectopically active Nav channels at a site of peripheral nerve damage engenders allodynia (centrally mediated) is the finding that full g_{Na} blockade (both fast-mode and slow-mode) at peripherally damaged membranes slows the onset of neuropathic pain. Liposomes containing saxitoxin administered percutaneously gave a localized long-term nerve block; silencing damage-site Navs delayed the onset of the downstream consequences (Shankarappa et al., 2012).

15. THRESHOLD TRACKING, PERIPHERAL NEUROPATHIES, CANCER CHEMOTHERAPY

The impaired AP triggering and propagation of peripheral neuropathies can be monitored by computerized threshold tracking procedures for transdermal compound AP recordings (Kanai et al., 2006). Results indicate, for diverse membrane injury scenarios (diabetic neuropathy, carpal tunnel, motor neuron disease), that node of Ranvier persistent sodium current increases with membrane injury (Fig. 6D). While increases might reflect regenerative processes, a degenerative explanation, in the form of left-shifted g_{Na} (fast- and/or slow-mode) acting as a subthreshold persistent current, could, we suggest, also be a factor.

For example, via sensory neuron damage, the cancer chemotherapeutic compound, oxaliplatin, can cause hyperexcitability (repetitive compound muscle APs, neuromyotonic discharges), propagation infidelities, and depolarizing block (numbness, paresthesia) (Lehky, Leonard, Wilson, Grem, & Floeter, 2004). Threshold tracking (Park et al., 2009) detects signs of oxaliplatin neurotoxicity before overt symptoms appear. This sick cell process aligns with classic cases of increasingly overactive Nav channels (Fig. 1A).

Clinical benefit could be expected from noninvasive early detection of neuropathic states (via threshold tracking) if appropriate Nav inhibitors were then administered to stanch progressively worsening membrane damage. The interpretation of threshold tracking data sets would benefit from node of Ranvier sawtooth-clamp (Morris, Boucher, et al., 2012) data that could establish contributions of fast/slow modes to nodal "g_{Na}-persistent" (see Fig. 6B and C) and, as discussed in the previous section, whether severe injury produces left-shift in either.

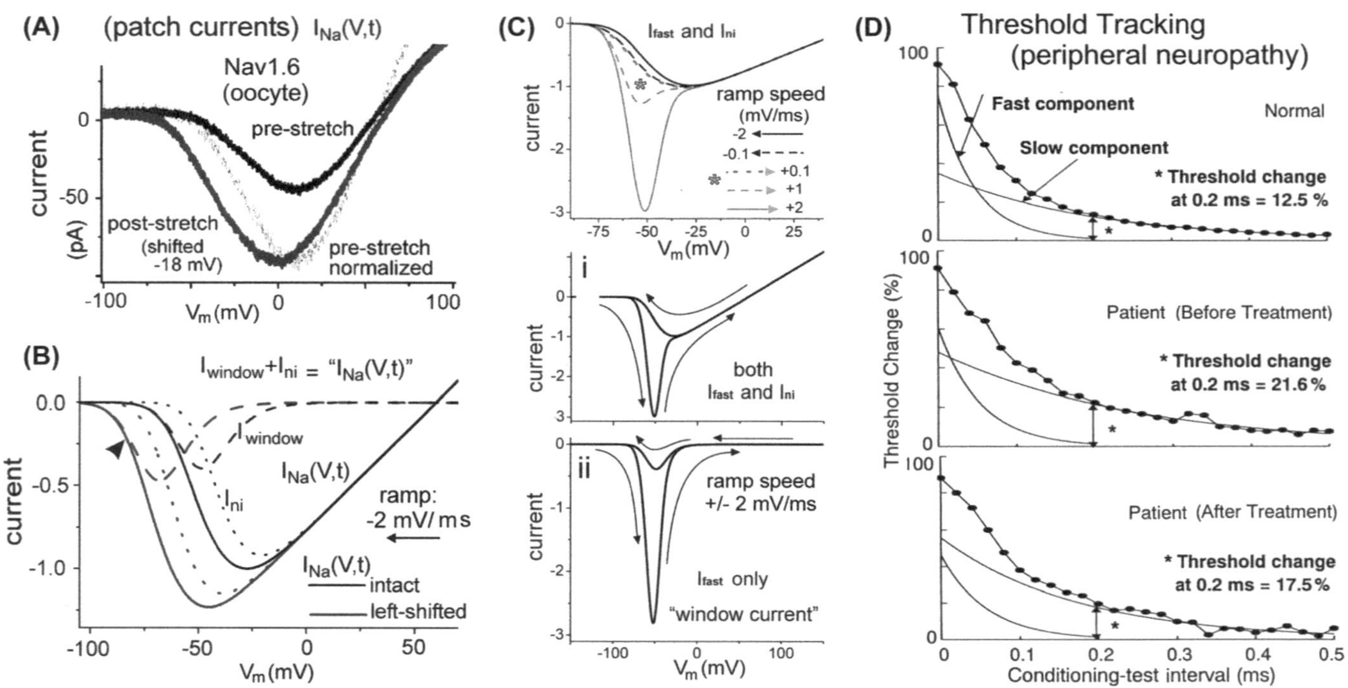

Figure 6 ***Persistent sodium current in practice and theory.*** (A) Recombinant persistent I_{Na} before and after patch damage; a slow hyperpolarizing V-ramp was used (not the standard slow depolarizing ramp). (B and C) Shows expectation from a hyperpolarizing ramp for an HH-transient (fast) and noninactivating Nav channels at a <1% ratio (1.00:0.0075) and the expectations when sawtooth ramp is used at various speeds; reporting such data and subjecting it to kinetic modeling would help clarify the relative contributions of slow and fast modes to ramp currents, especially if step voltage clamp was also included in the protocols. (D) Illustrates threshold tracking data, one of many

situations for which a more thorough knowledge of Nav-modal contributions (slow/fast) is needed to facilitate the development of appropriate therapeutic approaches. The raw data (*dots*) are fitted to two exponentials, a fast component associated to passive membrane properties, and a slow component which depends on persistent Na^+ currents. *(A) Modified from Wang, J. A., Lin, W., Morris, T., Banderali, U., Juranka, P. F., & Morris, C. E. (2009). Membrane trauma and Na^+ leak from Nav1.6 channels. American* Journal of Physiology and Cell Physiology, 297, *C823–C834, (B and C) Modified from Morris, C. E., Boucher, P. A., & Joós, B. (2012). Left-shifted Nav channels in injured bilayer: primary targets for neuroprotective Nav antagonists?* Frontiers in Pharmacology, 3, *19 and explained therein, and (D) Modified from Figure 2 in Isose, S., Misawa, S., Sakurai, K., Kanai, K., Shibuya, K., Sekiguchi, Y., … Kuwabara, S. (2010). Mexiletine suppresses nodal persistent sodium currents in sensory axons of patients with neuropathic pain.* Clinical Neurophysiology, 121, *719–724.*

16. WHEN A Nav CHANNEL LEFT-SHIFTS DUE TO MEMBRANE DAMAGE DOES THIS AFFECT BOTH FAST AND SLOW MODES?

Using recombinant Nav1.2, Nav1.4, or Nav1.6 (αβ) in oocyte patches, biophysical aspects of these fast/slow-mode issues could be substantially addressed prior even to examining modes for channels in their native settings. The modes arise from one population of Nav channels accessing different conformational states (Lenkey et al., 2011). Presently, it is unknown if membrane damage that left-shifts fast-mode g_{Na} perforce renders slow-mode g_{Na} left-shifted, and if so, by what amount. Also unknown: are the V-insensitive intermode transitions (see simplified scheme, Figure 1 of Morris, Boucher, et al., 2012) sensitive to membrane damage? The dramatic mode-shift response of αNav1.4 to membrane damage (Shcherbatko et al., 1999; Tabarean et al., 1999) highlights the importance of answering this question, but doing so where the relative contribution of slow-mode, while consequential enough for good signal-to-noise, is not considered anomalous (eg, αβNav1.4 in oocytes). Combined saw-tooth plus step clamp protocols (Morris, Boucher, et al., 2012) would be needed to simultaneously monitor slow and fast modes before, during, and after episodes of pipette suction. To reiterate, important unknowns regarding g_{Na} in acutely injured membrane include (1) does slow-mode left-shift directly echo fast-mode left-shift (and if not, what happens?), (2) are relative weights of the two modes constant or changeable with acute damage, (3) do intermode transitions respond to damage?

17. BLEB-DAMAGED T-TUBULAR MEMBRANES

A hallmark of bleb damage is loss of the actin-spectrin membrane skeleton (Zhang, Gao, Popov, Wen, & Hamill, 2000). Since Nav channel localization at excitable membranes depends dynamically on multiple elements of the spectrin skeleton (Bennett & Lorenzo, 2013; Ho et al., 2014) and since perturbing the lipid composition and architecture of T-tubules alters many conductances (O'Connell et al., 2015), we suggest T-tubular blebbing as possible source of dysfunctional I_{Na} in "sick" muscle cells. T-tubular Nav channels, though not extensively studied, contribute to cardiac (Westenbroek et al., 2013) as well as to skeletal muscle (DiFranco & Vergara, 2011) excitation–contraction coupling.

Bleb-type injury to T-tubules' grid-like arrangement is readily evident (as dilated invaginations) by confocal microscopy (Krolenko & Lucy, 2001; Moench, Meekhof, Cheng, & Lopatin, 2013). Osmotic swell/shrink perturbations, for instance, cause T-tubules to "remodel" (dilate), a pathological process analogous to the vacuole-like dilation (VLD) invaginations of neurons and myotubes (Morris & Homann, 2001; and see Figure 1 in Gauthier, Masters, & Sheetz, 2012). While membrane skeleton filaments (F-actin, β-spectrin) realign to coat some VLDs within minutes (Morris & Homann, 2001), much of the perturbed membrane remains; essentially it is invaginated membrane skeleton-free bleb membrane. Nav channels in such membranes continue to gate (Milton & Caldwell, 1990) in response to depolarizing steps and, if left-shifted, they could be a source of ectopic excitation until the dilated invagination pinched off from the surface.

Voltage-dye monitoring of T-tubular Nav channel activity (Nav1.4, presumably) is possible (DiFranco & Vergara, 2011), so it might be possible to optically monitor T-tubular Nav activity under normal conditions and during the course of increasing bleb damage, in the form of the strikingly visible (by fluorescence microscopy) fatigue-induced vacuolar dilation of the T-tubules (Lännergren, Bruton, & Westerblad, 2000).

18. PROPAGATION INFIDELITIES AND BLOCK DUE TO Nav-CLS

The basic consequences of Nav-CLS injury were explored with regard to node-to-node propagation of APs (Boucher et al., 2012). In a simple myelinated axon (modeled without pumps), mild Nav-CLS (ie, hyperexcitable regime, Fig. 3) is benign. Across the injured node, standard 1:1 throughput of incoming APs is achieved. However, with injury severe enough to elicit spontaneous firing (the model axon "resonant" frequency is ~70 Hz), faithful (1:1) throughput changes no longer pertain. The ratio, 1:>1, at first reflects the ectopic-rate firing at the injured node but as Nav-CLS injury worsens, the downstream intact nodes cannot follow; abruptly, at some Nav-CLS value, they begin firing at lower frequencies, albeit in an approximately integral fashion. Instead of 1:1 firing moves to 2:1, a "period-doubling" pattern reflecting intact-node refractoriness. Once Nav-CLS reaches "depolarizing block regime" severity, propagation fails beyond the injured node.

19. MILDLY INJURED EXCITABLE MEMBRANE? DO NOT DISTURB

A mildly injured cell, to have reasonable prospects for recovery, must have functional pumps. Adding pumps to a propagating-axon model allowed Lachance, Longtin, Morris, Yu, and Joos (2014) to address a key clinical issue for mild injuries: should having to cope, metabolically speaking, with normal excitatory inputs render propagation-competent but still-quiescent injured axons (hyperexcitable regime) vulnerable to further Nav activity–induced impairment? This was an "incipient diffuse axonal injury scenario" probed via a 10-node myelinated axon model with mild Nav-CLS.

What emerged was that with no increase in Nav-CLS values, the fully restabilized mildly damaged nodes (V_m and other parameters not varying) can go from quiescence to becoming ectopic signal generators if regular incoming AP traffic causes sufficient ion homeostatic stress. What happens is that spike thresholds move to more hyperpolarized values as per Fig. 5C. Such changes could contribute to acquired sodium channelopathies such as epileptic phenomena, diabetic neuropathies, and the neuropathic amplification of normally benign sensory inputs (allodynia). Further, they would be exacerbating the primary injury by fostering excitotoxically increased Nav-CLS (the Fig. 4A scenario).

Another phenomenon was noted: namely input spike patterns whose frequencies exceed the ectopic frequency can propagate with good fidelity through an ectopically firing site. This "propagation window" is robust, occurring despite Gaussian noise, large temporal jitter, and the presence of consecutive ectopic nodes.

20. Nav-CLS MODELING OF ACUTE MILD TRAUMATIC BRAIN INJURY

Volman and Ng (2013) implemented Nav-CLS (with pumps) for model myelinated axons whose diameters and internodes are typical of brain white matter (particularly corpus callosum). The regime studied was mild enough to avoid repetitive spike activity. As implemented (at single or at multiple nonadjacent nodes), mild CLS yielded reduced spike amplitudes and complex effects (slowing, failure) on spike propagation, consistent with mild traumatic brain injury studies (von Reyn et al., 2012) showing functional changes in axonal signal conduction. Results, which were also fully consistent with data from fluid percussion injury studies (Baker et al.,

2002; Reeves, Phillips, & Povlishock, 2005), could help explain the trauma intensity–dependent reduction of compound AP amplitudes.

21. CONCLUSION

Nav-CLS goes a long way in explaining diverse acquired sodium channelopathies associated with membrane damage. The modeled consequences of Nav-CLS for ongoing excitotoxic damage to injured cells show why fast-mode inhibition is needed until damaged excitable membrane can be reprocessed. By definition, Nav-CLS does not address Nav slow mode, and it will be important to establish whether left-shift of slow mode is a factor in dysfunctional excitability. Sawtooth ramp clamp is readily implemented but seldom used. When used in conjunction with step protocols, it would substantially augment information garnered while monitoring Nav activity in damaged membranes. Because membrane damage intensity can be imposed and controlled to a reasonable degree in cell-attached oocyte patches (this is untested for Nav channels in cell lines), we suggest oocytes as the heterologous expression system of choice for working out if and how Nav-CLS and slow-mode Nav gating changes coexist and interact in the damaged membranes of sick excitable cells. Finally, we suggest that future research will need characterize in detail the bilayer-structural changes associated with bleb damage and with membrane repair in excitable cells and in the heterologous systems in which Navs and other VGCs are studied. The paucity of information about plasma membrane structure in sick excitable cells is quite remarkable.

REFERENCES

Baker, A. J., Phan, N., Moulton, R. J., Fehlings, M. G., Yucel, Y., Zhao, M., … Tian, G. F. (2002). Attenuation of the electrophysiological function of the corpus callosum after fluid percussion injury in the rat. *Journal of Neurotrauma, 19*, 587–599.

Banderali, U., Juranka, P. F., Clark, R. B., Giles, W. R., & Morris, C. E. (2010). Impaired stretch modulation in potentially lethal cardiac sodium channel mutants. *Channels (Austin), 4*, 12–21.

Bennett, V., & Lorenzo, D. N. (2013). Spectrin- and ankyrin-based membrane domains and the evolution of vertebrates. *Current Topics in Membranes, 72*, 1–37.

Beyder, A., Rae, J. L., Bernard, C., Strege, P. R., Sachs, F., & Farrugia, G. (2010). Mechanosensitivity of Nav1.5, a voltage-sensitive sodium channel. *Journal of Physiology (Paris), 588*, 4969–4985.

Bosmans, F., Milescu, M., & Swartz, K. J. (2011). Palmitoylation influences the function and pharmacology of sodium channels. *Proceedings of the National Academy of Sciences of the United States of America, 108*, 20213–20218.

Boucher, P. A., Joós, B., & Morris, C. E. (2012). Coupled left-shift of Nav channels: modeling the Na^+-loading and dysfunctional excitability of damaged axons. *Journal of Computational Neuroscience, 33*, 301–319.

Boulbitch, A., Simson, R., Simson, D. A., Merkel, R., Häckl, W., Bärmann, M., & Sackmann, E. (2000). Shape instability of a biomembrane driven by a local softening of the underlying actin cortex. *Physical Review E, 62*, 3974–3985.

Capes, D. L., Goldschen-Ohm, M. P., Arcisio-Miranda, M., Bezanilla, F., & Chanda, B. (2013). Domain voltage-sensor movement is both sufficient and rate limiting for fast inactivation in sodium channels. *Journal of General Physiology, 142*, 101–112.

Celes, M. R., Malvestio, L. M., Suadicani, S. O., Prado, C. M., Figueiredo, M. J., Campos, E. C., … Rossi, M. A. (2013). Disruption of calcium homeostasis in cardiomyocytes underlies cardiac structural and functional changes in severe sepsis. *PLoS One, 8*, e68809.

Chen, L. Q., Santarelli, V., Horn, R., & Kallen, R. G. (1996). A unique role for the S4 segment of domain 4 in the inactivation of sodium channels. *Journal of General Physiology, 108*, 549–556.

Conti, F., Inoue, I., Kukita, F., & Stuhmer, W. (1984). Pressure dependence of sodium gating currents in the squid giant axon. *European Biophysical Journal, 11*, 137–147.

Costigan, M., Scholz, J., & Woolf, C. J. (2009). Neuropathic pain: a maladaptive response of the nervous system to damage. *Annual Review of Neuroscience, 32*, 1.

DiFranco, M., & Vergara, J. L. (2011). The Na conductance in the sarcolemma and the transverse tubular system membranes of mammalian skeletal muscle fibers. *Journal of General Physiology, 138*, 393–419.

Douglas, H. A., Callaway, J. K., Sword, J., Kirov, S. A., & Andrew, R. D. (2011). Potent inhibition of anoxic depolarization by the sodium channel blocker dibucaine. *Journal of Neurophysiology, 105*, 1482–1494.

Filatov, G. N., & Rich, M. M. (2004). Hyperpolarized shifts in the voltage dependence of fast inactivation of Nav1.4 and Nav1.5 in a rat model of critical illness myopathy. *Journal of Physiology (Paris), 559*, 813–820.

Finol-Urdaneta, R. K., McArthur, J. R., Juranka, P. F., French, R. J., & Morris, C. E. (2010). Modulation of KvAP unitary conductance and gating by 1-alkanols and other surface active agents. *Biophysical Journal, 98*, 762–772.

Friedrich, O. (2008). Critical illness myopathy: sepsis-mediated failure of the peripheral nervous system. *European Journal of Anaesthesiology Supplement, 42*, 73–82.

Gauthier, N. C., Masters, T. A., & Sheetz, M. P. (2012). Mechanical feedback between membrane tension and dynamics. *Trends in Cellular Biology, 22*, 527–535.

Gil, Z., Silberberg, S. D., & Magleby, K. L. (1999). Voltage-induced membrane displacement in patch pipettes activates mechanosensitive channels. *Proceedings of the National Academy of Sciences of the United States of America, 96*, 14594–14599.

Haeseler, G., Foadi, N., Wiegand, E., Ahrens, J., Krampfl, K., Dengler, R., & Leuwer, M. (2008). Endotoxin reduces availability of voltage-gated human skeletal muscle sodium channels at depolarized membrane potentials. *Critical Care Medicine, 36*, 1239–1247.

Hamill, O. P., & McBride, D. W., Jr. (1992). Rapid adaptation of single mechanosensitive channels in *Xenopus* oocytes. *Proceedings of the National Academy of Sciences of the United States of America, 89*, 7462–7466.

Hinman, J. D., Rasband, M. N., & Carmichael, S. T. (2013). Remodeling of the axon initial segment after focal cortical and white matter stroke. *Stroke: A Journal of Cerebral Circulation, 44*, 182–189.

Hirn, C., Shapovalov, G., Petermann, O., Roulet, E., & Ruegg, U. T. (2008). Nav1.4 deregulation in dystrophic skeletal muscle leads to Na^+ overload and enhanced cell death. *Journal of General Physiology, 132*, 199–208.

Ho, T. S., Zollinger, D. R., Chang, K. J., Xu, M., Cooper, E. C., Stankewich, M. C., ... Rasband, M. N. (2014). A hierarchy of ankyrin-spectrin complexes clusters sodium channels at nodes of Ranvier. *Nature Neuroscience, 17*, 1664–1672.
Hodgkin, A. L. (1948). The local electric changes associated with repetitive action in a non-medullated axon. *Journal of Physiology, 107*, 165–181.
Hodgkin, A. L., & Huxley, A. F. (1952). A quantitative description of membrane current and its application to conduction and excitation in nerve. *Journal of Physiology, 117*, 500–544.
Isose, S., Misawa, S., Sakurai, K., Kanai, K., Shibuya, K., Sekiguchi, Y., ... Kuwabara, S. (2010). Mexiletine suppresses nodal persistent sodium currents in sensory axons of patients with neuropathic pain. *Clinical Neurophysiology, 121*, 719–724.
Iwata, A., Stys, P. K., Wolf, J. A., Chen, X. H., Taylor, A. G., Meaney, D. F., & Smith, D. H. (2004). Traumatic axonal injury induces proteolytic cleavage of the voltage gated sodium channels modulated by tetrodotoxin and protease inhibitors. *Journal of Neuroscience, 24*, 4605–4613.
Jin, X. (2014). Membrane resealing as a promising strategy for early treatment of neurotrauma. *Neural Regenerative Research, 9*, 1876–1877.
Jung, H. J., Lee, J. Y., Kim, S. H., Eu, Y. J., Shin, S. Y., Milescu, M., ... Kim, J. I. (2005). Solution structure and lipid membrane partitioning of VSTx1, an inhibitor of the KvAP potassium channel. *Biochemistry, 44*, 6015–6023.
Kanai, K., Kuwabara, S., Misawa, S., Tamura, N., Ogawara, K., Nakata, M., ... Bostock, H. (2006). Altered axonal excitability properties in amyotrophic lateral sclerosis: impaired potassium channel function related to disease stage. *Brain: A Journal of Neurology, 129*, 953–962.
Kilinc, D., Gallo, G., & Barbee, K. A. (2008). Mechanically-induced membrane poration causes axonal beading and localized cytoskeletal damage. *Experimental Neurology, 212*, 422–430.
Kinnun, J. J., Mallikarjunaiah, K. J., Petrache, H. I., & Brown, M. F. (2015). Elastic deformation and area per lipid of membranes: atomistic view from solid-state deuterium NMR spectroscopy. *Biochimica et Biophysica Acta, 1848*, 246–259.
Klint, J. K., Senff, S., Rupasinghe, D. B., Er, S. Y., Herzig, V., Nicholson, G. M., & King, G. F. (2012). Spider-venom peptides that target voltage-gated sodium channels: pharmacological tools and potential therapeutic leads. *Toxicon: Official Journal of the International Society on Toxinology, 60*, 478–491.
Kovalsky, Y., Amir, R., & Devor, M. (2009). Simulation in sensory neurons reveals a key role for delayed Na^+ current in subthreshold oscillations and ectopic discharge: implications for neuropathic pain. *Journal of Neurophysiology, 102*, 1430–1442.
Krolenko, S. A., & Lucy, J. A. (2001). Reversible vacuolation of T-tubules in skeletal muscle: mechanisms and implications for cell biology. *International Revue of Cytology, 202*, 243–298.
Lachance, M., Longtin, A., Morris, C. E., Yu, N., & Joós, B. (2014). Stimulation-induced ectopicity and propagation windows in model damaged axons. *Journal of Computational Neuroscience, 37*, 523–531.
Lacroix, J. J., Campos, F. V., Frezza, L., & Bezanilla, F. (2013). Molecular bases for the asynchronous activation of sodium and potassium channels required for nerve impulse generation. *Neuron, 79*, 651–657.
Laitko, U., & Morris, C. E. (2004). Membrane tension accelerates rate-limiting voltage dependent activation and slow inactivation steps in a Shaker channel. *Journal of General Physiology, 123*, 135–154.
Lännergren, J., Bruton, J. D., & Westerblad, H. (2000). Vacuole formation in fatigued skeletal muscle fibres from frog and mouse: effects of extracellular lactate. *Journal of Physiology, 526*, 597–611.

Latronico, N., & Bolton, C. F. (2011). Critical illness polyneuropathy and myopathy: a major cause of muscle weakness and paralysis. *Lancet Neurology, 10*, 931–941.

Lehky, T. J., Leonard, G. D., Wilson, R. H., Grem, J. L., & Floeter, M. K. (2004). Oxaliplatin-induced neurotoxicity: acute hyperexcitability and chronic neuropathy. *Muscle & Nerve, 29*, 387–392.

Lenkey, N., Karoly, R., Epresi, N., Vizi, E., & Mike, A. (2011). Binding of sodium channel inhibitors to hyperpolarized and depolarized conformations of the channel. *Neuropharmacology, 60*, 191–200.

Lenkey, N., Karoly, R., Lukacs, P., Vizi, E. S., Sunesen, M., Fodor, L., & Mike, A. (2010). Classification of drugs based on properties of sodium channel inhibition: a comparative automated patch-clamp study. *PLoS One, 5*, e15568.

Lin, M. J., Fine, M., Lu, J. Y., Hofmann, S. L., Frazier, G., & Hilgemann, D. W. (2013). Massive palmitoylation-dependent endocytosis during reoxygenation of anoxic cardiac muscle. *eLife, 2*, e01295.

Lin, W., Laitko, U., Juranka, P. F., & Morris, C. E. (2007). Dual stretch responses of mHCN2 pacemaker channels: accelerated activation, accelerated deactivation. *Biophysical Journal, 92*, 1559–1572.

Manno, S., Takakuwa, Y., & Mohandas, N. (2002). Identification of a functional role for lipid asymmetry in biological membranes: phosphatidylserine-skeletal protein interactions modulate membrane stability. *Proceedings of the National Academy of Sciences of the United States of America, 99*, 1943–1948.

Martindale, J. J., & Metzger, J. M. (2014). Uncoupling of increased cellular oxidative stress and myocardial ischemia reperfusion injury by directed sarcolemma stabilization. *Journal of Molecular and Cellular Cardiology, 67*, 26–37.

Maxwell, W. L. (1996). Histopathological changes at central nodes of Ranvier after stretch-injury. *Microscopy Research and Technique, 34*, 522–535.

Merkies, I. S., Faber, C. G., & Lauria, G. (2015). Advances in diagnostics and outcome measures in peripheral neuropathies. *Neuroscience Letters, 596*, 3–13.

Methfessel, C., Witzemann, V., Takahashi, T., Mishina, M., Numa, S., & Sakmann, B. (1986). Patch clamp measurements on *Xenopus laevis* oocytes: currents through endogenous channels and implanted acetylcholine receptor and sodium channels. *Pflügers Archiv, 407*, 577–588.

Mihailescu, M., Krepkiy, D., Milescu, M., Gawrisch, K., Swartz, K. J., & White, S. (2014). Structural interactions of a voltage sensor toxin with lipid membranes. *Proceedings of the National Academy of Sciences of the United States of America, 111*, E5463–E5470.

Milescu, M., Bosmans, F., Lee, S., Alabi, A. A., Kim, J. I., & Swartz, K. J. (2009). Interactions between lipids and voltage sensor paddles detected with tarantula toxins. *Nature Structural & Molecular Biology, 16*, 1080–1085.

Milton, R. L., & Caldwell, J. H. (1990). Na current in membrane blebs: implications for channel mobility and patch clamp recording. *Journal of Neuroscience, 10*, 885–893.

Moench, I., Meekhof, K. E., Cheng, L. F., & Lopatin, A. N. (2013). Resolution of hyposmotic stress in isolated mouse ventricular myocytes causes sealing of t-tubules. *Experimental Physiology, 98*, 1164–1177.

Moran, O., Tammaro, P., Nizzari, M., & Conti, F. (2000). Functional properties of sodium channels do not depend on the cytoskeleton integrity. *Biochemical and Biophysical Research Communications, 276*, 204–209.

Morris, C. E. (2001). Mechanoprotection of the plasma membrane in neurons and other non-erythroid cells by the spectrin-based membrane skeleton. *Cellular & Molecular Biology Letters, 6*, 703–720.

Morris, C. E. (2011a). Why are so many ion channels mechanosensitive? In N. Sperelakis (Ed.), *Cell physiology source book* (4th ed., pp. 493–505). Elsevier.

Morris, C. E. (2011b). Voltage-gated channel mechanosensitivity: fact or friction? *Frontiers in Physiology, 2*, 25.

Morris, C. E. (2011c). Pacemaker, potassium, calcium, sodium: stretch modulation of the voltage-gated channels. In P. Kohl, F. Sachs, & M. Franz (Eds.), *Cardiac mechano-electric coupling and arrhythmias: From pipette to patient* (2nd ed., pp. 43−49). Elsevier Saunders.

Morris, C. E., Boucher, P. A., & Joós, B. (2012). Left-shifted Nav channels in injured bilayer: primary targets for neuroprotective Nav antagonists? *Frontiers in Pharmacology, 3*, 19.

Morris, C. E., & Homann, U. (2001). Cell surface area regulation and membrane tension. *Journal of Membrane Biology, 179*, 79−102.

Morris, C. E., & Horn, R. (1991). Failure to elicit neuronal macroscopic mechanosensitive currents anticipated by single-channel studies. *Science, 251*, 1246−1249.

Morris, C. E., & Juranka, P. F. (2007a). Nav channel mechanosensitivity: activation and inactivation accelerate reversibly with stretch. *Biophysical Journal, 93*, 822−833.

Morris, C. E., & Juranka, P. F. (2007b). Lipid stress at play: mechanosensitivity of voltage-gated channels. *Current Topics in Membranes, 59*, 297−338.

Morris, C. E., Juranka, P. F., & Joós, B. (2012). Perturbed voltage-gated channel activity in perturbed bilayers: implications for ectopic arrhythmias arising from damaged membrane. *Progress in Biophysics and Molecular Biology, 110*, 245−256.

Morris, C. E., Juranka, P. F., Lin, W., Morris, T. J., & Laitko, U. (2006). Studying the mechanosensitivity of voltage-gated channels using oocyte patches. *Methods in Molecular Biology, 322*, 315−329.

Morris, C. E., & Lecar, H. (1981). Voltage oscillations in the barnacle giant muscle fiber. *Biophysical Journal, 35*, 193−213.

Morris, C. E., Prikryl, E. A., & Joós, B. (2015). Mechanosensitive gating of Kv channels. *PLoS One, 10*, e0118335.

Nagoshi, N., Nakashima, H., & Fehlings, M. G. (2015). Riluzole as a neuroprotective drug for spinal cord injury: from bench to bedside. *Molecules: A Journal of Synthetic Chemistry and Natural Product Chemistry, 20*, 7775−7789.

Novak, K. R., Nardelli, P., Cope, T. C., Filatov, G., Glass, J. D., Khan, J., & Rich, M. M. (2009). Inactivation of sodium channels underlies reversible neuropathy during critical illness in rats. *Journal of Clinical Investigation, 119*, 1150−1158.

O'Connell, R. P., Musa, H., Gomez, M. S., Avula, U. M., Herron, T. J., Kalifa, J., & Anumonwo, J. M. (2015). Free fatty acid effects on the atrial myocardium: membrane ionic currents are remodeled by the disruption of t-tubular architecture. *PLoS One, 10*, e0133052.

Park, S. B., Lin, C. S., Krishnan, A. V., Goldstein, D., Friedlander, M. L., & Kiernan, M. C. (2009). Oxaliplatin-induced neurotoxicity: changes in axonal excitability precede development of neuropathy. *Brain: A Journal of Neurology, 132*, 2712−2723.

Pathak, M., Kurtz, L., Tombola, F., & Isacoff, E. (2005). The cooperative voltage sensor motion that gates a potassium channel. *Journal of General Physiology, 125*, 57−69.

Picot, L., Chevalier, S., Mezghani-Abdelmoula, S., Merieau, A., Lesouhaitier, O., Leroux, P., … Feuilloley, M. G. (2003). Cytotoxic effects of the lipopolysaccharide from Pseudomonas fluorescens on neurons and glial cells. *Microbial Pathogenesis, 35*, 95−106.

Prescott, S. A., De Koninck, Y., & Sejnowski, T. J. (2008). Biophysical basis for three distinct dynamical mechanisms of action potential initiation. *PLoS Computational Biology, 4*, e1000198.

Rad, I., Mobasheri, H., Najafi, F., & Rezaei, M. (2014). Efficient repairing effect of PEG based tri-block copolymer on mechanically damaged PC12 cells and isolated spinal cord. *Journal of Materials Science: Materials in Medicine, 25*, 1539−1551.

Rawicz, W., Smith, B. A., McIntosh, T. J., Simon, S. A., & Evans, E. (2008). Elasticity, strength, and water permeability of bilayers that contain raft microdomain-forming lipids. *Biophysical Journal, 94*, 4725–4736.

Reeves, T. M., Phillips, L. L., & Povlishock, J. T. (2005). Myelinated and unmyelinated axons of the corpus callosum differ in vulnerability and functional recovery following traumatic brain injury. *Experimental Neurology, 196*, 126–137.

von Reyn, C. R., Mott, R. E., Siman, R., Smith, D. H., & Meaney, D. F. (2012). Mechanisms of calpain mediated proteolysis of voltage gated sodium channel α-subunits following in vitro dynamic stretch injury. *Journal of Neurochemistry, 121*, 793–805.

Rognoni, A., Barbieri, L., Cavallino, C., Bacchini, S., Veia, A., Degiovanni, A., … Bongo, A. S. (2013). Ranolazine: effects on ischemic heart. *Recent Patents on Cardiovascular Drug Discovery, 8*, 197–203.

Rusinova, R., Hobart, E. A., Koeppe, R. E., 2nd, & Andersen, O. S. (2013). Phosphoinositides alter lipid bilayer properties. *Journal of General Physiology, 141*, 673–690.

Schmidt, D., & MacKinnon, R. (2008). Voltage-dependent K^+ channel gating and voltage sensor toxin sensitivity depend on the mechanical state of the lipid membrane. *Proceedings of the National Academy of Sciences of the United States of America, 105*, 19276–19281.

Schmidt, D., del Mármol, J., & MacKinnon, R. (2013). Mechanistic basis for low threshold mechanosensitivity in voltage-dependent K^+ channel. *Proceedings of the National Academy of Sciences of the United States of America, 109*, 10352–10357.

Sezgin, E., Gutmann, T., Buhl, T., Dirkx, R., Grzybek, M., Coskun, Ü., … Schwille, P. (2015). Adaptive lipid packing and bioactivity in membrane domains. *PLoS One, 10*, e0123930.

Shankarappa, S. A., Tsui, J. H., Kim, K. N., Reznor, G., Dohlman, J. C., Langer, R., & Kohane, D. S. (2012). Prolonged nerve blockade delays the onset of neuropathic pain. *Proceedings of the National Academy of Sciences of the United States of America, 109*, 17555–17560.

Shcherbatko, A., Ono, F., Mandel, G., & Brehm, P. (1999). Voltage-dependent sodium channel function is regulated through membrane mechanics. *Biophysical Journal, 77*, 1945–1959.

Sheetz, M. P., Sable, J. E., & Döbereiner, H. G. (2006). Continuous membrane cytoskeleton adhesion requires continuous accommodation to lipid and cytoskeleton dynamics. *Annual Review of Biophysics and Biomolecular Structure, 35*, 417–434.

Shelat, P. B., Plant, L. D., Wang, J. C., Lee, E., & Marks, J. D. (2013). The membrane-active tri-block copolymer pluronic F-68 profoundly rescues rat hippocampal neurons from oxygen-glucose deprivation-induced death through early inhibition of apoptosis. *Journal of Neuroscience, 33*, 12287–12299.

Small, D. L., & Morris, C. E. (1994). Delayed activation of single mechanosensitive channels in Lymnaea neurons. *American Journal of Physiology, 267*, C598–C606.

Song, X. S., Huang, Z. J., & Song, X. J. (2009). Thiamine suppresses thermal hyperalgesia, inhibits hyperexcitability, and lessens alterations of sodium currents in injured, dorsal root ganglion neurons in rats. *Anesthesiology, 110*, 387–400.

Song, Y., Li, H. M., Xie, R. G., Yue, Z. F., Song, X. J., Hu, S. J., & Xing, J. L. (2012). Evoked bursting in injured Aβ dorsal root ganglion neurons: a mechanism underlying tactile allodynia. *Pain, 153*, 657–665.

Suchyna, T. M., Markin, V. S., & Sachs, F. (2009). Biophysics and structure of the patch and the gigaseal. *Biophysical Journal, 97*, 738–747.

Sun, G. C., Werkman, T. R., & Wadman, W. J. (2006). Kinetic changes and modulation by carbamazepine on voltage-gated sodium channels in rat CA1 neurons after epilepsy. *Acta Pharmacological Sinica, 12*, 1537–1546.

Tabarean, I. V., Juranka, P., & Morris, C. E. (1999). Membrane stretch affects gating modes of a skeletal muscle sodium channel. *Biophysical Journal, 77*, 758–774.

Taddese, A., & Bean, B. P. (2002). Subthreshold sodium current from rapidly inactivating sodium channels drives spontaneous firing of tuberomammillary neurons. *Neuron, 33*, 587–600.

Ursell, T., Agrawal, A., & Phillips, R. (2011). Lipid bilayer mechanics in a pipette with glass-bilayer adhesion. *Biophysical Journal, 101*, 1913–1920.

Volman, V., & Ng, L. J. (2013). Computer modeling of mild axonal injury: implications for axonal signal transmission. *Neural Computation, 25*, 2646–2681.

Wagner, S., Maier, L. S., & Bers, D. M. (2015). Role of sodium and calcium dysregulation in tachyarrhythmias in sudden cardiac death. *Circulation Research, 116*, 1956–1970.

Wan, X., Juranka, P., & Morris, C. E. (1999). Activation of mechanosensitive currents in traumatized membrane. *American Journal of Physiology, 276*, C318–C327.

Wang, J. A., Lin, W., Morris, T., Banderali, U., Juranka, P. F., & Morris, C. E. (2009). Membrane trauma and Na^+ leak from Nav1.6 channels. *American Journal of Physiology and Cell Physiology, 297*, C823–C834.

Wang, X., Xie, W., Zhang, Y., Lin, P., Han, L., Han, P., ... Cheng, H. (2010). Cardioprotection of ischemia/reperfusion injury by cholesterol-dependent MG53-mediated membrane repair. *Circulation Research, 107*, 76–83.

Wang, Y. Y., Wen, Z. H., Duan, J. H., Zhu, J. L., Wang, W. T., Dong, H., ... Hu, S. J. (2011). Noise enhances subthreshold oscillations in injured primary sensory neurons. *Neurosignals, 19*, 54–62.

Westenbroek, R. E., Bischoff, S., Fu, Y., Maier, S. K., Catterall, W. A., & Scheuer, T. (2013). Localization of sodium channel subtypes in mouse ventricular myocytes using quantitative immunocytochemistry. *Journal of Molecular and Cellular Cardiology, 64*, 69–78.

White, S. H., Brisson, C. D., & Andrew, R. D. (2012). Examining protection from anoxic depolarization by the drugs dibucaine and carbetapentane using whole cell recording from CA1 neurons. *Journal of Neurophysiology, 107*, 2083–2095.

Xie, W., Strong, J. A., Kim, D., Shahrestani, S., & Zhang, J. M. (2012). Bursting activity in myelinated sensory neurons plays a key role in pain behavior induced by localized inflammation of the rat sensory ganglion. *Neuroscience, 206*, 212–223.

Yifrach, O., & MacKinnon, R. (2002). Energetics of pore opening in a voltage-gated K(+) channel. *Cell, 111*, 231–239.

Yu, N., Morris, C. E., Joós, B., & Longtin, A. (2012). Spontaneous excitation patterns computed for axons with injury-like impairments of sodium channels and Na/K pumps. *PLoS Computational Biology, 8*, e1002664.

Zhang, Y., Bekku, Y., Dzhashiashvili, Y., Armenti, S., Meng, X., Sasaki, Y., ... Salzer, J. L. (2012). Assembly and maintenance of nodes of ranvier rely on distinct sources of proteins and targeting mechanisms. *Neuron, 73*, 92–107.

Zhang, Y., Gao, F., Popov, V. L., Wen, J. W., & Hamill, O. P. (2000). Mechanically gated channel activity in cytoskeleton-deficient plasma membrane blebs and vesicles from *Xenopus* oocytes. *Journal of Physiology (Paris), 523*, 117–130.

Zhou, X., Luo, Y. C., Ji, W. J., Zhang, L., Dong, Y., Ge, L., ... Li, Y. M. (2013). Modulation of mononuclear phagocyte inflammatory response by liposome-encapsulated voltage gated sodium channel inhibitor ameliorates myocardial ischemia/reperfusion injury in rats. *PLoS One, 8*, e74390.

Zou, S., Chisholm, R., Tauskela, J. S., Mealing, G. A., Johnston, L. J., & Morris, C. E. (2013). Force spectroscopy measurements show that cortical neurons exposed to excitotoxic agonists stiffen before showing evidence of bleb damage. *PLoS One, 8*, e73499.

CHAPTER EIGHTEEN

Unusual Voltage-Gated Sodium Currents as Targets for Pain

C. Barbosa and T.R. Cummins[1]
Indiana University School of Medicine, Indianapolis, IN, United States
[1]Corresponding author: E-mail: trcummin@iupui.edu

Contents

Abstract

Pain is a serious health problem that impacts the lives of many individuals. Hyperexcitability of peripheral sensory neurons contributes to both acute and chronic pain syndromes. Because voltage-gated sodium currents are crucial to the transmission of electrical signals in peripheral sensory neurons, the channels that underlie these

Current Topics in Membranes, Volume 78
ISSN 1063-5823
http://dx.doi.org/10.1016/bs.ctm.2015.12.005

currents are attractive targets for pain therapeutics. Sodium currents and channels in peripheral sensory neurons are complex. Multiple-channel isoforms contribute to the macroscopic currents in nociceptive sensory neurons. These different isoforms exhibit substantial variations in their kinetics and pharmacology. Furthermore, sodium current complexity is enhanced by an array of interacting proteins that can substantially modify the properties of voltage-gated sodium channels. Resurgent sodium currents, atypical currents that can enhance recovery from inactivation and neuronal firing, are increasingly being recognized as playing potentially important roles in sensory neuron hyperexcitability and pain sensations. Here we discuss unusual sodium channels and currents that have been identified in nociceptive sensory neurons, describe what is known about the molecular determinants of the complex sodium currents in these neurons. Finally, we provide an overview of therapeutic strategies to target voltage-gated sodium currents in nociceptive neurons.

1. OVERVIEW

Pain is one of the most prevalent reasons for which people in the United States seek medical care. According to the National Center for Health Statistics, around 76.2 million people experience persistent pain (Institute of Medicine Report, 2011). A third of people who experience pain find that it's so significant that it impacts their daily routine (Portenoy, Ugarte, Fuller, & Haas, 2004). One of the main problems faced in pain management is that the available therapeutics either are not effective or have unwanted side effects (Institute of Medicine Report, 2011). Such a high incidence of pain and inadequate relief creates a need for the development of new therapeutics.

In a linear perspective, peripheral sensory neurons are the first step in the initiation of pain sensations. In mammals, these neurons are pseudo unipolar neurons, possessing one axon that sends branches into two areas: one to the periphery (skin, muscle, and tissue) and the other to the central nervous system (CNS) (spinal cord or brainstem). Increased firing of peripheral sensory neurons in dorsal root ganglia (DRG) and trigeminal ganglia is a common mechanism in pain pathologies (Baranauskas & Nistri, 1998; Cummins & Rush, 2007; Liu & Wood, 2011; Millan, 1999). Voltage-gated ion channels contribute to the excitability of sensory neurons. In particular, voltage-gated sodium channels (VGSCs or Nav) are classically viewed to generate the rising phase of the action potential (AP). There is strong evidence from genetic and pain model studies that support that VGSCs are functionally important for pain sensations (Baker & Wood, 2001; Cummins, Dib-Hajj, Black, & Waxman, 2000; Cummins, Sheets, & Waxman, 2007; Dib-Hajj, Cummins, Black, & Waxman, 2007; Waxman, Dib-Hajj, Cummins, & Black, 1999).

For example, in humans and mice loss-of-function of VGSC isoform, Nav1.7, is associated with congenital indifference to pain. Furthermore, alteration to the activity of other VGSC isoforms such as Nav1.3, Nav1.6, Nav1.8, and Nav1.9 has also been associated with pain pathologies (Deuis et al., 2013; Dib-Hajj, Cummins, Black, & Waxman, 2010; Xie, Strong, Ye, Mao, & Zhang, 2013; Xie, Strong, & Zhang, 2015). This evidence suggests that inhibition of specific VGSC isoforms or selective blockade of altered activity may provide new therapeutic strategies for pain.

Specific patterns or modes of sodium current activity are also likely to be important in nociception and the transmission of pain sensations. Pronounced slowly inactivating and persistent currents (generated by Nav1.8 and Nav1.9 channels, respectively) are readily observed in many peripheral neurons in addition to the classic fast-activating and fast-inactivating currents observed in most excitable cells. Another novel candidate target for altered activity in VGSC that has recently emerged is resurgent sodium current. Resurgent sodium currents, an atypical VGSC current, are thought to increase the firing frequency of neurons by providing a depolarizing drive at repolarizing voltages (Akemann & Knopfel, 2006; Jarecki, Piekarz, Jackson, & Cummins, 2010; Khaliq, Gouwens, & Raman, 2003). For example, increased resurgent currents in sensory neurons have been implicated in paroxysmal extreme pain disorder (PEPD), sea anemone toxin ATX-II-induced pain, and oxaliplatin acute cooling–aggravated painful neuropathy (Dib-Hajj et al., 2008; Jarecki et al., 2010; Jarecki, Sheets, Jackson, & Cummins, 2008; Klinger et al., 2012; Sheets, Jackson, Waxman, Dib-Hajj, & Cummins, 2007; Sittl et al., 2012; Tan, Piekarz, et al., 2014; Tan, Priest, et al., 2014). Under normal physiological conditions, resurgent currents are observed in a subpopulation of DRG neurons where the major contributing isoform is Nav1.6 (Cummins, Dib-Hajj, Herzog, & Waxman, 2005). Nav1.6 has recently been implicated in inflammatory and neuropathic pain models since knockdown of Nav1.6 blocks the development of persistent pain and increased activity of sensory neurons (Xie et al., 2015). Therefore, in normal physiology and pathophysiological states, resurgent currents may prove to be important modulators of painful sensations and therapeutic targets.

In this chapter we will review the expression of VGSC in sensory neurons. We will then focus on currents with unusual inactivation properties and atypical VGSC currents (ie, resurgent currents). We will discuss recent findings regarding pain sensations suggesting the involvement of Nav1.6 and Nav1.9 (in addition to the more established roles of Nav1.7 and Nav1.8) in pain sensations. Significant lessons have been gained from studying VGSC disease mutations and pain activity in transgenic animals, and therefore key studies relating to these efforts will also be highlighted. Lastly, we will discuss how classic and atypical VGSC currents may prove to be important targets in the development of new therapeutics for pain management.

2. SODIUM CHANNELS IN PERIPHERAL SENSORY NEURONS

2.1 Alpha subunit

Peripheral sensory neurons, at least in mammals, differ from neurons in the CNS in a number of ways, including morphology and physiology (Devor, 1999). Although it has been written that with VGSCs "one is impressed more with the similarity of function than with the differences" (Hille, 1984), this is clearly not the case with sodium currents in peripheral sensory neurons where sodium currents exhibit broad diversity. While sodium currents in the CNS are for the most part highly sensitive to the guanidinium toxin tetrodotoxin (TTX; IC50s near 10 nM), distinct components of the complex currents that can be elicited in peripheral dorsal root ganglion and trigeminal sensory neurons are roughly 10,000-fold less sensitive. The kinetic diversity of sodium currents in peripheral sensory neurons is also greater then in the CNS. While the activation and inactivation kinetics of the voltage-gated sodium currents observed in different CNS neuron populations typically vary by only about 2-fold, the kinetics of currents in peripheral sensory neurons differ by as much as 20-fold (Dib-Hajj, Black, Cummins, & Waxman, 2002; Roy & Narahashi, 1992). In particular, nociceptive peripheral neurons express fast, slow, and persistent sodium currents (Caffrey, Eng, Black, Waxman, & Kocsis, 1992; Rush, Cummins, & Waxman, 2007). The fast gating current is generated by multiple TTX-sensitive (TTXS) isoforms including Nav1.1, Nav1.3, Nav1.6, and Nav1.7. TTXS currents can generate small persistent currents, sometimes as a result of window currents that arise from an overlap between the voltage dependence of activation and steady-state inactivation and sometimes from incomplete fast inactivation (Crill, 1996). However, the large slowly inactivating and persistent currents observed in peripheral sensory neurons (Cummins et al., 1999; Elliott & Elliott, 1993; Habib, Wood, & Cox, 2015) are resistant to TTX (TTXR) and are generated by Nav1.8 and Nav1.9. Although Nav1.8 and Nav1.9 channels are not exclusively expressed in nociceptive neurons, they are highly and preferentially expressed in nociceptive neurons. These isoforms generate unusual sodium currents that could be therapeutic targets in pain pathophysiologies.

VGSCs are transmembrane proteins that selectively mediate the inward flow of sodium ions that generate the rapid upstroke of the AP (Hodgkin & Huxley, 1952). The principal subunit of VGSC is the alpha (α)-subunit. Nine α-subunit genes that encode VGSC isoforms (Nav1.1–Nav1.9) have been identified in mammals, and an individual α-subunit is sufficient to form a functional channel. The overall structure of the nine α-subunit isoforms is highly conserved, containing four homologous domains (DI–DIV) each with six transmembrane segments (Fig. 1A). The α-subunit folds into a tridimensional structure in which S5/6 segments within each domain form an aqueous pore selectively permeable to sodium ions. Segment 4 within each domain is sensitive to changes in membrane potential, and movement produces conformational changes in the protein structure (Catterall, Goldin, & Waxman, 2005; Noda et al., 1984; Payandeh,

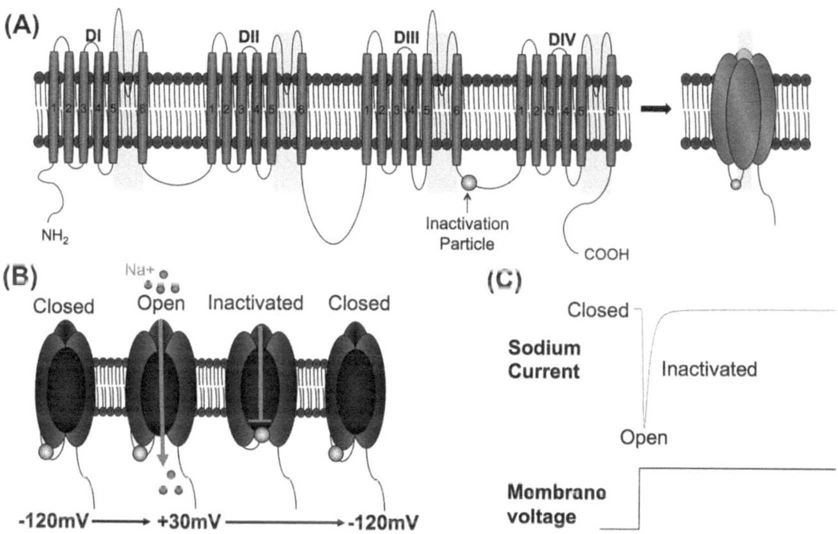

Figure 1 (A) Linear representation of voltage-gated sodium channel (VGSC) α-subunit. Regions highlighted in blue (light gray in print verions) represent the regions between S5 and S6 that yield the aqueous pore selectively permeable to sodium. *Reproduced with permission from (Barbosa, C., Tan, Z. Y., Wang, R., Xie, W., Strong, J. A., Vasko, M. R.,... Cummins, T.R. (2015). Navβ4 regulates resurgent currents and excitability in sensory neurons. Molecular Pain, in press).* (B) Cross-sectional representation of VGSC conformational changes in response to changes in membrane potential. (C) Representative trace of whole-cell patch-clamp recording of sodium current upon a depolarizing voltage step stimulus and corresponding state transitions from closed–open-inactive states.

Scheuer, Zheng, & Catterall, 2011). In this manner, VGSCs change between conformations that are either permissive of sodium ions (open state) or nonconducting (inactive states and closed states). Classically, sodium currents are generated when the membrane is depolarized, and channels become activated or open. Sodium currents are then terminated by channel inactivation (Fig. 1B and C). Once inactivated, VGSCs become refractory and require repolarization of the membrane to recover before they are able to open again (Armstrong & Bezanilla, 1974). Among the nine isoforms, Nav1.8 and Nav1.9 show the most diversity in terms of properties and sequence with 50–60% amino acid identity compared to the other VGSCs. In contrast, Nav1.1, Nav1.2, Nav1.3, Nav1.6, and Nav1.7 are 75–95% identical in their amino acid sequences (Goldin et al., 2000).

As mentioned, nine mammalian VGSC α-subunit isoforms are often divided into two major groups based on their sensitivity to inhibition by TTX into TTXS and TTXR. VGSCs that are TTXS have fast kinetics, whereas TTXR channels have slower or persistent kinetics (Caffrey et al., 1992; Dib-Hajj et al., 2002; Elliott & Elliott, 1993). Expression of the VGSC α-subunits can be developmentally regulated and tissue specific. For example, expression of Nav1.3 and Nav1.5 can be detected in embryonic DRG neurons but are downregulated to none or low levels in adult neurons (Beckh, Noda, Lubbert, & Numa, 1989; Kerr et al., 2007; Renganathan, Dib-Hajj, & Waxman, 2002). Expression of Nav1.4 is not observed in DRG neurons at any stage, and its expression is restricted mainly to skeletal muscle tissue (Trimmer et al., 1989). Mature DRG neurons can express a combination of VGSC channels that include the TTXS isoforms, Nav1.1, Nav1.6, Nav1.7, and TTXR isoforms, Nav1.8 and Nav1.9 (Amaya et al., 2000; Caffrey et al., 1992; Dib-Hajj et al., 2002; Dib-Hajj, Tyrrell, et al., 1999; Ho & O'Leary, 2011). Expression of VGSC α-subunits varies between different classes of peripheral sensory neurons. Three classes of sensory fibers (C, Aδ, and Aα/β) are often defined based on conduction velocity, soma diameter, and activation threshold (Table 1). These different types of sensory fibers are recruited depending on the peripheral sensory input. Fibers that are recruited when noxious (and potentially painful) stimulus is presented are termed nociceptors. Nociceptors are generally classified as high-threshold neurons (Beaulieu & International Association for the Study of Pain, 2010). While nociceptors are classically identified as fibers in the C/Aδ class, there is substantial evidence that supports that a subset of Aβ fibers also likely function

Table 1 Peripheral sensory neurons classification

Fiber Class	**Aα/β**	**Aδ**	**C**
Soma diameter	<35 um	25–35 um	>25 um
	Large	Medium	Small
Myelination	Thick	Thin	None
Conduction	35–75 m/s	5–30 m/s	0.5–2 m/s
Threshold	Low–high	Low–high	Low–high
VGSC (predominant α-subunits)	Nav1.1, Nav1.6, Nav1.7 (Unique subpopulation expresses Nav1.8)		Nav1.7 Nav1.8 Nav1.9

as nociceptors (Djouhri & Lawson, 2004). In C-fiber neurons, the predominant VGSCs expressed are the TTXS VGSC Nav1.7 and the TTXR VGSCs Nav1.8 and Nav1.9 (Cummins et al., 1999; Djouhri, Fang, et al., 2003; Elliott & Elliott, 1993; Ho & O'Leary, 2011). In contrast, A-fiber neurons predominantly express the TTXS VGSCs Nav1.1, Nav1.6, and (to a lesser extent) Nav1.7 with the exception of a discrete subpopulation that also expresses Nav1.8 at high levels (Ho & O'Leary, 2011; Ramachandra, McGrew, Baxter, Howard, & Elmslie, 2013; Shields et al., 2012).

2.2 Auxiliary proteins

2.2.1 Beta-subunits

Auxiliary subunits can associate in multiple ways with VGSCs resulting in a variety of heteromultimers that can impact VGSC properties. Beta (β)-subunits are the most commonly known (Isom, 2001; Patino & Isom, 2010). To date four β-subunits (Navβ1–β4) have been identified. These subunits share some common features which include a single transmembrane domain, an intracellular C-terminal domain, and an extracellular N-terminal domain (Isom, 2001). β-subunits can associate with α-subunits noncovalently (Navβ1, β3) or covalently (Navβ2, β4) (Catterall et al., 2005). All β-subunits are expressed in normal DRG neurons although distribution can vary between specific subclasses (Chahine & O'Leary, 2011). For example, Navβ4 (the main subunit associated with resurgent currents generation) mRNA and protein expression has been reported to be present at higher levels in medium-large population than in small diameter neurons (Barbosa et al., 2015; Ho, Zhao, Malinowski, Chahine, & O'Leary, 2012; Klinger et al., 2012). In a similar manner, Navβ1 has been reported to be more abundant in medium-large than in small diameter, and Navβ3 is more abundant in small diameter neurons than in medium-large diameter DRG neurons. Navβ2 is ubiquitously expressed throughout the different size classes (Takahashi et al., 2003).

β-subunits can alter the expression and biophysical properties of the channel such as voltage dependence of activation and inactivation, rate of inactivation and inactivation, peak current density, persistent current, and resurgent current (O'Malley & Isom, 2015). These effects can vary depending on the α-subunit expression and cell background. For example, in small diameter neurons deletion of Navβ2 decreased the expression of TTXS channels and increased their rate of activation and inactivation. However, TTXR channels were not significantly modulated (Lopez-Santiago et al., 2006). In contrast, Navβ1 deletion decreased the expression of TTXR Nav1.9 and shifted the voltage dependence of inactivation for TTXS channels to positive potentials, which resulted in hyperexcitability of small diameter neurons (Lopez-Santiago, Brackenbury, Chen, & Isom, 2011). There is good evidence to support the potential role of β-subunits in pain sensations. For example, Navβ2 null mice exhibit reduced pain behavior responses after spared nerve injury, a neuropathic pain model (Lopez-Santiago et al., 2006). Furthermore, upregulation of Navβ3 in sensory neurons has been associated with increased pain responses in different animal models (Shah et al., 2001, 2000; Takahashi et al., 2003). The role of Navβ4 in pain sensations has not been fully explored, but one possibility is that Navβ4 modulates pain sensations through regulation of VGSC's resurgent currents (Barbosa et al., 2015; Klinger et al., 2012). Additionally, β-subunits can act as cell adhesion molecules (Isom & Catterall, 1996). As such, β-subunits may also be important for neuronal outgrowth. However, this potential role and implication to pain sensations has not been extensively studied in sensory neurons.

2.2.2 Fibroblast growth factor homologous factors

Numerous other proteins have been identified that reportedly interact with the VGSC α-subunits, including calmodulin, ankyrins, multiple protein kinases, Nedd4-2, CRMP2, syntrophin, contactin, annexin II light chain, and KIF5 (to name a few). One family of auxiliary proteins to VGSC that is particularly interesting is the fibroblast growth factor homologous factors family (FHFs). FHFs are expressed mainly in excitable cells, and contrary to their homologue counterparts are neither secreted nor do they activate FGF receptors (Goldfarb, 2005). To date four genes have been identified in vertebrate species (FHF1–4). Each gene has multiple variants with distinct amino terminus regions that result from alternative splicing and variable transcription initiation sites (Goldfarb, 2005; Munoz-Sanjuan, Smallwood, & Nathans, 2000). Two commonly identified variants are variant A (FHFA) and variant B (FHFB). FHFBs are cytosolic; in contrast, FHFAs contain a bipartite nuclear localization signal that, depending on the cell background, is active or silent (Smallwood et al., 1996). The FHF superfamily shares a conserved surface domain capable of binding to the C-terminus of Navs (Goetz et al., 2009). FHF variants can differentially modulate VGSC gating and surface expression. In particular, FHF's binding to the C-terminus of VGSC can shift the voltage dependence of inactivation. This effect has been observed consistently independent of the type of isoform or variant. Furthermore, FHFs are proposed to modulate resurgent current generation in excitable cells (Yan, Pablo, Wang, & Pitt, 2014).

Mature DRG neurons have been reported to express FHF1 and FHF2 (Hubert et al., 2008; Rush, Wittmack, et al., 2006; Wittmack et al., 2004). FHF1 is present mainly as variant B (FHF1B) as FHF1A is downregulated in adult animals. FHF1 expression is found in membrane, cytosol, and nucleus; therefore the FHF1A nuclear signal is predicted to be active in DRG neurons. Liu et al. (2001) identified FHF1B's possible interaction with the C-terminus of Nav1.9. However, this interaction is not predicted to occur in sensory neurons since both proteins are not generally expressed together in the same population of DRG neurons (Hubert et al., 2008; Liu, Dib-Hajj, & Waxman, 2001). FHF2 variants colocalize with Nav1.6 in different subpopulations of DRG neurons: FHF2A colocalizes with Nav1.6 in C-fibers, whereas FHF2B colocalizes with Nav1.6 in A-fibers (Rush, Wittmack, et al., 2006; Wittmack et al., 2004). The FHF2A nuclear localization signal is predicted to be silent since expression is detected at the membrane and in the cytosol. FHF4 and FHF3 expression have not been extensively studied in sensory neurons (Goldfarb, 2005).

Overall, VGSC subunits can form heteromultimeric complexes with a diverse combination of α-subunits and auxiliary proteins. The formation of the complexes depends on expression, colocalization, and interaction between the subunits in a given cell type. Furthermore, as VGSC activity is modulated by interactions with auxiliary subunits, FHFs and other interacting proteins have the potential to be molecular regulators of excitability and pain sensations.

3. INACTIVATION AND RESURGENT CURRENT CONTRIBUTION TO EXCITABILITY OF SENSORY NEURONS

3.1 Intrinsic mechanisms of inactivation

Many of the auxiliary proteins that have been studied have been implicated as modulators of VGSC inactivation. Inactivation is key to terminating sodium current from VGSCs, thus it is a contributing factor to AP waveforms and firing frequency modulation. There are several mechanisms by which VGSCs intrinsically inactivate, and major mechanisms include fast inactivation and slow inactivation (Goldin, 2003; Ulbricht, 2005). Fast inactivation occurs within milliseconds of channel activation, as the intracellular loop between the DIII-DIV folds (inactivation particle shown in Fig. 1) and occludes the cytoplasmic face of the pore (West et al., 1992). Interestingly, the DIII-DIV intracellular loop is almost completely conserved among the nine different isoforms, including Nav1.9 despite its dramatically slower kinetics. VGSCs can undergo both closed-state and open-state fast inactivation, but it is not entirely clear if the mechanisms for these two are identical. Alterations in fast inactivation are believed to be important modulators of nociceptor activity and pain sensations. Indeed, local anesthetics such as lidocaine that target sodium channel activity in peripheral fibers are believed to act largely by enhancing fast inactivation of VGSCs in nociceptors (Sheets, Jarecki, & Cummins, 2011). In contrast to fast inactivation, slow inactivation is a gradual process that occurs on the timescale of seconds to minutes through conformational changes of the channel pore (Goldin, 2003; Ulbricht, 2005). It is less clear to what extent slow inactivation contributes to alterations in nociception. Slow inactivation of Nav1.8 seems to modulate the duration of the AP firing in nociceptive neurons in response to sustained stimulation (Blair & Bean, 2003). Interestingly, slow inactivation of TTXS channels is enhanced by cold, which could contribute to the ability of tissue cooling to reduce some types of pain sensations. Conversely, a lack of cold sensitivity for Nav1.8 slow inactivation is believed to be crucial to cold-induced pain sensations (Zimmermann et al., 2007). Several of the Nav1.7 erythromelalgia mutations that have been identified in patients with severe burning pain alter slow inactivation of Nav1.7 channels (Cheng et al., 2011; Cummins, Dib-Hajj, & Waxman, 2004; Harty et al., 2006; Sheets et al., 2007), but while some mutations impair slow inactivation others enhance it; the overall contribution of alterations to slow inactivation to erythromelalgia pain phenotypes is unclear.

3.2 Alternative mechanisms

3.2.1 Open-channel block

In addition to classic fast and slow inactivation, there are atypical mechanisms for inactivation. Open-channel block is a competing mechanism with classic fast inactivation (Lewis & Raman, 2013). Although binding of the open-channel blocker initially terminates sodium current, rapid unbinding of a blocker at intermediate voltages (eg, −40 mV) allows sodium current to be uncharacteristically generated in periods where the channel is usually refractory. This mechanism is known to produce atypical currents known as resurgent sodium currents. The present model for resurgent current generation proposes that the open-channel blocker binds at positive voltages and further unbinds upon membrane repolarization (Fig. 2A and B) (Cannon & Bean, 2010; Grieco, Malhotra, Chen, Isom, & Raman, 2005; Raman & Bean, 2001). Open-channel block bypasses fast inactivation, enabling the channel to recover faster and have a shorter refractory period (Grieco et al., 2005). Therefore, resurgent currents are predicted to increase neuronal firing frequency through a more rapid channel recovery in conjunction with the depolarizing contribution in the downward phase of the AP (Akemann & Knopfel, 2006; Cruz et al., 2011; Jarecki et al., 2010; Khaliq et al., 2003; Patel, Barbosa, Xiao, & Cummins, 2015). There are various examples that support that there is a positive correlation between DRG neuronal activity and resurgent currents; these will be discussed later on.

Evidence from studies on CNS neurons (Bant & Raman, 2010; Miyazaki et al., 2014), and more recently from DRG sensory neurons (Barbosa et al., 2015), strongly suggests that the open-channel blocker underlying resurgent

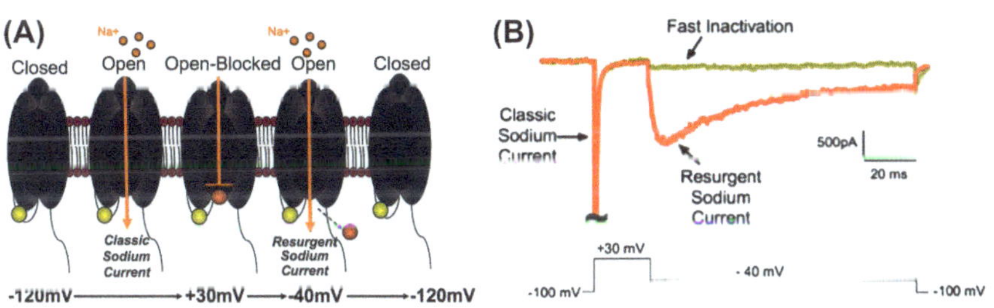

Figure 2 (A) Simplified Model of Resurgent Current State Transitions. (B) Representative traces of a dorsal root ganglia neuron with resurgent sodium current (red) and no resurgent sodium current (yellow) due to channel fast inactivation with respective voltage command protocol.

currents is part of the Navβ4 subunit. The Navβ4 subunit was initially focused on because the cytoplasmic C-terminus contains positively charged and hydrophobic residues that would have the necessary properties to act as the voltage-dependent open-channel blocker (Grieco et al., 2005; Yu et al., 2003). Based on this observation, a synthetic β4 peptide was designed from the C-terminus of Navβ4 (β4 peptide: KKLITFILKKTREKKKECLV; (Grieco et al., 2005; Lewis & Raman, 2011)). In DRG neurons, incorporation of the β4 peptide into the intracellular recording solution evokes resurgent currents in neurons that do not endogenously generate them (Klinger et al., 2012). In neurons that endogenously generate resurgent currents, the synthetic peptide increased resurgent current amplitude (Tan, Piekarz, et al., 2014).

Two types of resurgent currents have been identified in DRG neurons and are classified based on their kinetics (fast and slow; see Fig. 3A and B). Nav1.6 is the main contributing isoform to fast resurgent currents (Cummins et al., 2005). Under normal conditions fast resurgent currents are observed in a percentage of medium-large diameter neurons but not in small diameter neurons (Cummins et al., 2005; Jarecki et al., 2010). In contrast, slow resurgent currents are mediated by Nav1.8 and are mainly observed in a percentage of small-medium diameter neurons (Tan, Piekarz, et al., 2014). The voltage dependence of resurgent currents is different between these two

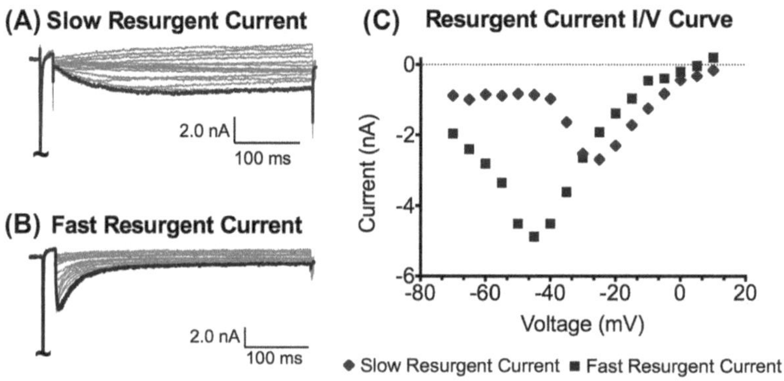

Figure 3 Slow and Fast Resurgent Currents have different kinetic and voltage properties. (A) Representative trace of slow resurgent current and (B) fast resurgent current obtained from dorsal root ganglia neurons. Peak current amplitude for slow resurgent current is −25 mV (red (light gray in print verions) trace) and −45 mV for fast resurgent current (blue (dark gray in print verions) trace). (C) Current−Voltage (I/V) curve for slow (red (light gray in print verions)) and fast resurgent current (blue (dark gray in print verions)).

currents; fast resurgent currents peak at −45 mV, whereas slow resurgent currents peak at −25 mV (Fig. 3C), likely reflecting differences in binding of the open-channel blocker to the pores of Nav1.6 and Nav1.8. The Navβ4 subunit is the major determinant of fast resurgent currents generated by Nav1.6 (Barbosa et al., 2015). Knockdown of Navβ4 in rat DRG neurons substantially reduced fast resurgent currents. By contrast, overexpression of Navβ4 increased fast resurgent currents and sensory neuron excitability (Barbosa et al., 2015). It is unclear if Navβ4 is the major determinant of the slow resurgent current generated by Nav1.8 channels, but this seems to be a strong possibility.

3.2.2 Long-term inactivation

Another alternate extrinsic mechanism of inactivation is long-term inactivation, which is mediated by FHFA variants through a sequence of amino acid residues in the N-terminus domain. This pathway is also a distinct and competing mechanism to fast inactivation (Fig. 4). For FHF2A, the long-term inactivation particle is proposed to be 20 residues long composed by amino acid residues: AAAIASSLIRQKRQARERE (Dover, Solinas, D'Angelo, & Goldfarb, 2010). This sequence is highly conserved between all FHFA isoforms. Binding of the long-term inactivation particle to the α-subunit prolongs the recovery period of the VGSC from 10–20 to 100–400 ms (Dover et al., 2010). In essence, when a VGSC undergoes long-term inactivation, the channel remains in an inactivated state for a longer period relative to classic fast inactivation, and thus the refractory period is prolonged. For example, in a DRG-derived cell line (ND7/23),

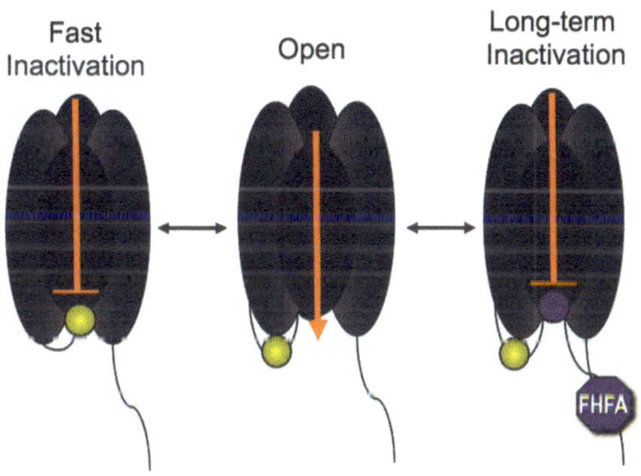

Figure 4 Fast Inactivation and Long-term Inactivation as competing mechanisms.

FHF2A expression causes accumulation of channel inactivation at all frequency stimulations due to slowing of recovery from inactivation (Rush, Wittmack, et al., 2006). Current clamp simulations incorporating long-term inactivation parameters suggest that this mechanism decreases firing frequency (Dover et al., 2010). Interestingly, Goldfarb (2012) proposed that open-channel block by Navβ4 and FHFA long-term inactivation compete not only with fast inactivation but also with each other. Inclusion of the FHF2A long-term inactivation peptide with the β4 peptide in the recording pipette (at equimolar concentrations) decreased evoked resurgent currents in Neuro2A cells (Venkatesan, Liu, & Goldfarb, 2014). It is plausible that in DRG neurons, these mechanisms (ie, fast inactivation, long-term inactivation, and open-channel block) may compete with each other to modulate excitability and, ultimately, pain sensations. Long-term inactivation induced by FHFAs is a fairly recent finding, and there are no reported studies in DRG neurons. It will be interesting to see if long-term inactivation can impact the activity of all sensory neuron VGSCs and if long-term inactivation is impaired or reduced in conditions where resurgent currents are enhanced.

4. CLASSIC SODIUM CURRENTS ROLE IN PAIN

VGSCs are crucial to the transmission of electrical impulses in nerve fibers. Because Nav1.7, Nav1.8, and Nav1.9 are preferentially expressed in peripheral neurons and highly expressed in typical nociceptor neurons, a great deal of attention has focused on these three isoforms. Studies of animal models of pain, transgenic mice and human disease mutations have confirmed that these three isoforms play important roles in pain mechanisms and therefore are potentially important targets for pain therapeutics. The relevant literature has been the subject of multiple reviews (Cummins et al., 2007; Dib-Hajj et al., 2010; Habib et al., 2015); therefore the discussion of the role of classic sodium currents in pain mechanisms here will be relatively brief.

4.1 Nav1.8

Fairly soon after the development of tight-seal whole-cell patch-clamp techniques, it was noted that peripheral sensory neurons expressed unusual TTXR sodium currents with kinetics that were slower than TTXS currents typically observed in neurons (Kostyuk, Veselovsky, & Tsyndrenko, 1981).

The voltage dependence of activation and inactivation for these TTXR currents was more depolarized than that of TTXS currents. However, the α-subunit underlying these slow TTXR currents, Nav1.8 (also called SNS), was not cloned until the mid 1990s (Akopian, Sivilotti, & Wood, 1996). Although Nav1.8 currents are TTXR and slower than TTXS currents, they fulfill a classic sodium channel role, contributing in a major fashion to the upstroke of APs in sensory neurons (Blair & Bean, 2002; Renganathan, Cummins, & Waxman, 2001). The depolarized voltage dependence of activation and inactivation is likely to help nociceptive neurons to function under compromised conditions (such as following tissue injury) when intact pain sensations serve an important defensive role. Multiple studies indicated that Nav1.8 currents are substantially altered in animal models of nerve injury and inflammation (Cummins & Waxman, 1997; Dib-Hajj et al., 2010; Dib-Hajj, Fjell, et al., 1999; Gold, Reichling, Shuster, & Levine, 1996; Okuse et al., 1997; Tanaka et al., 1998). Subsequent studies knocking-down Nav1.8 (Khasar, Gold, & Levine, 1998; Lai et al., 2002) or using transgenic mice (Akopian et al., 1999; Zimmermann et al., 2007) indicated that Nav1.8 plays an important role in pain sensations, particularly in inflammatory, cold, and mechanical pain. A number of mutations in *SCN10A* (the gene that encodes Nav1.8) have been identified in humans with painful neuropathies (Faber et al., 2012; Han et al., 2014), and several of these have been shown to increase sensory neuron excitability, helping validate the importance of Nav1.8 in pain.

4.2 Nav1.7

Nav1.7 channels are heavily expressed in nociceptive neurons (Djouhri, Newton, et al., 2003). Initially it was unclear if Nav1.7 would be a good target for pain therapeutics. Firstly, Nav1.7 exhibits high homology to other TTXS VGSCs in terms of its sequence, which suggested that it could be difficult to selectively target. Secondly, sensory neurons express multiple TTXS α-subunits (Felts, Yokoyama, Dib-Hajj, Black, & Waxman, 1997), raising concern that there was built-in redundancy. Thirdly, in some electrophysiological studies very little evidence of TTXS current was found in populations of putative nociceptive neurons (Cardenas, Cardenas, & Scroggs, 2001; McLean, Bennett, & Thomas, 1988), raising a concern that TTXS currents were not major contributors to the transmission of nociceptive information. Therefore, while it was reported that Nav1.7 had some intriguing properties compared to other TTXS VGSCs (Cummins, Howe, & Waxman, 1998), it received little attention until human mutations in

Nav1.7 were identified. Gain-of-function mutations in Nav1.7 were identified in patients with severe neuropathic pain sensations (Fertleman et al., 2006; Fischer & Waxman, 2010; Waxman, 2007; Yang et al., 2004). PEPD mutations predominantly impair fast inactivation, slowing the decay of the activated sodium currents and shifting the voltage dependence of steady-state inactivation to more positive potentials (Dib-Hajj et al., 2008; Fertleman et al., 2006; Jarecki et al., 2008). These PEPD mutations tend to increase AP firing in sensory neurons. By contrast, erythromelalgia mutations typically do not impair fast inactivation but rather shift the voltage dependence of activation in the negative direction (Cummins et al., 2004; Sheets et al., 2007). This shift in activation is thought to reduce the depolarization necessary to reach threshold for AP generation in sensory neurons. Interestingly, while PEPD-associated pain is manifested in ocular and rectal regions, erythromelalgia mutations cause pain predominantly in the hands and feet. Although both sets of mutations (PEPD and erythromelalgia) are presumably present in all tissues that express Nav1.7 channels, the mutations have very distinct impact on pain sensations. The mechanism behind the region-specific effects is not clear but may involve differential expression of other channels and/or auxiliary proteins. Indeed, it has been demonstrated that the impact of erythromelalgia mutations on hyperexcitability depends, at least in part, on coexpression of Nav1.8 channels (Rush, Dib-Hajj, et al., 2006). Conversely loss-of-function mutations in Nav1.7 are seemingly associated with complete insensitivity to pain (Cox et al., 2006; Goldberg et al., 2007). However, it is unclear if Nav1.7 is crucial to all pain sensations, especially in adults. Loss of Nav1.7 from birth could predispose individuals to a complete inability to sense pain throughout life. Interestingly, animals with loss of Nav1.7 in sensory neurons do not exhibit a complete insensitivity to all pain conditions (Minett et al., 2014). This raises the possibility that specific pain conditions, such as cold-induced pain or cancer chemotherapy—induced pain, are independent of Nav1.7 expression. From a therapeutic perspective it is important to consider that loss of Nav1.7 from birth may have distinct consequences compared to pharmacologic inhibition of Nav1.7 in mature individuals.

4.3 Nav1.3

Nav1.3 channels are TTXS and, as with other TTXS channels, are fast-activating and fast-inactivating, albeit with a propensity to recover from inactivation more rapidly than Nav1.7 (Cummins et al., 2001). Nav1.3 channels are not highly expressed in adult DRG sensory neurons, but because expression of these channels increases following nerve injury (Waxman, Kocsis, & Black, 1994), it has been proposed that Nav1.3 could contribute to abnormal sensory neuron hyperexcitability. Data from Nav1.3 knockout mice indicated that Nav1.3 is not important for the development of nerve injury—related pain (Nassar et al., 2006). However, several studies have implicated Nav1.3 in neuropathic pain using viral knockdown techniques (Samad et al., 2013; Tan, Samad, Dib-Hajj, & Waxman, 2015). Nav1.3 may also modulate pain sensations by playing a role in spinal cord neurons (Hains, Saab, Klein, Craner, & Waxman, 2004).

5. ATYPICAL SODIUM CURRENTS AND PAIN

5.1 Resurgent currents

5.1.1 Paroxysmal extreme pain disorder and resurgent currents

Studies of human VGSCs mutations are the first reported example of the potential contribution of resurgent currents to pain pathologies. Inherited and de novo Nav1.7 mutations identified in patients that suffer from PEPD result in functional channels with altered inactivation properties and increased resurgent currents (Fig. 5A and B) (Choi et al., 2011; Jarecki et al., 2010; Jarecki et al., 2008; Jarecki, Sheets, Xiao, Jackson, & Cummins, 2009; Theile, Jarecki, Piekarz, & Cummins, 2011). Recombinant expression of Nav1.7-PEPD mutant channels in sensory neurons results in hyperexcitability, which is thought to underlie the painful sensations (Dib-Hajj et al., 2008).

Nav1.7 is an isoform that normally does not frequently generate robust resurgent currents in DRG neurons. However, PEPD mutations theoretically increase the probability of binding of the open-channel blocker due to delayed inactivation (ie, voltage dependence of inactivation shifted to positive potentials and in some cases accompanied by slowed rate of inactivation). DRG simulation studies of the Nav1.7-I1462T PEPD mutant showed that altered gating properties result in a decrease in the magnitude of stimulus needed to elicit an AP (rheobase). Interestingly, when increased resurgent currents were added into the model there was a substantial increase in the number of evoked APs (Fig. 5C). Therefore, resurgent currents were identified as a key factor that contributes to the increased firing frequency (Jarecki et al., 2010). The combined changes in rheobase and firing frequency are thought to contribute to painful sensations in PEPD patients.

PEPD is lifelong disease with no current cures. The pain attacks are extreme, episodic, localized (eyes, jaw or rectum, limbs), or diffuse (Elmslie, Wilson, & Rossiter, 1996; Fertleman & Ferrie, 2006). The onset of the

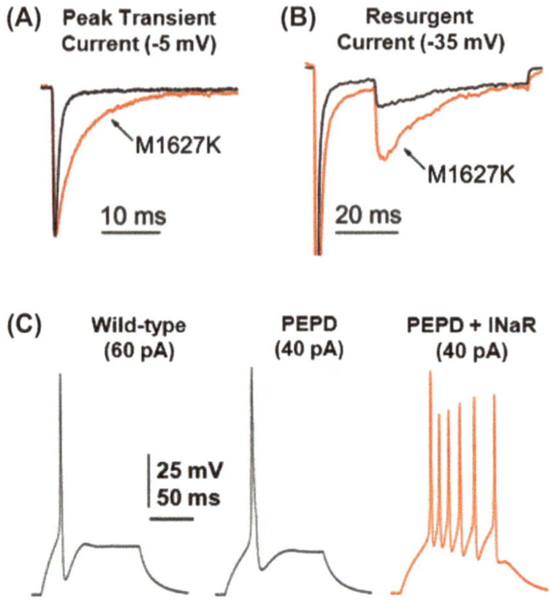

Figure 5 Comparison of (A) classic transient sodium current and (B) resurgent sodium current for wild-type Nav1.7 (black) and Nav1.7-M1627K (red) in HEK293 cells. (C) Computer simulation of APs elicited with wild-type Nav1.7 channels and Nav1.7-PEPD channels with and without incorporation resurgent current parameters. The PEPD mutation reduces threshold but the addition of resurgent currents increase AP burst firing. *PEPD*, paroxysmal extreme pain disorder.

episodes is often triggered by different factors including emotional distress, spicy food, and slight increases in temperature. One of the current treatments for PEPD is carbamazepine; however, it's only partially effective at high doses (Fertleman et al., 2007). Carbamazepine's classic mechanism of action is through stabilization of the VGSC-inactivated state. Theile and Cummins showed that carbamazepine is not selective for Nav1.7-mediated resurgent currents but partially inhibits them at high concentrations (Theile & Cummins, 2011). A more effective drug for this disorder may arise from selective targeting or more potent inhibition of resurgent currents in conjunction with stabilization of inactivated states.

5.1.2 Noninherited (acquired) pain pathologies and resurgent currents

Through a similar mechanism as seen with PEPD, increased resurgent currents may contribute to increased electrical activity of sensory neurons in noninherited (acquired) pain pathologies. Three main examples of this are described as follows. The first example is ATX-II-induced pain. ATX-II is a site-3 neurotoxin isolated from the venom of sea anemone (*Anemonia sulcata*; (Alsen, 1983)). Human contact with *A. sulcata* often results in pain with varying intensities depending on the area of contact (Abody & Klein-Kremer, 2006; Maretic & Russell, 1983). Intradermal injection of ATX-II toxin into humans induces pain and itch that resembles those experienced by divers, suggesting this component of the venom contributes to the pain sensations (Klinger et al., 2012). Using the differential nerve block approach, A-type sensory fibers were identified to mediate the pain and itch sensations. In congruence with this observation, ATX-II increased TTXS resurgent and persistent currents in medium-large diameter neurons (associated with A-type fibers) but not small diameter neurons (Klinger et al., 2012). Increased resurgent and persistent currents potentially increase the electrical activity in A-fibers. ATX-II's mechanism of action is likely due to direct interaction of the toxin with VGSCs. As a site-3 toxin, ATX-II binds to the extracellular face of VGSCs and restricts DIVS4 voltage sensor movement (Moran, Gordon, & Gurevitz, 2009). DIVS4 voltage sensor movement is important for the channel to change conformations that favor binding of the inactivation particle (Chahine et al., 1994; Ji, George, Horn, & Barchi, 1996; Kuhn & Greeff, 1999). Thus ATX-II indirectly impairs the VGSC fast inactivation process and potentially increases the binding of the open-channel blocker. Studies with ATX-II and β4 peptide support the hypothesis that the open-channel blocker has greater accessibility when DIVS4 sensor deployment is impaired (Lewis & Raman, 2013). Other site-3 toxins from scorpions, sea anemones, and spiders may also induce painful sensations through a similar mechanism (Stevens, Peigneur, & Tytgat, 2011).

The second example is oxaliplatin-induced acute painful neuropathy. Oxaliplatin is an anticancer drug used mainly to treat colorectal cancer. The acute stage of the neuropathy arises shortly after infusion and is observed with high incidence (Saif & Reardon, 2005). Sensory, and in some cases motor, symptoms are aggravated or triggered by cold temperatures (Argyriou, Polychronopoulos, Iconomou, Chroni, & Kalofonos, 2008). Human A-type sensory fibers acutely exposed to oxaliplatin exhibit an increase in electrical activity that is significantly enhanced at cold temperatures (Sittl et al., 2012). Surprisingly no increase was observed in C-type fibers. Thus increased electrical activity in A-type sensory fibers is likely mediating the cold-aggravated painful sensations. Chemotherapeutic-induced pain does not seem to depend on classic nociceptors and sensory selective isoforms, Nav1.7, Nav1.8, or Nav1.9 (Deuis et al., 2013; Minett et al., 2014; Sittl et al., 2012). Studies with transgenic mice, selective inhibitors, and selective activators of Nav1.6 suggest that it is a key factor in mediating oxaliplatin-induced increases in electrical activity (Deuis et al., 2013; Sittl et al., 2012). For example, fibers from heterozygous knockout Nav1.6 mice exhibited a dampened cold-dependent increase in electrical activity to oxaliplatin exposure, whereas homozygous knockout Nav1.6 mice fibers exhibited no increase at all (Sittl et al., 2012). At a behavioral level, Deuis and colleagues showed reversal of cold-triggered pain-associated behaviors in an oxaliplatin mouse model by selectively inhibiting Nav1.6 with GIIIA toxin. Consistent with these findings, Nav1.6 activity is altered in a cold-dependent manner upon exposure of oxaliplatin. In particular, resurgent currents and persistent currents increase as temperature decreases suggesting that this mechanism contributes to the cold-aggravated and/or cold-triggered episodes. The increase in resurgent current and persistent current is theoretically due to impaired fast inactivation, as evidenced by slower inactivation kinetics observed in medium-large diameter neurons exposed to oxaliplatin (Sittl et al., 2012). It is unknown if oxaliplatin alters inactivation by direct interaction with VGSCs or through an indirect pathway. While other mechanisms may also contribute to the acute painful neuropathy, Nav1.6 may prove to be a key target that would enable the patient to withstand higher doses or more frequent treatments of oxaliplatin (Adelsberger et al., 2000; Deuis et al., 2013; Grolleau et al., 2001; Sittl et al., 2012; Webster, Brain, Wilson, Grem, & Vincent, 2005).

The third example we will discuss is increased resurgent currents under inflammatory conditions. Inflammation due to injury or tissue damage may lead to pain pathologies (Dray, 1995). While the initial pain sensations are protective, once the injury is resolved, pain may persist. This type of persistent pain is considered pathological. Several studies have shown that inflammatory mediators alter the firing properties of sensory neurons, which can result in hyperexcitability and contribute to persistent pain (Hucho & Levine, 2007; Richardson & Vasko, 2002). These mediators can increase VGSC's expression and/or alter the gating properties promoting increased electrical activity (Black, Liu, Tanaka, Cummins, & Waxman, 2004; Cummins & Rush, 2007; Lai, Porreca, Hunter, & Gold, 2004; Tanaka et al., 1998). For example, as previously demonstrated (Tan, Piekarz, et al., 2014) and as simulated in Fig. 6, inflammatory mediators can increase fast and slow resurgent currents in sensory neurons by ~75%. Furthermore, current clamp experiments showed that the firing frequency of DRG neurons was increased when slow resurgent currents were enhanced by

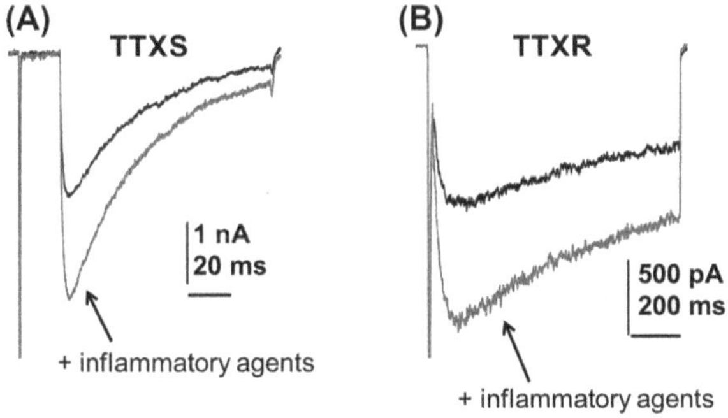

Figure 6 Increase of resurgent currents by inflammatory mediators in dorsal root ganglia neurons. Traces illustrate the relative magnitude of changes in TTXS (toxin tetrodotoxin sensitive) (A) and TTXR (toxin tetrodotoxin resistant) (B) resurgent currents before (black traces) and after (red (light gray in print verions) traces) exposure to inflammatory mediators.

injection of β4 peptide. In these experiments TTXS currents were inhibited, providing proof of concept that slow resurgent currents can increase firing frequency of neurons (Tan, Piekarz, et al., 2014). Slow and fast resurgent currents likely contribute to hyperexcitability in inflammatory conditions.

There is evidence indicating that posttranslational modifications can lead to enhanced resurgent currents (Grieco, Afshari, & Raman, 2002). Studies with Nav1.7 suggest that inflammation-induced activation of protein kinase C (PKC) pathways (Chattopadhyay, Mata, & Fink, 2008; Tan, Priest, et al., 2014; Vijayaragavan, Boutjdir, & Chahine, 2004) can lead to enhanced resurgent currents. In human embryonic kidney (HEK) cells, phosphorylation of Nav1.7 at PKC site S1479 near the inactivation particle resulted in increased resurgent currents (Tan, Priest, et al., 2014). Results from S1479 phosphorylation mimetics suggest that this is likely due to slowing of fast inactivation kinetics. Interestingly, this site is conserved in other isoforms (Nav1.1−Nav1.9), which may also be modulated by PKC. However, because modulation of VGSC is cell background dependent and isoform dependent, further studies should explore if activation of these pathways results in the same effects as those seen in HEK cells with Nav1.7. If they do, this may be an important target to decrease firing frequency and potentially pain sensations in inflammatory conditions.

5.2 Other unusual currents

5.2.1 Nav1.6

Nav1.6 is an isoform that is not classically associated with painful sensations, in part because its expression is not, by any measure, restricted to sensory neurons (Burgess et al., 1995; Schaller & Caldwell, 2000; Schaller, Krzemien, Yarowsky, Krueger, & Caldwell, 1995). Mutations identified in Nav1.6 are associated with pathologies in which brain, heart, and motor functions are altered (Blanchard et al., 2015; Burgess et al., 1995; Noujaim et al., 2012; Ohba et al., 2014). As such, until recently it has not been studied in as much detail in the context of pain as other isoforms. We have discussed contemporary findings that suggest the potential role of this isoform in oxaliplatin acute painful neuropathy. Other studies, which employed localized knockdown of Nav1.6, suggest that it may also be a potential target in inflammatory and neuropathic pain pathologies. For example, the role of Nav1.6 in chemogenic inflammation of the DRG, a proposed low back pain model, has been studied (Strong, Xie, Bataille, & Zhang, 2013; Xie et al., 2006). In early stages of chemogenic inflammation (postoperative day 3), increased Nav1.6 expression was positively correlated with increased spontaneous activity and firing (Xie et al., 2013). Localized injection (near the DRG) of Nav1.6 small inhibitory RNAs (siRNA) in parallel with induction of chemogenic inflammation prevented the development of persistent pain-associated behaviors. Furthermore, knockdown of Nav1.6 in sensory neurons abolished early-stage inflammation-mediated increases in electrical activity. While resurgent and persistent currents were not examined in this study, the increased propensity of Nav1.6 to generate these currents potentially accounts for the increased electrical activity. Indeed, we have recently shown in DRG neurons that overexpression of Navβ4 with recombinant Nav1.6 channels is sufficient to upregulate Nav1.6 resurgent currents and induce Nav1.6-dependent increases in spontaneous and evoked AP firing (Barbosa et al., 2015). It is nevertheless unclear if inflammation directly increases Navβ4 expression. However, as previously discussed, under inflammatory conditions TTXS resurgent currents may be enhanced through kinase activation pathways. Overall, these studies suggest that Nav1.6-mediated activity in sensory neurons in early stages of inflammation is important for the development of persistent pain.

The role of Nav1.6 was also investigated in two rat models of neuropathic pain (spinal nerve ligation and chronic constriction injury). Localized knockdown (as previously described) of Nav1.6 prior to nerve ligation or injury greatly reduced mechanical but not cold-dependent pain-associated behaviors (Xie et al., 2015). In a similar manner, in DRG neurons Nav1.6 knockdown reduced the spontaneous firing as well as the increased firing frequency that is commonly observed in these models (Song, Hu, Greenquist, Zhang, & LaMotte, 1999; Zhang, Song, &

LaMotte, 1999). Interestingly, Nav1.6 knockdown also decreased sympathetic sprouting into DRG neurons (Xie et al., 2015). The contribution of sympathetic sprouting to pain sensations is not fully understood. Studies suggest that sympathetic sprouting is stimulated by DRG's ectopic activity (Xie, Strong, Meij, Zhang, & Yu, 2005). Abnormal sympathetic contacts in turn stimulate DRG neurons, forming a cycle that maintains chronic pain sensations (Loh & Nathan, 1978; Xie, Strong, Li, & Zhang, 2007; Xie, Strong, & Zhang, 2010). This hypothesis is supported by the fact that in these models, sympathetic sprouting was highly colocalized to DRG neurons that expressed Nav1.6 (Xie et al., 2015). One possibility is that Nav1.6 mediates initial changes that may later contribute to the maintenance of chronic pain sensation, and thus targeting Nav1.6 activity (such as resurgent currents) in early stages may be a viable strategy to prevent the development of neuropathic pain.

The role of Nav1.6 in other pain pathologies is less clear. For example, the role of Nav1.6 in painful diabetic neuropathy has been controversial. In animal model studies of type 1 diabetic neuropathy (early stage), where increased pain-associated behavior was confirmed, Nav1.6 levels in DRG neurons have been reported to be both upregulated (Craner, Klein, Renganathan, Black, & Waxman, 2002) and downregulated (Hong, Morrow, Paulson, Isom, & Wiley, 2004). In a type 2 diabetes animal model, Nav1.6 is upregulated in early stages of diabetic neuropathy, which persisted in the later stages of the disease. In this model, changes in pain-associated behavior were positively correlated with increased Nav1.6 mRNA levels in DRG neurons (Ren et al., 2012). In a trigeminal pain model, Nav1.6 accumulates proximal to the site of injury particularly at the remodeled nodes (Henry, Freking, Johnson, & Levinson, 2007). Further studies should provide a better understanding of the role of Nav1.6 in painful diabetic neuropathy (type 1 or type 2) and trigeminal pain.

5.2.2 Nav1.9

In the late 1990s a second putative TTXR sodium channel expressed in sensory neurons (Nav1.9) was cloned (Dib-Hajj, Tyrrell, Black, & Waxman, 1998). It was soon after discovered that in addition to the slowly inactivating TTXR current, sensory neurons also expressed a previously unnoticed persistent TTXR current and it was proposed that this was generated by Nav1.9 channels (Cummins et al., 1999). At first it was unclear if Nav1.9 did indeed generate a TTXR sodium current in sensory neurons because recombinant Nav1.9 channels are extremely difficult to express; however, studies with Nav1.9-null mice (Amaya et al., 2006; Priest et al., 2005) and recombinant Nav1.9 (Vanoye, Kunic, Ehring, & George, 2013) helped confirm the original hypothesis. Nav1.9 currents are considered atypical and unusual in several aspects. The activation and inactivation kinetics of these currents are 10–20 times slower than that of classic TTXS currents (Cummins et al., 1999; Herzog, Cummins, & Waxman, 2001). Nav1.9 channels can activate at membrane potentials where other VGSCs are quiescent. Nav1.9 currents are subject to substantial modulation, exhibiting rapid rundown in whole-cell recordings (Leffler, Herzog, Dib-Hajj, Waxman, & Cummins, 2005) and modulation by G proteins (Baker, Chandra, Ding, Waxman, & Wood, 2003; Herzog et al., 2001; Ostman, Nassar, Wood, & Baker, 2008; Rush & Waxman, 2004). Mutations in *SCN11A* have been identified in humans with abnormal pain sensations. Interestingly, putative gain-of-function mutations have been identified in patients with enhanced pain sensations (Huang et al., 2014; Zhang et al., 2013) as well as patients with loss of pain sensations (Leipold et al., 2013). One hypothesis is that while slight enhancement of Nav1.9 channel activity might lead to pain, sufficient overactivation of Nav1.9 channels might lead to depolarization block and loss of pain sensations. If this is valid, it may indicate that both inhibitors and activators of Nav1.9 could have therapeutic potential.

6. THERAPEUTIC STRATEGIES

VGSCs in peripheral sensory neurons are attractive targets for treating pain. Studies of genetic mutations in VGSCs have validated Nav1.7, Nav1.8, and Nav1.9 as playing important roles in nociception and pain. Animal models of pain and human investigations have implicated Nav1.3 and Nav1.6 as also being potential targets. With a wealth of molecular targets, the question is how to develop drugs that effectively inhibit VGSCs in peripheral sensory neurons without eliciting adverse effects by inhibiting CNS and muscle sodium channels. Indeed, this is a major limitation of many of the drugs targeting VGSCs (eg, lidocaine, phenytoin) that are currently used in the clinic (Mulroy, 2002; Yanagidate & Strichartz, 2007). One appealing strategy is to develop isoform-specific blockers that target Nav1.7, Nav1.8, or Nav1.9. Many animal toxins have been identified that preferentially inhibit specific VGSC isoforms (Billen, Bosmans, & Tytgat, 2008; Bulaj et al., 2006; Terlau & Olivera, 2004; Wilson et al., 2011), but it is unclear if any of these toxins can be developed into successful therapeutics. Because Nav1.8 and Nav1.9 are less conserved compared to other VGSCs, these isoforms may seem the most promising for the development of isoform-specific small molecule inhibitors. Development of Nav1.9 inhibitors has been hindered by difficulties in expressing recombinant channels, although some investigators have developed clever chimeras that may partially alleviate this problem (Gilchrist & Bosmans, 2012; Goral, Leipold, Nematian-Ardestani, & Heinemann, 2015). Initial reports indicated that Nav1.8 small molecule inhibitors were feasible (Jarvis et al., 2007); however, the selectivity of the reported compound has been questioned (Han, Jiang, Xiao, Cao, & Wang, 2015; Stone, Kim, Yamauchi, Ruiz-Velasco, & Kaufman, 2013). Recent evidence indicates that Nav1.7 selective blockers are indeed feasible despite the strong sequence homology to other isoforms (Lee et al., 2014; McCormack et al., 2013; Sun et al., 2014). It is interesting to note that while many of the classic drugs that target VGSCs interact with pore residues (Ragsdale, McPhee, Scheuer, & Catterall, 1996), many of the animal toxins and recently described small molecules target the voltage sensors or other regions of the channels that are less conserved among isoforms than the pore is. The ability to target unique regions of specific isoforms is intuitively appealing and potentially quite powerful.

In addition to targeting specific isoforms, other strategies are being explored. Identifying sodium channel inhibitors that do not cross the blood–brain barrier may limit adverse effects. Alternatively, targeting specific patterns of sodium channel activity involved in sensory neuron hyperexcitability is a promising strategy. Indeed, classic local anesthetics and anticonvulsants are useful, at least in part, because they preferentially enhance inactivation of VGSCs (Bean, Cohen, & Tsien, 1983; Hille, 1977). These drugs are believed to mainly inhibit sodium channels in neurons that are either more active or that might be chronically depolarized. Specifically it appears that drugs like lidocaine bind to and enhance the fast-inactivated state and have a lower affinity for slow-inactivated channels (Sheets et al., 2011). However, enhancing slow inactivation of sensory neuron VGSCs might also help reduce hyperexcitability in nociceptive neurons. Slow inactivation can take seconds to minutes to occur. Because nociceptive neurons tend to have a depolarized resting membrane potential (eg, −50 mV) compared to that of many CNS neurons (eg, −65 mV), they may be more susceptible to modulators that enhance slow inactivation. Several different agents have been reported (Errington, Coyne, Stohr, Selve, & Lees, 2006; Hildebrand et al., 2011;

Jo & Bean, 2011) that seem to preferentially target slow inactivation of VGSCs. Interestingly, Hildebrand reported that a novel small organic compound that enhanced Nav1.7 and Nav1.8 slow inactivation could reverse thermal hyperalgesia and tactile allodynia associated with a rat model of neuropathic pain. In addition to targeting VGSCs, this compound also enhances slow inactivation of T-type calcium channels. Bladen, Zamponi, and colleagues have shown that several different VGSC inhibitors have activity against T-type calcium channels (Bladen, Hamid, Souza, & Zamponi, 2014; Bladen & Zamponi, 2012; Hildebrand et al., 2011), suggesting that binding sites and structural motifs are likely conserved between these related channel types. T-type calcium channels have been implicated as playing important roles in pain mechanisms (Todorovic & Jevtovic-Todorovic, 2011), and thus drugs that target T-type calcium channels and VGSCs in nociceptive neurons could have synergistic advantages as pain therapeutics (Bladen & Zamponi, 2012).

Resurgent sodium currents represent a type of activity that could potentially be targeted. Although carbamazepine, which is often at least partially effective in patients with PEPD pain, does not preferentially target resurgent currents, anandamide seems to preferentially target resurgent currents over peak transient Nav1.7 currents (Theile & Cummins, 2011), providing limited proof of concept. Theoretically, any agent that limits the ability of the open-channel blocker (most likely the β4-subunit) to interact with the α-subunit pore or enhances the rate of classic sodium channel inactivation would reduce resurgent current amplitudes. Compounds that modulate activity of the domain IV voltage sensor are appealing as this voltage sensor is selectively involved in determining open-channel inactivation (Xiao, Blumenthal, & Cummins, 2014). As the tail of FHFA variants likely competes with the resurgent current blocker (Goldfarb, 2012), enhancing long-term inactivation should reduce nociceptor excitability, in part by reducing resurgent current amplitudes.

Voltage-gated sodium currents in nociceptive neurons are complex, involving multiple isoforms, numerous configurations, a plethora of modifier proteins, and a variety of gating modes that present a wealth of targets for potentially reducing aberrant AP firing. Increased knowledge of VGSC structure–function relationships should aid the development of analgesic compounds that might selectively target specific channel isoforms or modulate unusual and atypical patterns of sodium channel activity underlying problematic pain sensations.

REFERENCES

Abody, Z., & Klein-Kremer, A. (2006). *Anemonia sulcata* sting. *Harefuah, 145*(10), 736–737, 782.

Adelsberger, H., Quasthoff, S., Grosskreutz, J., Lepier, A., Eckel, F., & Lersch, C. (2000). The chemotherapeutic oxaliplatin alters voltage-gated Na(+) channel kinetics on rat sensory neurons. *European Journal of Pharmacology, 406*(1), 25–32. pii:S0014-2999(00)00667-1.

Akemann, W., & Knopfel, T. (2006). Interaction of Kv3 potassium channels and resurgent sodium current influences the rate of spontaneous firing of Purkinje neurons. *Journal of Neuroscience, 26*(17), 4602–4612. http://dx.doi.org/10.1523/JNEUROSCI.5204-05.2006. pii:26/17/4602.

Akopian, A. N., Sivilotti, L., & Wood, J. N. (1996). A tetrodotoxin-resistant voltage-gated sodium channel expressed by sensory neurons. *Nature, 379*(6562), 257–262.

Akopian, A. N., Souslova, V., England, S., Okuse, K., Ogata, N., Ure, J. … Wood, J. N. (1999). The tetrodotoxin-resistant sodium channel SNS has a specialized function in pain pathways. *Nature Neuroscience, 2*(6), 541–548.

Alsen, C. (1983). Biological significance of peptides from *Anemonia sulcata*. *Federation Proceedings, 42*(1), 101–108.

Amaya, F., Decosterd, I., Samad, T. A., Plumpton, C., Tate, S., Mannion, R. J. … Woolf, C. J. (2000). Diversity of expression of the sensory neuron-specific TTX-resistant voltage-gated sodium ion channels SNS and SNS2 (Research Support, Non-U.S. Gov't Research Support, U.S. Gov't, P.H.S.) *Molecular and Cellular Neuroscience, 15*(4), 331–342. http://dx.doi.org/10.1006/mcne.1999.0828.

Amaya, F., Wang, H., Costigan, M., Allchorne, A. J., Hatcher, J. P., Egerton, J. … Woolf, C. J. (2006). The voltage-gated sodium channel Na(v)1.9 is an effector of peripheral inflammatory pain hypersensitivity. *Journal of Neuroscience, 26*(50), 12852–12860.

Argyriou, A. A., Polychronopoulos, P., Iconomou, G., Chroni, E., & Kalofonos, H. P. (2008). A review on oxaliplatin-induced peripheral nerve damage. *Cancer Treatment Reviews, 34*(4), 368–377. http://dx.doi.org/10.1016/j.ctrv.2008.01.003.

Armstrong, C. M., & Bezanilla, F. (1974). Charge movement associated with the opening and closing of the activation gates of the Na channels. *Journal of General Physiology, 63*(5), 533–552.

Baker, M. D., Chandra, S. Y., Ding, Y., Waxman, S. G., & Wood, J. N. (2003). GTP-induced tetrodotoxin-resistant Na^+ current regulates excitability in mouse and rat small diameter sensory neurones. *Journal of Physiology, 548*(Pt 2), 373–382.

Baker, M. D., & Wood, J. N. (2001). Involvement of Na^+ channels in pain pathways. *Trends in Pharmacological Sciences, 22*(1), 27–31. pii:S0165-6147(00)01585-6.

Bant, J. S., & Raman, I. M. (2010). Control of transient, resurgent, and persistent current by open-channel block by Na channel beta4 in cultured cerebellar granule neurons. *Proceedings of the National Academy of Sciences of the United States of America, 107*(27), 12357–12362. http://dx.doi.org/10.1073/pnas.1005633107.

Baranauskas, G., & Nistri, A. (1998). Sensitization of pain pathways in the spinal cord: cellular mechanisms. *Progress in Neurobiology, 54*(3), 349–365.

Barbosa, C., Tan, Z. Y., Wang, R., Xie, W., Strong, J. A., Vasko, M. R. … Cummins, T. R. (2015). Navβ4 regulates resurgent currents and excitability in sensory neurons. *Molecular Pain*. in press.

Bean, B. P., Cohen, C. J., & Tsien, R. W. (1983). Lidocaine block of cardiac sodium channels. *Journal of General Physiology, 81*(5), 613–642.

Beaulieu, P., & International Association for the Study of Pain. (2010). *Pharmacology of pain*. Seattle, WA: IASP Press.

Beckh, S., Noda, M., Lubbert, H., & Numa, S. (1989). Differential regulation of three sodium channel messenger RNAs in the rat central nervous system during development. *EMBO Journal, 8*(12), 3611–3616.

Billen, B., Bosmans, F., & Tytgat, J. (2008). Animal peptides targeting voltage-activated sodium channels. *Current Pharmaceutical Design, 14*(24), 2492–2502.

Black, J. A., Liu, S., Tanaka, M., Cummins, T. R., & Waxman, S. G. (2004). Changes in the expression of tetrodotoxin-sensitive sodium channels within dorsal root ganglia neurons in inflammatory pain. *Pain, 108*(3), 237–247.

Bladen, C., Hamid, J., Souza, I. A., & Zamponi, G. W. (2014). Block of T-type calcium channels by protoxins I and II. *Molecular Brain,* 7, 36. http://dx.doi.org/10.1186/1756-6606-7-36 1756-6606-7-36.

Bladen, C., & Zamponi, G. W. (2012). Common mechanisms of drug interactions with sodium and T-type calcium channels. *Molecular Pharmacology, 82*(3), 481–487. http://dx.doi.org/10.1124/mol.112.079715 mol.112.079715.

Blair, N. T., & Bean, B. P. (2002). Roles of tetrodotoxin (TTX)-sensitive Na^+ current, TTX-resistant Na^+ current, and Ca^{2+} current in the action potentials of nociceptive sensory neurons. *Journal of Neuroscience, 22*(23), 10277–10290. pii:22/23/10277.

Blair, N. T., & Bean, B. P. (2003). Role of tetrodotoxin-resistant Na+ current slow inactivation in adaptation of action potential firing in small-diameter dorsal root ganglion neurons. *Journal of Neuroscience, 23*(32), 10338–10350. pii:23/32/10338.

Blanchard, M. G., Willemsen, M. H., Walker, J. B., Dib-Hajj, S. D., Waxman, S. G., Jongmans, M. C.… Kamsteeg, E. J. (2015). De novo gain-of-function and loss-of-function mutations of SCN8A in patients with intellectual disabilities and epilepsy. *Journal of Medical Genetics, 52*(5), 330–337. http://dx.doi.org/10.1136/jmedgenet-2014-102813.

Bulaj, G., Zhang, M. M., Green, B. R., Fiedler, B., Layer, R. T., Wei, S.… Olivera, B. M. (2006). Synthetic muO-conotoxin MrVIB blocks TTX-resistant sodium channel NaV1.8 and has a long-lasting analgesic activity. *Biochemistry, 45*(23), 7404–7414.

Burgess, D. L., Kohrman, D. C., Galt, J., Plummer, N. W., Jones, J. M., Spear, B., & Meisler, M. H. (1995). Mutation of a new sodium channel gene, Scn8a, in the mouse mutant 'motor endplate disease'. *Nature Genetics, 10*(4), 461–465. http://dx.doi.org/10.1038/ng0895-461.

Caffrey, J. M., Eng, D. L., Black, J. A., Waxman, S. G., & Kocsis, J. D. (1992). Three types of sodium channels in adult rat dorsal root ganglion neurons. *Brain Research, 592*(1–2), 283–297.

Cannon, S. C., & Bean, B. P. (2010). Sodium channels gone wild: resurgent current from neuronal and muscle channelopathies. *Journal of Clinical Investigation, 120*(1), 80–83. http://dx.doi.org/10.1172/JCI41340.

Cardenas, L. M., Cardenas, C. G., & Scroggs, R. S. (2001). 5HT increases excitability of nociceptor-like rat dorsal root ganglion neurons via cAMP-coupled TTX-resistant Na(+) channels. *Journal of Neurophysiology, 86*(1), 241–248.

Catterall, W. A., Goldin, A. L., & Waxman, S. G. (2005). International Union of Pharmacology. XLVII. Nomenclature and structure-function relationships of voltage-gated sodium channels. *Pharmacological Reviews, 57*(4), 397–409.

Chahine, M., George, A. L., Jr., Zhou, M., Ji, S., Sun, W., Barchi, R. L., & Horn, R. (1994). Sodium channel mutations in paramyotonia congenita uncouple inactivation from activation. *Neuron, 12*(2), 281–294.

Chahine, M., & O'Leary, M. E. (2011). Regulatory role of voltage-gated Na channel beta subunits in sensory neurons. *Frontiers in Pharmacology, 2*, 70. http://dx.doi.org/10.3389/fphar.2011.00070.

Chattopadhyay, M., Mata, M., & Fink, D. J. (2008). Continuous delta-opioid receptor activation reduces neuronal voltage-gated sodium channel (NaV1.7) levels through

activation of protein kinase C in painful diabetic neuropathy. *Journal of Neuroscience, 28*(26), 6652–6658. http://dx.doi.org/10.1523/JNEUROSCI.5530-07.2008.
Cheng, X., Dib-Hajj, S. D., Tyrrell, L., Te Morsche, R. H., Drenth, J. P., & Waxman, S. G. (2011). Deletion mutation of sodium channel Na(V)1.7 in inherited erythromelalgia: enhanced slow inactivation modulates dorsal root ganglion neuron hyperexcitability. *Brain, 134*(Pt 7), 1972–1986. http://dx.doi.org/10.1093/brain/awr143.
Choi, J. S., Boralevi, F., Brissaud, O., Sanchez-Martin, J., Te Morsche, R. H., Dib-Hajj, S. D.... Waxman, S. G. (2011). Paroxysmal extreme pain disorder: a molecular lesion of peripheral neurons. *Nature Reviews Neurology,* 7(1), 51–55. http://dx.doi.org/10.1038/nrneurol.2010.162.
Cox, J. J., Reimann, F., Nicholas, A. K., Thornton, G., Roberts, E., Springell, K.... Woods, C. G. (2006). An SCN9A channelopathy causes congenital inability to experience pain. *Nature, 444*(7121), 894–898.
Craner, M. J., Klein, J. P., Renganathan, M., Black, J. A., & Waxman, S. G. (2002). Changes of sodium channel expression in experimental painful diabetic neuropathy. *Annals of Neurology, 52*(6), 786–792. http://dx.doi.org/10.1002/ana.10364.
Crill, W. E. (1996). Persistent sodium current in mammalian central neurons. *Annual Review of Physiology, 58*, 349–362.
Cruz, J. S., Silva, D. F., Ribeiro, L. A., Araujo, I. G., Magalhaes, N., Medeiros, A.... Oliveira, F. A. (2011). Resurgent Na^+ current: a new avenue to neuronal excitability control. *Life Sciences, 89*(15–16), 564–569. http://dx.doi.org/10.1016/j.lfs.2011.05.016.
Cummins, T. R., Aglieco, F., Renganathan, M., Herzog, R. I., Dib-Hajj, S. D., & Waxman, S. G. (2001). Nav1.3 sodium channels: rapid repriming and slow closed-state inactivation display quantitative differences after expression in a mammalian cell line and in spinal sensory neurons. *Journal of Neuroscience, 21*(16), 5952–5961.
Cummins, T. R., Dib-Hajj, S. D., Black, J. A., Akopian, A. N., Wood, J. N., & Waxman, S. G. (1999). A novel persistent tetrodotoxin-resistant sodium current in SNS-null and wild-type small primary sensory neurons. *Journal of Neuroscience, 19*(24), RC43.
Cummins, T. R., Dib-Hajj, S. D., Black, J. A., & Waxman, S. G. (2000). Sodium channels and the molecular pathophysiology of pain. *Progress in Brain Research, 129*, 3–19.
Cummins, T. R., Dib-Hajj, S. D., Herzog, R. I., & Waxman, S. G. (2005). Nav1.6 channels generate resurgent sodium currents in spinal sensory neurons. *FEBS Letters, 579*(10), 2166–2170.
Cummins, T. R., Dib-Hajj, S. D., & Waxman, S. G. (2004). Electrophysiological properties of mutant Nav1.7 sodium channels in a painful inherited neuropathy. *Journal of Neuroscience, 24*(38), 8232–8236.
Cummins, T. R., Howe, J. R., & Waxman, S. G. (1998). Slow closed-state inactivation: a novel mechanism underlying ramp currents in cells expressing the hNE/PN1 sodium channel. *Journal of Neuroscience, 18*(23), 9607–9619.
Cummins, T. R., & Rush, A. M. (2007). Voltage-gated sodium channel blockers for the treatment of neuropathic pain. *Expert Review of Neurotherapeutics,* 7(11), 1597–1612.
Cummins, T. R., Sheets, P. L., & Waxman, S. G. (2007). The roles of sodium channels in nociception: Implications for mechanisms of pain. *Pain, 131*(3), 243–257. S0304-3959(07)00430-7.
Cummins, T. R., & Waxman, S. G. (1997). Downregulation of tetrodotoxin-resistant sodium currents and upregulation of a rapidly repriming tetrodotoxin-sensitive sodium current in small spinal sensory neurons after nerve injury. *Journal of Neuroscience, 17*(10), 3503–3514.
Deuis, J. R., Zimmermann, K., Romanovsky, A. A., Possani, L. D., Cabot, P. J., Lewis, R. J., & Vetter, I. (2013). An animal model of oxaliplatin-induced cold allodynia reveals a

crucial role for Nav1.6 in peripheral pain pathways. *Pain, 154*(9), 1749–1757. http://dx.doi.org/10.1016/j.pain.2013.05.032. pii:S0304-3959(13)00271-6.

Devor, M. (1999). Unexplained peculiarities of the dorsal root ganglion. *Pain*, (Suppl. 6), S27–S35.

Dib-Hajj, S., Black, J. A., Cummins, T. R., & Waxman, S. G. (2002). NaN/Nav1.9: a sodium channel with unique properties. *Trends in Neurosciences, 25*(5), 253–259.

Dib-Hajj, S. D., Cummins, T. R., Black, J. A., & Waxman, S. G. (2007). From genes to pain: Na v. 1.7 and human pain disorders. *Trends in Neurosciences, 30*(11), 555–563.

Dib-Hajj, S. D., Cummins, T. R., Black, J. A., & Waxman, S. G. (2010). Sodium channels in normal and pathological pain. *Annual Review of Neuroscience, 33*, 325–347. http://dx.doi.org/10.1146/annurev-neuro-060909-153234.

Dib-Hajj, S. D., Estacion, M., Jarecki, B. W., Tyrrell, L., Fischer, T. Z., Lawden, M.... Waxman, S. G. (2008). Paroxysmal extreme pain disorder M1627K mutation in human Nav1.7 renders DRG neurons hyperexcitable. *Molecular Pain, 4*, 37.

Dib-Hajj, S. D., Fjell, J., Cummins, T. R., Zheng, Z., Fried, K., LaMotte, R.... Waxman, S. G. (1999). Plasticity of sodium channel expression in DRG neurons in the chronic constriction injury model of neuropathic pain. *Pain, 83*(3), 591–600.

Dib-Hajj, S. D., Tyrrell, L., Black, J. A., & Waxman, S. G. (1998). NaN, a novel voltage-gated Na channel, is expressed preferentially in peripheral sensory neurons and down-regulated after axotomy. *Proceedings of the National Academy of Sciences of the United States of America, 95*(15), 8963–8968.

Dib-Hajj, S. D., Tyrrell, L., Cummins, T. R., Black, J. A., Wood, P. M., & Waxman, S. G. (1999). Two tetrodotoxin-resistant sodium channels in human dorsal root ganglion neurons. *FEBS Letters, 462*(1–2), 117–120.

Djouhri, L., Fang, X., Okuse, K., Wood, J. N., Berry, C. M., & Lawson, S. N. (2003). The TTX-resistant sodium channel Nav1.8 (SNS/PN3): expression and correlation with membrane properties in rat nociceptive primary afferent neurons. *Journal of Physiology, 550*(Pt 3), 739–752.

Djouhri, L., & Lawson, S. N. (2004). Abeta-fiber nociceptive primary afferent neurons: a review of incidence and properties in relation to other afferent A-fiber neurons in mammals. *Brain Research Brain Research Reviews, 46*(2), 131–145. http://dx.doi.org/10.1016/j.brainresrev.2004.07.015.

Djouhri, L., Newton, R., Levinson, S. R., Berry, C. M., Carruthers, B., & Lawson, S. N. (2003). Sensory and electrophysiological properties of guinea-pig sensory neurones expressing Nav 1.7 (PN1) Na^+ channel alpha subunit protein. *Journal of Physiology, 546*(Pt 2), 565–576.

Dover, K., Solinas, S., D'Angelo, E., & Goldfarb, M. (2010). Long-term inactivation particle for voltage-gated sodium channels. *Journal of Physiology, 588*(Pt 19), 3695–3711. http://dx.doi.org/10.1113/jphysiol.2010.192559.

Dray, A. (1995). Inflammatory mediators of pain. *British Journal of Anaesthesia, 75*(2), 125–131.

Elliott, A. A., & Elliott, J. R. (1993). Characterization of TTX-sensitive and TTX-resistant sodium currents in small cells from adult rat dorsal root ganglia. *Journal of Physiology, 463*, 39–56.

Elmslie, F. V., Wilson, J., & Rossiter, M. A. (1996). Familial rectal pain: is it under-diagnosed? *Journal of the Royal Society of Medicine, 89*(5), 290P–291P.

Errington, A. C., Coyne, L., Stohr, T., Selve, N., & Lees, G. (2006). Seeking a mechanism of action for the novel anticonvulsant lacosamide. *Neuropharmacology, 50*(8), 1016–1029.

Faber, C. G., Lauria, G., Merkies, I. S., Cheng, X., Han, C., Ahn, H. S.... Waxman, S. G. (2012). Gain-of-function Nav1.8 mutations in painful neuropathy. *Proceedings of the National Academy of Sciences of the United States of America, 109*(47), 19444–19449. http://dx.doi.org/10.1073/pnas.1216080109.

Felts, P. A., Yokoyama, S., Dib-Hajj, S., Black, J. A., & Waxman, S. G. (1997). Sodium channel alpha-subunit mRNAs I, II, III, NaG, Na6 and hNE (PN1): different expression patterns in developing rat nervous system. *Brain Research Molecular Brain Research, 45*(1), 71–82.

Fertleman, C. R., Baker, M. D., Parker, K. A., Moffatt, S., Elmslie, F. V., Abrahamsen, B. ... Rees, M. (2006). SCN9A mutations in paroxysmal extreme pain disorder: allelic variants underlie distinct channel defects and phenotypes. *Neuron, 52*(5), 767–774.

Fertleman, C. R., & Ferrie, C. D. (2006). What's in a name—familial rectal pain syndrome becomes paroxysmal extreme pain disorder. *Journal of Neurology, Neurosurgery and Psychiatry, 77*(11), 1294–1295. http://dx.doi.org/10.1136/jnnp.2006.089664.

Fertleman, C. R., Ferrie, C. D., Aicardi, J., Bednarek, N. A., Eeg-Olofsson, O., Elmslie, F. V. ... Stephenson, J. B. (2007). Paroxysmal extreme pain disorder (previously familial rectal pain syndrome). *Neurology, 69*(6), 586–595.

Fischer, T. Z., & Waxman, S. G. (2010). Familial pain syndromes from mutations of the NaV1.7 sodium channel. *Annals of the New York Academy of Sciences, 1184*, 196–207. http://dx.doi.org/10.1111/j.1749-6632.2009.05110.x.

Gilchrist, J., & Bosmans, F. (2012). Animal toxins can alter the function of Nav1.8 and Nav1.9. *Toxins (Basel), 4*(8), 620–632. http://dx.doi.org/10.3390/toxins4080620.

Goetz, R., Dover, K., Laezza, F., Shtraizent, N., Huang, X., Tchetchik, D. ... Mohammadi, M. (2009). Crystal structure of a fibroblast growth factor homologous factor (FHF) defines a conserved surface on FHFs for binding and modulation of voltage-gated sodium channels. *Journal of Biological Chemistry, 284*(26), 17883–17896. http://dx.doi.org/10.1074/jbc.M109.001842.

Gold, M. S., Reichling, D. B., Shuster, M. J., & Levine, J. D. (1996). Hyperalgesic agents increase a tetrodotoxin-resistant Na^+ current in nociceptors. *Proceedings of the National Academy of Sciences of the United States of America, 93*(3), 1108–1112.

Goldberg, Y. P., MacFarlane, J., MacDonald, M. L., Thompson, J., Dube, M. P., Mattice, M. ... Hayden, M. R. (2007). Loss-of-function mutations in the Nav1.7 gene underlie congenital indifference to pain in multiple human populations. *Clinical Genetics, 71*(4), 311–319.

Goldfarb, M. (2005). Fibroblast growth factor homologous factors: evolution, structure, and function. *Cytokine & Growth Factor Reviews, 16*(2), 215–220. pii:S1359-6101(05)00012-2.

Goldfarb, M. (2012). Voltage-gated sodium channel-associated proteins and alternative mechanisms of inactivation and block. *Cellular and Molecular Life Sciences, 69*(7), 1067–1076. http://dx.doi.org/10.1007/s00018-011-0832-1.

Goldin, A. L. (2003). Mechanisms of sodium channel inactivation. *Current Opinion in Neurobiology, 13*(3), 284–290. pii:S0959438803000655.

Goldin, A. L., Barchi, R. L., Caldwell, J. H., Hofmann, F., Howe, J. R., Hunter, J. C. ... Catterall, W. A. (2000). Nomenclature of voltage-gated sodium channels. *Neuron, 28*(2), 365–368.

Goral, R. O., Leipold, E., Nematian-Ardestani, E., & Heinemann, S. H. (2015). Heterologous expression of Na1.9 chimeras in various cell systems. *Pflugers Archiv*. http://dx.doi.org/10.1007/s00424-015-1709-1.

Grieco, T. M., Afshari, F. S., & Raman, I. M. (2002). A role for phosphorylation in the maintenance of resurgent sodium current in cerebellar purkinje neurons. *Journal of Neuroscience, 22*(8), 3100–3107.

Grieco, T. M., Malhotra, J. D., Chen, C., Isom, L. L., & Raman, I. M. (2005). Open-channel block by the cytoplasmic tail of sodium channel beta4 as a mechanism for resurgent sodium current. *Neuron, 45*(2), 233–244.

Grolleau, F., Gamelin, L., Boisdron-Celle, M., Lapied, B., Pelhate, M., & Gamelin, E. (2001). A possible explanation for a neurotoxic effect of the anticancer agent oxaliplatin on neuronal voltage-gated sodium channels. *Journal of Neurophysiology, 85*(5), 2293–2297.

Habib, A. M., Wood, J. N., & Cox, J. J. (2015). Sodium channels and pain. *Handbook of Experimental Pharmacology, 227*, 39–56. http://dx.doi.org/10.1007/978-3-662-46450-2_3.

Hains, B. C., Saab, C. Y., Klein, J. P., Craner, M. J., & Waxman, S. G. (2004). Altered sodium channel expression in second-order spinal sensory neurons contributes to pain after peripheral nerve injury. *Journal of Neuroscience, 24*(20), 4832–4839.

Han, C., Vasylyev, D., Macala, L. J., Gerrits, M. M., Hoeijmakers, J. G., Bekelaar, K. J. … Waxman, S. G. (2014). The G1662S NaV1.8 mutation in small fibre neuropathy: impaired inactivation underlying DRG neuron hyperexcitability. *Journal of Neurology, Neurosurgery and Psychiatry, 85*(5), 499–505. http://dx.doi.org/10.1136/jnnp-2013-306095.

Han, Z., Jiang, Y., Xiao, F., Cao, K., & Wang, D. W. (2015). The effects of A-803467 on cardiac Nav1.5 channels. *European Journal of Pharmacology, 754*, 52–60. http://dx.doi.org/10.1016/j.ejphar.2015.02.019.

Harty, T. P., Dib-Hajj, S. D., Tyrrell, L., Blackman, R., Hisama, F. M., Rose, J. B., & Waxman, S. G. (2006). Na(V)1.7 mutant A863P in erythromelalgia: effects of altered activation and steady-state inactivation on excitability of nociceptive dorsal root ganglion neurons. *Journal of Neuroscience, 26*(48), 12566–12575.

Henry, M. A., Freking, A. R., Johnson, L. R., & Levinson, S. R. (2007). Sodium channel Nav1.6 accumulates at the site of infraorbital nerve injury (Research Support, N.I.H., Extramural) *BMC Neuroscience, 8*, 56. http://dx.doi.org/10.1186/1471-2202-8-56.

Herzog, R. I., Cummins, T. R., & Waxman, S. G. (2001). Persistent TTX-resistant Na+ current affects resting potential and response to depolarization in simulated spinal sensory neurons. *Journal of Neurophysiology, 86*(3), 1351–1364.

Hildebrand, M. E., Smith, P. L., Bladen, C., Eduljee, C., Xie, J. Y., Chen, L. … Snutch, T. P. (2011). A novel slow-inactivation-specific ion channel modulator attenuates neuropathic pain. *Pain, 152*(4), 833–843. http://dx.doi.org/10.1016/j.pain.2010.12.035. pii:S0304-3959(10)00801-8.

Hille, B. (1977). Local anesthetics: hydrophilic and hydrophobic pathways for the drug-receptor reaction. *Journal of General Physiology, 69*(4), 497–515.

Hille, B. (1984). *Ionic channels of excitable membranes* (1st ed.). Sunderland, MA: Sinauer Associates, Inc.

Ho, C., & O'Leary, M. E. (2011). Single-cell analysis of sodium channel expression in dorsal root ganglion neurons. *Molecular and Cellular Neuroscience, 46*(1), 159–166. http://dx.doi.org/10.1016/j.mcn.2010.08.017.

Ho, C., Zhao, J., Malinowski, S., Chahine, M., & O'Leary, M. E. (2012). Differential expression of sodium channel beta subunits in dorsal root ganglion sensory neurons. *Journal of Biological Chemistry, 287*(18), 15044–15053. http://dx.doi.org/10.1074/jbc.M111.333740.

Hodgkin, A. L., & Huxley, A. F. (1952). A quantitative description of membrane current and its application to conduction and excitation in nerve. *Journal of Physiology, 117*(4), 500–544.

Hong, S., Morrow, T. J., Paulson, P. E., Isom, L. L., & Wiley, J. W. (2004). Early painful diabetic neuropathy is associated with differential changes in tetrodotoxin-sensitive and -resistant sodium channels in dorsal root ganglion neurons in the rat. *Journal of Biological Chemistry, 279*(28), 29341–29350.

Huang, J., Han, C., Estacion, M., Vasylyev, D., Hoeijmakers, J. G., Gerrits, M. M. … Waxman, S. G. (2014). Gain-of-function mutations in sodium channel Na(v)1.9 in painful neuropathy. *Brain, 137*(Pt 6), 1627–1642. http://dx.doi.org/10.1093/brain/awu079.

Hubert, T., Bourane, S., Venteo, S., Mechaly, I., Puech, S., Valmier, J. ... Fichard-Carroll, A. (2008). Fibroblast growth factor homologous factor 1 (FHF1) is expressed in a subpopulation of calcitonin gene-related peptide-positive nociceptive neurons in the murine dorsal root ganglia. *Journal of Comparative Neurology, 507*(4), 1588–1601. http://dx.doi.org/10.1002/cne.21631.

Hucho, T., & Levine, J. D. (2007). Signaling pathways in sensitization: toward a nociceptor cell biology. *Neuron, 55*(3), 365–376. http://dx.doi.org/10.1016/j.neuron.2007.07.008.

Institute of Medicine Report. (2011). *Relieving pain in America: A blueprint for transforming prevention, care, education, and research.* Washington, DC.

Isom, L. L. (2001). Sodium channel beta subunits: anything but auxiliary. *Neuroscientist, 7*(1), 42–54.

Isom, L. L., & Catterall, W. A. (1996). Na^+ channel subunits and Ig domains. *Nature, 383*(6598), 307–308.

Jarecki, B. W., Piekarz, A. D., Jackson, J. O., 2nd, & Cummins, T. R. (2010). Human voltage-gated sodium channel mutations that cause inherited neuronal and muscle channelopathies increase resurgent sodium currents. *Journal of Clinical Investigation, 120*(1), 369–378. http://dx.doi.org/10.1172/JCI40801.

Jarecki, B. W., Sheets, P. L., Jackson, J. O., 2nd, & Cummins, T. R. (2008). Paroxysmal extreme pain disorder mutations within the D3/S4-S5 linker of Nav1.7 cause moderate destabilization of fast inactivation. *Journal of Physiology, 586*(Pt 17), 4137–4153.

Jarecki, B. W., Sheets, P. L., Xiao, Y., Jackson, J. O., 2nd, & Cummins, T. R. (2009). Alternative splicing of Na(V)1.7 exon 5 increases the impact of the painful PEPD mutant channel I1461T. *Channels (Austin), 3*(4), 259–267. pii:9341.

Jarvis, M. F., Honore, P., Shieh, C. C., Chapman, M., Joshi, S., Zhang, X. F. ... Krafte, D. S. (2007). A-803467, a potent and selective Nav1.8 sodium channel blocker, attenuates neuropathic and inflammatory pain in the rat. *Proceedings of the National Academy of Sciences of the United States of America, 104*(20), 8520–8525.

Ji, S., George, A. L., Jr., Horn, R., & Barchi, R. L. (1996). Paramyotonia congenita mutations reveal different roles for segments S3 and S4 of domain D4 in hSkM1 sodium channel gating. *Journal of General Physiology, 107*(2), 183–194.

Jo, S., & Bean, B. P. (2011). Inhibition of neuronal voltage-gated sodium channels by brilliant blue G. *Molecular Pharmacology, 80*(2), 247–257. http://dx.doi.org/10.1124/mol.110.070276. pii:mol.110.070276.

Kerr, N. C., Gao, Z., Holmes, F. E., Hobson, S. A., Hancox, J. C., Wynick, D., & James, A. F. (2007). The sodium channel Nav1.5a is the predominant isoform expressed in adult mouse dorsal root ganglia and exhibits distinct inactivation properties from the full-length Nav1.5 channel. *Molecular and Cellular Neuroscience, 35*(2), 283–291. http://dx.doi.org/10.1016/j.mcn.2007.03.002.

Khaliq, Z. M., Gouwens, N. W., & Raman, I. M. (2003). The contribution of resurgent sodium current to high-frequency firing in Purkinje neurons: an experimental and modeling study. *Journal of Neuroscience, 23*(12), 4899–4912.

Khasar, S. G., Gold, M. S., & Levine, J. D. (1998). A tetrodotoxin-resistant sodium current mediates inflammatory pain in the rat. *Neuroscience Letters, 256*(1), 17–20.

Klinger, A. B., Eberhardt, M., Link, A. S., Namer, B., Kutsche, L. K., Schuy, E. T. ... Lampert, A. (2012). Sea-anemone toxin ATX-II elicits A-fiber-dependent pain and enhances resurgent and persistent sodium currents in large sensory neurons. *Molecular Pain, 8*, 69. http://dx.doi.org/10.1186/1744-8069-8-69. pii: 1744-8069-8-69.

Kostyuk, P. G., Veselovsky, N. S., & Tsyndrenko, A. Y. (1981). Ionic currents in the somatic membrane of rat dorsal root ganglion neurons-I. Sodium currents. *Neuroscience, 6*(12), 2423–2430.

Kuhn, F. J., & Greeff, N. G. (1999). Movement of voltage sensor S4 in domain 4 is tightly coupled to sodium channel fast inactivation and gating charge immobilization. *Journal of General Physiology, 114*(2), 167–183.

Lai, J., Gold, M. S., Kim, C. S., Bian, D., Ossipov, M. H., Hunter, J. C., & Porreca, F. (2002). Inhibition of neuropathic pain by decreased expression of the tetrodotoxin-resistant sodium channel, NaV1.8. *Pain, 95*(1–2), 143–152.

Lai, J., Porreca, F., Hunter, J. C., & Gold, M. S. (2004). Voltage-gated sodium channels and hyperalgesia. *Annual Review of Pharmacology and Toxicology, 44*, 371–397.

Lee, J. H., Park, C. K., Chen, G., Han, Q., Xie, R. G., Liu, T. ... Lee, S. Y. (2014). A monoclonal antibody that targets a NaV1.7 channel voltage sensor for pain and itch relief. *Cell, 157*(6), 1393–1404. http://dx.doi.org/10.1016/j.cell.2014.03.064.

Leffler, A., Herzog, R. I., Dib-Hajj, S. D., Waxman, S. G., & Cummins, T. R. (2005). Pharmacological properties of neuronal TTX-resistant sodium channels and the role of a critical serine pore residue. *Pflugers Archiv, 451*(3), 454–463.

Leipold, E., Liebmann, L., Korenke, G. C., Heinrich, T., Giesselmann, S., Baets, J. ... Kurth, I. (2013). A de novo gain-of-function mutation in SCN11A causes loss of pain perception. *Nature Genetics, 45*(11), 1399–1404. http://dx.doi.org/10.1038/ng.2767.

Lewis, A. H., & Raman, I. M. (2011). Cross-species conservation of open-channel block by Na channel beta4 peptides reveals structural features required for resurgent Na current. *Journal of Neuroscience, 31*(32), 11527–11536. http://dx.doi.org/10.1523/JNEUROSCI.1428-11.2011.

Lewis, A. H., & Raman, I. M. (2013). Interactions among DIV voltage-sensor movement, fast inactivation, and resurgent Na current induced by the NaVbeta4 open-channel blocking peptide. *Journal of General Physiology, 142*(3), 191–206. http://dx.doi.org/10.1085/jgp.201310984.

Liu, C., Dib-Hajj, S. D., & Waxman, S. G. (2001). Fibroblast growth factor homologous factor 1B binds to the C terminus of the tetrodotoxin-resistant sodium channel rNav1.9a (NaN). *Journal of Biological Chemistry, 276*(22), 18925–18933. http://dx.doi.org/10.1074/jbc.M101606200.

Liu, M., & Wood, J. N. (2011). The roles of sodium channels in nociception: implications for mechanisms of neuropathic pain. *Pain Medicine, 12*(Suppl. 3), S93–S99. http://dx.doi.org/10.1111/j.1526-4637.2011.01158.x.

Loh, L., & Nathan, P. W. (1978). Painful peripheral states and sympathetic blocks. *Journal of Neurology, Neurosurgery and Psychiatry, 41*(7), 664–671.

Lopez-Santiago, L. F., Brackenbury, W. J., Chen, C., & Isom, L. L. (2011). Na^+ channel Scn1b gene regulates dorsal root ganglion nociceptor excitability in vivo. *Journal of Biological Chemistry, 286*(26), 22913–22923. http://dx.doi.org/10.1074/jbc.M111.242370.

Lopez-Santiago, L. F., Pertin, M., Morisod, X., Chen, C., Hong, S., Wiley, J. ... Isom, L. L. (2006). Sodium channel beta2 subunits regulate tetrodotoxin-sensitive sodium channels in small dorsal root ganglion neurons and modulate the response to pain. *Journal of Neuroscience, 26*(30), 7984–7994.

Maretic, Z., & Russell, F. E. (1983). Stings by the sea anemone *Anemonia sulcata* in the Adriatic Sea. *American Journal of Tropical Medicine and Hygiene, 32*(4), 891–896.

McCormack, K., Santos, S., Chapman, M. L., Krafte, D. S., Marron, B. E., West, C. W. ... Castle, N. A. (2013). Voltage sensor interaction site for selective small molecule inhibitors of voltage-gated sodium channels. *Proceedings of the National Academy of Sciences of the United States of America, 110*(29), E2724–E2732. http://dx.doi.org/10.1073/pnas.1220844110.

MeLean, M. J., Bennett, P. B., & Thomas, R. M. (1988). Subtypes of dorsal root ganglion neurons based on different inward currents as measured by whole-cell voltage clamp. *Molecular and Cellular Biochemistry, 80*(1–2), 95–107.

Millan, M. J. (1999). The induction of pain: an integrative review. *Progress in Neurobiology, 57*(1), 1–164.

Minett, M. S., Falk, S., Santana-Varela, S., Bogdanov, Y. D., Nassar, M. A., Heegaard, A. M., & Wood, J. N. (2014). Pain without nociceptors? Nav1.7-independent pain mechanisms. *Cell Reports, 6*(2), 301–312. http://dx.doi.org/10.1016/j.celrep.2013.12.033.

Miyazaki, H., Oyama, F., Inoue, R., Aosaki, T., Abe, T., Kiyonari, H.… Nukina, N. (2014). Singular localization of sodium channel beta4 subunit in unmyelinated fibres and its role in the striatum. *Nature Communications, 5*, 5525. http://dx.doi.org/10.1038/ncomms6525.

Moran, Y., Gordon, D., & Gurevitz, M. (2009). Sea anemone toxins affecting voltage-gated sodium channels—molecular and evolutionary features. *Toxicon, 54*(8), 1089–1101. http://dx.doi.org/10.1016/j.toxicon.2009.02.028.

Mulroy, M. F. (2002). Systemic toxicity and cardiotoxicity from local anesthetics: incidence and preventive measures. *Regional Anesthesia and Pain Medicine, 27*(6), 556–561. pii: S1098733902000779.

Munoz-Sanjuan, I., Smallwood, P. M., & Nathans, J. (2000). Isoform diversity among fibroblast growth factor homologous factors is generated by alternative promoter usage and differential splicing. *Journal of Biological Chemistry, 275*(4), 2589–2597.

Nassar, M. A., Baker, M. D., Levato, A., Ingram, R., Mallucci, G., McMahon, S. B., & Wood, J. N. (2006). Nerve injury induces robust allodynia and ectopic discharges in Nav1.3 null mutant mice. *Molecular Pain, 2*, 33.

Noda, M., Shimizu, S., Tanabe, T., Takai, T., Kayano, T., Ikeda, T.… Numa, S. (1984). Primary structure of *Electrophorus electricus* sodium channel deduced from cDNA sequence. *Nature, 312*(5990), 121–127.

Noujaim, S. F., Kaur, K., Milstein, M., Jones, J. M., Furspan, P., Jiang, D.… Jalife, J. (2012). A null mutation of the neuronal sodium channel NaV1.6 disrupts action potential propagation and excitation-contraction coupling in the mouse heart. *FASEB Journal, 26*(1), 63–72. http://dx.doi.org/10.1096/fj.10-179770.

Ohba, C., Kato, M., Takahashi, S., Lerman-Sagie, T., Lev, D., Terashima, H.… Saitsu, H. (2014). Early onset epileptic encephalopathy caused by de novo SCN8A mutations. *Epilepsia, 55*(7), 994–1000. http://dx.doi.org/10.1111/epi.12668.

Okuse, K., Chaplan, S. R., McMahon, S. B., Luo, Z. D., Calcutt, N. A., Scott, B. P.… Wood, J. N. (1997). Regulation of expression of the sensory neuron-specific sodium channel SNS in inflammatory and neuropathic pain. *Molecular and Cellular Neuroscience, 10*(3–4), 196–207.

O'Malley, H. A., & Isom, L. L. (2015). Sodium channel beta subunits: emerging targets in channelopathies. *Annual Review of Physiology, 77*, 481–504. http://dx.doi.org/10.1146/annurev-physiol-021014-071846.

Ostman, J. A., Nassar, M. A., Wood, J. N., & Baker, M. D. (2008). GTP up-regulated persistent Na^+ current and enhanced nociceptor excitability require NaV1.9. *Journal of Physiology, 586*(4), 1077–1087.

Patel, R. R., Barbosa, C., Xiao, Y., & Cummins, T. R. (2015). Human Nav1.6 channels generate larger resurgent currents than human Nav1.1 channels, but the Navbeta4 peptide does not protect either isoform from use-dependent reduction. *PLoS One, 10*(7), e0133485. http://dx.doi.org/10.1371/journal.pone.0133485.

Patino, G. A., & Isom, L. L. (2010). Electrophysiology and beyond: multiple roles of Na^+ channel beta subunits in development and disease. *Neuroscience Letters, 486*(2), 53–59. http://dx.doi.org/10.1016/j.neulet.2010.06.050.

Payandeh, J., Scheuer, T., Zheng, N., & Catterall, W. A. (2011). The crystal structure of a voltage-gated sodium channel. *Nature, 475*(7356), 353–358. http://dx.doi.org/10.1038/nature10238.

Portenoy, R. K., Ugarte, C., Fuller, I., & Haas, G. (2004). Population-based survey of pain in the United States: differences among white, African American, and Hispanic subjects. *Journal of Pain, 5*(6), 317–328. http://dx.doi.org/10.1016/j.jpain.2004.05.005.

Priest, B. T., Murphy, B. A., Lindia, J. A., Diaz, C., Abbadie, C., Ritter, A. M.… Martin, W. J. (2005). Contribution of the tetrodotoxin-resistant voltage-gated sodium channel NaV1.9 to sensory transmission and nociceptive behavior. *Proceedings of the National Academy of Sciences of the United States of America, 102*(26), 9382–9387.

Ragsdale, D. S., McPhee, J. C., Scheuer, T., & Catterall, W. A. (1996). Common molecular determinants of local anesthetic, antiarrhythmic, and anticonvulsant block of voltage-gated Na+ channels. *Proceedings of the National Academy of Sciences of the United States of America, 93*(17), 9270–9275.

Ramachandra, R., McGrew, S. Y., Baxter, J. C., Howard, J. R., & Elmslie, K. S. (2013). NaV1.8 channels are expressed in large, as well as small, diameter sensory afferent neurons (Research Support, N.I.H., Extramural) *Channels (Austin), 7*(1), 34–37. http://dx.doi.org/10.4161/chan.22445.

Raman, I. M., & Bean, B. P. (2001). Inactivation and recovery of sodium currents in cerebellar Purkinje neurons: evidence for two mechanisms. *Biophysical Journal, 80*(2), 729–737.

Ren, Y. S., Qian, N. S., Tang, Y., Liao, Y. H., Yang, Y. L., Dou, K. F., & Toi, M. (2012). Sodium channel Nav1.6 is up-regulated in the dorsal root ganglia in a mouse model of type 2 diabetes. *Brain Research Bulletin, 87*(2–3), 244–249. http://dx.doi.org/10.1016/j.brainresbull.2011.10.015.

Renganathan, M., Cummins, T. R., & Waxman, S. G. (2001). Contribution of Na(v)1.8 sodium channels to action potential electrogenesis in DRG neurons. *Journal of Neurophysiology, 86*(2), 629–640.

Renganathan, M., Dib-Hajj, S., & Waxman, S. G. (2002). Na(v)1.5 underlies the 'third TTX-R sodium current' in rat small DRG neurons. *Brain Research Molecular Brain Research, 106*(1–2), 70–82.

Richardson, J. D., & Vasko, M. R. (2002). Cellular mechanisms of neurogenic inflammation. *Journal of Pharmacology and Experimental Therapeutics, 302*(3), 839–845. http://dx.doi.org/10.1124/jpet.102.032797.

Roy, M. L., & Narahashi, T. (1992). Differential properties of tetrodotoxin-sensitive and tetrodotoxin-resistant sodium channels in rat dorsal root ganglion neurons. *Journal of Neuroscience, 12*(6), 2104–2111.

Rush, A. M., Cummins, T. R., & Waxman, S. G. (2007). Multiple sodium channels and their roles in electrogenesis within dorsal root ganglion neurons. *Journal of Physiology, 579*(Pt 1), 1–14.

Rush, A. M., Dib-Hajj, S. D., Liu, S., Cummins, T. R., Black, J. A., & Waxman, S. G. (2006). A single sodium channel mutation produces hyper- or hypoexcitability in different types of neurons. *Proceedings of the National Academy of Sciences of the United States of America, 103*(21), 8245–8250.

Rush, A. M., & Waxman, S. G. (2004). PGE2 increases the tetrodotoxin-resistant Nav1.9 sodium current in mouse DRG neurons via G-proteins. *Brain Research, 1023*(2), 264–271.

Rush, A. M., Wittmack, E. K., Tyrrell, L., Black, J. A., Dib-Hajj, S. D., & Waxman, S. G. (2006). Differential modulation of sodium channel Na(v)1.6 by two members of the fibroblast growth factor homologous factor 2 subfamily. *European Journal of Neuroscience, 23*(10), 2551–2562.

Saif, M. W., & Reardon, J. (2005). Management of oxaliplatin-induced peripheral neuropathy. *Therapeutics and Clinical Risk Management, 1*(4), 249–258.

Samad, O. A., Tan, A. M., Cheng, X., Foster, E., Dib-Hajj, S. D., & Waxman, S. G. (2013). Virus-mediated shRNA knockdown of Na(v)1.3 in rat dorsal root ganglion attenuates nerve injury-induced neuropathic pain. *Molecular Therapy, 21*(1), 49–56. http://dx.doi.org/10.1038/mt.2012.169.

Schaller, K. L., & Caldwell, J. H. (2000). Developmental and regional expression of sodium channel isoform NaCh6 in the rat central nervous system. *Journal of Comparative Neurology, 420*(1), 84–97.

Schaller, K. L., Krzemien, D. M., Yarowsky, P. J., Krueger, B. K., & Caldwell, J. H. (1995). A novel, abundant sodium channel expressed in neurons and glia. *Journal of Neuroscience, 15*(5 Pt 1), 3231–3242.

Shah, B. S., Gonzalez, M. I., Bramwell, S., Pinnock, R. D., Lee, K., & Dixon, A. K. (2001). Beta3, a novel auxiliary subunit for the voltage gated sodium channel is upregulated in sensory neurones following streptozocin induced diabetic neuropathy in rat. *Neuroscience Letters, 309*(1), 1–4.

Shah, B. S., Stevens, E. B., Gonzalez, M. I., Bramwell, S., Pinnock, R. D., Lee, K., & Dixon, A. K. (2000). beta3, a novel auxiliary subunit for the voltage-gated sodium channel, is expressed preferentially in sensory neurons and is upregulated in the chronic constriction injury model of neuropathic pain. *European Journal of Neuroscience, 12*(11), 3985–3990.

Sheets, P. L., Jackson, J. O., 2nd, Waxman, S. G., Dib-Hajj, S. D., & Cummins, T. R. (2007). A Nav1.7 channel mutation associated with hereditary erythromelalgia contributes to neuronal hyperexcitability and displays reduced lidocaine sensitivity. *Journal of Physiology, 581*(Pt 3), 1019–1031.

Sheets, P. L., Jarecki, B. W., & Cummins, T. R. (2011). Lidocaine reduces the transition to slow inactivation in Na(v)1.7 voltage-gated sodium channels. *British Journal of Pharmacology, 164*(2b), 719–730. http://dx.doi.org/10.1111/j.1476-5381.2011.01209.x.

Shields, S. D., Ahn, H. S., Yang, Y., Han, C., Seal, R. P., Wood, J. N. … Dib-Hajj, S. D. (2012). Nav1.8 expression is not restricted to nociceptors in mouse peripheral nervous system. *Pain, 153*(10), 2017–2030. http://dx.doi.org/10.1016/j.pain.2012.04.022 (Research Support, U.S. Gov't, Non-P.H.S.).

Sittl, R., Lampert, A., Huth, T., Schuy, E. T., Link, A. S., Fleckenstein, J. … Carr, R. W. (2012). Anticancer drug oxaliplatin induces acute cooling-aggravated neuropathy via sodium channel subtype Na(V)1.6-resurgent and persistent current. *Proceedings of the National Academy of Sciences of the United States of America, 109*(17), 6704–6709. http://dx.doi.org/10.1073/pnas.1118058109.

Smallwood, P. M., Munoz-Sanjuan, I., Tong, P., Macke, J. P., Hendry, S. H., Gilbert, D. J. … Nathans, J. (1996). Fibroblast growth factor (FGF) homologous factors: new members of the FGF family implicated in nervous system development. *Proceedings of the National Academy of Sciences of the United States of America, 93*(18), 9850–9857.

Song, X. J., Hu, S. J., Greenquist, K. W., Zhang, J. M., & LaMotte, R. H. (1999). Mechanical and thermal hyperalgesia and ectopic neuronal discharge after chronic compression of dorsal root ganglia. *Journal of Neurophysiology, 82*(6), 3347–3358.

Stevens, M., Peigneur, S., & Tytgat, J. (2011). Neurotoxins and their binding areas on voltage-gated sodium channels. *Frontiers in Pharmacology, 2*, 71. http://dx.doi.org/10.3389/fphar.2011.00071.

Stone, A. J., Kim, J. S., Yamauchi, K., Ruiz-Velasco, V., & Kaufman, M. P. (2013). Attenuation of autonomic reflexes by A803467 may not be solely caused by blockade of NaV 1.8 channels. *Neuroscience Letters, 543*, 177–182. http://dx.doi.org/10.1016/j.neulet.2013.03.015. pii:S0304-3940(13)00220-6.

Strong, J. A., Xie, W., Bataille, F. J., & Zhang, J. M. (2013). Preclinical studies of low back pain. *Molecular Pain, 9*, 17. http://dx.doi.org/10.1186/1744-8069-9-17.

Sun, S., Jia, Q., Zenova, A. Y., Chafeev, M., Zhang, Z., Lin, S. … Oballa, R. M. (2014). The discovery of benzenesulfonamide-based potent and selective inhibitors of voltage-gated sodium channel Na(v)1.7. *Bioorganic & Medicinal Chemistry Letters, 24*(18), 4397–4401. http://dx.doi.org/10.1016/j.bmcl.2014.08.017.

Takahashi, N., Kikuchi, S., Dai, Y., Kobayashi, K., Fukuoka, T., & Noguchi, K. (2003). Expression of auxiliary beta subunits of sodium channels in primary afferent neurons and the effect of nerve injury. *Neuroscience, 121*(2), 441–450.

Tan, A. M., Samad, O. A., Dib-Hajj, S. D., & Waxman, S. G. (2015). Virus-mediated knockdown of Nav1.3 in dorsal root ganglia of STZ-induced diabetic rats alleviates tactile allodynia. *Molecular Medicine*. http://dx.doi.org/10.2119/molmed.2015.00063. pii:molmed.2015.00063.

Tan, Z. Y., Piekarz, A. D., Priest, B. T., Knopp, K. L., Krajewski, J. L., McDermott, J. S. … Cummins, T. R. (2014). Tetrodotoxin-resistant sodium channels in sensory neurons generate slow resurgent currents that are enhanced by inflammatory mediators. *Journal of Neuroscience, 34*(21), 7190–7197. http://dx.doi.org/10.1523/JNEUROSCI.5011-13.2014.

Tan, Z. Y., Priest, B. T., Krajewski, J. L., Knopp, K. L., Nisenbaum, E. S., & Cummins, T. R. (2014). Protein kinase C enhances human sodium channel hNav1.7 resurgent currents via a serine residue in the domain III-IV linker. *FEBS Letters, 588*(21), 3964–3969. http://dx.doi.org/10.1016/j.febslet.2014.09.011.

Tanaka, M., Cummins, T. R., Ishikawa, K., Dib-Hajj, S. D., Black, J. A., & Waxman, S. G. (1998). SNS Na+ channel expression increases in dorsal root ganglion neurons in the carrageenan inflammatory pain model. *NeuroReport, 9*(6), 967–972.

Terlau, H., & Olivera, B. M. (2004). Conus venoms: a rich source of novel ion channel-targeted peptides. *Physiological Reviews, 84*(1), 41–68.

Theile, J. W., & Cummins, T. R. (2011). Inhibition of Nav{beta}4 peptide-mediated resurgent sodium currents in Nav1.7 channels by carbamazepine, riluzole and anandamide. *Molecular Pharmacology*. pii:mol.111.072751.

Theile, J. W., Jarecki, B. W., Piekarz, A. D., & Cummins, T. R. (2011). Nav1.7 mutations associated with paroxysmal extreme pain disorder, but not erythromelalgia, enhance Navbeta4 peptide-mediated resurgent sodium currents. *Journal of Physiology, 589*(Pt 3), 597–608. pii:jphysiol.2010.200915.

Todorovic, S. M., & Jevtovic-Todorovic, V. (2011). T-type voltage-gated calcium channels as targets for the development of novel pain therapies. *British Journal of Pharmacology, 163*(3), 484–495. http://dx.doi.org/10.1111/j.1476-5381.2011.01256.x.

Trimmer, J. S., Cooperman, S. S., Tomiko, S. A., Zhou, J. Y., Crean, S. M., Boyle, M. B. … Sigworth, M. J. (1989). Primary structure and functional expression of a mammalian skeletal muscle sodium channel. *Neuron, 3*(1), 33–49.

Ulbricht, W. (2005). Sodium channel inactivation: molecular determinants and modulation. *Physiological Reviews, 85*(4), 1271–1301. pii:85/4/1271.

Vanoye, C. G., Kunic, J. D., Ehring, G. R., & George, A. L., Jr. (2013). Mechanism of sodium channel NaV1.9 potentiation by G-protein signaling. *Journal of General Physiology, 141*(2), 193–202. http://dx.doi.org/10.1085/jgp.201210919.

Venkatesan, K., Liu, Y., & Goldfarb, M. (2014). Fast-onset long-term open-state block of sodium channels by a-type FHFs mediates classical spike accommodation in hippocampal pyramidal neurons. *Journal of Neuroscience, 34*(48), 16126–16139. http://dx.doi.org/10.1523/JNEUROSCI.1271-14.2014.

Vijayaragavan, K., Boutjdir, M., & Chahine, M. (2004). Modulation of Nav1.7 and Nav1.8 peripheral nerve sodium channels by protein kinase A and protein kinase C. *Journal of Neurophysiology, 91*(4), 1556–1569. http://dx.doi.org/10.1152/jn.00676.2003.

Waxman, S. G. (2007). Nav1.7, its mutations, and the syndromes that they cause. *Neurology, 69*(6), 505–507. pii:69/6/505.

Waxman, S. G., Dib-Hajj, S., Cummins, T. R., & Black, J. A. (1999). Sodium channels and pain. *Proceedings of the National Academy of Sciences of the United States of America, 96*(14), 7635–7639.

Waxman, S. G., Kocsis, J. D., & Black, J. A. (1994). Type III sodium channel mRNA is expressed in embryonic but not adult spinal sensory neurons, and is reexpressed following axotomy. *Journal of Neurophysiology, 72*(1), 466–470.

Webster, R. G., Brain, K. L., Wilson, R. H., Grem, J. L., & Vincent, A. (2005). Oxaliplatin induces hyperexcitability at motor and autonomic neuromuscular junctions through effects on voltage-gated sodium channels. *British Journal of Pharmacology, 146*(7), 1027–1039. pii:0706407.

West, J. W., Patton, D. E., Scheuer, T., Wang, Y., Goldin, A. L., & Catterall, W. A. (1992). A cluster of hydrophobic amino acid residues required for fast Na(+)-channel inactivation. *Proceedings of the National Academy of Sciences of the United States of America, 89*(22), 10910–10914.

Wilson, M. J., Yoshikami, D., Azam, L., Gajewiak, J., Olivera, B. M., Bulaj, G., & Zhang, M. M. (2011). mu-Conotoxins that differentially block sodium channels NaV1.1 through 1.8 identify those responsible for action potentials in sciatic nerve. *Proceedings of the National Academy of Sciences of the United States of America, 108*(25), 10302–10307. http://dx.doi.org/10.1073/pnas.1107027108.

Wittmack, E. K., Rush, A. M., Craner, M. J., Goldfarb, M., Waxman, S. G., & Dib-Hajj, S. D. (2004). Fibroblast growth factor homologous factor 2B: association with Nav1.6 and selective colocalization at nodes of Ranvier of dorsal root axons. *Journal of Neuroscience, 24*(30), 6765–6775. http://dx.doi.org/10.1523/JNEUROSCI.1628-04.2004.

Xiao, Y., Blumenthal, K., & Cummins, T. R. (2014). Gating-pore currents demonstrate selective and specific modulation of individual sodium channel voltage-sensors by biological toxins. *Molecular Pharmacology, 86*(2), 159–167. http://dx.doi.org/10.1124/mol.114.092338.

Xie, W., Strong, J. A., Li, H., & Zhang, J. M. (2007). Sympathetic sprouting near sensory neurons after nerve injury occurs preferentially on spontaneously active cells and is reduced by early nerve block. *Journal of Neurophysiology, 97*(1), 492–502. http://dx.doi.org/10.1152/jn.00899.2006.

Xie, W., Strong, J. A., Meij, J. T., Zhang, J. M., & Yu, L. (2005). Neuropathic pain: early spontaneous afferent activity is the trigger. *Pain, 116*(3), 243–256. http://dx.doi.org/10.1016/j.pain.2005.04.017.

Xie, W., Strong, J. A., Ye, L., Mao, J. X., & Zhang, J. M. (2013). Knockdown of sodium channel NaV1.6 blocks mechanical pain and abnormal bursting activity of afferent neurons in inflamed sensory ganglia. *Pain, 154*(8), 1170–1180. http://dx.doi.org/10.1016/j.pain.2013.02.027.

Xie, W., Strong, J. A., & Zhang, J. M. (2010). Increased excitability and spontaneous activity of rat sensory neurons following in vitro stimulation of sympathetic fiber sprouts in the isolated dorsal root ganglion. *Pain, 151*(2), 447–459. http://dx.doi.org/10.1016/j.pain.2010.08.006.

Xie, W., Strong, J. A., & Zhang, J. M. (2015). Local knockdown of the Na1.6 sodium channel reduces pain behaviors, sensory neuron excitability, and sympathetic sprouting in rat models of neuropathic pain. *Neuroscience*. http://dx.doi.org/10.1016/j.neuroscience.2015.02.010.

Xie, W. R., Deng, H., Li, H., Bowen, T. L., Strong, J. A., & Zhang, J. M. (2006). Robust increase of cutaneous sensitivity, cytokine production and sympathetic sprouting in rats with localized inflammatory irritation of the spinal ganglia. *Neuroscience, 142*(3), 809–822. http://dx.doi.org/10.1016/j.neuroscience.2006.06.045.

Yan, H., Pablo, J. L., Wang, C., & Pitt, G. S. (2014). FGF14 modulates resurgent sodium current in mouse cerebellar Purkinje neurons. *Elife, 3*, e04193. http://dx.doi.org/10.7554/eLife.04193.

Yanagidate, F., & Strichartz, G. R. (2007). Local anesthetics. *Handbook of Experimental Pharmacology*, (177), 95–127.

Yang, Y., Wang, Y., Li, S., Xu, Z., Li, H., Ma, L. ... Shen, Y. (2004). Mutations in SCN9A, encoding a sodium channel alpha subunit, in patients with primary erythermalgia. *Journal of Medical Genetics, 41*(3), 171–174.

Yu, F. H., Westenbroek, R. E., Silos-Santiago, I., McCormick, K. A., Lawson, D., Ge, P. ... Curtis, R. (2003). Sodium channel beta4, a new disulfide-linked auxiliary subunit with similarity to beta2. *Journal of Neuroscience, 23*(20), 7577–7585.

Zhang, J. M., Song, X. J., & LaMotte, R. H. (1999). Enhanced excitability of sensory neurons in rats with cutaneous hyperalgesia produced by chronic compression of the dorsal root ganglion. *Journal of Neurophysiology, 82*(6), 3359–3366.

Zhang, X. Y., Wen, J., Yang, W., Wang, C., Gao, L., Zheng, L. H. ... Liu, J. Y. (2013). Gain-of-function mutations in SCN11A cause familial episodic pain. *American Journal of Human Genetics, 93*(5), 957–966. http://dx.doi.org/10.1016/j.ajhg.2013.09.016.

Zimmermann, K., Leffler, A., Babes, A., Cendan, C. M., Carr, R. W., Kobayashi, J. ... Reeh, P. W. (2007). Sensory neuron sodium channel Nav1.8 is essential for pain at low temperatures. *Nature, 447*(7146), 855–858.

Index

'*Note:* Page numbers followed by "f" indicate figures and "t" indicate tables.'

D

K

L

O

P

T

Kolibri 360 GmbH
Ferdinand-Jühlke-Straße 7
99095 Erfurt, Deutschland
produktsicherheit@kolibri360.de